Fire and Life Safety in Health Care Facilities

IJ
2007

Fire and Life Safety in Health Care Facilities

Second Edition

Marvin J. Fischer
Thomas W. Gardner
Burton R. Klein
James K. Lathrop

NFPA®

National Fire Protection Association
Quincy, Massachusetts

Product Manager: Brad Gray
Editorial-Production Services: SPI Publisher Services
Composition: SPI Publisher Services
Cover Design: Cameron Inc.
Manufacturing Manager: Ellen Glisker
Printer: R.R. Donnelley

Copyright © 2006
National Fire Protection Association, Inc.
One Batterymarch Park
Quincy, Massachusetts 02169-7471

All rights reserved. No part of the material protected by this copyright notice may be reproduced or utilized in any form without acknowledgment of the copyright owner nor may it be used in any form for resale without written permission from the copyright owner.

Notice Concerning Liability: Publication of this work is for the purpose of circulating information and opinion among those concerned for fire and life safety and related subjects. While every effort has been made to achieve a work of high quality, neither the NFPA nor the authors and contributors to this work guarantee the accuracy or completeness of or assume any liability in connection with the information and opinions contained in this work. The NFPA and the authors and contributors shall in no event be liable for any personal injury, property, or other damages of any nature whatsoever, whether special, indirect, consequential, or compensatory, directly or indirectly resulting from the publication, use of or reliance upon this work.

This work is published with the understanding that the NFPA and the authors and contributors to this work are supplying information and opinion but are not attempting to render engineering or other professional services. If such services are required, the assistance of an appropriate professional should be sought.

The following are registered trademarks of the National Fire Protection Association:

Life Safety Code® and *101®*
National Fire Alarm Code® and *NFPA 72®*
National Electrical Code® and *NEC®*

NFPA No.: FLS06
ISBN-13: 978-0-87765-744-6
ISBN-10: 0-87765-744-0
Library of Congress Control No.: 2005930710

Printed in the United States of America
06 07 08 09 10 5 4 3 2 1

To our families, without whose patience,
perseverance, and understanding this book
could not have been written.

Contents

Preface — xv

Part I Introduction — 1

1 Health Care Facility Codes and Standards — 3
1.1 Fire and Life Safety Codes and Standards — 4
1.2 NFPA Codes and Standards That Apply to Health Care Facilities — 7
1.3 Creating and Revising Codes and Standards — 10
1.4 Cross-Referencing in Codes and Standards — 11
1.5 Authorities Having Jurisdiction — 12
1.6 Conflicts and Duplications Among Codes and Standards — 13
1.7 Resolving Code Conflicts — 14

2 Overview of Health Care Facility Organization — 16
2.1 Health Care Facility Categories — 16
2.2 Building Occupancies in Health Care Facilities — 20
2.3 Typical Areas Within Health Care Facilities — 22
2.4 Typical Health Care Facility Departments — 27

3 Inspections and Surveys — 29
3.1 JCAHO Standards and Accreditation — 30
3.2 Preparing for Surveys and Inspections — 30
3.3 When an Inspector or Surveyor Is on Site — 32
3.4 Record Keeping — 34

Part II Electrical Systems and Equipment — 35

4 Electrical Hazards in Health Care Facilities — 37
4.1 Types of Electrical Hazards in Health Care Facilities — 37
4.2 Avoiding Electrical Hazards — 39

5 Electrical Supply at the Entrance to the Facility — 44
5.1 Utility Service Arrangements and Reliability — 45
5.2 Service Layout and Equipment — 50

6 Distribution Systems — 59

- 6.1 Wire and Cable — 59
- 6.2 Raceways — 64
- 6.3 Wiring Methods — 68
- 6.4 Panelboards, Switchboards, and Motor Control Centers — 69
- 6.5 System Protection and Coordination — 70
- 6.6 Grounding — 76
- 6.7 Circuiting — 79

7 Electrical System Devices and Lighting — 83

- 7.1 Receptacles — 83
- 7.2 Switches — 87
- 7.3 Ground-Fault Circuit Interrupters — 89
- 7.4 Lighting — 91

8 Essential Electrical Systems — 97

- 8.1 Codes and Standards — 97
- 8.2 Types of Essential Electrical Systems — 98
- 8.3 Wiring of Essential Electrical Systems — 101
- 8.4 Emergency Generators — 101
- 8.5 Switching and Load Shedding — 103
- 8.6 Energy Converters — 106
- 8.7 Alarms, Indicators, and Safety Devices — 109
- 8.8 Testing, Maintenance, and Record Keeping — 110

9 Isolated Power Supplies and Other Special Systems — 115

- 9.1 Isolated Power Systems — 115
- 9.2 Alarms — 121
- 9.3 Communication Systems — 123

10 Requirements for Specific Areas and Facilities — 127

- 10.1 Hospitals, Nursing Homes, Limited Care Facilities, and Other Health Care Facilities — 127
- 10.2 General Care Areas — 128
- 10.3 Critical Care Areas — 129
- 10.4 Wet Locations — 130
- 10.5 Inhalation Anesthetizing Locations — 131
- 10.6 Operating and Labor and Delivery Suites — 132
- 10.7 Radiology — 133
- 10.8 Laboratories — 134
- 10.9 Hazardous Locations — 135
- 10.10 Psychiatric Units and Hospitals — 136
- 10.11 Emergency Departments — 138
- 10.12 Hyperbaric and Hypobaric Areas and Facilities — 139
- 10.13 Information System Areas — 140

Contents **ix**

11	Medical Equipment	142
11.1	Equipment Manufacturer Requirements	143
11.2	Equipment Purchase Order Specifications	148
11.3	Facility Equipment Testing	151
11.4	Equipment Installation and Grounding	156
11.5	Medical Equipment Training	157

12	Testing, Maintenance, and Record Keeping	160
12.1	Testing Electrical Systems	161
12.2	Maintaining Electrical Systems	163
12.3	Record Keeping	168

Frequently Asked Questions	172
Checklist	176

Part III Building Construction and Life Safety   179

13	Building and Life Safety Codes	181
13.1	Building Codes in the United States	181
13.2	NFPA *101*®, *Life Safety Code*®	185
13.3	Fire Prevention Codes	186
13.4	AIA *Guidelines for Design and Construction of Hospitals and Health Care Facilities*	187
13.5	Occupancy Classifications	187
13.6	Multiple, Mixed, and Separated Occupancies	195
13.7	New versus Existing	199

14	Construction Types	203
14.1	Defend-in-Place versus Evacuation	204
14.2	Fire Resistance	205
14.3	Number of Stories and Heights and Areas	206
14.4	Construction Types	208
14.5	Construction Type Consistency and Conversions	212
14.6	Determining Construction Type	214

15	Compartmentation	218
15.1	Barrier Continuity	218
15.2	Fire Resistance of Barriers	223
15.3	Types of Fire Barriers	223
15.4	Barrier Openings and Penetrations	229
15.5	Barrier Sealing and Penetration Protection	238

x Contents

16	**Egress Components**	242
16.1	Means of Egress—Definitions	243
16.2	Protecting Exits and Exit Enclosures	244
16.3	Components of Means of Egress	254

17	**Principles of Egress**	279
17.1	Egress Capacity	279
17.2	Number of Exits	285
17.3	Travel Distance	288
17.4	Arrangement of Means of Egress	289
17.5	Accessible Means of Egress for the Mobility Impaired	302
17.6	Illumination of Egress and Emergency Lighting	303
17.7	Marking the Means of Egress	305

18	**Fire Protection Features**	309
18.1	Vertical Opening Protection	309
18.2	Protection of Concealed Spaces	315
18.3	Smoke Barriers	316
18.4	Special Hazard Protection	327
18.5	Interior Finish	332
18.6	Contents and Furnishings	336
18.7	Fire Alarm Systems	337
18.8	Extinguishing Equipment	338

19	**Building Service Equipment**	339
19.1	Utilities	339
19.2	Heating, Ventilating, and Air-Conditioning Equipment	343
19.3	Smoke Control	347
19.4	Elevators, Escalators, and Vertical Conveyors	348
19.5	Rubbish Chutes, Incinerators, and Laundry Chutes	349
19.6	Cooking Equipment	351

Frequently Asked Questions	355
Checklist	361

Part IV Fire Protection Systems 369

20	**Importance of Fire Protection Systems**	371
20.1	Codes and Standards for Active Fire Protection Systems	372

21	**Automatic Fire Sprinkler Systems**	375
21.1	Sprinkler Facts and Myths	376
21.2	History of Automatic Sprinklers	377

21.3	System Activation and Controls	378
21.4	System Types	380
21.5	Design Considerations for New Construction and Retrofits	387
21.6	General Requirements for Inspection, Testing, and Maintenance	396
21.7	Inspection of Sprinkler Systems	398
21.8	Testing of Sprinkler Systems	402
21.9	Sprinkler Maintenance	409

22 Standpipe Systems 413

22.1	History of Standpipes in Health Care Facilities	413
22.2	Standpipe System Requirements	416
22.3	Standpipe System Types	416
22.4	Standpipe System Classes	417
22.5	Design Considerations	417
22.6	Inspection, Testing, and Maintenance	419

23 Fire Pumps 431

23.1	Location Requirements	431
23.2	Fire Pump Types	432
23.3	Pump Drivers	433
23.4	Design Considerations	438
23.5	Inspection, Testing, and Maintenance	441

24 Kitchen Hood Fire-Extinguishing Systems 450

24.1	Methods of Control	451
24.2	UL 300–Compliant Suppression Systems	452
24.3	System Types	453
24.4	System Design and Modification	456
24.5	Inspection and Maintenance	459
24.6	Testing	462

25 Specialized Suppression Systems 463

25.1	Clean Fire Suppression Agents	463
25.2	Gaseous Fire Suppression Agents	464
25.3	Water Mist Extinguishing Systems	469

26 Portable Fire Extinguishers 474

26.1	Buildings Where Required	474
26.2	Classification of Fire Types and Extinguisher Ratings	474
26.3	Portable Fire Extinguisher Selection and Placement	475
26.4	Extinguisher Inspection, Maintenance, and Recharging	483
26.5	Employee Training	486

27 Fire Detection and Alarm Systems — 488

- 27.1 System Types — 489
- 27.2 Fire Alarm Control Panels — 491
- 27.3 Alarm Devices and Appliances — 494
- 27.4 Specialized Features for Health Care Fire Alarm Systems — 498
- 27.5 Inspection, Testing, and Maintenance — 499
- 27.6 Procedure in Case of a Fire — 501

Frequently Asked Questions — 509

Checklist — 511

Part V Gas and Vacuum Systems and Equipment — 515

28 History and Hazards of Piped Gas and Vacuum Systems — 517

- 28.1 Piped Gas Systems — 517
- 28.2 Piped Vacuum Systems — 520
- 28.3 Disposal of Waste Anesthetic Gas — 522

29 Patient Piped Gas Systems — 524

- 29.1 System Levels — 525
- 29.2 Design Considerations — 526
- 29.3 System Sources — 527
- 29.4 Gas Distribution — 539
- 29.5 End of the Line — 547
- 29.6 Monitors and Alarms — 552
- 29.7 System Installation, Testing, and Verification — 556
- 29.8 Staff Training — 559

30 Patient Piped Vacuum Systems — 561

- 30.1 Piped Vacuum Sources (Pumps) — 562
- 30.2 Vacuum Distribution — 565
- 30.3 End of the Line — 567
- 30.4 Monitors and Alarms — 569
- 30.5 System Performance Criteria — 569
- 30.6 System Renovations — 571
- 30.7 System Installation, Testing, and Verification — 571
- 30.8 Staff Training — 572

31 Other Gas/Vacuum Systems in Health Care Facilities — 573

- 31.1 Waste Anesthetic Gas Disposal — 573
- 31.2 Instrument Air Piped Gas Systems — 576
- 31.3 Non-Patient Piped Gas Systems — 576
- 31.4 Non-Patient Piped Vacuum Systems — 580

32	Patient Gas and Vacuum Equipment	583
32.1	Patient Gas Equipment	583
32.2	Patient Vacuum Equipment	588

Frequently Asked Questions	589
Checklist	592

Part VI Communication Among Health Care Facility Departments — 595

33	Fire and Life Safety Communications in Health Care Facilities	597
33.1	Communicating with Administrators	598
33.2	Communicating with Various Departments	600
33.3	Communicating with Clinical Departments	603
33.4	Communicating with Support Services	603
33.5	Communicating with Outside Agencies and Groups	604
33.6	Meeting JCAHO Communications Standards	605

34	Record Keeping	606
34.1	Record-Keeping Requirements	606
34.2	Equipment Manuals	607
34.3	Equipment Records	607
34.4	Keeping and Storing Records	609

35	Construction Projects	610
35.1	Project Planning	610
35.2	Construction	611
35.3	Progress Reports	611
35.4	System Testing	612
35.5	Walk-Throughs	613
35.6	Shakedown Period	613

36	Health Care Facility Training	615
36.1	Training for All Personnel	615
36.2	In-Service Training and Continuing Education for the Engineering Staff	617

Selected NFPA Definitions	619
References	633
Index	641

Preface

Health care facility designers, engineers, and administrators must be knowledgeable about a myriad of codes, standards, and guidelines to promote fire and life safety in hospitals, nursing homes, limited and ambulatory care facilities, clinics, medical and dental offices, and other types of buildings. With this second edition of *Fire and Life Safety in Health Care Facilities,* NFPA brings together a vast amount of information on a range of codes, standards, and other key documents that address fire and life safety in these structures. In addition to addressing the numerous NFPA codes and standards that are enforced in many jurisdictions around the country, this book addresses the codes, standards, and guidelines generated by the American Institute of Architects (AIA), the American National Standards Institute (ANSI), the American Society for Testing and Materials (ASTM), the Institute of Electrical and Electronic Engineers (IEEE), the Joint Commission on Accreditation of Health Care Organizations (JCAHO), and other major code-making groups. The codes and standards set forth by the major national and international building code organizations are also discussed. The information and tools presented in this book provide guidance on how to comply with these codes and standards and, foremost, how to prepare for inspections and surveys conducted by many authorities, including JCAHO, the U.S. Occupational Safety and Health Administration (OSHA), local building officials and fire marshals, and other authorities having jurisdiction.

The systems addressed in this book include electrical, building construction, life safety, fire protection, and gas/vacuum systems. The specific parts of this book are as follows:

- **Part I**—Introduction (introduces health care facility codes and standards, health care facility organization, and the inspection and survey process)
- **Part II**—Electrical Systems and Equipment
- **Part III**—Building Construction and Life Safety (for various occupancies typically associated with health care facilities)
- **Part IV**—Fire Protection Systems
- **Part V**—Gas and Vacuum Systems and Equipment
- **Part VI**—Communication Among Health Care Facility Departments (including in-house interactions as well as those with outside agencies, record keeping, and training)

Each part describes relevant aspects of the system at hand and the requirements in the codes and standards related to materials, design, installation, use, testing, maintenance, inspections, and other system features. Complex issues surrounding compliance are discussed, and suggestions are presented for making the job of compliance easier.

Special features are included throughout the book that can assist the reader in understanding code compliance. These features include case studies of actual circumstances, frequently asked questions (FAQs) that address the most common concerns, and checklists of tasks that, if completed successfully, will help a facility pass inspections and receive or maintain certifications. Definitions extracted from NFPA codes, as well as references, are included at the end of the book.

This book was written to assist health care facility architects, designers, planners, engineers, contractors, maintenance personnel, and administrators in understanding the codes, standards, and guidelines related to achieving fire and life safety in health care facilities and designing and running these facilities accordingly. Readers with nontechnical backgrounds will become familiar with these systems and key aspects of code compliance. Technical readers and others more familiar with the systems and codes will become more informed about common concerns and problems and will be able to apply the information to actual facility design, engineering, and management.

This book is intended to serve as an operational manual to enhance fire and life safety in health care facilities and promote effective communications regarding safety, from the design process to system use, preventive maintenance, testing, and repairs. This book is not a design handbook; thus, it does not address all elements of design for the various systems.

The text in this book is based on the editions of documents listed in the References section at the back of the book. For facilities built before the editions listed, readers need to learn what editions were in effect or being enforced at the time of construction, installation, or renovation.

This is the second edition of *Fire and Life Safety in Health Care Facilities*. As such, new editions of documents have been referenced and technical changes in the codes and standards covering health care facilities have been noted.

Acknowledgments

The authors wish to thank the many individuals who provided valuable input to this book from the inception of the first edition. These include Mark Allen, Beacon Medical Products; Robert Bornstein, Youville Hospital; John Bruner, M.D. (retired), Massachusetts General Hospital; David Lees, M.D., Georgetown University Medical Center; David Mohile, Medical Engineering Services; Gary Slack, P.E., Healthcare Engineering Consultants; Brad Nicodemus, B.S.N., M.S.; and the administrative and engineering staffs at Brookdale University Hospital and Medical Center, Brooklyn, New York.

Very special thanks are due to the technical staff at NFPA who again took time from their busy schedules to provide invaluable comments, suggestions, and advice on the content of the second edition of this book. To all of the following, a sincere expression of our appreciation: Robert Benedetti, Dennis Berry, Richard Bielen, Guy

Colonna, Mark Conroy, Ron Coté, Christian Dubay, Mark Earley, Dana Haagensen, David Hague, Gregory Harrington, James Lake, Theodore Lemoff, Milosh Puchovsky, Jeffrey Sargent, Joseph Sheehan, and Robert Solomon.

Finally, the authors wish to thank Brad Gray. He was more than a project manager for this book—the book was his idea. He put the team together to write it, got us started, and kept us on the right track.

About the Authors

Marvin J. Fischer, P.E., FASHE, FACHE

Marvin J. Fischer is a private consultant for health care facilities and the legal profession. He began his affiliation with NFPA in 1971, serving as chair of the Technical Correlating Committee for Health Care Facilities for almost 20 years. During this period, he was involved in the committee's work of consolidating 13 separate documents to create NFPA 99, *Standard for Health Care Facilities*. He has also served as chair of both the Technical Committee on Essential Electrical Systems (NFPA 76A, *Essential Electrical Systems in Health Care Facilities*, which has been incorporated into NFPA 99) and the Technical Committee on Emergency Power Supplies (NFPA 110, *Standard for Emergency and Standby Power Systems*, and NFPA 111, *Standard on Stored Electrical Energy Emergency and Standby Power Systems*). In 1992, he received the NFPA Committee Service Award and is now a life member of NFPA.

From 1972 to 1996, Fischer was Vice President for Facilities Planning and Engineering Services at Brookdale University Hospital and Medical Center. In this capacity, he was responsible for the departments of facilities planning, construction, plant operation and maintenance, biomedical engineering, communications, security, safety, fire protection, and environmental engineering. Previously, he was chief engineer and an associate with a consulting engineering firm in New York City.

Fischer has lectured at many symposiums on health care facility codes, design, construction, and safety and has held an adjunct appointment as a lecturer in the graduate program for Healthcare Administration at the State University of New York at Stony Brook. He has also served on code-making committees for many organizations and for the state of New York. He was appointed the representative of the American Society for Healthcare Engineering to the International Federation of Hospital Engineering for 1982 and 1983. The American Society for Healthcare Engineering presented him with the Engineer of the Year award in 1985.

Fischer's publications include *Designing Electrical Systems for Hospitals*, third edition (McGraw-Hill, 1985), and *Pocket Guide to the National Electrical Code*, eighth edition (Prentice Hall, 2005). Over the years, he has been a contributing author to many other health care publications.

Fischer is a licensed professional engineer, a certified clinical engineer (Emeritus), and a Fellow in both the American Society for Healthcare Engineering and the American College of Healthcare Executives. He holds a bachelor's degree in electrical engineering (B.E.E.) and a master's degree in business administration (M.B.A.).

Thomas W. Gardner, P.E.

Thomas W. Gardner is the Managing Director of the Atlanta and Miami offices of Schirmer Engineering Corporation. He has had a long association with NFPA. He served as the editor of the sixth and seventh editions of NFPA's *Health Care Facilities Handbook,* and he is currently a member of the Technical Correlating Committee on Health Care Facilities (NFPA 99, *Standard for Health Care Facilities*) and the Technical Committee on Health Care Facilities of NFPA *101®, Life Safety Code®.* Gardner is also a member of the executive board of the NFPA Health Care Section (HCS) and has served on the HCS Codes and Standards Review Committee since 1990.

Gardner has practiced a broad spectrum of fire protection engineering specialties, including life safety analysis, code consultation, fire resistance, fire modeling, JCAHO Statements of Condition, and system design. He has designed automatic sprinkler, standpipe, fire pump, foam fire protection, clean agent fire protection, and fire detection/alarm systems. From 1989 to 2003, he was the staff consultant representing the American Health Care Association (AHCA) providing on-call fire protection consulting to AHCA members, investigating fire losses, and representing AHCA in all fire protection matters involving NFPA and the model code organizations.

Gardner has also worked for the U.S. Fire Administration, Grinnell Fire Protection Systems, and two other consulting engineering firms. In addition, he has been a member of the volunteer fire service for 30 years and has served with three fire departments. He is currently a member and former captain of the Manassas Volunteer Fire Company in Virginia and a former volunteer fire marshal in Hyattsville, Maryland.

Gardner is a full member of the Society of Fire Protection Engineers (SFPE), a past chair of SFPE's Engineering Education Committee, past president of the SFPE Chesapeake Chapter, and current instructor of their *Sprinkler Design for Engineers* class. He is a registered professional engineer (fire protection engineering) in Alabama, Georgia, Louisiana, Maryland, Mississippi, Ohio, South Carolina, and Virginia. He completed his B.S. degree in general science from Fordham University in 1980 before earning a B.S. degree in fire protection engineering from the University of Maryland in 1986.

Burton R. Klein, P.E.

Burton Klein is president of Burton Klein Associates, a consulting firm specializing in health care facility fire and electrical safety issues and in seminars on NFPA 99, *Standard for Health Care Facilities.* During his 20-year tenure as Chief Health Care Fire Protection Engineer at NFPA, he served as staff liaison to the Technical Correlating Committee on Health Care Facilities and to the various technical committees involved in the development of NFPA 99.

Klein also served as staff liaison to the committees responsible for NFPA 110, *Standard for Emergency and Standby Power Systems*; NFPA 37, *Standard for the Installation and Use of Stationary Combustion Engines and Gas Turbines*; NFPA 53, *Recommended Practice on Materials, Equipment, and Systems Used in Oxygen-Enriched Atmospheres*; NFPA 170, *Standard for Fire Safety and Emergency Symbols*; and NFPA

115, *Standard for Laser Fire Protection*. He was also the executive secretary to NFPA's 3000-member Health Care Section and publisher of the Health Care Section's quarterly newsletter.

Previously, Klein was a medical electronics engineer at Tufts-New England Medical Center in Boston, where he created a medical electronics department, as well as participated in the creation and teaching of the first hospital-affiliated medical electronics technician program (between Tufts and Franklin Institute of Boston).

In addition to writing and editing the first five editions of NFPA's *Health Care Facilities Handbook,* Klein's publications include *The Healthcare Electrician* (coauthored with Dan Chisholm and Ron Smidt, Motor & Generator Institute, 1998), *Health Care Facility Planning and Construction* (coauthored with John Platt, American Institute of Architects, Van Nostrand Reinhold, 1984), *Introduction to Medical Electronics for Medical and Technical Personnel,* second edition (Tab Books, 1975), *Health Care Fire, Electrical and Life Safety Compendium* (Burton Klein Associates, 1999), ASHE *Electrical Standard Compendium* (American Society for Healthcare Engineering, 1999), and ASHE *Fire, Electrical & Life Safety Compendium* (American Society for Healthcare Engineering, 2002).

Klein is a member of the National Fire Protection Association and the American Society for Healthcare Engineering. He is also a member of the Technical Committee on Electrical Systems and the Technical Committee on Piping Systems (both in the NFPA 99 Project). He received a B.S. degree in electrical engineering from Tufts University in 1965, and an M.S. degree in engineering from The George Washington University in 1970.

James K. Lathrop, FSPE

James K. Lathrop is vice president of the firm Koffel Associates, Inc., a fire protection and code consulting firm. He works from their Connecticut office. He has had a long relationship with NFPA, as an employee for almost 18 years, first working for NFPA's Fire Investigations Department and later for the *Life Safety Code* Field Services Department, where he served as Chief Life Safety Engineer. During this period, he served as staff liaison and secretary to the Committee on Safety to Life (NFPA *101*) and, at one time or another, on all of its various subcommittees. Since leaving NFPA, he remains very active with the association. Lathrop is the former chair for the Committee on Residential Occupancies for the *Life Safety Code* and for *NFPA 5000®*, *Building Construction and Safety Code®*. Currently he chairs NFPA's Technical Committee on Residential Occupancies of the *Life Safety Code* and the Pyrotechnics Committee for NFPA 1123 and NFPA 1124 and serves on several other NFPA committees as well.

Lathrop has been an NFPA *Life Safety Code* seminar instructor since 1980 and has lectured on the *Code* around the world. He has also contributed to the technical content of numerous NFPA *Life Safety Code* seminar programs and is author of the current NFPA *Life Safety Code* seminars. Lathrop has also served as NFPA staff representative to the Board for the Coordination of Model Codes and to all three model building code groups. He continues this activity with ICC's *International Building Code* and now with the *Building Construction and Safety Code*.

Lathrop is the author or editor of numerous publications on life safety and health care facilities, including the second, third, fourth, and fifth editions of NFPA's *Life Safety Code® Handbook,* the first and second editions of *The Statement of Conditions Manual: A Comprehensive Guide to JCAHO Compliance* (Opus Communications), and the 1997 and 2000 editions of *Life Safety Code® Field Guide for Health Care Facilities* (Opus Communications). He has also been a contributing author to NFPA's *Fire Protection Handbook* and *Fire and Life Safety Inspection Manual.*

Lathrop is a full member of the Society of Fire Protection Engineering (SFPE). He has served in the fire services, both volunteer and paid, for 40 years and is currently assistant chief with the Niantic Connecticut Fire Department, where he started in 1964. He holds a B.S. degree in fire protection engineering from the University of Maryland.

Part I
Introduction

Understanding and promoting fire and life safety in health care facilities covers a lot of territory. Health care facilities contain a vast array of systems and equipment, not just those required to operate any type of facility but also those highly specialized and vital for the care of patients—people with a range of ambulatory conditions and abilities to care for their own needs. Health care facility architects, planners, and designers, as well as managers, operators, maintenance personnel, and clinical staff, have a huge responsibility for ensuring the safety of the patients in their care, as well as of visitors to the facility. Health care facilities must, therefore, meet a unique series of challenges for preventing electrical hazards, incorporating safety measures during construction and renovation, preventing fires, providing protection if fires occur, promoting life safety, and providing for the safe use of gas and vacuum systems. For each of these activities, codes and standards must be met that stipulate what materials must be used and how systems and equipment must be designed, installed, tested, operated, maintained, and repaired. Additional codes and standards cover staff training and essential record keeping.

Being knowledgeable about the myriad of relevant codes and standards is one major aspect of promoting safety in health care facilities. To aid in accomplishing that goal, Chapter 1, "Health Care Facility Codes and Standards," provides a list of the various relevant codes and standards, discusses the concept of multiple authorities having jurisdiction over a particular facility or system within it, and gives some suggestions for dealing with conflicting codes. Chapter 2, "Overview of Health Care Facility Organization," defines and differentiates the various institutions that cluster under the umbrella term *health care facility* and discusses how requirements vary from one type of facility to another. Chapter 3, "Inspections and Surveys," concentrates on how health care facilities can prepare successfully for code compliance inspections by authorities having jurisdiction and how they can prepare for surveys conducted by the Joint Commission on Accreditation of Healthcare Organizations (JCAHO), to receive JCAHO accreditation and perhaps additional sources of funding.

1
Health Care Facility Codes and Standards

Most professionals involved in designing, installing, using, and maintaining the electrical, medical, gas/vacuum, fire protection, and communication systems and equipment found in health care facilities are well aware of the numerous codes and standards that apply to this work. They are also well aware that codes and standards must be adhered to in incorporating life safety features into the structures themselves. However, being aware of the codes and standards is a far cry from knowing which specific codes and standards apply to a particular facility or, perhaps more importantly, how to comply with the codes and standards adopted by a particular legislative body. Yet this knowledge is essential for providing safe facilities and preparing for inspections and surveys that affirm facility safety and identify where improvements are needed.

NFPA defines a *code* as a standard that is an extensive compilation of provisions covering broad subject matter or that is suitable for adoption into law independently of other codes and standards. Codes typically detail minimum requirements with which a facility must comply. Most health care facilities must comply with numerous codes covering all aspects of building construction and the systems and equipment within the structure, many of which deal directly with fire and life safety.

NFPA defines a *standard* as a document, the main text of which contains only mandatory provisions using the word "shall" to indicate requirements and which is in a form generally suitable for mandatory reference by another standard or code or for adoption into law. Nonmandatory provisions shall be located in an appendix, annex, footnote, or fine-print note and are not to be considered part of a standard. Several applicable standards in this case cover manufacturing, inspecting, testing, and maintaining fire protection equipment, systems, and building components; installing, using, and maintaining emergency and standby power systems; testing the fire endurance of building construction and materials; and determining outlet connections for medical gas cylinders, to name a few. Many codes incorporate references to specific standards that must be followed for code compliance and vice versa. The codes and standards that a particular facility must use depend on several factors, such as the city, county, and state in which the facility is located; the date the building was constructed or renovated; the date the original system was installed; and the particular *authorities having jurisdiction* (AHJ) that are involved in the project (i.e., organizations, offices, or individuals responsible for enforcing the requirements of a code or standard or for approving equipment, materials, an installation, or a procedure).

Many professional societies and organizations have prepared voluntary codes, standards, and recommended practices that address fire and life safety in health care facilities. One example is IEEE Std 602-1996, *Recommended Practice for Electric Systems in Health Care Facilities*. Another example is the American Institute of Architects' (AIA's) *Guidelines for Design and Construction of Hospital and Health Care Facilities—White Book*. This standard is used as a reference code and standard by the federal government, many states, and the Joint Commission on Accreditation of Healthcare Organizations (JCAHO). It also has been adopted into law in many cases. Also, government agencies at the federal, state, and municipal levels have either written their own regulatory codes and standards or have adopted the voluntary codes and standards of other organizations—or in some cases they have done both. Although many of the codes and standards are widely adopted nationwide, each governmental jurisdiction has the authority to adopt different codes and standards into law and to amend existing codes and standards. Depending on the rules and regulations adopted by a particular locale, each health care facility must comply with a unique set of codes and standards. As a result, countless sets of rules governing health care facilities are in effect across the country.

To add to this confusion, more than one edition of a particular code or standard may apply to a particular facility, more than one authority can have jurisdiction over a facility, the codes and standards these authorities enforce might conflict, and the facility managers may not be sure which code or standard to follow. It is clear that those responsible for designing, building, and maintaining health care facilities must either be knowledgeable about the applicable codes or standards or have consultants who are.

This chapter describes the major organizations that write fire and life safety codes and summarizes many of the codes that, if adopted into law, can affect health care facilities. Some of the common problems and conflicts that facilities managers encounter when attempting to comply with codes and standards and how to address these problems are also addressed.

1.1 Fire and Life Safety Codes and Standards

The primary voluntary organizations that prepare fire and life safety codes and standards and the most applicable codes and standards they produce for health care facilities follow.

- **American Institute of Architects.** AIA is a professional society that promotes a "more humane built environment and a higher standard of professionalism for architects" (AIA, 2000). With assistance from the U.S. Department of Health and Human Services, AIA has prepared *Guidelines for Design and Construction of Hospital and Health Care Facilities,* commonly referred to as the *Guidelines*. These guidelines address minimum engineering design criteria for plumbing, medical gas, electrical, heating, ventilating, and air conditioning systems for hospitals; nursing facilities; freestanding psychiatric facilities; outpatient and rehabilitation facilities; mobile, transportable, and relocatable units; hospice care; assisted living; and adult day care facilities. AIA's *Guidelines,* although only guidelines, must be followed if the federal government has some jurisdiction during the construction of the facility or areas. As noted previously, many states and other organizations have adopted these guidelines into their codes and standards.

- **American National Standards Institute (ANSI).** ANSI is a nonprofit organization that administers and coordinates a private-sector voluntary standardization system. The institute itself does not develop the standards; rather, it establishes consensus for each standard among qualified groups. ANSI codes and standards have no legal status unless a legislative body adopts them into law. However, ANSI approves standards issued by other organizations if the standards were developed following principles of openness and due process and if consensus was achieved among those directly and materially affected by the standard. NFPA is an ANSI-accredited codes and standards organization, and all NFPA codes and standards are ANSI accredited.

- **American Society for Testing and Materials (ASTM).** ASTM develops standard test methods, specifications, practices, guides, classifications, and terminology in areas such as metals, paints, plastics, textiles, construction, medical services and devices, computerized systems, electronics, and many others. ASTM standards serve as a basis for manufacturing goods, procuring products, and generating some regulations. Some ASTM standards that relate to health care facilities are as follows:
  - ASTM E 84, which is the same as NFPA 255, *Standard Test Method for Surface Burning Characteristics of Building Materials*
  - ASTM E 119a, which is the same as NFPA 251, *Standard Test Methods for Fire Tests of Building Construction and Materials*
  - ASTM E 814, *Standard Test Methods for Fire Tests of Through-Penetration Fire Stops*
  - ASTM E 1537, *Standard Test Method for Fire Testing of Upholstered Furniture*
  - ASTM E 1590, *Standard Test Method for Fire Testing of Mattresses*

- **Compressed Gas Association (CGA).** CGA develops standards and recommendations for safe and environmentally responsible practices related to the manufacture, storage, transportation, distribution, and use of industrial gases, cryogenic liquids, and related products. Some of these codes and standards are as follows:
  - CGA P-2, *Characteristics and Safe Handling of Medical Gases*
  - CGA P-2.6, *Transfilling of Liquid Oxygen Used for Respiration*
  - CGA P-2.7, *Guide for the Safe Storage, Handling, and Use of Portable Liquid Oxygen Systems in Healthcare Facilities*

- **Gypsum Association (GA).** GA is an international trade association that represents gypsum board manufacturers and serves builders, contractors, architects, code officials, educational institutions, and the general public by providing technical information and assistance. One product used by health care facilities is GA-600, *Fire Resistance Design Manual*. This manual provides design, construction, and inspection information related to a wide range of fire-resistant gypsum systems, classified according to use and fire resistance.

- **Institute of Electrical and Electronic Engineers (IEEE).** IEEE writes rules often considered by professionals in the field to be recommended practices for the industry. These practices and guidelines address electrical systems in general as well as electrical systems particular to health care facilities. One of the better-known practices specific to health care facilities is IEEE Std 602, *Recommended Practice for Electric Systems in Health Care Facilities*. This recommended practice is commonly referred to as the *White Book*.

- **Joint Commission on Accreditation of Healthcare Organizations.** JCAHO has written performance standards for many types of health care facilities such as hospitals, pathology and clinical laboratories, and facilities that provide assisted living, home care, long-term care, behavioral care, and ambulatory care services, the most extensive being those for hospitals. JCAHO standards are followed on a voluntary basis. Facilities can request JCAHO accreditation, which is based on passing an extensive survey conducted by JCAHO surveyors. Although JCAHO accreditation and compliance are voluntary, it is often in a facility's best interest to obtain this accreditation. It is usually accepted by the federal government as evidence of compliance with Medicare and Medicaid regulations. As a result, most institutions that can do obtain JCAHO accreditation. The standards and the type of survey are continually being changed and updated. It is important that the facilities engineer or manager be aware of the latest edition and type of survey. The standard that is most important to the facilities engineer and manager is the *Management of the Environment of Care*.

- **National Fire Protection Association (NFPA).** NFPA is perhaps the most familiar of the organizations that write codes and standards. NFPA has as its mission "to reduce the worldwide burden of fire and other hazards on the quality of life by providing and advocating scientifically based consensus codes and standards, research, training, and education." NFPA publishes approximately 300 codes, standards, guides, and recommended practices, about 60 of which are relevant to health care facilities. These codes and standards have no legal status unless adopted as law by a legislative body having jurisdiction, such as the federal government or a municipal or state governing body. Numerous legislative authorities having jurisdiction around the country and around the world, however, have adopted NFPA codes.

    NFPA codes and standards consist of chapters and one or more appendixes or annexes. The chapters of the document include only mandatory provisions. The appendixes or annexes provide additional information that further explains the associated text or provides reference to other documents, all of this on an informatory basis. The appendixes or annexes are not part of the required provisions of the document. Any agency can make an annex or appendix mandatory by legislative act.

    NFPA codes and standards applicable to health care facilities are listed in Section 1.2.

- **Underwriters Laboratories (UL).** UL is a product safety testing and certification organization that writes close to 750 standards and evaluates more than 18,000

product types. Some UL standards and standards documents relevant to health care facilities are as follows:
- UL 9, *Standard for Fire Tests of Window Assemblies*
- UL 10B, *Standard for Fire Tests of Door Assemblies*
- UL 263, *Standard for Fire Tests of Building Construction and Materials*
- UL 498, *Standard for Attachment Plugs and Receptacles*
- UL 924, *Standard for Safety Emergency Lighting and Power Equipment*
- *UL Fire Resistance Directory*

The following, while not complete, is a list of some other testing agencies that are frequently used: Electrical Testing Laboratories (ETL), Omega Point Laboratories (OPL), and Warnock Hersey, all three of which are now part of Intertek Testing Services (ITL); Southwestern Research Institute (SwRI); and Factory Mutual (FM).

1.2 NFPA Codes and Standards That Apply to Health Care Facilities

Of the NFPA codes and standards that apply to health care facilities, some are more prevalent and widely used than others, as listed in Section 1.2.1. Other codes that can apply to health care facilities, if adopted, are listed in Section 1.2.2. Because this book is not a code or standard, no edition dates are specified here. The mandatory code or standard that references the following documents must specify the edition being referenced.

1.2.1 Major NFPA Codes and Standards Relevant to Health Care Facilities

- NFPA 1, *Uniform Fire Code*™, contains general fire safety regulations for code enforcement and administration.
- NFPA 13, *Standard for the Installation of Sprinkler Systems,* addresses the design and installation of automatic fire sprinkler systems and exposure protection sprinkler systems, including the character and adequacy of water supplies and the selection of sprinklers, fittings, piping, valves, and all materials and accessories.
- NFPA 25, *Standard for the Inspection, Testing, and Maintenance of Water-Based Fire Protection Systems,* provides information and requirements for inspections, testing, maintenance, and the handling of impairments of systems utilizing water as a method of extinguishment. These include sprinkler systems, standpipe and hose systems, fire service piping and appurtenances, fire pumps, water storage tanks, fixed water spray systems, foam–water systems, valves, and allied equipment.
- NFPA 70, *National Electrical Code*® *(NEC*®*),* details the requirements for the practical safeguarding of persons and property from hazards arising from the use of electricity. The *NEC* applies to all types of structures, although it includes specific requirements for electrical construction and installation in health care facilities. This code is the most commonly used electrical code, almost universally adopted by the governing legislatures in all jurisdictions in the United States.

- *NFPA 72®, National Fire Alarm Code®,* covers the application, installation, location, performance, and maintenance of fire alarm systems and their components. Although titled as a code, it is really a standard, as it tells how, not when.
- NFPA 99, *Standard for Health Care Facilities,* details the requirements for the use, performance, maintenance, and testing of electrical, gas/vacuum, and environmental systems and equipment in health care facilities; manufacturers' requirements; and requirements for specific types of facilities. Although the information is not part of the requirements, the standard also provides guidance on the safe use of high-frequency electricity in health care facilities.
- *NFPA 101®, Life Safety Code®,* is a widely accepted, consensus-based code for protecting occupants in new and existing structures from fire. As stated in the *Code,* its scope includes those construction, protection, and occupancy features necessary to minimize danger to life from fire, including smoke, fumes, or panic. The *Code* also establishes minimum criteria for the design of egress facilities so as to permit prompt escape of occupants from buildings or, where desirable, into safe areas within buildings. Other provisions included in the *Code* address protective systems, building services, maintenance activities, and other safeguards for adequate egress time or protection for people exposed to fire. The occupancies covered (i.e., a particular use or intended use for a building or portion of a building) that may apply to health care facilities include assembly, educational, day care, health care, ambulatory health care, detention and correctional, some residential categories, business, industrial, and storage.

1.2.2 Additional NFPA Codes and Standards Relevant to Health Care Facilities

A number of other codes and standards may be applicable to a particular health care facility, as follows:

- NFPA 10, *Standard for Portable Fire Extinguishers*
- NFPA 11, *Standard for Low-, Medium-, and High-Expansion Foam*
- NFPA 12, *Standard on Carbon Dioxide Extinguishing Systems*
- NFPA 12A, *Standard on Halon 1301 Fire Extinguishing Systems*
- NFPA 14, *Standard for the Installation of Standpipe and Hose Systems*
- NFPA 17, *Standard for Dry Chemical Extinguishing Systems*
- NFPA 17A, *Standard for Wet Chemical Extinguishing Systems*
- NFPA 20, *Standard for the Installation of Stationary Pumps for Fire Protection*
- NFPA 22, *Standard for Water Tanks for Private Fire Protection*
- NFPA 24, *Standard for the Installation of Private Fire Service Mains and Their Appurtenances*
- NFPA 30, *Flammable and Combustible Liquids Code*
- NFPA 31, *Standard for the Installation of Oil-Burning Equipment*
- NFPA 33, *Standard for Spray Application Using Flammable or Combustible Materials*
- NFPA 37, *Standard for the Installation and Use of Stationary Combustion Engines and Gas Turbines*
- NFPA 45, *Standard on Fire Protection for Laboratories Using Chemicals*

- NFPA 51B, *Standard for Fire Prevention During Welding, Cutting, and Other Hot Work*
- NFPA 53, *Recommended Practice on Materials, Equipment, and Systems Used in Oxygen-Enriched Atmospheres*
- NFPA 54, *National Fuel Gas Code*
- NFPA 55, *Standard for the Storage, Use, and Handling of Compressed Gases and Cryogenic Fluids in Portable and Stationary Containers, Cylinders, and Tanks*
- NFPA 58, *Liquified Petroleum Gas Code*
- NFPA 70E, *Standard for Electrical Safety in the Workplace*
- NFPA 75, *Standard for the Protection of Information Technology Equipment*
- NFPA 80, *Standard for Fire Doors and Fire Windows*
- NFPA 80A, *Recommended Practice for Protection of Buildings from Exterior Fire Exposures*
- NFPA 82, *Standard on Incinerators and Waste and Linen Handling Systems and Equipment*
- NFPA 85, *Boiler and Combustion Systems Hazards Code*
- NFPA 88A, *Standard for Parking Structures*
- NFPA 90A, *Standard for the Installation of Air-Conditioning and Ventilating Systems*
- NFPA 90B, *Standard for the Installation of Warm Air Heating and Air-Conditioning Systems*
- NFPA 91, *Standard for Exhaust Systems for Air Conveying of Vapors, Gases, Mists, and Noncombustible Particulate Solids*
- NFPA 92A, *Standard for Smoke-Control Systems Utilizing Barriers and Pressure Differences*
- NFPA 92B, *Standard for Smoke Management Systems in Malls, Atria, and Large Spaces*
- NFPA 96, *Standard for Ventilation Control and Fire Protection of Commercial Cooking Operations*
- NFPA 97, *Standard Glossary of Terms Relating to Chimneys, Vents, and Heat-Producing Appliances*
- NFPA 99B, *Standard for Hypobaric Facilities*
- NFPA 101A, *Guide on Alternative Approaches to Life Safety*
- NFPA 105, *Standard for the Installation of Smoke Door Assemblies*
- NFPA 110, *Standard for Emergency and Standby Power Systems*
- NFPA 111, *Standard on Stored Electrical Energy Emergency and Standby Power Systems*
- NFPA 115, *Standard for Laser Fire Protection*
- NFPA 170, *Standard for Fire Safety and Emergency Symbols*
- NFPA 204, *Standard for Smoke and Heat Venting*
- NFPA 211, *Standard for Chimneys, Fireplaces, Vents, and Solid Fuel–Burning Appliances*
- NFPA 214, *Standard on Water-Cooling Towers*
- NFPA 220, *Standard on Types of Building Construction*
- NFPA 221, *Standard for High Challenge Fire Walls, Fire Walls, and Fire Barrier Walls*

- NFPA 232, *Standard for the Protection of Records*
- NFPA 241, *Standard for Safeguarding Construction, Alteration, and Demolition Operations*
- NFPA 251, *Standard Methods of Tests of Fire Resistance of Building Construction and Materials* (also known as ASTM E 119a)
- NFPA 252, *Standard Methods of Fire Tests of Door Assemblies*
- NFPA 253, *Standard Method of Test for Critical Radiant Flux of Floor Covering Systems Using a Radiant Heat Energy Source*
- NFPA 255, *Standard Method of Test of Surface Burning Characteristics of Building Materials* (also known as ASTM E 84)
- NFPA 257, *Standard on Fire Test for Window and Glass Block Assemblies*
- NFPA 259, *Standard Test Method for Potential Heat of Building Materials*
- NFPA 265, *Standard Methods of Fire Tests for Evaluating Room Fire Growth Contribution of Textile, Coverings on Full Height Panels and Walls*
- NFPA 286, *Standard Methods of Fire Tests for Evaluating Contribution of Wall and Ceiling Interior Finish to Room Fire Growth*
- NFPA 291, *Recommended Practice for Fire Flow Testing and Marking of Hydrants*
- NFPA 418, *Standard for Heliports*
- NFPA 424, *Guide for Airport/Community Emergency Planning*
- NFPA 600, *Standard on Industrial Fire Brigades*
- NFPA 701, *Standard Methods of Fire Tests for Flame Propagation of Textiles and Films*
- NFPA 704, *Standard System for the Identification of the Hazards of Materials for Emergency Response*
- NFPA 780, *Standard for the Installation of Lightning Protection Systems*
- NFPA 801, *Standard for Fire Protection for Facilities Handling Radioactive Materials*
- NFPA 1600, *Standard on Disaster/Emergency Management and Business Continuity Programs*
- NFPA 1961, *Standard on Fire Hose*
- NFPA 1962, *Standard for the Inspection, Care, and Use of Fire Hose, Couplings, and Nozzles and the Service Testing of Fire Hose.*
- NFPA 2001, *Standard on Clean Agent Fire Extinguishing Systems*
- NFPA 5000, *Building Construction and Safety Code*

The American Society for Healthcare Engineering publishes a *Fire, Electrical and Life Safety Compendium* that lists many of the preceding documents and the sections where material relevant to a topic can be found.

1.3 Creating and Revising Codes and Standards

When faced with the often difficult task of having to comply with a complicated code, a facility manager may find it helpful to know the process that goes into generating a code. Many of the code-making organizations formulate their codes via committees made up of volunteer professionals in their respective fields.

NFPA has a specific set of rules governing how codes and standards are written and revised. During each revision cycle for an NFPA document (a cycle is the time period between revisions, usually 3 to 5 years), thousands of hours of labor, not just a day or two of hearings, are dedicated to the task of formulating the document. Each document is written by a group of volunteer committees under the NFPA consensus development process. For example, currently 19 technical committees (Code Making Panels) and one correlating committee write the *NEC*. Eight technical committees and one technical correlating committee write NFPA 99, *Standard for Health Care Facilities*. For the *Life Safety Code,* each of the 13 technical committees is responsible for addressing an occupancy, a series of occupancies (such as the Technical Committee on Health Care Occupancies), or a technical area of expertise (such as the Technical Committee on Means of Egress). There is also a technical correlating committee. By employing this multicommittee approach, the committees can tap experts in each field without the disadvantage of having to deal with large, unwieldy committees.

NFPA regulations dictate that each committee must be balanced by interest, with no more than one-third of a committee representing any single interest group (i.e., insurance, user, manufacturer, consumer, enforcer, labor, installer/maintainer, research/testing, and special expert). As such, the Technical Committee on Health Care Occupancies, for example, will have members representing the AIA, the American Hospital Association (AHA), the American Health Care Association (AHCA), the U.S. Department of Veterans Affairs (USDVA), the U.S. Department of Health and Human Services (USDHHS), the International Fire Marshals Association (IFMA), and many more groups.

Because NFPA, JCAHO, and other standards organizations revise their documents and accreditation manuals according to different time schedules, the documents can, in a short time, contain outdated references to other codes and standards (e.g., an organization revises its Code A, which includes a reference to the latest edition, by year, to Code B. A year later, the organization revises Code B. Thus, Code A now carries an outdated reference). Also, when legislatures adopt codes or standards as law, they cannot legally specify the "latest edition" of that code or standard but must instead adopt a year-specific document and then readopt the document into the law when it is revised.

It is important to be aware of the edition of each code and standard that is being enforced by each authority having jurisdiction. Checking the latest edition of each code helps facility managers keep abreast of all code changes and helps determine if compliance is required or if existing facilities are exempt from meeting newer standards. For example, a facility built under the auspices of the 1997 *Life Safety Code* may not have to comply with all the new requirements in the 2000 edition of the *Life Safety Code,* although some retrofits may be necessary under the requirements for existing buildings.

1.4 Cross-Referencing in Codes and Standards

In some cases, a published code or standard cross-references other codes and standards. For example, if federal funds are used during construction, the federal government requires compliance with the AIA *Guidelines,* which, in turn, contains references to NFPA codes and other standards. NFPA 99 references NFPA 50 on bulk oxygen containers. On occasion, other organizations, such as ANSI or AIA, adopt IEEE rules and include IEEE standards within their own documents. Many codes, such as the *National Electrical Code,* are cross-referenced by such codes as the IEEE *National Electrical Safety Code* and apply to all types of structures and occupancies, including health care facilities.

The federal government requires compliance with the AIA *Guidelines* for a facility to receive federal funding for construction. Also, the federal government accepts the accreditation of JCAHO as proof that an institution meets certain requirements. This acceptance, therefore, eliminates the need for the government to conduct separate surveys or inspections.

1.5 Authorities Having Jurisdiction

The existence of codes and standards does not independently ensure facility compliance. It is the role of federal, state, and local governments to adopt codes and standards into laws and regulations with which facilities must comply and the role of authorities having jurisdiction to enforce these codes and standards. NFPA's definition of the term *authority having jurisdiction* is "the organization, office, or individual responsible for enforcing the requirements of a code or standard, or for approving equipment, materials, an installation, or a procedure."

NFPA documents use the term *authority having jurisdiction* in a broad manner because jurisdictions and approval agencies vary, as do their responsibilities. Where public safety is primary, the AHJ may be any one of the following federal, state, local, or other regional department or person:

- Fire chief or fire marshal
- Chief of a fire prevention bureau
- Labor department or health department
- Building official
- Electrical inspector
- JCAHO surveyor
- Any number of pertinent federal agencies
- Another statutory authority

For insurance purposes, an insurance inspection department, a rating bureau, or another insurance company representative may be the AHJ. In many circumstances, the property owner or his or her designated agent assumes the role of the AHJ; at government installations, the commanding officer or departmental official may be the AHJ.

At some level, facilities engineers and administrators and general contractors for new construction may be AHJs. For example, every time a person in one of these roles accepts the installation of a fire door or wall covering or approves an emergency plan, that person is in fact "approving" an item required by the code and is, therefore, acting as an AHJ.

Many have argued that JCAHO should not be considered an AHJ. This opinion would be a mistake, as it is important that JCAHO surveyors have the authority to approve the various structures, systems, and equipment within a facility, especially alternative methods and materials under the equivalency provisions of most codes.

The term *AHJ* is not common to the major model building codes in the United States. Most of these codes use the term *code official* or *building official* to refer to the AHJ. The basic concept, however, is the same, in that a public official is enforcing the building code.

It may not always be clear to facility administrators, engineers, or architects which authority has jurisdiction over a facility or system. In many cases, one or more persons or groups may share the duties of the AHJ. For example, the fire marshal may have jurisdiction over certain items, such as fire alarms or fire pumps, whereas other responsibilities are shared or divided among building officials, electrical inspectors, or municipal engineers or planners.

A confusing circumstance can occur when one AHJ accepts a particular form of compliance but another AHJ finds that compliance unacceptable. Because AHJs are responsible for interpreting the requirements and how to apply the codes and standards, when more than one AHJ is involved in a project, they may have different interpretations and requirements for health care facilities.

1.6 Conflicts and Duplications Among Codes and Standards

In addition to not always knowing who all of their AHJs are, facilities may also be confused about which codes they must follow. With so many codes written on the same subject and revised at different times, some of the codes inevitably overlap and conflict, and it might not be clear which *edition* of a code must be followed, such as when most of a building must comply with the code in effect at the time of its construction but a renovated area must comply with the code in effect at the time the newer work was done.

An example of various AHJs applying different editions of the same document to one facility is when JCAHO enforces the 2000 edition of the *Life Safety Code*, whereas the state government enforces an earlier edition of the same document. A hypothetical example of a code conflict is when Code A mandates that an item be mounted at a height of 30 to 34 in. (0.762 to 0.863 m) and Code B prohibits the installation of such a device between 30 and 34 in. The following case study illustrates another example of a code conflict.

Code Conflict

Here is one example of a situation in which a code conflict was avoided. The code conflict was related to patient room door closers. Several years ago, NFPA's Health Care Subcommittee of the Committee on Safety to Life (now referred to as the Technical Committee on Health Care Occupancies) discussed including a provision in NFPA *101* to prohibit self-closers on patient room doors. The committee had

(Continued)

several reasons for prohibiting self-closers and believed this prohibition should be in the *Code*. At that time, however, most of the major model building codes required closers on patient room doors. If the prohibition of self-closers had been accepted, a true conflict would have arisen or automatic closers would have had to be used. As a result, the subcommittee did not accept the prohibition. Since that time, changes have been made in the model building codes to allow the use of doors without closers in health care facilities under certain conditions. Refer to Chapter 13 for more information on building codes.

Another example of conflicting codes is that not all legislative bodies have adopted the *National Electrical Code,* which is a widely adopted code. Some municipalities have adopted the *NEC* with certain changes, and others follow their own electrical code, which may contain requirements different from those of the *NEC*. As a result, conflicts can arise between a state that has adopted the *NEC* and a local government that has issued its own electrical code. Another example of a conflict is that the state health department may require all health care facilities to conform to the state's code (often the *NEC*), *as well as* all local codes, but these codes conflict.

In many cases, what is referred to as a conflict may in fact be only an additional requirement. For example, if Code A requires smoke detection in all patient sleeping rooms and Code B does not, no conflict exists because smoke detection devices can be installed to comply with Code A without their installation affecting compliance with Code B. According to Code B, the devices may be considered unnecessary but are not a source of conflict.

1.7 Resolving Code Conflicts

Facilities personnel can take several steps to avoid some of the problems associated with code or standard conflicts and having multiple AHJs. It is recommended that those responsible for complying with the regulations know the specific edition of each applicable document being enforced by their AHJs. This information can often be obtained by calling the AHJ, but the facility engineer should always ask for such information in writing as well. Copies of this documentation, which should be on site at all times, will be helpful when inspectors are on site and attempt to apply a different code or edition (either because they are not aware of the correct code or edition applicable to that facility or because they wish to use another code with which they are more familiar). Facility managers should also ensure that all applicable AHJs review projects as early as possible.

Because jurisdictions update requirements periodically, facility engineers should reconfirm the applicable requirements on a regular basis to keep their information current. If facility managers cannot determine applicable regulatory information, they should seek professional assistance to protect the facility from costly misapplication of code or standard requirements.

Conflicts between the codes enforced by more than one AHJ can be fully resolved only through discussions with the AHJs, who are in a position to make exceptions to the rules and grant waivers. When codes or standards cover the same material, the enforcing agencies involved usually require a facility to follow the more stringent of the two codes. Determining which code is the more stringent one, however, can pose its own set of problems, as illustrated in the following case study.

> **Determining the More Stringent Code**
> One of the electrical codes used by one city for a health care facility restricted the branch circuits to 15-A (ampere) loads. The state, which adopted the *NEC,* allowed health care facilities to use 15- or 20-A branch circuits. The state health department also required conformance with all local laws, which in this case would permit only 15-A branch circuits. In this case, the more stringent regulation would be the 15-A restriction.

Another case study shows one example of how a code conflict was resolved.

> **Resolving a Code Conflict**
> A federal-level code (which has since been changed) required operating suites to have twist-lock receptacles, whereas the state applied the newly revised edition of the applicable code, which did not require twist-lock receptacles. The conflict was resolved when the AHJ that required the twist-lock receptacles decided that either twist-lock or regular receptacles would be acceptable. The federal agency later adapted the newly revised edition and solved the conflict.

Although complying with the most stringent code is the most common way of solving differences, it often results in little or no increase in life safety at significant expense, because *both are considered safe*. An alternative way to solve conflicts, which should be discussed with the AHJs involved, is to allow the more specific code to prevail. For example, the *Life Safety Code* could prevail in situations where the conflicts relate to egress issues, and the building code could prevail when the conflicts relate to construction issues (refer to Part III). Fire alarm requirements, sprinkler thresholds, and similar items could then be based on the more stringent code—and the application of the *best* of the two codes with little duplication.

As demonstrated, a major responsibility for health care facility designers, planners, engineers, and administrators is to determine a facility's AHJs, establish an active working relationship with them, and become familiar with all the codes and standards enforced by each AHJ that apply to their facility. It is also essential to learn the areas where jurisdiction among AHJs overlaps, how the codes themselves overlap and perhaps conflict, and how to resolve all conflicts.

In addition to contributing to the design of a facility, knowing what each AHJ considers compliance can contribute to formulating the flow of activities related to installing, using, testing, and maintaining all systems and equipment, activities that must constantly take place within health care facilities. Implementing these activities as smoothly as possible not only facilitates the process of preparing for an inspection but also assists in providing a safer facility for patients, staff, and visitors.

2

Overview of Health Care Facility Organization

All types of health care facilities and the various areas and departments found within these facilities need to incorporate a range of electrical, gas and vacuum, life safety, and fire protection features set forth in the codes and standards to reduce hazards and create a safe environment for patients, staff, and visitors. The National Fire Protection Association (NFPA), American Institute of Architects (AIA), and the Joint Commission on Accreditation of Healthcare Organizations (JCAHO) are three code-making organizations that have created codes and standards for health care facilities and specific areas within such facilities. These groups categorize health care facilities in a number of ways, not so much according to the "name on the door" but according to the activities that take place or are intended to take place within a facility or the specialized areas within a facility. The nature of health care and the unique circumstances that occur within these facilities make the type of safety measures required for each type of facility and area unique.

Administrators, clinical department staff, building engineers, support service personnel, and other staff are all affected by a building's safety measures and play a role in maintaining the safety of a facility. Each department has its responsibilities, which can include approving or not approving new equipment purchases or facility repairs and renovations; assisting with the design of a facility's safety features; installing, testing, and maintaining equipment; training staff; using equipment properly; and handling emergency situations.

This chapter describes types of health care facilities addressed in the safety codes, as well as areas and departments in large health care facilities for which the codes include fire and life safety requirements. Roles for key personnel who maintain facility safety are also addressed.

2.1 Health Care Facility Categories

According to NFPA 99, *Standard for Health Care Facilities,* health care facilities are buildings or portions of buildings in which medical, dental, psychiatric, nursing, obstetrical, or surgical care is provided. Health care facilities include, but are not limited to, hospitals, nursing homes, limited care facilities, clinics, medical and dental offices, and ambulatory care centers, whether permanent or movable. Because of changes in the medical care industry, the technical committee responsible for NFPA

99 has modified the categorization of health care facilities listed in NFPA 99. Over the years, however, the various code-making organizations have not developed a common set of names for the various types of health care facilities. These differences can be seen by comparing the names used in NFPA documents to those used by AIA and JCAHO.

NFPA 99 includes requirements for hospitals, nursing homes, limited care facilities, hyperbaric facilities, freestanding birthing centers, home care, and other health care facilities. Other health care facilities cover those facilities not specifically covered in designated chapters and include such facilities as medical and dental offices, ambulatory care centers, and all others not specifically covered. AIA's *Guidelines for the Design and Construction of Hospital and Health Care Facilities* includes minimum program, space, engineering, and equipment guidelines for all clinical and support areas of hospitals, nursing facilities, psychiatric hospitals, outpatient facilities, and rehabilitation facilities. JCAHO accredits hospitals, managed care organizations, pathology and clinical laboratories, and organizations that provide assisted living, home care, long-term care, behavioral health care, and ambulatory care services. Article 517 in NFPA 70, *National Electrical Code®* (*NEC®*), refers simply to *health care facilities with a definition included*. NFPA 101®, *Life Safety Code®*, is categorized according to building occupancies, or the ways in which a building or an area within a facility are used or intended to be used. The main types of facilities are described next. Building occupancies are described in Section 2.2 of this chapter.

2.1.1 Hospitals

It may be obvious to the public what a *hospital* is, but the code-making bodies have several definitions for this type of facility. The most common definitions are found in NFPA 99 and NFPA *101*, which are similar.

NFPA *101* and NFPA 99 define a *hospital* as a building or portion thereof used on a 24-hour basis for the medical, psychiatric, obstetrical, or surgical care of four or more inpatients. Whenever NFPA *101* or NFPA 99 uses the term *hospital*, it includes general hospitals, psychiatric hospitals, mental hospitals, tuberculosis hospitals, children's hospitals, and any such facilities providing inpatient care.

2.1.2 Nursing Homes

The definition of *nursing home* is basically the same in NFPA *101* and NFPA 99 and is as follows: a building or portion of a building used on a 24-hour basis for the housing and nursing care of four or more persons who, because of mental or physical incapacity, might be unable to provide for their own needs and safety without the assistance of another person. According to the codes, nursing homes include nursing and convalescent homes, skilled nursing facilities, intermediate care facilities, and infirmaries in homes for the aged.

2.1.3 Limited Care Facilities

The definition for a *limited care facility* is the same in NFPA 99 and NFPA *101*. A limited care facility is a building or portion of a building used on a 24-hour basis for the housing of four or more persons who are incapable of self-preservation because of age, physical limitation resulting from accident or illness, or mental limitations such as mental retardation or developmental disabilities, mental illness, or chemical dependency.

2.1.4 Psychiatric Hospitals

There are two locations of concern with respect to the treatment of psychiatric patients. The first location includes areas within a general hospital environment. The second location is a hospital devoted only to the treatment of psychiatric patients. In either case, the patients fall into three general categories. They may be inpatients requiring overnight accommodations and all of the ancillary facilities. They may be day patients who come only for treatment and then go home. They may also be emergency patients who require immediate treatment and then an evaluation as to whether they should be admitted as inpatients or sent home and required to return for additional treatment.

The added electrical, gas/vacuum, life safety, and fire protection safety features and work procedures that must be incorporated into psychiatric hospitals and psychiatric areas within general hospitals take account of the mental and emotional conditions of these patients. AIA has separate guidelines for psychiatric hospitals. Various federal and state construction codes also often require all areas set aside for the specific care of psychiatric patients to include special safeguards such as lockable fire alarms, ceiling tiles that cannot be removed, and tamper-proof electrical outlets, windows, and window screens.

2.1.5 Other Health Care Facilities

In an effort to categorize the requirements for other health care facilities in terms of the procedures performed within a facility rather than in terms of the title of the facility, beginning with the 1999 edition of NFPA 99, facilities other than hospitals, nursing homes, or limited care facilities were grouped into a new chapter titled "Other Health Care Facilities." This chapter addresses safety requirements for facilities, or portions thereof, that provide diagnostic and treatment services to patients in health care facilities other than hospitals, nursing homes, limited care facilities, home care, hyperbaric facilities, and freestanding birthing centers. Some examples of facilities that are covered under this new chapter are ambulatory health care centers, clinics, and medical and dental offices. Although not explicitly stated, a major distinguishing characteristic of "other health care facilities" is that there are no provisions for treating or housing patients for more than 24 hours (e.g., there are no sleeping beds).

2.1.6 Freestanding Birthing Centers

The NFPA Technical Correlating Committee on Health Care Facilities added requirements for freestanding birthing centers to the 1999 edition of NFPA 99 in the form of a new chapter. Before the addition of these codes, members of the NFPA Technical Correlating Committee for Health Care Facilities and representatives from birthing center associations had many discussions as to whether or not birthing centers were health care facilities. In many cases, authorities having jurisdiction were not sure what codes and standards were applicable. Conformance requirements varied in different parts of the country.

What these groups determined was that birthing centers vary in size, structure, and location. Some centers are small, rural, converted one-family wood frame houses, whereas others are either contiguous to or are part of a hospital in an urban setting. Even though birthing centers screen their patients to lower the risk of complicated births, they do concede that the unexpected can happen. Many of these birthing centers have also stated that their facilities are not staffed by MDs or RNs and do not provide "medical care."

The Technical Correlating Committee concluded its research and discussions by invoking requirements for the safe use of electrical and gas equipment, and for electrical, gas, and vacuum systems if used for the delivery and care of infants in freestanding birthing centers.

It is likely that these requirements will change over time as new information is received and the practices at these facilities change. It is advisable that facility maintenance personnel keep abreast of the changes made to the standards and check with federal, state, and local jurisdictions, as some have already written codes covering these types of facilities.

2.1.7 Hyperbaric and Hypobaric Facilities

NFPA 99 defines *hyperbaric* as pressures above atmospheric pressure. Hyperbaric areas are used to treat conditions related to burns, decompression sickness (i.e., the "bends"), some diabetic conditions, and other medical conditions for which increased oxygen saturation is deemed appropriate. These chambers can contain atmospheres at higher than normal atmospheric pressure, a partial oxygen content greater than 0.21 atm absolute, or both. The code classifies these chambers as follows:

- Class A—Human, multiple occupancy
- Class B—Human, single occupancy
- Class C—Animal, no human occupancy

Hypobaric chambers are areas where pressures are below atmospheric pressure. With the advent of space travel, chambers to simulate pressures below atmospheric pressure were built and used. When NFPA determined that hypobaric chambers were no longer used for medical purposes, it separated the requirements for hyperbaric and hypobaric chambers, with hyperbaric facilities remaining in NFPA 99 and hypobaric facilities being placed in a separate document, NFPA 99B, *Standard for Hypobaric Facilities*.

2.1.8 Home Care

As medical care is changing, patient home care is increasing. It is not out of the question today for patients to use medical equipment at home that was once limited to hospital use. As most homes and private living quarters are not constructed to conform to health care facility requirements, the use of medical equipment at home can create hazards, particularly in terms of the electrical system and oxygen use. Information presented in the 2005 edition of NFPA 99 on electrical and gas equipment for home care may provide some guidance.

2.2 Building Occupancies in Health Care Facilities

The *Life Safety Code* categorizes buildings according to occupancies, that is, how a facility or the areas within a facility are used or are planned to be used, rather than according to procedures. Typical occupancies within a health care facility include health care, ambulatory health care, business, industrial, and storage occupancies. The *Life Safety Code* includes requirements for how to incorporate fire and life safety measures into the structure of each occupancy, so that each area provides fire and life safety for occupants. Refer to Chapter 13 for more information on building occupancies.

2.2.1 Health Care Occupancies

The *Life Safety Code* defines a *health care occupancy* as an occupancy used for purposes of medical or other treatment or care of four or more occupants where such occupants are mostly incapable of self-preservation because of age, physical or mental disability, or because of security measures not under the occupants' control. The vast majority of health care facilities, including hospitals, nursing homes, and similar facilities, are categorized as health care occupancies, although the facilities themselves may contain many other occupancies, such as business and storage occupancies.

2.2.2 Ambulatory Health Care Occupancies

Ambulatory health care occupancies are buildings or portions of buildings used to provide services or treatment simultaneously to four or more patients that provide, on an outpatient basis, one or more of the following: (1) treatment for patients that renders the patients incapable of taking action for self-preservation under emergency conditions without the assistance of others (2) anesthesia that renders the patients incapable of taking action for self-preservation under emergency conditions without the assistance of others (3) emergency or urgent care for patients who, because of the nature of their injury or illness, are incapable of taking action for self-preservation under emergency conditions without the assistance of others. Beginning with the 2000 edition of the *Life Safety Code,* ambulatory health care occupancies are addressed in their own chapter.

2.2.3 Detention and Correctional Occupancies

Detention and *correctional occupancies* include facilities that provide housing for four or more individuals who are mostly incapable of self-preservation because of security measures not under the occupants' control. Health care facilities typically include this occupancy when they are part of a correctional facility, such as a prison hospital, or when a general hospital has a detention wing or section. Most psychiatric hospitals are not considered to be detention and correctional occupancies or facilities, but they sometimes can be, as in the case of facilities for the criminally insane. Having a few secure rooms in an emergency department does not normally qualify as a detention and correctional occupancy.

2.2.4 Residential Occupancies

Residential occupancies provide sleeping accommodations other than for health care or for detention and correctional purposes. Under the *Life Safety Code,* there are five different residential occupancies: one- and two-family dwellings, lodging and rooming houses, hotels and dormitories, apartment buildings, and board and care facilities. In some circumstances, health care facilities own, manage, and are responsible for residential occupancies, such as when housing medical students or staff or relatives of inpatients, or when they own such buildings for investment purposes. Staff sleeping areas, such as on-call rooms, are also residential occupancies.

2.2.5 Business Occupancies

A *business occupancy* is any building or portion thereof used for keeping accounts and records and transacting business other than the sale of products. Typical examples of business occupancies in health care facilities are administrative buildings or areas in medical office buildings that are not ambulatory health care occupancies (i.e., do not have four or more patients rendered incapable of self-preservation). Outside of health care occupancies, business occupancies are probably the most common occupancies in health care facilities.

2.2.6 Assembly Occupancies

Assembly occupancies are buildings or portions thereof used for gathering 50 or more people for the purposes of amusement, eating, drinking, worship, deliberation, or similar functions. Assembly occupancies in health care facilities could include the cafeteria, conference rooms, and auditoriums. Most hospital chapels are too small to be categorized as assembly occupancies.

2.2.7 Day Care Occupancies

Day care occupancies are buildings or portions thereof used for the care, maintenance, and supervision of four or more clients on less than a 24-hour basis. Clients can be adults as well as children. Day care facilities can operate 24 hours per day, but no one client can receive care for 24 hours. Designers and managers of staff day care centers located in health care facilities should pay careful attention to these requirements.

2.2.8 Industrial Occupancies

Industrial occupancies are buildings or portions thereof wherein anything is made, manufactured, processed, assembled, mixed, packaged, finished, decorated, or repaired. In a health care facility, this area could be a boiler plant, maintenance shop, or similar area or building.

2.2.9 Storage Occupancies

Storage occupancies are areas where anything is stored. Examples in health care facilities include parking garages, central warehouses, or large storage areas within a facility. Small storage rooms and closets do not warrant a separate occupancy classification.

2.2.10 Other Occupancies

The *Life Safety Code* categorizes several other areas into occupancies that are allowed in health care facilities but are not typically found. *Mercantile occupancies* include buildings or portions thereof used for the sale of products on display, such as grocery or department stores. Typical gift shops in health care facilities are usually not large enough to be categorized as mercantile occupancies, although shopping malls, as found in some facilities, are large enough. *Educational occupancies* are not typically found in health care facilities either, because the *Life Safety Code* limits educational occupancies to those providing education up to the twelfth grade for six or more persons, and health care facility classrooms are most often used for adult education. In pediatric areas of a hospital, the location where the children are given education instruction by the local board of education usually does not have more than six persons at a time. There are cases, however, where true educational classrooms are provided in health care facilities and the requirements of the *Life Safety Code* for educational occupancies must be observed.

2.3 Typical Areas Within Health Care Facilities

Many of the code-making organizations have further categorized health care facilities into specific areas, based on the types of procedures and activities that take place within these areas and the typical equipment used. NFPA 99 includes requirements for general patient care areas, critical patient care areas, wet locations, anesthetizing locations, labor and delivery suites, and other specialized areas. NFPA 99 states that the governing body of each facility determines the classification of each area (general care area, critical care area, and wet location).

Although it might be tempting to downgrade the classification of a specific area to save these costs, the procedures performed in a specific area must determine its proper classification and the protective systems and equipment that must be installed. For example, if major surgery is sometimes performed in a room, but patients are more routinely brought into this room for less intensive planned surgical procedures, the room must still conform to the requirements of an operating room. Patient safety is the first consideration.

Although it is important not to underclassify an area, it is equally as important economically not to overclassify an area. For example, special equipment does not need to be installed in operating rooms where procedures that would constitute a wet condition are never intended to be performed, provided other operating rooms are classified to operate this way and will always be available when needed. Each facility must determine the equipment and systems most useful to its needs. No matter how costly it is to build or operate an area, these areas are inspected or surveyed based on the chosen designations. Thus, to prepare for and pass inspections, the area must comply with the classification it has been assigned by the governing body.

2.3.1 Patient Care Areas

Although the definitions of such terms as *patient care areas, patient bed locations,* and *patient care vicinity* vary slightly between NFPA 70 (the *NEC*) and NFPA 99, it is useful for facility designers and engineers to become familiar with these terms. *Patient care areas*, as defined in NFPA 99, include any portion of a "health care facility" wherein patients are intended to be examined or treated. *Patient care vicinities* are spaces within a location intended for examination and treatment that extend 6 ft (1.8 m) beyond the normal location of the bed, chair, table, treadmill, or other device that supports the patient during examination and treatment. Patient care vicinities also extend to 7 ft 6 in. (2.3 m) above the floor. A *patient bed location* is the location of a patient sleeping bed, or the bed or procedure table of a critical care area.

According to the *NEC* and NFPA 99, patient care areas are further categorized as general care or critical care areas. Either category can also be classified as a wet location if it also meets the definition in NFPA 99 for a "wet location." The governing board of the hospital or its designees are responsible for categorizing these areas. Conducting medical procedures in areas not classified to handle such procedures places staff and patients at risk of serious electrical shock.

2.3.1.1 General Care Areas

The definition of a *general care area* differs editorially between the *NEC* (Article 517) and NFPA 99. Both documents consider patient bedrooms, examining rooms, treatment rooms, clinics, and similar areas to be general care areas where patients can come into contact with ordinary appliances such as nurse call systems, electric beds, examining lamps, telephones, and entertainment devices. The *NEC* adds that in general care areas, patients may also come into contact with electrical medical devices, including heating pads, electrocardiographs, drainage pumps, monitors, and other equipment.

2.3.1.2 Critical Care Areas

A *critical care area*, which is another type of patient care area, is one where invasive procedures are intended to be performed and patients are to be connected to line-operated patient-care electrical appliances. According to NFPA 99, critical care areas include intensive care units, coronary care units, angiography laboratories, cardiac catheterization laboratories, operating and delivery rooms, postanesthesia recovery rooms, and emergency rooms. A definition of *invasive procedure* has been added in the 2005 edition of NFPA 99:

> Any procedure that penetrates the protective surfaces of a patient's body (i.e., skin, mucous membrane, cornea) and that is performed with an aseptic field (procedural site). [Not included in this category are placement of peripheral intravenous needles or catheters used to administer fluids and/or medications, gastrointestinal endoscopies (i.e., sigmoidoscopies), insertion of urethral catheters, and other similar procedures.]

2.3.2 Wet Locations

NFPA 99, 2005 edition, defines *wet locations* as follows:

> The area in a patient care area where a procedure is performed that is normally subject to wet conditions while patients are present including standing fluids on the floor or drenching of the work area, either of which condition is intimate to the patient or staff.

It is important to note that the definition refers to patient care areas only, that is, areas where patients are examined and treated. The definition of *wet location* in Article 100 of the *National Electrical Code* was originally intended for non–health care patient settings, such as kitchens, laundry rooms, and bathrooms.

The obvious reason for special precautions and requirements in wet locations is to minimize the hazard of electric shock (see Part II). What is not as obvious is how to determine which areas of a facility fall into this category with respect to patient care areas. The classification of a "patient care" wet location by the governing body depends entirely on the procedures that are intended to be performed within the room and whether these procedures will create wet conditions for patients and staff. The following case study provides an example of how to delineate wet locations.

Delineating Wet Locations

When planning a new operating suite, one institution divided the rooms into three types of areas. One area was designated for simple procedures that would not necessarily fall under the definition of a wet location. The second area, designed for major surgery, was classified as a wet location. The third area was planned as a swing room that could be used for any type of surgery. In this case, too, the room was classified as a wet location. This delineation led to substantial savings for the facility, because only two of the three areas had to conform to the requirements for a wet location.

2.3.3 Inhalation Anesthetizing Locations

The *NEC* and NFPA 99 define an *anesthetizing location* in a similar way, as follows: "any area of a health care facility that has been designated to be used for the administration of nonflammable inhalation anesthetic agents in the course of examination or treatment, including the use of such agents for relative analgesia." NFPA 99 also addresses the use of flammable inhalation anesthetic agents.

Inhalation anesthetizing locations that are designated for use of *nonflammable* anesthetizing agents only are classified as *other-than-hazardous (classified)* locations. *Flammable* anesthetizing agents have been virtually eliminated from use in the United States, however. In 1996, the Technical Committee on Anesthesia Services decided to eliminate the associated requirements from NFPA 99 because no committee member was aware of any facility in the United States that administered flammable anesthetics, or any medical school that provided instruction on how to administer flammable anesthetics. Because committee members were aware that flammable anesthetizing agents might still be used outside the country, information on the necessary precautions was transferred to a newly created annex in NFPA 99, to be retained for technical informational purposes only. This information was moved to Annex E starting in the 2002 edition of NFPA 99.

According to the *NEC*, any location in which flammable anesthetics are employed is considered to be a Class 1, Division 1, hazardous location from the floor level to a point 5 ft (1.52 m) above the floor. An area or room in which flammable anesthetics are stored is considered a Class 1, Division 1, hazardous location from the floor to the ceiling.

2.3.4 Operating and Labor and Delivery Suites

Governing bodies designate *operating* and *labor and delivery suites* as wet locations or inhalation anesthetizing locations, depending on the types of procedures that are intended to take place within these areas. In some instances, depending on the anesthetizing agent or other chemical that might be used, these areas might be categorized as hazardous locations. Each room must comply with the requirements of the classification it has received.

2.3.5 Birthing Rooms

In many institutions, *birthing rooms* have replaced labor rooms and delivery suites and can act as labor, delivery, and postpartum rooms. In some cases, birthing rooms are located in hospitals, and in other cases, they are located in freestanding birthing centers. Procedures taking place in the latter facilities would not be included as "medical procedures" if no medical or nursing personnel were present.

The governing body of the facility must determine any special classifications for birthing rooms, depending on what procedures or agents will be used in a particular room. It is advisable to consult with the authorities having jurisdiction regarding the classification of a particular birthing area. A birthing room can look much like a home and less institutional, but it still must meet all safety requirements.

2.3.6 Psychiatric Areas

As mentioned previously, many general hospitals contain areas set aside for the specific care of psychiatric patients. Refer to Section 2.1.4 for more information on the delineation of psychiatric areas and some safeguards that must be taken.

2.3.7 Radiology Areas

Radiology covers the use of x-ray, magnetic resonance imaging, positron emission tomography, computed tomography scan, angiography, mammography, and other similar procedures used for diagnosis and treatment. In many cases these areas are known as imaging centers. Although the codes do not categorize these areas with a separate designation, NFPA 99 [Sections 13.4.1.2.6.2, 20.3.2.1.2(1), E.4.6.3, 2005 edition] and the *NEC* (Article 517, Part V of the 2005 edition) include specifications for installing x-ray equipment.

2.3.8 Laboratories

Laboratories are buildings, spaces, rooms, or groups of rooms intended to involve investigation, diagnosis, or treatment activities and procedures in which flammable, combustible, or oxidizing materials are to be used. One of the major problems for facilities engineers responsible for designing, installing, or maintaining laboratories is determining which code has jurisdiction. NFPA 45, *Standard on Fire Protection for Laboratories Using Chemicals,* typically has primary jurisdiction over all laboratories, including those in health care facilities. Chapter 11, "Laboratories," in NFPA 99 has additional and more stringent requirements than NFPA 45 for laboratories located in health care facilities that house inpatients, or that perform procedures that make outpatients incapable of self-preservation in an emergency. Charts in Annex C (Section C.11) of NFPA 99 should be reviewed.

2.3.9 Hazardous Locations

The most obvious health care facility areas that are classified as *hazardous* are those in which flammable agents are used. Clinical and research laboratories in which certain chemical agents are used are classified as hazardous locations. In some instances, just a specific area within a laboratory might receive this classification. Less obvious hazardous locations include alcohol storage units and gas storage tank areas. (See Section 2.2.10 for the *Life Safety Code* definition of hazardous area.) It thus becomes necessary to learn what agents are intended to be used or stored in a particular area.

2.3.10 Emergency Departments

Emergency departments can fall into many categories, from small general care emergency rooms to major trauma centers. The type of emergency department present in a facility dictates the space used, the procedures performed, and the type and complexity of the equipment used.

Major trauma centers can contain operating rooms, cardiac care rooms, radiology departments, hazardous material detoxification rooms, pediatric areas, and specialty areas such as ophthalmology, dentistry, obstetrics, and so forth. The same codes and standards that apply to these areas in the main facility apply to the corresponding areas in the emergency department.

2.3.11 Information Systems

Information system areas generally are areas in which computerized data processing takes place. With the use of computerized diagnostic and treatment equipment; financial, medical, and engineering record-keeping systems; and nurses' station equipment, it is clear that the use of computers in health care facilities has increased dramatically. Some information systems are small scale and local, such as those used in a specific piece of equipment or in a single department. Other information systems are networked facility-wide, such as administrative information relevant to all staff. Other information systems are shared among departments. A common example of a shared system is when the medical department must provide the billing department with information on the specific procedures performed on various patients, the amount of medication used, and the like.

With so many reports and data generated for patient care, some health care facility information systems are devoted strictly to patient care information. For example, newer computerized telemetry systems allow facility staff, physicians, and nurses to enter information into the patient chart at a bedside computer using a small transmitting device. These systems eliminate the need for staff to carry the chart over from the nurses' station. Permanent copies can be printed out at a later time. Another type of patient care information system is a portable computer carried from bedside to bedside. The data can be transmitted into the permanent records either manually or automatically. In some cases this equipment is wireless and the information is transferred immediately to the patient records.

Many health care facilities have departments dedicated to information systems management. These facilities typically have large computer equipment rooms with central data processing systems that interconnect the entire facility. In some cases, the central system stores and processes information from individual departments. For example, many facilities place the maintenance and preventive maintenance programs on the facility's main system, which can then be accessed by both the facilities engineering department and the clinical engineering department.

2.4 Typical Health Care Facility Departments

Although a local dental office typically has a simple organizational structure, large health care facilities usually have more complex administrative levels. Each department plays a role in providing a safe facility, from the administrators and financial officers, who make major decisions on equipment, renovations, and repairs, to support service personnel, who must properly use plugs and outlets and report equipment malfunctions.

2.4.1 Administration

Through the chief executive officer (CEO) and the chief operating officer (COO), the administration department is responsible for the ultimate functioning of the entire health care facility. Major decisions regarding equipment purchases, adding staff, and approving or not approving renovations and repairs, to name a few, are all made by administrators. Although organizational structures and functional designations vary among facilities, in many institutions, department heads and directors of service report to various administrators, who in turn report to the COO, who reports to the CEO. The size of the institution determines the number of administrative layers that exist.

2.4.2 Clinical Departments

The primary role of the clinical departments—medical, surgical, neonatal, pediatric, psychiatric, dental, nursing—is to attend to the needs of the patient. To accomplish this task, all clinical personnel must be thoroughly trained on how to properly operate medical equipment and some aspects of specialized electrical power systems (i.e., essential, emergency, and isolated power supplies; see Chapters 8 and 9), handle equipment malfunctions, and report equipment problems. Some members of the clinical staff typically are part of a facilities emergency preparedness team, which JCAHO-accredited facilities must coordinate and maintain (see Chapter 36). In this capacity, clinical staff assist patients, other staff, and visitors during emergency situations and provide emergency care, emergency clinical procedures, and necessary movement within a building, as well as communicate with outside emergency workers and officers.

2.4.3 Engineering Departments

In some institutions, the engineering function is divided into two departments, facilities engineering and clinical engineering (the latter also known as medical engineering or biomedical engineering), whereas in others, both functions are within one department. Facilities engineers play a major role in ensuring the safety of a facility on a daily basis. They are frequently involved in the design stage of a facility, providing input on including the proper safety measures. They are involved in installing, testing, and maintaining all equipment and systems and in preparing facilities for inspections and surveys. They also are often involved in training personnel on the safe use of equipment and systems.

The role of clinical engineers is to ensure the safe installation and use of all medical equipment. They should be involved in purchase decisions and in testing, installing, and maintaining the equipment. They also generally train clinical staff on the proper use of medical equipment. In some cases they assist in research and design equipment for special purposes.

Some facilities differentiate responsibilities at the point where building utility systems interface with, or make connection to, medically related equipment. This could be electrical receptacles and junction boxes, gas outlets and vacuum inlets, or water connections.

2.4.4 Other Departments

Other health care facility departments include information management, environmental services, security, and food service. Personnel in these departments play a role in operating equipment properly and reporting repair needs (lightbulbs, broken sockets, and the like). Information management personnel must safely use much electronic equipment associated with a vast amount of wires and cords (see Chapter 10). Food service personnel must operate all kitchen equipment properly and must know how to operate kitchen-hood fire protection systems (see Chapter 24). Security guards and telephone operators must know the different alarms connected to areas within their command and know how to respond to any of them quickly.

3
Inspections and Surveys

To make sure facilities are in compliance with their codes and standards, authorities having jurisdiction (AHJs) routinely inspect health care facilities. Some of the AHJs that a facility might encounter are federal, state, county, and local agencies Although not required to do so, many facilities also elect to meet the standards set by the Joint Commission on Accreditation of Healthcare Organizations (JCAHO) and undergo its surveys to receive JCAHO accreditation. Every aspect of a facility related to safety, including its equipment, systems, and records, can be checked, and often is, by such organizations as the U.S. Occupational Safety and Health Administration (OSHA) and by state and local health departments. The level of detail of the inspections varies widely, depending on the inspecting agency, the size of the facility, and the type of inspection.

The frequency with which surveys and inspections take place also varies. Many AHJs, such as local fire marshals and insurance industry inspectors, visit facilities once per year. JCAHO surveys laboratories every 2 years and other health care facilities every 3 years and conducts interim inspections as well, from 6 months to 30 months after the biennial or triennial inspections have been completed. Although JCAHO used to provide about 30 days' notice of upcoming biennial or triennial surveys, this is no longer being done and surveys are now unannounced. Other AHJs do likewise. Many inspectors and surveyors make unannounced visits. OSHA inspectors typically make such visits. Although JCAHO surveyors in the past provided 24-hour notice of interim surveys, since adopting the Random Unannounced Survey Policy, they are now arriving at facilities totally unannounced for those surveys as well.

Thus, all health care facilities must be prepared to undergo an inspection (or survey, if JCAHO accredited) and, if not in compliance with applicable codes and standards, must take steps to bring the systems, equipment, and records into compliance. Facilities managed in such a way that compliance is incorporated into the daily routine are more apt to pass unannounced surveys or inspections. Preparing for announced events is usually hectic (i.e., it can be very taxing on an engineering department's normal schedule as well as that of the entire health care facility) and often takes more than 30 days.

Preparing for health care facility inspections has similarities around the country. Because of the wide range of codes and standards in place, however, and the numerous authorities having jurisdiction that establish their own methods of conducting inspections, there is great diversity regarding what each facility specifically must do to prepare for such events. JCAHO standards and accreditation are nationwide, and the organization conducts surveys in a consistent way across the country for each type of facility, depending on a facility's scope of services. Thus, knowing how to prepare for a JCAHO survey can assist facility engineers and managers in preparing for inspections in general and can bring a facility into compliance and ready for any inspection. Reviewing JCAHO's survey procedures and expectations, as well as typical steps followed during other types of inspection, can assist facility engineers and managers in preparing for all these important events. Many seminars are given to instruct on how to prepare a facility for a JCAHO survey.

3.1 JCAHO Standards and Accreditation

JCAHO standards address a facility's level of performance in specific areas, both clinical and nonclinical. These standards set achievable levels of performance that can be expected for various activities, systems, and building features, all of which relate to providing high-quality and safe health care. JCAHO provides evaluation and accreditation services for more than 15,000 health care organizations in the United States based on these performance standards, including the following types of facilities (JCAHO, 2005):

- General, psychiatric, children's, rehabilitation, and critical access hospitals
- Health care networks
- Home care organizations
- Nursing homes and other long-term care facilities
- Behavioral health care facilities
- Ambulatory care facilities
- Clinical laboratories

From the onset of facility planning, health care facility planners, designers, and managers should base system installations, operations, testing, maintenance, management, and tracking and staff training on meeting JCAHO standards and the content of surveys, as well as all other codes and standards that must be met. Having JCAHO accreditation can assist a facility in many ways.

JCAHO provides a list of the health care facilities it has accredited. The organization also makes in-depth individual performance reports available for many accredited organizations surveyed after January 1, 1996, which provide details on a facility's performance and how it compares with similar organizations. JCAHO accreditation may thus also enhance community confidence in a facility and medical staff recruitment.

3.2 Preparing for Surveys and Inspections

The steps taken to prepare for an inspection or survey depend on the type of survey or inspection being performed, the purpose of the survey or inspection, and the authority having jurisdiction that is performing the inspection or survey. The easiest way to determine what is expected of a facility is to call the appropriate authority and request any available relevant documentation. Another source is the local or state health care engineering society. The American Society for Healthcare Engineering (ASHE) may be another source of information, although it usually does not provide information for specific localities. For some facility engineering departments, compliance with the applicable NFPA standards is sufficient.

JCAHO publishes accreditation manuals for each type of health care facility it accredits to assist a facility in bringing it into compliance with JCAHO standards. These manuals contain the standards, the intent of the standards, and the scoring methods and guidelines used by the surveyors. They also include worksheets and checklists that help the facility understand what JCAHO expects of the facility and what an inspector will likely look at during an inspection. The requirements of JCAHO's entire manual should be familiar to the appropriate engineering staff. Some of the suggestions made by JCAHO for preparing for surveys are discussed here (JCAHO, 2000).

Facilities that wish to be newly accredited should allow 9 to 12 months of preparation *before* the survey date and be in compliance with the standards for at least 4 months before an initial survey. During this time, facilities should take the following steps:

- Carefully review the standards
- Conduct a self-assessment of current processes
- Make improvements where needed
- Develop new policies or processes, as necessary
- Train staff accordingly

Facilities being resurveyed must show a 12-month "track record" of compliance. The following tasks can assist facilities in preparing for initial and triennial surveys:

- Read the applicable JCAHO accreditation manual and standards to determine their relevance. Facilities being resurveyed must also make sure a facility is in compliance with any standards that have become effective since the last survey was conducted. JCAHO surveyors look for organization-wide approaches to meeting the standards; thus, compliance should not be limited to specific departments or disciplines. Because the standards and the type of survey are continually being changed and updated, ensure that the latest standards are being used and become familiar with the type of survey being conducted.
- Attend seminars and meetings sponsored by professional associations to understand the standards better. Network with colleagues who have recently gone through the accreditation process or call counterparts in other organizations. A newer way to gain assistance is to obtain it online from professional association bulletin boards and chat rooms. These activities can be ongoing.
- Develop programs to educate staff about how to comply with standards and new systems. JCAHO surveyors interview staff members to determine their level of understanding about a facility's processes.

- Use JCAHO guidelines to conduct a mock survey and identify and document any areas that are not in full compliance. During the mock survey, review policies and procedures, audit records, and interview staff. It is helpful to conduct mock surveys regularly to help judge a facility's efforts at making improvements. Facilities that do not have time or expertise to conduct mock surveys can hire consultants to conduct these surveys. Review the results of the mock survey with appropriate staff, develop a plan to correct the problems, and establish realistic priorities and a schedule for making improvements. Improving a facility's existing methods and processes on an ongoing basis can prevent problems from becoming serious.
- Review expectations with staff. Discussing what is likely to take place during the survey can help relieve some anxieties about the survey process.

For laboratories, which must be resurveyed every 2 years and show a 2-year track record of compliance with the requirements for proficiency testing, preparing for a survey can begin with meeting the recommendations of the most recent survey and either submitting a written progress report or undergoing a focused survey to show compliance. In addition to the steps listed previously, other steps that laboratory managers can take to prepare for surveys are as follows (JCAHO, 2000).

- Make sure the following activities are up to date: in-service training, continuing education, staff orientation, proficiency testing, and preventive maintenance for laboratory equipment. During this time, focus on maintaining skills and staff performance levels and discuss how to evaluate the services provided by contractor laboratories. Testing policies and procedures must be strictly followed.
- Continually gather required information on all lab services and certificates, which must be available during the survey.
- Maintain key documents that the surveyor may be interested in reviewing, including general and quality control policies, procedures, specimen collection manuals, departmental records for patients, and applicable licenses. Independent laboratories should also post a notice about public information interviews at all laboratory locations and inform communities about the process through classified ads or other means.

3.3 When an Inspector or Surveyor Is on Site

JCAHO surveyors and other inspectors evaluate a facility's compliance against its codes and standards that are applicable to that facility. For each code or standard, the inspector or surveyor evaluates a facility based on a series of questions regarding the standard and a physical examination of the records and physical facility and observations.

The results of JCAHO surveys indicate the level of a facility's actual performance, not what a facility is capable of. That is, a facility may have to show the surveyor the actual physical plant and the records of all tests, maintenance, and preventive maintenance of the electrical, gas/vacuum, and fire protection systems; medical equipment; and building life safety features. With the new type of survey, patient tracking, the surveyor will look at the physical plant and documentation as a product of the patient care rather than as a separate item. Larger facilities (generally more than 200 beds) will have a specific JCAHO surveyor who will evaluate the facility's compliance with the *Life Safety Code*. Many other inspectors evaluate similar criteria.

Being aware of what typically takes place when an inspector or surveyor arrives on site may help a facility prepare for such an event and perhaps even ease some tensions regarding the process. Note, however, that the length of time an inspector or surveyor reviews a facility varies widely, depending on the schedule of the inspector, what is being inspected, and the size of the facility. Inspectors can be on site from several hours to a week or more.

The first event that usually takes place is that the inspector requests to meet with the highest-ranking official of the facility or with his or her representative to explain the reason and intentions for the visit.

If this is a JCAHO survey, it will begin with an opening conference on the morning of the first day of the survey. This will introduce the survey team to the key members of the health care facility, along with the tentative survey schedule. The CEO determines who attends the conference on behalf of the facility. This conference will generally be limited to 30 minutes.

If the inspection is specific to a facility problem, the appropriate facility personnel will then become involved. If this is a JCAHO survey, after the opening conference the survey team might review certain specific documents that will orient them to the way the hospital functions. These documents focus on the hospital's performance and include committee minutes; reports of measurement and assessment activities; reports to the medical staff, hospital committees, and the governing body and the bylaws; planning documents; and other evidence of performance. The latest type of survey then follows patient tracking. The surveyors follow a patient's care from the time that patient entered the hospital to the time of discharge. While doing this tracking, the surveyor will come into contact with medical equipment as well as the facility. The surveyors talk to patients and staff and are exposed to all facets of the facility. In this way, all of the records might be reviewed. Inspectors also often ask to review the safety management plans for fire protection, hazard management, security, infection control, utilities, essential electrical systems, and equipment. Thus, during an actual survey, all records and system documents must be readily available for surveyors, and the latest copies of all safety management plans must be "on the shelf" at all times.

Most AHJs have different methods of conducting an inspection. The inspection (other than JCAHO) is usually for a specific part of the facility or area of concern such as by a fire marshal, electrical or plumbing inspector, or OSHA. Usually most state department of health inspections are all-inclusive, similar to JCAHO. Therefore, an inspector might only want to inspect a facility and not look at records, or vice versa. How the inspector proceeds depends on the type of inspection and the inspecting authority.

After a walk-through, a surveyor or inspector may wish to review some records again, based on any questions raised during the site visit. At the close of an inspection or survey, the inspector usually provides information regarding any issues uncovered during the site visit. JCAHO surveyors provide an exit conference with administrators and department heads. At the exit conference the staff has the opportunity, if they wish, to provide any additional information that they think will help to avoid receiving a citation. Any remaining questions might be handled either by facility personnel promising to provide information or the inspector calling back for additional information. A written report is always provided, although it can take some time to receive. If the facility believes that any of the JCAHO surveyor findings are wrong, it must be worked out at this time.

3.4 Record Keeping

Keeping accurate records of all tests, preventive maintenance, and maintenance tasks that take place throughout the year is essential to be prepared for unannounced, as well as impending, surveys or inspections. To comply with JCAHO and other record-keeping requirements, a facility must develop and maintain such documents as management plans, minutes of meetings, maintenance records, test logs, and use and inspection records for many types of equipment. Facilities must also keep diagrams of various in-house systems.

When keeping records, no harm is done in indicating that a malfunction was found during a routine test or preventive maintenance procedure. The important thing is to document when and how it was corrected. Refer to Chapters 12 and 34 for more details on record keeping.

Part II
Electrical Systems and Equipment

Electrical systems in health care facilities have been expanding with the ever-changing needs of the medical and allied health care professions. New types of equipment are found in almost all clinical departments, and the need for many facilities to upgrade to accommodate not just new types of equipment but a greater amount of equipment places great demands on a facility's electrical service and systems. Systems specific to health care facilities, such as essential electrical systems and isolated power systems, are associated with unique requirements and hazards.

As in all facilities, the proper functioning of electrical systems depends on items such as wire and cable, switchboards and panelboards, grounding and disconnect devices, and receptacles and switches. And as in all facilities, the electrical system is often taken for granted and assumed to be both reliable and safe. In health care facilities, though, staff and patients count on the electrical service not only for turning on a standard light but also for operating sensitive medical equipment and systems that assist in both diagnosis and treatment and that often sustain life itself. Additionally, this reliable source of electric power is needed for the proper functioning of the communications and signaling systems that also help to provide a safe and secure environment.

Part II of *Fire and Life Safety in Health Care Facilities* focuses on the electrical systems and equipment in health care facilities and issues specific to the installation, use, testing, maintenance, and record keeping of electrical systems in such facilities. All chapters cover general and specific requirements for these systems in accordance with the relevant codes and standards. Specific topics are as follows:

Chapter 4: The importance of the continuity of electrical service, systems, and equipment in health care facilities and the hazards that are present

Chapter 5: Electrical supply at the entrance to health care facilities, including the service arrangement, equipment, and service equipment layout; transformers; and voltage and current conditioners

Chapter 6: Components of the electrical distribution systems—wire and cable, raceways, wiring methods, switchboards, panelboards, and motor control centers—as well as system protection and coordination, grounding, and circuiting

Chapter 7: Electrical system devices and lighting, such as receptacles, switches, and ground-fault circuit interrupters, in addition to lighting of specific areas, energy conservation, and maintenance

Chapter 8: Essential electrical systems, including system types and wiring; emergency generators; switching and load shedding; energy converters; alarms, indicators, and safety devices; and testing, maintenance, and record keeping

Chapter 9: Isolated power supplies and other special systems, including alarms and communication systems

Chapter 10: Requirements for specific areas and facilities, such as general care areas, critical care areas, wet locations, inhalation anesthetizing locations, operating and labor and delivery room suites, radiology, laboratories, hazardous locations, psychiatric units, emergency departments, hyperbaric and hypobaric facilities and areas, and information system areas

Chapter 11: Medical equipment manufacturing, performance standards, and testing requirements; guidelines for purchase order specifications; facility equipment testing; ongoing equipment tests; equipment installation and grounding; and medical equipment training

Chapter 12: Testing of service switches and panelboards, wiring, disconnect devices, loads, and patient care areas; general electrical system maintenance; and record keeping

4

Electrical Hazards in Health Care Facilities

In today's world, everyone assumes that the source of electricity will be safe and reliable. We also expect the delivery of power to resume quickly when it does fail, so as not to disrupt the typical activities of our daily lives. In health care facilities, however, the safety and sometimes the very lives of patients depend on power being available at all times and, in some cases, without interruption at all. Additionally, the electricity must be clean of transients and harmonics (power quality) and delivered at the proper voltage for the proper operation of the equipment it is powering. Equipment must also be installed in a safe manner so that it will not cause harm to the user by means of a fire, shock, or thermal damage.

To say the least, providing such safe, reliable, and uninterrupted service is extremely important. It is also, most often, very complex, because of the many unique features these systems must incorporate to meet these objectives. A myriad of codes and standards provide guidance for incorporating these features to create highly dependable and safe electrical service in health care facilities. These codes address all phases of service, including system design, installation, maintenance, testing, and use. Equally as important as meeting the requirements of the codes by providing reliable and safe service is the facility planners' addressing facility-specific resources, needs, and constraints. To service certain types or volumes of equipment, more power may be needed in a specific area of a facility than the amount specified in the codes, or a facility might have to limit an activity in a certain area because of conflicts between available service and certain equipment specifications. Meeting all of these requirements and taking actions above and beyond the codes aim to accomplish one thing: the minimization of electrical hazards in health care facilities.

4.1 Types of Electrical Hazards in Health Care Facilities

According to NFPA 99, *Standard for Health Care Facilities*, electrical hazards in health care facilities include loss (interruption) of power, electric shock, fire and explosion, and thermal injury. These hazards are caused by human error, defects or deterioration in equipment (mechanical failure), or acts of nature. Human error entails improper design, improper installation, or improper use of electrical components and equipment. Mechanical defects include defective wiring, a defective component, deteriorating insulation, and the like. Acts of nature are events such as earthquakes, hurricanes, tornadoes, floods, explosions, and storms.

4.1.1 Lack of Continuity of Electrical Service

Most pieces of medical and diagnostic equipment are sophisticated devices that need a reliable source of electric power for tests to be completed in a timely fashion with accurate results. Without a reliable source of power, results of tests such as magnetic resonance imaging (MRI) and positron emission tomography (PET), both of which are used, for example, to diagnose brain dysfunctions; angiography, which is used to diagnose heart and vascular conditions; and sonography, which helps diagnose a range of medical conditions, are often unclear and full of artifacts. Supplying low voltage to older pieces of medical equipment that may not have built-in voltage regulation features can also cause inaccurate laboratory results, such as unclear X-rays or X-rays with a large number of artifacts. Transients in a line (i.e., surges in voltage that can last from a fraction of a cycle to a few cycles), created either on or off site by starting up large motors, arcing grounds, or switching loads, capacitors, or faults, for example, can also cause malfunctions in medical and office equipment. Many laboratory devices perform numerous tests that, once started, cannot be interrupted.

Treatment areas in health care facilities also need reliable sources of power. Modern surgery requires sophisticated electrically powered equipment with no interruption of power. The electrical equipment used in cardiology, for example, must be relied upon to sustain life by artificially circulating blood, stimulating and regulating heart action, and preventing suffocation by removing excess body fluids. Many other types of electrical life support and life safety equipment are used throughout hospitals, especially in patient rooms. Cardiac monitors, respirators, suction pumps, and feeding pumps, all of which may serve one patient simultaneously, require uninterrupted electric power to avoid harm to the patient.

Data processing and information systems in health care facilities require uninterruptible electrical power, not just for frequent access to patient records but also to function properly. Even a momentary disruption of power can cause errors or malfunctions (see Chapter 10). An uninterruptible power supply (UPS) ensures that there will be enough power for a sufficient amount of time for an orderly shutdown of the system or until emergency generators restore power. Additionally, health care facilities need properly functioning signaling and communications systems to support the accurate and timely diagnosis and treatment of patients, as well as to provide a safe environment for patients, staff, and visitors. Signaling systems include fire and medical gas alarms. The communication system can be a simple nurse call system or a complicated voice and radio paging system.

4.1.2 Shock and Thermal Injury Hazards

Examples of potential shock hazards in health care facilities are numerous. Operating rooms, critical care areas, laboratories, and physical therapy areas can contain many exposed, electrically conductive surfaces that increase the risk of injury by electrical shock and thermal injury to patients and staff. Patients are vulnerable to electric shock as a result of the use of catheters, electronic treatment and diagnostic equipment, and similar devices that are connected directly to the patient and in some cases invade the body. An excess amount of electrical current flowing through a conductor or device causes that wire or device to become hot, which can cause thermal injury if it comes into contact with the patient or staff. Wet locations (i.e., patient care areas, such as some operating rooms, that are normally subject to wet conditions while patients are present) also pose significant electrical shock hazards to patients and staff.

4.1.3 Fire and Explosion

Electrical fires and explosions can be caused by malfunctions or improper use or installation of the electrical system. Overcurrent devices, ground fault circuit interrupters, and grounding systems are some of the means that are used to prevent this. Annex B in NFPA 99, 2005 edition, contains some information on the nature of hazards and electrical hazards.

4.2 Avoiding Electrical Hazards

There are three basic aspects to providing a safe electrical environment in health care facilities. They include the following:

- Properly designing and installing all systems. This means that the facility should conform not only to the code but also to the needs of the patients and staff. It is essential to acquire input from appropriate personnel and consultants during all phases of any project—design and construction, purchase, installation, and testing.
- Properly maintaining and testing the facility and equipment, not only to pass inspections or surveys but also to ensure that the facility and equipment are safe at all times
- Properly educating staff through in-service courses on the proper use of the equipment and on all facets of the electrical distribution system

It may not be possible to avoid hazards caused by acts of nature, but facilities can be prepared to handle hazards should they arise. Some key points to keep in mind are as follows.

4.2.1 System Design, Installation, and Use

Creating and using electrical systems that avoid all potential hazards involves designing, installing, maintaining, and using electrical systems that provide and distribute the necessary amounts of power for the required size loads without causing excessive voltage drops, transients, ground faults, or other system shortcomings. These aims should be factored into the design of new facilities, as well as when conducting electrical renovations that will place additional loads on existing systems. Some of the main factors to consider when designing, installing, and using electrical systems in health care facilities to meet the codes and prevent hazards are as follows.

4.2.1.1 Providing the Proper Amount of Power

To ensure that the electrical supply delivers the proper amount of power at the proper voltage, the panelboard serving a load should be as close as possible to the load, which will shorten the branch circuit run and thus reduce voltage drops (see the following case study). Care must also be taken to locate panelboards and all other equipment that generates interference away from sophisticated diagnostic and treatment equipment. This precaution will ensure that the electromagnetic fields generated in any alternating current circuit will not affect the medical equipment. Making sure that motors are not circuited to the lighting panelboards can minimize transients and the effect of starting and stopping the motors on branch circuits.

> **Designing for Variations in Voltage**
>
> In one facility, the settings of the transformers for the X-ray equipment were such that increased loads on the feeders caused additional small drops in voltage and unclear X-rays. Since it was not possible for the facilities engineer to keep changing the settings on the transformers, he rewired the system to correct the problem. The low-quality results of the X-rays could have been avoided had the original design of the facility accounted for varying voltage regulation by different devices.

Additional circuits are frequently needed in health care facilities to accommodate modern electrically powered health care equipment. Thus, when designing a new installation, some spare capacity should be included in the load calculations to provide for future expansion or change in service. Running wire to an existing panelboard is much easier than installing a new panelboard. Before adding load, it is advisable to check the load on the panelboard each time spare capacity is used. Spare circuit breakers or spaces in the panelboard do not ensure that feeder capacity is available or that the capacity of a main circuit breaker on the panelboard will not be exceeded.

The design must be such that electrical power is delivered to the areas where it is needed. The minimum amount of circuiting required by the codes, however, may not be adequate for the proper functioning of the area. If certain areas will be used to treat patients using equipment that consumes large amounts of electricity, such as respirators, special circuits with adequate power for this equipment must be installed in these areas (see the following case studies). Additionally, designers must address how to provide emergency power. Refer to Chapter 8 for a discussion of essential electrical systems.

> **Planning for Power Needs**
>
> At one institution, the use of a greater number of respirators was required on the medical floors. Depending on what other loads were using these circuits at any given time, the addition of the large electrical load demanded by the respirators caused some of the circuit breakers to open on overload. It became necessary to group patients requiring respirators in one area of each floor and to add additional power and circuits in these rooms to accommodate the respirators.

> **Improper Shortcut Electrical Design**
>
> In a large facility, the circuiting of the corridor receptacles was originally designed to take into account the number and location of cleaning machines that would be used simultaneously. Not all these receptacles were powered from the same circuit. In an effort to save money, a contractor connected all the corridor receptacles on one floor to the same circuit breaker, which was code compliant. When more than one machine was used at the same time, however, circuit breakers opened. Additional circuits were subsequently added.

Many hospitals, as well as other types of health care facilities, have "swing rooms," which can be used for various types of treatments. To avoid electrical hazards in these areas, the maximum amount of power and circuiting that may eventually be needed should be installed initially.

Note should be taken of seasonal problems. In some areas, during the summer months the demand on the electrical utility company is high. To compensate for this while using existing equipment, the utility company will reduce the voltage supply by 1 to 3 percent in some instances and higher for brief periods. The facility should have ample warning from the utility to make whatever provisions are necessary to compensate for the reduced voltage. This can include turning on the emergency generator system for critical areas.

4.2.1.2 Isolating Faults

The electrical system must be designed so that an electrical fault anywhere in the system can be isolated and power lost to only a small portion of the system affected. To do so, a system must include proper short-circuit coordination (also called selective coordination; see Chapter 6). The distribution system must also be broken down into smaller feeders, keeping in mind the cost of accomplishing this. The larger the area that a feeder serves, the more people and equipment are affected when that feeder is out of service. The same is true of a branch circuit. It may be less expensive to load up every circuit to its maximum, but it will cause greater interruption of service during a local outage.

4.2.1.3 Safeguarding from Shock

The electrical system must also be designed with the proper safeguards so that patients and staff are safe from electric shock. One safeguard from shock is to electrically insulate energized parts and conductive surfaces that a person may come in contact with. The most common safeguard is to properly ground metallic surfaces that can become energized. To ground these surfaces properly, a low-impedance return path for faults or short circuits should be provided. This low-impedance return path will ensure that fault currents are of sufficient magnitude to operate overcurrent devices (i.e., open circuits and blow fuses) and thus de-energize the grounded object. Grounding also ensures that surfaces likely to become energized and come into contact with a person remain at a minimal voltage. (Refer to Chapter 6 for a more detailed discussion of grounding.) Staff training on the proper use of equipment is another means of preventing electrical shocks.

4.2.1.4 *Providing Other Electrical Safeguards*

Lighting must also be designed, installed, and used in ways that avoid hazards. In addition to making sure that health care professionals have enough light to diagnose certain conditions and conduct complicated procedures, lighting must be installed in ways that avoid contact with patient and staff skin and with materials such as bedding. Lighting is discussed in more detail in Chapters 7 and 17, on lighting and life safety, respectively.

Some jurisdictions require a licensed electrician to supervise the installation of electrical systems, to make sure they are installed properly. Although system designers or contractors typically believe that their installations meet the requirements of the code, all systems should nevertheless be tested immediately after installation and periodically thereafter. The systems should also be maintained properly to ensure that they continue to function properly and provide for the needs of the patients, staff, and visitors. Maintenance and testing of electrical systems are discussed in more detail in Chapter 12. Special needs for essential electrical systems are discussed in Chapter 8.

To reduce the number of hazards caused by human error, all personnel who use the system and equipment must be properly trained. For example, staff should be trained on how to use the receptacles that are connected to the essential electrical system. All devices that must have continuous power should be plugged into these distinctively marked receptacles. In many cases, staff members mistakenly believe that these receptacles are powered only when the generators are on and therefore hardly ever use them. Another example of a human error that creates a hazard is when staff members connect many pieces of equipment in one area without discussing this power load with the engineering staff.

4.2.2 Avoiding Mechanical Failures

Although medical and electrical equipment is subject to manufacturers' testing, equipment can sometimes malfunction due to damage incurred in shipping. Sometimes equipment designs do not adequately minimize leakage currents or prevent shocks or do not meet the particular codes and standards for the jurisdiction within which a facility is located. For these reasons, all electrical and medical equipment must be tested before being used in a facility.

Even after a facility tests and approves electrical and medical equipment for use, equipment in health care facilities can face extreme environmental stress. Wet locations not only pose significant electrical hazards to patients and staff in that they provide a direct low-impedance path to ground, but they also subject equipment to a wet environment. Damp, hot, or humid conditions can adversely affect equipment that is not manufactured to meet these conditions. When medical devices are transported around a facility, particularly those transported frequently and at great pace, they are subjected to mechanical stresses. All these stresses can lead to mechanical failures. To minimize equipment and system malfunctions and failures, it is necessary to adhere to the schedules for equipment testing, inspection, and preventative maintenance stipulated by the codes. Refer to Chapter 12 for a discussion of testing, inspection, and maintenance.

4.2.3 Dealing with Acts of Nature

Earthquakes, hurricanes, tornadoes, floods, explosions, storms, and other acts of nature can cause interruptions of power supplies. When acts of nature strike, proper advance planning can prevent a disaster. All health care facilities should have a disaster plan. The staff should be thoroughly familiar with the procedures to follow when there is a loss of water, electrical power, medical gas systems, and the like. To provide for continuity of electrical service, an emergency generator or an uninterruptible power supply may be required. In certain health care facilities where this is not required, it certainly is suggested. Refer to Chapter 8 for more details on emergency power in health care facilities.

5

Electrical Supply at the Entrance to the Facility

No utility company can guarantee 100 percent reliability and continuity of electrical service. Designing a reliable electrical system for a health care facility requires a reliable power supply from the utility, a reliable backup generator and essential electrical system, a distribution system with well-coordinated protective equipment, proper voltage at the point of use, low harmonic content, and receptacles that are properly placed, grounded, and connected to the normal or essential electrical system.

The first step in providing reliable service is to make the most advantageous service arrangements for a facility's anticipated size of load and estimated budget. In addition to investigating the options that are available from the utility, facility planners must also investigate such factors as a utility's service record and geographic constraints. Due to economies of scale, larger health care facilities usually have more leverage in determining service arrangements with utility companies than smaller facilities. Yet active coordination between utility companies and planners of all sizes of facilities is essential for developing electrical service arrangements that provide the most reliable service with the safest installation possible under any given set of circumstances. Although health care facility personnel usually do not work directly with the utility company to design the electrical service supply, it may be useful for facilities engineers involved in renovation projects to be aware of the pros and cons of the most common service options.

Once the service arrangements are agreed upon, facility designers can plan the service layout and determine the electrical equipment that can be safely and reliably used in a facility. It is essential to have equipment that can disconnect electrical service when and where necessary and protect the system from overcurrents (i.e., overloads) and damaging line-to-ground faults. Allocating enough space for major pieces of equipment is also critical. Making sure that the system includes the ability to adjust voltage and currents increases system reliability as well. The codes and standards address all of these considerations to some degree. Designing the service arrangements and layout according to code provides greater assurance that the facility will be reliable and safe. At times, it may be necessary to design the service arrangement according to more stringent requirements for greater reliability.

Chapter 5. Electrical Supply at the Entrance to the Facility

5.1 Utility Service Arrangements and Reliability

Industry-wide, utilities can offer a wide variety of service arrangements. The service available in a particular region, however, depends on a particular utility's infrastructure. Its number of substations, transformers, power lines, and aboveground and underground equipment, as well as its connection to the overall electrical grid, all factor into what a utility can provide to any facility.

Some types of service are more reliable than others, depending on the availability of backup power and how it is connected to the facility. Other factors to consider in determining the reliability of an electric service include the utility company's record of service interruptions, the stability of the geographic location of both the utility and the facility, and whether the facility will be large enough to receive primary voltage and maintain its own electrical supply equipment. In most cases, the greater the reliability of the proposed service arrangements, the higher the price for the service. In other circumstances, the desired service arrangement may not be available or will be available only after paying a surcharge.

IEEE's *Recommended Practice for Electric Systems in Health Care Facilities* (IEEE Std. 602, 1996) includes a detailed description of the various service arrangements available. Figures 3.14, 3.17, 3.18, and 3.19 there show different types of electrical supply arrangements available from utility companies. These arrangements are discussed here.

5.1.1 Types of Service Available

Typical electric service in the United States is either dedicated or mixed-load service. Dedicated service occurs when one substation provides power to only one facility. Dedicated supplies are usually more reliable and stable than those from substations that service more than one facility at a time, because the one facility is not affected by conditions created by all the other facilities being supplied. If the dedicated supply should fail, however, the facility being supplied would experience a loss of power.

It is more common to have mixed-load service, in which one substation supplies power to more than one facility. Service can be either a radial or network arrangement. A radial arrangement is one in which power is sent from a single substation on a single circuit to the loads. In a secondary network arrangement, the secondaries of the utility company transformers are tied together so that if one transformer fails, other transformers pick up the load.

5.1.1.1 Radial Service

A straightforward but not highly reliable radial arrangement occurs when a utility provides a single service drop or service lateral from the street main to the service entrance conductors. The electricity can be received from only one substation, generating station, or transformer at a time, rather than from two different and separate sources. If electrical service to the one power source is interrupted, all service is lost to the facility until the utility transfers the load to another power source. Similarly, any disturbance on this single radial line will be transferred to the facility. This arrangement is usually provided when the load is very small. Figure 5.1 shows a typical radial system.

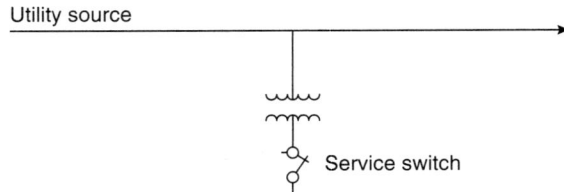

Figure 5.1 Single radial distribution system.

Greater reliability is obtained when the utility company provides a radial system that has two feeders entering the facility from two different power sources. In this arrangement, one of the radial feeders typically supplies the entire service through one transformer or a group of transformers, and the second radial feeder connects to the same transformers through a switching arrangement. Only one radial feeder is connected to the transformers at any given time. Should one of the feeders fail, the load is transferred to the second feeder by use of either manual or automatic switches. Figure 5.2 shows a distribution system in which the facility is supplied from two radial feeders.

When the load is large enough, usually considered to be 750 kVA (kilovolt amperes) or more, a more reliable double-ended radial system can be arranged. In this case, the utility uses two radial feeders, each of which supplies only part of the load through two main service switches (either fused switches or circuit breakers). The two radial feeders are connected on the load side of the main switches by means of a tie switch (i.e., a device, such as a circuit breaker, that makes sure two lines do not feed into each other or that can divert the power to keep the electricity going in the desired direction). Under normal circumstances, the tie switch is in the open position. Should one of the feeders lose power, the main switch for the feeder without power opens and the tie switch closes, transferring the entire load to a single feeder. This switching arrangement can be manual or automatic. The protectors are arranged so that only two of three switches can be closed at one time. Figure 5.3 shows a double-ended arrangement.

5.1.1.2 Network Supplies

One of the most reliable service arrangements is a *secondary network supply* that provides power from a number of sources, usually to many customers. Two or more trans-

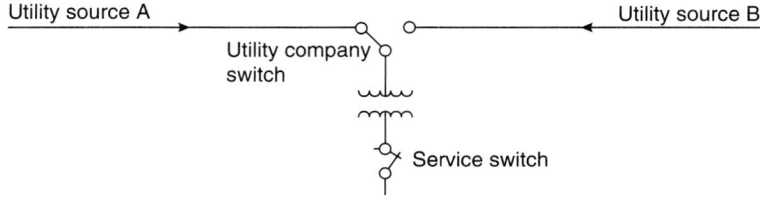

Figure 5.2 Radial distribution system with two feeder lines. (Note the utility company's switch to enable power to be supplied from either distribution line.)

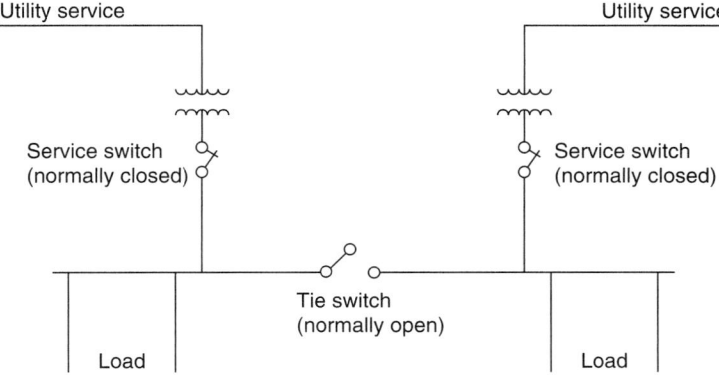

Figure 5.3 Double-ended radial distribution system.

formers that operate with their secondaries tied together through network protectors supply the secondary network. Primary feeders, often from separate substations, feed each transformer. Network protectors installed on each transformer secondary open like circuit breakers when power flows in reverse to ensure that electricity does not feed back into the primary grid through one of the transformers. A primary network could also be used when the utility company ties primary feeders together through network protectors to supply transformers for secondary voltage.

Spot networks are highly reliable secondary networks that provide power to only one customer from two or three substations that contain many transformers. In this type of arrangement, the utility company's secondary transformers are connected through network protectors to a common bus, and all of a facility's service switches are connected to that common bus. If either a primary feeder or a secondary transformer is out of service, the entire load feeds through the remaining transformers to the common bus. Figure 5.4 shows a secondary network arrangement.

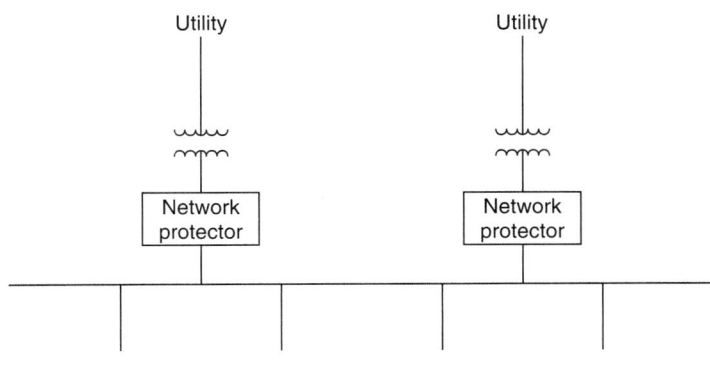

Figure 5.4 Seconday network distribution system.

There are other types of service arrangements. These are usually one of the types just discussed with slight variations depending on the load and the utility company.

5.1.1.3 Service Continuity with Cogeneration Plants

Some health care facilities receive electricity from cogeneration plants. One type of cogeneration plant uses one source of energy to generate the primary electricity and the heat energy generated as a by-product of the main power generation process as another energy supply. Cogeneration plants can be stand-alone facilities that are not connected to the utility company or plants that run in parallel with and are connected to the utility company.

For health care facilities that receive power from stand-alone cogeneration plants, either the utility company or on-site emergency generators must provide power during emergency situations. A number of different arrangements can provide emergency power to facilities that receive power from cogeneration plants that are connected to a utility. In the cogeneration plant run by the health care facility, fossil fuel is typically used to generate steam for heat and to run a generator to produce electricity. Cogeneration can also be done using fuel to run a motor generator set and the by-product used for heating purposes. To ensure continuity of service and the proper design and installation of protection devices, facility personnel must closely plan and coordinate the cogeneration service arrangements with the utility. (Refer to Chapter 8 on essential electrical systems for more details on providing emergency power and cogeneration.)

5.1.2 Primary and Secondary Voltage Supplies

Facility planners are responsible for determining the anticipated loads the facility is likely to draw for the normal and backup electrical systems. The calculation considers many factors, including square footage, the motor loads, and the type of fuel used for heating, which affects the amount of electricity used. One decision based on this estimation is whether the facility is interested in receiving the electric service at primary voltage or secondary voltage from the utility. A primary voltage service is an option most often taken by very large facilities or facilities with campus-type settings.

Primary voltage is the term used to describe electricity that is provided at a voltage higher than 600 V (volts). This level of electricity must be "stepped down" in order for a facility to make use of the power. *Secondary voltage* is voltage below 600 V. This electricity can most often be used at its incoming level and does not need to be stepped down. If the secondary service voltage is 277/480 V, then transformers will have to be used to step this amount down to 120 V, where this voltage is required. Most facilities elect to take secondary voltage service from a utility company, the most common configuration being 120/208 or 277/480 V, three-phase, four-wire, "Y"-configuration service. The utility company will usually determine what will be supplied, although the facility can make a specific request. A lot depends on the size of the facility and the electrical load. A utility can supply secondary voltage service from either a radial system or a network system. Smaller facilities may obtain service at 110/220 V, single phase, three wire.

Facilities that elect to receive high (primary) voltage receive power at more than 600 V. This is usually on the order of 4000 V up to 13,000 V nominal. Such facilities install, operate, and maintain their own high-voltage transformers, switches, and other equipment necessary for stepping the primary voltage down to a secondary voltage that can be used throughout the institution. A facility that manages its own electrical supply equipment has the same options for configuring its service arrangement and secondary transformers as the utility company has.

Larger facilities may also elect to take two services, one at each voltage. In cases where different voltages and systems are available, the facility can likely negotiate favorable system arrangements. In general, facilities that receive primary voltage usually are charged lower rates.

5.1.3 Utility Company Service Record

When making service arrangements for new construction or renovations, investigating the past history of the utility company provides information on the company's frequency of electrical outages and any steps the company might have taken to reduce its number of outages. Greater care should be taken when coordinating a service arrangement with a company that has a record of frequent outages. Less complicated arrangements can be made with companies that have infrequent outages and provide extremely reliable service.

It is also beneficial to investigate which types of electrical system configurations may cause disruptions in service and under what circumstances. The distribution lines of some utilities are subject to transients, which can be transmitted to the health care facility. Another disruption in service may occur during peak energy consumption times, when the utility lowers its supply voltage to satisfy these peak loads. (Note that both transients and excessive voltage drops also can be generated at the facility, causing malfunctions of sensitive medical and office equipment, including radiological devices, computers, and sensitive monitoring equipment.)

5.1.4 Geographic Factors

Geographic location should also be taken into account when planning for the continuity of electrical service. In areas that are prone to floods, earthquakes, tornadoes, or other acts of nature, factors such as the location of the service equipment, its protection from the elements, and contingency arrangements for emergency power should be scrutinized. In some instances equipment should be placed on the rooftop, in others on the lowest level, preferably underground, to provide greater reliability. In disaster-prone areas, even well-planned and potentially reliable service arrangements may not ensure the highest continuity of electrical service (see case study).

> **Floodplain Service**
>
> To compensate for locating a facility on the banks of a river that had a history of flooding, the service equipment was placed on the roof of the building instead of in the basement. The utility company brought the electric service conductors to the roof using conduits encased in a 2 inch concrete envelope and placed the transformers next to the rooftop service entrance. According to NFPA 70, *National Electrical Code®*
> *(Continued)*

(*NEC*®), 2005 edition, Section 230.6, the service conductors in the concrete envelope were considered to be outside the building, which provided the necessary safety to run the service conductors inside the building.

5.1.5 Communications

To enhance the continuity of operations, it may be advantageous for the health care facility engineer to establish open communications with the electric utility company. For example, there may be times, especially during the summer months, when increased loads on the utility lead to a reduced system voltage being supplied to the facility or voluntary load shedding by the facility to protect the integrity of the utility's entire electrical system. With advanced notice from the utility company, a system engineer can provide for a utility's drop in voltage, which can avoid interference with the functioning of a facility, or use an emergency generator to relieve the load on the utility. Similarly, electric utility companies can provide portable generators, which can be useful when in-house emergency generators malfunction. This type of planning and mutual cooperation between the utility and the facility can lead to smoother operations overall.

5.2 Service Layout and Equipment

Determining how all of the equipment and supplies will be laid out safely and reliably is like fitting a puzzle together. How to connect from the service to the building, locate major pieces of equipment, situate distribution lines, and locate protective devices are just a few of the factors that must be considered in laying out the electrical system. The codes and standards cover basic requirements.

The *NEC* (Article 230) delineates requirements for laying out and installing the service equipment for all types of facilities regardless of voltage. Additional source material can be found in the NFPA *National Electrical Code Handbook*. The codes address how to make connections from overhead service drops, underground service laterals, and overhead service locations; to locations that have service of 600 V or less; and where the service is more than 15,000 V. These requirements provide criteria for insulation; size, rate, and clearances for conductors; attachment of equipment to the facility; wiring and cabling; service equipment; the disconnecting means; overcurrent protection; and other system features. Article 110 addresses general workspace requirements.

Ground-fault protection (*NEC* Articles 215, 230, 240, and 517 have some of the requirements) is another major factor in laying out electrical equipment in health care facilities because it protects equipment from being damaged by line-to-ground fault currents. It is also necessary to design the system with short-circuit coordination, which involves selecting the overcurrent devices and preserving the electrical supply to as many areas as possible during a fault. The short-circuit availability is the amount of current that would flow at the service equipment if a three-phase fault occurred at the service equipment. Information on short-circuit availability can be obtained from the utility company.

Major pieces of equipment critical to facility design are transformers and voltage and current conditioners. Article 450 of the *NEC* addresses the installation of all types of transformers and transformer vaults, and Articles 455 and 460 cover phase converters and capacitors, two main types of current conditioners. Although most of these topics are discussed in more detail in Chapters 6 and 7 of this book, they are briefly discussed here to provide an overview of the key elements in determining system layout.

5.2.1 Number of Services

Most jurisdictions allow a facility to receive power from only one service, although they may permit additional services to supply electricity for fire pumps, emergency systems, legally required standby systems, optional standby systems, and parallel power production. Some jurisdictions allow additional services, such as when a facility is too large for one utility service or when a facility uses different service voltages. As an additional example, the utility's primary power plant may provide a facility's main service, and a cogeneration plant may service its emergency generator or fire alarm system. Some jurisdictions require the fire alarm system to be on a separate service if a facility has no emergency generator, or at a minimum, be connected to the incoming service on the line side on the main service switches. Authorities having jurisdictions often allow large medical centers with multi-building complexes to receive power from more than one service, provided that each building is served by only one service. It is important to check with all authorities having jurisdiction as to what codes are followed and what exceptions are allowed.

Because each service has its own meter arrangement, facilities that receive electricity from more than one service may not reap the advantages of multi-tiered billing (i.e., when the utility charges the highest rate for the first level of kilowatts used and then succeedingly lower rates for higher levels of kilowatts used). Both consumption and demand charges for each building are charged using multi-tiered billing. In some instances, such as when a facility is under one ownership and management, a utility company may permit concurrent billing, in which case the utility will add the consumption and demand for each building and produce one bill. This billing arrangement is usually allowed if each service uses the same voltage. Here again, proper working relationships with the utility company can be advantageous.

5.2.2 Disconnects

A *disconnect* is a device or a group of devices that can disconnect conductors of a circuit from their source of supply. For emergency situations, all systems must provide a means to disconnect simultaneously all ungrounded service conductors entering a building. There also must be a means to disconnect the grounded conductor from the premises wiring in the service equipment. This requirement is usually accomplished by removing the grounded conductors from the ground bus.

According to the *NEC* (Section 230.70), the disconnecting means must be located either outside the building or inside the building at the closest point to where the service entrance conductors enter the building. This location should be easily accessible to qualified maintenance personnel so that they can quickly operate the disconnects in an emergency. The service conductors are considered to be located outside the building if they are encased in a minimum of 50 mm (2 in.) of concrete or brick that is situated within the building or under at least 2 inches of concrete located under a building. The service conductors in the facility situated on or in a floodplain (see the previous case study), where the utility company brought the service to the roof of the building through a concrete chase, were considered to be outside the building.

The *NEC* permits each service to have a maximum of six service disconnects. The disconnecting means must be grouped together and marked to indicate the load being served and to facilitate operation of the correct switch during an emergency. If a facility already has six service disconnect switches but is undergoing renovations that require additional service switches, two or more existing switches can be grouped together and fed from a new service switch. The new switch will be the service switch, and the switches being fed by the new switch will become distribution switches. For these facilities, it would be advantageous to have concurrent billing for all of the service switches to reduce overall power costs.

5.2.3 Overcurrent Protection

Each ungrounded service conductor must also have overcurrent protection to protect the feeders from overloading or short circuiting, which could cause a fire or thermal damage. When a switch is used as a disconnect, overcurrent protection is usually accomplished by installing a fuse. Properly rated circuit breakers also can serve as disconnecting devices as well as provide overcurrent protection.

5.2.4 Workspace Requirements

Space is usually at a premium in health care facilities. All too often the electrical service room or the electrical closets become ancillary storage areas. The *NEC* (Article 110) specifies the types of layouts that constitute clear work spaces around electrical equipment, to maintain order and provide ready access for maintenance, servicing, and quick operation when necessary. The code specifies a range of required work space depths for services of 600 V or less from 900 mm (3 ft) to 1.2 m (4 ft), depending on different conditions of exposed live parts. There are also requirements for working distance in front of and above the service equipment. Naturally, the requirements change for voltages above 600 V.

Some jurisdictions not only require clear working space around electrical equipment but also prohibit any other wires, including telephone and television wiring, to pass around the electrical equipment or through the room that houses the equipment. The AIA *Guidelines for Design and Construction of Hospital and Health Care Facilities* (2001 edition, Section 7.32, "Electrical Standards for Hospitals") requires that the main switchboards be located in areas that do not include plumbing or mechanical equipment and that they be readily accessible to authorized personnel (and only to authorized personnel). The switchboards must also be located in dry, ventilated spaces and be free of corrosive or explosive materials and gases. Figure 5.5 shows a service equipment room at a large hospital.

Chapter 5. Electrical Supply at the Entrance to the Facility

Figure 5.5 Typical service equipment room with multiple bolted pressure service switches.

5.2.5 Ground-Fault Protection

Equipment is protected from damaging line-to-ground fault currents by devices that open all ungrounded conductors of a ground-faulted circuit. These are necessary, as many line-to-ground faults will generate damaging currents, but the magnitude will not be large enough to open an ordinary circuit breaker or fuse in time to prevent damage from occurring. Ground-fault protection for service equipment must be installed if the following conditions are present: (1) the service voltage is more than 150 V to ground but not more than 600 V phase to phase; (2) the service is a solidly grounded "Y" configuration; and (3) all service disconnects are rated 1000 A or more. Ground fault protection must be provided for feeders with the same conditions except where it is provided for the service disconnects. In this case the 1000 A disconnect requirement would refer to the feeder disconnect. The *NEC* (Article 517) requires that if the service of a health care facility has ground-fault protection, it must be installed on the next level of distribution toward the load (i.e., the next level of feeder disconnecting means after the service switch). This requirement holds true if there is no service ground-fault protection but there is protection to a feeder as described previously. This requirement for ground-fault protection, its ramifications, and solutions to layout challenges are discussed in more detail in Chapters 6 and 7.

5.2.6 Transformers

Transformers serve many functions related to supplying power and can either be part of or situated near the service equipment or be located throughout the building.

Transformers located near the service entrance usually serve to step down the utility's supply voltage to the secondary voltage necessary for use by the facility. Utility companies typically own and maintain the transformers located at facilities with a secondary voltage supply. If the supply voltage is at a primary voltage, the facility typically owns the transformer(s). In larger institutions and multi-building campuses, transformers serve to increase or decrease the voltage for distribution throughout the facility.

Figure 5.6 shows an electric closet with step-down transformers. These transformers step the voltage down from 480 V (which is the voltage used for the distribution system throughout the facility) to 120/208 V for utilization at lights, receptacles, and so on. One transformer is for the normal distribution system and the other is for the essential electrical system. Transformers that supply *isolated power* (i.e., an ungrounded electrical system that reduces the hazards of electric current flow in certain areas) are addressed in more detail in Chapter 9.

Article 450 of the *NEC* addresses transformer and transformer vault requirements. Part I of Article 450 includes general requirements for transformers in various types of arrangements, such as where loads can be connected, how to protect and control tie circuits, and how to protect from overcurrents. Part II of Article 450 provides specific requirements for different types of transformers, including dry-type transformers installed indoors, dry-type transformers installed outdoors, nonflammable fluid-insulated transformers, askarel-insulated transformers, and oil-insulated transformers. If a facility has any one of these types of transformers, it is useful for the operating personnel to become familiar with the applicable parts of the *NEC* and other relevant codes and standards for maintaining a safe installation.

Part III of Article 450 lists the requirements for transformer vaults. Transformer vaults must be located where they can be vented to the outside without using flues or

Figure 5.6 Electric closet with panelboards and step-down transformers.

ducts where possible. The system design must plan for proper ventilation at all times, because as the load increases on the transformer, the heat emitted also increases. Facilities engineers and maintenance personnel should be aware that if heat builds up in a transformer room that is not ventilated properly, transformer ratings diminish. Additionally, Article 450 of the *NEC* stipulates that the walls, floors, and roofs of transformer vaults be constructed of material that has a 3-hour fire rating, and that pipes and ducts that are not connected to the electrical installation do not pass through the transformer vault.

5.2.7 Voltage and Current Conditioners

Voltage and current conditioners are devices that remove potentially harmful variations in the voltage or current, such as transients, spikes, harmonics, and reactive current, to protect equipment, avoid hazards, and use power more efficiently and effectively. Voltage and current conditioners come in various forms; some of the more common ones are phase converters, capacitors, voltage regulators, and power conditioners.

5.2.7.1 Phase Converters

A phase converter is an electrical device that converts single-phase power to three-phase power, enabling the use of devices that require three-phase power in areas where only single-phase power is available. Because utility companies typically provide facilities with the power phase delineated in the service arrangement, most health care facilities usually purchase equipment that already uses the correct phase. The *NEC* (Article 455) addresses the requirements for the installation, use, and maintenance of phase converters in terms of conductor capacity, overcurrent protection, disconnecting means, and other criteria. In the rare instance where phase converters are used, it is necessary to comply with the requirements of the *NEC*.

5.2.7.2 Capacitors

Operating equipment such as motors, transformers, and fluorescent light ballasts places inductive and capacitive reactance on the electrical circuit. The result is that the utility company must supply both *real* and *imaginary* power. The real power, called *watts*, provides the energy to do the mechanical work, and the imaginary power, called the *inductive* or *reactive power* (VAR, volt amperes reactive), provides the power for the reactive component of the load. The vector sum of the two is called *apparent power*, which is the power that the utility company supplies. The *power factor* is the ratio of the real power to the apparent power. A unity power factor (a value of 1) means that all of the power is being used for the work and there is no reactive power. As the reactive power increases, the value of the power factor goes down. The utility company usually charges only for real power (using a watt-hour meter). If a utility objects to a facility having a low power factor, it might install a meter at the facility that measures the reactive power (kVAR, or kilovolt ampere reactive power) and charge for a low power factor. Utility companies and electrical engineers can often suggest ways to reduce the reactive power.

Health care facilities often use capacitors as surge protectors and also to improve the power factor of the system. (To protect personnel working on the unit, the residual voltage of the capacitor must be reduced to 50 V within 5 min of the capacitor being disconnected from the circuit.) The *NEC* (Article 460) covers capacitor installation and use, means of discharging stored energy, capacitor conductor ampacity and overcurrent protection, motor overload devices, disconnecting means, and capacitor grounding and marking. Figure 5.7 shows power-factor correction capacitors with discharge resistors.

Figure 5.7 Power factor correction capacitors with discharge resistors.
Source: NEC Handbook, 2005, Exhibit 460.2.

5.2.7.3 Voltage Regulators

Voltage regulators are often used when, for example, a piece of older equipment, or the system itself, can tolerate only small variances in voltage. Some equipment comes with built-in voltage regulation devices. In situations where external voltage regulators are required, the characteristics of the voltage regulators must match the needs of the equipment. It is also necessary to match the operating speed of the voltage regulator with the requirements of the equipment for the voltage regulator to be effective.

5.2.7.4 Power Conditioners

Power conditioners usually combine voltage and current regulation into one device.

With the greater use of equipment that produces nonstandard waveform pulses, such as computers and other solid-state devices, facilities engineers must pay attention to the harmonic content of a facility's distribution system. It is not the purpose here to provide a complete discussion of harmonics. Suffice it to say that a harmonic waveform, which is a multiple of the base frequency (60 Hz), is present in a facility's distribution system. Harmonics appear as current and generate heat in conductors, overcurrent devices, and distribution equipment, which can affect all types of equipment connected to the system. Although it is not necessary for system personnel to understand fully what harmonic waveforms are, it is useful to know that as a result of harmonics, a calculated load might show that a circuit is loaded to 70 percent of its capacity, while the load-generating heat, which affects performance and overcurrent devices, actually is much higher. This heat could damage equipment or expose the system to unnecessary power outages.

Figure 5.8 A clamp-on ammeter that uses true rms measurements.
Source: NEC Handbook, 2005, Exhibit 310.5; photo courtesy of Fluke Corp.

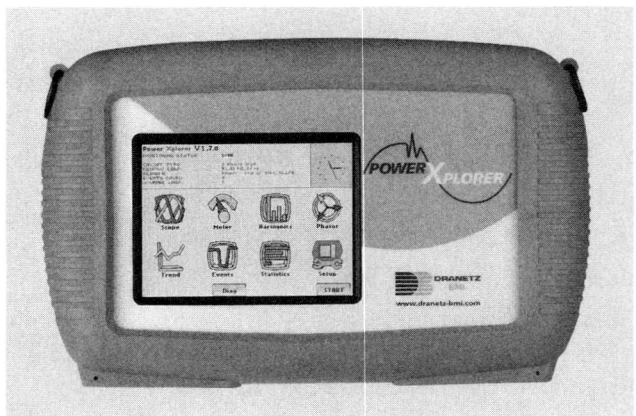

Figure 5.9 A portable tool for such tasks as diagnostic power analysis including harmonic distortion.
Source: NEC Handbook, 2005, Exhibit 310.6; photo courtesy of Dranetz-BMI.

Measurement of this type of waveform requires a special meter. The ordinary average-responding ammeter that is usually carried by the electrical staff does not give a true reading. A true root-mean-square (rms) meter should be used if a facility engineer suspects that a high degree of harmonic content is present in the system. The use of isolation transformers, harmonic filters, and delta-connected transformers can suppress or at least control harmonic currents. Figure 5.8 shows an ammeter that uses true rms measurements. Figure 5.9 shows a portable tool that is used for such tasks as diagnostic power analysis, including harmonic distortion.

6

Distribution Systems

Electrical distribution systems in health care facilities must not only distribute power to equipment common to all buildings, they must also provide power to preserve lives and provide for a safe environment for all who come into contact with the system through the use of medical equipment. Electrical distribution systems start at the main service switch and end at the receptacles, user devices, and local switches. Specific components of electrical distribution systems include wire and cable, raceways, switchboards, panelboards, motor control centers, system protection and coordination equipment, and grounding and circuiting devices.

Most distribution systems in health care facilities must comply with NFPA 70, *National Electrical Code*® (*NEC*®), either by direct reference or by cross-reference in another code. For example, NFPA 99, *Standard for Health Care Facilities*, states that the wiring and installation of electrical equipment in health care facilities must meet the specific requirements of the *NEC*. Also, the American Institute of Architects' (AIA) *Guidelines for the Design and Construction of Hospital and Health Care Facilities* state that all electrical material and equipment must be installed in compliance with applicable sections of the *NEC* and NFPA 99.

6.1 Wire and Cable

Wire and cable are the veins and arteries of the distribution system. There are numerous types of wire and cable. A wire is the metal conductor through which electricity flows (sometimes called a conductor). If there is insulation around a wire, it is called an insulated wire or insulated conductor. A cable is an assembly of one or more insulated conductors with a jacket or cover around the assembly. The *NEC*, 2005 edition (Article 300), contains the requirements for wiring methods and materials, generally for voltages of 600 V (volts) or less. Article 310 lists the requirements for many conductor and cable types.

6.1.1 Conductor and Cable Types

The *NEC* lists copper, copper-clad aluminum, and aluminum as permitted materials for conductors. There is a restriction in both the *NEC* and NFPA 99, however, that in patient care areas, the installed grounding conductor must be an insulated copper conductor. Aluminum and aluminum-clad conductors generally have lower ampere ratings than copper conductors of the same size. Thus, to maintain the required ampacity when using aluminum conductors, it may be necessary to use larger conductors and possibly larger conduits than when using copper conductors.

Additional precautions must be taken when using aluminum or aluminum-clad conductors. Proper terminating devices must be used. Torque wrenches should be used to tighten screws and bolts to prevent the aluminum from creeping (thinning and stretching). The conductor must be cleaned with a proper cleaning compound and brush before making the connection to prevent the aluminum from oxidizing before the connection is made.

The *NEC* (Section 310.8) differentiates conductor and cable types according to the location in which the conductors can be used (i.e., dry locations, dry and damp locations, and wet locations). All insulated conductors and cables noted in the *NEC* can be used in dry locations. Locations where different insulated conductors can be used are listed in Section 310.8. The conductor application and insulations are listed in Table 310.13 of the *NEC*.

New types of insulation are regularly being introduced by the manufacturers. Rather than list all of the conductor and insulations types, suffice it to say that the facilities managers should be familiar with these insulated conductors.

6.1.2 Conductor Size

The physical size of the insulated conductor is another important criterion. When installed in a raceway, conductors of size 8 AWG and larger must be stranded. Conductors sizes 1/0 AWG and larger are permitted to be run in parallel, but the conductors must be the same size and length, have the same type of conductor material and insulation, and be terminated the same way. The *NEC* (Table 310.5) lists the minimum size conductors allowed for different conductor voltage ratings.

The percentage of fill and derating of the conductor are important as well. A greater number of conductors with thinner insulation can maintain the same percentage of fill in a conduit that has fewer conductors with thicker insulation. This factor becomes important when planning renovations and determining if additional conductors can be installed in an existing conduit or if completely new conduit and conductors must be installed.

6.1.3 Conductor Insulation

Generally, all conductors must be insulated. The various types of materials allowed as insulating compounds are as follows:

- Rubber, designated with the letter R
- Neoprene® (RHW), not made of rubber
- Thermoplastic compound (T)
- Thermosetting compound (X in the insulation designation)
- Moisture-resistant compounds
- Heat-resistant compounds
- Combinations of moisture- and heat-resistant compounds
- Compounds with other characteristics

Refer to Table 310.13 in the *NEC* for allowed conductor insulation types and thicknesses.

6.1.4 Operating Temperature

The operating temperature of conduits affects the number of conductors allowed within them. Conductors cannot be used in situations where the operating temperature exceeds the temperature for which the insulation was designed. This situation can occur where there is a high ambient temperature, such as the boiler room, or where more than three current-carrying conductors are used in a single conduit. In cases such as these, it might be necessary to lower the maximum allowable current.

Special connecting devices must be installed when using aluminum as a conductor, and special care must be taken during the installation: Loose aluminum joints can heat up and eventually arc, creating a fire hazard. Because aluminum has a tendency to creep with pressure and oxidizes almost immediately upon exposure to the air, installers should use a torque wrench to tighten all connectors and ensure that the pressure is correct and not excessive. Before connections are made, the conductors must first be cleaned to remove the oxidation. Then they must immediately be coated with a compound to prevent any further oxidation. If aluminum conductors or bus bars are used, they should be properly inspected and maintained, such as by periodically conducting thermal investigations of all panelboards and other connection points and preparing inspection reports. This information will let the facility engineer know when some of the aluminum has crept and which joints may be slightly loose and heating up. The additional precautions that must be taken when using aluminum rather than copper conductors may offset the lower cost of using aluminum.

Thermal testing is recommended for all installations, regardless of whether they are copper or aluminum. Inspections and testing are covered in more detail in Chapter 12.

6.1.5 Conductor Ampacity

Current ampacity is the current, in amperes (A), that a conductor can carry continuously under its conditions of use without exceeding its temperature rating. The *NEC* now permits two methods for determining conductor ampacity. The first method uses a formula to calculate the ampacity, the results of which must be verified by an engineer. The second method of determining conductor ampacity is to select the ampacity from tables presented in the *NEC* (Tables 310.16 through 310.21 and 310.67 through 310.86; Annex B, Tables B.310.1 through B.310.10) that list the allowable ampacities for various combinations of conductors based on calculations as well as information about the calculation of ampacities. Section 374.6 discusses ampacity of conductors in gutters.

A correction factor to the ampacity tables must be applied when a raceway includes more than three current-carrying conductors. This can occur either upon initial installation or during a renovation to increase the number of circuits leading to an area. Additionally, a correction factor must be used when the conductors are located in areas having different ambient temperatures than those listed in the ampacity tables (40°C, 104°F). The *NEC* ampacity tables include the correction factors.

6.1.6 Conductor Marking

All conductors and cables must be properly marked to inform system users of the types of wire or cable present. Markings must include the rated voltage, the type, the manufacturer's name, the size designation, which will be in either AWG size or circular mil area depending on the size of the conductor, and a notation that indicates if the cable assembly has a smaller neutral conductor than the other conductors. The marking is usually placed on the surface of the conductor insulation. Marker tape is used for metal-clad, multiconductor cables. Insulated or covered grounded conductors must be identified in accordance with Section 200.6 of the *NEC*. Ungrounded conductors must be a color other than white, natural gray, or green.

6.1.7 Cable Types

The *NEC* (Articles 320 through 340) lists the various types of cable permitted. For each type of cable, the code lists the definition, permitted locations, installation and construction specifications, required markings, allowable fittings, and other specific requirements. Although mineral-insulated (MI), armored (AC), and metal-clad (MC) are the types most commonly used, other options are available. All options are described.

6.1.7.1 Type FCC Cable

Type *FCC cable* is flat conductor cable, designed for use under carpet squares. It has the advantage of facilitating flexibility, since the carpet squares can easily be lifted to change the cable.

The *NEC* prohibits the use of FCC cable in any part of a hospital building. This means that it cannot be used in the administrative portion or any other portion of the hospital building, regardless of the occupancy classification. Refer to Chapter 13 on classification of occupancies in the "Life Safety" section.

The *NEC* has limitations and restrictions as to where this type of cable can be used. It is strongly suggested that the *Code* should be consulted and a thorough understanding be had before using this cable in a health care facility.

6.1.7.2 Type MI Cable

Type *MI cable* is mineral-insulated, metal-sheathed cable. It consists of one or more copper conductors insulated with a highly compressed refractory mineral and covered with a copper or steel alloy metal sheath. The only place where MI cable *cannot* be used is where the cable may be exposed to destructive corrosive conditions.

6.1.7.3 Type AC Cable

Type *AC cable* is armored cable, also known as BX cable. AC cable, which is one of the most often used cable assemblies, is fabricated of insulated conductors in a flexible metallic enclosure with a bonding strip that can be used as a ground. Underwriters Laboratories lists AC copper cable in sizes No. 14 through No. 1 and AC aluminum and copper-clad aluminum cable in sizes No. 12 through No. 1. AC cable is recognized as an equipment-grounding conductor unless otherwise required. There is also a bonding strip in the cable. Consult the *NEC* to find out where this cable can be used. Figure 6.1 shows an example of type AC cable.

Chapter 6. Distribution Systems

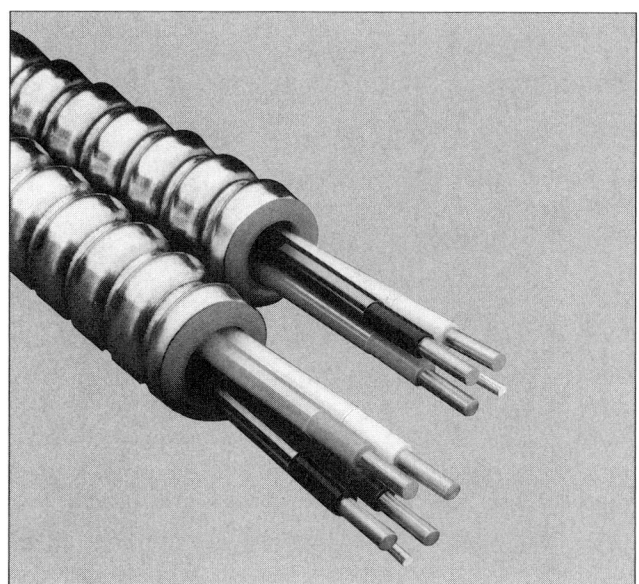

Figure 6.1 An example of type AC cable.
Source: NEC Handbook, 2005, Exhibit 320.1; photo courtesy of AFC Cable Systems, Inc.

6.1.7.4 Type MC Cable

Type *MC cable* is metal-clad cable. It is factory assembled and consists of insulated circuit conductors within an armor of interlocking tape made of smooth or corrugated metal. MC copper cable is manufactured in size No. 18 or larger; MC aluminum or copper-clad aluminum cable is manufactured in size No. 12 or larger. Although MC cable can be used in most health care facility locations, it may not be used in corrosive or destructive areas unless the metal sheath has additional protection for these destructive environments. Figure 6.2 shows examples of MC cable.

6.1.7.5 Types NM, NMC, and NMS Cable

Types *NM, NMC*, and *NMS cable* are nonmetallic-sheathed cables, factory assembled with two or more insulated conductors and an outer sheath of flame-retardant, moisture-resistant, nonmetallic material. There are many restrictions limiting the use of this cable in health care facilities.

6.1.7.6 Types SE and USE Cable

Types *SE* and *USE cable* are service entrance cables. They are assembled cables that are used primarily at the entrance to the facility. USE cable can be used underground. Again, the reader should consult the *NEC* to determine where this cable can be used at a health care facility.

6.1.7.7 Type UF Cable

Type *UF cable* is an underground feeder and branch circuit cable. Consult the *NEC* to determine where this type of cable can be used in a health care facility.

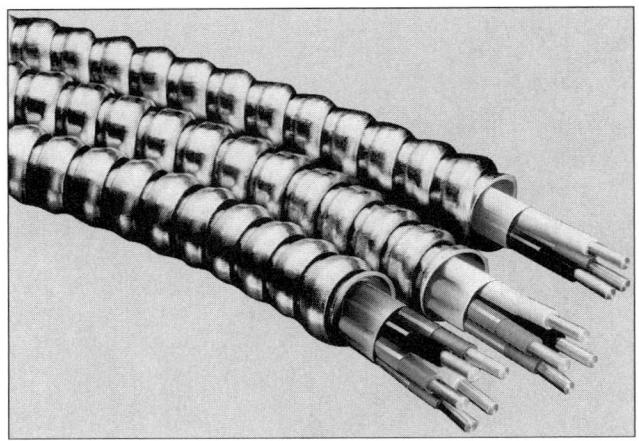

Figure 6.2 Some examples of type MC cable.
Source: NEC Handbook, 2005, Exhibit 330.1; photo courtesy of AFC Cable Systems, Inc.

6.2 Raceways

Raceways are enclosed metal or nonmetallic channels designed to hold wires, cables, or busbars and to serve additional functions as allowed by code. Raceways provide physical protection for the conductors and, when metallic, provide a grounding path. Raceways are most often made of steel, aluminum, or plastic and can be either rigid or flexible, either heavy or thin walled, and, as in the case of cellular concrete floor raceways, part of the structure. It is useful for facility engineers to be familiar with the many types of available raceways shown in the following list, especially when planning a renovation. The corresponding *NEC* (2005 edition) article numbers indicate the articles that stipulate requirements for the use, size, number of conductors, fittings, supports, grounding, bends, and splices and taps for each type of raceway:

- Electrical nonmetallic tubing (Article 362)
- Intermediate metal conduit (Article 342)
- Rigid metal conduit (Article 344)
- Rigid nonmetallic conduit (Article 352)
- Electrical metallic tubing (Article 358)
- Flexible metallic tubing (Article 360)
- Flexible metal conduit (Article 348)
- Liquid-tight flexible metal conduit (Article 350)
- Surface metal raceways (Article 386)
- Surface nonmetallic raceways (Article 388)
- Underfloor raceways (Article 390)
- Cellular metal floor raceways (Article 374)
- Cellular concrete floor raceways (Article 372)

- Metal wireways (Article 376)
- Nonmetallic wireways (Article 378)
- Busways (Article 368)
- High density polyethylene conduit (Article 353); there are limitations as to where this can be used
- Nonmetallic underground conduit with conductors (Article 354); there are limitations at to where this can be used
- Liquidtight flexible nonmetallic conduit (Article 356)
- Strut-type channel raceway (Article 384)
- Auxiliary gutters (Article 366)
- Cablebus (Article 370)
- Cable trays (Article 392)

Code requirements, economic factors, and safety concerns are the predominant factors that determine the choice of a specific raceway. In some cases where the raceway is metallic, the raceway is considered a grounding means. It is helpful for facility engineers to be familiar with the different types of raceways that are available, especially when planning a renovation.

6.2.1 Rigid Metallic Raceways

The most commonly used metal conduit is *rigid metal conduit* made of a ferrous material such as steel or a nonferrous material such as aluminum. These raceways can be used in all atmospheric conditions and occupancies. Rigid metallic raceways can be made of a corrosion-resistant material, such as coated galvanized iron, for use in wet locations. They can also be treated with a corrosive-resistant paint or covered with enamel for installation in a corrosive atmosphere, such as a battery room or some laboratories. Rigid metal conduit cannot be used in a trade size smaller than metric designator 15 (trade size ½) or larger than metric designator 155 (trade size 6).

Intermediate metal conduit (IMC) is a listed metal raceway similar to rigid metal conduit except that the wall of the IMC conduit is thinner. IMC raceways, which are made only of steel, can be used in all atmospheric conditions and atmospheres. When necessary, proper protection or coatings can be provided for corrosive areas and wet locations.

Electrical metallic tubing (EMT) is a rigid conduit made of either ferrous or nonferrous material and has the thinnest construction of the three rigid metallic raceways. It is sometimes known as *thinwall* conduit. EMT is a popular choice for conduit since it is very light, easy to handle, and the least costly of the three rigid metallic raceways. Many use restrictions apply to this type of conduit. It may not be used where it will be exposed to severe physical damage, where protected from corrosion solely by enamel, in moisture, in concrete under certain conditions, or in hazardous (classified) locations. It also cannot be used to support most fixtures or other equipment.

6.2.2 Flexible Metallic Raceways

The most commonly used flexible metallic raceway is *flexible metal conduit (FMC)*. FMC conduit has a circular cross section and is made of an interlocked metallic strip that is helically wound and formed. The FMC raceway can be used in all exposed and concealed locations. It cannot be used in wet locations or hoistways unless specifically approved for those uses, in hoistways unless otherwise permitted, in storage battery rooms, or where it might be subject to physical damage. FMC also cannot be used underground or embedded in poured concrete or aggregate, where it would be exposed to materials that would have a deteriorating effect, or in any hazardous (classified) locations, except where specifically permitted by code. Generally, this type of conduit can be used as a grounding means and usually must include an equipment grounding conductor unless the run is less than 6 ft (1.82 m).

Liquid-tight flexible metal conduit is similar to flexible metal conduit except that this type of raceway has a liquid-tight, nonmetallic, sunlight-resistant jacket over the flexible conduit. It is usually used in wet locations or where protection from vapors is required. Generally, the same rules and restrictions apply to liquid-tight flexible metal conduit as to flexible metal conduit.

As its name implies, *flexible metallic tubing* is both metallic and flexible. It also has a circular cross section and is liquid-tight without having a nonmetallic jacket. It can be used in dry locations, where concealed, in accessible locations, and for system voltages of 1000 V maximum. Its use restrictions are similar to those for FMC conduit.

6.2.3 Rigid Nonmetallic Raceways

Rigid nonmetallic raceways are made of nonmetallic materials, some of the more common being polyvinyl chloride (PVC), high-density polyethylene, fiberglass, and fiberglass-reinforced epoxy. Each material is listed as being suitable for specific permitted uses. However, this type of conduit cannot be used in (1) hazardous (classified) locations, unless specifically permitted; (2) to support fixtures; (3) where subject to physical damage; (4) in theaters; or (5) where ambient temperatures could be higher than 50°C (122°F) or higher than the temperature allowed for the conductors.

6.2.4 Flexible Nonmetallic Raceways

Electrical nonmetallic tubing, type ENT, is a pliable corrugated tubing with a circular cross section that is made of PVC. ENT is moisture and flame resistant, as well as resistant to chemical atmospheres. It can be bent by hand without the use of other devices. ENT is prohibited from use in hazardous (classified) locations, to support fixtures and other equipment, in locations subject to temperatures higher than 50°C (122°F), and in other areas. Note that in some jurisdictions, the fire department might request information on where nonmetallic raceways are used because this material may give off toxic fumes during a fire.

6.2.5 Other Raceways

Other types of raceways can be used in health care facilities, including cable trays, surface metallic and nonmetallic raceways, underfloor raceways, cellular metal floor raceways, and cellular concrete raceways. The most common types installed in health care facilities are surface metal raceways and surface nonmetallic raceways.

Surface metal and nonmetallic raceways are convenient to use where an exposed raceway is not objectionable, such as over laboratory benches or in some offices. They come in many sizes and finishes that can make the raceway look like a piece of molding or blend in with the wall or ceiling. Special fittings can connect these raceways to surface-mounted boxes or lighting fixtures, and they may come with receptacles and contain branch circuit wiring. Raceways fitted in this manner are called multioutlet assemblies and are common installations in laboratories, repair shops, and clerical areas, where many receptacles are needed for equipment. Surface raceways can be rewired easily, and additional outlets can be connected after the original installation is in place. Figure 6.3 shows an example of a surface nonmetallic raceway extending from an existing receptacle outlet.

Another type of raceway used to bring power to areas, but not through a wall, is an underfloor raceway. This type of raceway can be installed under the floor or in a concrete floor flush with the top of the floor. In the latter case, the raceway sometimes comes equipped with a removable cover. Underfloor raceways installed in concrete must be in place when the concrete is poured. Cellular metal floor raceways must also be installed before the concrete floor is poured. The cellular openings in precast cellular concrete slabs can be aligned to form wireways. The decision to use this flooring material and underfloor raceways must be made during the planning phase of a new building or renovation.

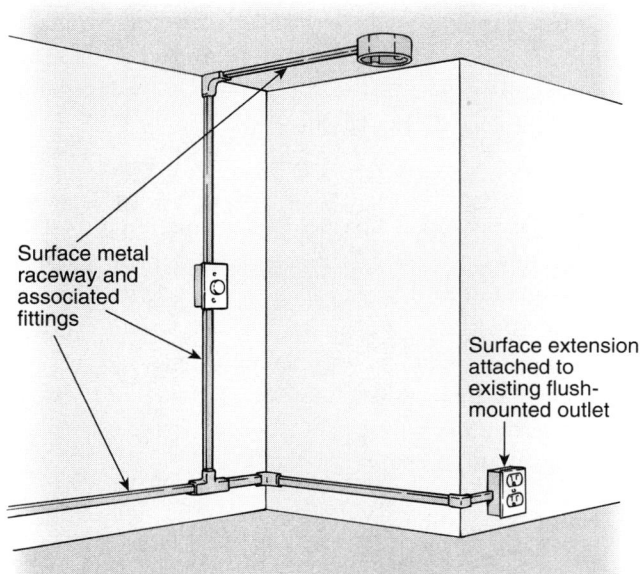

Figure 6.3 An example of a surface nonmetallic raceway extending from an existing receptacle outlet.
Source: NEC Handbook, 2005, Exhibit 386.1; photo courtesy of The Wiremold Co.

6.3 Wiring Methods

Electrical continuity of metal raceways and enclosures is important for the continuity of a grounding system if the raceway is part of that system. The entire system has to be electrically bonded to ensure this continuity. As defined by the *NEC*, bonding is the permanent joining of metallic parts to form an electrically conductive path that ensures electrical continuity and the capacity of the system to conduct any likely current safely. The entire electrical system of a health care facility must be electrically bonded to ensure this electrical continuity and provide an effective low-impedance path to dissipate fault currents. Wiring methods must comply with Article 300 of the *NEC*.

6.3.1 Installing Conductors in Raceways

The number of conductors that can be installed in a single raceway is limited by how much and how fast the raceway can dissipate the heat generated by the current flow and how well the conductors can be installed and removed without damaging them. The number of conductors that are permitted in a particular type of raceway is determined by the percentage of fill allowed in that raceway. The total number of each type of conductor that can be installed in a particular raceway is calculated based on the following criteria: the percent cross section of the raceway needed for the conductor; the dimensions and percent area of the raceway; and the dimensions of the insulated conductors and fixture wire. The *NEC* (Chapter 9, Tables 1, 4, and 5) provides the necessary information to compute this calculation; Annex C of the *NEC* lists the results of the computation. Note that if the calculation shows that the total fill can include more than three current-carrying conductors and the raceway is filled to the allowable capacity, it may be necessary to apply a derating factor to the wire. Although it is not mandatory to use the calculations in the *NEC* Annex C, using them eliminates the need to make these complicated calculations.

6.3.2 Adding Power

When additional power is needed at a particular location, it may be possible to replace the original wires within the original conduit. Either larger wires or a larger number of wires with smaller insulation may be installed in the original conduit to provide for more than one circuit. Replacing wires in an original conduit can usually be accomplished under the following conditions: the conduit is rigid; it is possible to install new conductors with insulation that has a total cross-sectional area less than the original conductors; and the maximum amount of fill allowed is not exceeded and the proper derating factors are taken into account.

6.3.3 Sealing Partition Penetrations

All too often, a 6-in.- (15.24-cm-) diameter hole is made in a fire or smoke partition for a 3-in.- (7.62-cm-) diameter conduit. To prevent the spread of fire or other products of combustion through parts of the electrical installation, it is essential to make the size of the penetration through these partitions appropriate to the size of the conduit. The *NEC* requires that electrical installations not substantially increase the potential for the spread of fire or products of combustion. NFPA *101*®, *Life Safety Code*®, 2006 edition, the Joint Commission on Accreditation of Healthcare Organizations (JCAHO), and all other design and safety manuals also address this issue. JCAHO surveyors usually examine the penetrations made by the electrical installation through these walls and partitions. (Refer to Chapter 15 for more information on sealing penetrations in fire and smoke partitions.)

To maintain NFPA *101* fire and smoke resistance ratings, openings around the electrical penetrations must be sealed with an approved material. Many compounds are available for this purpose. Additional requirements related to sealing the wall penetration are found in NFPA *101* (Sections 8.3.5 and 8.5.5), which address construction and compartmentation and smoke barriers, respectively. The facility engineer is responsible for ensuring that all of these penetrations are the proper size and are properly sealed, whether the work was conducted in house or by an outside contractor.

6.3.4 Installing Wiring in Ducts or Plenums

Additional care must be taken when installing wiring in manufactured ducts or plenums. For example, it can be very tempting to use a hung ceiling, when available, for the installation of additional wiring. In many cases, however, the area between the hung ceiling and the structural floor above is used as a return-air plenum, in which case the types of cable and conduit permitted are restricted. There also may be restrictions on the equipment, such as lighting fixtures, permitted in or partially in the return-air plenum. Equipment installed in these areas must have metal enclosures, and the entire assembly must be fire rated for this type of use. Wiring through air-handling areas present under raised floors in information technology areas is governed by the *NEC* (Article 645) with various restrictions.

6.4 Panelboards, Switchboards, and Motor Control Centers

According to the *NEC*, a panelboard is a single panel or a group of panel units, designed for assembly in the form of a single panel including buses and automatic overcurrent devices, and equipped with or without switches for controlling lights, heat, or power circuits. Panelboards are placed in a cabinet or cutout box or are placed against a wall or partition and are accessible only from the front. A switchboard is a large single panel, frame, or assembly of panels on which switches, overcurrent devices, other protective devices, and instruments are mounted on the face, back, or both. Switchboards are usually accessible from both the front and the rear, and are not intended to be stored in cabinets. Motor control centers are assemblies of one or more enclosed sections having a common power bus that contain primarily motor control units. Collectively, this equipment provides a convenient and safe means for controlling the distribution of the power that feeds equipment and devices.

6.4.1 Panelboard Devices

Although devices are always being added to the electrical system, each lighting cabinet or appliance branch circuit panelboard is allowed to have a total of 42 overcurrent device poles. The configuration may be 42 single-pole devices, 14 three-pole devices, or any combination in between. Many panelboards can accommodate more than 42 poles, considering empty space on the board, and it can be very tempting to fill the available space without regard to the maximum number of pole devices allowed. Doing so, however, might lead to concerns about compromised safety and reliability due to excessive heat generation. The requirements for this equipment are addressed in the *NEC* (Articles 408 and 430).

6.4.2 Equipment Location and Access

It is essential to locate panelboards, switchboards, motor control centers, and other wiring equipment in convenient places that are readily available but not obtrusive. They should also be situated in ways that align vertically in multistory structures to minimize the horizontal feeder and branch circuit runs, and as close as possible to the loads they are serving, to decrease the chance for voltage drops and other losses.

The equipment must be locked and tamperproof but also easily accessible by *authorized* personnel. Since continuity of electric service is critical in health care facilities, hospitals in particular, it might be tempting for a facilities engineer to allow all staff working on a patient floor to have access to the panelboards. For example, when a circuit breaker opens, someone can immediately go to the panelboard and turn it back on, without having to wait for maintenance personnel to arrive. The problem with allowing all staff to have access to the panelboards, however, is that general staff usually are not properly trained in handling panelboards. They potentially could turn the circuit breaker back on when there is a short circuit or overload that could cause damage and possibly a fire. It is usually recommended that only trained personnel have access to the panelboards.

Another factor to consider when locating electrical supply equipment is the effect that this equipment might have on other electronic equipment, such as highly sophisticated electronic diagnostic and treatment equipment. Electromagnetic interference is a by-product of alternating current and normally is not strong enough to interfere with these sophisticated medical devices. Electrostatic interference, however, can be caused by arcing of switches and relays. Thus, to avoid electrostatic interference, electronic diagnostic and treatment devices should not be located in proximity to switchboards, panelboards, or motor control centers. If it is not possible to locate this equipment elsewhere, then the medical electronic equipment should be provided with proper shielding. Many new devices come equipped with electromagnetic or electrostatic shielding or both. It may also be possible to retrofit shields onto equipment that is already installed.

As noted previously for electrical service and transformer rooms, equipment rooms and areas should *not* be used as storage space. Clear working space around the equipment is required to ensure proper access at all times. Figure 6.4 shows the clear space in a switchboard room.

6.5 System Protection and Coordination

Switches, fuses, circuit breakers, short-circuit coordination, and ground-fault circuit interrupters provide overcurrent and short-circuit protection to the electrical distribu-

Chapter 6. Distribution Systems **71**

Figure 6.4 Switchboard room in a large hospital. (Note the clear working space around the equipment and the rubber safety mats in front of the equipment.)

tion system—the conductors and the equipment. For each ungrounded conductor, an overcurrent device must be located at the point where the conductor receives its supply. Feeder taps are permitted without overcurrent protection under certain conditions.

Most of the requirements for this equipment are not unique to health care facilities. The *NEC* (Article 240) covers general requirements for overcurrent protection switches, fuses, and circuit breakers, including allowable sizes, ampacity, ampere rating, voltage, location, enclosures, disconnecting means, marking, replacement, interchangeability, installation, and additional criteria. The *NEC* and NFPA 99 both address the requirements for ground-fault circuit interrupters for specific areas of use. Special requirements for protecting electrical systems in health care facilities are covered in detail in Chapter 7.

6.5.1 Switches and Fuses

A *switch* is a device that provides a means to connect or disconnect load conductors directly to and from a power source. A *fuse* is an overcurrent protective device with a part that opens a circuit when it is heated and severed by an overcurrent passing through it. A disconnecting means is a device or group of devices by which the conductors of a circuit can be disconnected from their source of power. Switches are used primarily as disconnecting devices and are manufactured either as fused or nonfused disconnects, depending on whether overcurrent protection is needed.

Switches come with from one to three poles, can be single or double throw, and can be operated manually or by means of an electromagnet. The most common type of switch used in hospitals is the knife-blade type. A bolted-pressure-type switch is usually used for capacity requirements more than 800 A. A fuse-puller-type switch and fuse installation can be used in service entrances under certain conditions. In these cases, the fuse holder pulls out of the circuit, so the device also serves as a switch or a disconnect. Switches used for the facility's service entrance must be rated for that type of use. Fuse types include plug fuses, Edison-base fuses, type S fuses, which have limited use, and cartridge fuses. Cartridge fuses come in different classes, such as current limiting, short-circuit interrupting, and the like. Figure 6.5 shows a main switch and distribution panelboard.

When disconnecting devices are required, such as for motors, appliances, and services, the *NEC* stipulates what size switches or fuses can be used, where the devices must be located, criteria for on/off switches, and so on. In many cases, these types of switches can be locked in the open position. This setting ensures that the switch cannot be closed accidentally when a circuit is being maintained and the switch

Figure 6.5 Main switch and adjacent distribution switchboard.

is out of sight of the maintenance personnel. When overcurrent and disconnection is required, a switch-and-fuse combination is used, during which the switch is always on the line side (supply side) of the fuse. When overcurrent protection is not required, unfused disconnect devices can be used.

6.5.1.1 Spare Fuses

Obviously, blown fuses must be replaced. To ensure immediate restoration of a circuit, a supply of spare fuses should be kept on hand. It is essential to have spares of the larger-type fuses used in the service switch and distribution panels because they are usually not readily available from outside sources. The spare supply should also include at least two of each type and size of fuse that the system uses. In situations where three-phase circuits use fuses as overcurrent protection, it is good practice to have at least six spare fuses of each size and type that are used. This will ensure that if the first replacement should blow, a second set is available as soon as the trouble on the circuit is corrected. It is good practice to replace all spare fuses as soon as they are used to maintain the emergency supply. All too often, a fuse blows when a replacement cannot be readily obtained, although the affected circuit must be restored immediately for a life-saving procedure.

6.5.1.2 Time-Delay Characteristics

Fuses used with motor circuits should have some time-delay characteristics. The time-delay characteristic of a fuse can be determined by looking at the characteristic time–current curve of the fuse. Generally, the larger the current that is flowing through the fuse, the sooner the fuse will blow and open. This type of characteristic is known as an inverse time–current relationship. The operator should make sure that the amount of starting current for the motor and the time it takes for the motor to return to its normal running current will not cause the fuse to open. Naturally, if the fuse is a replacement for an existing fuse that has been used successfully for some time, then an exact replacement may be used. One problem that may exist, however, is that if loads have been added to the circuit over time, the starting current may no longer be tolerated as it was when the circuit was originally put into use. With the starting current of the motor, the added loads now place the current level high enough to blow the fuse upon starting the motor. When conducting renovations, one should know the type and size of load that was initially on the circuit, the size and type of load that will be added, and how the total will be affected by the characteristics of the fuses being used.

6.5.1.3 Single Phasing

One problem with fuses being used for overcurrent protection is that each fuse protects only the single phase that it is on, so that three fuses are required for three-phase circuits. When a circuit is overloaded, a fuse opens only the overloaded phase. If only one fuse opens, this may mean that two of the phases remain powered, providing single-phase operation of a three-phase motor. Thus, when using fuses, single-phase protection should be applied for three-phase motors.

6.5.2 Circuit Breakers

Circuit breakers are devices that can both open and close a circuit manually, and open it automatically when a predetermined level of overcurrent occurs. Circuit breakers are considered to be both disconnects and overcurrent devices.

Circuit breakers come in single-, two-, or three-pole configurations. Multiple-pole circuit breakers open all phases, even if the overload or fault is on only one phase. A single-pole circuit breaker can be used as a multiple-pole circuit breaker if it has a handle tie to ensure that all poles operate simultaneously. Circuit breakers are made with either nonadjustable or adjustable trip devices so that the long-time pickup setting can be adjusted.

Any circuit breaker used in an enclosure increases the temperature inside the enclosure to above the ambient temperature outside the enclosure, due to the heat generated by the current flow in the panelboard. This is especially true in large panelboards and for circuit breakers mounted on the upper half of the panelboard. To compensate for this excess heat, circuit breakers have been rated for use in various locations and conditions and with consideration of the ambient temperature. Circuit breakers used in enclosures must therefore be derated if uncompensated or must be enclosure compensated.

Uncompensated circuit breakers are calibrated in 25°C (77°F) open air. *Enclosure-compensated thermal-magnetic circuit breakers* are calibrated in an enclosure with the ambient temperature outside the enclosure at 25°C (77°F); these circuit breakers will not have to be derated when used inside an enclosure. *Ambient-compensated circuit breakers* are rated at a much higher ambient temperature. This type of circuit breaker is used in high-temperature areas such as boiler rooms.

The most common circuit breakers used are uncompensated thermal-magnetic circuit breakers for use in open air. The most widely used circuit breaker in hospitals is the *molded-case thermal-magnetic circuit breaker*. In this type of device, the thermal aspect of the device protects the circuit from small overcurrents such as overloads. When there is a high current overload, such as when there are short-circuit conditions, the magnet part of the device reacts to open a circuit with a much quicker reaction than the thermal reaction time.

If circuit breakers are opening because of overloaded conditions, the facility engineer should check the actual load. It may be that the load does not exceed the circuit parameters but that the circuit breaker type must be changed to compensate for the extra heat dissipated in the panelboard.

While tripping of the circuit breaker usually requires only resetting the circuit breaker to put the circuit back in operation, sometimes the circuit breaker will be damaged or its ratings will be compromised. This is especially true if it has already tripped many times. This is another reason why it is good practice to keep spare circuit breakers of all ratings and types that are being used in a facility.

Like fuses, circuit breakers also have characteristic time–current curves, which must be referenced when designing or replacing a circuit breaker if the original conditions of the circuit have changed. The use of these curves is discussed in the following section.

Also available are electronic and microprocessor-based tripping units. These are used on low-voltage circuit breakers and are usually a part of the tripping element.

6.5.3 Selective Coordination

Selective coordination occurs when the system opens only the closest overcurrent device on the supply side of a fault. All other devices must remain closed. Short-circuit coordination preserves the electrical supply to as many areas as possible during a fault.

The operating parameters for all overcurrent devices have characteristic curves that indicate a time–current relationship, the amount of time needed for the device to open for a specific amount of current. Usually, the higher the current, the shorter the time needed. To limit the outage to the area downstream from a fault, the overcurrent devices that are the farthest away from the electrical supply must open more quickly than the overcurrent devices closest to the source of supply for the same amount of short-circuit current. For example, if a fault occurs at a piece of equipment or any place on a branch circuit, the branch circuit overcurrent device should open before the feeder overcurrent device. Similarly, if the fault is on a feeder, that feeder's overcurrent device should open before the main service overcurrent device.

Selective coordination among overcurrent devices should be established during the design stage of a new or renovated facility or repair project. The facilities maintenance manager should oversee short-circuit coordination and take the necessary precautions during renovations or when replacing an overcurrent device.

6.5.4 Ground-Fault Protection

Ground-fault protection occurs when a disconnecting means (i.e., a ground-fault circuit interrupter) opens all ungrounded conductors of a faulted circuit to protect equipment from damaging line-to-ground fault currents. Protection is provided at current levels that are lower than those that trigger overcurrent devices to operate. *Ground-fault circuit interrupters* (GFCIs) are devices that protect personnel from being shocked. They operate by deenergizing a circuit (or a portion of a circuit) within a certain period of time when a current to ground is higher than a predetermined level. This current level is less than the level required to operate the overcurrent device on the supply circuit.

Ground-fault coordination is necessary when ground-fault interrupters are installed in the distribution system. Both the *NEC* and NFPA 99 state general requirements for ground-fault protection at service switches and specific ground-fault requirements for health care facilities. For a health care facility, when ground-fault protection is required at the service switch or feeder disconnecting means as stated in the *NEC*, it is also required in all next feeder disconnecting means downstream toward the load (i.e., one step down from the service switch). If there is selectivity among the various stages of ground-fault protection and a ground fault occurs, then only the feeder that services the fault will shut down. There are three exceptions to this requirement: The additional level of protection does not need to be installed on the load side of an essential electrical system transfer switch or between an on-site generator and the essential electrical system transfer switch, or on an electrical system that is not a solidly grounded wye system with greater than 150 V to ground but not exceeding 600 V phase to phase. Chapter 8 discusses essential electrical systems in detail.

One specific requirement for ground-fault protection [NEC, Section 517.17(C)] is that there must be a six-cycle minimum separation between the two stages of ground-fault protection devices. Also, the trip setting and time required for the devices to operate should be considered when coordinating a system. For additional protection, ground-fault interrupters can be installed farther downstream as well. The only considerations are the cost of the added protection and the possibility of nuisance tripping, a problem that has been greatly reduced in recent years.

6.6 Grounding

The possibility of electric shock exists in any building with an electrical distribution system. In health care facilities, particularly hospitals, nursing homes, and surgical centers, the shock hazard is greater than in other facilities because of the greater amount of electrical equipment used for patient care. To avoid electrical shock to patients and staff, it is essential to properly ground all electrical systems and equipment in health care facilities, particularly where invasive procedures are performed on patients or electronic equipment is used for diagnosis and treatment. Ungrounded systems are used in special cases to conform to codes and for special systems. They therefore do not have system grounding. Ungrounded systems, which do require equipment cases to be grounded, are discussed in Chapter 9.

The term *grounding* usually refers to both system and equipment grounding. System grounding controls system overvoltages and provides fault protection by intentionally connecting an electrical system conductor and the ground. Equipment grounding is the electrical connection to the ground of all non-current-carrying metal parts of the distribution system and other equipment.

When renovating or maintaining a system or installing equipment, care must be taken not to interrupt the integrity of the grounding system. The resistance of the grounding circuits must always remain low enough to provide a low-impedance path to ground when there is a ground fault. In that way, all of the fault current will be directed to ground. Requirements such as the size of conductors, the size of ground rods, and so forth are stated in Article 250 of the *NEC*. Periodic testing of the integrity of the grounding system and its resistance should be part of the preventive maintenance schedule of all facilities.

Grounding requirements fall under three categories: general, special locations, and quiet grounding.

6.6.1 General Requirements for Grounding

Even if special grounding precautions are taken, patients may not be protected if the general grounding system in the facility is compromised. Anyone involved in designing, installing, or maintaining electrical distribution systems should become familiar with the essential grounding requirements for all types of structures as well as those applicable to health care facilities. The *NEC* (Article 250) provides general requirements for grounding, as well as requirements and methods for grounding circuits and electrical systems overall, including electrode systems and conductors; enclosures, raceways, and service cables; and equipment. The *NEC* also cover the bonding of metal parts to form electrically conductive paths that ensure electrical continuity and safety. Although the *NEC* was completely revised in 1999, for convenience, an Appendix or Annex is provided in the editions after 1999 that shows cross-references between the 1999 and later editions.

6.6.2 Grounding Patient Care and Other Areas

The *NEC* and NFPA 99 include additional grounding requirements for protecting patients in health care facilities. For example, the grounding terminals of all receptacles in patient care areas must be grounded by insulated copper conductors. The same requirement is applicable to all non-current-carrying conductive surfaces of fixed electrical equipment that are likely to become energized, are subject to personal contact, and are operating at more than 100 V. The copper conductor must be installed in a metal raceway and be connected to the grounding terminal on the receptacle and then to the grounding bus in the panelboard where the circuit originates. Metal raceways that qualify as an equipment grounding path must be used for wiring in patient care areas. Note that according to the *NEC* (Section 517.13), metal raceways do not need to be used in patient care areas where metallic cable armor or sheath assemblies are used, and they are an acceptable grounding return path.

In 1996, NFPA 99 dropped the requirements for testing the impedance and voltage for existing facilities. A facility should still provide an equivalent safe environment. These requirements are discussed in more detail in Chapter 12, on testing, maintenance, and record keeping.

Panelboards for both the normal power source (i.e., the main electrical supply) and the essential electrical system power source typically supply patient care areas with electricity. During normal service, however, it is possible for a patient or staff member to come into contact with the ground of both sources of power, since equipment often is connected to both sources. To avoid having ground currents present and the possibility of an electrical shock, albeit of small electrical potential, both grounding systems should be interconnected to ensure that the ground potentials of both systems are the same.

At one time, equipotential grounding systems were required in critical care areas and other patient treatment areas. In these areas, the space around each patient bed or table where a procedure took place had an equipotential ground bus, and all equipment was grounded to this point. This point, now called a patient-equipment grounding point by both the *NEC* and NFPA 99, is no longer required, although it is permitted. Consideration should be given to using this type of grounding system in areas such as coronary care units, intensive care units, operating rooms, and parts of the emergency department, especially trauma centers. The added stringency of the older code provided a higher level of ground protection in these areas, where much equipment, some of it invasive, is used on patients.

Reliable grounding of patient equipment at the patient-equipment grounding point can be accomplished with a row of grounding jacks connected together. This is usually done in an intensive care unit (ICU) or coronary care unit (CCU). Each piece of equipment has a separate grounding wire plugged into these jacks. The disadvantage of this arrangement is that each piece of equipment has an additional conductor coming from it, which could congest the area around a patient. Figure 6.6 shows a row of grounding jacks under the wall unit at the head of a patient in an ICU. In some cases a properly designed and maintained grounding system will be sufficient. The facility engineer, in consultation with the staff, is responsible for determining which grounding system to install.

Special requirements for grounding anesthetizing locations and wet areas are discussed in Chapter 10, on requirements for specific areas.

6.6.3 Quiet Grounding

The definition of a quiet ground that is listed in NFPA 99 is a system of grounding conductors insulated from portions of the conventional grounding of the power system, which interconnects the grounds of electric appliances for the purpose of improving immunity to electromagnetic noise. Quiet grounds are most often used when a ground circuit that is not subject to disturbances is needed to operate sensitive equipment and some computers properly, as well as to ensure the safety of the people who come into contact with this equipment or system.

When quiet grounds are used, a separate ground conductor is run from the equipment to the ground bus in the panelboard that supplies the circuit, similar to the arrangement in patient care areas. If this arrangement is not sufficient to eliminate any transients that get into the grounding system, it may be necessary to install a separate ground conductor, either from the distribution panel or from the service area to the

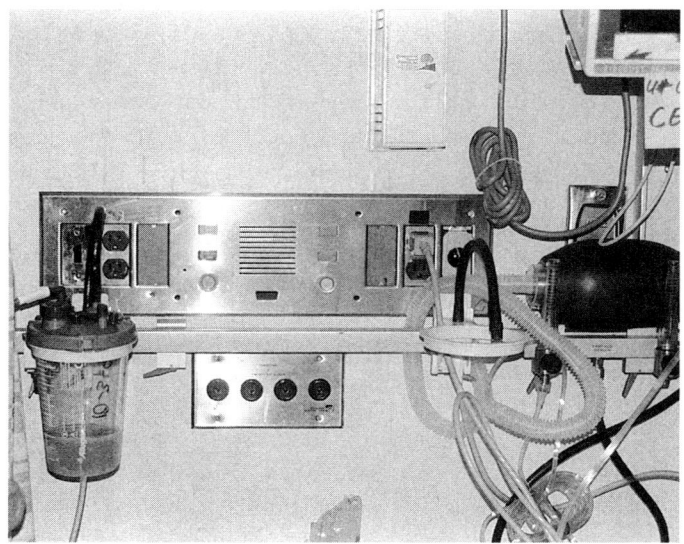

Figure 6.6 Headwall in an ICU. (Note that grounding jacks are located under wall unit.)

local panelboard. Installation should follow the manufacturers' requirements; if the requirements are stringent, manufacturers typically provide technical help.

6.7 Circuiting

The circuiting or layout of the conductors is the heart of any distribution system. It consists not only of the distribution system layout (i.e., the number of feeders and the distribution of the load on each feeder), but also the location of the disconnects. Unless an institution is in the midst of designing a new building or renovating an existing one, the facility engineer has very little input into how the electrical distribution system is laid out.

6.7.1 General Requirements for Circuiting

One important criterion for designing and renovating the circuiting of a distribution system is that it must be flexible enough to address the changing needs of a facility, as well as maintain the expected level of system reliability. For example, as the needs of medical care change, power requirements in different locations of a facility often change. A flexible circuiting system allows work to be performed without electrical outages.

One way to maintain flexibility of the circuiting of the distribution system is to provide actual spare capacity, not just in the main system but also in the various branches of the system being compartmentalized. It often appears that space is available on panelboards and wires that are not loaded to their maximum allowed ratings. However, when the facility engineer tries to make use of this spare capacity, he finds that the spaces cannot be used because the panelboard has the maximum amount of poles allowed by the code. Alternatively, additional load cannot be placed on the conductors due to the derating of the overcurrent devices protecting them as well as the derating factors necessary for the cable. When panelboards and wire have no additional capacity, the facility will have to take other measures to either increase system capacity or eliminate the need for the added load. Spare capacity should be real, not just a paper notation. (See the following case study.)

Inflexible Circuiting

A high-rise building had taps at every floor in electrical closets and used a bus duct system for the vertical feeders. For economic reasons, however, the only disconnects provided were at the main distribution board where the feeder started, and not at each tap on each floor. When the needs of the facility changed, additional power was needed on two of the patient floors. Enough spare capacity was available in the feeders for the floors, but since the voltage being used on these feeders was 480/277 V, the electrical contractor did not want to work on the tap with the connections energized.

The only way to work on the system was during a planned power outage. The outage took place at sunrise, which was early enough not to interfere with patient care

(Continued)

and to allow enough light in each patient room to minimize patient anxiety. The emergency generators were turned on to provide emergency power. However, all other power was turned off for the period of time needed to install the new feeder taps. Overtime staff was present in the event of an emergency with the patients. The entire process, which was costly and time consuming, could have been avoided had disconnects been provided at each floor, adding to the flexibility of the circuiting that was installed.

The circuiting of the distribution system should provide for shutdowns of small areas in the event of a short circuit or overload, as well as to accommodate maintenance and renovation activities. To provide for these shutdowns, additional feeders, branch circuits, and overcurrent devices must be installed. These added materials, of course, raise the installation, testing, and maintenance costs. Facility engineers and those with overall budget responsibilities must weigh this added cost against the potential loss of power in large areas in the event of an outage.

To prevent transients due to motor starting current from occurring on lines that supply medical equipment or light fixtures, motors should be powered by branch circuits from panelboards that are separate from the lighting or appliance branch circuits. This arrangement, too, becomes an economic decision. It may not be desirable to spend additional money for extra branch circuits or panelboards for a few small motors. However, due to the starting and stopping of these motors, the effect of the transient spikes on the medical equipment (i.e., possible wrong diagnostic results depending on the equipment) and lighting (if they were placed on the same panelboards) must be measured against the additional cost.

6.7.2 Specific Requirements for Circuiting

The specific requirements for circuiting a health care institution primarily relate to the needs of special equipment such as magnetic resonance imaging, information system areas, and laboratories. System parameters must be decided on an individual basis due to the specific requirements for each type of equipment and health care occupancy area and the effect they may have on other equipment and parts of the distribution system.

In most cases, these special areas require separate feeders with quiet grounds and usually large blocks of power. If necessary, the voltage for the feeder of a 120/208-V system can be stepped up to 277/480 V. If a 277/480 V system is used, then smaller conductors can be used, in which case the voltage would be stepped down at the load. Some equipment requires the higher voltage. Where ground-fault circuit interrupters are used, it is essential to verify that the electrical system meets the voltage and power requirements of commercially available ground-fault circuit interrupter equipment.

There are several specific requirements for circuiting patient care areas. The *NEC* states that in general care areas, each patient bed location must have at least two branch circuits, one supplied from the normal distribution system and one supplied from the emergency system. Only one normal branch circuit panelboard can supply a patient location. If emergency power is required, then these branch circuits can be served from more than one emergency branch circuit distribution panel. Critical care areas are required to have at least two branch circuits, a minimum of one served from a single automatic transfer switch and a minimum of one supplied from the normal distribution system. This requirement ensures that there will be some power for critical care patient bed locations in all power conditions. One exception to this rule allows branch circuits serving only special-purpose receptacles or equipment in critical care areas to be served by other panelboards. This same exception is true of general care areas. Another exception is that locations served from two separate transfer switches on the emergency system are not required to have circuits from the normal system (NEC).

The following case studies describe incidences in which the patient care area was not properly circuited.

Circuiting Patient Care Areas
In one extreme case, an entire patient care area was connected to the essential electrical system. During a generator test, the transfer switch did not transfer back to normal power. To work on the switch, its power supply had to be turned off. All power connected to the essential electrical system was lost during the repair, and temporary circuits had to be installed to maintain power during this time.

As seen in the following case study, certain circuit arrangements could have eliminated the problem faced by the previously mentioned facility and facilitated the testing of the emergency power system without turning off all power or jeopardizing patient care.

Testing the Circuiting System
One facility had a circuiting and service arrangement that enabled the facility engineer to test the essential electrical system in its entirety while the rest of the institution maintained its normal power. This service was designed so that one service switch would be dedicated to the loads that would be placed on the essential electrical system. During system testing, that service switch or the test switches on the transfer switches could be opened to simulate a loss of electrical service, while the rest of the institution remained with normal power. The test could be conducted at any time; therefore, the actual load that could be on the generator was connected.

Obviously, the areas that had both normal power and emergency power supplying it did not encounter any inconvenience other than the 5- to 10-second periods without power on the emergency circuits. If a test was desired to simulate a total power failure to check the reaction of the staff and patients, then all of the service switches could be opened. This institution had the usual type of arrangement in other parts of the facility, which faced losses of power during the test.

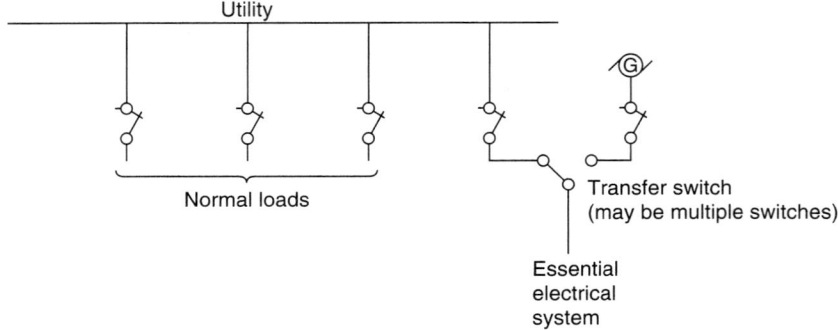

Figure 6.7 Distribution system with one service switch feeding the essential electrical system.

Figure 6.7 is a riser diagram of a system in which one service switch controls all of the normal power to the essential electrical system.

Chapter 8 discusses the essential and emergency power systems in more detail.

7

Electrical System Devices and Lighting

Electrical devices in health care facilities are much like those in other facilities. The main devices include receptacles, switches, and ground-fault circuit interrupters (GFCIs), which serve to provide safe and reliable systems. The lighting of health care facilities is also the same as for other facilities. Safe levels of lighting are required by the various codes in many areas throughout health care facilities, and comfortable levels of lighting are provided in accordance with acceptable standards. Lighting in health care facilities does serve several unique functions, however. It must assist patients in becoming oriented to day and night and must provide illumination to enable staff to perform complicated medical procedures. It also must facilitate diagnosis, treatment, and life safety.

7.1 Receptacles

As defined in NFPA 70, *National Electrical Code® (NEC®)*, 2005 edition, a receptacle is a contact device installed at the outlet for the connection of an attachment plug. A single receptacle is a single contact device with no other contact device on the same yoke, whereas a multiple receptacle provides two or more contact devices on the same yoke.

7.1.1 Receptacle Types

Receptacles come in various configurations, with different numbers of poles and different current and voltage ratings, and are designed for specific purposes. For example, locking receptacles are often installed in ceilings, such as in operating rooms. The blades of the plug are inserted into the receptacle and the plug is then turned and locked in place, preventing the cord from falling out or being pulled out.

The *NEC* (Article 210) includes some requirements for receptacles in terms of how they will be grounded, their required ampere ratings, and maximum loads. Tables 210.21(B)(2) and (B)(3) in the *NEC* indicate maximum allowed cord- and plug-connected loads to receptacles and ratings for various size circuits for the types of receptacles and plugs currently manufactured. The *National Electrical Code Handbook* has two charts in Article 406 that show the receptacle and plug configuration for various types and sizes of nonlocking and locking plugs and receptacles. These charts are

83

useful references, not only when replacing common-type receptacles but also when adding specific receptacles often needed during repairs or to support newer equipment. Various other articles in the *NEC* list the requirements for receptacles.

The *NEC* requires any receptacle used on a 15- and 20-A branch circuit to be of the grounding type. The American Institute of Architects' (AIA's) *Guidelines for Design and Construction of Hospital and Health Care Facilities* further require patient rooms in most types of health care facilities to be equipped with duplex grounded receptacles. Any ungrounded receptacles still in service in older facilities should be replaced with the grounded type. This may necessitate installing ground wires where none presently exist.

The AIA *Guidelines* further stipulate the types of receptacles that should be installed in pediatric and newborn intensive care units, trauma and resuscitation rooms, and emergency department examination and treatment rooms. While many facilities are not required to comply with these guidelines, they do provide an excellent reference. Both the *NEC* and the *Guidelines* require pediatric areas to have tamper-resistant receptacles or listed tamper-resistant covers for receptacles. The *Guidelines* also require the use of tamper-resistant receptacles in psychiatric units.

Another *NEC* requirement (Article 517) is that receptacles located in patient bed locations must be listed as *hospital grade*. These receptacles are manufactured for use in areas where equipment is constantly being plugged in and removed, and they should be used for as much equipment as practicable. Commonly referred to as green-dot receptacles, hospital-grade receptacles are usually identified with a green dot on the face of the receptacle. Although some top-of-the-line specification-grade receptacles might pass the same, if not greater, rigors as some hospital-grade receptacles, some lesser-quality specification-grade receptacles do not meet the standards of hospital-grade receptacles. Before 1990, the *NEC* required only critical care bed locations to have hospital-grade receptacles. A three-prong grounding plug should always be used where a grounding conductor is part of the cord assembly.

7.1.2 Receptacle Specifications for Various Locations

Most codes stipulate how many receptacles should be installed in the various areas of health care facilities. The *NEC* and NFPA 99, *Standard for Health Care Facilities*, require four receptacles to be installed in general care patient bed locations and six receptacles to be installed in critical care patient bed locations. The AIA *Guidelines* require one duplex receptacle on each wall of a patient room, as well as in other locations in health care facilities. The *Guidelines* also specify additional requirements for the location and use of the receptacles in patient bed locations, such as at the foot of the bed for televisions.

The AIA *Guidelines*, as well as many state and local codes, also specify the location for corridor receptacles. In most cases, these receptacles provide electricity to operate cleaning machines and maintenance equipment. The facilities engineer should coordinate with the maintenance and housekeeping supervisors regarding the types of equipment they use. Cleaning department personnel often use "cheaters" (i.e., gadgets that change the configuration of the receptacle to fit the plug on the piece of equipment; this is done by having the plug fit the receptacle on the wall and on the other end of the cord connect to a receptacle that matches the equipment that is to be plugged in). If such is the case, the receptacle should be changed. The use of cheaters should be prohibited, because it is potentially hazardous to plug in a piece of equipment that is incompatible with the electrical ratings of the branch circuit to which the equipment is connected or the provisions for grounding and overcurrent protection.

Although the codes do not prohibit or require the installation of receptacles in *bathrooms* or *toilets*, there is much controversy among design and facilities engineers as to whether they should be installed here. These receptacles, which are installed for the convenience of patients using electric razors, hair dryers, and other personal equipment, can present a shock hazard. All receptacles that are installed in bathrooms and toilets must be connected to a GFCI, which interrupts the power when a ground fault exceeds a certain level. The decision to install receptacles in bathrooms and toilets must be made by the facility designers, as well as the staff. GFCIs are discussed in more detail in Section 7.3.

Receptacles in other *wet locations*, such as kitchens, some laboratories and operating rooms, or where hydrotherapy takes place, must also be equipped with ground-fault interruption protection if interruption of power under fault conditions can be tolerated. (If interruption of power cannot be tolerated in these areas, they must be served by an isolated power supply—see Chapter 9.)

7.1.3 Adding Receptacles

Although many state and local codes require the installation of at least a minimal number of receptacles, sometimes the minimum number stated in the codes may not be sufficient for the proper functioning of the room. Very few authorities having jurisdiction, however, permit the use of extension cords or cube taps other than for an emergency, and then only until other accommodations for the load can be made. Both of these types of devices can cause overloads or fire and shock hazards. Extension cords are also often stepped on or rolled over when moving equipment around in patient rooms and can become overheated or frayed. Thus, in older facilities, additional receptacles may need to be installed to accommodate new equipment or new procedures.

If there are financial constraints to installing more than the required minimum number of receptacles, the facility engineer can gather statistics on the number of cube taps and extension cords that are being used, the number of times the extension cords have become frayed and been repaired, and related information. These data may persuade the decision makers to allow the installation of additional receptacles. *Usually Joint Commission Accreditation of Healthcare Organizations (JCAHO) surveyors and other inspectors do not favor the use of cube taps, especially in patient care areas.*

Adding receptacles can be accomplished when renovating or upgrading the electrical system in a room or an area. Figure 7.1 shows the wall behind a patient bed equipped with surface-mounted receptacles that were added during a renovation.

7.1.4 Safety Precautions

To ensure system safety and reliability, all the wires in an institution must have the proper polarity. Polarity checks are accomplished by testing the wires at each

86 Part II. Electrical Systems and Equipment

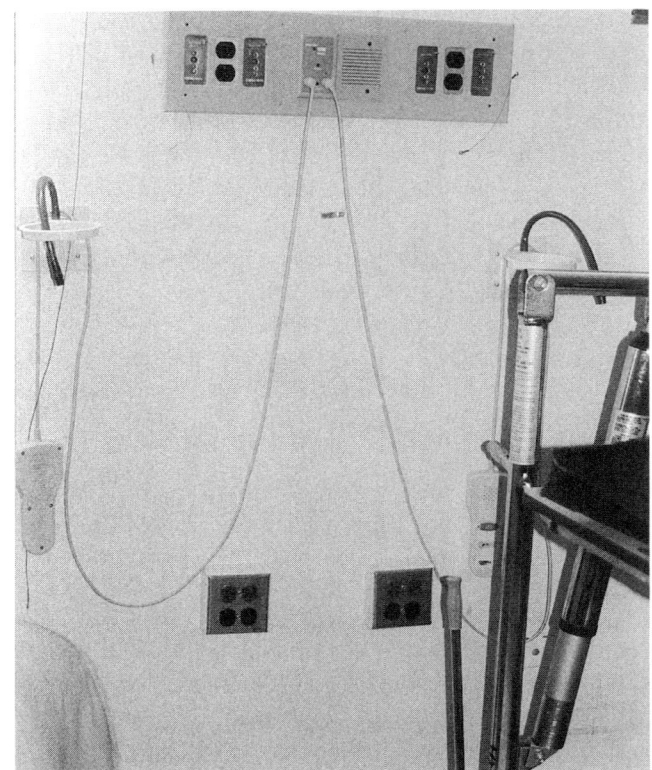

Figure 7.1 Surface-mounted receptacles added during a renovation.

receptacle, either when conducting acceptance tests on new installations or while doing preventive maintenance. Because electrical inspectors typically do not conduct polarity checks, facilities engineers must do so. While checking polarity of the wires, the stability of the receptacle and outlet box also should be checked to make sure they are firmly attached to the structure and will not move when plugs are inserted or removed. The cover plate should *not* be the item that secures the receptacle and box.

Facilities engineers must also ensure that all 15-A and 20-A convenience duplex receptacles are installed and function properly, which may not always be the case. See the following case study.

Faulty Construction of a Duplex Receptacle

In a new facility, when checking out the electrical system before occupancy, the staff found that it was difficult to insert plugs into both of the outlets of a duplex receptacle. This receptacle had been wired around one of the screw terminals, with the other

(Continued)

screw terminal not tightly closed. The problem did not occur when the receptacle was back-wired. Further investigation of the problem revealed that in the portion of the receptacle where the plug was forced in, contact was made on only one point of the plug blades, not on the length of the plug blades. It appeared that the configuration of the receptacle's plastic case allowed the copper inside the receptacle that led from one outlet to the other to deform when one plug was inserted, making it necessary to force the second plug into the other part of the duplex receptacle. This action, in turn, deformed the blade of the second plug. This condition was remedied by changing all of the receptacles to another brand that did not exhibit this problem.

The type of problem noted in this case study could go undetected if only one plug is used in a duplex receptacle or if the staff does not report any difficulty when inserting two plugs into a duplex receptacle. This condition can be easily checked, however, when doing an initial inspection or when performing preventive maintenance in a room, just by inserting two new plugs into each duplex receptacle. If one plug must be forced, the receptacle should be checked further. When performing the test, one must make sure that the blades of the receptacle do not become deformed and are always parallel with each other.

Note that, effective December 1997, Underwriters Laboratories changed its test procedure (UL 498, *Standard for Attachment plugs and Receptacles*) to ensure that all products would be free of this problem. To be listed by Underwriters Laboratories, all receptacles must meet the new test.

Another obvious safety precaution that should be part of every facility's safety program is that staff members should be taught how to remove plugs from receptacles properly. All too often, the equipment's cord is pulled on to remove a plug from a low-wall receptacle. Even if the cord is knotted, pulling on it will stress the conductors and insulation and eventually cause the cord to fray, which could create a fire and shock hazard.

Another safety measure is that plugs should always match the receptacles in which they will be used. When purchasing equipment, determining the configuration of the plug that will be used for that equipment can save a large amount of retrofitting after the equipment is brought into the facility. The compatibility of the plug to the receptacles can be determined when either the engineering department or the clinical engineering department inspects the equipment's safety and performance.

In terms of maintenance, every facility should have in stock at least one of every type of receptacle used in the facility. The number of receptacles used and the frequency of repair will determine the actual number of receptacles to keep in inventory.

7.2 Switches

Many types of switches are used in health care facilities. Whatever type is installed, it is important to make sure that each switch is rated for the type of service for which it will be used. It is also important to install switches to conform to convention. That is, make sure "up is on and down is off!"

7.2.1 Switch Types

The most common switch used in all facilities to control lighting or provide power to equipment is the *snap switch*. Snap switches are also used to control motors, such as those in small exhaust fans. Because snap switches are not always readily accessible to the patients, more accessible *pull-chain switches* are used to control patient reading lights. Because these switches are used rather frequently and often require high maintenance, facility engineers should always maintain a spare supply of pull cords and pull-cord switches. The use of silent or quiet-type switches should be considered for all patient bed locations, if not for the entire facility, as a courtesy to resting patients while hospital staff are taking care of other patients during nighttime hours. *Three-way switches* should be used when an area has two entrances and it is desirable to control the lights or motors from both locations.

Health care facilities use many types of time switches. One type of time switch uses a clock to regulate the on–off sequence and control the lighting of certain areas. Another type of time switch activates as soon as it detects motion and turns off after a preset time in which it does not detect motion. This type of time switch is often used to control automatic door openers and the lighting in some areas and offices. Switches that are activated when a beam of light is broken can also control automatic door openers. Switches controlled with photocells are often used to activate outdoor lighting. Photocells must be maintained and kept free of dirt to operate properly. Some common special-purpose switches include *temperature-controlled switches*, such as thermostats and break glass switches, which are used for emergency shutoff.

Circuit breakers with the proper rating can be used as switches. However, to avoid turning off the power in areas that should retain power at all times, only a few trained personnel should have access to the panelboard where circuit breakers are used. Refer to the case study.

Circuit Breakers Used as Lighting Controls

One facility used circuit breakers to control lighting in an office, and all personnel had access to the panelboard. The last person to leave the area was responsible for turning off the lights. Unfortunately, even though the circuit breakers were labeled, many times when the lights were turned off, equipment such as facsimile machines and answering machines that were to remain on also were turned off. This could easily happen in other areas.

7.2.2 Switching Systems

A switching system with relays (i.e., devices that can control the switches in the same or a different circuit) is sometimes installed to control many circuits that are all used for the same purpose. For example, this type of arrangement is often used where large areas of lighting composed of many circuits are controlled by one switch. In this configuration, the switch controls the relay, which in turn opens and closes all of the circuits to which the relays are connected. This arrangement can be accomplished at a panelboard that has a split bus, which is commonly called a split-bus panelboard, with the relay controlling power to a portion of the panelboard. The relay circuit can be a low-voltage type.

7.3 Ground-Fault Circuit Interrupters

The shock hazard present to patients, visitors, and staff in health care facilities is always of great concern due to the potential for line-to-ground faults. As mentioned in the preceding discussion on receptacles in wet locations, areas that can tolerate a loss of power can be equipped with ground-fault circuit interrupters.

Both the *NEC* and NFPA 99 define GFCIs as devices that interrupt power to a load when there is a fault to ground. The power interruption occurs at a predetermined value of the ground-fault current. The *NEC Handbook* (Section 210.8) notes that GFCIs operate on currents of 5 mA (milliamperes), and states that listing standards permit a differential of 4 to 6 mA. NFPA 99 requires GFCIs to interrupt power in wet locations when the ground-fault current exceeds 6 mA.

Facilities engineers should be aware that GFCIs protect personnel from shock hazards only when there is a fault to ground, not where the contact is phase to neutral or phase to phase. This type of protection is desired because the fault to ground through a patient or staff member can be low enough in current so as not to open an overcurrent device but still large enough to create a shock hazard. For patients who have indwelling catheters and devices connected to the skin, these low values of current can even cause death.

7.3.1 GFCIs in Wet Locations

While many institutions prohibit the use of hair dryers, electric razors, and similar devices in patient bathrooms, not all patients, visitors, or staff adhere to these rules. There are cases when a malfunctioning piece of equipment caused the GFCI to open the circuit while the operator was not aware that there was ground leakage. In some instances, the shape of the plug will cause the test switch to operate and open the circuit.

Both the *NEC* and NFPA 99 require the use of GFCIs that limit the possible ground-fault current caused by low-value first faults in wet locations. The GFCI can be accomplished by means of a receptacle or a circuit breaker with an integral GFCI, a portable plug-in-type GFCI, attachment plugs with integral GFCIs, or any other listed system. The site-specific situation and requirements will determine which GFCI or combination of GFCIs should be used and at what locations.

7.3.2 GFCI Installations

There are various ways to install GFCIs that not only meet the codes but also provide the desired level of protection above the code. The method used to obtain the least interruption of power is to equip each receptacle with its own integral GFCI, which protects only that one receptacle. When there is a ground fault in the equipment plugged into a receptacle with its own integral GFCI, it will be the only receptacle to lose power. Although this type of GFCI protection is the most costly and time consuming to install and to test, faults are easy to locate when they do occur. These GFCI receptacles contain their own test and reset switches. Figure 7.2 shows a receptacle with integral GFCI protection.

Integral GFCIs can also be arranged so that they protect receptacles downstream from the GFCI installation. While this is a less expensive arrangement than installing integral GFCIs on each receptacle, all the receptacles downstream from the GFCI installation will lose power when there is a ground fault on a downstream receptacle. (Downstream is away from the source of supply. One GFCI receptacle can be placed so that it will protect other receptacles downstream from it.) It will also be more difficult to locate the fault because all the receptacles downstream that are being used must have their loads tested to determine which one caused the trouble. It is helpful to mark distinctively all of the receptacles protected by this type of GFCI installation, to indicate exactly where their GFCI controls are located.

Entire circuits also can be protected by installing integral GFCIs on the circuit breakers. This installation basically has the same drawbacks as having one GFCI controlling other receptacles downstream—when there is a fault at one part of the circuit, all receptacles on that circuit will lose power, and it can be a challenge to determine the location of the ground fault. The same marking recommendations apply as well; it is helpful to mark each receptacle on the circuit distinctively, with an indication of the GFCI protection, the location of the circuit breaker (i.e., the panelboard number and exact physical location), and the number of the circuit breaker in that panelboard.

Figure 7.2 A 15-A duplex receptacle with an integral GFCI that also protects downstream loads. *Source: NEC Handbook,* 2005, Exhibit 210.8; photo courtesy of Pass & Seymour/Legrand®.

Facility engineers should know that there is a requirement for installing temporary GFCIs during construction, renovation, or repairs to protect construction workers. This requirement is for all temporary installations. Use of portable GFCIs and attachment plugs with GFCI protection are two options to consider. The *NEC* (Section 590.6) addresses the use of GFCIs for temporary installations.

7.3.3 GFCI Testing

Regardless of its location, each GFCI should be tested on a regular basis. Where individual GFCIs control only one receptacle, the facility engineer might instruct some staff members on how to reset a receptacle if one should open. Most people today know how to do this, as GFCI receptacles have for some time been required in the bathrooms of homes, hotels, and motels. The important fact to impress upon staff is that after the faulty equipment is unplugged and the receptacle is reset, the equipment itself must be thoroughly checked for damage before it is plugged in again.

7.4 Lighting

While this book is not a design manual, knowledge about some of the basic principles of lighting can assist the facilities engineer in making lighting recommendations for a range of projects, from repainting a wall to making major renovations.

The amount of overall light present in a room is the result of a combination of the types and number of fixtures used; the amount of natural light present; the light sources; the colors used for the floor, walls, and ceiling; and the finish used on the floor, walls, and ceiling (i.e., smooth and shiny tiles or nonreflective carpeting). Naturally, the lighter the colors and the more reflective the surfaces, the greater the amount of light reflected from the surfaces. Other aspects of lighting are brightness, brightness ratio, and glare factor. That is, if a person is concentrating on a brightly lit object and the surrounding area is much darker, the person requires a period of adjustment when shifting his attention from one area to another.

Most areas in health care facilities are illuminated by using a combination of natural and artificial lighting. State codes often require patient rooms to have windows, which help to orient patients in terms of the time of day or night. It is also not unusual for people to want an office with a view to the outside.

Although artificial light comes from a variety of sources, the most common sources are incandescent bulbs and fluorescent tubes. Other sources include high-intensity, metallic, and pressurized lights. Many of these light sources require ballasts, and in some cases the fixtures must warm up for a few minutes before providing full illumination. A fixture that requires a warm-up period would not be desirable in high-security areas or areas that lead to means of egress. Few people would appreciate having to wait in the dark for a few minutes until a fixture or ballast warms up after a power failure, while the emergency power source was able to restore power elsewhere in the facility within 10 seconds of the power loss.

7.4.1 Lighting Codes and Standards

The Illuminating Engineering Society of North America (IESNA) recommends lighting levels for different areas and functions in health care facilities. Refer to the *IESNA Lighting Handbook* (formerly consisting of *Applications Volume* and *Reference Volume*) for specific information on lighting in various locations of a health care facility. Many codes, standards, and recommended practices reference these IESNA publications.

The AIA *Guidelines* include generalized lighting requirements, the most detailed being for hospitals. These guidelines reference the *IESNA Handbook* for requirements on general and emergency lighting and NFPA *101*®, *Life Safety Code*®, for requirements on illuminating the means of egress and exit ways. The *NEC* includes many general requirements for the construction and installation of fixtures and the calculation of the electrical load for fixtures but no codes for required levels of lighting or fixture types or restrictions on such conditions as glare and brightness.

7.4.2 Lighting Specific Areas

Many areas within health care facilities require special consideration when designing, installing, using, and maintaining lighting fixtures and wiring.

7.4.2.1 Patient Bedrooms

One of the most important areas in any health care facility is the patient bedrooms. Most codes include specific lighting requirements for patient rooms to ensure the well-being and safety of patients. In addition to natural lighting, patient bedrooms require general lighting and examination lighting, as well as reading lights and nightlights.

General lighting is required to assist the staff in maintaining the room and treating patients, and to make it safer for visitors to move about. When designing general lighting for patient rooms, it is useful to keep in mind that many patients lie in bed all day facing the ceiling and often do not appreciate looking at glaring lights. Patient bedrooms also should have nightlights that are bright enough for the staff to access the room and check on patients but not so bright as to disturb the patients' sleep.

Each patient bed area also should have an examination light. Some examination lights use light sources that direct the light to a small area but that give off a lot of heat. When using these lights, care must be taken so that the light source does not dry up a wound while an examination is taking place. It is imperative that hot areas of the fixtures do not come into contact with patients staff members, or any items that might burn.

The AIA *Guidelines* recommend that care should be taken to prevent burns to patients. As an added precaution, the light source should be protected. This can be accomplished with a diffuser or a lens. In many installations, flexible arms are used on reading lights. Care must be taken to ensure that these light sources as well do not come into contact with bed linen, and that patients do not fall asleep with the light on and then come into contact with the fixture. As mentioned in the preceding section on switches, reading light controls should be accessible to the patient, both for convenience and safety. In most cases, other than in psychiatric areas, pull chains are used to control reading lights. Figure 7.3 illustrates a patient bed lamp.

Chapter 7. Electrical System Devices and Lighting

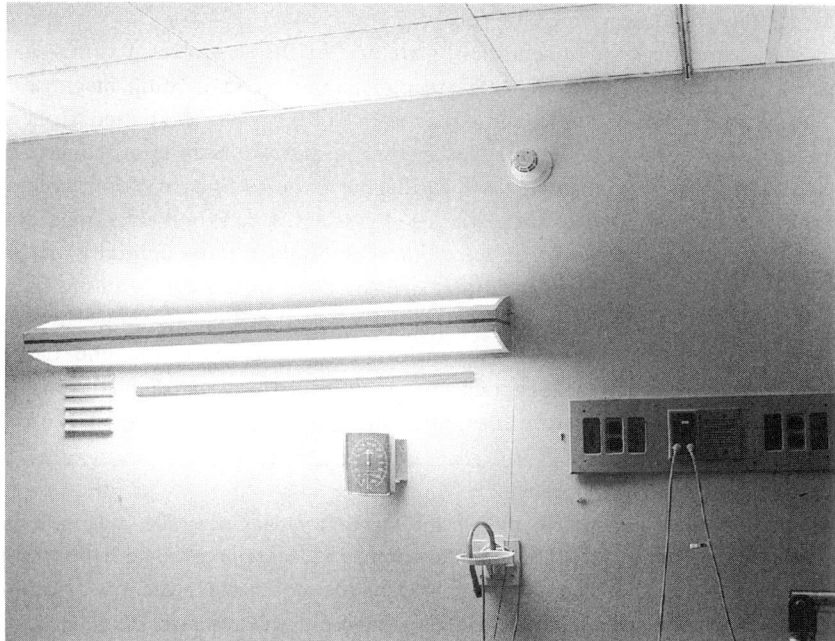

Figure 7.3 Patient over-the-bed lamp, with pull chain for convenience and safety.

7.4.2.2 Corridors

Corridors in nursing units require lighting that remains on at all times. The lights should be able to be dimmed at night for the convenience of the patients but still be adequate for the staff to move about and provide service. One way to provide this type of lighting control is to connect the lighting fixtures that will remain on for 24 hours directly to the circuit breaker panel and the essential electrical system. In this way, these lights will not be able to be turned off accidentally and will provide light during an electrical power outage without anyone having to switch on the lights at any time. The remaining light fixtures, which can be turned on and off, can be connected to switches at the nurses' stations.

7.4.2.3 Elderly Care Areas

Special care must be taken to provide for the safety of the elderly when lighting elderly patient care areas. Shadows and drastic changes in lighting levels can create a walking hazard for the elderly (as well as for people who have visual impairments), who might interpret a change in the level of illumination as a change in the level of the floor beneath them. In general, elderly care areas require a slightly higher level of illumination and a low level of contrast and brightness of the lighting levels between adjoining areas. Care must be taken in selecting fixture types that reduce extremes in lighting levels and shadows, especially considering the economic shift toward using less expensive task lights that create more extreme lighting differentials.

7.4.2.4 Diagnostic and Examination Areas

As in patient care areas, color rendition, glare and brightness ratios, fixtures on movable arms, and high-intensity lighting are considerations in lighting diagnostic and examination rooms. For example, some diagnoses are based on the patient's perceived skin tone or color. This is especially true in the newborn nursery and neonatal intensive care units. Therefore, light fixtures in diagnostic areas must not emit colors that affect the skin tone or color of the patient being examined. When selecting lights for diagnostic areas, keep in mind that the color of the object being lighted is not only a result of the color of the object itself but also the color of the light emitted from the fixture and the color of the surrounding walls, floor, and ceiling. As shown in the case study, certain manufacturers of very sensitive electronic diagnostic and treatment equipment require the use of incandescent lamps in fixtures that light the diagnostic areas in which their equipment resides, prohibiting the use of fluorescent lamps.

Fluorescent Lamp Interference

New equipment in a cardiac catheterization laboratory was controlled by a remote control unit similar to ones used for television sets. The remote would not operate properly, however, until the fluorescent lamps in the room were replaced with incandescent ones. After this situation was discovered, the manufacturer distributed an addendum to its product literature that addressed the environmental requirements necessary for the proper functioning of their equipment.

Radio-frequency interference (RFI) filters must sometimes be used where fluorescent lamps are used. These filters, which block out radio-frequency waves that can cause interference to the proper functioning of certain electronic equipment, can be obtained either as shields for the lens of the fixture or already equipped around fluorescent light ballasts. Keeping records about the location of these special installations assists facility engineers in installing proper replacement parts when necessary.

7.4.2.5 Areas with Isolated Power Supplies

In all areas equipped with an isolated power supply, which is an ungrounded source of power (see Chapter 9), care must be taken to keep the total leakage current from the wiring, lighting, and equipment less than the amount that will trigger the leakage current alarms of the isolated power supply. It should also be low enough to allow for the installation of future equipment. Any fluorescent lighting fixtures installed in these areas may need to have low-leakage ballasts.

7.4.2.6 Means of Egress

Egress lighting includes exit lights, as well as lights in the egress passage. Refer to Chapter 17, section on life safety, for a detailed discussion of the requirements and recommendations for illuminating the means of egress.

7.4.2.7 Outdoor Areas

Outdoor lighting helps to provide security by promoting the safe movement about the grounds of a facility. Outdoor lights are often controlled by means of time switches or photocells. Photocells require periodic cleaning for proper functioning. In high-security areas where surveillance cameras are used, lighting levels must be coordinated with the requirements of the cameras.

7.4.3 Emergency Lighting

Adequate lighting should be provided during power failures for diagnosis and treatment, security and safety in moving about, and to help keep patients calm in times of emergency. Refer to Chapter 8 on essential electrical systems for details regarding emergency lighting.

7.4.4 Energy Conservation

Newer facilities are presently designed in accordance with local and national energy codes, for environmental and economic reasons. Lighting levels of the past, however, were generally higher than they are today, based on past design trends and the level of lighting that was deemed acceptable. As a result, lighting levels in some older facilities (when the fixtures are clean) are much higher than is necessary to meet today's standards. By learning present lighting needs and existing levels of use and selecting a variety of new fixtures and retrofit equipment, efficient lighting systems that conserve energy can be created.

The first step toward energy conservation is to conduct a lighting survey. The survey should include, at a minimum, collecting information on the general illumination level in each room or area; how the lights are controlled; the types of fixtures that are used in each room; the number, size, and type of lamps and lenses present in each fixture; the type of ballasts used in fluorescent fixtures; whether any special reflectors are used in the fixtures; and the maintenance that is performed on the fixtures. After the facilities engineer inventories the lights and controls that a facility uses, he or she can replace or retrofit them as necessary to reduce energy usage and long-term costs.

Another step that can be taken to reduce energy usage is to replace lamps with high wattage-to-lumen-output ratio, such as incandescent lamps, with low-ratio lamps, such as fluorescent lamps. Fluorescent lights can also be retrofitted to reduce the number of lamps. Lighting controls that allow lights in unoccupied areas to be turned off are another important energy-saving device. Either time-controlled switches or switches that are sensitive to movement and occupancy can be installed for this type of control. The only time lights should be left on in unoccupied areas is for security purposes.

Many manufacturers today make high-reflectance reflectors that can be installed in existing fixtures to raise the lighting levels. Using a combination of reflectors and a reduced number of lamps in certain instances can result in equal or higher lighting levels and less energy consumption than was provided by the original fixtures. The facilities engineer has two choices when retrofitting a four-lamp, two-ballast fixture to three lamps. One of the ballasts can be removed and replaced with a single-lamp ballast, or a phantom tube can be used in place of the removed tube. Note, however, that the reflectors of all fixtures may not provide the desired results. It may be useful to see a demonstration of a few typical fixtures before purchasing this equipment; record the lighting levels before and after installing them, and clean the lens and fixture before and after the retrofit. These steps can lead to some excellent results, as well as avoid a potential disaster. (Refer to the case study.)

Improper Lighting Retrofit

A large institution retrofitted many corridor fixtures in a number of buildings with reflectors. It also replaced one working tube with one phantom tube in each fixture, so that only three tubes worked in four-tube fixtures, and only one tube worked in two-tube fixtures.

The institution tested only one type of fixture before purchasing the reflectors and phantom tubes. When the work was completed, the results were as expected in areas that used the same type of fixture that was tested. In the long-term care facility, however, where lighting levels are so important to the elderly, a different type of fixture was used than the ones in the rest of the institution, and the lighting levels obtained were lower than they were before the retrofit. The original tube configurations had to be reinstalled. Many arguments ensued, and the distributor of the reflector was threatened with lawsuits concerning stated and implied guarantees.

Incidentally, when receiving complaints concerning low lighting (and low temperatures), which may happen after a lighting retrofit, facility engineers should investigate room colors. The perception of low lighting may not be from the level of lighting per se but from the colors used in the area.

7.4.5 Lighting Maintenance

A proper maintenance program should be set up for the lighting system. Dirt on the lenses or on the reflectors inside the lighting fixtures reduces lighting levels. Facility personnel should keep records concerning the use and maintenance of special fixtures, such as those used in operating rooms, to help evaluate the performance of the fixtures and make decisions regarding how to replace them when necessary.

Balancing lighting requirements with budgetary constraints can be a challenge. To assist in this process, the first step is to make sure that proper lighting controls are installed to facilitate turning off lights when not in use. Second, it is important to make sure that all lighting that is installed is as maintenance free as possible. Special tools or large amounts of time should not be required for periodically cleaning the fixtures and their lenses or replacing lamps. High-maintenance equipment should not be purchased if at all possible. For example, lighting fixtures on brackets require high maintenance. They frequently break and are notorious for banging into and damaging walls and wall finishes.

8

Essential Electrical Systems

Essential electrical systems (EESs) in health care facilities basically are systems of alternate power that are designed to ensure the continuity of electrical service to specific areas and pieces of equipment when there is a disruption in the normal source of power [NFPA 70, *National Electrical Code*® (*NEC*®); NFPA 99, *Standard for Health Care Facilities*]. Authorities having jurisdiction (AHJs) may highly *recommend* that industries such as chemical processing, research, and data processing have emergency sources of power but *require* that health care facilities have such electrical systems. The type of essential electrical system required depends on the type of facility and the medical procedures performed.

Essential electrical systems include emergency sources of power and their distribution systems and equipment and means to transfer power between normal and emergency sources. The systems also include alarms, indicators, and safety devices to alert staff of any system malfunctions. As with all aspects of electrical systems in health care facilities, essential electrical systems require testing and maintenance. It is important to keep in mind that emergency power systems are typically a subset of essential electrical systems; the two systems are not interchangeable when referring to health care facilities.

8.1 Codes and Standards

In addition to state and local authorities, the *NEC* (mainly Chapter 7 and Article 517); NFPA standards 99, *101*®, *Life Safety Code*®, 110, *Standard for Emergency and Standby Power Systems*, and 111, *Standard on Electrical Energy Emergency and Standby Power Systems*; the Joint Commission on Accreditation of Healthcare Organizations (JCAHO's) comprehensive accreditation manuals for various types of health care facilities; and the American Institute of Architects' (AIA's) *Guidelines for Design and Construction of Hospital and Health Care Facilities* include requirements for essential electrical systems.

NFPA 110, which addresses *emergency power supply systems* (EPSSs) and *standby power systems*, and NFPA 111, which covers stored emergency power supply systems (SEPSSs) and *standby power systems*, typically have jurisdiction over the generators, emergency power sources, transfer switches, ancillary equipment, and transfer and operating times needed for the safe and reliable operation of essential electrical systems. Standby power simply refers to the alternate power supply used when the normal power system fails, such as a generator or battery system. Stored electrical energy refers to power provided by central battery units. NFPA 110 and NFPA 111 delineate *types*, *classes*, and *levels* of EPSSs and SEPSSs, based primarily on the urgency of a facility to restore power and the length of time the emergency source has to function.

NFPA 99 stipulates requirements for essential electrical systems based on the types of activities that take place within a facility and the arrangement of the normal and alternate power supplies. NFPA 99 references information from NFPA 110 and NFPA 111 for coverage of the requirements for the alternate sources of power required for each type of EES. The *NEC* references NFPA 110 and NFPA 111 with respect to construction and installation of EESs in hospitals, nursing homes, and limited care facilities. Since many of the requirements listed in the *NEC* are also listed in NFPA 99 and vice versa, it is important to note the scope of each document. NFPA *101*, includes all regulations for emergency lighting of means of egress. JCAHO manuals and the AIA *Guidelines* generally refer to or accept conformance to the NFPA codes and standards as evidence of compliance. Anyone responsible for an EES should have a thorough knowledge of these lengthy and detailed codes.

8.2 Types of Essential Electrical Systems

NFPA 99 stipulates requirements for three types of essential electrical systems in health care facilities, simply referred to as Type 1, Type 2, and Type 3, and states in which facility each type of system must be used. The codes require that the type of EES selected should be determined by the extensiveness of the medical procedures being performed in a facility. That is, facilities that perform surgery and other invasive procedures and use life support equipment and general anesthesia must use the most extensive type of essential electrical system, whereas facilities that perform simple procedures that do not require life support equipment and general anesthesia can use a less extensive type of EES. The three types are classified according to many items, including classifications given in NFPA 110, the breakdown of loads the systems serve, the distribution system, the source of power, and so on. The codes further stipulate requirements concerning time delays for powering up equipment, the restoration of power after the normal supply has been restored, and what to do if the emergency power source fails either during start-up or after operation has started.

8.2.1 Type 1 Essential Electrical Systems

Type 1 essential electrical systems are typically used in facilities where extensive medical procedures take place and reliable service is absolutely necessary to preserve lives. This type of EES consists of two subsystems. One subsystem is the equipment system used for such functions as heating critical care and sometimes patient care areas and operating clinical air compressors and vacuum pumps, some air systems, elevators, and other medical care and nonmedical equipment. The second subsystem is the emergency system that provides electricity to patient care areas, life support systems, anesthetizing areas, other loads essential for medical care, limited lighting, some alarms, and other equipment essential for life safety when normal electrical service is interrupted.

The emergency system of Type 1 EESs is further divided into two branches, the *life safety branch* and the *critical branch*. NFPA 99 delineates the items that are *required* and those that are *permitted* on each branch. The items that must be connected to the life safety branch include egress and exit sign lighting, fire and other alarms, communication systems, selected equipment at the generator site, elevators, and automatically operated egress doors. The code allows auxiliary functions on fire alarm systems that comply with *NFPA 72®*, *National Fire Alarm Code®*, to be connected to the life safety branch.

NFPA 99 is a little more liberal with the critical branch. This branch must supply power for task illumination, fixed equipment, selected receptacles, and selected power circuits that serve anesthetizing locations; areas with isolated power supplies (see Chapter 9); some patient care and treatment areas and laboratories; nurse call systems; blood, bone, and tissue banks; telephone equipment; and task illumination, receptacles, and selected power circuits for a list of specified areas. The code also allows additional task illumination, receptacles, and selected power circuits to be connected to the critical branch for "effective facility operation."

Type 1 EESs require two sources of power—a normal source, which supplies the entire facility and generally is the utility company itself, and an alternate source. When the normal source of power is the utility company, the alternate source must be an on-site generator. When the normal source of power is an on-site generator, the alternate source can be either the utility company or a second on-site generator.

The classification of an NFPA 99 Type 1 essential electrical system must be classified Type 10, Class X, Level 1, per NFPA 110. Type 10 means that power must be restored within 10 seconds of failure of the normal source of power (actually, the time the EPSS permits the load terminals of the transfer switch between the normal and alternate power systems to be without acceptable electrical power). Class X means that the emergency source of power must be available for the amount of time required by code. The Level 1 designation is for systems that have the most stringent equipment performance requirements.

8.2.2 Type 2 Essential Electrical Systems

Type 2 systems are used in facilities that provide less critical forms of care and as such have less stringent requirements. The two subsystems of Type 2 essential electrical systems include an emergency system and a critical system. The items permitted on the emergency system of a Type 2 EES are similar to those permitted on the life safety branch of a Type 1 system. It must supply power for illuminating the means of egress; exit signs and lights; fire alarms; communication systems; enough lighting in dining and recreation areas to have at least 5 ft-candles of illumination to exit ways; task illumination and selected receptacles at the generator; and elevator cab lights, controls, and communications. The requirements for Type 2 critical systems are similar to a combination of a Type 1 equipment system and critical branch of the emergency system. Type 2 critical systems illuminate patient care areas, ventilate infectious isolation rooms, and operate major apparatus and associated control systems, smoke control and stair pressurization systems, kitchen hood supply, and exhaust fans, if necessary during a fire, and heating equipment for general patient rooms. Some of the items on the critical system must be connected for nondelayed automatic connection, while others are permitted to be connected for delayed automatic connection or manual connection. Type 2 systems have the same classification requirements as Type 1 systems (Type 10, Class X, Level 1, per NFPA 110).

8.2.3 Type 3 Essential Electrical Systems

The Type 3 essential electrical system has the least stringent requirements of the three EESs used in health care facilities. Type 3 systems must be capable of supplying a limited amount of lighting and power considered essential for life safety and the orderly cessation of procedures when normal service is interrupted. Note that medical procedures do not have to be completed—they just need to come to an orderly cessation. Type 3 systems are typically used in facilities where no patients or residents are being sustained on electrical life support equipment or where no general anesthesia is used.

The alternate source of power for Type 3 systems can be a generator, a battery system, or a self-contained battery that is integral to the equipment. These sources must conform to the same generator, battery, and distribution requirements as for Type 1 systems. Most often, Type 3 facilities provide emergency power through the use of an automatic battery-powered system that complies with the provisions of NFPA *101*, NFPA 111, and the *NEC*. Type 3 EESs are classified as Type 10, Class X, Level 2, per NFPA 110. The Level 2 classification per NFPA 110 applies to applications where failure of the EPSS to perform is less critical to human life and safety.

8.2.4 EESs for Specific Facilities

The most significant difference in the requirements between hospitals and nursing homes is that hospitals must have Type 1 EESs and nursing homes do not need to use Type 1 systems but can use Type 2 essential electrical systems. If a nursing home does not admit patients that require electrical life support equipment or provide surgical treatment that requires general anesthesia and has an automatic battery-powered system or equipment that provides lighting in accordance with the *Life Safety Code* and the *NEC*, then the facility can use self-contained battery-powered units that are effective for at least 1½ hours per NFPA 99, in lieu of having a Type 2 essential electrical system. With changes in insurance reimbursement policies, many patients are being moved from hospital beds to nursing home beds but still require electrically powered life support equipment. These facilities are seen as *long-term care facilities* (a JCAHO delineation). It is important for the safety of all patients in the facility that the proper essential electrical system be provided.

Limited care facilities are required to have a Type 2 essential electrical system. The same exception that applies to nursing homes concerning the use of battery-powered units applies to limited care facilities.

The main requirement for "other" facilities is that the essential electrical system can be a Type 3 system. However, if the facility uses electrical life support equipment or if critical care areas are present, then the facility must be equipped with a Type 1 system.

Small independent "other" health care facilities usually do not have an engineering or maintenance staff that is familiar with all of the health care facilities codes and standards. In many cases, the owners or administrators of the facilities are not aware of these requirements. In many jurisdictions, there are no health care requirements for private offices. A typical example is the dental office. If the power fails while a dentist is performing an extraction after the sun has set (i.e., when there is no natural light), many jurisdictions have no requirement for a light to come on automatically so that the dentist can bring the procedure to an orderly cessation and ensure that the health of the patient is not threatened. This problem can be avoided simply by installing a fluorescent fixture that includes an integral battery unit. Most patient treatment rooms already have a fluorescent fixture in the center of the ceiling.

If the facility requires a JCAHO accreditation or a state certification, there is then some degree of assurance that a safe environment will be provided. If there are no requirements, then it is usually a matter of conscience to provide safety measures.

8.3 Wiring of Essential Electrical Systems

The wiring of emergency/life safety and critical branches of Type 1 and 2 essential electrical systems must be installed separately from the wiring of the normal system circuits, although the wiring of the equipment system does not have to be separated as such. The emergency system wiring also requires mechanical protection. The cover plates or the receptacles themselves must be of a distinctive color or have a distinctive marking so that they can easily be identified. Red faceplates and receptacles usually signify that these devices are connected to the essential electrical system. In some cases, labeling strips placed on the normal-color faceplates and receptacles indicate that these devices are part of the essential electrical system. Care must be taken during routine inspections and preventive maintenance checks to make sure that the labels have not accidentally come off or been removed. Figure 8.1 shows the typical wiring arrangement for an EES in a hospital.

8.4 Emergency Generators

Although emergency generators were used almost exclusively in the past as alternate sources of power when normal sources failed, today's codes, influenced by the energy conservation movement, permit generators to be used for several other purposes. These include controlling peak demands and internal voltage, providing load relief for the utility company, and providing a source of cogenerated power. There are two restrictions on the use of generators in these purposes (NFPA 99, 2005 edition, Sections 4.4.1.1.7.1 and 4.4.1.1.7.2) The first is that the essential electrical system must have more than one generator installed such that with the largest single generator out of service, the demands of the essential electrical system will be met. The second is that a single generator set can be used provided that its use will not decrease the mean period between service overhauls to less than 3 years.

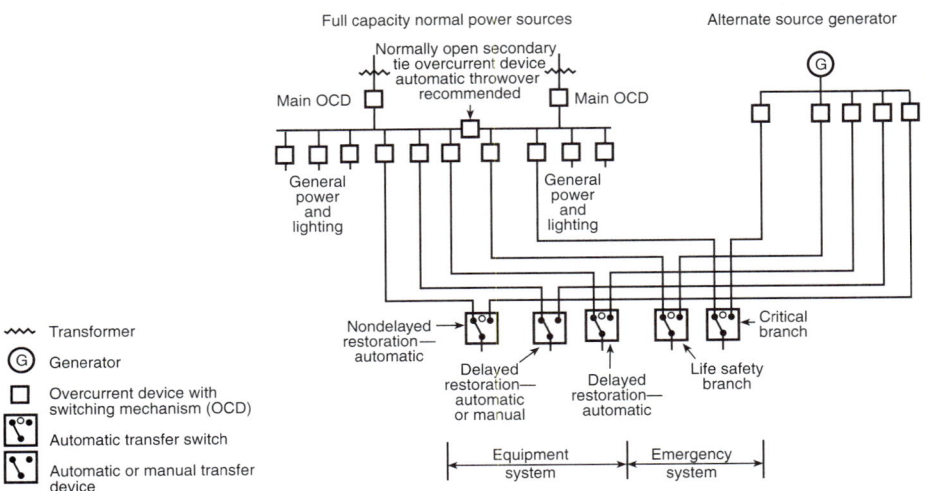

Figure 8.1 Typical wiring arrangement for the essential electrical system in a hospital. *Source:* NFPA 99, 2005, Figure C.4.1.

NFPA 99 requires emergency generators to be in a service room dedicated to the generating equipment and separated from the rest of the building with a fire-rated partition. It is imperative to ventilate generator rooms properly. When determining adequate ventilation for a generator room, the facilities engineer must keep in mind that during a power failure, the generators could likely run for longer than the monthly 30-minute testing period. (Power outages after storms have been known to last longer than 48 hours.) Generators that have run for long periods in inadequately ventilated rooms have been known to fail. During these emergency situations, it is often difficult, if not impossible, for personnel to attend to the generators due to excessive heat in the room.

The lighting in the generator room is also important. The NEC and NFPA 99 both require that the lighting for the rooms housing the generator sets for Type 1 and Type 2 systems be on the emergency generators. NFPA 110 requires that the room that houses the generator in Type 1 and Type 2 systems must be equipped with battery-powered lighting. If the generator cannot be started during a power failure, light is needed to perform necessary maintenance.

A word of caution: *Do not overload the generator*. The essential electrical system is not intended to supply power to the entire facility. It may be convenient, although not required, to add certain loads to the generator. If care is not taken, the generator will overload and stop running. Very few generators are tested to run for a long period of time at the peak loads similar to what would take place during a power failure.

If load-shedding circuits are used, there are restrictions as to what loads can be shed automatically. For example, generator load-shed circuits must not shed life safety branch loads, critical care area loads, medical air compressors or surgical vacuum pumps, fuel or jockey pumps, or other generator accessories. (Refer to the following section for more details on load shedding.)

NFPA 110 and NFPA 99 include requirements for load pickup, maintenance of room and jacket water temperature, ventilating air, cranking batteries, compressed-air starting devices, safety devices, alarm annunciators, automatic transfer switches, and bypass isolation switches.

8.5 Switching and Load Shedding

Switching the loads between the normal and alternate sources of power occurs during a loss of the primary source of power and when testing the essential electrical systems for emergency preparedness, which in itself can be disruptive to facility operations. Load shedding, which reduces the load on a generator, is used to prevent the generator from becoming overloaded.

8.5.1 Transfer Switches

Transfer switches connect the electrical load to the emergency source of power when the normal source of power fails and then returns the load back to the normal source of power when normal power is restored. The switch looks like a large, sophisticated double-throw switch. For any system other than a battery-powered system, the transfer switch must use electrical power to operate from each of its two positions (normal and emergency) and must be held in each position mechanically. Figure 8.2 is a photograph of a small transfer switch with the cover removed.

In the event of an actual emergency or a planned system test, the automatic starting operation is accomplished by means of undervoltage-sensing devices that are installed on either the generator or the generator control panel. When the voltage falls below the minimum operating voltage of the system, the transfer switch activates the generator. This value is usually set at the factory. Voltage- and frequency-sensing equipment monitor the emergency power source. When the generator is up to speed and voltage, the transfer takes place.

To prevent the alternate-source generator from starting in the event of momentary power dips and interruptions in the normal source of power, automatic transfer switches are equipped with time delays for the starting cycle. Time delays are also added to the engine shutdown cycle to allow for added cooling. Regardless of any time delay built into the system, according to NFPA 110 and NFPA 99, health care facilities must have the alternate source of power running at proper speed (frequency) and voltage and the load transferred in 10 seconds. Similarly, when the normal source of power is restored, the transfer switch is set with a delay to allow for any instability in the normal source at the moment of restoration. This value is also usually preset at the factory but can be changed in the field.

To reduce excessive motor-starting currents that can be present when power is switched to the emergency source, certain motors are allowed to be started with a delayed automatic transfer switch. In this case, the transfer is automatic, but after a predetermined delay.

Some systems require certain loads to be transferred manually, which requires personnel to be available to perform the switching operation. Unfortunately, when there is

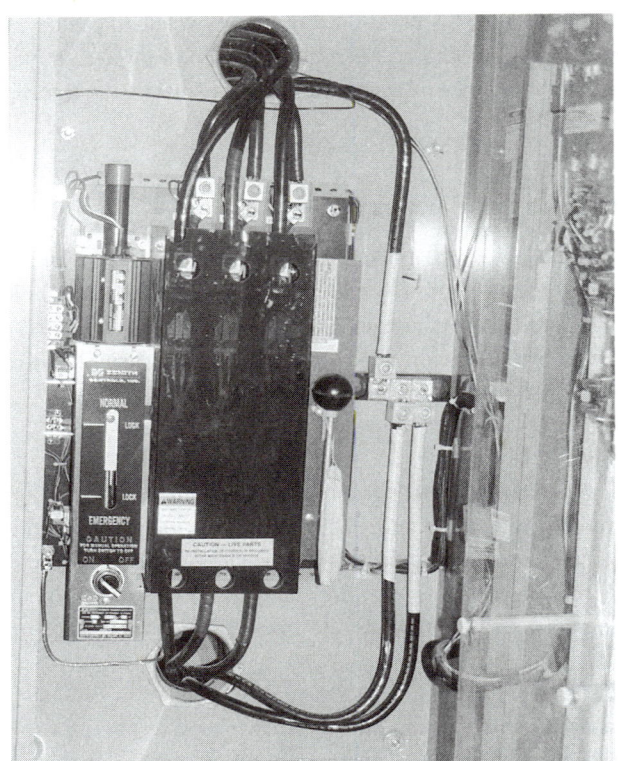

Figure 8.2 Small transfer switch with the cover removed.

a loss of electrical power, there usually are not enough staff present to handle all of the tasks required in emergency situations, even for power outages that occur during less busy times.

Some older installations may still use as a transfer switch two circuit breakers installed in tandem and interlocked with electric controls, such that one circuit breaker is locked in the open position and the other circuit breaker is locked in the closed position. Basically, each circuit breaker holds one position of the double-throw switch. The case study that follows shows a problem with this type of arrangement. This type of arrangement is not permitted by the codes now.

Automatic transfer switches rarely fail, but when they do, the results can be drastic if there is no bypass isolation switch. This type of switch is a manually operated device used to bypass the transfer switch to connect the load directly to an alternate power source and disconnect the transfer switch. While bypass isolation switches are not required by the code, they are permitted and, if installed, must be code compliant. These switches are a worthwhile investment because they allow the transfer switch to be isolated. This isolation makes it easier to conduct preventive maintenance activities without having to interrupt the power and allows the transfer switch to be bypassed in the event of its failure during normal operations without loss of power to the facility. The situation described in the following case study could have been attended to more quickly and efficiently had a bypass isolation switch been installed.

Emergency Power System Transfer Switch

The transfer switch of an emergency power system in a large medical center was configured with two circuit breakers installed in tandem, interlocked with electric controls. One of the switches was locked in the open position and the other was locked in the closed position. After conducting a test of the essential electrical system and attempting to transfer power back to the normal system, the circuit breaker connected to the emergency source opened, but the circuit breaker connected to the normal source failed and got stuck in the open position. Since this occurred during a test and not an actual emergency, the normal source of power was available by use of temporary wiring, but the facility did not have power available for all the loads connected to the essential electrical system. Fortunately, only a small area of the facility was involved, and someone familiar with the system immediately restored the generators and the emergency source of power.

Because the emergency system was not equipped with a bypass isolation switch, the staff was unable to repair the transfer switch. While the generators were kept operating, power had to be brought temporarily from an area that was not served by this transfer switch. After this source of power was connected, the generator and power to the broken transfer switch were shut down and maintenance was performed.

Each transfer switch must have a test switch that can simulate a failure of the normal source of power. This test switch can be used for the required testing of the essential electrical systems. (For information on testing the EES, refer to Section 8.8 in this chapter.) The operation of a transfer switch can be either automatic or manual. If there is a specific problem with an automatic transfer switch, the facility engineer should contact the manufacturer.

While the requirements for the transfer switch seem less restrictive in Type 2 essential electrical systems than in Type 1 EESs, basically these switches must transfer automatically with and without delays, have a means of manual control, and the like.

8.5.2 Load Shedding

Another way to retain an emergency source of power is to use a load-shedding system. Load-shedding systems reduce the load on the generator to prevent it from becoming overloaded. They are used primarily at facilities with multigenerator installations, where the generators are run in parallel and are connected to a common bus, in the event of a malfunction of one or more of the generators. Reducing part of the load keeps at least some of the emergency power available to the essential electrical system.

Load-shedding systems function at the start-up of the alternate system, as well as during system operation. As the generators start up and are placed on line, the loads are brought on line on a priority basis. The entire system must be functioning within the 10-second requirement.

Generators running in parallel typically provide greater generating capacity and spare capacity to service the loads than systems in which each generator is an independent source of power with its own transfer switch and load. That is, when all of the generators are run in parallel and are connected to a common bus, all of the combined capacity and the spare capacity of each generator can be applied to the total load. When one of the generators malfunctions, loads of the least priority can be shed, while the other generators continue to supply power to the rest of the system. With independent generators, when each generator is connected to a separate load and one generator fails, power is lost to that load regardless of priority. Similarly, the spare capacity of each generator may be too small individually to supply power to an additional load. The advantages of greater generator and spare capacity lead to a system with greater integrity overall.

NFPA 110 stipulates requirements for load-shedding systems that use two or more generators connected in parallel. These requirements address what loads may be shed during the normal course of operation. Each facility should have a load-shedding priority that suits its needs and functions to handle malfunctioning generators.

Installations with generators that run in parallel require more equipment and safety devices than systems in which each generator is connected to a separate load. The former requires not only load-shedding equipment but also paralleling equipment. For a system to function properly when the generators are run in parallel, each generator must produce the same voltage and frequency. Equipment must be installed to ensure that a new generator is synchronized with the generator(s) already connected to the common bus before it, too, is connected to the bus. Figure 8.3 is a photograph of synchronizing equipment.

Generators connected to a common bus must also produce the same voltage and frequency when they are running. A generator running in parallel with other generators will act like a motor if its voltage or frequency is lower than the rest of the system. Instead of the generator producing power, it acts as a motor and takes power from the other generators. The synchronizing equipment is equipped with safety devices to prevent this situation. If one of the generators goes out of synchronization, it must be removed from the system.

Based on the numerous equipment requirements for load-shedding systems, these installations are more costly than transfer-switch installations and require additional capability to operate. The performance of this arrangement usually justifies the extra cost and effort, however.

8.6 Energy Converters

Energy converters simply convert one form of energy to another. A battery unit converts the chemical energy of the battery to electrical energy to illuminate the lights. In the case of an engine generator set, the chemical energy of the fuel is converted to mechanical energy by the diesel motor, which drives the generator. The generator in

Figure 8.3 Synchronizing equipment used for generators running in parallel.

turn converts the mechanical energy of the motor to electrical energy. Generally, only two types of energy converters are used in an essential electrical systems—a rotating electric generator with a prime mover and a stored energy system. Type 1 and Type 2 EESs (Level 1 emergency power supply systems per NFPA 110) require a rotating generator driven by a prime mover. Type 3 EESs (Level 2 emergency power supply systems) can have either a generator with a prime mover or a stored energy system.

8.6.1 Stored Energy Systems

Stored energy systems can be as simple as a battery lighting unit or energy conversion equipment. They consist of an uninterrupted power supply, a central battery system, or a motor generator, powered by a stored electrical energy source. The stored system is equipped with a transfer switch that monitors the power source of the preferred and alternate loads and all necessary control equipment (NFPA 111). The types, classes, and levels for stored emergency power supply systems are the same as those for EPSS covered under NFPA 110. NFPA 111 states the requirements for an SEPSS, some of which are alarms; installation; types of batteries used; heating, ventilating, and air conditioning (HVAC) and humidity controls; and ventilation requirements.

8.6.2 Rotating Electric Generators

The generator set at a health care facility usually consists of an electric generator driven by a motor. Figure 8.4 is a photograph showing a multigenerator installation. The sources of fuel for the motor permitted by NFPA 110 include liquid petroleum, liquid petroleum gas, and natural or synthetic gas, the most common being diesel fuel. The fuel supply for a Level 1 EPSS cannot be used for any other purpose. One exception to this requirement is that fuel from an enclosed tank may be used for other purposes if the tank's fuel level indicators can ensure that the required amount of fuel for the emergency generator will not be used. NFPA 110 includes requirements for sensing low levels in the fuel supply.

Engine generator sets can be installed outdoors, if within a completely enclosed housing. Figure 8.5 is a photograph of an outdoor engine generator set installation.

8.6.3 Cooling Systems for Energy Converters

Every prime mover requires a cooling system. Level 1 EPSSs can have a unit-mounted radiator and fan, a remote radiator, or a liquid-to-liquid heat exchanger. If a unit-mounted radiator system is used, the room housing the engine generator set must be properly ventilated because all of the heat given off by the engine is transmitted to the generator room. If the temperature gets too high, it will be impossible for personnel to work in the room and the engine may overheat. Provisions should be made to exhaust

Figure 8.4 Multigenerator installation.

Chapter 8. Essential Electrical Systems 109

Figure 8.5 Outdoor engine generator set enclosed in a housing.

this heat for at least 48 hours following operation of the generator set. The room housing the generator set should be checked after that time to make sure it is at a proper temperature. All too often the temperature is checked only during the monthly 30-minute test, and no one checks to see what the temperature of the room would be after many hours of operation.

Cooling systems for generators are the same regardless of system type.

8.7 Alarms, Indicators, and Safety Devices

Alarms and indicators for the generator sets alert system operators about a host of equipment conditions, including water and engine temperatures; lube oil, air, and hydraulic pressures; speed; coolant levels; the load on the emergency power supply; battery voltage; battery charger failure; and other criteria. NFPA 110, 2005 edition, includes a table [Table 5.6.5.2] that lists the various required safety indicators for Level 1 and Level 2 EPSSs.

Alarms and indicators must be located at or near the engine generator set location so that they can be observed during the operation of the equipment. Engineering department personnel should be trained on how to observe all of the alarms and indicating devices to ensure that the system is operating properly and, most important, what to do if one of the generators malfunctions and an alarm goes off. Particularly in health care facilities with Type 1 essential electrical systems, emergencies cannot wait for personnel to arrive at the scene to get the system operational after a malfunction. Usually, when alarms are set off during Monday-through-Friday day shift, sufficient staff are available to handle the malfunction. But problems can also arise during evening or weekend shifts or during holidays, so staff working all shifts must be properly trained. If no one is on site who is trained in the proper operation of the essential electrical system, someone should be on call and immediately available.

By code, alarm annunciators must also be located outside the generator room, at a remote location. One of the locations, either the generator equipment room or the remote location, should have someone on duty 24 hours a day, 7 days a week. A typical remote location is the telephone operators' room, since this is usually one location where someone is on duty at all times. Locating alarms in the telephone operators' room can create a problem for the telephone operators, however, especially during an emergency. It is very difficult to observe fire alarms, medical gas alarms, engine generator alarms and indicators, to name a few, while answering telephones and possibly operating a paging system. Observing alarms and indicators becomes even more difficult for telephone operators if the emergency occurs in the middle of the night, when fewer operators are on duty (for short periods just after midnight, only one person may be at the switchboard). For these reasons, if the operators are to be made responsible for monitoring the remote equipment alarms, they must be well trained and dedicated. All the requirements for alarms and indications can be found in NFPA 99, NFPA 110, and NFPA 111.

In addition to requirements for alarms and indicators, the codes specify requirements for safety devices for the generator sets. Internal combustion engines serving generator sets must be equipped with sensors and warning devices such as ones that indicate engine temperature, oil pressure, coolant levels, overcranking, and so on. A complete list is given in NFPA 110 and NFPA 99.

8.8 Testing, Maintenance, and Record Keeping

Testing and maintenance requirements for Types 1, 2, and 3 essential electrical systems are similar, except that when stored energy power sources are used, as is more typical for Type 3 EESs, they must comply with the provisions of NFPA 111. As for all health care facility systems, facilities engineers must maintain records of all tests of the essential electrical system.

8.8.1 Operational Inspection and Testing of EPSSs

NFPA 110 contains comprehensive requirements for operational inspection and testing of emergency power supply systems; most of the requirements in NFPA 99 have been extracted from this code. JCAHO also usually accepts compliance with NFPA codes and standards as equivalent to complying with their requirements, except for some specific requirements, such as what constitutes a full-load test, the frequency of testing, and what must be recorded. The *NEC* refers to NFPA 99 for performance, maintenance, and testing requirements for essential electrical systems in hospitals, nursing homes, and limited care facilities.

8.8.2 Inspecting, Load Testing, and Record Keeping for Generator Sets

NFPA 110 requires the generator sets in each Level 1 and Level 2 emergency power supply system and its components to be visually inspected weekly and exercised under loaded conditions for at least 30 minutes at least monthly. NFPA 99 states that generator sets must be tested for 30 minutes under loaded conditions 12 times a year, with the time between tests at least 20 days but not longer than 40 days. This 20-day window was provided to give facility operators flexibility in testing their systems. At certain facilities, even a 10-second interruption of power might cause problems when a very delicate medical procedure is taking place. With this window, the system test can be undertaken during a less precarious time.

The amount of load to use during a test has been subject to controversy based on changing as well as conflicting requirements among the code-writing organizations. The organizations have frequently revised the amount of load that must be used to test the essential electrical system, and current codes vary. NFPA 110 requires facilities to test diesel generator sets at loads that maintain operating temperature and at least 30 percent of the nameplate rating, at loadings that maintain the minimum exhaust gas temperatures recommended by the manufacturer, or, if the 30 percent load criteria cannot be met, then it is to be operated until the water temperature and the oil pressure have stabilized and the test is then teminated before the 30-minute time period. (The requirement for the exhaust gas temperature relates to a condition known as wet stacking. If the load is too small, there might be unburned fuel in the exhaust system, which can be seen as black smoke during the test.) JCAHO for where now basically requiring these same loading conditions. NFPA 110 also has provisions for where a diesel-powered EPS installation cannot meet the preceding criteria. If it cannot meet these criteria, then it must be exercised monthly with the available load, and exercised annually with supplemental loads at 25 percent of nameplate rating for 30 minutes, followed by 50 percent of nameplate rating for 30 minutes, followed by 75 percent of nameplate rating for 60 minutes for a total of 2 continuous hours.

All too often, a system functions for the 30 minutes of a load test but would be overloaded if it had to function for longer periods. Periodically connecting a load bank and testing the generator under a full-rated load ensures that the generator can still operate at its nameplate ratings. However, there are no longer any requirements in any of the codes to conduct a full-load test of the generator other than at the time of the initial acceptance of the equipment. Therefore, systems are usually never tested under the largest loads that could be present during an electrical power failure, such as those that occur in the middle of the day during weekdays.

Although testing under these circumstances should not put the facility under risk, it could place it in legal jeopardy if there is a failure in the system and a patient is injured during the test. Facilities must follow some procedures to ensure that the generators will not be overloaded if a power failure occurs during times of peak loading. Doing any test with a load larger than that required by the codes is an issue that must be discussed by all key players involved, including the legal staff, and decided by each facility individually.

Keeping records of the entire system and the loads, both the connected loads and the highest possible loads based on current levels of equipment, can ensure that overloads on the generator do not occur during a test or during a loss of power. To determine a facility's peak loads, the facilities engineer can take four readings annually—day and nighttime readings for peak summer and winter months. Records should indicate the dates and totals for these peak loads. After the initial peak loads have been determined, the records should then indicate the added load from each new piece of equipment and its circuit and feeder, as well as diminished loads from equipment that has been removed. The ideal way to do this is to use recording equipment. If this cannot be done, then instantaneous readings must be used. Multiple instantaneous readings should be taken to ensure that the peak was noted. This is discussed in Chapter 12.

Annex A of NFPA 110 includes a sample operation and testing log for emergency power supply systems along with a suggested operation and testing procedure. Figure 8.6 shows a sample operation and testing log for rotating equipment (NFPA 110). This is Figure A.8.4.1(a) in NFPA 110, 2005 edition.

8.8.3 Battery Tests

The batteries used for cranking the motor generator set should be visually inspected at least weekly to determine if there is any corrosion at the terminals, fluid levels are proper, the battery case is leaking, and the like.

Essential electrical systems powered by battery units require different preventive maintenance than essential electrical systems powered by generators. Battery units must be inspected periodically to ensure that the internal transfer switch functions when the power is lost, the lamps are still lighting, and the battery is fully charged. Where the battery is not sealed, liquid levels must be checked. Most units come with a test switch. Another testing method is simply to open the circuit breaker that distributes power to the circuit the battery units are on. Most units come with manufacturer's instructions as to what tests are needed and how to make sure the units are fully charged and functional. The requirements for maintenance and testing of stored energy (battery) emergency power supply systems are found in NFPA 111, Chapter 6.

8.8.4 Feeder and Circuit Breaker Inspections and Tests

NFPA 99 requires the main and feeder circuit breakers to be inspected annually and exercised in accordance with manufacturer's recommendations. Exercising these circuit breakers requires proper planning and great care to avoid problems associated with the lack of power in health care facilities. The power should be off for a minimal amount of time. Establishing lines of emergency communication so that staff in the areas affected can communicate with the personnel doing the test is essential. Notifying all staff who will be affected by the test well in advance of the procedure will allow enough time for them to make alternate plans for patient care during this brief period. The time of day or night selected for conducting this test should be when the facility will be minimally affected.

8.8.5 System Documentation

NFPA 110 requires manufacturers to supply at least two sets of instruction manuals for all of the major components of the emergency power supply system. One set should

	Operation and Testing Log												
	Performed by												
	Date												
Item*	Fill in Appropriate Readings												
1. Maintenance schedule													
2. RTM													
3. Power fail													
4. T/D start													
5. Crank time													
6. Transfer													
7. (a) ac voltage													
(b) Hz													
(c) ac amperage													
8. (a) Oil pressure													
(b) dc amperage													
9. (a) Oil pressure													
(b) dc amperage													
(c) W/A temp.													
10. Restore normal													
11. (a) Oil pressure													
(b) dc amperage													
(c) W/A temp.													
(d) ac voltage													
(e) Hz													
(f) ac amperage													
12. T/D retransfer													
13. T/D stop													
14. Auto mode													
Comments													

* See Suggested Operation and Testing Procedures for explanation of items.

© 2005 National Fire Protection Association

Figure 8.6 Sample operation and testing log.
Source: NFPA 110, 2005, Figure A.8.4.1(a).

be kept near the equipment and one set should be kept in a secure location. These manuals should contain a detailed explanation of the operation of the system, instructions for routine maintenance, instructions for repairs, and illustrated and schematic drawings of the system. It is not enough just to keep these manuals on hand; the personnel who will be conducting the tests and maintenance activities should know where the manuals are located and what information they contain. The manufacturer must also provide a parts list and part numbers. The facility should keep these spare parts on hand at all times. The spare parts that a facility keeps on hand must take into consideration the geographic location of the institution and the availability of spare parts from outside sources.

Annex A of NFPA 110, 2005 edition, includes a maintenance log and guide for emergency power supply systems. NFPA 99 also includes a maintenance guide intended to assist administrative, supervisory, and operating personnel in establishing and evaluating maintenance programs. Refer to Chapter 12 for more details on system maintenance.

9

Isolated Power Supplies and Other Special Systems

In addition to the essential electrical systems, health care facilities have many other specialized systems that ensure safety and alert staff of possible hazards. Because many specialized medical procedures are invasive in nature and the environmental conditions in which they take place are more extreme in terms of temperature and wet conditions, they require specialized ungrounded and uninterrupted electrical service. These services are used in areas where ground-fault circuit protection is required but interruption of service cannot be tolerated. It is also essential to have sophisticated alarms systems, not only to alert people in the event of fires but also to signal possible security breaches or unsafe environmental conditions in such unique areas as blood and tissue banks and narcotics cabinets. And it can be argued that not many types of facilities have such a vital need for sophisticated communications system. All of these specialized systems have their own set of requirements aimed at providing a safe environment.

9.1 Isolated Power Systems

With few exceptions, all electrical systems in the United States must be grounded. When a ground fault occurs in a grounded system, the fault current follows a return path through ground that causes the overcurrent device (circuit breaker or fuse) on that circuit to open, remove the electrical source, and eliminate the dangerous condition. Figure 9.1 shows a simplified basic grounded system. The neutral conductor is grounded at the service entrance to the facility. If a person should come in contact with the phase leg (hot wire) and become grounded (i.e., come into contact with earth ground or a grounded device), the fault current would flow through the person. The time it takes for the circuit breaker to open would depend on the amount of current that was flowing. Figure 9.2 shows this situation.

Several common conditions often require some form of ground-fault circuit interruption. When an interruption of power cannot be tolerated, an isolated power supply can be used. These conditions involve the possibility that the following situations might be present:

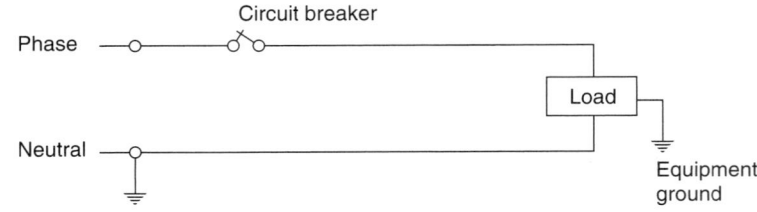

Figure 9.1 Simple grounded electrical system.

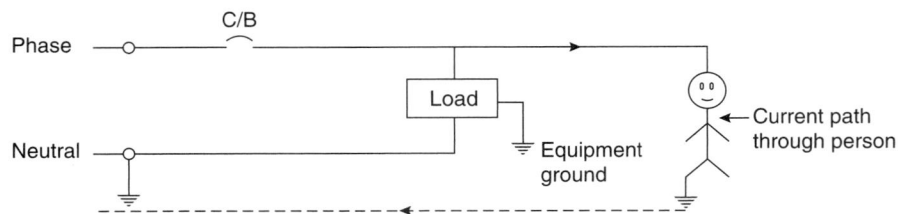

Figure 9.2 Simple grounded electrical system with a person in contact with ground and an energized conductor.

- There is a wet location, whereby standing fluids are on the floor and patients and staff are frequently wet with prepping solutions, bodily fluids, and other conductive fluids that could provide a path to ground.
- Line-operated devices are being used that are normally connected to and can induce a current in the patient's body. Note that the electrical resistance of the patient's body with these internal electrical devices is lower for several reasons. One reason is that the patient's skin no longer provides electrical resistance. Another reason is that the patient is vulnerable, often unconscious, and unable to care for himself.
- Flammable anesthetics, which can ignite if sparks and arcing occur, are being used.

The way an ungrounded electrical system works is that when a ground fault occurs, there is no return path for the current to complete its circuit. This is illustrated in Figure 9.3.

Terminals A and B are on the secondary side of the isolation transformer. Since neither point is grounded, there is no path for the current to return to point B and hence no current would flow through the person. Note that the equipment case is grounded but does not enter into the normal conductive path for the current

However, no electrical system, even an ungrounded one, can remain totally ungrounded. This is because capacitive leakage (capacitive coupling) occurs between wires that carry alternating current. The result is equivalent to placing a capacitor between the two conductors. When this capacitive coupling is of very high impedance, the value of the leakage current is limited and the circuit still acts as if it were

Chapter 9. Isolated Power Supplies and Other Special Systems   **117**

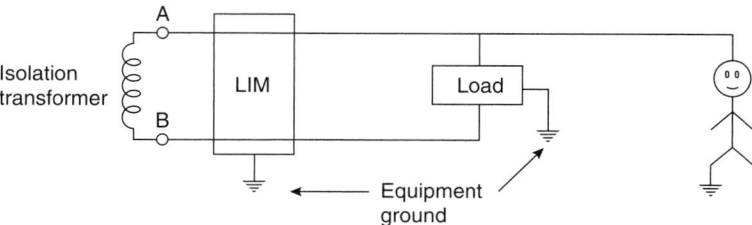

Figure 9.3 Ungrounded electrical system with a grounded patient, in which current cannot flow to point B through the patient.

completely ungrounded. Should the impedance of the coupling fall below a certain point, however, the circuit acts as if it were grounded. Moreover, if a second ground fault occurs under these circumstances, it would be the same as if a ground fault in a grounded system occurred, which could be harmful or even lethal to the patient. Figure 9.4 illustrates an ungrounded system with the capacitive coupling to ground and a patient grounded. Having the patient grounded is required for the proper functioning of certain medical devices. The amount of current that can flow through the patient is determined by the value of the impedance of the capacitive coupling and the resistance of the patient.

Because an isolated power system (IPS) must retain power to serve the critical medical procedure that is taking place and prevent circuit breakers from opening, all IPSs are equipped with *line isolation monitors* (LIMs). An alarm device is included in the LIM, which lets the staff know when the leakage current to ground has reached a level at which it can become dangerous or at which the system can be considered to be grounded. When the LIM sounds, the staff have the opportunity to either bring the procedure to an orderly cessation, examine the patient area and equipment to determine where the grounding or leakage is occurring, or replace questionable equipment. Figure 9.5 shows a line isolation monitor used in an operating room.

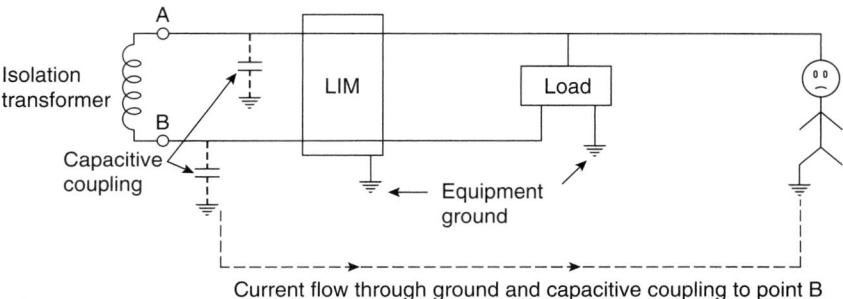

Figure 9.4 Ungrounded electrical system with a grounded patient, in which current will flow through the patient to point B.

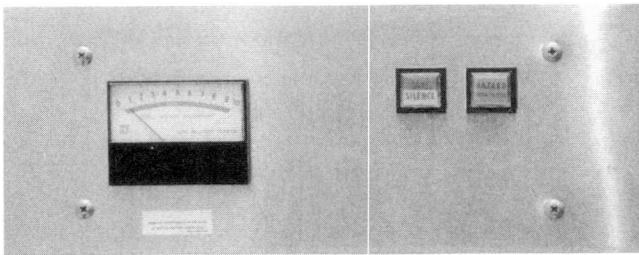

Figure 9.5 A line isolation monitor in an operating room.
Source: Health Care Facilities Handbook, 2005, Exhibit 4.2.

In the early years of isolated power systems (i.e., the 1940s), the term *leakage current* was used to describe the current that flowed between the ungrounded conductors and ground. That term has now been changed to *hazard current*, which means the total current that would flow through a low impedance if it were connected between either one of the isolated conductors and ground. The various hazard currents include fault hazard currents, monitor hazard currents, and total hazard current. A *fault hazard current* is the hazard current that flows with all devices connected except the line isolation monitor. A *monitor hazard current* is the hazard current that flows from the line isolation monitor by itself. The *total hazard current* is the hazard current that flows with all devices including the line isolation monitor connected.

9.1.1 IPS Codes and Standards

According to NFPA 70, *National Electric Code*® (*NEC*®), 2005 edition, isolated power systems can be used in wet locations to satisfy the requirement to limit the ground fault current in those areas where an interruption of power cannot be tolerated. NFPA 99, *Standard for Health Care Facilities*, 2005 edition (Annex E), requires IPSs to be installed in hazardous (classified) anesthetizing locations, such as where flammable anesthetics are used. NFPA 99 does not require isolated power systems in patient care areas other than in specific cases, but they are permitted as long as the installation conforms to all of the requirements of NFPA 99 and Article 517 of the *NEC*. Annex A of NFPA 99 and the *Health Care Facilities Handbook* both contain explanatory information and commentary about isolated power supplies. Each facility must also consult with its authority having jurisdiction for individual requirements for isolated power systems.

9.1.2 IPS Equipment

An IPS is composed of a shielded isolation transformer or its equivalent, a LIM, circuit breakers, power receptacles, a test switch, and grounding jacks. The transformer's electrostatic shield is located between the primary and secondary windings of the transformer. This device serves to reduce capacitive coupling, which more thoroughly isolates the secondary from ground. IPS equipment is usually contained within one unit and is available in single-phase or three-phase designs. Initially, this equipment was available only with single-phase output, but recent demand for the use of three-phase equipment in areas served by IPSs has led to the availability of three-phase IPS equipment. Good design practice dictates that separate systems should be used for single- and three-phase equipment.

Line isolation monitors are usually equipped with a meter, an audio alarm device, red and green indicator lights, and an alarm-silencing switch. The meter indicates the amount of hazard current that is flowing. If the total hazard current exceeds 5 mA, the visual and audible alarm must be activated. The LIM alarms shall not activate for a fault hazard of less than 3.7 mA. The alarm cautions staff that a ground-fault condition exists and assists them in determining whether to stop the procedure in progress. The lights indicate whether the system is in a normal state or an alarm state, and the silencing switch turns the audible alarm on and off. The silencing switch must reset automatically when the fault condition is corrected and the system reverts back to the normal monitoring state.

Although today's LIMs continuously monitor and indicate the value of the current that might exist between each conductor and ground, some of the early line isolation monitors were static monitors that measured only the leakage current between one of the ungrounded conductors and ground. They also contained ground-fault detectors, not line isolation monitors. (If an institution still uses these older types of monitor, they should consider replacing all of them with the type that is now required by the codes.)

When isolated power systems were initially used, the allowable value for total hazard current on systems that were not equipped with alarms was 2 mA. As medical care has evolved, the allowable hazard current value has been raised to 5 mA, which allows for the use of additional equipment without causing any harm to the patient or staff. The staff and engineering department should be aware of the types of systems that are installed in their facilities.

9.1.3 IPS Installation, Testing, and Maintenance

The codes contain many requirements for the installation, maintenance, and testing of IPSs. When installing an IPS, the impedance, both capacitive and resistive, to ground of either conductor must exceed 200,000 ohms, which will prevent leakage current from flowing in excessive amounts. NFPA 99 outlines a test procedure to check this value. To meet this requirement, the size of the isolation transformer may have to be limited to 10 kVA (kilovolt amperes), and conductor insulation with low leakage values should be used.

The entire IPS must be periodically tested, and all test records must be kept. The facilities engineer should ascertain the frequency of the testing as required by the various authorities having jurisdiction as well as by the Joint Commission on Accreditation of Healthcare Organizations (JCAHO). JCAHO still allows testing every 6 months on most equipment but now allows the facility to make up its own maintenance schedule based on many other factors (JCAHO *Environment of Care Handbook*, 2nd edition). NFPA 99 requires the LIM circuit to be tested upon installation and at least at intervals of not more than 1 month by actuating the LIM test switch. If the LIM circuit has automated self-test and self-calibration capabilities, the test can be performed at intervals of not more than 12 months. After every repair or renovation to the electrical distribution system, the LIM circuit has to be tested as if it were initially installed. This test involves grounding each line of the energized distribution system through a specified resistance that activates the visual and audible alarms. The facilities engineer must investigate all areas where leakage might exist based on test results.

All staff members working in a location where an isolated power system is used should be thoroughly familiar with the operation of these systems. Not only is it essential to keep the leakage current to a low value, it is also important for staff to know what actions to take when an LIM alarm goes off, rather than continuing with the procedure. Obviously, there are times when a procedure cannot be stopped, but all pertinent staff must know how to eliminate the source of the problem and be aware of the potential hazards if they do not do so. As soon as possible after an LIM goes off, the incident should be reported to the department responsible for maintaining the equipment and the room where the procedure is taking place. It is helpful when reporting a problem to be able to provide the symptoms of the problem and a possible explanation.

It is also important for maintenance personnel to use the correct terminology when performing maintenance and interpreting test results in terms of the various hazard currents that exist—fault hazard currents, monitor hazard currents, and total hazard currents.

Finally, Section 8.5.5.1 of NFPA 99, 2005 edition, states the following requirement:

> Personnel concerned for the application or maintenance of electrical appliances shall be trained on the risks associated with their use.

The following case study shows how important it is for staff to understand how isolated power supplies work and that care must be taken by appropriate personnel to review all new equipment before purchase.

Cumulative Hazard Currents

One large medical center needed to purchase replacements for the patient monitoring system used for a suite of 12 operating rooms. A new system from a reputable manufacturer had a feature that would have allowed any one of the monitors in a particular room to act as a master monitor for all the suites. Even during surgery, with the press of a button, an anesthesiologist working in one operating room could have observed the patient indications on a monitor in any of the other rooms. This would have allowed the director of anesthesiology, for example, to work in one operating room and be available for consultation at all times, while observing the status of the patients in the other rooms, if necessary.

Careful analysis of the equipment by the staff during the review period before its purchase revealed a drawback to this system, however. There obviously would have to be a connection between the monitors to allow the signal to flow between them. A close examination of the circuitry revealed that the way these connections would have been made would have caused the leakage current in each of the operating rooms to be added to each other. The resultant total hazard current of the 12 interconnected operating rooms could have potentially exceeded the set-point limit, and all of the

(Continued)

line isolation monitors alarms would have gone off simultaneously. It would have appeared that all the rooms had a fault condition, and there would be no indication of which room had a fault. In the event of an actual fault, it would have been impossible to determine where the fault existed.

The manufacturer's representatives had not been aware that this type of situation could exist, nor did they understand the code requirements, but they confirmed that the analysis of the potential problem was correct. The company had never considered the cumulative effect of all 12 operating rooms being in use simultaneously. Care must be taken by relevant personnel to review all equipment before purchase.

9.2 Alarms

Alarms are used for many purposes throughout health care facilities. They alert staff about fires, life safety concerns, security issues, and the improper operation of life safety devices. They also alert personnel when blood bank and food refrigeration-unit temperatures go beyond set limits, and when narcotics cabinets are opened inappropriately.

Care should be taken that the proper staff person attends to the various parts of the alarm systems and that that person is specially trained for the task. The facilities engineer typically is responsible for maintaining the electrical supply necessary for all of the alarms systems. The engineering department is often responsible for the blood bank alarms, refrigeration alarms, narcotics alarms, and security alarms.

9.2.1 Fire Alarms

Article 760 in the *NEC* contains requirements for installing fire alarm system wiring and equipment. *NFPA 72®*, *National Fire Alarm Code®*, also provides requirements for installing fire alarm systems, as well as for monitoring these systems for integrity, supplying electrical power, and providing continuity of service.

Most jurisdictions require the fire alarm system in certain health care facilities to be connected to the local fire department or, if that arrangement is not permitted, to a central alarm station, which would then transmit the signal to the fire department. These connections are usually accomplished over telephone lines.

Larger facilities usually have a safety officer on staff who is responsible for the maintenance and testing of the fire alarm system. In smaller institutions, the engineering department may be responsible for these tasks. Another common arrangement is for an outside contractor to conduct the required semiannual testing and maintenance of fire alarm systems and to be on call for emergency repairs. Hospital personnel conduct the fire drills and, if necessary, make an emergency repair while waiting for the outside contractor to respond to a call. Fire alarms are covered in more detail in Chapter 27.

9.2.2 Medical Gas/Vacuum Alarms

NFPA 99 (Chapter 4) specifies that fire and medical gas/vacuum alarms and alerting systems that are required for the piping of nonflammable medical gases must be connected to the essential electrical system of the facility, including Type 1 and Type 2 essential electrical systems. If the essential electrical system does not have a generator, such as in some Type 3 systems, then the supply for the fire alarms system should be connected on the line side (i.e., ahead) of the main service switch of the facility. The medical gas/vacuum systems are usually under the jurisdiction of a facility's mechanical or plumbing department. Refer to Chapters 8, 29, and 30 for more details on essential electrical systems and medical gas/vacuum alarms, respectively.

9.2.3 Blood Bank Alarms

The blood bank alarm system is basically a refrigeration alarm. If the temperature in the blood bank refrigerator increases or decreases below a predetermined point, or if there is a loss of power at the blood bank refrigerator, then an alarm is set off. The blood bank refrigerator must be connected to the essential electrical system and would receive power from the generator in the event of a major power outage. It would not receive power if a local or branch circuit failed, however.

9.2.4 Narcotics Alarms

Narcotics alarms are usually placed on all cabinets that store narcotics. These cabinets often are located in the medical preparation areas of nurses' stations, satellite pharmacies, or the main pharmacy. When the door to the cabinet is opened, an indicator light located somewhere in the surrounding area is usually activated to indicate to the staff that someone is at the narcotics storage area. This indicator is usually local to the medical preparation area in nurses' stations, since the narcotics cabinet is usually opened and closed many times during normal hours. Also, there is usually a staff person within earshot of the station. Some facilities are required to install an alarm that has a remote indicator on the main narcotics cabinet in the main pharmacy. Under this arrangement, if the main pharmacy is closed for a period of time or left unattended, the remote alarm can be activated, in effect keeping the narcotics storage cabinet under supervision.

9.2.5 Refrigeration Alarms

Refrigeration alarms are usually installed on the refrigerators and freezers in the main kitchen, if there is one. If the temperature should exceed a low or high set point, an alarm is activated. These alarms can be installed on any refrigerator. As with narcotics alarms, refrigerators in the medicine preparation area in nurses' stations are usually equipped with local alarms. Remote alarms are not necessary because these refrigerators are constantly being opened and closed, and typically, a supervisory staff member must check the temperature readings periodically as required by the local health department's sanitation codes (not addressed in this book).

9.2.6 Security Alarms

There are two types of security alarms: alarms that indicate unauthorized access and alarms that summon immediate help. Alarms that are used to summon help are usually installed in areas such as the cashier, pharmacy, emergency room, certain psychiatric areas, and the warehouse. Staff might request that these alarms be installed in other locations as well. Access-type alarms are located on doors where access is restricted to certain personnel or to all personnel at certain times of the day. Access to restricted areas can be made with keys, punch code systems, or card access. Card access systems with alarms can be expensive but have many advantages. One card can be used for attendance and time card information as well as access.

9.2.7 Alarm Indicators

Most alarm systems require alarm indicators to be located at the equipment itself, with remote indicators placed in an area that is staffed 24 hours per day, 7 days per week. In most facilities, this location is often the telephone operators' room. Aside from fire and medical gas alarms, the alarms for emergency generators, refrigeration units, blood banks, and security systems are usually found in the telephone operators' room. A typical problem associated with locating many of the remote alarm indicators there, however, is that there often is not enough wall space to lay out the annunciators properly, in addition to the communication systems equipment (i.e., voice and wireless paging and telephone switchboard equipment) that is normally located in this room. Sometimes, there are too many alarms. Both of these situations can lead to confusion when multiple alarms sound or light up.

9.3 Communication Systems

The basic function of a communication system is to convey information from one point to another, often for the care and safety of patients, staff, and visitors. Conveying information can be accomplished by using voice, tones, or writing. Many systems in a health care facility can be categorized as communication systems, including the alarm indicators and annunciators, as discussed previously. After all, alarm indicators and anunciators communicate a normal or abnormal condition in the systems to which they are connected.

9.3.1 Types of Communication Systems

Some of the other familiar communication systems found in health care facilities are as follows:

- Voice paging and radio paging
- Nurse call
- Patient physiological monitoring
- Central dictation
- Data processing
- Telephones – wired and wireless
- Televisions
- Intercoms
- Teletypewriters
- Physician in-out registers
- Central clocks
- Background music
- Security communications

Many of these systems are used in emergency conditions to convey important information at a rapid pace to the patients and staff who are involved. Most communication systems use low voltage.

9.3.1.1 Paging Systems

There are two general types of paging systems, overhead voice and radio/wireless. Most facilities install an overhead system so the entire staff throughout the facility can hear messages. This system is usually accessed through one person such as the telephone operator. The radio paging (wireless) system can be accessed two ways. The first is through one location and person, such as the telephone operators' room. The second way is either from any telephone or from selected telephones. Radio/wireless systems can be owned and operated by the facility or rented, in which case the owner of the system is responsible for maintaining them.

9.3.1.2 Nurse Call Systems

The features of nurse call systems vary from simple visual signaling systems that indicate that a patient or staff member needs attention to complete audiovisual systems that provide both signal and voice communications between stations. Some of the more sophisticated systems include physiological monitoring and television signals.

Patient controls on nurse call systems are a point at which a patient can potentially receive an electric shock, especially if liquids have accidentally spilled on the controls. For this reason, most nurse call systems use low-voltage patient controls. These controls should not be overlooked when planning a new system or replacing an old one. When maintaining an existing system, it is important to ensure that no hazardous conditions, such as frayed wires, exist.

9.3.1.3 Patient Physiological Monitoring Systems

Patient physiological monitoring systems, such as cardiac output or vital-signs units, can transmit data from the patient to a central point by means of dedicated wires through the nurse call system or, in some cases, through wireless radio transmission. In many cases, ambulatory cardiac patients must be monitored while being treated, which can be accomplished through radio transmission systems. Since the output of these systems is necessary for the continued treatment and life support of the patient, the equipment should be connected to the essential electrical system.

9.3.1.4 Central Dictation Systems

Health care facility staff use central dictation systems to communicate patient information to be entered into the permanent patient record along with the patient chart. These systems can be configured as a separate set of wires between the various transmission points (the patient's bedside, the nurses' station, etc.) and the receiving station, usually in the medical records department. The system can also be installed as part of the telephone system. Newer systems can use hand-carried computers or other wireless systems.

9.3.2 Installation Requirements for Communication Systems

Many codes address the installation of communication systems Articles 720 and 725 of the *NEC*, 2005 edition, govern communication systems that use low voltage (i.e., less than 50 V) and Class 1, Class 2, and Class 3 remote control, signaling, and power-limited circuits, respectively. Article 800 in the *NEC* addresses communication circuits. The codes frequently state that the communication equipment being used should be listed for that purpose by a third-party testing organization, such as Underwriters Laboratories (UL), or another approved testing agency, such as Electrical Testing Laboratories (ETL).

There are many requirements for the separation of communication wire and cables from other power, lighting, and special wiring systems, both inside and outside a building. These are all addressed in Article 800 of the *NEC*.

When installing new wire and cabling for communication systems, there can be a tendency just to lay the wires and cables in the hung ceiling without adding any ties, dressing, or other devices to make the wires neat and accessible. One of the major concerns about installing communication wires in conduits or raceways is the fire rating and toxicity rating of the insulation on these wires and cables. Many fire departments are concerned that the insulation used on communication cables and wires may give off toxic fumes when exposed to fire. The *NEC* (Section 800.179) allows just nine types of cable to be used as communication cable. The code also requires communication wires and cables to have a voltage rating of not less than 300 V.

In general, space is a problem when installing wire and cables. Many jurisdictions prohibit wire or cable, other than light and power distribution cables, running through electric closets. Wire and cable also should not run through wet areas. One institution that was having a private telephone system installed discovered that the installation company had run the telephone vertical riser cable through the janitors' closets next to the slop sink, because the closet was accessible, convenient, and shortened the runs. Obviously, this type of installation must be avoided.

Communication systems that use standard voltage must comply with the applicable *NEC* requirements rather than those for low-voltage systems.

9.3.3 Backup Communication Systems

JCAHO requires each accredited health care facility to have communication backup plans, primarily for the paging, telephone, and nurse call systems, to facilitate communications between staff and patients and between staff members during an emergency or disaster. Not many institutions, however, have access to a second paging, telephone, or nurse call system if the primary one fails. Nonetheless, if a backup system is not available, staff members must in some way continue to communicate with patients and emergency workers. The following case study illustrates two innovative ways to allow communications to continue during an emergency.

> **Communication System Backups**
>
> When the nurse call system in one area of a hospital failed, the staff handed out bells to the patients—the kind that rings when shaken, such as a "dinner bell," and the kind that rings when a knob is depressed, such as those often situated on front desks with signs that say, "Ring for service." While primitive, these bells filled the void until the communication system could be repaired.
>
> In another example, a large medical facility had a number of telephone lines installed that bypassed the hospital's main switchboard. These lines had a series of telephone numbers different from the hospital series. The telephones for these lines were strategically located (at each nurses' station, the security department, administration, etc.) and were locked until needed. If the telephone switchboard failed, these telephones, as well as the coin telephones, could be used for outside communications. A list of these telephone numbers was kept with the disaster plan of the institution.

Now, with the use of cell phones, there is an almost infinite number of possibilities for communication during normal or emergency operation.

9.3.4 Testing and Maintenance

Depending on the size of the facility and its organization as set forth by its governing body and administration, purchasing, operating, and maintaining the communication system can fall under the jurisdiction of the communications department, the engineering department, or the clinical engineering department. Often, the engineering department is responsible for testing and maintaining some, if not all, of the facility's communication systems. JCAHO and many other governmental agencies have requirements for testing and maintaining communication systems. For information pertaining to these requirements for testing and maintaining electrical systems in health care facilities, refer to Chapter 12.

10

Requirements for Specific Areas and Facilities

Health care facilities include a number of different types of environments, from general care areas, where patients are examined and treated, to areas where complex procedures take place. Each type of area presents unique challenges for providing safe and reliable electrical service for patients and staff. Some of the more challenging areas are as follows:

- Wet locations, where fluids come in contact with both the floor and the staff
- Critical care areas, inhalation anesthetizing locations, and operating and delivery room suites, where electrical life support equipment is used in patients internally and a multitude of equipment all needs power simultaneously
- Radiology, laboratories, hyperbaric chambers, and computer areas, where the equipment typically is extremely sensitive
- Hazardous locations, where explosive and toxic materials may be in use
- Psychiatric units, where mentally incapacitated patients must be cared for
- Emergency departments, where any combination of the above can take place at one time

Areas within health care facilities that are designed or maintained inappropriately and not according to code can lead to serious consequences, such as electric shock or burn, and even death, for patients and staff.

10.1 Hospitals, Nursing Homes, Limited Care Facilities, and Other Health Care Facilities

Hospital facilities present the largest concern for electrical safety, as most patients stay overnight, are subjected to tests and treatment that put them at risk for electrical shock, and are often unable to care for themselves. NFPA 99, *Standard for Health Care Facilities*, 2005 edition (Chapter 13), includes the following general requirement for hospitals:

> It shall be the responsibility of the hospital to provide an environment that is reasonably safe from the shock and burn hazards attendant with the use of electricity in patient care areas. [NFPA 99–05: 13.2.3]

127

The requirements for hospitals are so numerous that they have been discussed in all chapters in this book, as well as noted throughout NFPA 99.

Although the electrical requirements for nursing homes and limited care facilities are not discussed in detail in Chapters 14, 17, and 18, of NFPA 99, 2005 edition, the code does contain in these chapters the requirements for essential electrical systems (EESs), as discussed in Chapter 8 of this book. Briefly, nursing homes and limited care facilities must comply with all of the general codes and standards for health care facilities that are applicable for the service provided. In general, nursing homes must conform to a Type 2 essential electrical system. Under certain circumstances, a battery powered system is permitted. Limited care facilities have the same requirements as a nursing home. Other health care facilities are required to have a Type 3 EES. If the facility uses electrical life support equipment or contains critical care areas, then a Type 1 system must be used. If the facility is an ambulatory health care center, then the electrical distribution system for patient care areas must conform to the requirements of Chapter 4 of NFPA 99. Hospitals must have Type 1 EESs. Essential electrical systems are discussed in Chapter 8 of this book.

NFPA 99 (Chapter 21) states that the electrical systems and equipment in freestanding birthing centers must comply with the applicable sections of Chapter 4 of the same code in terms of electrical distribution, essential electrical systems, grounding, equipment accessories, equipment testing, maintenance and repairs, and record keeping. As more experience is gained as to what problems are encountered, this section of the code will most likely be expanded.

As with freestanding birthing centers, the section in the code for home health care is brief. It states that electrical equipment used in the home for health care should conform to the applicable standards found in Chapter 8 of NFPA 99. The equipment supplier is responsible for this conformance, instructing the user on how to operate the equipment safely and maintain the equipment. In cases where the equipment supplier is a health care facility, the facilities engineer or the clinical engineer is responsible for testing and maintaining the equipment. These tasks are conducted in a manner similar to how equipment that is used on patients staying in a health care facility are tested and maintained.

It is likely that these requirements, too, will change as new information is received and home care practices are revised. Again, it is advisable for maintenance personnel to keep abreast of changes made to the standard and to check with federal, state, and local jurisdictions, as some have already written codes covering these types of facilities.

10.2 General Care Areas

The focus of the electrical requirements in general care areas is on receptacles. Both NFPA 70, *National Electrical Code® (NEC®)*, and NFPA 99 require a minimum of four receptacles (or two duplex receptacles) at patient bed locations in general care areas. In consultation with the medical staff, the facilities engineer must determine the areas that should be equipped with more than the minimum number of receptacles, based on the number of medical devices that potentially can be used in that area at one time. General care areas can be equipped with many medical devices, and it is important to provide enough receptacles for these devices without the use of extension cords or cube taps.

The American Institute of Architects (AIA) *Guidelines for Design and Construction of Hospital and Health Care Facilities* (2001) requires at least one receptacle at each side at the head of each bed, one for a television set if used, one on every other wall, and one for each motorized bed. Receptacles may be omitted on exterior walls where construction or room configuration makes it impractical.

The *NEC* requires each receptacle in general care areas to be grounded by means of an insulated copper equipment-grounding conductor. Although each receptacle does not have to be grounded to the reference grounding point, as a practical matter they usually are. By definition in NFPA 99, the reference grounding point is a terminal bus that is the equipment grounding bus, or an extension of the equipment grounding bus, and a convenient collection point for installed grounding wires or other bonding wires where used. The grounding of the receptacle sets up a redundant grounding system, since the conduit is also usually bonded and therefore grounded to the ground bus of the panelboard. All receptacles are required to be listed as hospital grade and identified as such. These are the receptacles that have the green dot.

10.3 Critical Care Areas

Both the *NEC* and NFPA 99 require that each patient bed location be supplied by at least two branch circuits, one from the essential electrical system and one from the normal service. These rules ensure that if the essential electrical system malfunctions, the normal system will be able to provide service. The second service can be either the normal service or stem from a second critical branch circuit transfer switch. Although it is more common for the second service to stem from the normal power supply service rather than a second critical branch circuit transfer switch, either arrangement is permitted.

The *NEC* requires all receptacles in critical care areas to be listed as *hospital grade* and identified as such. These receptacles are produced to be more durable than the more common *specification-grade* receptacles so as to be able to support and withstand the frequent use (and often misuse) they face in critical care areas. These receptacles are usually identified with a green dot on the face of the receptacle and are thus commonly referred to as *green-dot receptacles*. Refer to Chapter 7 for more information on hospital-grade receptacles.

The *NEC* and NFPA 99 require each receptacle in critical care areas to be grounded to the reference-grounding point by means of an insulated copper equipment-grounding conductor, setting up a redundant grounding system, as mentioned previously. Figure 10.1 shows a typical method of wiring a receptacle to meet these code requirements.

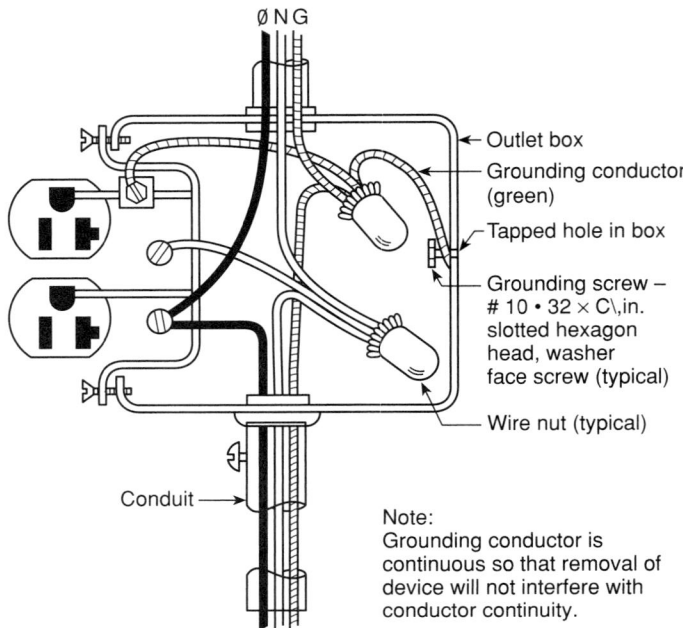

Figure 10.1 Typical method of wiring a receptacle to meet the criteria for grounding of receptacles in critical care areas.
Source: Health Care Facilities Handbook, 2005, Exhibit 4.1.

Both the *NEC* and NFPA 99 require each critical care area patient bed location to have a minimum of six receptacles. These receptacles can be either single or duplex types. The AIA *Guidelines for the Design and Construction of Hospital and Health Care Facilities* require a minimum of seven receptacles at the head of each bed, crib, or bassinet in hospital critical care areas.

The facilities engineer must consult with the staff to determine the exact number of receptacles needed in patient bed locations and the rooms as a whole. Keep in mind that critical care areas are most often used for patients who require special treatment and equipment. The area around the bed or treatment area is often surrounded by many pieces of medical equipment and associated tubes and electrical wires, which can limit access to the patient. Long wires and extension cords can create a tripping hazard and exacerbate what is likely an already trying situation. When a staff member trips on a cord or tube, two people are affected—the staff member and the patient.

10.4 Wet Locations

As defined in Chapter 2 of this book, wet locations are patient care areas normally subject to wet conditions while patients are present. The electrical requirements for wet locations relate to protecting the area from ground faults and minimizing the hazard of electric shock. The main requirement is to use an electrical distribution system that limits the first ground-fault current to a low value. When the electrically dependent activities that take place in a wet location cannot tolerate an interruption of power, limiting the ground-fault current can be accomplished by using an *isolated power system*. This type of system is an ungrounded system designed to keep the current below a level that could harm the patient by alerting staff (by means of alarms and indicators) when the collective leakage current from the circuits and all of the line-operated devices connected to the patient attain a preset level. Isolated power systems are discussed in detail in Chapter 9. When a power interruption in the wet location can be tolerated, a ground-fault circuit interrupter can be used to limit the ground-fault current, provided that the device interrupts the circuit when the fault current is larger than 5 mA (milliamperes). Listing agencies usually permit a trip point between 4 and 6 mA. Authorities having jurisdiction may waive these requirements for existing construction.

10.5 Inhalation Anesthetizing Locations

The electrical requirements for anesthetizing locations pertain primarily to areas where nonflammable anesthetics are used, although the *NEC* also provides information on safety measures to follow where flammable anesthetics are used. The distribution systems in nonflammable inhalation anesthetizing locations must be run in metal raceways with green grounding wires, which must be the same size as the other conductors run in the raceway. All devices connected to the distribution system must be grounded to the raceway at the device. Additionally, the *NEC* requires that one or more battery-powered emergency lighting units be provided.

Both the *NEC* and NFPA 99 permit the installation of a grounded electrical distribution system in anesthetizing locations, provided the area is not classified as a hazardous location and the location complies with all other code requirements. However, if any line-powered electrical equipment is used in these locations and the equipment induces an electric current into a patient's body, the output circuit should be isolated from ground. Since this type of medical equipment is used, most facilities equip some operating suites with isolated power supplies, although this depends on the procedures being performed. Annex B of NFPA 99 deals with shock hazards.

Another requirement for anesthetizing locations is that precautionary signs must be posted in each area indicating whether the area is designed for use of flammable or nonflammable anesthetics. Sample wording for such signs is presented in Annex C of NFPA 99.

In the rare instance that a flammable anesthetizing agent is used, all of the requirements of the *NEC* that pertain to hazardous locations (Articles 500 through 505) would be enforced. Precautions must be followed in locations where flammable anesthetics are used: The production of static electricity and the ensuing spark could cause an explosion. Precautionary measures for avoiding these risks include installing conductive flooring and requiring personnel to wear shoes that discharge static electricity before it creates a potentially destructive spark. This type of equipment also serves to prevent a number of other potential problems, including the discharge of static electricity through sensitive components of electrical equipment, electrostatic cling that can impair efficiency, or the involuntary reaction of personnel to electrostatic discharges.

If flammable anesthetic agents are prohibited in an area, the use of conductive flooring and conductive shoes are not necessary. Likewise, in areas where flammable anesthetics are no longer used, the conductive flooring can be disconnected from the ground system, which in effect creates a regular floor. It also eliminates the need for the facility to test the floor regularly for conductivity and for staff working in the room to wear conductive boots or check their conductivity before entering the area. These reduced measures can lead to large financial savings for a facility. If the floor is disconnected from the grounding system, signs should be posted indicating that this has been done.

Although flammable anesthetics are no longer used in hospitals, there are still reports of fires in operating suites apparently caused by some form of electrical usage. Investigations concluded that many of the fires are caused by negligence of the staff while conducting medical procedures, and that these fires could have been avoided. As an outcome of these incidences and investigations, NFPA added to NFPA 99 (Chapter 13) a series of electrical safeguards to follow while operations are taking place in anesthetizing locations. These requirements cover physical safeguards, scheduled inspections, replacement of defective electrical equipment, personnel training about these safeguards, electrical safety equipment testing, new equipment approval, and isolation of line-powered equipment that induces current into a patient's body.

10.6 Operating and Labor and Delivery Suites

As mentioned, most operating suites are usually equipped with isolated power supplies as they can be classified as wet locations, inhalation anesthetizing locations, or, depending on the anesthetizing agent or other chemical that might be used, hazardous locations. The type of procedure and the governing body determine the classification of the area. If the room is used for different types of procedures, some of which can cause the area to be classified as a wet location, then it will have to have an isolated power supply. Refer to Chapter 2 for more information on area definitions and Chapter 9 for details on the electrical requirements for isolated power systems.

The AIA *Guidelines* require each operating and delivery room suite to have at least six receptacles convenient to the head of the bed and a total of 8 duplex or 16 simplex receptacles in the room. In addition, operating rooms must be equipped with specialized receptacles with appropriate configurations to accommodate special equipment such as portable X-ray machines. To eliminate transforming static electricity (and shocks) to patients in operating and delivery suites, it is essential to ground all electrical outlets and equipment properly.

Several electrical configurations can accommodate the physiological monitoring equipment used in operating and delivery rooms. Some equipment systems interconnect all of the monitoring units in an entire operating suite, which allows staff to use any one of the units to monitor activities taking place in the other rooms. When using this circuitry, care must be taken to prevent the leakage currents typically present in each room from adding together and exceeding the value for tripping the line isolation monitor of the isolated power system that alerts of excessive leakage current. Recall the case study in Chapter 9, which illustrates why this is so.

10.7 Radiology

NFPA 99 includes very few electrical requirements for the radiology department. Radiology areas categorized as wet or hazardous, and associated equipment used in these areas, must comply with the requirements for those classifications. The *NEC*, 2005 edition (Part V of Article 517), includes specific requirements that address a number of concerns for X-ray installations, including supply conductor ratings, supply circuit connection, control circuit conductors, disconnecting means, overcurrent protection, equipment installations, transformers and capacitors, installation of high-tension X-ray cables, and guarding and grounding.

10.7.1 Branch Circuit Conductors

One specific *NEC* requirement (Section 517.73) for radiology areas is that branch circuit conductors must be rated for at least 50 percent of the momentary rating or 100 percent of the long-term rating, whichever is greater. The facilities engineer should be familiar with the characteristics of the equipment in question to determine the existing load on a branch circuit or feeder. If the equipment presents a momentary large load, such as when an X-ray unit is used, then the continuous loading is small—and the effect of the medical device on the total continuous load, and thus all conductors and overcurrent devices, is small.

If the loading on the branch circuit conductor is momentary but in rapid succession for a long period of time, such as when a CAT scan procedure is conducted, then the load has the effect of being continuous. This type of continuous load in turn affects the continuous loading on all conductors, overcurrent devices, and generators. All generators are capable of handling momentary overloads for finite values of time. Long-time overloads, however, that approach being continuous can cause the generators to shut down on overload if they are not sized for this type of load.

If requested, some equipment manufacturers provide information on wire sizes, length of conductors, and other criteria for the supply power to radiology areas. They may, however, give this information out very cautiously and, if necessary, err on the side of increasing wire sizes and overcurrent device trip ratings. Manufacturers typically recommend wire sizes for each conductor and the maximum length of the conductor that can be used before the wire size must be increased.

Because the length and size of the conductor obviously affect the voltage drop, and there are various voltages used in radiology equipment, voltage transformers are often installed in radiology areas. Manufacturers should be consulted as to the optimal location for this equipment, to avoid interference with other nearby medical devices. The manufacturer usually insists that all of the power and control cables for radiology equipment be located in some form of a wireway or raceway. Loose exposed cables should be kept to a minimum, both for safety purposes and to aid the housekeeping department in keeping a clean environment.

10.7.2 Radiology Equipment Installation

Radiology equipment manufacturers often prepare the installation plan and work directly with the facility engineer during the installation of the radiology equipment. Usually the manufacturer's instructions state that the installation must meet all local codes, although the manufacturer may not be familiar with the specific local electrical codes that apply. The facilities engineer is ultimately responsible for ensuring that the equipment meets all applicable codes, fits within the allocated space, and does not affect the surrounding areas by creating electromagnetic interference.

Another factor to consider when installing radiology equipment is that the lighting of the area must not affect the operation of equipment that uses remote controls or controllers. Likewise, the operation of the equipment must not affect the setup in the surrounding areas. Refer to the case study in Chapter 7 about the situation where the fluorescent lighting affected the remote control of some radiological devices. This manufacturer required incandescent lighting to be installed in the room where remote-controlled equipment was used but did not list this requirement in its equipment specifications. For the installation of any equipment, it is prudent to determine if there are installation requirements other than those provided in the manufacturer's written materials and specifications.

In general, recessed film illuminators (i.e., backlights mounted on the wall for viewing X-rays and other diagnostic images) are safe to use because they have no sharp protruding edges. This feature will be most appreciated in exam and treatment rooms, where space might be at a premium.

10.8 Laboratories

NFPA 45, *Standard on Fire Protection for Laboratories Using Chemicals*, has primary jurisdiction over laboratories in general, including laboratories located in health care facilities. NFPA 99 sometimes requires more stringent rules for laboratories in health care facilities compared to other types of facilities. Although very few requirements in NFPA 99, 2005 edition (Section 8.5.2.5 and Chapter 11), address the electrical system in laboratories per se, this code does include more stringent requirements for health care facility laboratories in terms of laboratory structure, equipment, fire protection, emergency showers, the storage and use of flammable and combustible liquids and gas cylinders, the transfer of gases, construction details, and maintenance and inspection considerations. The type of work conducted in the laboratory and the chemicals used determine which sections of the *NEC* should be applied.

In terms of electrical requirements, laboratories must have an adequate supply of receptacles with a certain degree of flexibility built into their location. (Power cords lying all over a laboratory floor is a disaster waiting to happen.) Flexibility is often accomplished through the use of multioutlet assemblies, an underfloor duct system where receptacles can be installed at any future time, or properly located power poles. Multioutlet assemblies are surface-mounted raceways connected to surface-mounted boxes or light fixtures (refer to Chapter 6). All laboratory receptacles should be of a durable grade and be properly grounded and bonded.

Much of the newer equipment used in health care facility laboratories is computer driven, with very specific and stringent voltage requirements. As a result, voltage fluctuations as well as transients must be kept to a minimum, and it may be necessary for a laboratory to be equipped with a separate branch circuit with a quiet ground (see Chapter 6). Here again, the facilities engineer should work closely with the equipment manufacturer and the laboratory director to determine the optimal electrical arrangement for the laboratory.

The results of some laboratory tests might be seriously affected by the 10-second loss of power that can occur when the normal power fails and the emergency generators have not yet come on line. To accommodate this need for uninterrupted power, some equipment is manufactured with an *uninterruptable power supply* (UPS) within the equipment. These units have varying amounts of time capacity, usually a few minutes, that allow either the generator to come on line or the test procedure to be brought to an orderly cessation (not necessarily to completion). If the equipment cannot be obtained with this feature and uninterruptable power is required, the facilities engineer must install a separate UPS unit.

By the nature of the work conducted in laboratories, proper lighting is required for both the safety of the personnel moving about the laboratory and the proper operation of the equipment being used. Additionally, determining the results of some diagnostic culture tests depends on lighting that does not affect color rendition. Refer to Chapter 7 for more details on lighting.

Another important safety feature in the laboratory is proper ventilation. The electrical staff is responsible for bringing the proper amount of energy at the proper voltage to the ventilation units and ensuring that the units start and run properly.

10.9 Hazardous Locations

The *NEC* (Articles 500 through 506 and 510) stipulates very strict, lengthy, and specific electrical requirements for hazardous (classified) locations in any facility, based on the types of materials in use and their varying degrees of hazard. Depending on the hazardous location classification, the *NEC* requires such items as explosion-proof receptacles and switches, sealing of all conduit and boxes, special lighting fixtures, and so on.

Communication among the departments and staff about all procedures and chemicals to be used in all locations within the facility is essential—first to determine which areas should be classified as hazardous, and then to maintain safety, as illustrated in the following case study. For example, when a member of the professional staff wishes to use a new substance or procedure in a hazardous location, he should notify the engineering department to ensure code compliance in the area, as well as safety for the staff and patients. The facilities engineer should also emphasize to the administrators and staff the importance of not using any unauthorized material.

> **Improper Use of Hydrogen Gas**
> A physician in a cardiac catheterization laboratory of a large teaching hospital used hydrogen gas in a certain procedure. Before trying it out, however, he failed to notify any of the administrative or engineering staff. While the physician was aware that hydrogen was a highly flammable gas and handled it very carefully, he was not aware of other precautions to take, such as to conduct the procedure in a laboratory that was classified as hazardous and was therefore constructed and wired appropriately. Luckily, an accident did not occur before the physician discontinued the procedure.

10.10 Psychiatric Units and Hospitals

This section addresses the electrical requirements for psychiatric units within general hospitals as well as hospitals devoted to psychiatric care. All electrical requirements for psychiatric areas and hospitals are designed to prevent patients from harming themselves, either by accident or by self-infliction, to prevent harm to staff members and visitors, and to protect all people in the event of an emergency. Psychiatric patients can be very inventive when it comes to bypassing the precautions and special equipment that are often used in these areas.

All of the NFPA 99 requirements that apply to hospitals also apply to psychiatric hospitals. The AIA *Guidelines for Design and Construction of Hospital and Health Care Facilities* (2001) contains a separate chapter on psychiatric hospitals. State and local authorities may impose additional electrical requirements, such as those related to the total number of receptacles required, where the receptacles must be located, the type of lighting fixtures that can be used, and maintenance and repair procedures.

Based on the AIA *Guidelines*, the electrical requirements for psychiatric areas within general hospitals are basically the same as for other patient care areas with respect to the protection of the patient. The requirements for the emergency room may vary in specific areas set aside for psychiatric patients, but not if the patients are closely supervised in the emergency room.

The AIA *Guidelines* require all receptacles in psychiatric hospitals and psychiatric areas within a general hospital to be tamperproof or equipped with a ground fault circuit interrupter, particularly where patients are likely to come into contact with these devices. The tamperproof receptacles can be similar to the type used in pediatric areas, that is, constructed in such a way that when an object, such as a paper clip, is inserted into one slot of the receptacle, the other slot of the receptacle is energized. To make a full connection, an object must be inserted into both slots. In some receptacles, something also must be inserted into the ground connection to activate the circuit.

At one time, the use of plastic covers that inserted into the receptacle to cover them was allowed. Because patients could remove them too easily, these covers are no longer accepted as adequate protection in many jurisdictions. Receptacle cover plates should be secured with tamperproof screws that can be removed only with special tools. State and local authorities may require various numbers of receptacles in patient rooms or head walls and may regulate the manner in which they are concealed. The AIA *Guidelines* require grounded duplex receptacles to be installed in patient rooms. There has to be one receptacle at each side of the head of each bed and one on every other wall. There is an exception for exterior walls.

The lighting fixtures in psychiatric areas and hospitals should also be tamperproof. The location of the lighting fixtures may be determined by the location of furniture in the room or the procedures that are being performed in the room. Having a light shine directly into a patient's eyes should be avoided. The lens that is used on the lighting fixture should be unbreakable if possible. All fixtures should have some type of protection over the bulb or tube to protect it from being damaged by the patient. Controls for lighting in psychiatric units should be limited to snap switches or key switches. The locations for the controls may be dictated by local codes. Pull cords should not be used, because they can be used to commit suicide or to inflict harm on another patient or staff member. If the patient care area is a locked unit, the area may need special electrical security and communication equipment that the facilities engineer must install and maintain.

The use of emergency power in psychiatric areas should be discussed with the staff and administration. The AIA *Guidelines* require that nursing facilities must comply with NFPA 99; NFPA *101*®, *Life Safety Code*®; and NFPA 110, *Standard for Emergency and Standby Power Systems*. During a power failure, it is particularly important to keep psychiatric patients calm, which may require the use of additional emergency power.

Facilities engineers must make special provisions when installing new equipment or performing or scheduling maintenance or repairs in psychiatric hospitals or areas. A general rule to follow is to discuss work with the staff before its implementation. It is particularly important to consult with the psychiatric staff at all times, before, during, and after the repair or maintenance work is conducted. The work should be scheduled as far in advance as possible to allow both the maintenance and psychiatric staffs to make all necessary arrangements. If variances are needed, such as locked fire exits that open electrically after an alarm, it may be helpful to consult with the authority having jurisdiction.

Installation of special equipment for use in psychiatric areas and associated code requirements also should be discussed with the staff. For example, staff may want a hidden buzzer to be installed in an interview room to alert staff outside the room of a problem occurring within the room, or staff may wish to take other precautions. The reverse can also be true; the staff may want to make the installation look less institutional and request that some code requirements not be put into place, such as added screening on the windows. Any waivers of the requirements must be discussed with the staff and the authorities having jurisdiction, and all paperwork must be readily available for inspectors.

When conducting any work, it may be advisable to have a person from the psychiatric staff at the repair site, if many patients will also be there. The staff should conduct the work at least in pairs, even if it is a one-person job. One worker should make the repair, while the other worker watches to make sure that a patient does not take any tools or equipment or get too close to the operation. If security cannot be maintained, the engineering and psychiatric staffs will have to cooperate to keep the area of the repair patient free. All tools and materials must be secured at all times.

For security reasons and to prevent patients from eloping, many doors and equipment that are usually unlocked and required to be unlocked in hospitals, including fire exits and fire alarm boxes, should be or are required by the staff to be locked in psychiatric hospitals and psychiatric areas in general hospitals. This usually requires that the authority having jurisdiction approve the method of locking and unlocking. Some methods used are specially marked keys that must be carried by all staff members, or electrically controlled locks.

10.11 Emergency Departments

Major trauma centers almost resemble mini-hospitals. Most trauma centers contain major operating rooms, cardiac care rooms, radiology departments, hazardous material detoxification rooms, pediatric areas, and specialty areas such as ophthalmology, dentistry, obstetrics, and so on. These areas can include wet locations, anesthetizing locations, hazardous locations, and the like. The same electrical codes and standards that apply to these areas in the main facility apply to these areas in the emergency department.

Emergency departments typically are large users of electrical power and equipment. NFPA 99 requires the emergency power system to supply electricity to task lights, receptacles, and power circuits in selected emergency room treatment areas. Choosing the areas to be connected to the emergency system can be difficult, due to uncertainty regarding how the areas will most often be used. In many cases, the emergency department serves as the control area when the disaster plan for the facility is put into effect. If the emergency area is a major trauma center, it might be best to place the entire emergency department on emergency power, except for some administrative areas. In an emergency, administrative functions can take place from the nurses' station.

The AIA *Guidelines* include requirements for emergency service. Although these requirements are not necessarily enforceable in every facility, they are certainly worth looking at. For example, they include a requirement for a well-illuminated entrance. Emergency rooms must also be able to maintain radio contact and other means of communications among the emergency department staff, other relevant staff members, emergency response teams, and local police and fire departments. Because emergency departments are high-maintenance areas by nature, it is important for the facilities engineer to ensure that maintenance staff monitor these areas carefully.

Emergency departments are continually evolving. Therefore, when renovating or planning a new emergency department, provisions should be made to accommodate future needs with spare capacity and the ability to connect the area to the spare capacity without major shutdowns. It is one thing to have the power available at the perimeter of the department and another to be able to bring the power to a location within the department.

10.12 Hyperbaric and Hypobaric Areas and Facilities

It should be obvious that in hyperbaric areas, which have increased atmospheric pressure and increased partial oxygen content, certain electrical system requirements must be met to avoid serious hazards. All applicable requirements of the *NEC* must be followed. Likewise, the equipment must comply with the requirements of the medical equipment requirements of NFPA 99. (Refer to Chapter 8 for information on electrical requirements for medical equipment.)

Chapter 20 of NFPA 99 (2005 edition) contains the requirements for hyperbaric facilities and chambers. One main electrical requirement for hyperbaric chambers in NFPA 99 is that all hyperbaric facilities for humans must have electric service from two independent sources. If the hyperbaric facility is based in a health care facility and uses a prime mover generator set as one source of power, then the set is classified as the emergency system and conforms to Chapter 4 of NFPA 99. Another requirement is that only electrical wiring or equipment that is absolutely necessary for the safe operation of the chamber and the care of the patient can be located inside a chamber. The wiring and equipment must be rated for this type of installation, and the wiring to the chamber must have two sources of power.

If the unit is installed in a health care facility, electrical power circuits within the chamber must be supplied from an ungrounded system. In the event of a fire in a hyperbaric area, nonessential electrical equipment should be turned off as much as possible before extinguishing the fire. Other requirements for hyperbaric chambers address the wiring methods, lighting, low-voltage equipment, patient-related electrical appliances, battery-operated devices, cord-connected devices, and other aspects of grounding, all of which are very important. Figure 10.2 shows a Class B hyperbaric chamber (single human occupancy) located within a health care facility.

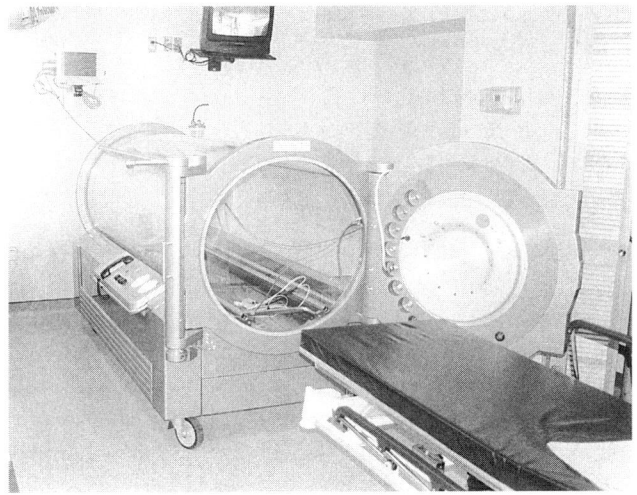

Figure 10.2 Class B (single-occupancy) movable hyperbaric chamber.

Personnel responsible for installing and maintaining hyperbaric chambers must work closely with the operating personnel to ensure that the chamber meets all of the requirements for installation and use. All electrical circuits in hyperbaric chamber areas must be tested before chamber pressurization, and specialized training and experience are necessary to operate hyperbaric systems. Proper preventive maintenance and repairs are essential and should be performed only by qualified personnel. Careful records of all such work must be kept.

While hypobaric chambers are no longer used in health care facilities, health care facility personnel may find it useful to know that electrical requirements for hypobaric chambers closely resemble those for hyperbaric chambers. NFPA 99B, *Standard for Hypobaric Facilities*, has all of the requirements. As with hyperbaric chambers, personnel involved in the operation of hypobaric chambers must be properly trained and experienced.

10.13 Information System Areas

The requirements for installing information system equipment in health care facilities are found in two main codes. The *NEC* (Article 645) is devoted to information technology equipment. NFPA 75, *Standard for the Protection of Information Technology Equipment*, contains construction and fire protection requirements for data processing equipment and all relevant electrical requirements. Both codes address equipment and wiring within information technology equipment rooms and wiring outside the rooms. These codes frequently cross-reference each other.

10.13.1 Cable Installation and Marking

In many information technology equipment rooms, the interconnecting cable is installed under a raised floor. Requirements for this type of installation relate to the type of cable that can be used, the construction of the floor, the fire protection and detection systems employed, and the like. The *NEC* includes several codes and standards for the following types of cable that are connected to the system but are installed outside the equipment room:

- Article 725—signaling circuits
- Article 770—fiber-optic circuits
- Article 760—fire alarm systems
- Article 800—communication circuits

Refer to Chapter 6 for a listing of the *NEC* codes that address wiring methods and cables.

If it is not possible to avoid having cables lying exposed on the floor, the cables should at least be grouped as much as possible to avoid accidents and aid the housekeeping services for the room. Cables should be protected to ensure that the insulation is not damaged, and they should be properly supported. The location of the room, equipment in the room, and cable within and outside the room should be selected with features that ensure that there will be no interference with other electronic medical devices in the facility.

Planning for future data processing electrical needs is important. As the use of data processing expands, so will the need for cabling to transport the information (until it is all sent over the airwaves). Allowing space for additional cable runs throughout the facility and making sure that all cables are properly labeled periodically throughout the run will accommodate future needs.

10.13.2 Information System Area Disconnects

One electrical requirement for information technology equipment rooms is that two main disconnects must be installed in these areas. One disconnect protects all of the electronic equipment installed. The second disconnect serves the dedicated heating, ventilating, and air-conditioning (HVAC) system, which should disconnect power to the HVAC, as well as close all fire and smoke dampers, when necessary. One switch can be used for both functions. The control for these disconnects must be at the exit door.

10.13.3 Uninterruptible Power Supplies

Health care facility data processing and information systems operators may require an uninterruptible power supply to function properly during interruptions of power, as momentary disruptions of power can cause errors or malfunctions. A UPS will provide power for an orderly shutdown of data processing equipment and procedures when the primary power supply has failed. The UPS ensures that there will be enough power for an adequate amount of time for the system to shut down properly or until the emergency generator comes on. The emergency generators can take up to 10 seconds before they restore power (see Chapter 8).

According to the *NEC* (Section 645.11), the supply and output circuits of a UPS must have the same two disconnecting means as stated previously. The disconnecting means must also disconnect the battery from its load. Refer to the *NEC* for several exceptions to these requirements.

10.13.4 Grounding

There are stringent requirements for grounding the equipment in information system rooms, with some exceptions for double-insulated equipment. The manufacturers of equipment installed in the main data processing area usually provide an installation booklet that describes these requirements. Many equipment sales representatives have no formal engineering training, however, and cannot provide information on how to install equipment. It may be feasible, for example, to use less costly and more convenient installation options than those suggested by the manufacturer or sales representative. Facilities engineers or other staff members who have experience or training in installing equipment should question the manufacturer's suggestions when necessary. Unfortunately, some manufacturers approve the use of only their own installation method, and the facility must use that method to avoid nullification of any warranties or guarantees.

11

Medical Equipment

The hazards associated with using medical equipment for patient diagnosis and treatment have always been a concern. Most medical equipment developed during the past few decades has been electronic or has exposed patients to electricity in some form. As a matter of course, today's patients are subjected to electrical currents in the microampere range, such as through various indwelling catheters, monitoring leads, and other pathways. Hydrotherapy that requires the use of electrical devices is just one example of a medical treatment that requires extra precaution. Questions are frequently raised about the effect of electrical shock, the amount of current that constitutes a safe range for electrical shock, steps that equipment manufacturers can take to protect the patient, and the type of preventive and regular maintenance that facilities can perform to keep equipment safe for patient use.

At one time, these concerns were addressed in NFPA 76B, *Standard on the Safe Use of Electricity in Patient Areas of Health Care Facilities*, 1980 edition. These requirements, with additions written later, were incorporated into various Chapters of NFPA 99, *Standard for Health Care Facilities*. Law-making bodies and other authorities having jurisdiction (AHJs) most often refer to the requirements of NFPA 99 when addressing medical equipment, although some local and state authorities may also issue regulations that pertain to medical equipment. Additional requirements for medical equipment are noted in Joint Commission on Accreditation of Healthcare Organizations (JCAHO) accreditation manuals.

Two other organizations that publish recommended practices for medical equipment are the Institute of Electric and Electronics Engineers (IEEE) and the Association for the Advancement of Medical Instrumentation (AAMI). IEEE's *Recommended Practice for Electric Systems in Health Care Facilities* (1996) devotes an entire chapter to medical equipment and instrumentation; while this book is not considered a code or standard, it is an excellent source for information on this topic. AAMI writes many recommendations for specific pieces of equipment. Those responsible for testing and maintaining medical equipment in health care facilities, whether in-house staff or outside contractors, should be familiar with IEEE and AAMI recommendations. Although these recommendations may not have any legal status, they make excellent reference guides.

The requirements and guidelines for medical equipment cover manufacturer and facility responsibilities. Manufacturers must develop equipment that meets specifications set forth in the codes and must conduct numerous equipment tests before any device can be sold. Facility purchasing agents must make sure to order equipment that meets these code requirements, as well as other conditions deemed necessary by the facilities engineer and the authority having jurisdiction. Clinical engineers must test all equipment before any unit is installed or used, and they must install, test, maintain, and repair the equipment on an ongoing basis. Depending on the institution, staff from the biomedical engineering, clinical engineering, or facilities engineering department are responsible for conducting this work, or else the work is contracted to outside specialists. In each institution, the department responsible for testing and maintaining equipment is responsible for conducting staff training programs on the proper and safe use of medical equipment.

11.1 Equipment Manufacturer Requirements

Chapter 10 of NFPA 99, 2005 edition, includes a list of manufacturers' requirements that covers the performance, maintenance, and testing of equipment used in health care facilities to promote the manufacture of safe equipment.

11.1.1 Performance Standards

Manufacturing performance standards address wiring and circuitry, the power cord and plug, materials approved for use, voltage requirements, and the like. Many manufacturer equipment requirements in NFPA 99 cover the cord and plug arrangement, including the material and gauge of the power cord, its grounding conductor, different types of cord sets, cord and plug strain relief, the plug grounding prong, and the manner in which the wire is connected to the plug. These specifications are especially important for portable equipment that often undergoes strain due to frequent plugging and unplugging from a receptacle.

One specific requirement (NFPA 99, 2005 edition, Section 8.4.1.2.1.1) is that all cord-connected electrically powered appliances used in the patient care vicinity that are not double insulated must have a three-wire power cord and a three-pin grounding-type plug. The power cords of an appliance that doesn't require or have a grounding conductor must not have a grounding-type plug. The cords for line-voltage equipment used in anesthetizing locations, such as portable lamps or portable electric appliances operating at more than 12 V (volts) between conductors, shall be continuous (i.e., without having switches from the appliance to the attachment plug) and of a type designated for *extra-hard usage*. The cord must also be long enough to reach its proper receptacle from any position where the portable device will be used.

Another basic requirement (NFPA 99, Section 10.2.3.3) is that all power cords should be accessible for replacement, unless the manufacturer does not want the user to make this repair. In this case, it is suggested that provisions be made at the time of purchase for performing this type of repair. The manufacturer should provide information about where authorized repairs can be made, shipping arrangements, and turnaround time.

Another important but unfortunately often neglected NFPA 99 manufacturer requirement concerns wiring and switches. The wiring within appliances equipped with a power cord should be installed to minimize accidental contact between energized wires and the case of the equipment. This can be accomplished by connecting the grounding conductor to the external metal case by means of a terminal or bolt. For units that use a primary control switch, and most do, the equipment switch should disconnect all primary power conductors, including the neutral conductor, but not the grounding conductor. (Recall that the neutral conductor and the grounding conductor are not the same; the neutral conductor is grounded at the main service switch.) Refer to Chapter 6 for further discussion of grounding in the distribution system.

Several manufacturer requirements aim to prevent the equipment patient leads from becoming energized by the normal electrical supply, thereby ensuring that the current transferred to the patient will be below a value that will cause discomfort or harm. For example, the plugs and jacks for patient leads must not be interchangeable with the normal power-supply plugs and receptacles. To further avoid interchanging patient leads with the normal power supply, manufacturers must index all patient-lead receptacles. Because the values of current believed to cause hazards have changed over the years, it is essential for the facilities engineer to keep abreast of any changes made to the code in this regard.

NFPA 99 (Section 10.2.5) requires all appliances to operate within the line-voltage variations prescribed in American National Standards Institute's (ANSI's) C84.1, *Electric Power Systems and Equipment—Voltage Ratings*. Utility companies often lower voltage when their systems become overloaded, most often during hot summer months when the use of air conditioning strains utility company distribution systems. Excessive voltage drops can occur within health care facilities equipped with distribution systems that have not been designed to account for this. Many medical devices are equipped with a voltage-regulated power supply in the front end of each unit, to ensure that the voltage supply to the internal workings of the equipment is kept within the required narrow ANSI range and to meet specific design requirements that impose more stringent limits on voltage variation.

Most medical electronic equipment today has some form of computer chip within the circuitry for controlling or programming the equipment. However, when there is a loss of power and subsequent resumption of power (either through the start-up of an emergency generator or the restoration of normal power), this type of equipment may not automatically revert to normal operations. To avoid this type of problem, manufacturers must configure equipment to either continue normal operations after such a loss of power or default into a nonhazardous or start-up status. If equipment is not configured to go into start-up status, it should include both an audible and a visual indication of the status it is in.

Other NFPA 99 design and manufacturing requirements (Chapter 10, Section 10.2.6) for patient care–related electrical appliances can be paraphrased as follows:

- **Thermal hazards.** Exposed surfaces cannot get too hot.
- **Toxic materials.** Surfaces that contact patients shall be free of materials that commonly cause toxic reactions.

- **Chemical agents.** Equipment must be designed so that the replenishment of any chemical agents used for its operation occurs without spillage.
- **Electromagnetic compatibility.** Equipment must be able to operate in a radio-frequency electromagnetic environment (although cell phones are restricted from use in selected health care facility areas, to prevent interference with some equipment).
- **Operation with essential electrical system.** The equipment must be able to operate normally when energized by an essential electrical system.
- **Programmable appliances.** If there is a loss of power, the program of the equipment cannot be lost.
- **Fire and explosion hazards.** Equipment must be constructed with noncombustible materials and must be impermeable to liquids and gases.
- **Electrical equipment in oxygen-enriched atmospheres.** Equipment must comply with rules for gas equipment and hyperbaric facilities.

There are other manufacturer requirements listed in Chapter 10.

11.1.2 Manufacturer Testing Requirements

Chapter 10 of NFPA 99 requires medical equipment manufacturers to perform the following five leakage current tests. These tests are similar to the tests that must be conducted at the health care facility after it receives the equipment (see Sections 11.3.1 through 11.5, this chapter).

1. *With all the ground conductors open at the end nearest the power receptacle, measure the current from an exposed conductive surface of the appliance to ground.* Figure 11.1 shows an acceptable test circuit for measuring leakage current from exposed conductive surfaces. If no exposed conductive surface is present, then the manufacturer must simulate one by having bare metal foil, 3.9 in. by 7.8 in. (10 cm by 20 cm), come into contact with the surface. In a cord-connected appliance

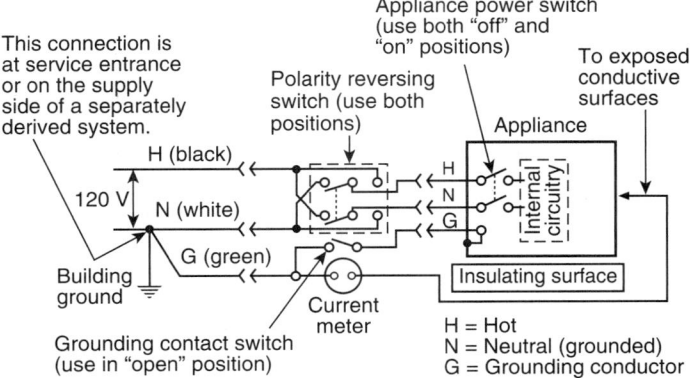

Figure 11.1 Manufacturers' test circuit for measuring leakage current from exposed conductive surfaces.
Source: NFPA 99, 2005, Figure 10.2.13.4.1.

intended for use in a patient care vicinity, the leakage current cannot exceed 300 µA (microamperes). In a permanently wired device, the leakage current cannot exceed 5.0 mA (milliamperes).

2. *Test lead to ground with a nonisolated input.* The leakage current should not exceed 100 µA. Figure 11.2 depicts an acceptable manufacturers' test circuit that can be used for measuring leakage current between patient leads and ground (nonisolated).
3. *Conduct the preceding test with an isolated input.* Figure 11.3 illustrates an acceptable manufacturers' test circuit for measuring leakage current between patient leads and ground (isolated). The leakage current must not exceed 10 µA with the ground intact and 50 µA with the ground open.
4. *Test the isolation between each of the patient leads and ground with an isolated input.* Figure 11.4 shows an acceptable manufacturers' test circuit that can be used for measuring the electrical isolation of isolated patient leads. The leakage current cannot exceed 50 µA at the end of the patient leads and 25 µA at the apparatus terminals.
5. *Test the leakage current between any of the patient leads with both a nonisolated input and an isolated input.* Figure 11.5 depicts an acceptable manufacturers' test circuit that can be used for this test. The leakage current with a nonisolated input cannot exceed 50 µA. The leakage current with an isolated input cannot exceed 10 µA with the ground intact and 50 µA with the ground open.

11.1.3 Manufacturer Documentation

Manufacturers must supply operator, maintenance, and repair manuals with all equipment. According to NFPA 99, the manuals must contain the following information:

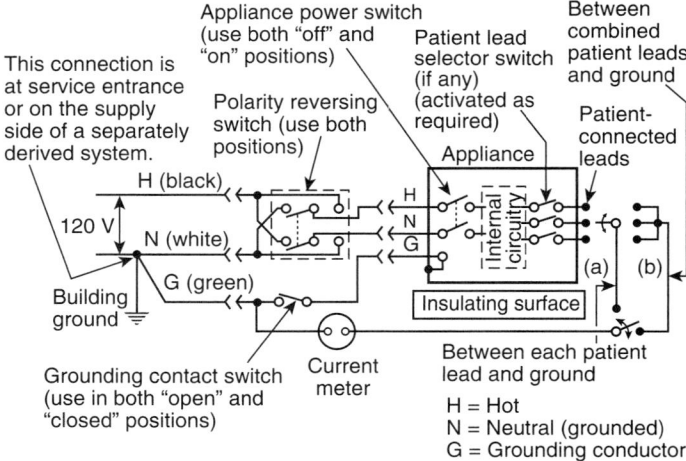

Figure 11.2 Manufacturers' test circuit for measuring leakage current between patient leads and ground (nonisolated).
Source: NFPA 99, 2005, Figure 10.2.13.5.1.4.

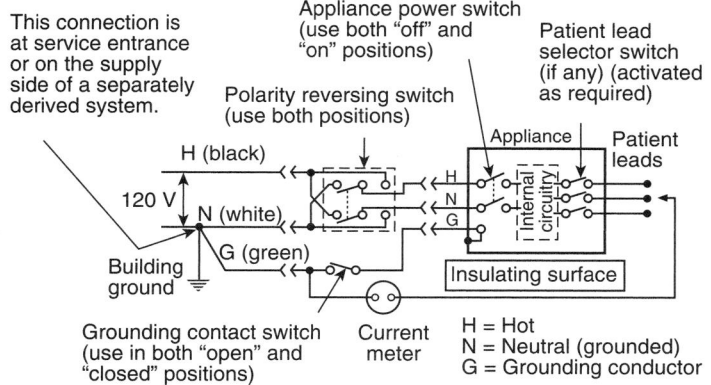

Figure 11.3 Manufacturers' test circuit for measuring leakage current between patient leads and ground (isolated).
Source: NFPA 99, 2005, Figure 10.2.13.5.2.3.

- Illustrations that show the locations of the controls
- Explanations of the function of each control
- Illustrations of proper connections to the patient and other equipment
- Step-by-step procedures for proper use of the appliance
- Safety considerations in application and in servicing
- Difficulties that might be encountered and care that must be taken if the appliance is used on a patient simultaneously with other electrical appliances
- Schematics, wiring diagrams, mechanical layouts, parts lists, and other pertinent data for the appliance as shipped
- Functional description of the circuit

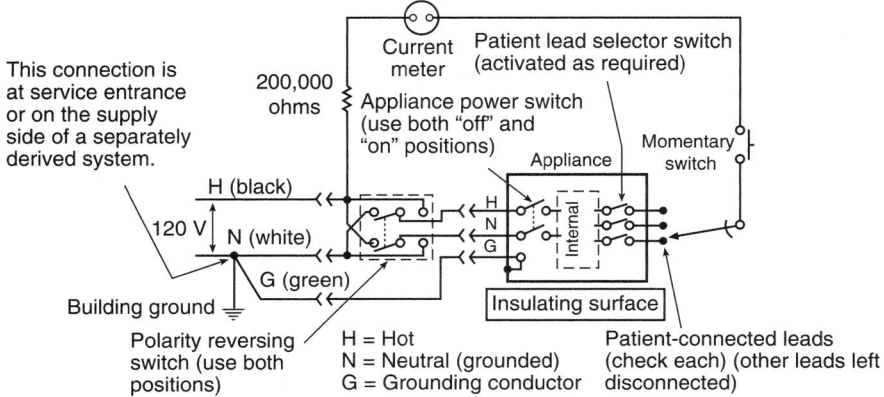

Figure 11.4 Manufacturers' test circuit for measuring electrical isolation of isolated patient leads.
Source: NFPA 99, 2005, Figure 10.2.13.5.3.2.

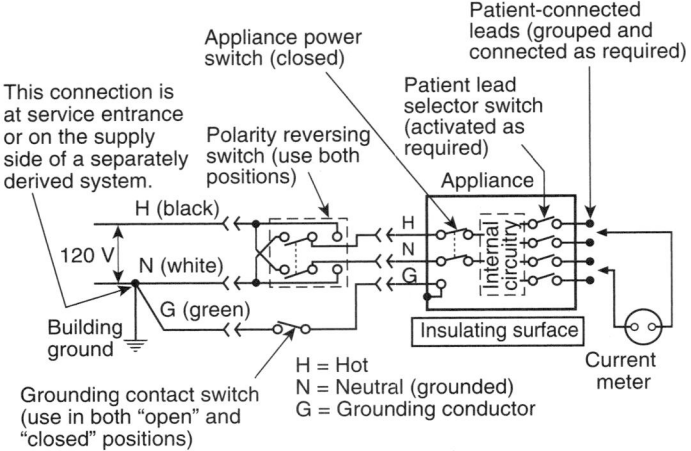

Figure 11.5 Manufacturers' test circuit for measuring leakage current between patient leads (nonisolated and isolated).
Source: NFPA 99, 2005, Figure 10.2.13.5.4.1.

- Electrical supply requirements (voltage, frequency, amperes, and watts), heat dissipation, weight, dimensions, output current, output voltage, and other pertinent data
- Performance specifications of the appliance for the applicable limits of the electrical supply voltage
- Technical performance specifications including design levels of leakage current
- Instructions for unpacking (readily available upon opening), inspecting, installing, adjusting, and aligning
- Comprehensive preventive and corrective maintenance and repair procedures

11.2 Equipment Purchase Order Specifications

Medical personnel, facilities engineers, purchasing agents, and administrative staff all have a say in purchasing medical equipment, and most likely there is not always agreement on what equipment should be acquired. Most parties can agree, however, that whatever equipment is purchased, it must meet manufacturing specifications as well as any additional requirements set forth by the facility and by the AHJ. If the purchase specifications are not the same as the manufacturing specifications, then facilities engineers can provide purchasing agents with information on any special conditions that must be included in a facility's purchase orders, such as particular voltage requirements or safety measures for equipment that will be used in hazardous locations.

NFPA 99 (Chapter 8) lists the conditions of purchase that facility procurement authorities must include on purchase orders for new equipment. These conditions cover equipment listing and certification, manufacturer test data, special conditions of use, unusual environmental conditions, the type of electric power system that will energize the equipment, the nature of the overcurrent devices, and the use of auxiliary emergency power. Some of these conditions become "boilerplate" items that are included in every purchase order. The conditions required by code, as well as other typical conditions with which equipment manufacturers must comply, are as follows:

- *Include a statement that the equipment is in compliance with NFPA 99.* If manufacturers are reticent to make such a broad statement, the purchase order should include comparable requirements that ensure compliance.
- *Include appliance listing or certification, if any.* For all new equipment orders, a facility or the authority having jurisdiction must require manufacturers to include a copy of the equipment's certification or listing from a third-party testing laboratory. However, manufacturers may not yet have submitted equipment that is either a proprietary item or is new on the market to a testing agency for listing.
- *Provide equipment that operates under special conditions of use (such as in anesthetizing or other locations with special hazards) as stated in the purchase order.* If the piece of equipment is slated for use in a hazardous area, it should be listed for that purpose. In some instances, the facility may have to describe the potential hazard to the manufacturer. For example, foreign companies may not be fully aware of medical equipment requirements in the United States. The case study illustrates such an example.

Equipment Specifications for Special Conditions

A lithotriptor is a piece of equipment that requires patients to be wholly or partially immersed in water within an electrical device. The location where lithotripography takes place is considered to be a wet location (some of the newer types may not be considered wet locations), which requires ground-fault interruption (see Chapter 7). The manufacturer of the first such piece of equipment that was installed in this state was not American and was unaware of NFPA 70, *National Electrical Code®* (NEC®), 1999 edition, requirements for this type of equipment or location. The company thus provided equipment with a voltage rating of 480/277 V, which no listed ground-fault interrupter manufactured in the United States could accommodate. Eventually the equipment had to be wired to afford the proper protection for the patient.

- *Provide equipment that operates under unusual environmental conditions (such as high humidity, moisture, salt spray, etc.) as stated.* If the equipment will be used in an area with any extreme or unusual environmental condition, the facility may require the manufacturer to apply additional treatments to the equipment's components to avoid malfunctions. In addition, some equipment may cause interference with existing equipment, such as with the lighting. Recall the case study in Chapter 7 that described how fluorescent lights interfered with equipment remote controls.
- Although it is not specified as a special condition, it is also important for hospital staff to be aware of the physical size of the equipment before it is ordered and of where it will be installed. Sufficient space should be provided for normal testing

and maintenance. If the device is portable, area personnel should determine where it will be stored and how it will be transported and set up for use.
- *Provide equipment that operates using the type of electric power system (i.e., grounded or isolated) that will energize the appliance, and overcurrent devices, auxiliary emergency power, and so forth, that will be present.* All facilities must include in each purchase order pertinent information regarding the characteristics of the power system under which the new equipment will operate, so that the manufacturer can provide equipment that meets the facility's requirements. For example, equipment manufacturers must be informed about whether a piece of equipment will be used for procedures that cannot tolerate an interruption of power. If such is the case, the company may be able to install an integral uninterrupted power supply (UPS) within the unit. If this is not possible, the facility may have to purchase an external power unit with the equipment.

Another electrical feature often overlooked when purchasing medical equipment is the compatibility of the equipment plug with the receptacles installed in areas where the equipment will be used. A simple characteristic such as this can cause problems as equipment is moved around the facility.

The case study illustrates the importance of informing equipment manufacturers of specific electrical system requirements.

Special System Characteristics

In one institution, the utility company always supplied higher than nominal voltage on the 460/265-V service. There was very little voltage drop in the distribution system, and the taps on the equipment transformers in the radiology department required a voltage setting on the high side. The utility company kept the voltage supply just above this setting except during the summer months, when it provided a slightly lower system voltage. During this time, the taps on the transformers needed to be changed to obtain optimal results from the equipment. Had the facility informed the manufacturer prior to purchasing the equipment of the need for the equipment to accommodate a range of voltages, the manufacturer could have installed a special tap arrangement.

- *Include manufacturers' test data, where pertinent.* As mentioned previously, all manufacturers must thoroughly test and calibrate their medical equipment before sending it to a facility. The health care facility department responsible for equipment testing and maintenance (see Section 11.3, this chapter) must ascertain from the manufacturer what tests have been performed on each piece of equipment, the test circuits that were used, and the results of each test. A list of the tests conducted by a manufacturer and of the test results helps health care staff understand the standards set by the manufacturer for its equipment. The data also set a baseline level of performance for the equipment. To save time, avoid in-house testing complications, and prevent equipment use hazards after the equipment has

been delivered, the facility should require the manufacturer to ship this information along with the equipment documentation. Note that in some cases, however, the manufacturer will not reveal information it feels is proprietary or a trade secret.
- *Provide additional documentation.* Each purchase order should include a requirement for the manufacturer to supply the proper documentation. The documentation should include a requisite number of instruction manuals, operation manuals, and maintenance manuals (in addition to the testing data mentioned previously). It is a good idea for facilities to require at least two sets of these documents. One set of manuals should be kept with the equipment. A second set should be kept at the location where the equipment is to be maintained and tested, usually the clinical engineering department. If there is no such department at the facility, then these manuals should be kept in another central location where the equipment is sent for periodic testing and maintenance.

The facility can also require additional items to be included in the manuals or in addenda to the manuals, as they deem necessary for a specific piece of equipment. For example, a facility can require a condensed version of operating instructions that can be permanently attached to a piece of equipment so as to be highly visible to the staff. The attachment of these types of instructions is required for equipment where there is a possibility that injury or death could result to the operator or a patient due to improper use of the equipment. During many surveys, JCAHO requires the staff to produce the operating instructions for such equipment. It is permissible to have a separate operating manual, but the maintenance manual must contain the operating instructions.

Note that if there is a conflict between what is written in a code for medical equipment and what the manufacturer recommends in the equipment documentation, the more stringent requirement should apply.

- *Provide training.* If the equipment is new to the market or to the staff, the purchase order should include a requirement for the manufacturer to supply in-service training on the proper use and maintenance of the equipment.

Although not included in purchase orders per se, facilities must also issue very specific conditions and standards for newly developed equipment on which manufacturers wish to conduct on-site trials.

11.3 Facility Equipment Testing

Health care facilities must ensure that all equipment performs in accordance with the manufacturer's specifications and the published data contained in the manuals that accompany the equipment, both before the equipment is installed and used (i.e., acceptance testing) and on an ongoing basis. This holds true for newly purchased equipment and equipment new to the market, as well as equipment that is undergoing on-site trials.

11.3.1 Acceptance Testing

Since malfunctioning medical equipment can severely harm a patient, it is not wise to assume that equipment packaged directly from the factory and labeled with a factory inspection certificate meets all operational specifications or that the equipment is safe to use on patients. Making sure equipment has not been damaged during shipping, double checking the tests conducted by the manufacturer, and conducting additional tests before equipment is used on a patient provide additional safety assurances that equipment meets or exceeds all stated purchase and manufacturer specifications. In the event that a system is purchased that consists of many individual pieces of equipment, each piece must be tested individually in addition to testing the system as a whole.

The clinical engineering department or staff who serve as clinical engineers typically perform these initial tests. If a facility does not have provisions for testing the equipment in house, an outside contractor must be hired to test equipment on a piece-by-piece basis.

As required by NFPA 99 (Chapter 8), acceptance tests must be performed on fixed and portable equipment for any chassis leakage current and on portable equipment for lead leakage current. The chassis leakage current of a permanently wired appliance must be tested before installing the appliance, while the equipment is temporarily insulated from ground. The leakage current from frame to ground of a permanently wired (fixed) appliance cannot exceed 5.0 mA with all grounds lifted. After the equipment is permanently installed, it is considered to be part of the building's wiring system and must be tested periodically according to the following criteria: The voltage measurements are made under no-fault conditions between a reference point and exposed fixed electrical equipment with conductive surfaces in a patient care vicinity (NFPA 99, 2005 edition, Section 4.3.3.1.3) with the limits of 500 mV (millivolts) in general care areas and 40 mV in critical care areas.

The chassis leakage current for cord-connected appliances cannot exceed 300 μA. The code allows two exceptions, one for existing equipment and one for special equipment where the leakage current is between 300 and 500 μA under special conditions. Figure 11.6 shows a test circuit for measuring chassis leakage current.

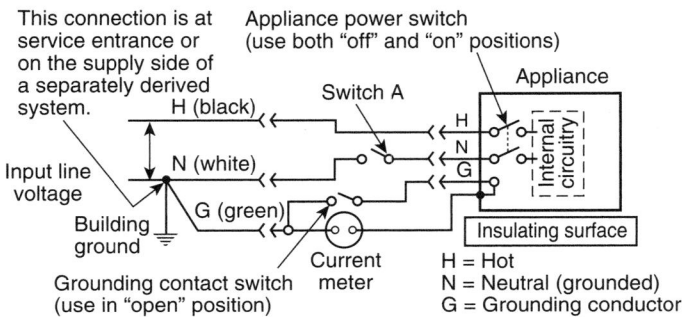

Figure 11.6 Test circuit for measuring chassis leakage current.
Source: NFPA 99, 2005, Figure 8.4.1.3.5.5.

Chapter 11. Medical Equipment

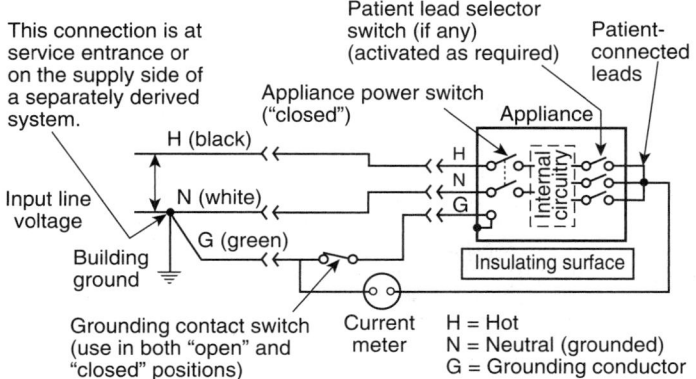

Figure 11.7 Test circuit for measuring leakage current between patient leads and ground (nonisolated).
Source: NFPA 99, 2005, Figure 8.4.1.3.6.1.

NFPA 99 includes four tests for checking lead leakage current on portable equipment:

1. *Test lead to ground with all of the patient leads connected together and the input nonisolated.* Figure 11.7 shows a test circuit for measuring leakage current between patient leads and ground (nonisolated). In the example test configuration shown in the figure, the power plug is connected normally and the device is on. The leakage current cannot exceed 100 µA for ground wire, open and closed.
2. *Test lead to ground for each patient lead with isolated leads.* For this test, the power plug is connected normally and the device is on, but the leakage current cannot be greater than 10 µA with the ground intact and 50 µA with the ground open. Figure 11.8 is an example of one test method configuration, showing a test circuit for measuring leakage current between patient leads and ground (isolated).

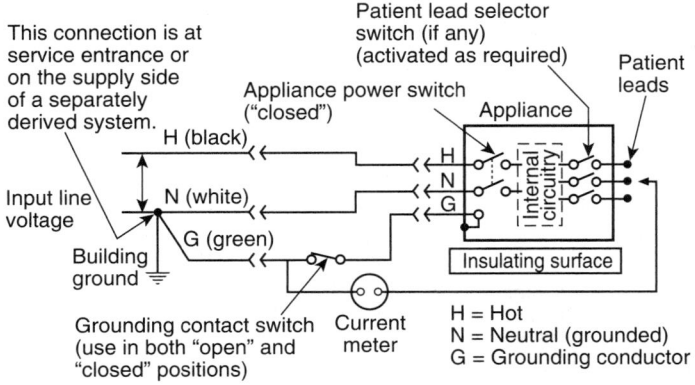

Figure 11.8 Test circuit for measuring leakage current between patient leads and ground (isolated).
Source: NFPA 99, 2005, Figure 8.4.1.3.6.2.

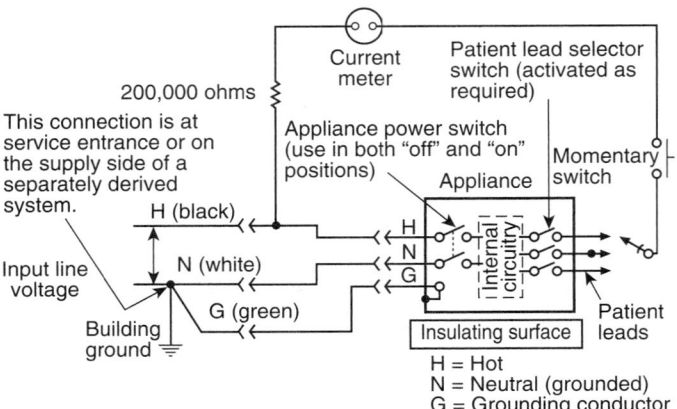

Figure 11.9 Test circuit for measuring the electrical isolation of isolated patient leads.
Source: NFPA 99, 2005, Figure 8.4.1.3.6.3.

3. *Test isolation with current driven into the leads.* Figure 11.9 indicates the test circuit to be used for measuring the electrical isolation of isolated patient leads. The leakage current cannot be greater that 50 μA.
4. *Test the leakage current between any two patient leads.* Figure 11.10 shows a test circuit that can be used for measuring leakage current between patient leads (nonisolated and isolated). The leakage current cannot exceed 50 μA for a nonisolated input with the ground wire open or intact. The leakage current cannot be greater than 10 μA with the ground intact and 50 μA with the ground open for an isolated input.

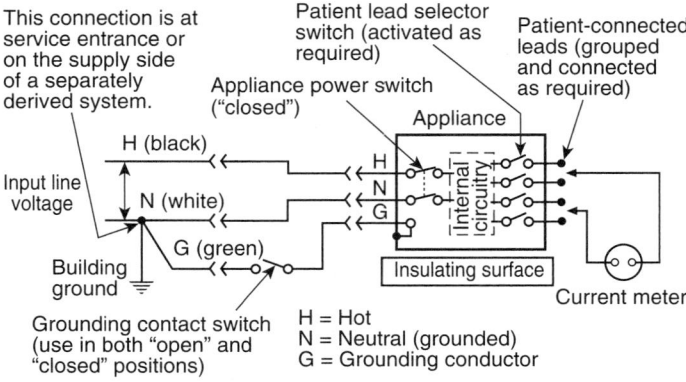

Figure 11.10 Test circuit for measuring leakage current between patient leads (nonisolated and isolated).
Source: NFPA 99, 2005, Figure 8.4.1.3.6.4.

11.3.2 Ongoing Equipment Tests

Each facility determines the equipment tests it will conduct and the intervals between them, although the tests and intervals must at least meet the minimal standards prescribed in the various codes. NFPA 99 (Chapter 8) requires appliances used in patient care areas to be tested when first put into use, after any repair or modification takes place, and then at intervals not exceeding the following times:

- General care areas—12 months
- Critical care areas—6 months
- Wet locations—6 months

JCAHO usually accepts testing at these intervals and meeting the requirements of NFPA 99 as evidence of compliance. Since this can change, it is best to become familiar with the latest JCAHO requirements.

JCAHO also requires each facility to implement a management plan that promotes the safe and effective use of medical equipment. Some of the items that the plan should address are:

- Purchasing, identifying, and inventorying equipment
- Assessing the risk of equipment use
- Managing recall notices
- Reporting incidents related to equipment use
- Monitoring reportable incidents due to equipment use
- Orienting and educating staff on how to use equipment, monitor activities regarding the use of equipment, handle equipment emergency procedures, and oversee the management plan for the medical equipment

The biomedical engineering department is typically responsible for overseeing the management plan as well as equipment management, testing, and maintenance. Facilities that do not have a biomedical engineering department or other staff qualified to do this work must use outside contractors for this work.

The ongoing testing and maintenance of the medical equipment takes two forms. One is general for all pieces of equipment (leakage tests, visual testing of cords and plugs, etc.). The other is specific to each piece of equipment (e.g., calibration). In many cases, the same computer programs that are used for the general electrical system and equipment in the facility can be used for medical equipment. The tests basically collect the same information, such as the name of the equipment, its identifying numbers, date of purchase, preventive maintenance schedule, dates of preventive maintenance, description of the maintenance performed, and history of preventive and general maintenance.

As with all tests, it is important to maintain proper records and documentation. Almost all of the AHJs will look for the tags on the equipment indicating the last date that preventive maintenance was performed. Inspectors may also look at the record books to ensure that preventive maintenance has been performed on a regular basis and that historical data have been recorded. Figure 11.11 shows a sample preventive maintenance work order. Figure 11.12 shows a sample work order for an unscheduled repair.

11.4 Equipment Installation and Grounding

As with the electrical distribution system, one of the most important requirements of medical equipment is for it to be properly grounded. NFPA 99 (Chapter 8) states that equipment that is permanently connected to the electrical supply must be grounded to the equipment grounding bus in the distribution panel by means of an insulated conductor run with the power conductors. It is important for the facility engineer to ensure that this grounding method is always intact, particularly in older facilities that have been renovated or where the initial installation did not have a separate grounding conductor. All cord-connected patient care equipment must have a three-wire power cord and three-pin grounding-type plug. The only exception to this rule that NFPA 99 allows is the use of two conductor cords for double-insulated appliances. Per NFPA 99

Community Regional Hospital

PREVENTIVE MAINTENANCE WORK ORDER

DATE 12/24/99

WORK ORDER NO 160128

TYPE — ☒ P — PREV MAINT

AREA B BIO-MEDICAL ENGIN

DESCRIPTION OF WORK   FUNCTION TEST

BRIEF DESCRIPTION   S I M U L A T O R S

FOR MAINTENANCE USE ONLY

EQUIPMENT NUMBER........................ B 3 2 8 0 4
TOTAL MAN HOURS (HHHH.H) 1.5
TOTAL PARTS COST (DOLLARS).............   0 0
DATE COMPLETED (MM-DD-YY)............. 0 1 0 5 0 0
WORKER INITIALS .... M J F
STANDARD WORK CODE .... * *
INSTRUCTION NUMBER (PM ONLY).......... 1
MATERIAL ON ORDER AS OF.................   N/A
PREVENTIVE MAINTENANCE OUTCOME ... ok

************ PM DATA ************
PM CYCLE-A MODEL NO-113A
         SERIAL NO-3110
ADDITIONAL INFORMATION BIO-MED ENG
                        (DMITRY)
         NO PM

☐ A—LOCKSMITH   ☐ B—PLUMBER   ☐ C—HVAC
☐ D—CARPENTER   ☐ E—ELECTRIC ☐ F—LABOR
☐ G—MECHANIC   ☐ H—PAINTER   ☐ I—SIGNS
☒ J—BIO MED   ☐ K—COMMUNICATION
☐ L—MASON   ☐ M—OTHER

— CHECK ONE OF THE ABOVE —

CHARGEABLE COST CENTER   1   835

BUILDING   C

FLOOR AND ROOM   B L 9 7

DATE OF REQUEST   0 1 1 0 0

SERVICE DEPT. APPROVAL

Figure 11.11 Sample preventive maintenance work order for the biomedical engineering department.

(Continued)

DESCRIPTION OF MATERIALS USED				
ITEM	QTY	UNIT COST	NET COST	
None	—	—	0.0	
		.	.	
		.	.	
		.	.	
		.	.	
		.	.	
		.	.	
		.	.	
		.	.	
		.	.	
		.	.	
		.	.	
		.	.	
		.	.	
		.	.	
TRANSFER TOTAL COST IN DOLLARS TO FRONT		TOTAL COST	0.0	

START TIME — 09:00   STOP TIME — 10:30
START TIME — __:__   STOP TIME — __:__
START TIME — __:__   STOP TIME — __:__
START TIME — __:__   STOP TIME — __:__
START TIME — __:__   STOP TIME — __:__
START TIME — __:__   STOP TIME — __:__
START TIME — __:__   STOP TIME — __:__

TOTAL TIME — ⌊__⌊__⌊1⌋.⌊5⌋

— — — — — TRANSFER TOTAL TIME TO FRONT — — — — —

COMMENTS

Completed preventive maintenance.

M.J.F.

Figure 11.11 Cont'd

(Section 8.4.2.2.2.1), portable equipment intended for laboratory use must be provided with an approved method to protect personnel against shock.

NFPA 99 permits the use of extension cords, provided the cords are tested and follow the code requirement for power cords.

11.5 Medical Equipment Training

Medical equipment operated by properly trained personnel is both safer and requires fewer repairs. All facilities should require new staff to attend safety orientation programs and all employees to attend annual in-service refresher classes. One example of an obvious, but seldom taught, procedure is that equipment users should unplug a device by grasping the plug and then pulling the cord out of the receptacle. It is common, however, for people to grab the cord at some convenient point, usually far from the plug, and then give the cord a quick yank. This type of action may eventually fray the cord and compromise the strain-relief device within the plug, thereby breaking the wire. When the cord is worn, there is a risk of a spark or short circuit causing a fire or electric shock to the user.

Part II. Electrical Systems and Equipment

It is also important to train and test all staff who will be using a piece of equipment on how to read, interpret, and understand the instructions attached to that equipment, especially for devices that are used to treat patients in an emergency. Staff also should be able to read and interpret the more detailed equipment instructions provided by the manufacturers. The case study shows the importance of being trained on using a new piece of equipment.

Community Regional Hospital

REPAIR WORK ORDER

WORK ORDER NO: F 291

AREA: B BIO-MED-ENG

DESCRIPTION OF WORK
GROUND RESISTANCE: N/A mΩ   WORST CASE LEAKAGE: N/A

PRINT NAME DEPT. HEAD: Joseph Jones
(SIGN) DEPARTMENT HEAD APPROVAL: J.J.

DESCRIPTION: REPLACE CORD & PLUG

FOR MAINTENANCE USE ONLY

EQUIPMENT NUMBER: B 3 2 8 0 4
TOTAL MAN HOURS (HHHH.H): 0.5
TOTAL PARTS COST (DOLLARS): 1.5
DATE COMPLETED (MM-DD-YY): 0 1 1 2 0 0
WORKER INITIALS: JOS
3rd PARTY REPAIR: ☐ YES ☐ NO
WARRANTY REPAIR: ☐ YES ☐ NO
MATERIAL ON ORDER AS OF:
CO. OR MANUFACTURERS NAME: ABC CO.
NAME OF EQUIPMENT: simulator
MODEL NO.: 113 A   SERIAL NO.: 3110
P.O. NO.: N/A

Lead Leakage Test (Worst Case)
N/A
II ___ uA   IV ___ uA
III ___ uA   V ___ uA

CHARGEABLE COST CENTER: 1 8 3 5

BUILDING: C

FLOOR AND ROOM: BL 9 7

DATE OF REQUEST: 0 1 1 1 0 0

SERVICE DEPT. APPROVAL: M.J.F.

Figure 11.12 Sample maintenance (repair) work order for the biomedical engineering department.

(Continued)

Equipment Training

A large medical center received new cardiac monitors for its operating suite. One evening soon after, a patient required an emergency procedure. The morning following the incident, the emergency staff informed the Clinical Engineering Department that none of the new units had worked properly, causing a delay in the procedure until one of the old units was reinstalled. An investigation determined that all of the monitors did indeed function properly, but the anesthesiologist did not understand the

(Continued)

DESCRIPTION OF MATERIALS USED				
ITEM	QTY	UNIT COST	NET COST	
Power cord and plug	1	15.00	15.00	

START TIME — 10 :30   STOP TIME — 11 :00
START TIME — ___:___   STOP TIME — ___:___
START TIME — ___:___   STOP TIME — ___:___
START TIME — ___:___   STOP TIME — ___:___
START TIME — ___:___   STOP TIME — ___:___
START TIME — ___:___   STOP TIME — ___:___
START TIME — ___:___   STOP TIME — ___:___

TOTAL TIME — |_|_|_0|.|5|

— — — — — TRANSFER TOTAL TIME TO FRONT — — — — —

COMMENTS
Cord frayed and plug damaged

TRANSFER TOTAL COST IN DOLLARS TO FRONT   TOTAL COST   15.00

Figure 11.12 Cont'd

instructions that were printed on the unit and had not been trained in how to operate it. Consequently, he made an error in setting the equipment controls and this prevented the monitoring screen from functioning properly. Instead of turning a few dials to correct the problem, the staff determined that the equipment did not function properly.

Refer to Chapter 36 for more details on staff training.

12

Testing, Maintenance, and Record Keeping

An integral part of operating safe electrical systems in all types of health care facilities is implementing well-organized testing, maintenance, and record keeping programs. Acceptance testing involves making sure all equipment and systems, whether newly installed as a result of construction, acquisition, or renovations or newly repaired, are in working order and are safe before being put into service. Tests are also conducted at set intervals as part of the preventive maintenance program to make sure that existing equipment continues to operate properly. The results of these tests may indicate that a piece of equipment must be replaced before it totally malfunctions. In addition to testing, maintenance programs involve conducting necessary repairs. Creating and maintaining accurate system records is essential for keeping relevant staff informed of the status of all systems and equipment, test results, repair history, system upgrades or changes, inventory needs, and similar matters.

While many facilities test, maintain, and keep records of their electrical systems as a matter of course, the National Fire Protection Association (NFPA), American Institute of Architects (AIA), and Joint Commission on Accreditation of Healthcare Organizations (JCAHO) all issue codes, standards, rules, and guidelines that require or recommend that these activities take place on a periodic basis. All of these activities, properly conducted, greatly assist in preparing for inspections and surveys. Not only do they offer written evidence of the myriad of tasks that have been conducted, but they also help to avoid the typical scramble to figure this out before an inspection occurs.

This chapter covers how to test and maintain the electrical distribution system and equipment in health care facilities. It also provides general information on how to conduct testing and maintenance in all facility departments. Refer to Chapters 8, 9, and 11 for specific information on testing and maintaining essential electrical systems, isolated power systems, and medical equipment, respectively. Refer to Parts III through V for information on testing other systems within a facility, including fire protection systems and gas/vacuum systems, and parts of a facility's structure.

12.1 Testing Electrical Systems

One of the most important and often neglected systems to test in health care facilities is the electrical distribution system. Because the electrical distribution system is so critical, acceptance testing is essential to prevent an interruption of power that can be caused by a faulty, untested system component that goes on line. Facility engineers should require all new equipment and work to be acceptance tested. If an outside contractor conducts the work, acceptance testing should be included in the construction contract.

Like acceptance testing, conducting tests at regular intervals may prevent the breakdown of a system that contains equipment that is failing but not yet completely malfunctioning. The engineering department is primarily responsible for conducting these tests, from the main service switch through the load at the receptacles and switches. The results from all tests and any corrective action should be noted and kept as a permanent record (see the discussion of record keeping in Section 12.3).

Testing requirements for the electrical distribution system in health care facilities are included in NFPA 99, *Standard for Health Care Facilities*, the AIA *Guidelines for Design and Construction of Hospital and Health Care Facilities*, and JCAHO *Comprehensive Accreditation Manual for Hospitals*. The AIA *Guidelines* state that the electrical installation and operation in each section of a facility must be tested to demonstrate that the equipment not only meets all code requirements (i.e., baseline safety requirements), but that it also is set up and functions properly for optimal safety, above and beyond what the codes require. JCAHO requires all utility systems, including electrical distribution systems, to be maintained, tested, and inspected regularly. JCAHO is in a transition stage and the requirements are changing. It is advisable to find out the latest requirements.

12.1.1 Testing Service Switches and Panelboards

Testing the main service switch of a new facility before the facility opens its doors for business can avoid the problems associated with a complete loss of power. If one switch or circuit breaker controls the entire essential electrical system, an acceptance test of this switch or circuit breaker can be conducted without affecting power to the rest of the facility. It is also one way to test simultaneously both the service switch and the distribution of power to the essential electrical system. Opening the one service switch that serves only the essential electrical system should simulate a power failure, and the generators should start. Since all the other service switches should remain closed, all of the normal power should be supplied to the institution.

None of the codes requires service switches to be acceptance tested, although it is a good idea to inspect them visually at regular intervals. Also, in place of an operational test, an infrared or thermographic test can be done. This type of test consists of taking pictures of each switch to indicate the location of any *hot spots*, which are areas of heat buildup that generally indicate a loose connection or another present or imminent malfunction.

Periodically, the entire distribution system should undergo thermographic testing. In conducting these thermographic tests, pictures of each switchboard and panelboard are taken to locate hot spots.

12.1.2 Testing the Wiring

The *megger* test is an acceptance test for wiring. This test determines whether the insulation resistance on the wiring system is effective and has not been compromised during the installation and pulling of the wire and cable. This test should be done on any new wiring before putting it in service. Since many variables affect the readings, the user should be familiar with the use of the instrument and follow the manufacturer's recommendations.

12.1.3 Testing Disconnecting Devices

Disconnecting devices should be checked to ensure that they open and close properly. These devices can be tested without having a load placed on them. A seal from an acceptable third-party testing laboratory implies that a device should function properly under load. However, operating disconnects at least once before use provides added assurance that the devices have been installed properly and without damage.

NFPA 70, *National Electrical Code*® (NEC®), 2005 edition, Section 230.95, requires ground-fault protection systems installed at the service switch and farther down the line in the electrical distribution system to be tested when they are first installed at the site. The tests must be conducted in accordance with the instructions provided by the manufacturer of the equipment. If a testing agency lists certain equipment, the agency promulgates test procedures for that equipment.

12.1.4 Testing Loads

Although it is important to make sure that the total load on each device is within a safe range, obtaining load readings can be tedious. The facility engineer must first accurately estimate the times of day and night that represent the peak times for each unit in question. He or she must then take four readings for each unit—summer day and night readings and winter day and night readings. Either clamp-on meters or recording meters are used to take these readings. Using a recording meter over a 24-hour period or longer enables the user to determine the peak readings for the day and night hours. When clamp-on meters are used, many readings must be taken over a short period of time to determine the peak reading.

12.1.5 Testing Patient Care Areas

Section 4.3.3 of NFPA 99 lists performance criteria and testing requirements for grounding systems and receptacles located in patient care areas. In addition to the requirements, the code specifies the test equipment that must be used. NFPA 99 (Annex A) includes a detailed explanation of various tests.

NFPA 99 requires the grounding system of newly constructed patient care areas to be acceptance tested but not periodically tested, although portions of the electrical system that have been altered or replaced must be tested after initial acceptance. If the grounding system in a patient care area has never been tested, however, or if records of any initial test cannot be found, it is advisable for facility engineers to conduct at least one test, where possible, to ensure that the facility is in compliance and that patients, staff, and visitors are safe.

NFPA 99 (Section 4.3.3.2) requires facilities to conduct four different tests on receptacles in patient care areas after their initial installation, replacement, or servicing of the devices. Testing receptacles should be conducted at regular intervals. The tests include a visual inspection, verification of the grounding circuit, a polarity check, and a test of the retention force of the grounding blade. While these tests may seem cumbersome, they can be conducted using one or two small, hand-held devices. Engineering department staff can perform the necessary tests on the receptacles and grounding system while they conduct the inspection of patient care areas required by the various agencies.

The grounding conductors for the receptacles and equipment installed in wet locations must also be tested to ensure that they are installed and maintained in accordance with the *NEC* and comply with the performance requirements of NFPA 99.

12.2 Maintaining Electrical Systems

The purpose of system maintenance is to check equipment to prevent it from malfunctioning and to repair equipment that is already not working properly. A maintenance program should be established for all systems and equipment that have been accepted and placed into operation. Preventive maintenance is required for compliance with various authorities and to maintain JCAHO certification of the electrical systems. At one time most of the preventive maintenance requirements had to take place on an annual basis or more frequently. JCAHO has been changing its requirements as to the frequency of preventive maintenance testing. Some AHJs may still require annual preventive maintenance checks. Each facility should check with its respective agencies. It is good policy to perform preventive maintenance testing and preventive maintenance even if not required. All personnel responsible for meeting JCAHO compliance should become familiar with the organization's preventive maintenance requirements as published in the latest JCAHO accreditation manual for the type of facility in question.

In many cases, facilities are not required to conduct specific preventive maintenance activities, or if they are, the authorities having jurisdiction may not have provided specific procedures to follow. In these cases, it is best to follow the manufacturer's recommendations and consult the applicable codes. If there is a conflict between the manufacturer's recommendations and the codes, the more stringent requirements should be followed.

12.2.1 General Maintenance Procedures

The initial stage of any preventive maintenance program is to prepare a complete list of all items in a system that require maintenance, including equipment, panelboards, switches, disconnecting devices, and the like. Based on the relevant codes, manufacturer requirements, and additional facility requirements, a list of all necessary maintenance tasks for each item can be generated. Many computer software programs are available to track the required tasks and schedules and inventory used, as well as record the work that has actually been conducted and when it was completed. Computerized maintenance and preventive maintenance systems are discussed later in this book, in Chapter 34, Section 34.3.2. The same programs also record the work done on a historical record of the equipment. The same program and computer system that is used for the electrical systems and equipment can be used for the medical equipment.

Properly trained in-house staff can do routine maintenance and repairs. There are times, however, when this function is contracted to outside contractors who specialize in this work on specific devices. If an outside contractor is hired to conduct routine maintenance or repair work, a facility staff member should be present to ensure that all work has been completed. The records of this work must be maintained with the other records of the facility's engineering or maintenance department.

Some jurisdictions require licensed electricians to perform work on the electrical system. It is highly recommended that at least two workers be present when major electrical components are being repaired or maintained, because working alone can be dangerous, as illustrated in the following case study.

Too Few Workers on Hand

While performing preventive maintenance on a 2000-A (ampere) service switch, one of the workers assigned to the job had to leave for a few minutes. The other worker started to remove the front panel himself, but it proved to be too heavy for him. He subsequently lost control of the panel, and it fell into the open switch across the ground bus and one of the phase buses. The short-circuit available current in this installation at the main service for a single-phase fault was greater than 125,000 A. The current was so high that some of the metal on the cover was vaporized before the circuit opened. Luckily, the magnetic forces did not push the front cover into the worker.

One of the problems often encountered when conducting maintenance work is that the maintenance personnel arrive at the site only to find that the proper parts and tools are not available. Well-thought-out work-order request sheets and protocols may encourage personnel to report problems accurately and to provide enough details about the maintenance or repair work to assist maintenance personnel in arriving at the job site prepared, so as to minimize the number of trips they must make to the central tool and supply area.

Figure 12.1 shows a sample preventive maintenance work-order request sheet from a medical center's plant engineering and maintenance department. The first page of the form contains general information, including the work order number, the date the order was generated, the physical location of the equipment (in this case, electrical panels), the skill required (in this case, an electrician), the equipment asset number, the estimated time required to conduct the repair, plus additional information that may be available. The second sheet contains specific instructions for all the work that should be performed and space for indicating inventory materials used and ammeter readings. The person performing the maintenance must sign the form. Figure 12.2 is a sample repair work-order form, which uses a format similar to the preventive maintenance work-order form.

Note that these two forms could be used by the biomedical engineering department for scheduling preventive maintenance and repair work for medical equipment, as described in Chapter 11. Similarly, the forms shown in Chapter 11 for the biomedical engineering department could be used by the plant engineering and maintenance departments. These forms also can be generated in advance by a computer program, so that an ample supply can be readily available upon request, which can assist in scheduling the work. Note that these forms are only samples of the many types that may be used.

Of course, to avoid long delays in completing any repairs, facilities engineers must keep an adequate supply of spare parts on hand, depending on the types of equipment used, the age of the equipment, the availability of parts from outside sources, the size

---

Community Regional Hospital Work Order
(Follow LockOut/TagOut Procedures)

4/26/2000   Page 1

WO:	43536	Date Orig:	8/23/2000
Type:	PM Preventive Maintenance	Priority:	3 Routine
Location:	ENERGY CENTER	Reference:	
Building:	EC		
Account:	0830 - PLANT ENG - MAINTENANCE		
Req Name:	N/A	Req Phone:	N/A
		Skill:	Electrician
Req Remarks:	PM Work Order for EPEDP50006 ENERGY CENTER, REFRIGERATION LEVEL		

Asset Data

Asset No:	EPEDP50006	Desc: ELECTRICAL PANELS	
Manuf:		Warranty:	
Model No:		Start:	
Serial No:		Expiration:	
Location:	EC	Asset Location:	ENERGY CENTER

Seq: 1   Task No: 77   Desc: EPEDP06—PANEL ELECTRICAL DISTRIBUTION

Skill: Electrician   Shut Down Required: No
Est Time: 1.01

Refer to p.2 for preventive maintenance work to be completed.

Figure 12.1 Sample preventive maintenance work order for the plant engineering and maintenance department.

(Continued)

Work Order
(Follow Lockout/Tagout Precedures)

Instructions: PANEL, ELECTRICAL DISTRIUBUTION
CHECK MAIN BREAKER (IF APPLICABLE).
    CHECK CONNECTIONS AT TERMINALS.
    CHECK CONTRACTS ON BARS (PLUG-IN TYPE).
    INSPECT WIRING FOR FRAYED INSULATION.
    CHECK CONTRACTS (IF ACCESSIBLE).
    CHECK MECHANISM OPERATION.
    CHECK FOR FREEDOM OF MOVEMENT.
    LUBRICATE MOBILE PARTS, PER MANUFACTURE'S RECOMMENDATIONS.
INSPECT BRANCH CIRCUITS.
    INSPECT WIRING FOR FRAYED INSULATION.
    TIGHTEN LUGS AS REQUIRED.
    CHECK BREAKERS (IF APPLICABLE).
    OPERATE TO ENSURE FREE MOVEMENT.
    CHECK FUSE(S), (IF APPLICABLE).
    CHECK SIZE AND TYPE.
    CHECK FOR SECURE MOUNTING.
CHECK CURRENT AND PHASE BALANCING.
    PHASE A   290   Amps
    PHASE B   286   Amps
    PHASE C   295   Amps
    RE-BALANCE LOAD AS REQUIRED.
INSPECT ENCLOSURE.
    CHECK DOOR, HINGES, AND LOCK FOR SECURE CLOSURE.
    CHECK FOR UNPLUGGED KNOCK-OUTS.
    INSPECT MECHANICAL GROUNDING.
ENSURE THAT PANELBOARD AND BREAKERS ARE CLEARLY IDENTIFIED.
NOTE: ENSURE THAT A MOMENTARY INTERRUPTION OF POWER SUPPLIED BY THIS PANEL WILL NOTE ADVERSELY AFFECT EQUIPMENT SERVED.
NOTE: WHEN INSPECTING ELECTRICAL EQUIPMENT, AN INFRARED DETECTOR IS RECOMMENDED TO REVEAL "HOT-SPOTS." THERMOGRAPHY IS A FAST AND SAFE MEHTOD OF INSPECTING ELECTRICAL EQUIPMENT.

Inventory Materials Used

Qty:
Description:
    None

Completed by: M.J.F.   #:   374   Date: 8/28/00 Hours: 1.25

Figure 12.1 cont'd

```
                    Community Regional Medical Hospital Work Order
                              (Follow LockOut/TagOut Procedures)

4/26/2000                                                                           Page 1

          WO:      134753                          Date Orig:    4/20/2000
         Type:     HVAC   HOT/COLD CALL            Priority:     3
                                                                 Routine
      Location:    Room 1302                       Reference:    PMAGGIO
      Building:    Ruth Pavilion
      Account:     0759 - PSYCH. ADMINISTRATION
     Req Name:     Dr. Fischer                     Req Phone:
                                                      Skill:     HVAC
   Req Remarks:    NON PM
                   HOT—Temperature in room 80°F
                              Inventory Materials Used
          Qty:
   Description:


      Comments:    Adjusted thermostat and reheat coil.
                   No p.2. No preventive maintenance required.
  Completed by:       M.J.F.        #:   1234    Date:  4/20/00   Hours:   0.75
```

Figure 12.2 Sample maintenance (repair) work order for the plant engineering and maintenance department.

of the facility, and the like. Maintaining an inventory system that indicates the lowest number in stock for each spare part provides useful information for reordering items. All applicable manuals and special tools must also be readily accessible at all times to maintain a facility adequately.

Having the maintenance worker sign the worksheet after performing preventive maintenance or repair work provides accountability for the task, as well as a way to check that the work was done and done properly. If the supervisor spot-checks the work and finds that something was not done properly, then he can provide feedback to the worker and the problem can be corrected.

12.2.2 Visual Inspections

In addition to the load testing described previously and general repair work, other maintenance tasks are visual inspections and general cleaning. Some of the items and areas that can easily be visually inspected periodically and cleaned on a routine basis are as follows:

- Panelboards—for dirt and dust in the gutters. The accumulation of dust in a switchboard or panelboard can cause arcing.

- Receptacles—for cracks and breaks that can occur after constant use
- Switches and receptacles—to see if the cover plates are intact and, in fact, are there. A switch or receptacle that functions without a cover plate can lead to shock or fire hazards.
- Lighting fixtures—to ensure that the lenses are intact, not damaged, and clean, the fixture is firmly fastened to the ceiling or wall, and all of the bulbs or tubes are working, especially the exit lights
- Extension cords, if they are used—to ensure that there are no breaks in the insulation
- Electrical service rooms and closets—to keep them free of debris and dust. Some jurisdictions mandate the maintenance of a clear working space around electrical equipment.
- Transformer rooms and vaults—to keep them free and clear of all materials not necessary to the installation or operation of the transformer. It is essential for the transformers to be accessible to qualified personnel for inspection and maintenance, and that transformer rooms not be used as ancillary storage rooms.

12.3 Record Keeping

The importance of creating and maintaining a complete and accurate set of records cannot be underestimated. Having up-to-date diagrams of each system, manufacturers' documentation of all equipment, a complete log of all preventive maintenance and repair work that has taken place, and the results of all tests makes the facilities engineer's job easier in a number of ways. The diagrams show the location and specifications of all system components. This assists in conducting tests and maintenance work as well as renovation projects. Equipment documents provide essential information for proper testing, installation, operation, use, and maintenance. The log of all testing and preventive maintenance and repair work provides key historical data, as well as support materials that can prove invaluable in the event of a legal action involving any item in question, particularly if proper maintenance and testing have been performed. All of these documents facilitate (i.e., make less frantic) the survey and inspection process, no matter which jurisdiction is due to arrive.

As with testing procedures, authorities having jurisdiction typically require facilities to keep written records of all maintenance work. Facilities can develop their own log forms or use one of the forms provided in the codes. These records are also always required during inspections and surveys. Therefore, all records should be kept for future reference.

Exception reporting, which usually indicates only items that for some reason were out of the ordinary or not normal, is not as thorough as actually making an entry of all items tested, inspected, and maintained—that is, what actually was observed and accomplished. With exception reporting, items that appear to be operating normally can be overlooked. An actual record of the work conducted presents a clear, indisputable record.

12.3.1 As-Built Drawings

One condition that should be included in all construction and renovation contracts is that the builders and contractors provide accurate as-built drawings of the electrical installation. Although it is common to hear how "impossible" it is to indicate the exact location of the wiring and devices on a drawing, a one-line drawing should be provided to the facilities engineer showing all of the branch circuits, the size of the wire and conduit, the type of conductor and insulation that was used, the starting point and termination of each branch circuit, and finally the panelboard schedules. The panelboard schedule should be an accurate record of the location of the loads on each circuit, not simply be marked "receptacles" or "lights." If these records do not exist for an existing facility, then provisions should be made to trace and record this information and to prepare a set of records. The contractor for any new construction or renovation should provide an as built electrical riser diagram as described below.

12.3.2 Riser Diagrams and Electrical Distribution System Logs

Each electrical distribution system also needs a complete and up-to-date riser diagram that replicates the distribution system on all floors and building locations. Multiple-building installations require individual riser diagrams for each building, with the interconnections shown. Riser diagrams should contain the following information:

- All service switches and their ratings. If a switch and fuse are used, then the rating of the switch and fuse should be noted, as well as the type of switch. Similarly, if the main disconnect device is a circuit breaker, the frame size, trip rating, and type of circuit breaker should be noted. The short-circuit interrupting ratings should be noted for the overcurrent devices.
- All switchboards and panelboards. Each should have a designation that should also be indicated on the switchboard or panelboard. The ratings should be noted, as well as the bus size and any main overcurrent device if installed.
- All motor control centers
- Any large individual loads, such as air-conditioning compressors, and so on
- All transformers, with indications of their primary and secondary voltages and kVA (kilovoltampere) ratings
- Any other system devices, with their ratings, such as voltage regulators, capacitors, and so on
- All of the wire, cable, and conduit interconnections between the components of the system. Noted next to the wire and cable should be the wire size and type, the conduit size and type, and the number of conductors in the conduit. In some cases, a designation can be given to the feeders.
- Information about what is connected to each switchboard, panelboard, and motor control center

As a complement to the riser diagram, the engineering department should have a book with one or more pages devoted to each switchboard or distribution board, panelboard, and motor control center as required. The information that should be noted for each switchboard, panelboard, and motor control center is as follows:

- Designation of the unit. This is the same designation that is on the riser diagram and is physically on the unit.
- The actual location of the unit. It is important that the exact location as to building, floor, and room designation be noted. If the unit is in a corridor, the location *of* the corridor (i.e., main building, west wing, third floor) and the location *in* the corridor (i.e., between rooms 125 and 127) should be noted.
- General information, such as voltage rating, number of phases, short-circuit withstand rating, and special constructions, such as split bus, timers, or relays, and so on
- Size and type of main disconnect and overcurrent devices, including number of phases, short-circuit current interrupting rating, and so on
- In the case of a motor control center, information about the starter and controller that is used with all of the necessary ratings, and so on
- A schedule, similar to the panelboard schedule found in the panelboard, with specific notations as to what each overcurrent devices it serves, such as specific receptacles, lights, special loads, and other devices and their locations
- A schedule of the maximum load on each overcurrent device. This should include to which phase the load is connected on a multiphase panelboard and the load taken at peak day and night both for summer and winter.
- The total load on each phase of the switchboard, and so on
- Spaces for additional loads that are added, as well as room for general comments

If records of schedules and maximum loads for overcurrent devices and total loads on each phase of the switchboard do not exist, it may seem overwhelming for a facility engineer to compile this information. These records are very important, however, to help prevent overloads to the distribution system that may be caused by the constantly changing use of space. The records will also be essential for planning the installation of additional equipment as well as expansions.

The riser diagram and system log should be updated whenever loads are added or removed or other system modifications are made. When all load information is properly recorded, the record becomes a valuable reference for safely making additions to the electrical system. The load values, the dates that additional loads were added to a circuit, and the dates that all readings were taken on the loads should be noted on the appropriate record sheets in the system log with the load values. This information is necessary because when a load is added to a panelboard, it not only affects a branch circuit, it also affects the load on all feeders back to the service switch. It is not enough to indicate the additional load on the affected panelboard only. This information is invaluable when a renovation takes place.

The designations shown on the riser diagram should match those on the individual pages of the record book. If the facility engineer is preparing this record for the first time, it is advisable to create the service switch records first and then work down to the local lighting and appliance panelboards.

12.3.3 Maintenance Logs

To further facilitate testing, system maintenance, and preparation for inspections and surveys, it is essential to keep complete and accurate records of all modifications that have been made to the system and any maintenance, preventive maintenance, and testing procedures that have been conducted. All too often these records do not exist, or if they do, they do not reflect the existing conditions. Specifically, it is important to document all repairs in terms of the work that has been done, the materials used, the length of time it took to complete the repair, and the number of people who conducted the work. The records should also indicate any temporary measures that were taken during the repair.

Surveyors and inspectors typically take a close look at the records of all tests, maintenance, and preventive maintenance activities. It is not uncommon to have malfunctions, and there is no harm in indicating any malfunction that was found during a routine test or preventive maintenance procedure. The important thing is to document when and how the malfunction was corrected.

Refer to Chapter 34 for a general discussion of record keeping.

Electrical Systems and Equipment Frequently Asked Questions

1. How often should the emergency generator be tested?

Refer to both NFPA 99, *Standards for Health Care Facilities,* 2005 edition, and NFPA 110, *Standard for Emergency and Standby Power Systems,* 2005 edition. NFPA 110 (Chapter 8) states the requirements for routine maintenance and operational testing. Section 8.4.2 states that diesel generator sets in service shall be exercised at least once monthly, for a minimum of 30 minutes using one of three acceptable methods. Other testing methods are listed in the event that these criteria cannot be met.

NFPA 99, 2005 edition (Sections 4.4.4.1.1.2, 4.5.4.1.1.2, and 4.6.4.1.1.2), also includes criteria for testing the generator. This code states that the generator must be tested 12 times a year, with testing intervals of not less than 20 or more than 40 days. This leeway allows for the postponement of a scheduled test in a hospital in the event that conditions arise that might be detrimental to a patient. Note that in Annex A of NFPA 110 the maintenance log shown indicates a weekly inspection for the prime mover. This should not be confused with an operational test, which involves actually starting and running the generator.

2. What are the conditions for the generator test?

The 2005 edition of NFPA 110 (Section 8.4.2) states that a diesel generator must be exercised for a minimum of 30 minutes under one of the following three conditions: (a) loading that maintains the minimum exhaust temperatures as recommended by the manufacturer, (b) operating temperature conditions and not less than 30 percent of the EPS nameplate kW rating, or (c) if the engine cannot be loaded as required in (b), operating the engine until the water temperature and the oil pressure have stabilized and then terminating the test before the 30-minute time period expires. If diesel-powered EPS installations do not meet one of these requirements, they must undergo monthly tests with prescribed loads and test durations as further described in NFPA 110. There are also similar requirements for a spark-ignited generator.

NFPA 99 further states that the essential electrical system is to be classified as Type 10 and therefore must be capable of supplying power within 10 seconds after loss of the normal power. Competent personnel must perform the scheduled tests. NFPA 99 refers to NFPA 110.

3. Is a bypass isolation switch required on the transfer switch of an essential electrical system?

While the 2005 editions of both NFPA 110 (Section 6.4.1) and NFPA 99 (Section 4.4.2.1.7) include the same statements regarding the bypass isolation switch, NFPA 99 extracts the information from NFPA 110, the document having jurisdiction. A bypass isolation switch is not required on the transfer switch; however, it is permissible, and if it is installed, it must conform to the requirements listed in both documents. The advantage of installing a bypass isolation switch is that it provides a means to maintain power to the load on the essential electrical system while the transfer switch is being repaired or maintained.

4. Where are isolated power supplies required?

These requirements are found in NFPA 99, 2005 edition (Section 4.3.2.2.8), and NFPA 70, *National Electrical Code®* (*NEC®*) Section 517.20. An isolated power supply (IPS) is not required in any location. However, an IPS can be used to satisfy the requirement that wet locations must use an electrical distribution system that limits the first ground-fault current to a low value (either by interrupting the circuit or not interrupting the circuit), thereby protecting the area from ground faults and minimizing the hazard of electric shock. IPSs can also be used in hazardous anesthetizing locations. Where it is acceptable to interrupt the circuit, as in bathrooms, a ground-fault circuit interrupter (GFCI) can be used. Where interruption of power is not acceptable, such as in operating rooms, an isolated power supply is used.

5. What are the requirements for wet locations in existing construction?

Electrical installations in wet locations in existing construction do not require GFCIs or IPSs, under certain conditions. One condition is when the authority having jurisdiction allows a designated individual at the hospital to inspect the installation according to an accepted written inspection procedure. The testing must ensure that the equipment-grounding conductors for 120 V, single-phase, 15 A and 20 A receptacles, equipment connected by cord and plug, and fixed electrical equipment are installed and maintained in accordance with the *NEC* and applicable provisions of NFPA 99. The procedure must include electrical continuity tests of equipment, grounding conductors, and connections. The test must be done when the equipment is first installed, when there is evidence of damage, after any repairs have been conducted, or at intervals not exceeding 6 months.

6. Must conductive flooring in an existing operating room be maintained if no flammable anesthetics are being used?

No. At one time this requirement existed, but the codes are now silent on this issue. Conductive flooring is not required in areas where only nonflammable anesthetics are being used. NFPA 99 (Annex C.13.1.4.1.1 and Annex E) has suggested material on conductive flooring, and Annex C.13.3 contains requirements for posting signs and other notices which state that only nonflammable anesthetics are permitted in that area.

7. Is ground-fault protection required at any other place on the distribution system if it is on the main service switch?

Yes. The *NEC* includes a requirement Article 517.17 that in health care facilities where ground-fault protection is provided for operating the service disconnecting means or feeder disconnecting means as specified in Article 250.95 or 215.10, an additional step of ground-fault protection must be provided in the next level of all feeder disconnecting means downstream toward the load. There are three exceptions to this requirement, dealing with essential electrical systems, onsite generation, and grounded wye systems that are not solidly grounded.

8. What are the grounding requirements for receptacles located in patient care areas?

NFPA 99 requires the installed grounding circuit to a power receptacle to be as reliable as an electrically continuous copper conductor of appropriate ampacity. The *NEC* [Article 517.13(b)] states that in patient care areas, all receptacles and noncurrent carrying conductive surfaces of fixed electrical equipment that operate at over 100 V are subject to personal contact, and are likely to become energized must be grounded by an insulated copper conductor. The conductor must be sized in accordance with Table 250.122 of the *NEC* and installed in metal raceways with the branch-circuit conductors supplying the receptacles or fixed equipment.

9. If the conductors to the receptacles in a patient care area are run in metal conduit that is grounded, is a separate grounding conductor also needed?

Yes. *NEC* [Article 517-13(a)] states that in addition to the grounding wire, all branch circuits serving patient care areas must be provided with a ground path for fault current by being installed in a metal raceway or cable assembly. As noted in the *National Electrical Code Handbook*, this provision will result in providing a redundant ground path in patient care areas.

10. Can Type FCC (flat conductor cable) be installed in a part of a hospital that is not classified as a *health care occupancy in accordance with NFPA 101, Life Safety Code*?

No. The *NEC* 2005 edition (Article 324.12) lists the uses that are not permitted for Type FCC cable. Subparagraph 4 states that the cable cannot be used in residential, school, and hospital buildings. The code does not refer to an occupancy classification. Therefore, as long as the area is within the hospital building, Type FCC cable is prohibited.

11. Are there prescribed intervals for testing medical equipment?

Yes. NFPA 99, 2005 edition (Section 8.5.2.1.2), states maximum intervals permitted for testing equipment in different patient care areas, although health care facility personnel responsible for testing medical equipment should become familiar with a number of exceptions. It is also important to contact the authorities having jurisdiction, including JCAHO, to ascertain what intervals they deem acceptable.

12. Are there any requirements that medical equipment manufacturers must follow?

Yes. Chapter 10 of the 2005 edition of NFPA 99 is devoted entirely to manufacturer requirements. At the present time, these requirements apply only to patient care-related electrical appliances. All equipment purchase orders should stipulate that the manufacturer must comply with these requirements. The person responsible for maintaining and testing medical equipment should be familiar with these requirements to further ensure compliance.

13. What code contains the requirements for conductor and overcurrent device size for X-ray installations?

Article 517, Part V, of the *NEC*, includes requirements for X-ray installations. Article 517.73 states the requirements for the rating of supply conductors and overcurrent protection. In many cases, the manufacturer recommends its own requirements. The manufacturer requirements should be compared to those contained in the code to ensure that the manufacturer meets the requirements of the *NEC*. Considerations should be made for voltage drop in installations where the conductors exceed the distances recommended by the manufacturer.

14. What are the prescribed tests for receptacles in patient care areas?

NFPA 99 (Section 4.3.3.2) contains the following requirements for these tests:

(a) *The physical integrity of each receptacle shall be confirmed by visual inspection.* The receptacle must be checked to ensure that the cover plate is on and the receptacle is not cracked.
(b) *The continuity of the grounding circuit in each electrical receptacle shall be verified.*
(c) *Correct polarity of the hot and neutral connections in each electrical receptacle shall be confirmed.*
(d) *The retention force of the grounding blade of each electrical receptacle (except locking-type receptacles) shall be not less than 4 oz (115 g).*

The tests for the preceding items (b), (c), and (d) can be accomplished by using a simple receptacle-testing device.

15. How long should records be kept for such items as generator test results and reports of preventive and regular maintenance?

This question is best answered by the health care facility administration in consultation with its legal advisors. Obviously, the records should be kept for at least a minimum of two JCAHO surveys (usually 6 years) to present to the surveyors. State and local requirements may apply as well. The questions of liability for injury and statute of limitations must also be addressed. It is not the intent here to give legal advice, which should instead be obtained by the facilities engineer through the administrative department.

Electrical Systems and Equipment Checklist

The following checklist can be used for evaluating an existing facility before a survey by JCAHO, or before an inspection by any authority having jurisdiction. This checklist is for the electrical systems and equipment section of this book. The reader is cautioned that this is not a complete list but one that touches on many of the major items. To ensure compliance, the reader should be familiar with all of the codes and standards that are applicable.

✓	Item	Inspection Activity	Comments
	1.	Verify that the branch circuits in patient bed locations are supplied from both the normal and essential electrical systems.	
	2.	Verify the grounding of receptacles in patient care areas.	
	3.	Test the grounding system in patient care areas.	
	4.	Verify that the receptacles in pediatric areas are tamperproof.	
	5.	Test all receptacles in patient care areas.	
	6.	Verify that all patient care areas have the required minimum number of receptacles.	
	7.	Verify that the requirements have been met for wet locations.	
	8.	Verify that the requirements for ground-fault circuit interruption have been met.	
	9.	Verify that the requirements for ground-fault protection have been met.	
	10.	Verify that the required tests have been performed on all isolated power systems and that all records of the tests are available.	
	11.	Verify that all testing has been performed on the essential electrical system and that records are available.	
	12.	Verify that the required preventive maintenance has been performed on all plant electrical equipment and that the records are available.	

(Continued)

✓	Item	Inspection Activity	Comments
	13.	Verify that the required preventive maintenance has been performed on all medical equipment and that the records are available.	
	14.	Verify that all medical equipment is labeled to indicate the date of the last test and inspection.	
	15.	Verify that fire and smoke partitions that were penetrated by electrical construction have been properly sealed.	
	16.	Verify that testing of the electrical distribution system and that the records are available, including an electrical riser diagram of the electrical distribution system.	
	17.	Ensure that all records of tests, inspections, and maintenance of the distribution system and equipment and medical equipment are up to date and available. Have available a copy of the system used to perform and track preventive maintenance and regular maintenance.	
	18.	Have available records indicating the continuing education of the staff in the respective engineering and maintenance departments.	

Part III
Building Construction and Life Safety

Most of the other five parts of this book deal with items or materials that go into health care buildings, such as electrical wiring and equipment (Part II), fire protection equipment (Part IV), and gas piping and equipment (Part V). Part III deals with the ways in which health care facility buildings themselves are made safe. General concepts related to safety and some specific aspects of building safety are included, as follows:

Chapter 13: NFPA *101*®, *Life Safety Code*®; NFPA *5000*®, *Building Construction and Safety Code*®; and other building codes, including the occupancy classifications, mixed and separated occupancies, and requirements for new versus existing facilities

Chapter 14: Building construction types, defend-in-place versus evacuation; fire resistance; stories, heights, and areas; construction type consistency and conversions; and determining construction types

Chapter 15: Building construction features such as compartmentation and barrier continuity, fire resistance of barriers, types of fire barriers, barrier openings and penetrations, and sealing and penetration protection

Chapter 16: Components of means of egress, protection of exits and exit access corridors, doors, stairs, horizontal exits, ramps, exit passageways, fire escape stairs and ladders, and accessible means of egress and areas of refuge

Chapter 17: Principles of egress, including egress capacity, occupant load, number of exits, and travel distance; arranging the means of egress; providing accessible means of egress for the mobility impaired; illuminating egress and emergency lighting; and marking means of egress

Chapter 18: Fire protection features, such as vertical opening protection, concealed spaces, smoke barriers, special hazard protection, interior finish, fire alarm systems and extinguishing equipment, and building contents and furnishings

Chapter 19: Building service equipment related to utilities; heating, ventilating, and air conditioning; cooking areas; and elevators, escalators, and vertical conveyers

Throughout this part, as needed, references are made to other applicable chapters of this book and to other applicable codes and standards. For example, the discussion of mandates and incentives for automatic sprinklers refers to Part IV, which covers fire protection systems in health care facilities. As with the other parts of this book, the discussions are limited to those areas of interest to health care facilities and the various occupancies associated with health care facilities.

This part does not include all specific code requirements, as would be found in a design manual. Those who are involved in designing, building, altering, or maintaining buildings must refer to these various codes, standards, and guides for more details.

13

Building and Life Safety Codes

Building codes are used to ensure the integrity of a wide variety of features in the newly built environment, including the stability of a building as well as its ventilation, light, sanitation, accessibility, and means of egress. The codes provide these requirements for a range of building types from small dwellings to large industrial structures. All levels of jurisdiction around the country adopt these codes or their own state code into laws and regulations to protect life, property, and public welfare.

This protection of property and public welfare is what separates most building codes from NFPA *101®, Life Safety Code®*. This code basically has one goal—to protect human life from fire and similar emergencies—and does not usually regulate building construction or include requirements for such safety features as ventilation and sanitation, building materials, and accessibility, as building codes do. Another major difference between the *Life Safety Code* and building codes is that the *Life Safety Code* provides separate requirements for both new construction and existing buildings, whereas building codes apply primarily to new construction.

Historic building catastrophes provide the background for understanding the rationale of today's building and life safety codes. How the many codes have evolved, are adopted, and are structured today provide insights into how best to understand and apply them for all types of structures. Guidance provided in this chapter can help sort out some of the confusing aspects of these codes, such as when one building or area therein must comply with two or more different building and life safety codes, and how to determine if a building must comply with codes for new or for existing structures.

13.1 Building Codes in the United States

Jurisdictions around the country comply with a variety of codes for construction, safety, fire protection, plumbing, mechanical features, and the like. Although there have been attempts to make building codes more uniform, some regional differences continue to exist. Furthermore, each state has a different set of rules on how county, city, and local governments can adopt and modify the codes to suit local needs.

13.1.1 Building Code History

Building codes have been around for centuries. Originally, authorities in larger cities wrote building regulations, usually after a major disaster such as a conflagration. During the nineteenth and early twentieth centuries, several major conflagrations virtually wiped out some cities in the United States, including such famous fires as the great fires of San Francisco (1849, 1906), Portland, Maine (1866), Chicago (1871), Boston (1872, 1889), Baltimore (1904), and Chelsea, Massachusetts (1908).

As a result of these disasters, the National Board of Fire Underwriters (subsequently called the American Insurance Association) developed what is considered to be the first model building code in the United States. It was considered to be a model code because, in order to have legal standing, authorities needed to adopt it. This document was referred to as the *National Building Code,* the first edition of which was published in 1905. The insurance industry used this document for many years as a basis for measuring local building codes for insurance rating purposes.

Several additional model building codes have been developed that have enhanced uniformity of codes around the country. In 1927, an agency more recently referred to as the International Conference of Building Officials (ICBO) published the *Uniform Building Code.* This code was a model building code, but was developed by building officials rather than an insurance organization. Subsequently, in 1945, the Southern Building Code Congress (SBCCI) developed the *Southern Standard Building Code* (renamed the *Standard Building Code*® with the 1976 edition). The last of the model building codes was developed by BOCA, which originally referred to Building Officials Conference of America but later for Building Officials and Code Administrators International, Inc. For years, the BOCA code was called the *Basic Building Code.* The 1984 edition of this code was referred to as the *BOCA*® *Basic/National Building Code,* reflecting the fact that the American Insurance Association was discontinuing its sponsorship of the *National Building Code* and transferring rights to the document to BOCA. In fact, the only part of the original document used by BOCA was the title, *National Building Code.* In 1987, the transformation was complete, and the name of the BOCA family of codes was renamed BOCA National, such as the *BOCA*® *National Building Code, BOCA*® *National Fire Prevention Code,* and so on.

For the most part, these building codes were developed as and remained regional codes. With a few exceptions, the ICBO Uniform Codes were used primarily west of the Mississippi River, the BOCA National Codes were used primarily in the Northeast, while the SBCCI Standard Codes were used primarily in the Southeast. In most cases, individual cities, towns, or counties initially adopted these codes. Beginning in the 1970s, many state governments adopted these model codes, albeit in most states with some degree of amendment and allowances for city, county, and local governments to make further amendments to the codes. (In some cases, the amendment document is thicker than the model code itself.) Yet many areas have not adopted any building regulations at all.

Today, almost every jurisdiction in the United States has its own set of building regulations, making the use of building codes regionalized and fragmented. For example, some states have a building code that is mandatory statewide, without exception. Other states have adopted a statewide building code but allow localities to be more stringent, while still other states allow localities to adopt either the state code or another code, and yet others leave building code adoption entirely up to the localities. In some instances, the statewide code applies in all areas except certain, usually larger, cities, which sometime have the authority to write their own codes. Texas has no statewide building code, and New York City and Chicago have their own building codes, but both are looking at a model code. With 50 different states, there are probably close to 50 different rules in place regarding the adoption and adaptation of building codes. So, even though model codes have reduced differences in building codes across the country, there is still a wide lack of uniformity in building codes nationwide.

13.1.2 Building Code Uniformity

Until 1994, there had been no concentrated effort to develop a single building code for the United States, although many groups, such as the American Institute of Architects (AIA), have encouraged having one. Building regulations are considered a local issue, and Congress has not empowered the federal government to perform duties outside the written powers granted in the Constitution for the building code arena.

For close to 20 years, one group, the Board for the Coordination of Model Codes (BCMC), had attempted to make the three major model building codes and the *Life Safety Code* more consistent. This board consisted of two representatives each from the three regional model building code groups (BOCA, ICBO, and SBCCI) plus the NFPA. The BCMC published numerous reports, including two reports regarding health care occupancies, one in 1984 and one in 1990, which have motivated each code-making organization to modify its code significantly with regard to the construction of health care facilities. Model codes promulgated after 1990, if not amended locally, are generally consistent with NFPA *101* with regard to life safety issues for health care construction. Under the model building codes, most health care occupancies are called an *institutional use group* or *use group I*. (See Section 13.5 of this chapter for more information on building occupancies.)

Another event that contributed to code consistency is that all three of the major regional building code groups organized the information presented in their respective codes in the same way. Thus all of the major model codes have similar items in the same chapters. For example, *means of egress* is addressed in Chapter 10 in all three of the codes. This code organization assisted architects, builders, and facility managers and engineers in locating and comparing the requirements within the codes.

To further assist in the process to formulate one model building code for use in the United States and elsewhere, the three major model building code organizations created the International Code Council (ICC). Using the resources of the three parent organizations, the ICC developed one set of model codes, the International Family of Codes, for use throughout the United States. The first edition of the *International Building Code* was the 2000 edition. Some of the companion documents to this code, such as the *International Plumbing Code* and the *International Mechanical Code*, existed before 2000. The BOCA National codes, ICBO Uniform codes, and SBCCI Standard codes are no longer being published.

13.1.3 Family of Codes

Each of the former regional code organizations, as well as the ICC, published more than a building code. A *family* of codes usually contains a building code, fire prevention code, mechanical code, plumbing code, and so forth. For example, there was a *Uniform Fire Prevention Code,* a *Uniform Mechanical Code,* a *Uniform Plumbing Code,* and so forth.

Table 13.1 NFPA's Emerging Family of Codes

Document	Title	Current Edition
NFPA 5000	*Building Construction and Safety Code*	2006
NFPA *101*	*Life Safety Code*	2006
NFPA 70	*National Electrical Code® (NEC®)*	2005
NFPA 1	*Uniform Fire Code™*	2006
	Uniform Plumbing Code[a]	2003
	Uniform Mechanical Code[a]	2003
ASHRAE 901	*Energy Standard for Buildings Except Low-Rise Residential Buildings*[b]	2004
ASHRAE 902	*Energy-Efficient Design of Low-Rise Residential Buildings*[b]	2004

[a]As developed by IAPMO (International Association of Plumbing and Mechanical Officials). Using full, open ANSI consensus procedures, similar to all other NFPA codes and standards.
[b]As developed by ASHRAE (American Society of Heating, Refrigeration and Air Conditioning Engineers) using full, open ANSI consensus procedures similar to all other NFPA codes and standards.

In 2000, the NFPA announced that it would write a complete family of codes, detailed in Table 13.1, for the built environment, including a building code. This process has resulted in two major model code sets for the built environment in the United States—NFPA and ICC. The NFPA documents are written using NFPA's approved American National Standards Institute (ANSI) consensus process.

The NFPA building code is *NFPA 5000®, Building Construction and Safety Code®*. As with any NFPA code or standard it is an ANSI document. It covers items traditionally found in a building code. One of the very significant advantages of *NFPA 5000* is that it is in total harmony with *NFPA 101*. Due to this, most of the discussions in this book regarding NFPA *101* for new construction and building rehabilitation also apply to *NFPA 5000*. For example, the terminology used to describe an occupancy type (e.g., health care, ambulatory health care) in *NFPA 5000* is identical to what can be found in NFPA *101*. Chapter 15 of *NFPA 5000* deals with rehabilitation of existing buildings and is harmonized with the requirements for "existing" occupancies in NFPA *101*. A specific example of this concerns the triggering criteria for renovations of smoke compartments in a health care occupancy that require retroactive installation of automatic sprinklers.

Due to the array of codes that must be followed, it is very important for facility designers, managers, and engineers to determine the following information:

- The building code adopted by the state, county, or city within which a facility exists or will be built
- The edition of that code currently being used in that jurisdiction, and the one used at the time the facility was built
- Whether there are any amendments to that code

13.2 NFPA 101®, *Life Safety Code*®

Like the building codes, the *Life Safety Code* was also founded after 1900, not only because of the significant number of conflagrations during this period but also because of the significant loss of life as a result of major fires. These fires included the Rhodes Opera House—170 killed (1903), the Iroquois Theater—602 killed (1903), the Lakeview Grammar School—175 killed (1908), and the fire that was the real start of NFPA's Committee on Safety to Life, the Triangle Shirtwaist Company Fire—145 killed (1911). NFPA's Committee on Safety to Life produced its first document on exit drills in 1913 and then followed with the *Building Exits Code* in 1927, which was renamed the *Life Safety Code* in 1966.

Specifically, the *Life Safety Code* provides for construction, protection, and occupancy features necessary to minimize danger to life from fire, including smoke, fumes, or panic, in all types of structures. This code also has building construction requirements for where the conditions of occupants make it difficult to evacuate, such as health care facilities and detention and correctional centers, and for where there may be slowed evacuation, such as in some board and care and day care facilities, and areas of assembly, due to the concentration of people who may be unfamiliar with their surroundings. Among other items, the *Life Safety Code* includes provisions for the following:

- Determining occupant loads
- Designing egress features so as to permit prompt escape of occupants from buildings, or, where desirable, into safe areas within buildings
- Designing and maintaining compartmentation to prevent the spread of fires to other areas of a building
- Designing and maintaining other operating features that enhance safety

Each type of building occupancy (residential, health care, business, etc.—see Section 13.5 of this chapter), as defined in the *Life Safety Code,* has its own set of requirements.

One of the main features of the *Life Safety Code* is that since its onset, it has provided separate requirements for new construction and for all existing structures, which provides all code users a comprehensive and practical safety code for use in all buildings. Since most people live and work in existing buildings, this is a very important feature, which has been upheld in the U.S. Federal Appeals Courts as reported in NFPA's *Fire Journal*® (Brannigan, 1981). The fact that the *Life Safety Code* deals only with life safety helps justify and facilitate its application to existing buildings.

Although 38 states require the use of the *Life Safety Code,* in one form or another statewide, almost all states require its use for health care facilities. After a series of multiple-fatality fires in health care facilities that occurred in the 1960s, the U.S. Congress passed a law requiring health care facilities to comply with the *Life Safety Code* to receive reimbursement under Medicare provisions. In essence, this action made it mandatory for all health care facilities to comply with the *Life Safety Code,* except for facilities that do not rely on Medicare or Medicaid funds and are located in one of the few states that do not require health care facilities to comply with the *Life Safety Code.* Even in such cases, a facility would be wise to comply voluntarily with the

Life Safety Code, due to the fact that it is an ANSI document and, as such, has been cited in civil litigation as an indication of a standard of care even if not actually adopted in a particular jurisdiction. In addition, even privately run facilities that do not participate in the Medicare/Medicaid program typically want to maintain a performance level that is certified by the Joint Commission on Accreditation of Healthcare Organizations (JCAHO). JCAHO accreditation surveys verify conformance with all provisions of NFPA *101.* Failure to comply may result in provisional accreditation or loss of accreditation.

It is important for architects to be aware of the fact that the *Life Safety Code* is mandated, as illustrated in the following case study.

Mandated *Life Safety Code*

In one instance, a large addition (several hundred thousand square feet) was built onto a hospital located in a state that does not use the *Life Safety Code.* That state's building code did not require the installation of sprinklers in patient sleeping rooms, and none were installed. With no such exception in the *Life Safety Code,* the building was considered nonsprinklered and had to meet all of the additional requirements applicable to a nonsprinklered building (see Chapter 21 of this book for more information on sprinklers). Had the sprinklers been installed in the patient sleeping rooms (a relatively easy task at the time of construction), numerous exceptions would have applied. If the addition had occurred after the adoption of the 1991 edition of the *Life Safety Code,* a major violation of the code (mandatory sprinklers) would have occurred.

13.3 Fire Prevention Codes

Each of the three former model code organizations wrote a model fire prevention code. Model fire prevention codes currently being used include NFPA 1, *Uniform Fire Code;* BOCA *National Fire Prevention Code;* Western Fire Chiefs Association (WFCA), *Uniform Fire Code™;* SBCCI, *Standard Fire Prevention Code;* and the ICC *International Fire Code.* The *National Fire Prevention Code* and the *Standard Fire Prevention Code* are no longer being promulgated and over time will disappear. The WFCA *Uniform Fire Code* has been merged with the former NFPA 1, *Fire Prevention Code,* and is now NFPA 1, *Uniform Fire Code.* Fire prevention codes commonly deal with such issues as fire reporting, fire lanes, hazardous materials (flammable liquids, flammable gases, dusts, etc.), housekeeping, control of ignition sources, operating features, and sometimes education (see Part IV of this book for more information on fire protection).

All the model fire codes reference numerous NFPA codes and standards, such as NFPA 10, *Standard for Portable Fire Extinguishers;* NFPA 13, *Standard for the Installation of Sprinkler Systems;* and *NFPA 72®, National Fire Alarm Code®.* NFPA 1, which was dramatically revamped from the late 1980s through the mid-1990s and has grown significantly in popularity, was written to take advantage of other NFPA codes and standards, as well. In NFPA 1, material that is commonly used during a fire prevention inspection is extracted from the referenced codes and standards and repeated in NFPA 1. As mentioned previously, in 2000 the NFPA *Fire Prevention Code* merged with the *Uniform Fire Code* to become NFPA 1, *Uniform Fire Code*.

Aside from referencing NFPA documents, there is very little commonality among these fire prevention codes. Some fire prevention codes deal primarily with existing buildings and equipment, others deal primarily with new construction and equipment, while still others deal with both.

As with building codes, jurisdictions must adopt model fire prevention codes for them to be enforceable. Many locations do not use a model fire prevention code but use a set of state regulations or laws instead. Thus, it is essential for facilities engineers and designers to check with each city, county, and state to determine what fire prevention code is used and enforced.

13.4 AIA *Guidelines for Design and Construction of Hospitals and Health Care Facilities*

A reference sometimes used in mandatory fashion for designing health care facilities is the AIA *Guidelines for Design and Construction of Hospitals and Health Care Facilities*. The *Guidelines* have evolved over the years; its preface provides excellent information on the history and use of the *Guidelines*.

The AIA *Guidelines* are written to work with, and include references to, NFPA *101* and other NFPA codes and standards. Items covered by the *Guidelines* include energy conservation, sites, equipment, construction, records, and detailed provisions for various health care occupancies. The provisions for general hospitals include details for items such as critical care units, surgical suites, emergency services, pharmacies, medical records, mechanical systems, and electrical standards, among many other items. The provisions for nursing facilities include such items as resident units, activities, rehabilitation therapy, personal services, dietary facilities, and the like. There are also provisions for outpatient facilities; rehabilitation facilities; psychiatric hospitals; mobile, transportable, and relocatable units; and hospice care. Needless to say, this document is "mandatory" for any health care design work.

13.5 Occupancy Classifications

Occupancy classification is extremely important when applying the *Life Safety Code* or any building code. It is not possible to apply a code properly if the proper occupancy classification has not been determined. The occupancy classification system in the codes, whether the *Life Safety Code* or a building code, is not the same, but it is very similar.

Occupancy classifications are detailed in the *Life Safety Code* (Chapter 6). Most occupancies have two chapters in the *Life Safety Code,* one for new construction and one for existing buildings. See Section 13.7 of this chapter for a discussion of the differences between the codes for new and existing structures and equipment.

The following discussion explains the *Life Safety Code,* 2006 edition, occupancies and how the occupancies relate to health care facilities. Some types of occupancies are not often found in health care facilities. These occupancies are briefly mentioned, however, to be comprehensive and in the event that facilities designers, planners, and managers may occasionally need to make use of this information. Where appropriate, material from the model building codes is included.

13.5.1 Assembly Occupancies

Chapters 12 and 13 of the *Life Safety Code* address assembly occupancies. *Assembly occupancies* are buildings or portions thereof used for the gathering together of 50 or more people for the purposes of amusement, eating, drinking, worship, deliberation, or similar functions. In a typical large health care facility, assembly occupancies would include the cafeteria, large conference rooms, and auditoriums. However, these are assembly occupancies only if they have the potential for an occupant load greater than 50. (Refer to Chapter 17 of this book for more information on calculating occupant load.) Most chapels are small enough not to be considered assembly occupancies, but if they have a capacity of 50 people or more, then they must meet assembly occupancy requirements.

NFPA 5000 simply defines a category for assembly occupancies while other building codes refer to these as an *A use group,* an *A occupancy,* or a similar designation. The category is then commonly broken down into various subcategories, such as A-1, A-2, A-3, and so on, depending on the specific building code. Explanations are usually found in the early chapters of the specific building code.

13.5.2 Educational Occupancies

Chapters 14 and 15 of the *Life Safety Code* limit educational occupancies to those providing education up through the twelfth grade for six or more persons. The *Life Safety Code* also requires educational occupancies to be used for education for 4 hours or more per day or more than 12 hours per week. Therefore, a classroom, even if used for education up to the twelfth grade, is not an educational occupancy if it is used for fewer than 4 hours per day and fewer than 12 hours per week. Most of the building codes have this same limitation.

Educational occupancies are not typically found in health care facilities; health care facility classrooms are typically used for adult education and are not considered to be educational occupancies. These areas could be considered to be business occupancies (see Section 13.5.9) if they hold fewer than 50 people, and assembly occupancies if they hold 50 people or more. However, there are times when there are classrooms for first through twelfth grades in a health care facility, and the requirements for educational occupancies must be used. There are several items in Chapters 14 and 15 of the *Life Safety Code* that are more stringent than those found in the requirements for health care.

Most model building codes define educational occupancies similarly to NFPA *101* and *NFPA 5000*. They are usually called *group E occupancies* or *uses* in the other codes.

13.5.3 Day Care Occupancies

Chapters 16 and 17 of the *Life Safety Code* address day care occupancies. In the 1997 *Life Safety Code,* day care occupancies were moved from the end of the educational chapters to their own chapters. Day care occupancies include adult and child day care and buildings or portions thereof used for the care, maintenance, and supervision of four or more clients on less than a 24-hour basis. This allows the day care facility to be open 24 hours per day, but no one client can receive care for 24 hours. The provisions of these chapters will be important to designers and managers of staff day care centers located in health care facilities.

Only NFPA *101* and *NFPA 5000* provide specific criteria for day care occupancies. Depending on the building code used and the age and number of children in attendance, day care facilities could be considered educational, institutional, or even residential.

13.5.4 Health Care Occupancies

Chapters 18 and 19 of the *Life Safety Code* address health care occupancies. The *Code* defines a health care occupancy as follows:

> An occupancy used for purposes of medical or other treatment or care of four or more persons where such occupants are mostly incapable of self-preservation due to age, physical or mental disability, or because of security measures not under the occupants' control. [101–06: 3.3.168.7]

The facilities regulated by Chapters 18 and 19 are further limited to health care facilities that provide sleeping accommodations. The "four or more" criterion is consistent with federal regulations for health care.

The vast majority of health care facilities, including hospitals, nursing homes, and similar facilities, are health care occupancies. Up to the 1997 edition of the *Life Safety Code,* ambulatory health care facilities, which are facilities that have four or more people incapable of self-preservation but do not have sleeping facilities, were also addressed in the same chapters as health care. With the 2000 edition of the *Life Safety Code,* the requirements for ambulatory facilities were moved to their own chapters. Typical health care facilities actually consist of many occupancies, with health care and ambulatory health care being just two of them. Multiple, mixed, and separated occupancies are discussed in Section 13.6.

Again, most model building codes define health care occupancies similarly to the *Life Safety Code* and *NFPA 5000*. One difference is that the other model building codes commonly use six or more people as the lower cutoff point for a health care occupancy. These occupancies are referred to as *I occupancies,* as either I-1 or I-2.

13.5.5 Ambulatory Health Care Occupancies

Chapters 20 and 21 of the *Life Safety Code* regulate ambulatory health care occupancies. The definition of an ambulatory heath care occupancy has fluctuated somewhat over the last dozen years. When using the *Code* it is important to determine what definition is being used. The most accurate definition of an ambulatory health care occupancy is

A building or portion thereof used to provide services or treatment simultaneously to four or more patients that provides, on an outpatient basis, one or more of the following: (1) Treatment for patients that renders the patients incapable of taking action for self-preservation under emergency conditions without the assistance of others; (2) Anesthesia that renders the patients incapable of taking action for self-preservation under emergency conditions without the assistance of others; (3) Emergency or urgent care for patients who, due to the nature of their injury or illness, are incapable of taking action for self-preservation under emergency conditions without the assistance of others. [101–06: 3.3.168.1]

Note that this definition does not refer to sleeping or 24-hour care. It provides a bridge between medical office buildings (business occupancies) and hospitals (health care occupancies). In fact, the provisions require that the facility comply with the requirements for business occupancies as modified by the provisions in ambulatory health care.

With the exception of *NFPA 5000*, most of the major model building codes do not have an occupancy comparable to ambulatory health care. The enforcing authority would need to evaluate whether use group B or I would apply.

13.5.6 Detention and Correctional Occupancies

Chapters 22 and 23 of the *Life Safety Code* address detention and correctional occupancies. These occupancies include facilities that provide housing for four or more (one or more starting with the 2006 edition of the Code) individuals who are mostly incapable of self-preservation because of security measures not under the occupants' control. Health care facilities typically include this occupancy under two conditions. One condition is a health care facility that is part of a correctional facility, such as a prison hospital. The second condition is a general hospital that has a detention wing or section. Most psychiatric hospitals are not considered to be detention and correctional occupancies or facilities, but they sometimes can be, such as facilities for the criminally insane. A couple of secure rooms in the emergency department would not normally qualify as a detention and correctional occupancy.

It should be recognized that the detention and correctional chapters of NFPA *101* were written using health care occupancies as a basis. Both types of facilities deal with protecting people who are not capable of self-preservation. For detention and correctional facilities, locking of egress doors is necessary. When it is necessary to lock egress doors in health care facilities, the chapters for detention and correctional facilities can be referenced, although it is not mandatory.

Other than in *NFPA 5000,* detention and correctional occupancies are referred to as *institutional occupancies* in most model building codes, the same as health care occupancies. However, a different subclass is used, such as I-2 or I-3.

13.5.7 Residential Occupancies

Chapters 24 through 35 of the *Life Safety Code* address residential occupancies. As an overall class, *residential occupancies* provide sleeping accommodations other than for health care or detention and correctional purposes. In the model building codes, other than *NFPA 5000,* residential occupancies are generally referred to as *R occupancies* or *uses* and are sometimes further subdivided into classes R-1, R-2, and so on. Under the *Life Safety Code,* there are five different residential occupancies. This is one area where there are some significant differences between the *Life Safety Code, NFPA 5000,* and the other model building codes.

All sleeping areas for non-patients must be evaluated as a residential occupancy. The *Life Safety Code* does not allow for *incidental residential,* and therefore the provisions for residential occupancies must be applied to any space having characteristics of residential use that is not considered a patient sleeping area. An example might be an area that permits overnight accommodations for family members, or an area where facility staff might retire to during a shift break.

13.5.7.1 *One- and Two-Family Dwellings*

Chapter 24 of the *Life Safety Code* addresses one- and two-family dwellings. The first, and most commonly the smallest, of the residential occupancies is the *one- and two-family dwelling.* As the name implies, this category consists of buildings or portions thereof that have a maximum of two residential units (kitchen, bath, and living accommodations). Typically health care facilities do not have this type of occupancy unless the facility provides this type of housing as a form of employee housing. However, no more than three people in a non-family unit are permitted to live in each dwelling unit. The *Life Safety Code* covers both new and existing one- and two-family dwellings in the same chapter. Except for smoke alarms and a secondary means of escape, the requirements for one- and two-family dwellings are minimal.

13.5.7.2 *Lodging and Rooming Houses*

Chapter 26 of the *Life Safety Code* addresses the next largest residential occupancy, the *lodging and rooming house.* This occupancy is essentially a small hotel or dormitory, limited to a maximum of 16 people without separate cooking facilities—in other words, not an apartment house. These buildings are often converted one- and two-family dwellings, in which more than three people who are not part of a family unit live. It should be noted that lodging and rooming house occupancies must house four or more people.

If a health care facility has sleeping facilities for residents or for on-call staff, that area is often categorized as a lodging or rooming house occupancy. This is important, as the smoke alarm requirements for lodging and rooming house occupancies are more stringent than for health care occupancies, and in some cases, installing or not installing corridor door closers and sprinklers could be an issue.

13.5.7.3 *Hotels and Dormitories*

Chapters 28 and 29 of the *Life Safety Code* address *hotels and dormitories,* the next step up from lodging and rooming houses. Hotels and dormitories provide sleeping accommodations for more that 16 people, again without separate cooking facilities. This categorization is a matter of size: areas that accommodate 16 or fewer people are lodging or rooming houses; areas that accommodate more than 16 people are hotels or dormitories. If a health care facility floor or wing accommodates more than 16 people in this way, the area falls under the requirements for hotels and dormitories.

The concept of hotels and dormitories is important to keep in mind for providing safety in health care facilities, because the smoke alarm and corridor wall requirements are more stringent for hotels and dormitories than for health care occupancies. The difference is largely attributable to the staff assistance that is present in a health care occupancy.

13.5.7.4 Apartment Buildings

Chapters 30 and 31 of the *Life Safety Code* address apartment buildings. If a building has more than two dwelling units in it, it is an *apartment building*. Although it is not totally unheard of for health care facilities to have apartment building residential occupancies, they typically do not. For example, a hospital might own an apartment building as an investment or for future space, or to house medical staff or students. Also, it might be owned as a way to house relatives of patients, although these occupancies generally would be categorized as hotels.

13.5.7.5 Board and Care Facilities

Chapters 32 and 33 of the *Life Safety Code* address the last of the residential occupancies, the *board and care facility*. Residential board and care occupancies are those where four or more residents are housed and receive personal care services. Small board and care facilities care for up to 16 people, and large board and care facilities care for more than 16 people. This category is generally a bridge between normal residential care and health care.

Not too long ago, it could have been said that health care facilities did not generally provide board and care facilities. Today, however, many communities throughout the country have board and care facilities for mentally handicapped or impaired residents. Also, facilities commonly referred to as assisted living facilities, where older citizens live and are watched over but do not receive acute medical care, must meet the requirements for residential board and care requirements. Since many health care facilities are associated with assisted living facilities, the provisions for residential board and care may be of great interest.

Residential board and care was probably the most controversial of the occupancies in the *Life Safety Code* because this category required facility designers and managers in conjunction with the code official to make a determination regarding the evacuation capabilities of the occupants as a group. Using terms such as *prompt, slow,* and *impractical*, the *Life Safety Code* established requirements for each type of board and care facility (i.e., small—16 or fewer, or large—more than 16). However, beginning with the 2003 edition of the *Code*, the evacuation capability determination was deleted for new construction. This now only applies to existing facilities, and even there, the authority having jurisdiction is given flexibility in its use.

When most people who live in a board and care facility remain constant in their evacuation capabilities, the classification system works. Unfortunately, many clients, such as the elderly, do not maintain a constant level of evacuation capability. If the facility is built for "prompt" or "slow" evacuation capabilities of the clients as a group and the capability becomes "impractical," over time the residents may have to be relocated to a nursing home facility, a very undesirable action for people in board and care facilities, particularly the elderly.

Both NFPA *101* and *NFPA 5000* provide substantial coverage for this occupancy type. The other model building codes do not do a good job of regulating this occupancy. The categorization is confusing and the evacuation capabilities of the residents are poorly handled.

13.5.8 Mercantile Occupancies

Chapters 36 and 37 of the *Life Safety Code* address the *mercantile occupancy*. This occupancy includes buildings or portions thereof used for the sale of products on display, such as grocery stores, department stores, and similar facilities. To qualify as mercantile, products or goods must be on display. A hospital or nursing home does, in fact, sell a service, but it does not have products on display. Typically, a gift shop in a health care facility would not be large enough to be categorized as a mercantile occupancy, as discussed further in Section 13.6. Thus, mercantile occupancies would generally not be of interest to a health care facility unless it held investment property. However, some hospitals actually have a shopping mall as part of the hospital. In this case, a mercantile occupancy classification would be appropriate. There are also cases where the gift shop is large enough that it can no longer be classified as incidental.

The model building codes, other than *NFPA 5000*, generally refer to mercantile occupancies as *M uses* or *occupancies*.

13.5.9 Business Occupancies

Chapters 38 and 39 of the *Life Safety Code* address business occupancies. Besides health care occupancies, *business occupancies* are probably the most common areas in health care facilities. A business occupancy is any building or portion thereof used for keeping accounts and records or for transacting business other than mercantile. Typical examples in a health care facility are administrative buildings or sections of buildings or medical office buildings (MOBs) that are not ambulatory health care occupancies (i.e., do not have four or more patients rendered incapable of self-preservation). Again, it is important to review the material on multiple occupancies, since it is very common for business occupancies to exist in conjunction with health care occupancies.

Business occupancies are generally referred to as *B occupancies* or *uses* by the model building codes other than *NFPA 5000*.

13.5.10 Industrial Occupancies

Chapter 40 of the *Life Safety Code* addresses industrial occupancies. It may seem that health care facility designers and administrators would be able to ignore the requirements for industrial occupancies, but this is not the case. *Industrial occupancies* are buildings or portions thereof wherein anything is made, manufactured, processed, assembled, mixed, packaged, finished, decorated, or repaired. In a health care facility, this area could be a boiler plant, maintenance shop, or similar area or building.

Although a small boiler room or repair shop would not normally qualify as a separate industrial occupancy, separate boiler buildings, attached or not, or large boiler or shop facilities would qualify as such. This occupancy has only one chapter in the *Life Safety Code* for both new and existing buildings.

In the model building codes other than *NFPA 5000*, industrial occupancies are usually referred to as *F occupancies* or *uses*. They are often broken down into F-1 and F-2 categories, depending on the types of hazards that might be present.

13.5.11 Storage Occupancies

Chapter 42 of the *Life Safety Code* addresses *storage occupancies*, which, as the name implies, are areas where anything is stored. Examples in health care facilities include parking garages, central warehouses, or large storage areas within a facility. Small storage rooms and closets do not warrant a separate occupancy classification. Similar to industrial occupancies, storage occupancies have only one chapter for both new and existing facilities in the *Life Safety Code*.

Storage occupancies are usually referred to as S *occupancies* or *uses* in the model building codes other than *NFPA 5000*. Similar to F occupancies, S occupancies are usually further divided into S-1 and S-2 categories, based on the types of hazards that may be present in the area.

13.5.12 Hazardous Occupancies

One occupancy type that the model building codes include, but that the *Life Safety Code* and *NFPA 5000* do not, is what is referred to as *H occupancies*. Areas categorized as such have detonation or deflagration potential; contain flammable liquids, flammable gases, oxidizers, or similar materials; contain health hazards such as toxics, corrosives, or radioactive materials; or involve semiconductor manufacturing. These areas are usually further categorized, based on the hazards present, into categories H-1, H-2, H-3, and so on. NFPA codes recognize that few occupancies are actually *hazardous* per se. However, almost all occupancies have some level or quantity of hazardous contents. Since NFPA *101* and *NFPA 5000* are occupancy centered, the codes regulate the unique issues associated with hazardous contents based on the occupancy: health care, storage, industrial, and so on.

In NFPA *101*, hazardous areas are handled first, based on their primary occupancy categorization. Buildings or areas that contain certain types and levels of hazards are then further subcategorized as containing a *low, ordinary,* or *high* hazard of contents, as defined in NFPA *101*, *Life Safety Code* (Section 6.2.2) and must comply with more stringent requirements (see Section 7.11). For example, a laboratory that a model building code may classify as H-3, NFPA *101* may classify as an industrial occupancy with high hazard of contents (NFPA *101,* Sections 40.1.5 and 40.3.2).

As per NFPA *101*, hazard of contents is the relative danger of the start and spread of fire, the danger of smoke or gases generated, and the danger of explosion or other occurrence that can potentially endanger the lives and safety of the occupants of the building or structure. Low-hazard contents have low combustibility with little or no risk of self-propagating fire occurring. Ordinary-hazard contents are those likely to burn with moderate rapidity or give off a considerable volume of smoke. High-hazard contents are those likely to burn with extreme rapidity or from which explosions are likely.

Depending on the nature and quantity of hazardous materials in a health care facility, the *H use* group could have a major impact on a health care renovation project. The requirements are quite stringent both in most model building codes and in the model fire prevention codes.

13.6 Multiple, Mixed, and Separated Occupancies

When two or more occupancies occur in the same building or area, which is a common occurrence, the *Life Safety Code* now calls this a *multiple occupancy building*. Knowing exactly what the occupancy requirements are and how to meet them can become difficult at times. There are two ways to handle multiple occupancy buildings. One way is to separate the occupancies. The second way to deal with a multiple occupancy building is to treat the area as a mixed occupancy and comply with the provisions that are more or most stringent of the occupancies involved. Not all codes allow this second option. Some model building codes, especially older versions, mandate separation of occupancies.

The *Life Safety Code* (Section 6.1.14) allows either option. Which course to take is up to the owner or designer, but the building layout often suggests which method to use. When using the *separated occupancies* method, the *Code* does not establish any specific fire resistance requirements to separate the occupancies in order for them to be considered "separated" for an existing building. For new construction the *Life Safety Code* contains an occupancy separation table, based on *NFPA 5000*, which must be met for separated occupancies. The rating that must be used is provided in a matrix that lists the requirements for the different occupancies on both the x and y axes of the matrix. The intersection of the ratings for the two occupancies gives the required fire resistance rating to separate those two occupancies. In the *Life Safety Code,* many occupancy chapters include additional provisions for multiple occupancies. For example, Sections 18.1.2 and 19.1.2 of the health care occupancy chapters in the 2006 edition detail specific fire resistance–rated separations (2-hour), as well as other restrictions on use and egress for treating occupancies as not mixed in both new and existing buildings.

Thus, the *Life Safety Code* allows for separating and protecting each occupancy in a building or treating them together as one occupancy and meeting the more stringent requirements of all occupancies involved. Although compliance for multiple-occupancy buildings or areas would seem straightforward, further analysis of the codes is usually required.

For example, Figure 13.1 shows a diagram of a multiple-occupancy area. In this figure, since the corridor serves as a common egress system for all the occupancies involved, it must be treated as a mixed occupancy, and the more stringent provisions of each occupancy must be met. However, this rule does not apply to all

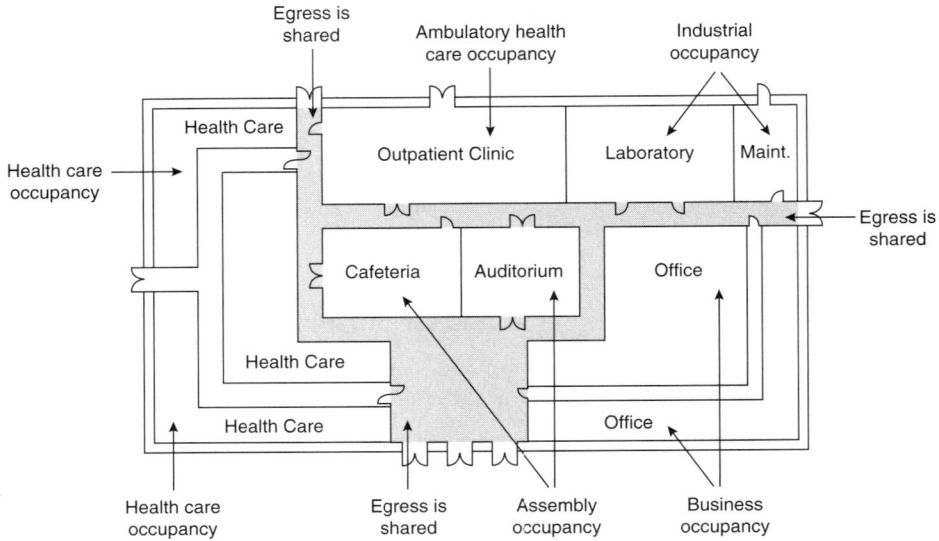

Figure 13.1 A multiple-occupancy area.
Source: Opus Communications and Koffel Associates, Inc., redrawn from *2000 Life Safety Code Workbook & Study Guide for Health Care Facilities*, 2003, Figure 3.2.

provisions. Since the cafeteria is an assembly occupancy and holds more than 100 people, all egress doors that serve that area require panic hardware or fire exit hardware, if a latch is provided. This does not mean, however, that the door from the maintenance area to the corridor or to the outside requires this type of hardware, because separate safeguards can be provided. Also, the health care area requires 8-ft-wide corridors; therefore, the corridors inside the health care area and those common corridors must be 8 ft wide. However, the corridors within the office area need not be 8 ft wide.

An important exception is provided in the *Life Safety Code* regarding multiple occupancies in that it allows minor incidental occupancies to be considered part of the predominate occupancy. The intent of the provision is to allow incidental storage areas, gift shops, minor incidental office areas, small maintenance areas, and similar spaces to exist without having to be classified as a separate or mixed occupancy. One of the most difficult aspects of this exception, though, is determining how much office area can be considered incidental. Obviously, the offices that are scattered around the building for nursing staff, department heads, and similar staff are minor and incidental. A whole administrative floor or wing is not. Anything in between requires sound judgment. This allowance has been revised over the last few code cycles, so it would be prudent for facility planners, managers, and engineers to review the edition that applies to their facility.

13.6.1 Health Care Facility Multiple Occupancies

The *Life Safety Code* offers two options for separating other occupancies from health care occupancies (Sections 18.1.2 and 19.1.2). Both options require the area to include a 2-hour fire resistance–rated separation. To determine how to handle a multiple occupancy in a health care facility, facility planners and managers must first determine if the other occupancy is or will be separated from the health care occupancy by a 2-hour fire resistance–rated separation. If not, then plans must be made to install the necessary separations if it is desirable to treat the areas as separate occupancies. For all areas that have or will have the proper fire resistance separations, it must then be determined how the space is being or will be used (i.e., what the other occupancies are). In the first option, as long as the 2-hour fire resistance criterion is met, any occupancy can be treated as a separate occupancy from the health care occupancy as long as the space is not being used for housing, treatment, or customary access by health care patients incapable of self-preservation. Thus, cafeterias, auditoriums, large boiler or storage areas, laboratories, and similar spaces can be treated as separate occupancies from health care occupancies if these areas have the proper fire resistance rating. This allowance does not mean that a cafeteria must prohibit patients from entering; it can allow access to patients, provided that they are capable of self-preservation.

The *Life Safety Code* has a special provision for when the occupancy being separated from the health care occupancy is a business or ambulatory health care occupancy (Sections 18.1.2.3 and 19.1.2.3, 2006 edition). This is the second option. In such cases, if that area meets the 2-hour separation requirement, it can be considered a separate occupancy if it is *primarily* intended to provide outpatient service and the facility does not provide services for four or more health care patients who are litter-borne (i.e., patients on litters or stretchers). If the patients are litter-borne, they must be limited to three or fewer in number. This allows a medical office building or ambulatory care facility that may be attached to a hospital to be considered a separate occupancy even though it may provide treatment to inpatients, as long as those inpatients are not litter-borne (or if litter-borne, they are three or fewer in number). Thus, the facility must primarily treat outpatients and a limited number of inpatients, but there is not a total prohibition on treating inpatients.

13.6.2 The Most Stringent Occupancy Requirements

The benefit of treating occupancies separately is that not all occupancies need to meet the most stringent requirements, which could create hardship for a facility. For example, some facilities would most likely find it difficult to have the cafeteria or assembly occupancies meet the requirements for health care occupancies and vice versa (the health care occupancy meet assembly requirements). Another difficult mixed occupancy is mixed residential and health care. The corridor requirements for residential occupancies are quite stringent.

A common mistake that health care facility planners make, however, is to assume that the requirements for health care occupancies are the most stringent and the ones that must be applied for areas that do not separate multiple occupancies. This is not always so when examined on an item-by-item basis. For example, depending on the requirements for *new* versus *existing* buildings and sprinklered versus nonsprinklered areas, the most stringent requirements for mixed occupancies may require installing self-closers on patient room doors (residential occupancies), introducing more stringent dead-end corridor provisions (assembly occupancies), and requiring corridor smoke detection and in-room smoke detection (residential occupancies). See Figure 13.2 for an example of separated occupancies. The figure shows three separate occupancies in the same building. A 2-hour separation is required for the health care occupancy. Egress through a separated occupancy is permitted only if via a horizontal exit. (See Section 16.3.3 of this book for more details on horizontal exits.)

13.6.3 Building Code Mixed Occupancies

Some of the model building codes treat multiple occupancies very similarly to the way the *Life Safety Code* and *NFPA 5000* do. One major difference, however, is that some model building codes do not permit the mixed occupancy concept at all, and separation is always required.

Some of the model building codes allow a certain percentage (e.g., 10 percent) of an area to be considered minor enough not to be called a different occupancy. For example, 10 percent of a hospital could be offices that do not comply with business occupancy requirements. The *Life Safety Code* does not have this 10 percent provision because the committee feels that this 10 percent provision may be too liberal in some cases. For example, a 500,000-ft^2 (46,450-m^2) warehouse could have 50,000 ft^2 (4645 m^2) of office area, which is too large an area not to meet the requirements for a business occupancy (i.e., a 500-person potential occupant load with no special protection). Conversely, the 10 percent provision may be too stringent in other situations, such as when a 100-ft^2 (9.3-m^2) office would be categorized as a business occupancy in a 900-ft^2 (83.6-m^2) storage building.

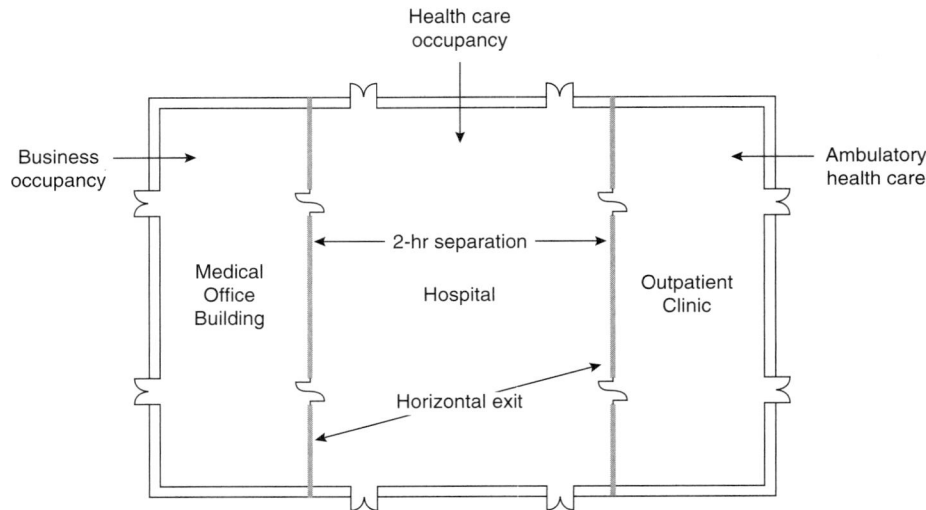

Figure 13.2 Separated occupancies.
Source: Opus Communications and Koffel Associates, Inc., redrawn from *2000 Life Safety Code Workbook & Study Guide for Health Care Facilities*, 2003, Figure 3.3.

13.7 New versus Existing

As mentioned previously, one of the major differences between the *Life Safety Code* and the model building codes is the *Life Safety Code*'s in-depth treatment of the requirements for both existing and new structures and equipment. The *Life Safety Code* has always covered both *new* and *existing*, but it was not until 1981 that it separated the occupancy requirements for new and existing into separate chapters. This separation makes it easier to use the *Life Safety Code*, as well as making the NFPA committee members who write the *Life Safety Code* more aware of when they add or revise requirements for existing structures. For example, the committees may add "a good idea" to the requirements for new structures but will wait until there is a well documented need before adding the new idea as a requirement for existing structures. As such, the differences between the requirements for new and existing structures are growing farther apart with each code cycle as more "good ideas" are added to *new* and more options for compliance are added to *existing*. It is therefore very important to understand the differences between the requirements for *new* and *existing* and how the code applies to each category.

The *Life Safety Code*'s definition of *existing* is basically something that is already in existence on the date when an authority puts into effect a specific edition of the *Life Safety Code* or that was erected or officially authorized prior to the date the *Life Safety Code* went into effect. There are several important issues related to this definition. The first issue relates to the date of adoption, which typically is not the date that NFPA issues the code. A facility's date of adoption of the *Life Safety Code* is generally based on how and when its authority having jurisdiction adopts the code. Some agencies adopt the *Life Safety Code* as soon as it is available; other agencies take many years. It is also not uncommon for jurisdictions to use very old editions of the *Life Safety Code*. For example, Subpart E of OSHA's *Means of Egress* regulation (29 CFR 1910.35–1910.40) was based on the 1970 edition of the *Life Safety Code* for decades. For many years the Centers for Medicare and Medicaid Services (CMS) used outdated editions of the code, some over 15 years out of date. JCAHO specifically puts the effective date of the edition of the *Life Safety Code* it is using on its Statements of Condition (SOC) forms. At present, the 2000 *Life Safety Code* is in effect at JCAHO (as well as by CMS). This edition is used to evaluate facilities as part of JCAHO accreditation (refer to Chapter 3 of this book).

It is also not uncommon for different agencies to apply different editions of the *Life Safety Code* to a facility that falls under more than one jurisdiction. It is helpful in such cases to try to obtain permission from the various agencies to use the most current edition that is referenced. The Committee on Safety to Life recognizes this dilemma and has added a note in the Annex of the code that emphasizes that newer editions of the *Life Safety Code* should be considered further refinements of prior editions. However, some authorities having jurisdiction may not allow the use of newer editions.

Once the effective date of the adoption is determined, figuring out whether a facility, building, or piece of equipment is *new* or *existing* is a straightforward process. If the building, for example, was in existence on the effective date of the *Life Safety Code,* it is considered to be *existing*. If it was not in existence at that time, it is considered to be *new*. The exception to this rule is for buildings that are officially approved prior to the effective date of the *Life Safety Code*. These structures and equipment are considered to be *existing* even though they are not yet built. For example, an architect working from the 2000 edition of the *Life Safety Code* had plans approved by the authority having jurisdiction. Before the building was actually built, however, the 2006 edition of the code went into effect. The building can continue to be built under the requirements for *new* for the 2000, not 2006, edition of the *Life Safety Code*. Any requirements for *existing* in the 2006 edition that are more stringent than those for *new* in the 2000 edition must be complied with, but this is a rare circumstance.

This rule basically means that every time a new edition of the *Life Safety Code* is adopted, all buildings that must comply with this code that were considered *new* become *existing*. For example, on March 1, 2003, JCAHO adopted the 2000 edition of the *Life Safety Code*. At that time, any health care facility building that was in existence (or for which plans had been approved) became *existing*. This shift can create a problem in some facilities because in many instances the differences between *new* and *existing* are significant, as illustrated in the following case study.

Complying with Codes for New versus Existing

The requirements for smoke barriers in a health care occupancy exemplify the classic problems related to compliance with the requirements for *new* versus *existing*. In new construction, smoke barriers are essentially required on all stories, but for existing conditions a smoke barrier is required only on patient sleeping floors, and then only when there are more than 30 patients. One hospital, built in 1978 under the 1976 edition of the *Life Safety Code,* had smoke barriers installed on all stories, complying with the conditions of *new* construction. Now, the building is *existing* and barriers are required only on patient sleeping floors having more than 30 patients. However, the *Life Safety Code* prohibits the facility from removing the barriers or stopping maintenance of them, even though they are not required for existing buildings. That is, the *Life Safety Code* says that any renovation or alteration (or installation of new equipment) must be done to meet the requirements for *new*. This is a very important concept in that it prevents facility managers from reducing the level of protection in a building when a new edition of the *Life Safety Code* is adopted. The facility does not have to upgrade to *new,* but it cannot downgrade to *existing*. If it does not meet the minimum requirements for *existing*, it must be upgraded.

In essence, the provisions for *existing* apply to existing conditions only. Likewise, the provisions for *new* can apply to any change that is being made.

It is also essential to determine carefully the requirements for areas categorized as hazardous. When a patient room is converted into a storage room, the area is considered to be a new hazardous area and must meet the requirements for *new,* not *existing*. A change in occupancy also can affect compliance with the codes for *new* versus *existing* prior to the 2006 edition. The *Life Safety Code* required that any time an occupancy is changed, the requirements for *new* must be met. Classic examples in the health care field of changes in occupancy are converting a one- or two-family dwelling into an office building (i.e., the area must meet the requirements for a new business occupancy), or converting a patient story or wing to offices (i.e., this area must meet the requirements for new business, which if it is nonsprinklered could cause significant corridor violations). With the 2006 edition, Chapter 43 provides specific provisions for change of occupancy, which may or may not require compliance with new.

Another case study shows some other considerations that must be made when converting occupancies.

Converting Occupancies

Due to downsizing, a small city hospital wanted to convert several patient rooms into storage areas. Since the rooms were not designed for enclosing a hazardous area, several problems arose. The first problem was to determine if the area was *new* or *existing*. Since the *hazardous areas* were new, the requirements for *new* needed to be applied. Since the rooms were to be used for storage and were over 100 ft² (9.3 m²) in area (see Chapter 18 of the 2006 edition of the *Life Safety Code*), the room needed to be both protected by sprinklers and separated by 1-hour walls with 45-minute self-closing, self-latching doors. Even if judged *existing*, either sprinkler protection with self-closing doors or 1-hr separation with 45-minute self-closing, self-latching doors would be required. Since most patient sleeping rooms do not meet any of these requirements, converting a patient sleeping room to storage is almost always going to require some renovation. In this case, the solution was to treat the entire story as another occupancy (i.e., there was 2-hour separation and the area was not used for patient sleeping, treatment, or customary access), which avoided the special hazard protection. In other words, rather than separating the room, the entire story (which was already sprinklered) was separated.

One problem in dealing with new versus existing is with renovations. In general, renovations must comply with the requirements for new construction. However, in the *Life Safety Code*, there are special provisions for renovations. In addition to the renovation itself complying with the requirements for new construction, if the renovation is a "major" renovation, the smoke compartment containing the renovation must be protected throughout with sprinklers meeting the requirements for new construction. This results in two problems. First, what is "major," and second, what to do if it is not major. Starting with the 2003 edition of the *Life Safety Code*, both of these questions have been answered. A "major" renovation is now defined as one exceeding 4500 ft² or 50 percent of the size of the smoke compartment, whichever is smaller. Work that is exclusively plumbing, mechanical, fire protection system, electrical, medical gas, or medical equipment is not included in the computation of the modification area within the smoke compartment. If it is a major rehabilitation, not only must the new work meet the *Code* requirements for new, but the smoke compartment must be protected by automatic sprinklers.

The 2003 edition of the *Code* also improved significantly how to do minor renovations in nonsprinklered smoke compartments. In the past one had to really dig and read between the lines to figure out what to do. Starting with the 2003 edition, a new section was added to Chapter 18 (18.4.3) specifically delineating the requirements for minor renovations in nonsprinklered smoke compartments.

With the introduction of *NFPA 5000*, the concept of "building rehabilitation" has been brought into the NFPA codes and standards. This takes what was simple "alterations, modernizations, and renovations" in the past into a broader subject of "building rehabilitation." Building rehabilitation addresses:

Repair: the patching, restoration, or painting of materials, elements, equipment, or fixtures for the purpose of maintaining such materials, elements, equipment, or fixtures in good or sound condition

Renovation: the change, strengthening, or addition of load-bearing elements; refinishing, replacement, bracing, strengthening, upgrading of existing materials; elements, equipment, and/or fixtures, without reconfiguration of spaces

Modification: reconfiguration of any space, the addition or elimination of any door or window, the reconfiguration or extension of any system, or the installation of any additional equipment

Reconstruction: reconfiguration of a space that affects an exit, or a corridor shared by more than a single tenant; or reconfiguration of space such that the rehabilitation work area is not permitted to be occupied because existing means of egress and fire protection systems, or their equivalent, are not in place or continuously maintained

Change of use: a change in the purpose or level of activity within a structure that involves a change in application of the requirements of the code

Change of occupancy classification: a change of use involving a change in the occupancy classification of a structure or portion of a structure

Addition: an increase in building area, aggregate floor area, height, or number of stories of a structure

Historic building: a building or facility deemed to have historical, architectural, or cultural significance by a local, regional, or national jurisdiction

This has significantly improved how to work with the code when dealing with work in existing buildings. This concept was incorporated into the *Life Safety Code* in the 2006 edition as Chapter 43.

14

Construction Types

Because health care occupancies deal with people who might be incapable of self-preservation, these facilities are often referred to as *defend-in-place occupancies*. Rather than evacuating the patients from the building, they are protected within the building, reducing the need to evacuate. Therefore, the building must be able to resist a fire long enough that its occupants will be safe within the structure until the fire is extinguished.

Many factors come into play to build a defend-in-place facility. Using specific fire-resistive materials is one aspect of accomplishing this task. How these materials are put together is another. Features such as the use of sprinklers, the number of stories, its compartmentation and use of smoke barriers, and how the openings in these barriers are handled are just a few of the many factors that contribute to the structure's ability to provide a viable defend-in-place facility.

With consideration of a building's materials and range of features and using a series of standardized tests, some code-making bodies have determined the fire resistance of various types of structures. These ratings are used to categorize buildings by construction type and are referenced within the codes to make sure that new and existing structures are built according to specified standards. That is, the codes specify how to construct new and maintain existing buildings and what materials to use to obtain a certain fire resistance.

In order to evaluate compliance with NFPA *101*®, *Life Safety Code*®, and to properly perform a JCAHO Statement of Conditions (SOC), architects, builders, and facility engineers and managers must know the construction type for all buildings containing health care, ambulatory health care, or assembly occupancies. It is also important to know how construction types are used by the *Life Safety Code* and in typical building codes and how to determine construction type for a particular structure. In addition to the *Life Safety Code*, many other codes and handbooks must be used, including the following:

- National Fire Protection Association
  - *Fire Protection Handbook*
  - *Life Safety Code*® *Handbook*
  - NFPA 80, *Standard for Fire Doors and Fire Windows*
  - NFPA 90A, *Standard for the Installation of Air-Conditioning and Ventilating Systems*

- NFPA 220, *Standard on Types of Building Construction*
- NFPA 221, *Standard for High Challenge Fire Walls, Fire Walls, and Fire Barrier Walls*
- NFPA 251, *Standard Methods of Tests of Fire Resistance of Building Construction and Materials*
- NFPA 252, *Standard Methods of Fire Tests of Door Assemblies*
- NFPA 257, *Standard on Fire Test for Window and Glass Block Assemblies*
- NFPA 259, *Standard Test Method for Potential Heat of Building Materials*
- NFPA 909, *Code for the Protection of Cultural Resource Properties—Museums, Libraries, and Places of Worship*
- American Society for Testing and Materials
  - ASTM E 84, *Standard Test Method for Surface Burning Characteristics of Building Materials*
  - ASTM E 119, *Standard Methods for Fire Tests of Building Construction Materials* (similar to NFPA 251)
  - ASTM E 814, *Standard Test Method for Fire Tests of Through-Penetration Fire Stops*
- Underwriters Laboratories
  - *UL Fire Resistance Directory*
  - UL 9, Standard for *Fire Tests of Window Assemblies* (similar to NFPA 257)
  - UL 10B, Standard for *Fire Tests of Door Assemblies* (similar to NFPA 252)
  - UL 263, Standard for *Fire Tests of Building Construction and Materials* (similar to NFPA 251)
- Gypsum Association
  - *Fire Resistance Design Manual*

(Refer to Chapter 1 for a more comprehensive list of codes and standards that are used to ensure fire and life safety in health care facilities.)

This chapter addresses fire resistance, construction types, and several other features related to determining construction type. How the code-making bodies deal with inconsistencies among the relevant codes and how to determine a building's construction type in spite of the inconsistencies are also addressed. One main factor in determining a building's capability to contain fire, its compartmentation, is addressed in Chapter 15.

14.1 Defend-in-Place versus Evacuation

The *Life Safety Code* recognizes that to protect people from fire, they can be moved out of harm's way (usually out of the building) or they can be defended in place. There are actually several levels of doing this:

- Defend a person where he is situated (i.e., defend-in-place). That is, people will not be evacuated from the room they are in. Such protection is accomplished by a combination of active and passive fire protection systems working in conjunction to limit the spread of a fire and its combustion products. An example is to leave the patient in his room with the door to the corridor closed. This relies on a proper

response from the facility staff and on the corridor walls and other features to keep fire and smoke away from the patient, depending on where the fire is in the building.
- Move a person to a safe area but not out of the building. There are many ways this can be accomplished, depending on the features provided in the building and where the fire is. The simplest is just to move the patient out to the corridor and close the room door, presuming the fire is in that patient's room. Another would be moving the patient past a smoke barrier or horizontal exit. Moving patients to another floor starts to become difficult unless they can move on their own. An example of where this approach might be used is in a high-rise medical office building, where people are moved from the fire floor and adjacent floors to lower floors in the building.
- The last, and least desirable approach for most nonambulatory patients, is building evacuation, which takes place when several fire protection measures are compromised. The decision to evacuate a health care facility during a fire should be made only as a last resort; due to the special needs of patients and residents, the environment exterior to the health care facility can pose a significant risk to their health. Although this is the most common option chosen for non–health care (or nondetention) occupancies, building evacuation can be very time consuming and problematic—even life threatening—for health care recipients.

In actuality, health care facilities, especially buildings containing health care occupancies, are supposed to be designed to provide all three options for protection. Features that relate to making a structure fire resistive are discussed in this chapter as well as in Chapter 15 on compartmentation and barriers. Features of buildings that allow safe egress either to safe areas within a structure or to the outdoors are discussed in Chapters 16 and 17.

14.2 Fire Resistance

Fire resistance relates to the time in minutes or hours that materials or assemblies have withstood a fire exposure when tested in accordance with the procedures specified in NFPA 251, also published as ASTM E 119 and UL 263. This test exposes the building element or system to a fire that is developed based on the standard *time–temperature curve*. It is important to understand that hourly fire resistance ratings based on this standard fire test are not necessarily the number of hours an element will last in a fire. An actual fire may be more or less severe than this curve. For example, at 2 hours this curve hits 1850°F (1009°C), which could easily be exceeded long before 2 hours, depending on the fuel package. These fire resistance ratings are found in the listings of various testing laboratories or in product guides. Figure 14.1 depicts the standard time–temperature curve used in NFPA 251.

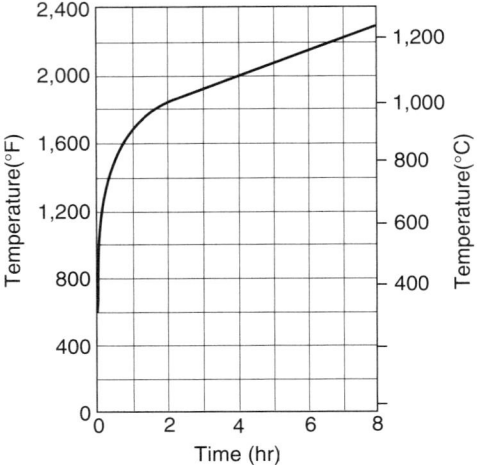

Figure 14.1 Standard time–temperature curve.
Source: NFPA 251, 1999 edition, Figure 201.1.

14.3 Number of Stories and Heights and Areas

Height is another factor that determines a building's construction type. Under the *Life Safety Code,* building height for health care occupancies is based on the number of stories counting up from the primary level of exit discharge. Where there is no primary level of exit discharge, the lowest story is considered to be the floor that is level with or above finished grade on the exterior wall line for 50 percent or more of its perimeter.

The heart of most building codes, including all major model building codes used in the United States, is the *tables of heights and areas.* These tables establish the maximum height and area allowed for each occupancy type based on building construction type. There are usually modifiers based on the use of sprinkler protection, types of building access, and clear space around the building. Unlike in the *Life Safety Code,* all occupancies are included in the heights and areas table of a building code.

NFPA *101* limits the total number of stories for buildings that contain a health care occupancy. Two tables in NFPA *101,* one for new construction and one for existing buildings, provide story limitations for various construction types. These tables are reproduced here as Tables 14.1 and 14.2, for new construction and existing buildings, respectively. In the tables, the Roman numeral designates the basic construction type.

Table 14.1 Construction Type Limitations for New Construction

Construction Type	1	2	3	4 or More
I (442)	X	X	X	X
I (332)	X	X	X	X
II (222)	X	X	X	X
II (111)	X	X	X	NP
II (000)	X	NP	NP	NP
III (211)	X	NP	NP	NP
III (200)	NP	NP	NP	NP
IV (2HH)	X	NP	NP	NP
V (111)	X	NP	NP	NP
V (000)	NP	NP	NP	NP

(Stories across columns 1–4)

X = Permitted type of construction
NP = Not permitted
Source: Life Safety Code, 2006 edition, Table 18.1.6.4.

Table 14.2 Construction Type Limitations for Existing Buildings

Construction Type	1	2	3	4 or More
I (442)	X	X	X	X
I (332)	X	X	X	X
II (222)	X	X	X	X
II (111)	X	X*	X*	NP
II (000)	X*	X*	NP	NP
III (211)	X*	X*	NP	NP
III (200)	X*	NP	NP	NP
IV (2HH)	X*	X*	NP	NP
V (111)	X*	X*	NP	NP
V (000)	X*	NP	NP	NP

(Stories across columns 1–4)

X = Permitted type of construction
NP = Not permitted
* Building requires automatic sprinkler protection. (See Section 19.3.5.1 of NFPA *101*, 2000.)
Source: Life Safety Code, 2006 edition, Table 19.1.6.4.

The three Arabic numbers in the parenthetical expression that follows the Roman numeral represent the hourly fire resistance rating for three building components, including the building exterior bearing walls, structural frame/columns/girders supporting more than one floor, and floor construction for the basic building types (i.e., Types I–V; see Section 14.4).

Note that all new health care construction requires sprinkler protection, regardless of building height or construction type. All the U.S. model building codes also require sprinkler protection in new health care construction. For existing health care occupancies, sprinklers may be required due to a combination of certain types of construction and building heights, or for certain allowances, such as corridor wall construction (refer to Chapters 15 and 16) and travel distance (see Chapter 17). In the 2006 edition of the *Life Safety Code*, all existing nursing homes, regardless of construction type, must be protected by automatic sprinkler protection. This change was promulgated following the multiple fatality nursing home fires in Hartford, Connecticut, and Nashville, Tennessee, in 2003.

14.4 Construction Types

Two main criteria for determining construction types are the type of material used (i.e., whether the material is combustible) and the degree of fire resistance of the construction. There are five traditional types of construction, as identified by different designators and depending on the code being used. The generic terms that were used in the past to describe traditional construction types are *fire resistive, noncombustible, ordinary, heavy timber,* and *wood frame*. In general, the codes no longer use these terms, but they do describe the basic types of construction using numerals, as discussed in the following subsections. Types of construction range from those built completely with heavily fire-resistive noncombustible materials to those that are built with nonrated combustible materials.

When a new or existing health care occupancy (or any of the other defend-in-place occupancies or slower-evacuating occupancies regulated by the *Life Safety Code*) requires compliance with a particular construction type, the *Life Safety Code* refers to NFPA 220 (Chapter 8). Table 14.3 is a reprint of NFPA 220, 2006 edition, Table 3.1, which is also included in Annex A of NFPA *101*, 2006 edition (Table A.8.2.1.2). This table provides fire resistance ratings required for various building components based on building construction types, as shown in Tables 14.1 and 14.2. It is necessary to refer to these tables to determine the required fire resistance of all elements.

14.4.1 Fire-Resistive Structures

Fire-resistive structures are intended to withstand significant fire exposure without losing structural integrity. The structures do not contribute fuel to the fire and tend to be made of noncombustible materials, such as concrete or steel with added fire protection, sometimes sprayed on. (Note that under the NFPA system, this category includes limited-combustible materials also; see Section 14.5.2 for more details.) The building codes go into detail as to combustible materials that are allowed in the construction of these buildings. Most structural elements of these buildings have at least a 2-hour fire resistance rating, with some elements being required to have fire resistance ratings as high as 4 hours.

Most of the codes refer to fire-resistive buildings as Type 1 or I construction. The degree of fire resistance in these structures is indicated in different ways. For example, NFPA *101,* which references NFPA 220, and *NFPA 5000*®, *Building Construction and*

Chapter 14. Construction Types **209**

Table 14.3 Fire Resistance Ratings for Type I Through Type V Construction (hours)

	Type I		Type II			Type III		Type IV	Type V	
	442	332	222	111	000	211	200	2HH	111	000
Exterior bearing walls[a]										
Supporting more than one floor,										
columns, or other bearing walls	4	3	2	1	0[b]	2	2	2	1	0[b]
Supporting one floor only	4	3	2	1	0[b]	2	2	2	1	0[b]
Supporting a roof only	4	3	1	1	0[b]	2	2	2	1	0[b]
Interior bearing walls										
Supporting more than one floor,										
columns, or other bearing walls	4	3	2	1	0	1	0	2	1	0
Supporting one floor only	3	2	2	1	0	1	0	1	1	0
Supporting roofs only	3	2	1	1	0	1	0	1	1	0
Columns										
Supporting more than one floor,										
columns, or other bearing walls	4	3	2	1	0	1	0	H	1	0
Supporting one floor only	3	2	2	1	0	1	0	H	1	0
Supporting roofs only	3	2	1	1	0	1	0	H	1	0
Beams, girders, trusses, and arches										
Supporting more than one floor,										
columns, or other bearing walls	4	3	2	1	0	1	0	H	1	0
Supporting one floor only	2	2	2	1	0	1	0	H	1	0
Supporting roofs only	2	2	1	1	0	1	0	H	1	0
Floor-ceiling assemblies	2	2	2	1	0	1	0	H	1	0
Roof-ceiling assemblies	2	1½	1	1	0	1	0	H	1	0
Interior nonbearing walls	0	0	0	0	0	0	0	0	0	0
Exterior nonbearing walls[c]	0[b]	0[b]	0[b]	0[b]	0[b]	0[b]	0[b]	0[b]	0[b]	0[b]

H = Heavy timber members (see text of NFPA 220 for requirements).
[a] See *NFPA 5000*, Section 7.3.2.1.
[b] See *NFPA 5000*, Section 7.3.
[c] See Sections 4.3.2.12, 4.4.2.3, and 4.5.6.8. of NFPA 220.
Source: NFPA 220, 2006, Table 4.4.1.

Safety Code®, use the terms Type I (442) or Type 1 (332) (see the discussion on NFPA 220 that follows), while the International Code Council's (ICC's) *International Building Code*® uses the terms Type 1A or Type 1B to indicate the degree of fire resistance.

14.4.2 Noncombustible Structures

All materials used in the structure of these buildings are noncombustible (the NFPA system includes limited-combustible materials also; see Section 14.5.2). This type of construction ranges from no fire resistance to significant fire resistance with 2 hours of protection. However, these structures, which are commonly built of steel with no or some fire resistance often added through encapsulation or by membrane protection, do not contribute fuel to the fire.

Most of the codes refer to this as Type 2 or II construction. NFPA *101* and *NFPA 5000* use the terms Type II (222), Type II (111), or Type II (000) to indicate the degree of fire resistance, while the ICC *International Building Code* uses the terms Type IIA or Type IIB to indicate the fire resistance ratings. Note that with Type II (000) construction, building collapse could occur fairly early in the fire. As with fire-resistive construction, the building codes go into detail as to combustible materials that are allowed in the construction of these buildings.

14.4.3 Ordinary Structures

Exterior walls of ordinary structures are made of noncombustible materials, usually brick or masonry block. The floors and roof are made of combustible (wood) construction and contribute fuel to the fire. This type of construction was very common around the turn of the last century and is typical of "Main Street U.S.A." From a life safety aspect, there is very little difference between ordinary construction and wood frame construction (see Section 14.4.5).

Ordinary structures can have either no fire resistance or 1 hour of fire resistance. Most of the codes refer to ordinary structures as Type 3 or III construction. NFPA *101* and *NFPA 5000* use the terms Type III (211) or Type III (200) to indicate the degree of fire resistance, while the ICC *International Building Code* uses the terms Type IIIA or Type IIIB.

14.4.4 Heavy Timber Structures

Heavy timber construction is also commonly referred to as mill construction or New England mill construction, as it was popular in the nineteenth and early twentieth centuries, especially in New England, for the construction of mills. Most of the codes refer to heavy timber construction as Type 4 or IV construction. NFPA 220 refers to heavy timber construction as Type IV (2HH) construction, while the ICC *International Building Code* uses the term Type 4 HT.

Exterior and interior bracing walls of heavy timber structures are made of noncombustible material, usually stone or brick. The floors and roof are made of wood construction but have very large dimensions. This is why the NFPA uses the term *HH designator,* and ICC *International Building Code* uses the term *HT designator* to refer to heavy timber, which has a degree, although unspecified, of fire resistance. What makes this type of structure different from ordinary construction is the timber sizes involved and the restrictions on concealed spaces. The following excerpts from NFPA 220 illustrate the requirements for the size of timber that can be used for heavy timber construction:

- Section 3.4.2—Wood columns supporting floor loads shall be not less than 8 in. (203 mm) in any dimension; wood columns supporting roof loads only shall be not less than 6 in. (152 mm) in the smallest dimension and not less than 8 in. (203 mm) in depth.
- Section 3.4.3—Wood beams and girders supporting floor loads shall be not less than 6 in. (152 mm) in width and not less than 10 in. (254 mm) in depth; wood

beams and girders and other roof framing, supporting roof loads only, shall be not less than 4 in. (102 mm) in width and not less than 6 in. (152 mm) in depth.
- Section 3.4.5—Floors shall be constructed of splined or tongued and grooved plank not less than 3 in. (76 mm) in thickness that is covered with 1-in. (25-mm) tongue and groove flooring, laid crosswise or diagonally to the plank.
- Section 3.4.6—Roof decks shall be constructed of splined or tongued and grooved plank not less than 2 in. (51 mm) in thickness.

Due to the large-geometry wood members, heavy timber construction lost popularity in the mid-1900s, but with the advent of "glue-lam" technology (in which smaller pieces of wood are actually "glued" together to form larger beams) it has become popular again.

Heavy timber construction is usually credited with having 1-hour fire resistance because of the size of the wood members. Due to the lack of concealed spaces in heavy timber construction and the size of the wood members used, these buildings are quite resistant to ignition. However, once these buildings get burning, the fire can be rather spectacular and of long duration. Figure 14.2 illustrates heavy timber construction.

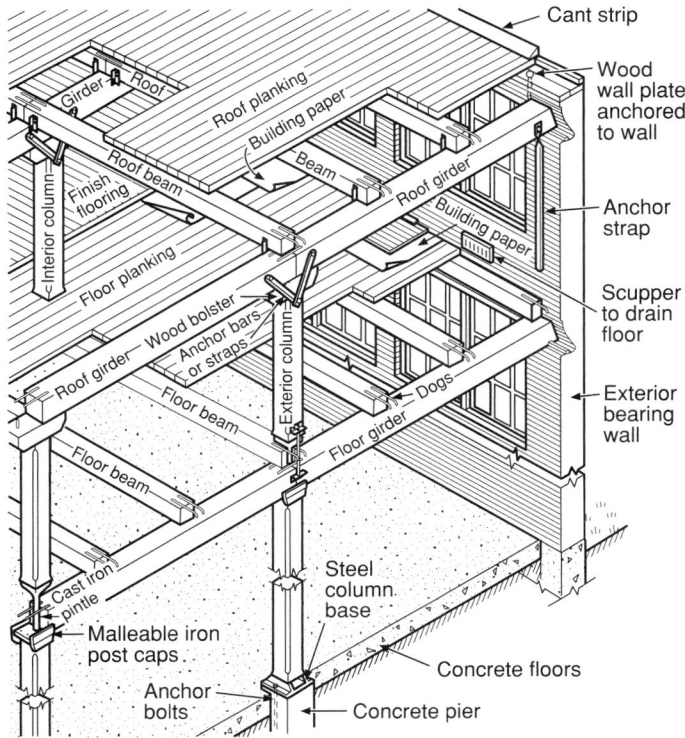

Figure 14.2 Components of heavy timber construction.
Source: Fire Protection Handbook, NFPA, 2003, Figure 12.2.27.

14.4.5 Wood Frame Structures

Wood frame construction is probably the most common type found in North America. Both walls and floors of wood frame structures are of wood construction, which can contribute fuel to the fire. This type of construction can have a rating of either no fire resistance or 1 hr of fire resistance.

Most of the codes refer to wood frame structures as Type 5 or V construction. NFPA *101* and *NFPA 5000* use the terms Type V (111) or Type V (000) to indicate the degree of fire resistance, while the ICC *International Building Code* uses the terms Type 5A or 5B. Figure 14.3 illustrates wood frame construction.

14.5 Construction Type Consistency and Conversions

Since the 1980s, much has been done to make the construction types allowed by NFPA codes and the major model building codes compatible for the requirements for health care occupancies. Typical building codes provide additional detail on materials that are allowed in each type of construction, such as to specifically allow the use of

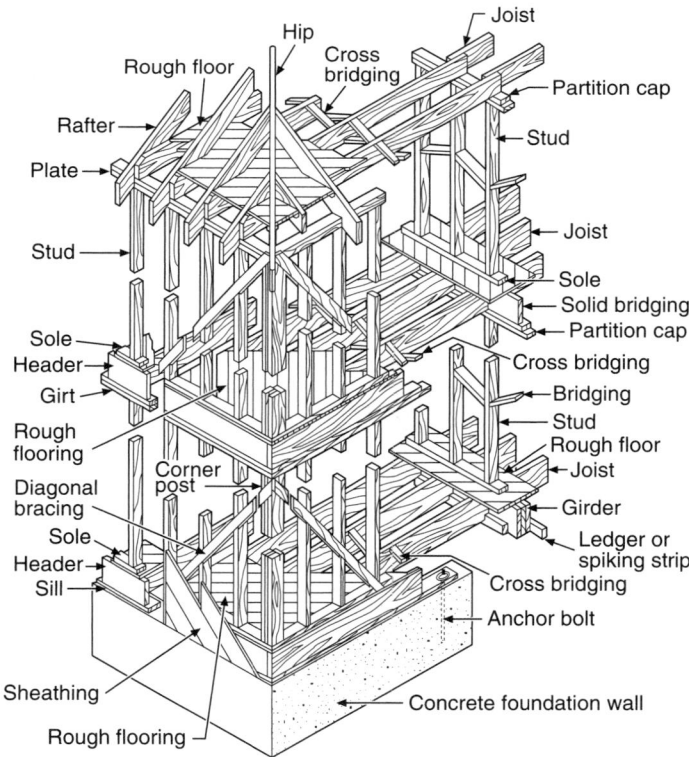

Figure 14.3 Components of wood frame platform construction.
Source: Fire Protection Handbook, NFPA, 2003, Figure 12.2.28.

fire retardant–treated wood or insulating plastics in buildings of noncombustible construction. This level of detail, along with subtle differences in the fire resistance required for various construction types, can make it difficult to design health care facilities and comply with the codes. State and local amendments to the model codes can sometimes make it difficult to comply with fire resistive requirements and building types, as well.

It is also important to note that many local codes are based on older editions of the model codes that were written before efforts were made to correlate the codes.

14.5.1 Fire Resistance Rating Conversions

As discussed previously, the various codes use slightly different symbols to indicate essentially the same type of construction. Table 14.4 provides a simple quick reference to which type of construction in each code is similar to the others and, more important, to NFPA 220. For example, Type 2B buildings under the *BOCA® National Building Code* are very close to Type II (111) buildings in NFPA 220, and therefore, NFPA *101*. Note that all but the *Standard Building Code®*, 1997 edition, use the same numbers, some Arabic, some Roman, with different methods of designating the fire resistance. The numbering system for the *Standard Building Code* does not follow this pattern. Although there are items unique to each code, Table 14.4 provides a general cross-reference. For example, NFPA 220 now does allow some partitions in Types I or II construction to be made of combustible materials (with some exceptions for fire retardant–treated lumber), and the model codes allow partitions to be made of fire-retardant lumber under certain conditions. Note that the terms have changed in some of the codes over the years, and some older plans may use other designations that were previously used in the codes—for example, the BOCA code used to have only four types of construction.

To determine the construction type of an existing structure with a particular fire resistance, the conversions given in Table 14.4 are usually sufficient as long as there is some verification by inspection. However, for determining the construction type for new construction, the project architect should be aware that the project must comply with the fire resistance ratings not only in the applicable building code but also in the *Life Safety Code*, which may include more detail than the building code or may be more stringent in other areas.

Table 14.4 Cross-Reference of Building Construction Types

NFPA 220	I (442)	I (332)	II (222)	II (111)	II (000)	III (211)	III (200)	IV (2HH)	V (111)	V (000)
IBC	—	IA	IB	IIA	IIB	IIIA	IIIB	IVHT	VA	VB
UBC	—	IFR	II FR	II 1-hr	II N	III 1-hr	III N	IV HT	V 1-hr	V-N
BNBC	1A	1B	2A	2B	2C	3A	3B	4	5A	5B
SBC	I	II	—	IV 1-hr	IV unp	V 1-hr	V unp	III	VI 1-hr	VI unp

Source: Life Safety Code Handbook, 2003, Table 8.1.
IBC = *International Building Code*; UBC = *Uniform Building Code*; BNBC = *BOCA National Building Code*; SBC = *Standard Building Code*.

14.5.2 Combustible, Noncombustible, and Limited-Combustible Materials

One major difference between NFPA 220, and therefore the *Life Safety Code* and *NFPA 5000*, and the other U.S. model building codes is that the other model building codes refer to combustible and noncombustible materials, whereas NFPA 220 refers to a third type, limited-combustible materials. The *Life Safety Code* defines a limited-combustible material as follows:

> A building construction material not complying with the definition of noncombustible (see 3.3.15.3) that, in the form in which it is used, has a potential heat value not exceeding 3500 Btu/lb (8141 kJ/kg), where tested in accordance with NFPA 259, *Standard Test Method for Potential Heat of Building Materials*, and includes either of the following:
> (1) Materials having a structural base of noncombustible material, with a surfacing not exceeding a thickness of 1/8 in. (3.2 mm) that has a flame spread index not greater than 50.
> (2) Materials, in the form and thickness used, having neither a flame spread index greater than 25 nor evidence of continued progressive combustion, and of such composition that surfaces that would be exposed by cutting through the material on any plane would have neither a flame spread index greater than 25 nor evidence of continued progressive combustion. [101–06: 3.3.150.2]

One common material that NFPA 220, *NFPA 5000*, and NFPA *101* consider to be limited combustible is gypsum board. In most cases, limited-combustible material is allowed wherever noncombustible is mandated, but there are exceptions.

NFPA 220 defines a noncombustible material as follows:

> A material that, in the form in which it is used and under the conditions anticipated, will not ignite, burn, support combustion, or release flammable vapors, when subjected to fire or heat. Materials that are reported as passing ASTM E 136, *Standard Test Method for Behavior of Materials in a Vertical Tube Furnace at 750°C*, shall be considered noncombustible materials. [101–06: 3.3.150.3]

Note that the *Life Safety Code* contains very stringent regulations for combustible construction and unprotected noncombustible construction. This is because patients may need to be protected in place.

14.6 Determining Construction Type

Several steps must be taken to determine the construction type of a particular building; some are straightforward, whereas others are more time consuming. If they are available, original building plans (often on the first page or soon thereafter) should indicate the construction type based on the code used at the time of construction. If the type is not already in NFPA 220 format, the next step is to convert it to this format using the conversion chart presented in Table 14.4. Then, because some of the model codes have modified the criteria for construction types over the years, the next step in determining the construction type of the building is to confirm that the type indicated is consistent with the type required by the code. Also, the construction type may have been affected if a building was not built according to the plans, or if changes were made to the structure over the years. For example, an asbestos removal program may have removed protection that was not replaced, or a renovation may have removed a ceiling that was providing some of the fire resistance for the floor or roof above. Thus, even if building plans are available that indicate the construction type, it will usually be necessary to inspect a facility fully in order to accurately determine the construction type of a structure.

If the construction type is not indicated on the plans, a thorough review of the plans along with field confirmation (i.e., inspection) can sometimes reveal the construction type. Reviewing the details on the drawings to try to determine construction materials and hourly fire resistance can assist greatly in this effort. In many instances, the designer refers to the use of design assemblies from documents such as Underwriters Laboratories' *UL Fire Resistance Directory*, or the Gypsum Association's *Fire Resistance Design Manual* directly on the plans. Even with this information on the plans, however, a field inspection must be conducted to establish that the building was built and maintained in accordance with the plans.

If plans are not available, it is necessary to conduct a thorough field inspection to document existing conditions and determine fire resistance. Again, documents such as the *UL Fire Resistance Directory* and the GA's *Fire Resistance Design Manual* are needed. Also, most U.S. model building codes and NFPA's *Fire Protection Handbook* provide typical assembly fire resistance ratings.

NFPA 909 contains extensive material to assist in determining construction (in Appendix L, "Guideline on Fire Ratings of Archaic Materials and Assemblies"). This material provides a methodology for evaluating building construction along with assistance in determining the fire resistance of materials that are no longer commonly used in construction. This is a valuable tool for anyone trying to determine the fire resistance of materials that are no longer commonly used, such as lath and plaster construction, terra cotta tile, and many other materials.

In many instances, facility personnel ask a consulting architect or engineer to make the determination about construction type. It is important to note that the *Life Safety Code* (but in this case, not *NFPA 5000*) treats building Types I (442), I (332), and II (222) the same. From the point of view of the *Life Safety Code*, and therefore JCAHO, there is no need to determine fire resistance ratings longer than 2 hours. The building code may require this determination, but the *Life Safety Code* does not. There have been numerous occasions on which a facility has gone to great efforts to determine whether a building was Type II (222) or Type I (332) when the effort was not needed.

14.6.1 Construction Types for Various Occupancies

It should also be noted that both ambulatory health care and assembly occupancies have requirements based on construction type that are different from those for health care occupancies. The requirements for ambulatory health care are quite minimal. Any type of construction recognized by NFPA 220 can be used, but if it is an unprotected type of construction [i.e., Type II (000), Type III (200), or Type V (000)] and the structure is two or more stories in height, then automatic sprinkler protection is required. As with health care facilities, this requirement is based on the number of stories in the building, starting at the primary level of exit discharge. The requirements for assembly occupancies differ significantly. The tables in the *Life Safety Code* chapters for new and existing assembly occupancies are based on the location of the assembly occupancy in the building in combination with the size of the assembly occupancy.

14.6.2 Multiple Construction Types

Many health care facilities have multiple construction types, particularly older facilities that have undergone numerous additions and renovations. To determine the construction type when multiple types are present, the overall building construction type is based on the least type involved. For example, if a building is basically Type I (332) but part of the building is Type II (000), then the whole building must be considered to be Type II (000). An exception is provided for wood roofing systems, supports, decking, or roofing. It is not uncommon to find buildings of Type I or Type II construction with a flat roof deck that has a wood peaked roof for water drainage. The *Life Safety Code* allows this type of construction, provided there is a substantial *floor* separating the peaked roof from the rest of the building and that this floor has a 2-hour fire resistance rating. The concept behind this regulation, and actual fires have shown it to be valid, is that the roof can burn off with little impact on the facility below. Beginning with the 2000 edition of the *Life Safety Code,* the code clarifies that the separation of construction types by walls that have at least a 2-hour fire resistance rating is valid; it has been a common practice but was not so stated in the code previously. In new construction, the wall separation must be a vertically aligned fire barrier wall in accordance with NFPA 221. Any previously approved separation between building construction types is not affected by this new provision.

There has also been some discussion among professionals in the field regarding the use of a 2-hour horizontal barrier to separate construction types. This type of barrier may be valid for situations such as a different type of roof construction, as well as other situations where the *lesser* type of construction (i.e., the less fire-resistive type) is above the better type. The *Life Safety Code* addresses this. However, professionals concur that 2-hour horizontal barriers should not be given general credit for separating construction types, especially when the lesser-type construction is below the more fire-resistive type. The following case study shows an example of how to deal with these types of multiple construction types.

Multiple Construction Types

The buildings in an older hospital complex that has undergone numerous additions are primarily of Type I construction, but they also contain wood structural members

(Continued)

in several areas, which have posed a problem. Without some form of plan, the buildings involved would have had to be reduced to Type III construction at best. The first location involving wood is in the roof framing. This area was easily handled by applying the exception in the *Life Safety Code* that allows the roof to contain wood members as long as the slab between the roof and the hospital is of at least 2-hour construction. In the other location, the wood structural members are on the fourth and fifth floors of a five-story building. It appears that the wood was originally intended to be a temporary bridge between two sections of the fourth and fifth floors, but was never removed during further additions and has remained an island of wood construction in a five-story Type I building. Since the barriers on all three connected sides of the wood portion are of 2-hour fire resistance rating and the floors below are also so rated, it was decided to treat this area as a separate building. Since there are no patient rooms (either treatment or sleeping) in the small portion of the building, it can also be treated as a business occupancy. Since it is separated by 2-hour walls and is not below any other construction type or occupancy, it presents little if any hazard to health care occupants.

15

Compartmentation

A major building feature in determining the safety of a structure is its compartmentation. Many different types and sizes of compartments are present in all types of structures. An entire building can be considered one compartment. Conversely, a small room categorized as a hazardous area can be a compartment. Health care facility compartments are typically divided vertically by floors and shaft enclosures and horizontally by room corridor walls, smoke barriers, and horizontal exits.

The basic concept of compartmentation is to minimize or prevent fire and smoke from spreading from one compartment to another, that is, to keep the fire and its products of combustion in the area in which it started.

Figure 15.1 illustrates this concept. Needless to say, compartmentation is an important concept in any occupancy, but even more so when dealing with defend-in-place occupancies, such as health care occupancies.

The building codes and NFPA *101*®, *Life Safety Code*®, deal with two different types of compartmentation—fire compartments and smoke compartments—and many buildings contain a combination of the two types. Installation of horizontal and vertical barriers form fire and smoke compartments that prevent spread of fires and resist passage of smoke, respectively. Typical horizontal fire and smoke barriers are floors, ceilings, and floor/ceiling assemblies. Typical vertical fire barriers are walls, also called fire walls, area separation walls, fire barrier walls, or fire partitions. Some walls are also considered to be vertical smoke barriers, also called smoke partitions. Smoke barriers and partitions are discussed in more detail in Chapter 18.

Barriers must meet a series of requirements based on how the compartments are used. Fire barriers must maintain a certain degree of fire resistance, while smoke barriers require only that the barrier resists the passage of smoke unless a fire rating is also specified. Openings in the barriers, including doors, windows, and vents, must be handled accordingly to meet the code in force and prevent the passage of fire and smoke. Another main concern in the codes is how to seal barrier openings and smaller penetrations.

15.1 Barrier Continuity

For a fire or smoke barrier to be effective, it must have vertical and horizontal continuity. For example, floors, ceilings, and floor/ceiling assemblies are effective barriers when they are continuous and extend horizontally to the extremities of the compartment. As such, these barriers have to extend either to the outside walls of the building

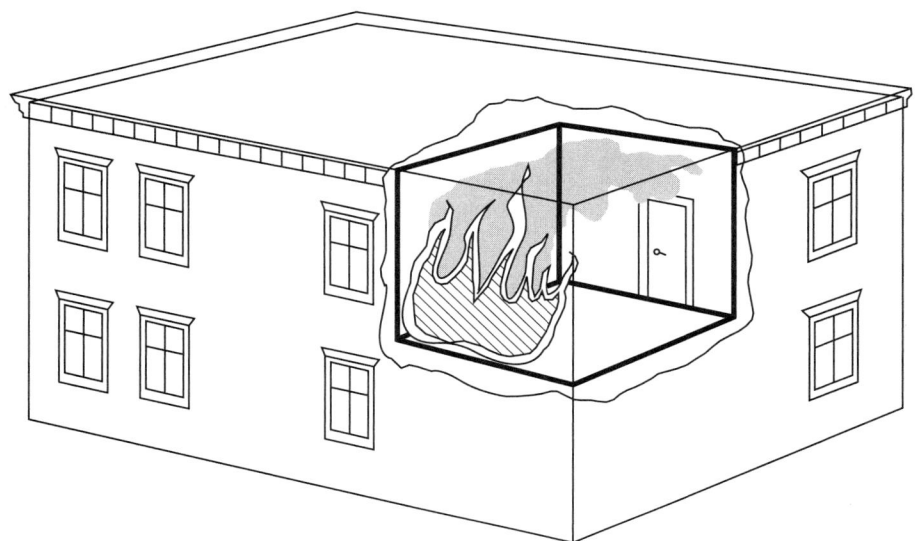

Figure 15.1 A fire compartment limits the spread of fire and smoke.

or, in the case of protecting a hazardous area, to the barriers around the hazardous area.

15.1.1 Horizontal Continuity

Walls that form the compartment must maintain horizontal continuity. These walls must extend horizontally from outside wall to outside wall, one fire or smoke barrier to another, or any combination thereof. Figure 15.2 illustrates this concept for fire barriers.

In Figure 15.2, wall A might be a building separation or occupancy separation wall. Note that this wall is continuous from outside wall to outside wall. Wall B might be an occupancy separation wall or smoke barrier or a wall around a hazardous area. This wall extends from the outside wall to another similar barrier. Wall A would have to be equivalent or better than wall B. The compartment formed by C is probably a hazardous area or a vertical opening. These barriers extend from barrier to barrier with no outside walls involved. Finally, D illustrates a corridor forrmed by barriers that run from an outside wall to a barrier to another barrier and back to the outside wall. In each case, the barrier must be continuous horizontally.

It seems obvious that, to be effective, these smoke and fire barriers must be continuous, because fire and smoke can easily reach around the end of an incomplete barrier. All too often, however, buildings contain barriers that terminate at either no barrier or at a barrier of lesser capability. The following case study illustrates a classic example of an improperly constructed barrier.

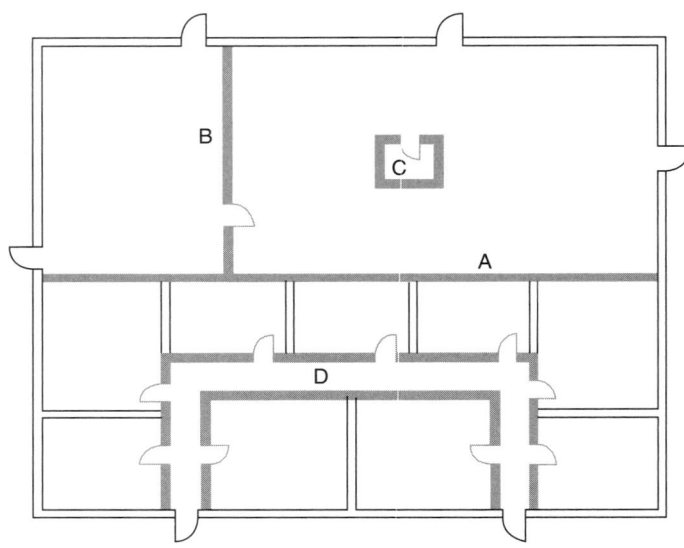

Figure 15.2 Typical fire barriers.
Source: Life Safety Code Handbook, 2003, Exhibit 8.2.

Improper Smoke Barrier

Hospital smoke barriers require a 1-hour fire rating and must extend from outside wall to outside wall. Part of this barrier in one hospital was a wall between two patient rooms. This wall was removed during a renovation, to form a small suite. The designer did not carefully research the building configuration and created a smoke barrier that terminated at a nonrated corridor wall with non–self-closing doors. Although the corridor walls were compliant for a corridor, they did not qualify as a smoke barrier. Smoke could easily travel around the end of this smoke barrier.

15.1.2 Vertical Continuity

Just as important as horizontal wall continuity is vertical continuity. Walls that form barriers must be continuous from the floor to the floor or roof above. In some cases, such as horizontal exits and building separation walls, this vertical continuity may be required to extend from the ground to the roof. In most cases, the barrier must extend through the ceiling to the floor deck or roof above. In general, one exception to creating vertical continuity through the ceiling is when the ceiling itself provides protection equivalent to the wall; that is, the ceiling itself has a fire resistance rating equal to or greater than that required for the wall.

When dealing with a barrier that only has to resist the passage of smoke, maintaining vertical continuity by using the ceiling is usually not difficult. In fact, corridor walls in sprinklered health care occupancies are specifically allowed to terminate at ceilings in most cases.

However, when dealing with barriers that must have a fire resistance rating, it is difficult to terminate the barrier at the ceiling, and problems are common. The reason is generally due to a misunderstanding of the difference between a ceiling and a floor/ceiling or roof/ceiling assembly. For a 1-hour fire barrier (or a smoke barrier required to have a 1-hour or ½-hour fire protection rating) to terminate at a ceiling, that ceiling, by itself, must have a 1-hour or ½-hour fire protection rating throughout the compartments formed. The typical "1-hour ceiling tile" does not in fact provide 1-hour fire resistance. Because the tile is part of a 1-hour assembly—either the floor/ceiling assembly or the roof/ceiling assembly—the ceiling in and of itself is not 1-hour rated. The ceiling provides a small part of the total rating. Figure 15.3 illustrates this point.

When dealing with such an assembly, any barrier required to have a fire resistance rating must extend into the ceiling space and be tight to the floor or roof above. When a barrier is required to have only a 1-hour fire protection rating, there is one ceiling assembly that would allow the barrier to terminate at the ceiling. The Gypsum Association's (GA's) *Fire Resistance Design Manual* identifies one assembly, GA No. FC 5406, in which the ceiling by itself provides 1 hour of fire resistance. This assembly consists of two layers of ⅝-in. type X gypsum wallboard. (See the GA design manual for further details.) Note that a single layer of ⅝-in. type X gypsum board does not provide the 1-hour protection needed to terminate 1-hour walls (except as discussed next). Unless specifically permitted by code—and the model building codes do in specific situations—fire-resistant barriers cannot terminate at the ceiling membrane of a fire-rated floor/ceiling assembly.

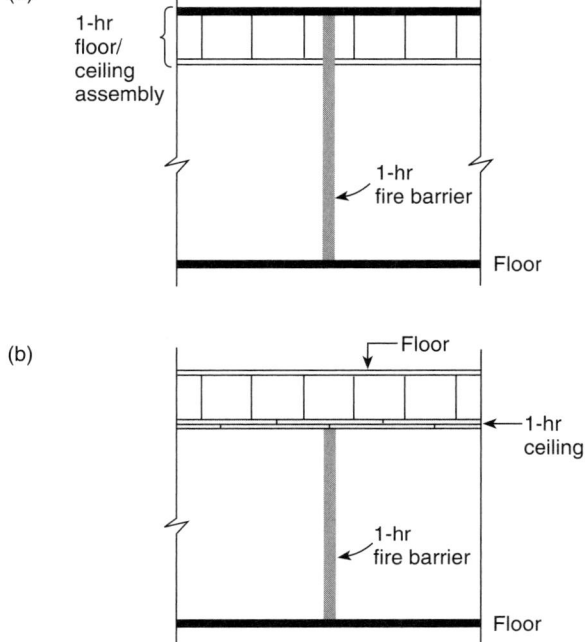

Figure 15.3 Fire barrier vertical continuity.
Source: Life Safety Code Handbook, 2003, Exhibit 8.3.

Above Ceiling Permits

The use of above ceiling work permits can help reduce the problem of unsealed penetrations above the ceiling, where most occur. Many hospitals have instituted this system with impressive results. One hospital makes enforcement of this policy very profitable. Any hospital employee seeing above ceiling work being done with no permit gets a $50 reward, which is billed back to the contractor or department doing the work. It doesn't take long for workers to realize they must have a permit when working above the ceiling.

There are cases, although few, in which the model building code may allow a barrier to terminate at a ceiling, whereas the *Life Safety Code* does not allow this. One typical problem is with corridor walls in non–health care occupancies. With some exceptions, the *Life Safety Code* requires 1-hour fire-rated corridor walls to extend to the floor or roof above, and most of the model building codes let this wall terminate at the ceiling of a 1-hour floor/ceiling or roof/ceiling assembly. For example, in the International Code Council's (ICC's) *International Building Code*®, a fire partition may terminate at the ceiling of a rated floor/ceiling or roof/ceiling system. Fire partitions in that code are used for certain corridor walls, tenant separation walls, and dwelling unit and guestroom separations. In a nonsprinklered new office building, the *Life Safety Code* usually requires 1-hour corridor walls. Under the *Life Safety Code,* these need to extend to the floor or roof above. Under the ICC *International Bulding Code* they are permitted, per the previous discussion, to terminate at the ceiling of a rated floor/ceiling or roof/ceiling assembly. The *Life Safety Code* places no requirements on the corridor walls of business buildings that are sprinkler protected or occupied by a single tenant, and therefore a wall built in compliance to the building code is sufficient. When using the *Life Safety Code* in conjunction with a model building code in creating a vertical smoke or fire barrier, the more stringent code must be followed.

Another problem with vertical continuity of barriers is ensuring that the wall is tight to the floor or roof. When dealing with a fluted steel deck or concrete construction with *pan decks* (concrete floor slabs that are poured over inverted pans), continuity to the deck is not always done properly. Just stuffing fiberglass into the voids between the wall and deck does not create the proper barrier. Materials are available that have been tested and listed to seal this gap. Underwriters Laboratories' *UL Fire Resistance Directory* as well as the directories of other testing laboratories list numerous such products. The following case study illustrates the problems of bringing walls with improper vertical continuity into compliance.

Improper Vertical Continuity

During a survey of a major hospital, it appeared that the walls around the exit enclosures did not extend into the voids above the wall where the wall met the fluted deck above. On closer examination, it was discovered that none of the walls, fire or smoke, in the entire building (a large addition to the complex) had been properly sealed at the deck. This resulted in a major project to get all the walls tight to the deck above. This was not only costly but difficult and disruptive. It would have been relatively simple to complete this work properly during construction.

15.2 Fire Resistance of Barriers

Different types of barriers provide different types of protection. For example, barriers that separate buildings are expected to provide more protection than barriers that form a corridor. As discussed previously, fire barriers are walls, floors, or floor/ceiling assemblies that are required to have some fire resistance rating. These ratings generally range from 30 minutes to 4 hours. Assemblies with ratings of under 1 hour are primarily for use in existing situations. There are cases where a 5-, 15-, or 20-minute thermal barrier is required, such as to protect a foamed plastic or a combustible wiring method. However, these barriers are not the same as the other fire barriers discussed in this chapter.

The fire resistance rating of the barrier is determined by test (NFPA 251, *Standard Methods of Tests of Fire Resistance of Building Construction and Materials*; ASTM E 119, *Standard Test Methods for Fire Tests of Building and Construction Materials*; UL 263, *Standard for Fire Tests of Building Construction and Materials*; or similar; see preceding discussion) or by calculation. The ratings are provided in numerous publications such as UL's *Fire Resistance Directory,* GA's *Fire Resistance Design Manual,* and others. NFPA 221 *Standard for High Challenge Fire Walls, Fire Walls, and Fire Barrier Walls,* has material on calculating fire resistance ratings. This information has changed over the years as documents available on this subject have been developed and gone out of date and out of print. Generically, the concept of calculating fire resistance is acceptable. Both *NFPA 5000®, Building Construction and Safety Code®,* and NFPA 221 specifically recognize both ASCE/SFPE 29 and ACI 216.1/TMS 0216.1 to be used for the calculation of fire resistance.

In the field, the assemblies must be constructed as shown in the directories. Some directories include extensive details that must be followed. A person who is thoroughly familiar with the materials and tests should be the only one allowed to alter these details, and then only after thorough evaluation. What might appear to be a logical step, although not included in the directions, has sometimes been shown by test not to be that logical.

Many times, the details required by the design are difficult to maintain in the field. An example of this is that many floor/ceiling or roof/ceiling assemblies require that the ceiling tiles be clipped down. Even if the clips are installed properly, they are often removed during maintenance or during installation of new wires or pipes and are rarely replaced. Another problem is when spray-on fireproofing is scraped off during the installation of pipes or tubes. When installing such equipment, the fireproofing must be repaired in accordance with manufacturer's instructions and the listings.

15.3 Types of Fire Barriers

One NFPA document of note that contains extensive information on fire walls and fire barrier walls is NFPA 221, first published in 1994. According to NFPA 221, *fire wall* and *fire barrier wall* are defined as follows:

- Fire wall—a wall separating buildings or subdividing a building to prevent the spread of fire and having a fire resistance rating and *structural stability* (emphasis added)
- Fire barrier wall—a wall, other than a fire wall, having a fire resistance rating

Under the new *International Building Code* there are three classifications of fire-resisting barriers, as follows:

1. Fire walls, which are similar to fire walls as defined by NFPA 221 and are required to have structural stability independent of collapse on either side
2. *Fire barriers,* which are similar to the fire barrier walls in NFPA 221 and NFPA *101*
3. Fire partitions, which can terminate at the ceiling of floor/ceiling or roof/ceiling assemblies and do not need to be supported on rated construction in many cases

The *Life Safety Code* uses the term *fire barrier* for both fire barriers and fire partitions but provides exceptions in some cases that result in walls that are similar to fire partitions under the *International Building Code*.

15.3.1 Building Separation Walls

Building separation walls are sometimes referred to as *area separation walls* or *fire walls*. These walls usually have the most stringent requirements of the fire barriers for normal buildings, with ratings from 2 to 4 hours. Traditionally addressed by the building codes, this type of wall is not referenced in many places in the *Life Safety Code*. However, since the *Life Safety Code* is often used without a building code, especially for existing buildings, the code does discuss these walls briefly (Chapter 8). The *Life Safety Code* states that for buildings or facilities that include additions or connected structures of different construction types, the rating and classification of the structure must be based either on separate buildings, if a 2-hour or greater, vertically aligned fire barrier wall exists between the portions of the building that is in accordance with NFPA 221, or, where there is no separation, the least fire-resistive type of construction of the connected portions. Structures built with previously approved separations are exempt from this rating requirement.

The *Life Safety Code* thus specifically allows a fire barrier wall that does not need to be structurally independent of both sides. Thus, even though the wall is separating buildings, or at least construction types, it is not what the building codes and NFPA 221 normally refer to as a fire wall or area separation wall. When the *Life Safety Code* is being used in conjunction with a building code, the building code will almost always be more stringent than the *Life Safety Code* in this regard.

The term *area separation wall*, as might be intuited, describes walls that are used to separate areas. These walls are usually needed when the size of the structure exceeds the area allowed based on the requirements for the structure's occupancy and type of construction. By installing one or more area separation walls, the structure is in fact broken down into multiple areas, each within the limits of the heights and areas tables. Separate buildings are not actually created, just separate areas within the same building. Another way to view this, especially when having to separate different types of construction, is that one structure is subdivided into multiple buildings—thus the rationale for the term *building separation walls*. The use of the term *fire wall* eliminates this potential confusion.

The structural stability of these building separation walls is important and can be difficult to design. The intent is that the area on either side of the wall should be able to collapse during a fire and not bring the wall down with the collapse. NFPA 221 provides assistance in this regard in the body of the text and related appendices.

In summary, under the *Life Safety Code,* for an area in an existing building to be evaluated as a separate building, a 2-hour, vertically aligned fire barrier wall or wall that was previously accepted for this purpose is required. The more stringent rules come into play only when required to conform to the building code, such as during renovations, alterations, or additions, or when the building code previously required the wall.

15.3.2 Horizontal Exit Walls

According to the *Life Safety Code* (Chapter 3), horizontal exits are defined as follows:

> A way of passage from one building to an area of refuge in another building on approximately the same level, or a way of passage through or around a fire barrier to an area of refuge on approximately the same level in the same building that affords safety from fire and smoke originating from the area of incidence and areas communicating therewith. [101–06: 3.3.70.1]

These exits are very useful in a health care occupancy because they not only eliminate the need for some of the normally required exit stairs but also provide for horizontal movement and protection. Figure 15.4 illustrates the use of a horizontal exit. Refer to Chapter 16 for more information on horizontal exits.

To comply with most codes, including the *Life Safety Code,* 2-hour fire barrier walls must form the walls used to create horizontal exits. They do not need to be structurally freestanding, but they must be supported by 2-hour construction and must penetrate any ceiling and continue to the floor or roof deck above. The *Life Safety Code* requires vertical continuity of these walls in that they must be continuous to the ground. The barrier is allowed to be omitted on any story below, provided the floor below the lowest level on which the barrier exists and all supporting members are 2-hour fire resistance rated and all exit stairs discharge directly to the outside. Figure 15.5 illustrates the vertical continuity of walls forming horizontal exits.

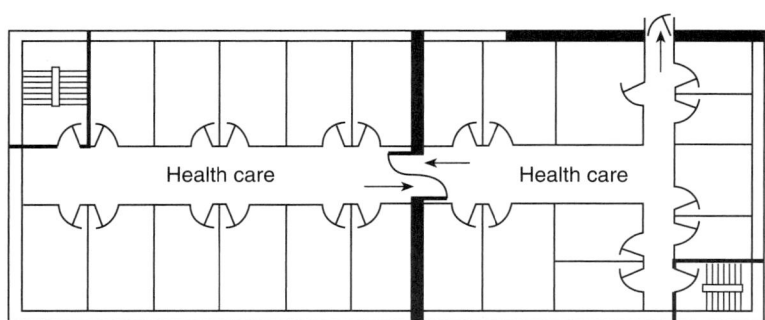

Figure 15.4 A horizontal exit, new construction.

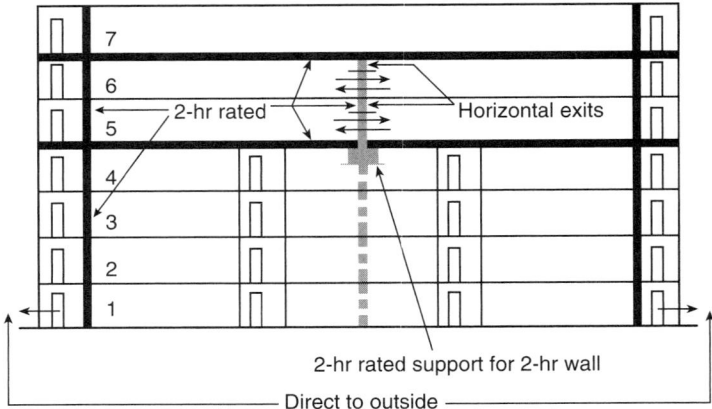

Figure 15.5 A building with horizontal exits on some floors.
Source: Life Safety Code Handbook, 2003, Exhibit 7.53.

The most common problems with horizontal exits in existing health care facilities tend to be with the doors, penetrations, and heating, ventilating, and air-conditioning (HVAC) dampers, as discussed in Section 15.4 in this chapter.

15.3.3 Exit Enclosures

Another common use of fire barriers is to create exit enclosures, both enclosed exit stairs and exit passageways. Figure 15.6 illustrates enclosed exit stairs and a typical use of an exit passageway. In this figure, the stairs are enclosed exit stairs, and there is an exit passageway connecting an exit stair to the outside. The need for the exit passageway is discussed in Chapter 16.

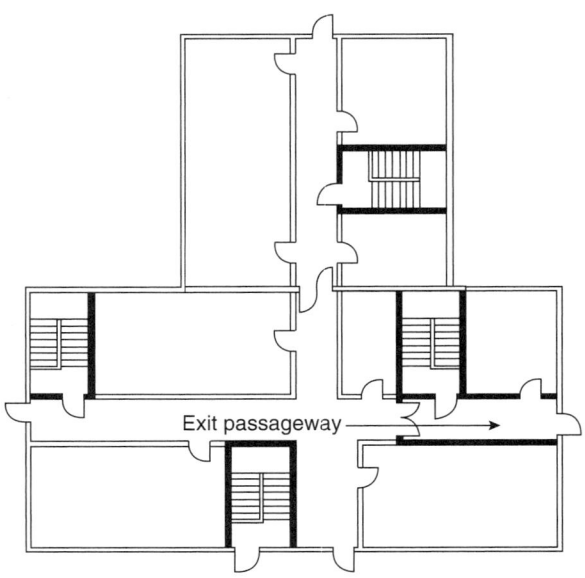

Figure 15.6 Examples of exit enclosures (indicated by heavy lines).
Source: Fire Protection Handbook, NFPA, 2003, Figure 4.3.6.

Exit enclosures must have either a 1- or 2-hour fire resistance rating, depending on the number of stories they serve. They are generally formed with fire barriers that are not required to be structurally freestanding but must be supported by structural elements having the same or greater fire resistance rating. Installing such barriers can be especially difficult in multistory non–fire-rated buildings [such as Type II (000)], which generally are not permitted for health care occupancies but could be used for business occupancies, such as medical office buildings, or industrial occupancies, such as service buildings. In these instances, an enclosure for an exit stair has to be supported by the ground, or the structural elements supporting the stair enclosure need to be either 1- or 2-hour rated, depending on the number of stories the stair serves.

The barriers forming exit enclosures are not permitted to terminate at ceiling membranes. As discussed in Chapter 16, there are some very stringent limitations on penetrations and openings into these enclosures. These limitations, along with door-related maintenance problems, tend to be some of the major problems health care occupancies face when fire barriers form exit enclosures.

15.3.4 Hazardous Area Separations

The *Life Safety Code* defines *hazardous areas* as those areas in structures or buildings that pose a degree of hazard greater than that normal to the general occupancy of a building or structure, such as those areas used for storing or using combustibles or flammables or heat-producing appliances. Depending on the occupancy, the nature of the hazard, and the presence or lack of sprinkler protection, hazardous areas, referred to by some building codes as *special use* or *incidental use* areas, may or may not require the installation of a fire barrier around them. If a fire barrier is required, it is generally only a 1-hour fire-rated barrier; however, under most model building codes there are some

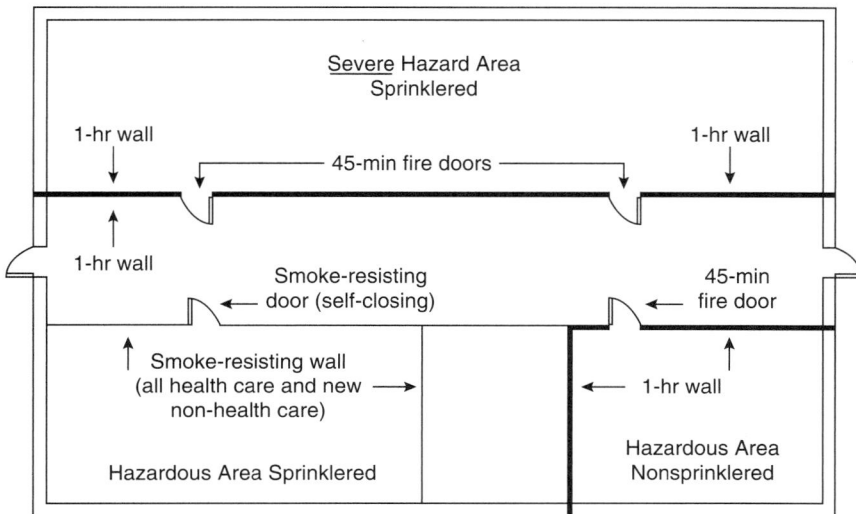

Figure 15.7 Protection of hazardous areas.
Source: Opus Communications and Koffel Associates, Inc., redrawn from 2000 *Life Safety Code Workbook & Study Guide for Health Care Facilities*, 2003, Figure 21-1.

requirements for 2-hour barriers. Figure 15.7 shows some of the concepts for protecting hazardous areas. Refer to Chapter 18 for more details on protecting hazardous areas.

When walls surrounding hazardous areas must have a fire resistance rating, they must go to the floor or roof above. A rare exception is when the ceiling of the hazardous area, in and of itself, can provide the same fire protection rating as required for the barrier. These walls, however, do not have to be supported on rated construction in nonrated buildings for two reasons, one practical and one technical. First, in nonrated construction, the floors and their supporting elements have no, or minimal, fire resistance. Since a hazardous area can occur almost anywhere in a building, it would be impractical to require the supporting elements to have a fire resistance rating. Second, hazardous areas are protected to keep the fire and smoke in those areas and away from the rest of the story. Usually, a fire on the floor below is not what is of concern for this protection. Therefore, not rating the supporting elements for the walls is a practical necessity justified from a technical viewpoint. The *International Building Code* provides a similar exception.

Determining whether a barrier is required to have a fire resistance rating or not and maintaining the doors within these barriers tend to be the problems health care facilities experience with fire barriers for hazardous areas.

15.3.5 Occupancy Separation Barriers

As discussed in Chapter 13, the *Life Safety Code* has relatively few requirements for occupancy separations for existing buildings. However, where specific fire-rated separations are required for existing buildings, such as for health care occupancies, the required rating is usually 2 hours of fire resistance. The U.S. model building codes, including *NFPA 5000*, and the *International Building Code* have extensive provisions for occupancy separations. In some of the outdated codes, occupancy separation is mandatory, and in others it is one of two options (the other option being meeting the most stringent codes for the occupancies involved in areas of mixed occupancies; see Chapter 13).

Fire barriers for occupancy separations in the building codes vary from 1 to 4 hours, depending on the code being used and the occupancies being separated. The *Life Safety Code* usually requires occupancy separations to extend to the floor or roof above. The 2-hour fire resistance–rated occupancy separations for health care occupancies are often used for horizontal exits. In such cases, the more stringent requirements for horizontal exits, as discussed in Section 15.3.2, must be met.

In health care occupancies, the most difficult part about occupancy separations is determining where the separations are required. Once identified, they tend to be used for horizontal exits also, and the typical problems are the same: penetrations, HVAC dampers, and maintaining the doors.

15.3.6 Corridor Separation

Fire barriers for corridor walls have probably the least stringent requirements of all the fire barriers in the *Life Safety Code* and some of the model building codes, but they are commonly violated nonetheless. For the most part, in the *Life Safety Code* each occupancy establishes its own requirements for corridors, but there are several common traits. Corridors, especially those used in health care occupancies, are discussed in more detail in Chapter 16. Figure 15.8 illustrates corridor requirements in general.

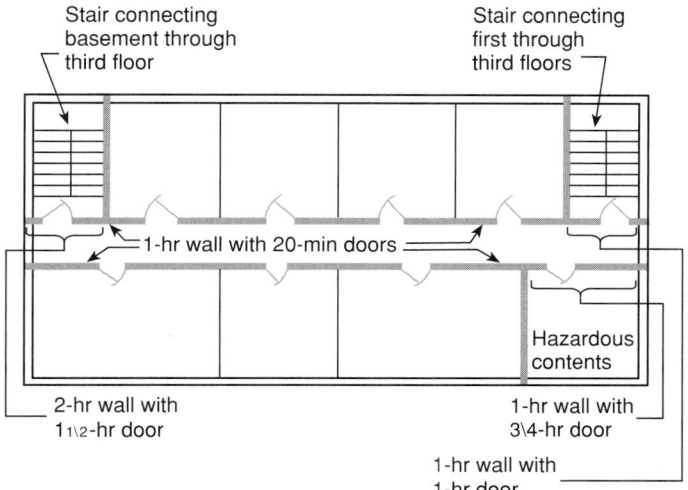

Figure 15.8 Protection of exit access corridors.
Source: Life Safety Code Handbook, 2003, Exhibit 7.4.

Corridor walls, when required to have a fire resistance rating, are usually 1-hour fire resistance rated. These walls do not need to be supported by elements having a fire resistance rating. Under the *Life Safety Code,* fire-rated corridor walls are required to extend to the floor or roof above. This is a major point, since some of the model building codes allow these walls to terminate at the ceiling membrane of a floor/ceiling or roof/ceiling assembly having a 1-hour fire resistance rating.

There tend to be numerous problems with corridor walls in health care occupancies, particularly in existing nonsprinklered facilities. These problems include having no wall where one is required, unprotected openings in the wall, doors that lack proper latches, doors with excessive undercuts, and penetrations that are not properly sealed (especially above the ceiling line). These problems are discussed in other sections; however, it is appropriate to emphasize a common finding. In nonsprinklered facilities, the corridor wall, with some minor uncommon exceptions, must extend to the floor or roof above. It is very typical for corridor walls in these situations to go up to the roof or floor but not tight to the deck. As mentioned, under vertical continuity, the wall must go tight to the floor or roof, and any gap must be sealed with material that has been tested and listed for such use.

15.4 Barrier Openings and Penetrations

Another consideration for regulating barriers is that, in addition to preventing the passage of smoke and fire, they also prevent the passage of people, goods, and building services. Due to the necessity to move people and goods, it is often necessary to have openings in barriers, yet openings create a significant code enforcement problem. Two different categories of openings in fire and smoke barriers are commonly discussed, *openings* and *penetrations*. Table 15.1 helps differentiate between the two terms.

Table 15.1 Barrier Openings and Penetrations

Openings	Penetrations
Doors	Pipes
Windows	Conduits
Access openings	Bus ducts
HVAC transfer openings	Tubes
HVAC ducts	HVAC ducts

Note that there is no real difference between barrier openings and penetrations other than that for penetrations, the object generally continues on each side of the barrier, and penetrations are usually (but not always) smaller than openings. The way that the openings are protected also differs. Generally, openings are protected with doors, panels (still considered to be a door), and dampers. With penetrations, the space around the penetrating item is firestopped but there is no protection within the penetrating item (such as with a door or damper). As shown in Table 15.1, HVAC ducts are considered to be both penetrations and openings.

For many types of barriers, the codes contain many restrictions on barrier openings. As discussed in Chapter 16, the codes place many limits on the openings allowed in exit enclosures. Similar restrictions occur for horizontal exits. The building codes also often limit the openings in fire walls. Therefore, the first issue that must be evaluated in terms of complying with the codes is whether an opening in the barrier is allowed. Once it has been determined that an opening is allowed, the next step is to determine what must be used to protect the opening. The most common types of openings in health care facilities are doors, access panels, windows, and HVAC ducts and grilles.

Fire doors and windows must be installed and maintained in accordance with NFPA 80, *Standard for Fire Doors and Fire Windows*. This is one of several NFPA documents that all health care facilities should have on hand. It describes in detail provisions for the following types of openings:

- Swinging doors with builders' hardware (typical pedestrian door)
- Swinging doors with fire door hardware (not common in health care occupancies, except possibly on a boiler room or similar area)
- Horizontal sliding doors (generally not used in health care occupancies—not permitted to go across the egress path except in certain limited conditions)
- Vertical sliding doors
- Rolling steel doors
- Special-purpose horizontal sliding accordion or folding doors (what the *Life Safety Code* refers to as sliding doors—permitted under limited conditions to go across the egress path)
- Hoistway doors for elevators and dumbwaiters
- Chute doors (common in health care occupancies for laundry and trash chutes)
- Fire shutters
- Access doors (access to equipment, shafts, or similar spaces)

- Installation of service counter doors (may be found at counters for pharmacy, medical records, radiology, or administration)
- Fire windows
- Glass blocks

15.4.1 Fire Doors

The following series of figures shows several examples of fire door openings:

- Figure 15.9 and 15.10—two examples of swinging doors with builder's hardware
- Figure 15.11—a swinging door with fire door hardware
- Figure 15.12—a horizontal sliding fire door with an inclined track closing device
- Figure 15.13—a vertical sliding fire door with closing device
- Figure 15.14—a rolling steel fire door

15.4.1.1 *Fire Door Fire Protection Ratings*

Fire doors, whether for people, products, access, or other reason, are required to have a *fire protection rating*. Note that the protectives used in barriers must have a fire protection rating, which is different from the fire resistance rating required for walls and floors. Fire protection ratings are based on the location of the opening (in an exit enclosure, hazardous area, corridor, vertical opening, etc.). The *Life Safety Code* establishes the requirements for fire protection ratings for each type of barrier. The requirements

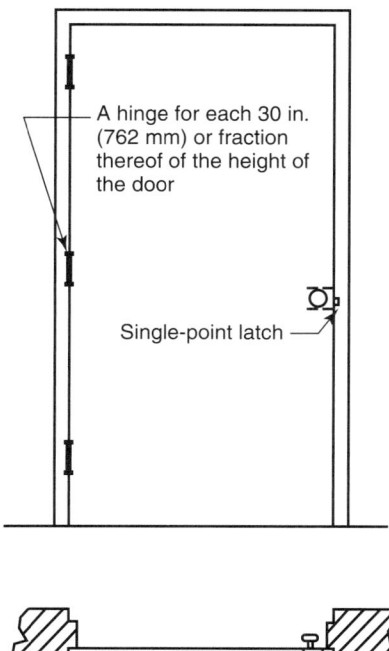

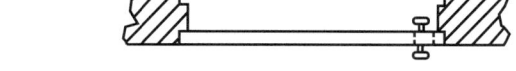

Figure 15.9 A single swinging door with builders' hardware (single-point latch; flush mounted). *Source:* NFPA 80, 1999, Figure B.20.

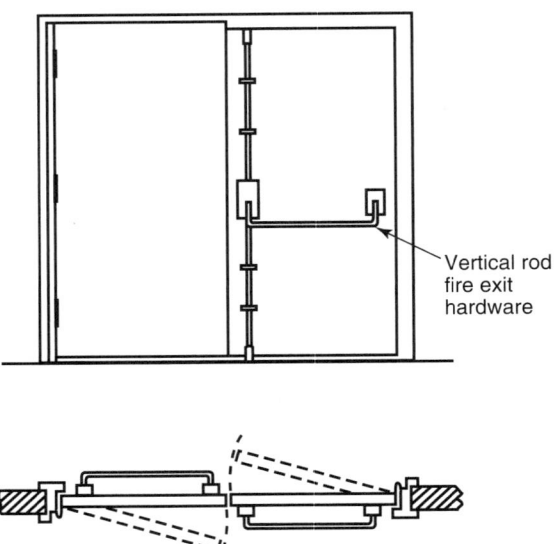

Figure 15.10 A double-egress door and frame with builders' hardware. *Source:* NFPA 80, 1999, Figure B.25.

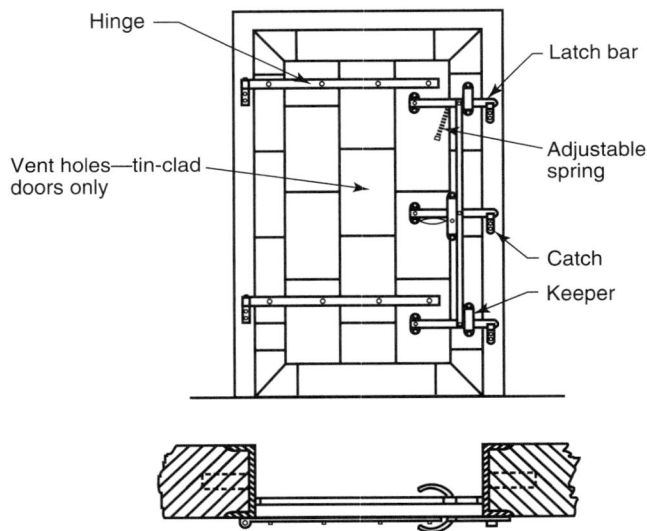

Figure 15.11 Fire door hardware on a single swinging door (flush mounted). *Source:* NFPA 80, 1999, Figure B.29.

are consistent with most of the modern U.S. model building codes. Table 15.2 summarizes requirements in the *Life Safety Code*. These ratings are based on tests conducted in accordance with NFPA 252, *Standard Methods of Fire Tests of Door Assemblies* (UL 10B), and NFPA 257, *Standard on Fire Test for Window and Glass Block Assemblies* (UL 9).

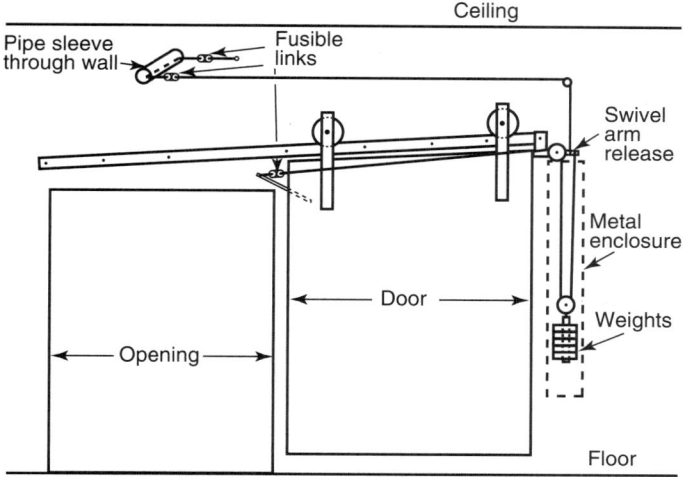

Note: Fusible links are needed on both sides of the wall.

Figure 15.12 An inclined track closing device for a horizontal sliding door.
Source: NFPA 80, 1999, Figure B.35.

Table 15.2 Minimum Fire Protection Ratings for Opening Protectives in Fire Resistance–Rated Assemblies

Component	Walls and Partitions (hr)	Fire Door Assemblies (hr)	Fire Window Assemblies (hr)
Elevator hoistways	2	1½	NP
	1	1	NP
Vertical shafts (including	2	1½	NP
stairways, exits, and	1	1	NP
refuse chutes)	½	⅓	NP
Fire barriers	2	1½	NP
	1	¾	¾
Horizontal exits	2	1½	NP
Exit access corridors[1]	1	⅓	¾
	½	⅓	⅓
Smoke barriers[1]	1	⅓	¾
Smoke partitions[1,2]	½	⅓	⅓

NP = not permitted.
[1] Fire doors are not required to have a hose stream test per NFPA 252.
[2] For residential board and care, see NFPA 101, 2006, Sections 32.2.3.1 and 33.2.3.1.
Source: NFPA 101, 2006, Table 8.3.4.2.

15.4.1.2 Fire Door Closing

In general, and with almost no exceptions, doors that must be fire rated by the *Life Safety Code* must be self-closing and self-latching. Doors can close automatically by smoke detection with some restrictions. Doors that must have a fire protection rating by the *Life Safety Code* may not be closed by the use of a fusible link (i.e., two pieces

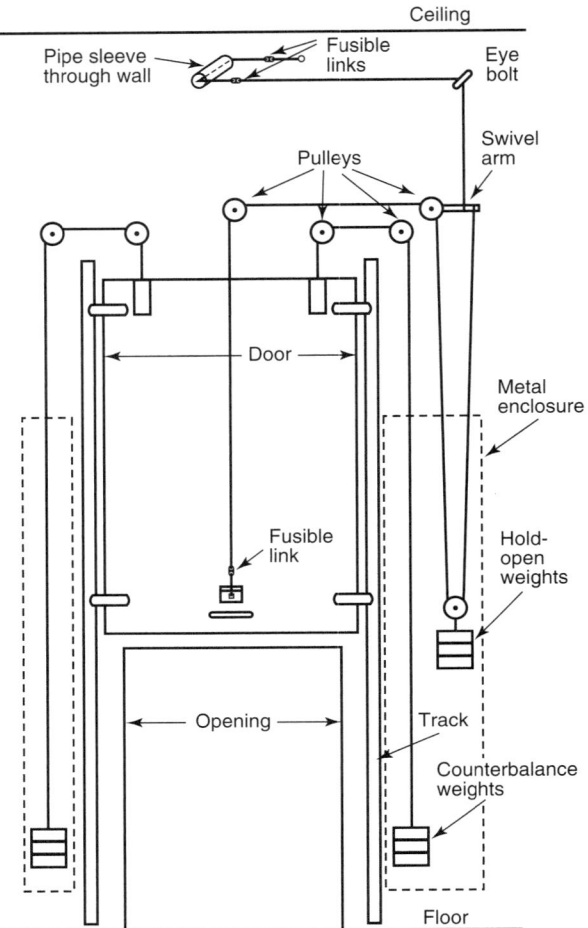

Note: Fusible links are needed on both sides of the wall.

Figure 15.13 Closing device for a vertically sliding fire door.
Source: NFPA 80, 1999, Figure B.46.

of metal held together by low-melting-point solder). This provision does not mean that all fusible links are prohibited, just that doors that must be self- or automatic closing, per the *Life Safety Code*, cannot have a fusible link. Other codes or insurance policies that require doors to be self- or automatic closing may allow these doors to have fusible links. It should be noted that the *Life Safety Code* does not prohibit fusible links from closing the doors at the bottom of trash and linen chutes.

15.4.1.3 Fire Door Labels

The normal way to determine the fire protection rating of a fire door is to look for the label of a recognized testing laboratory, which will provide the hourly rating of the door. If this label is missing, damaged, or painted over such that the rating is not legible,

Chapter 15. Compartmentation **235**

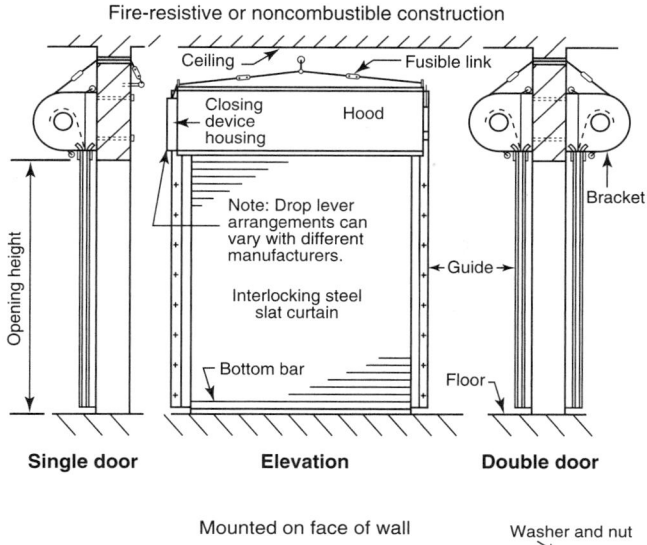

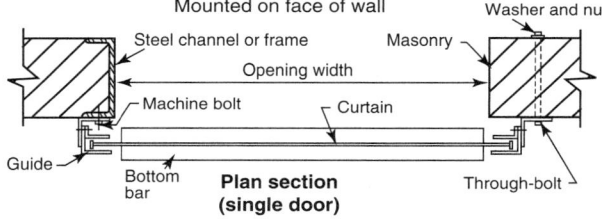

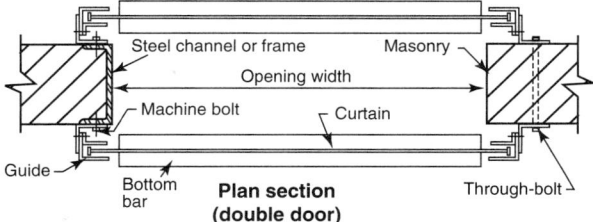

Note: Fusible links are needed on both sides of the wall.

Figure 15.14 Surface-mounted rolling steel door.
Source: Figure NFPA 80, 1999, Figure B.48.

there is no simple way to determine the rating. When the label is painted over, it will be necessary to clear the paint off the label, have a testing laboratory do a field examination of the door for relabeling, or assume that the door is not fire rated (as illustrated in the following case study) and replace the door.

Painted-Over Door Labels

It is not at all uncommon to find labels on fire doors painted over. It is tempting to say, "At least the doors are labeled." However, in one case when the labels were cleared

(Continued)

of paint, it was discovered that the labels stated that the doors were *not* fire rated doors. If the label is painted over, it has to be assumed that the door is not rated.

In addition to listing the fire protection rating, the label on the door will also include a letter designation. This designation is now somewhat archaic, although according to Appendix E of NFPA 80, the letter designation was one method used to classify the opening for which the fire door was considered suitable, as follows:

- **Class A**—Openings in fire walls and in walls that divide a single building into fire areas
- **Class B**—Openings in enclosures of vertical communications through buildings and in 2-hour-rated partitions providing horizontal fire separations
- **Class C**—Openings in walls or partitions between rooms and corridors having a fire resistance rating of 1 hour or less
- **Class D**—Openings in exterior walls subject to severe fire exposure from outside of the building
- **Class E**—Openings in exterior walls subject to moderate or light fire exposure from outside of the building

Another piece of information that may be on the label of a door is the temperature of the unexposed side of the door after the first 30 minutes of a fire test. Some codes require that doors used in exit stair enclosures have a limited rise in temperature during the first 30 minutes of the test. This limitation is usually 450°F (232°C). The *Life Safety Code* does not have this restriction.

15.4.1.4 Fire Door Glazing

Another fire door requirement relates to the size and type of glazing that can be used in a fire door (as also regulated by NFPA 80). Traditionally, glass must be wired glass with the limitations shown in Table 15.3.

Other materials are available today that could allow different glazing sizes that exceed 1296 in.^{2} (0.84 m^2) for all but the 3-hour fire door based on successful fire testing. The requirements in Table 15.3 should be viewed only as traditional allowances. See NFPA 80 for more details.

15.4.1.5 Fire Door Clearances

NFPA 80 also regulates the gaps between the doors and between the doors and the floor, according to the requirements presented in Table 15.4.

Table 15.3 Area Limitations on Fire Door Wired Glass Glazing

Fire Door Rating	Maximum Area of Glazing
½, ¾, or ¾ hr	1296 in.^{2} (0.84 m^2) with max. 54-in. (1.37-m) dimension
1, 1½, 3 hr	100 in.^{2} (0.065 m^2)

Table 15.4 Clearances Under the Bottoms of Doors

Clearance Between	Swinging Doors with Builders' Hardware		Swinging Doors with Fire Door Hardware		Horizontally Sliding Doors		Vertically Sliding Doors		Special-Purpose Horizontally Sliding Accordion or Folding Doors	
	in.	mm	in.	mm	in.	mm	in.	mm	in.	mm
Bottom of door and raised noncombustible sills	⅜	9.5	⅜	9.5	⅜	9.5	⅜	9.5	⅜	9.5
Floor where no sill exixts	¾	19.1	¾	19.1	¾	19.1	–	–	¾	19.1
Rigid floor title	⅝	15.9	–	–	–	–	–	–	–	–
Floor coverings	½	12.7	½	12.7	½	12.7	–	–	½	12.7

NFPA 80 regulations for floor coverings permit combustible floor coverings to extend through openings that must be protected by 1½-hour, 1-hour, or ¾-hour rated fire protection fire door assemblies without a sill where the coverings have a minimum critical radiant flux of 0.22 W/cm² in accordance with NFPA 253, *Standard Method of Test for Critical Radiant Flux of Floor Covering Systems Using a Radiant Heat Energy Source.*

An additional requirement for the clearance between the edge of the door on the pull side and the frame and the meeting edges of doors swinging in pairs on the pull side is that the clearance must not exceed ⅛ in. ± ¹⁄₁₆ in. (3.18 mm ± 1.59 mm) for steel doors and must not exceed ⅛ in. (3.18 mm) for wood doors.

According to NFPA 80, when factory-installed protective plates are present, they must be installed in accordance with the listing of the door. Also, field-installed protection plates must be labeled and installed in accordance with their listing. One exception to this rule is that labeling is not required where the top of the protection plate is not more than 16 in. (406 mm) above the bottom of the door. The *Life Safety Code* has some exceptions to the requirements for protective plates specifically for health care occupancies, as discussed in the sections for each specific area, such as for protecting hazardous areas in existing health care occupancies.

15.4.1.6 Health Care Facility Fire Door Problems

Several typical problems are associated with health care facility fire doors, including but not limited to the following:

- Labels are missing or painted over.
- Doors have holes from old hardware that are not properly protected.
- Closers are missing or disconnected.
- Closers are not properly adjusted (door must close totally and latch securely).
- Latches are missing or taped over or otherwise secured unlatched.
- Latches are not working properly.
- Vision panel is of an excessive size or the wrong glazing material was used.
- Nonrated kick plates extend up more than 16 in. (406 mm) from the bottom of the door.

- Pairs of doors have excessive gaps between them, missing or inoperative coordinators, or latch to each other but not to floor or frame.

15.4.2 Fire Windows

Fire windows are allowed in some fire-rated walls but not all. It is important to recognize two issues in this regard. First, fire windows and vision panels are not the same thing. Vision panels are allowed in fire-rated doors under the door provisions, not the window provisions. Second, not all glazing materials are considered windows. There are products available on the market that are tested and restricted as a wall, not as a fire window. As mentioned earlier, fire windows are tested in accordance with NFPA 257. Some glazing materials have been tested in accordance with NFPA 251. When so tested, the glazing material is considered part of the wall or another building element and is usually not restricted as a window. This is important because most codes have restrictions on opening protectives. Typical restrictions include

- Prohibitions in walls having a required fire resistance rating above 1 hour
- Prohibitions in walls used for most exit enclosures
- Limitations to a maximum of 25 percent of the wall area

15.4.3 Other Barrier Openings and Penetrations

As mentioned previously, HVAC ducts can be both an opening and a penetration. HVAC grilles and ductwork are regulated by NFPA 90A, *Standard for the Installation of Air-Conditioning and Ventilating Systems*. This is another "must have" of the NFPA family of documents, as it provides the requirements related to where and how fire dampers must be installed. There will be times, however, when the local mechanical code may be more stringent than NFPA 90A. One prominent example is for the installation of fire dampers in ductwork penetrating 1-hour fire-resistive barriers. Under NFPA 90A, ductwork passing through 1-hour fire-resistive barriers is not required to have fire dampers, while transfer openings are required to have these dampers. However, under most of the model mechanical codes, dampers are required in such ductwork, with some exceptions. Therefore, in new installations, it is important to be aware of what is required by both the local mechanical code, if there is one, and NFPA 90A. Figure 15.15 provides insights about many of the NFPA 90A requirements; refer to NFPA 90A for details.

15.5 Barrier Sealing and Penetration Protection

Regardless of the type of barrier used, smoke or fire, penetration protection is needed by properly protecting (i.e., firestopping) the space around the penetration. Although some trades involved in installing equipment for many years did not do anything at all to prevent the passage of fire and smoke, some trades did the best they could to firestop these penetrations by putting some type of material around the penetrating item. Sometimes this was no more than drywall tape and spackle or fiberglass insulation. More recently, materials have been developed and widely recognized and used

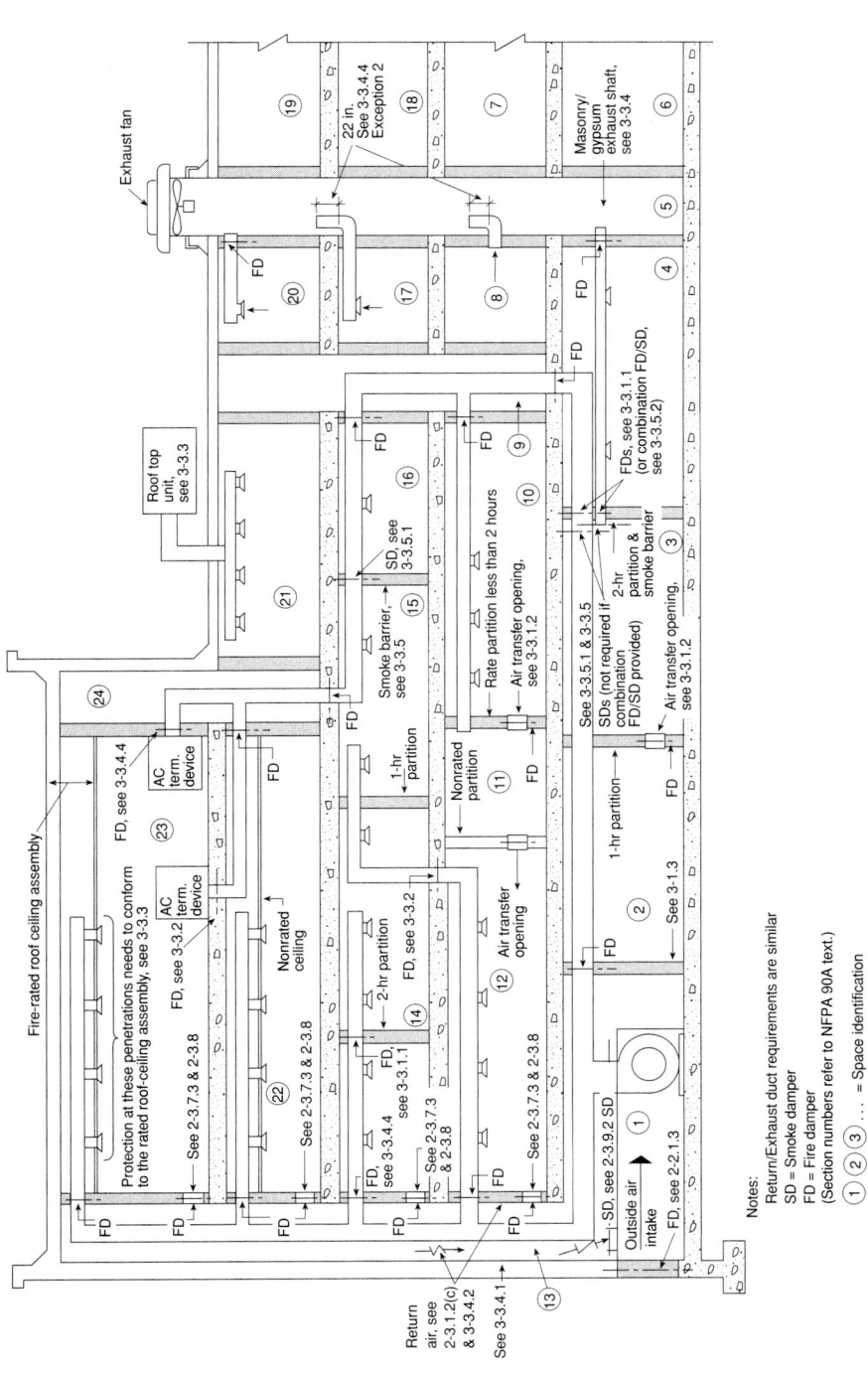

Figure 15.15 Application of penetration requirements for HVAC system.
Source: NFPA 90A, 2002, Figure A.5.3.

specifically for protecting penetration openings. Various testing laboratories, such as Factory Mutual, Intertek ETL SEMKO (which includes Warnock Hersey and the former Omega Point Laboratories), Southwest Research, Underwriters Laboratories, and others, use ASTM E 814, *Standard Test Method for Fire Tests of Through-Penetration Fire Stops*. In fact, the *UL Fire Resistance Directory* now comes in three volumes, with one entire volume dedicated to through-penetration firestop materials.

All the major building codes as well as the *Life Safety Code* address penetration protection. The *Life Safety Code* addresses penetration protection in the section relative to the wall that the penetration is in. The *Life Safety Code* has traditionally treated this issue in a performance manner. For example, the space between the penetrating item (e.g., pipes, conduits, bus ducts, cables, wires, air ducts, pneumatic tubes and ducts, and similar building service equipment that passes through fire barriers) and the fire barrier must be protected by filling the space with a material capable of maintaining the fire resistance of the barrier or otherwise protected by an approved device designed for this purpose. However, in newer editions of the *Code*, more specific requirements have been put in place. Due to this, there are provisions that allow previously approved installations to remain.

When the penetrating item uses a sleeve to penetrate the fire barrier, the sleeve must be set solidly in the fire barrier, and the space between the item and the sleeve must be filled with a material capable of maintaining the fire resistance of the fire barrier or otherwise protected by an approved device designed for this purpose. The *Life Safety Code* prohibits insulation and coverings for pipes and ducts to pass through the fire barrier unless the material can maintain the fire resistance of the fire barrier or is protected by an approved device. For designs that consider the transmission of vibration, any vibration isolation must be made on either side of the fire barrier or made by an approved device.

Some areas to consider when evaluating penetrations are as follows.

- For all barriers that must have a fire resistance rating or resist the passage of smoke, make sure that all penetrations are sealed. Many tradespeople are not as thorough as they should be about sealing the penetrations they make. Especially be aware when telecommunications, computer data, fire alarm, fire sprinkler, HVAC, medical gas piping, or electrical projects are being conducted. In some facilities, these activities are constantly taking place. Make sure the contract spells out who is responsible for sealing penetrations and what type of material must be used. Some facilities have a standing purchase order with a penetration sealant company to survey and seal barriers once or twice a year.
- Make sure that the proper material is used to seal the penetration and that it is installed properly. As mentioned previously, numerous products are available today to seal these penetrations. The acceptability of each product is based on many issues, including the material and thickness of the barrier penetrated, the material and size of the penetrating item, the size of the hole in the barrier, and the nature of the penetration (e.g., duct, pipe, conduit). The size of the hole in relation to the penetrating item can be a major issue. Contractors and employees should be

instructed to limit the hole size to that needed to pass the item through the wall. The smaller the space, the easier it is to seal.
- Train staff on the importance of these seals. Many penetrations are made by staff in their normal daily activities. Provide training programs for contractors, as well, to increase their awareness of the problem.

The most common problems related to these penetrations are that they lack any type of seal, the wrong type of seal has been used, or they have been sealed improperly. Another problem is unique to HVAC penetrations. It must be remembered that in many cases, smoke or fire dampers are needed where a duct or air transfer opening penetrates an assembly required to have a fire resistance rating or to resist the passage of smoke. Where a damper is installed, any material used to seal the area around the duct must not be a material that expands when heated, because this will cause the damper housing to deform and likely prevent the damper from working properly. Contact the manufacturer of the damper for information on sealing the opening around the damper.

16

Egress Components

Simply stated, a building's means of egress is the path one can take from any point in a building to the public street or sidewalk. People constantly egress buildings, but what most concerns code-making bodies and authorities having jurisdiction is ensuring safe egress during emergency situations. During these times in health care facilities, the means of egress must clearly direct all people in a building, including the mobility impaired, to safe areas either within the building or outside the building entirely.

A major portion of NFPA 101®, *Life Safety Code*®, is dedicated to means of egress. Since the *Life Safety Code* is an American National Standards Institute (ANSI) document, it is considered the "national" reference on means of egress in the United States. It is also used as a reference document in numerous countries. In addition, U.S. Occupational Safety and Health Administration (OSHA) regulations (29 *CFR* 1910, Sections 1910.35 to 1910.39, Subpart E) recognize the *Life Safety Code*. This regulation subpart addresses means of egress for the workplace. OSHA is on record as stating that facilities in compliance with the latest edition of the *Life Safety Code* should be considered in compliance with Subpart E of its regulations. This discussion therefore centers on the criteria set forth in the *Life Safety Code*. Note that although the major model building codes in the United States are mostly compatible with the *Life Safety Code*, there are differences, and compliance with a building code does not necessarily mean compliance with the *Life Safety Code*, as illustrated in the following case study.

Compliance Conflict

An on-site review of a new facility revealed improper height guards on all the stairs (36 in. versus the required 42-in. minimum). The architect had been informed that the facility had to comply with the *Life Safety Code* in addition to the local building code. Unfortunately, the architect assumed that complying with the local code, which was based on a model building code, would also ensure compliance with the *Life Safety Code*. Although much has been done over the last two decades to reduce differences among the codes, they still exist. Local and state amendments to the model codes sometimes make compliance even more difficult.

In addition to the *Life Safety Code*, numerous other sources of information on egress systems and people movement can be consulted when more information is needed or desired. Included are NFPA's *Fire Protection Handbook, The SFPE* (Society of Fire Protection Engineers) *Handbook of Fire Protection Engineering*—both of which have extensive references for additional reading—and NFPA's *Life Safety Code*® *Handbook*—a must-have document for every health care facility. It is not the intent here to provide an in-depth treatise on the subject of egress from buildings but to introduce key concepts related to egress and discuss areas that are problematic or typical in health care facilities, especially health care occupancies. These key issues include understanding the three elements of a means of egress (exit access, exit, and exit discharge), creating areas of refuge, protecting the means of egress, and designing and maintaining all egress components, such as doors, stairs, ramps, and horizontal exits. These features are discussed in this chapter. The principles of egress, such as how to determine egress capacity, occupant loads, clear widths, number of exits, common paths of travel, and the like, are discussed in Chapter 17.

16.1 Means of Egress—Definitions

The *Life Safety Code* defines *means of egress* as follows:

> A continuous and unobstructed way of travel from any point in a building or structure to a public way consisting of three separate and distinct parts: (1) the exit access, (2) the exit, and (3) the exit discharge. [101–06: 3.3.151]

When the term *means of egress* is used, it applies to all three parts. In some cases, one or more of the three components of means of egress is restricted. For example, when emergency lighting is required, the *Life Safety Code* does not necessarily require it in the exit discharge all the way to the public way, but only in designated areas of the exit discharge. Note that per the *Life Safety Code*, an accessible means of *egress is a path of travel, usable by a person with a severe mobility impairment, that leads to a pubic way, a horizontal exit, or an area of refuge.* Accessibility for the mobility impaired and areas of refuge are discussed in more detail later in this chapter and in Chapter 17.

The *Life Safety Code* defines a *public way* as follows:

> Any street, alley, or other similar parcel of land essentially open to the outside air deeded, dedicated, or otherwise permanently appropriated to the public for public use and having a clear width and height of not less than 10 ft (3050 mm). [101–06: 3.3.193]

Public ways usually refer to public streets or sidewalks. In a public way, a person can continually move away from the building the fire is in, without concern about smoke or impediment caused by fences or other barriers. As such, the definition of public way can cause some problems with large campus-like facilities (such as medical complexes, military reservations, universities, and large industrial facilities), since the streets and sidewalks are not actually "deeded, dedicated, or otherwise permanently appropriated to the public for public use." It is important in those cases to understand the concept of public way and not become overly technical. If the campus roads and sidewalks provide free movement, then it is logical to consider them a public way for the purposes of the *Code*. Should future construction affect this free movement, modifications to the area may be needed to ensure safety. This is normally not an issue with a true public way, but there are cases where public roads and sidewalks have been deeded over to private property. Therefore, even with true public ways, there is no absolute guarantee that they will remain so.

Exit is defined in the *Life Safety Code* as follows:

> That portion of a means of egress that is separated from all other spaces of the building or structure by construction or equipment as required to provide a protected way of travel to the exit discharge. [101–06: 3.3.70]

One problem with this definition is that the most common exit of all, the door to the outside at ground level, is not separated from the building. An annex note in the *Life Safety Code* clarifies this particular situation. (See Section 16.3.1 of this text for further discussion on doors.)

Exits are typically some combination of doors to the outside at ground level, properly enclosed stairs or ramps, exit passageways, horizontal exits, protected outside stairs, and, in limited cases, fire escape stairs. Exits typically found in health care facilities include doors to the outside at ground level, enclosed interior stairs (and smoke-proof enclosures), horizontal exits, exit passageways, protected outside stairs, and existing fire escape stairs (which are not allowed to serve a health care occupancy but can serve assembly, business, and industrial occupancies). Elevators can comply with exit requirements in some cases but not for uses typically associated with health care facilities.

Exit access is the path of travel from any point in a building to the exit, whereas *exit discharge* connects the exit to the public way. The vast majority of most buildings consists of exit access, including all spaces and corridors that can be occupied. The exit discharge is typically to outside courts and sidewalks leading to a public way; however, exit stairs sometimes can discharge inside a building, and then the travel from the stair to the door to the outside is interior exit discharge. Figure 16.1 provides examples of exit access, exit, and exit discharge.

16.2 Protecting Exits and Exit Enclosures

To comply with the requirement that the exit provide a protected way of travel to the exit discharge, all exit enclosures (i.e., enclosed exit stairs and exit passageways) must be constructed in ways that keep the exits as free from fire and smoke as is practical. As such, all exits must be protected with fire-protective-rated construction based on the number of stories served by the exit. Table 16.1 summarizes the fire resistance rating requirements for exit enclosures other than horizontal exits. (See Section 16.3.3 in this chapter for details on horizontal exits; refer to Chapter 15 for more on fire resistance ratings.)

16.2.1 Doors and Other Openings in Exit Enclosures

With regard to doors and other openings in exit enclosures, the biggest compliance problem in health care facilities tends to be doors that do not protect the exit enclosures properly. In some cases, the labels are missing or painted over so that the fire

Chapter 16. Egress Components

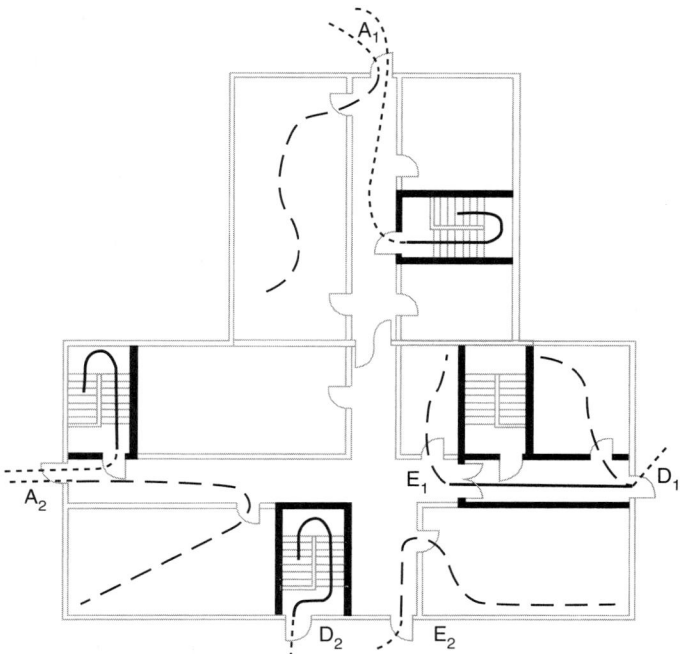

Figure 16.1 Examples of exit access, exit, and exit discharge. To the occupant of the building at the discharge level, the doors at A$_1$, A$_2$, E$_1$, and E$_2$ are exits, and the path denoted by dashes is the exit access. To the person emerging from the exit enclosure stairs, doors A$_1$ and A$_2$ and the paths denoted by dotted lines are the exit discharge. Doors D$_1$ and D$_2$ are exit discharge doors. Solid lines are within the exit.
Source: *Fire Protection Handbook*, NFPA, 2003, Figure 4.3.6.

resistance rating cannot be determined and verified. Also, many doors do not self-close and self-latch properly.

Since every opening and penetration in exit enclosures increases the chances for the leakage of smoke into an exit enclosure, the only openings allowed in an exit enclosure are doors leading from normally occupied areas and corridors and the doors that allow ultimate egress out. Heating, ventilating, and air-conditioning (HVAC) ducts, even if pro-

Table 16.1 Fire Resistance Rating Requirements for Exit Enclosures (excludes horizontal exits)

Exit	Walls (fire resistance rating)	Doors (fire protection rating)
New—serving ≤ 3 stories	1 hr	1 hr
New—serving ≥ 4 stories	2 hr	1½ hr
Existing exit—not high-rise	1 hr*	¾ hr*
Existing exit—high-rise	2 hr	1½ hr
Existing—building protected by sprinklers	1 hr*	¾ hr*

*For existing installations only, current protection cannot be reduced to this level if it exceeds this. Any alteration must meet the requirements for new construction.

vided with smoke dampers, transfer grilles, and panels, are not permitted, nor are doors leading from typically unoccupied areas (trash rooms, boiler rooms, mechanical equipment rooms, etc.). The only penetrations allowed in exit enclosures are as follows :

- Electrical conduit serving the stairway (not serving other portions of the building except for penetrations for fire alarm circuits where fire alarm circuits are installed in metallic conduit and penetrations are properly protected)
- Ductwork and equipment necessary for independent stair pressurization (not for HVAC systems serving the building)
- Water or steam piping necessary for the heating or cooling of the exit enclosure (not serving the rest of the building)
- Sprinkler piping and standpipes regardless of the area of service (sprinkler and standpipe risers and related valves can be in the exit enclosure)

Existing penetrations in exit enclosures, such as pipes and electrical conduit, can remain as long as the penetrations are properly sealed. Existing penetrations should be documented, and trade workers such as telephone and computer network installers and plumbers must be told that running material into exit enclosures is not tolerated. (See Section 15.5 in this book for information on penetrations.)

These provisions are often violated in both new and existing buildings. It is not at all uncommon to find panels covering access to concealed spaces or mechanical areas installed in exit enclosure walls. These openings are not permitted, regardless of the fire protection rating of the panel; they should be removed and the opening sealed the same as the wall. Access to these areas must be provided from outside the exit enclosure. Also, doors to normally unoccupied rooms such as mechanical equipment rooms, boiler rooms, telephone equipment rooms, and similar spaces are commonplace. This problem is often difficult to correct, because moving the door may be totally impractical. Adding a small vestibule on the room side of the door (in essence a small corridor) may be possible, but in many situations an equivalency is needed. (An equivalency is where one documents to the authority having jurisdiction that a certain arrangement, although not code compliant, provides an equivalent level of safety. This is different from a waiver, which is when an authority having jurisdiction modifies a code requirement due to difficulty in compliance.) More recent editions of the *Life Safety Code* allow existing fire protection–rated doors to interstitial spaces under certain conditions. See your edition of the *Code* for details.

16.2.2 Use of Exit Enclosures

Another problem related to protecting exits is how the exit enclosures are sometimes used. The *Life Safety Code* specifically prohibits an exit from being used for anything that could interfere with the use of the space as an exit. Therefore, it prohibits the storage of materials in exit enclosures; no equipment can be located in an exit enclosure other than that needed for the exit enclosure (i.e., exit signs, lights, heating or cooling equipment for the enclosure, sprinkler and standpipe equipment, and similar items). Storage of noncombustible items is not specifically prohibited unless it can interfere with egress. However, it is far easier to prohibit all storage in these areas because that eliminates the need to make a judgment of whether an item may interfere with egress or not. Storage under stairs is similarly prohibited, but there are specific provisions for enclosed storage. See the *Life Safety Code* for details.

16.2.3 Protection of Corridors

The *Life Safety Code* has general provisions for protecting corridors, as shown in Figure 16.2. In general, corridor provisions provide for a somewhat protected path of egress to the exit, although the corridor does not need to be protected as much as an exit. Another benefit of a corridor is that it helps contain the fire in the room of origin and protects people occupying other rooms. This is what might be described as the first level of defense. If the fire is in room A, everyone moves out of room A, and the door is closed, which keeps the fire and smoke in room A, protects the corridor for egress, and slows the movement of products of combustion into room B.

However, all of the occupancies, except assembly, include provisions that override the basic requirements and include their own provisions for corridor protection (as provided in Section 3.6 of each occupancy chapter). Due to these overrides, health care facilities must follow several different corridor provisions depending on the occupancy involved.

16.2.3.1 Health Care Occupancy Corridor Protection

The first thing to determine for a health care facility is whether a corridor wall is required. Health care occupancies clearly require *all* use areas to be separated from the corridor by a corridor wall. There are additional requirements for the wall in terms of fire rating, smoke resistance, and termination, and for the doors, air transfer openings, and miscellaneous openings.

Of course, there are exceptions to these requirements for separation to allow for function. Table 16.2 summarizes the exceptions that allow spaces to be open to the corridor. Although the exceptions may seem intimidating, exceptions based on the type

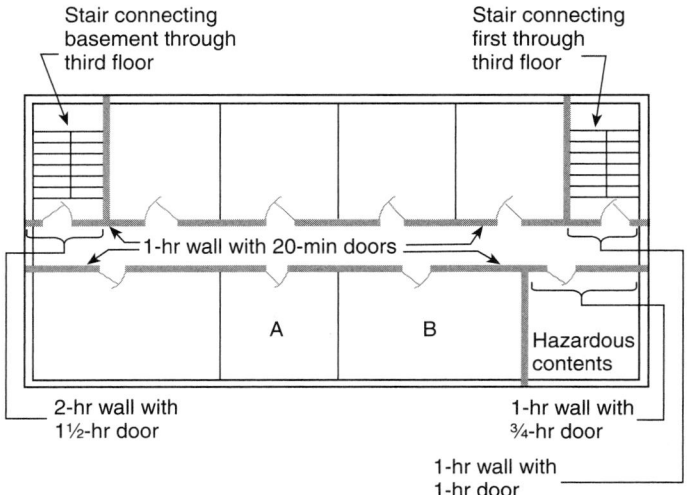

Figure 16.2 An example of protective features of exit access corridors.
Source: Life Safety Code Handbook, 2003, Exhibit 7.4.

Table 16.2 Exceptions to Requirements for a Corridor Wall

Space Open to Corridor[1]	Application	Use Limit[2]	Size Limit	Corridor Smoke Detection	Open Space Smoke Detection	Sprinkler Protection[3]
Any	New/existing	No sleeping or treatment	None	Yes, unless QRS[4]	Yes, unless direct observation	Yes, total smoke compartment
Waiting area	New/existing	Waiting	600 ft^2 (55.7 m^2) total per smoke compartment	No	Yes, unless direct observation	Yes, total smoke compartment
Nurse's station	New/existing	Nurses' station	No	No	No	New—yes Existing—no
Gift shops	New/existing	Nonhazardous area gift shop	See gift shop provisions	No	No	Yes—gift shop
Limited care facility	New/existing	Group meeting space	No	No	Yes, unless direct observation	Yes, total smoke compartment
Any	Existing	No sleeping or treatment	None	Yes	Yes	Yes, space only, or limit the contents
Waiting area	Existing	Waiting	600 ft^2 (55.7 m^2) per area	No	Yes	No
Limited-care facility	Existing	Group meeting space—staff supervision	1500 ft^2 (140 m^2)—1 per smoke compartment	No	Yes	No

[1]In no case can an area obstruct access to a required exit.
[2]In no case can a hazardous area be open to the corridor.
[3]See *Life Safety Code*, for details on extent of sprinkler protection and type of sprinklers required.
[4]QRS stands for quick response sprinklers.

of facility and the area under consideration can be ruled out. For example, when considering the requirements for hospitals, the requirements for limited care facilities can be eliminated. Likewise, areas such as nurses' stations and gift shops can be ruled out if they are not involved. This leaves only four options, as indicated by the categories in the four remaining rows on the table: *any* space in *new or existing* facilities; *waiting areas* in *new or existing* facilities, *any* spaces in *existing* buildings, and *waiting areas* in *existing* buildings. Waiting areas can comply with any of the requirements in these four remaining rows. If the space in question is not a waiting area, then only the requirements in the two remaining rows for any spaces apply. If the space in question is a smoke compartment that is not sprinklered, the only requirements that can be used are those listed for any area in an existing facility. Note that other than group meeting spaces in limited care facilities, open spaces to a corridor can never be used for patient sleeping or treatment. Hazardous areas can NEVER be open to a corridor. Please see the *Life Safety Code* for further details.

Figure 16.3 illustrates the requirements of the provisions allowed for any area in new and existing facilities (the first row of Table 16.2). Figure 16.4 illustrates the requirements of the provisions allowed for waiting areas in new and existing facilities (the second row of Table 16.2).

After determining where corridor walls are required, the construction of the walls and the doors must be evaluated. This construction depends on whether the smoke compartment is protected by sprinklers or not.

Sprinklered Construction

Of course, in new construction, including renovations and alterations, automatic sprinklers must protect the smoke compartment, allowing for quite simple requirements for walls in sprinklered compartments. These walls do not have to have a fire resistance rating, although the corridor walls and doors must resist the passage of smoke. The wall

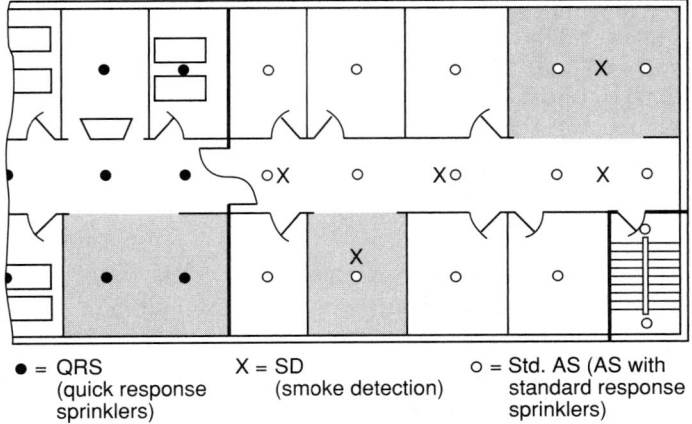

● = QRS (quick response sprinklers)   X = SD (smoke detection)   o = Std. AS (AS with standard response sprinklers)

Figure 16.3 Spaces of unlimited size that are open to the corridor.
Source: Life Safety Code Handbook, 2003, Exhibit 18/19.21.

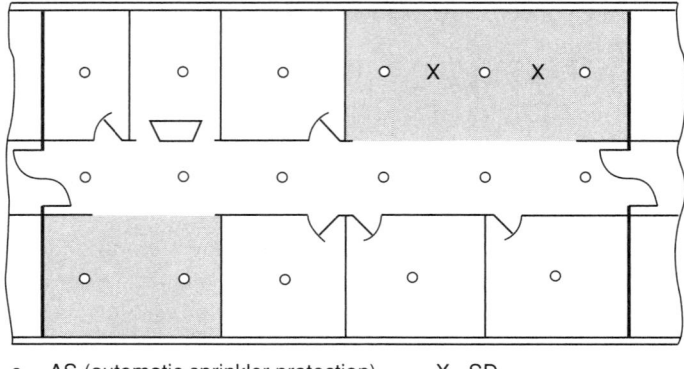

o = AS (automatic sprinkler protection)   X = SD

Figure 16.4 Waiting spaces of limited size that are open to the corridor.
Source: Life Safety Code Handbook, 2003, Exhibit 18/19.22.

can terminate at the ceiling, as long as the wall and the ceiling resist the passage of smoke. It is not the intent that the ceiling pass a "light test" (Does any light at all pass through the wall or ceiling?), only that it resist the passage of smoke. Some smoke may enter during a fire, but it will likely be limited in a sprinklered smoke compartment and will not likely have significant force to push through small openings after the very early stages of the fire.

Although drop-tile ceilings generally resist the passage of smoke, if the space above the ceiling serves as an air-handling plenum, then the ceiling will not likely resist the passage of smoke because of the air transfer louvers in the ceiling. In such cases, the corridor wall must continue to the floor or roof deck above. Other than that, the small ventilation openings in lights and similar objects are not considered to be a problem.

Corridor doors in new construction must be provided with a positive latch. Existing doors must have either a latch or a means for keeping the door closed against a 5-lb [22-N (newton)] force. Although the *Life Safety Code* does not prohibit existing roller latches, the Centers for Medicare and Medicaid Services (CMS) regulations do prohibit them. Door closers are not required unless they are required elsewhere, such as for exit stair doors or for doors to hazardous areas. Figures 16.5 and 16.6 illustrate the corridor wall and door requirements for a sprinklered smoke compartment.

Nonsprinklered Construction

As can be expected, the requirements for protecting corridors are more stringent in existing nonsprinklered smoke compartments. First, the wall must have at least a ½-hour fire resistance rating. This requirement is not difficult to meet, but it does restrict the use of items such as glass. A more significant requirement is that the wall must continue to the floor or roof deck above. A major problem in meeting this requirement for existing facilities is the presence of unprotected penetrations in corridor walls located above ceilings. Another common problem is making sure that the wall is tight

Chapter 16. Egress Components **251**

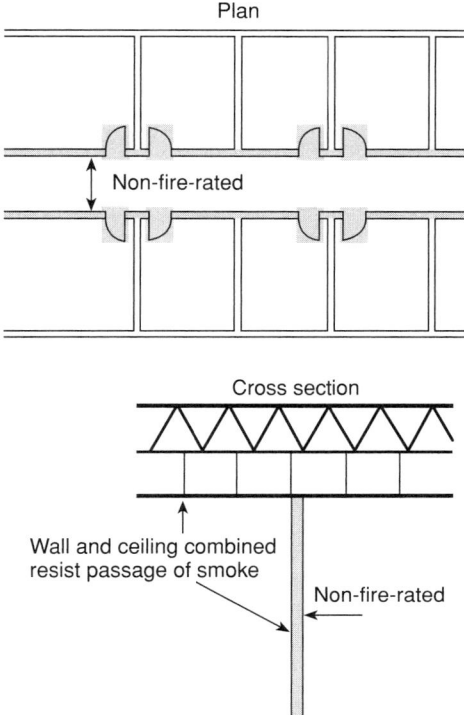

Figure 16.5 Corridor walls permitted in new health care occupancies and in existing sprinklered health care occupancies.
Source: Life Safety Code Handbook, 2003, Exhibit 18/19.23.

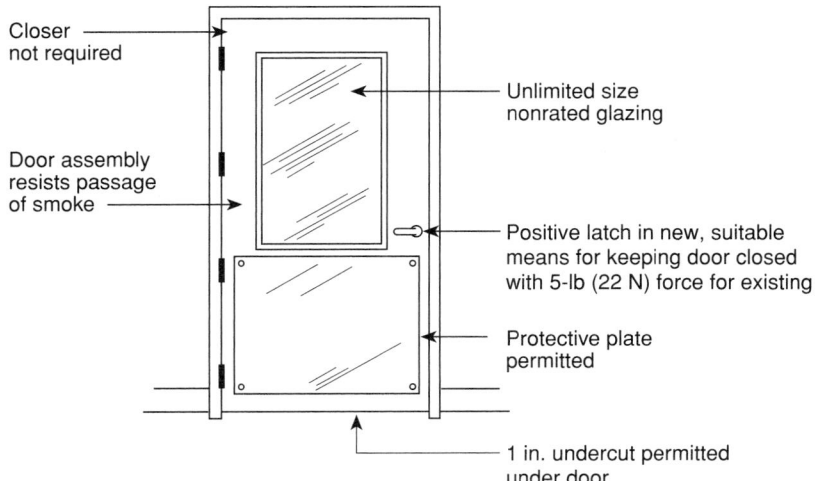

Figure 16.6 Requirements for corridor doors in new health care construction and in existing sprinklered health care occupancies.

to the deck above. Once the wall is fixed so that it continues to the floor or roof deck, it usually stays this way, but penetrations may continue to be a problem, as contractors and other employees may continually cause penetrations of the walls that are left unprotected. In most hospitals, the problem with penetrations in corridor walls above ceilings is so pervasive that the time needed to conduct a thorough Joint Commission on Accreditation of Healthcare Organizations (JCAHO) Statement of Conditions (SOC) survey is almost double for facilities that must have corridor walls that extend through the ceiling to the deck above versus those facilities where the corridor wall can terminate at the ceiling.

The door requirements for corridors in nonsprinklered facilities are similar to those for sprinklered corridors; however, these doors must resist the passage of fire for at least 20 minutes or be a 1¾-in. solid bonded wood core door. Although the *Life Safety Code* originally required these doors to be 20-minute fire doors, fire doors must have closers, and closers are exempt in health care occupancy corridors. Figures 16.7 and 16.8 illustrate corridor requirements in a nonsprinklered smoke compartment.

It is important to remember that any renovation or alteration must meet the requirements for *new* facilities, which are based on sprinkler protection. One major problem with meeting these requirements, however, occurs when making minor renovations to a nonsprinklered smoke compartment, since minor renovations are exempt

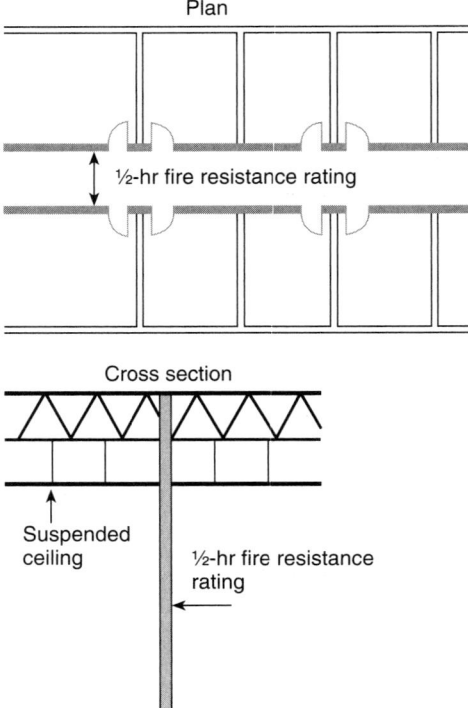

Figure 16.7 Corridor walls permitted in existing nonsprinklered smoke compartments.
Source: Life Safety Code Handbook, 2003, Exhibit 18/19.24.

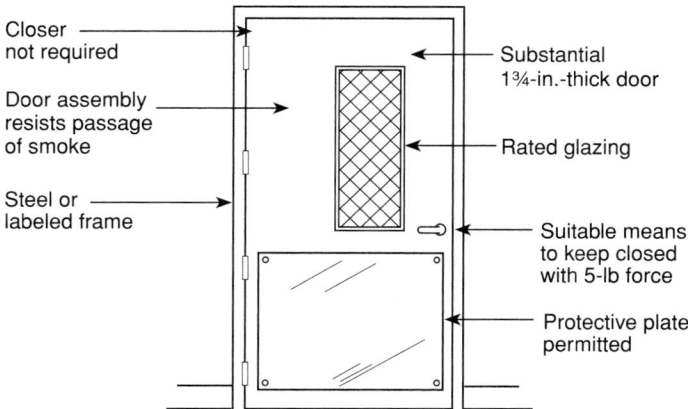

Figure 16.8 Requirements for corridor doors in existing nonsprinklered smoke compartments.

from meeting provisions that cover the sprinklered retrofits. Any corridor wall constructed in a nonsprinklered smoke compartment under the exception for sprinklers should be a 1-hour wall, tight to the deck above. In most cases where corridor walls are being replaced, however, the exception for sprinklers should not be exercised.

Dutch Doors

Another feature that is often found in error in corridor walls of health care occupancies is the way Dutch doors are arranged. In these situations, Dutch doors must be arranged as illustrated in Figure 16.9. Note that under the *Life Safety Code*, the top leaf self-latches into the bottom leaf and the bottom leaf self-latches into the frame. There must be an astragal between the two leaves (i.e., a projective strip on the edge between the top and bottom leaf that restricts the movement of smoke between the two leaves).

16.2.3.2 Corridor Protection in Other Occupancies

Table 16.3 provides an overview of the corridor protection requirements for other occupancies commonly found in a health care facility.

16.2.3.3 Corridor Protection in Mixed and Newly Created Occupancies

Protecting corridors involving mixed occupancies (see discussion in Section 13.6 of this book) and those involving newly created occupancies can present a significant problem. For example, if a former patient floor or wing were converted to offices, it would most likely be a new business occupancy. As such, unless it was a sprinklered building, 1-hour corridor walls with 20-minute self-closing doors might be required. Since this is not what the *Life Safety Code* requires for health care occupancies, the area most likely must be renovated to accommodate the more stringent provisions for the change in occupancy. Another example is where a corridor serves both health care

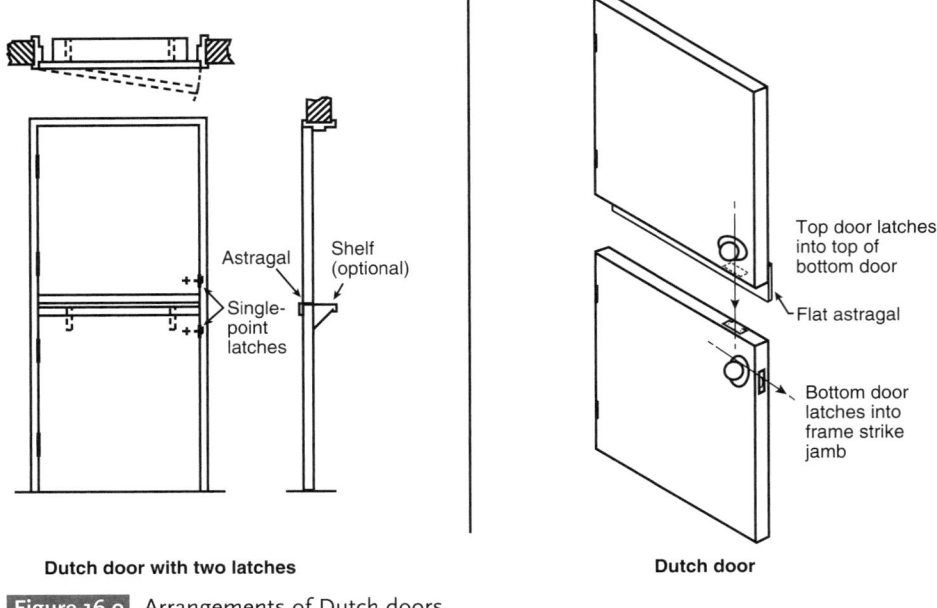

Figure 16.9 Arrangements of Dutch doors.
Source: NFPA 80, 1999, Figure B.26(A).

occupancies and dormitories (i.e., resident sleeping areas or doctors' ready rooms that accommodate more than 16 people). Since dormitories require doors that open into a corridor to be self-closing, this mixed occupancy would be very impractical in a health care facility.

16.3 Components of Means of Egress

The *Life Safety Code* establishes detailed requirements for different egress components, including doors, stairs (either interior or exterior), smoke-proof enclosures, horizontal exits, ramps, exit passageways, fire escape stairs, fire escape ladders, slide escapes, alternating-tread devices, areas of refuge, and elevators. Not all egress elements are allowed in all occupancies. For example, elevators, slide escapes, and fire escape stairs are not permitted to serve a health care occupancy. Additionally, some elements are allowed, but with restrictions. An example of such a restriction is that fire escape ladders and alternating-tread devices are not allowed in all locations in the building. Many devices have optional items such as delayed egress devices that can be used only when allowed in a particular occupancy. There can also be additional restrictions to the provisions for the general means of egress for specific occupancies (e.g., health care occupancies allow the use of delayed egress devices, but only one in each egress path).

Table 16.3 Corridor Protection Requirements for Other Occupancies

Occupancy	New Requirements	New Exceptions	Existing Requirements	Existing Exceptions
Assembly	1-hr 20-min self-closing door	NR–if AS NR–50% exits direct outside	NR	NA
Business and ambulatory health care	1-hr 20-min self-closing door	NR–if AS NR–if single tenant[1]	NR	NA
Industrial and storage	NR	NA	NR	NA
Dormitories	1-hr 20-min self-closing door	½-hr–If AS 20-min self-closing door	½-hr 1¾-in. self-closing door	Smoke resistant–if AS, closer req'd.
Lodging and rooming	Smoke resistant w/ self-closing door	Door closer not req'd. if AS	Smoke resistant w/ self-closing door	Door closer not req'd. if AS

[1] The single-tenant exception must be used very carefully. It is the intent of this exception that everyone works common hours and is familiar with his or her surroundings. This exception cannot be used in many medical office buildings.
NR = not required, AS = protected with an automatic sprinkler, and NA = not applicable

In summary, the provisions in the *Life Safety Code* for means of egress are as follows:

- Some provisions are mandatory for all occupancies.
- Some provisions are optional and allowed only when permitted by certain occupancies.
- Some provisions are optional and mandated only when required by certain occupancies.

What follows is an overview of the most common elements and issues related to means of egress in health care facilities. For details on egress elements see NFPA's *Life Safety Code Handbook*.

16.3.1 Doors

Needless to say, doors are the most common egress element, found in literally all buildings, and are listed as acceptable egress elements for all occupancies. Numerous provisions for doors as means of egress must be followed. The primary requirements relate to the following elements:

- Minimum width
- Type (swinging, sliding, rolling)
- Direction of swing
- Force needed to open
- Locks and latches
- Closers
- Special doors (power operated, revolving, turnstiles, etc.)

Refer to Chapter 15 of this book and NFPA 80, *Standard for Fire Doors and Fire Windows*, for more information on the fire protection rating of doors and the *Life Safety Code* or *Life Safety Code Handbook* for more details on the preceding elements.

16.3.1.1 Minimum Width

Minimum width is addressed in the Chapter 17 on means of egress, as well as in some of the chapters on specific occupancies. Unfortunately, measuring the width of a door is not always simple. There are basically two types of width: *net* or *clear width* and *gross* or *leaf width*. Each of these types of measurement has specific requirements that change at times, so it is important to refer to the specific edition of the *Life Safety Code* in use. In general, net or clear width is the actual clear usable width of the opening (e.g., how wide a box can get through the door) and is illustrated in Figure 6.10. Gross width or leaf width is the actual leaf size with no deductions for doorstops, hinge stile, or hardware. The general rule of the *Life Safety Code* is that net or clear width is used for determining the capacity of a door, whereas gross width or leaf width is used as a minimum width, unless clear width is specified. Table 16.4 summarizes the width required in most of the occupancies present in health care facilities. The 41½-in. (105-cm) dimension may appear odd for doors in new health care occupancies, but it is based on the clear width normally available from a 44-in. (112-cm) door. This door has been the type typically provided in health care occupancies and has

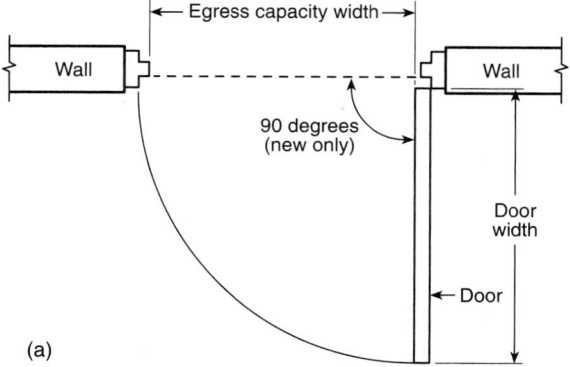

Figure 16.10 (a) Door width, egress capacity.
Source: Life Safety Code, 2006, Figure A.7.2.1.2.1(a).

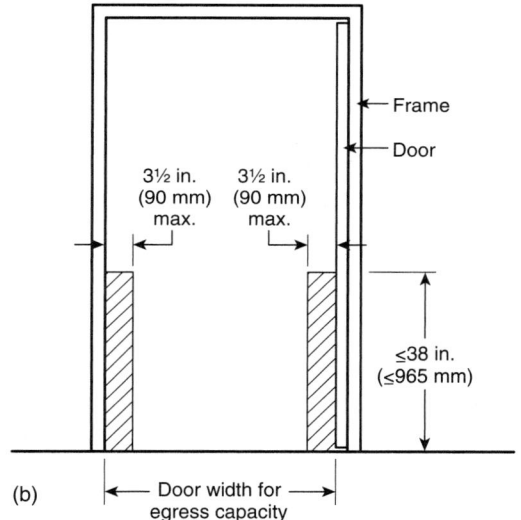

Figure 16.10 (b) Door width, egress capacity with permitted obstructions.
Source: Life Safety Code, 2006, Figure A.7.2.1.2.1(b).

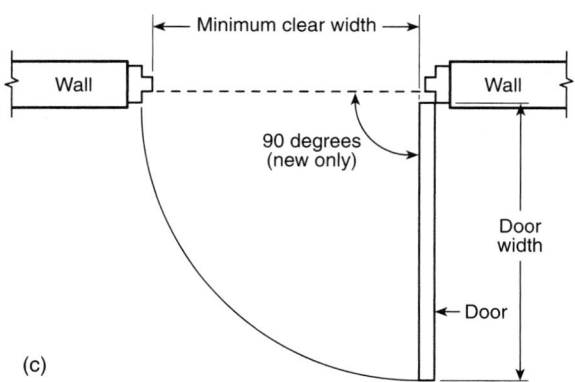

Figure 16.10 (c) Minimum clear door width.
Source: Life Safety Code, 2006, Figure A.7.2.1.2.3(a).

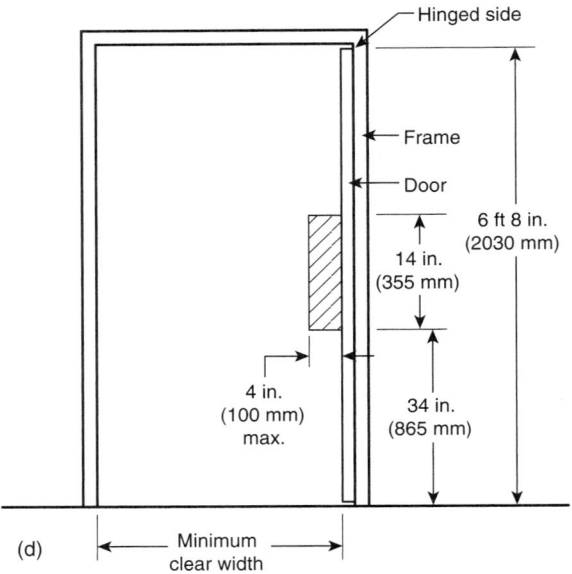

Figure 16.10 (d) Minimum clear door width with permitted obstructions. *Source: Life Safety Code, 2006, Figure A.7.2.1.2.3(b).*

proved satisfactory for egress purposes that might include the passage of gurneys, beds, and wheelchairs.

16.3.1.2 Type (Swinging, Sliding, Rolling)

In general, doors must be of the side-hinged or pivoted swinging type. In other words, rolling and sliding doors are not permitted; however, there are some exceptions. Table 16.5 summarizes some of the exceptions that might be applicable to health care facilities. A review of the table shows that in health care occupancies (or mixed occupancies

Table 16.4 Requirements for Minimum Door Widths, by Occupancy

Occupancy	New	Existing
Health care	41-½ in. (105 cm) clear (see code for exceptions)	32 in. (81 cm) clear or 34 in. (86 cm) leaf
Ambulatory health care	32 in. (81 cm) clear	32 in. (81 cm) clear or 34 in. (86 cm) leaf
Business, assembly	32 in. (81 cm) clear	28 in. (71 cm) leaf
Residential, means of egress	32 in. (81 cm) clear	28 in. (71 cm) leaf
Residential, means of escape	28 in. (71 cm) leaf 24 in. (61 cm) leaf for bathroom door	28 in. (71 cm) leaf 24 in. (61 cm) leaf for bathroom door

Table 16.5 Exceptions to Permitted Door Types

Type of Door	Restrictions	Occupancies	Application
Sliding	None	Detention, residential means of escape	New or existing
Security grilles	Numerous	Assembly, mercantile, business	New or existing
Horizontal sliding (special)	Numerous	All	New or existing
Any	Low or ordinary hazard contents, Occupant load ≤ 10	Private garages, business, industrial, storage (Health care ≤ 9)	New or existing
Sliding	Occupant load < 10	Health care	New or existing
Revolving	Numerous	All (not in egress) Assembly, hotels, business (in egress)[1]	New or existing
Horizontal sliding, vertical rolling	Fusible link operated plus other restrictions	Business, industrial, storage	Existing only

[1] Other occupancies permit this but are not typically associated with health care facilities.

involving a health care occupancy), side-hinged swinging doors must always be used, with the exception of special sliding doors. It should be noted that sliding doors that break away properly are considered side-hinged swinging doors. The *Life Safety Code* requires such doors to break away from any position (fully or partially closed). Beginning with the 2006 edition of the *Life Safety Code*, doors in health care facilities that serve fewer than 10 people do not have to be side-hinged swinging doors. In other words, sliding doors that do not break away are acceptable. This is especially important in suites. It allows sliding doors that do not break away and therefore do not need bottom tracks which are difficult to keep clean and maintain. It should also be noted that the special sliding doors allowed are very specific and have several limitations—these are not normal sliding doors. Figure 16.11 illustrates just one of the requirements for this type of door that make the door unusual.

16.3.1.3 Direction of Swing

Contrary to popular belief, there are only a few situations where the *Life Safety Code* requires doors to swing in the direction of egress:

- Where serving a room or area having an occupant load of 50 or more people
- Where used in an exit enclosure
- Where serving a high hazard contents area
- Where used in a horizontal exit

It is the first item that in most cases forces doors located across a corridor (i.e., a *cross-corridor* door) to swing in the direction of egress, forcing two doors to be installed,

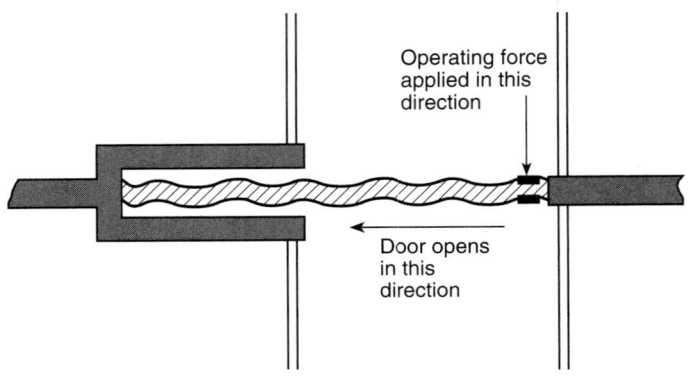

Figure 16.11 Horizontal sliding door operation.
Source: Life Safety Code Handbook, 2003, Exhibit 7.23.

one swinging one way and the second door swinging the other. [If people are coming from both directions in a corridor, one door must swing one way and one door must swing the other. Therefore, the corridor requires two doors. Stops on the door (required for other reasons) prevent one door from swinging in both directions.] Because of this, most smoke barrier cross-corridor doors (as well as horizontal exit cross-corridor doors; see Section 16.3.3 in this chapter) must consist of a pair of doors with one door swinging in each direction. In existing health care occupancies, both cross-corridor doors are allowed to swing in the same direction (i.e., one door swings in the *wrong* direction). When this is done, the doors most often require an astragal, or else close attention must be paid to the gap between the two doors. When astragals are provided and both doors swing in the same direction, a coordinator is needed to ensure that the door without the astragal projection closes before the door with the astragal does.

16.3.1.4 Force to Open

The purpose of the force-to-open requirement in the *Life Safety Code* is to ensure that anyone needing to use a door can open it. Doors in new facilities must be able to be opened with a force not exceeding 15 lb (67 N) to release the latch, 30 lb (133 N) to start opening the door, and 15 lb (67 N) to open the door. Interior doors without closers must open with a force not in excess of 5 lb (22 N). The force to open doors in existing facilities shall not exceed 50 lb (222 N). All forces are measured at the latch stile. For new doors this means that the doors should be relatively easy to open; however, existing doors could require significant force. For power-operated sliding or swinging doors that may be in the means of egress, the door must swing in the egress direction when a force not exceeding 50 lb (222 N) is applied. The application of force must be able to occur at any position in the door slide or swing.

16.3.1.5 Locks and Latches

Requirements for locks and latches are important and frequently violated areas of the *Life Safety Code*. Again, two areas must be addressed: the general provisions in the chapter on means of egress and specific provisions for the various occupancies, most importantly in this case, the provisions for health care occupancies.

Use of Keys, Tools, Special Knowledge, and Effort

The general rule is that all doors in the means of egress must be usable in the direction of egress without the use of *keys, tools, special knowledge,* or *effort*. Note: a lock that must be opened using an ID card requires the use of a key or tool, and a lock that must be opened using a code requires special knowledge. Doors with these types of locks in the direction of egress are acceptable only when the authority having jurisdiction accepts the door as equivalent.

Under limited conditions, double-cylinder locks (requiring a key for operation from the inside) are permitted for use in assembly, mercantile, and business occupancies. This would allow a key-operated double-cylinder lock on the front door into a cafeteria or auditorium or on the front door into a medical office building or administrative building. A door equipped with such a lock could not be in the egress path in health care occupancies unless it was there to meet the clinical needs of the patients (see the second following subsection titled "Special Locking Arrangements").

Number of Actions to Open

Opening the door must be accomplished with no more than one releasing operation. In other words, it is not acceptable for an egress door to have both a latch and a thumb-turn bolt. There are some exceptions to this provision as well, such as for meeting the clinical needs of psychiatric patients (see the next subsection).

An exception allows two (or, in existing installations, three) actions to open the door. However, this exception applies only to doors from individual living units or guest rooms of residential occupancies. For example, this would allow a latch, security bar, and deadbolt (limited to two actions for new doors) to be on the door of a ready room or doctors' sleeping room.

Special Locking Arrangements

Two special locking arrangements are recognized by the *Life Safety Code*. The first is a delayed egress lock and the second is an access control system. With the delayed egress lock, the individual arrives at the door and pushes on the door or a *panic bar* device to activate the system. The door stays locked for 15 to 30 seconds after being activated. During this time, an alarm at the door sounds, and the door releases after the time delay. The door releases immediately upon activation of sprinklers or a fire detection system. The locking mechanism must also be failsafe and release upon loss of power to the system. This system cannot be used unless either a complete sprinkler system or a complete fire detection system protects the building.

It is a major error to install this type of system in many buildings. Since the lock can easily be installed by a locksmith or by staff, it is often installed in buildings that are not protected by a complete automatic sprinkler system. In addition, some occupancies, such as health care, restrict the use of this lock so that not more than one of these locks is present in any single egress path. For example, delayed egress locks can be used to secure a pediatric unit. One lock can be installed on the door to the exit stair and one can be installed on the smoke barrier leading out of the compartment. However, the installation of one of these locks on the other stair is prohibited if travel is required through the smoke barrier to get to that stair, as this route would then have two such locks in it. Facility designers and engineers must be sensitive to how this type of lock might affect other people on the floor. If people are required to use the stair at the end of the pediatric wing for egress, then the delayed egress device could not be used to restrict access to the pediatric area. Figures 16.12 and 16.13 illustrate some acceptable arrangements using these devices. See the *Life Safety Code* for specific details on this system. The following case study illustrates another example of this restriction.

Delayed Egress Lock Restriction

While surveying a small hospital, it was noted that delayed egress locks were being installed for a psychiatric unit at the end of a wing. The wing was not sprinklered and was not fully detected. This violated one of the basic requirements for delayed egress locks. Further examination revealed that the exit in the psychiatric unit was required for egress for the rest of the story. This mandated that the occupants had to go through two delayed egress devices to reach the exit—a prohibited arrangement for a health care occupancy.

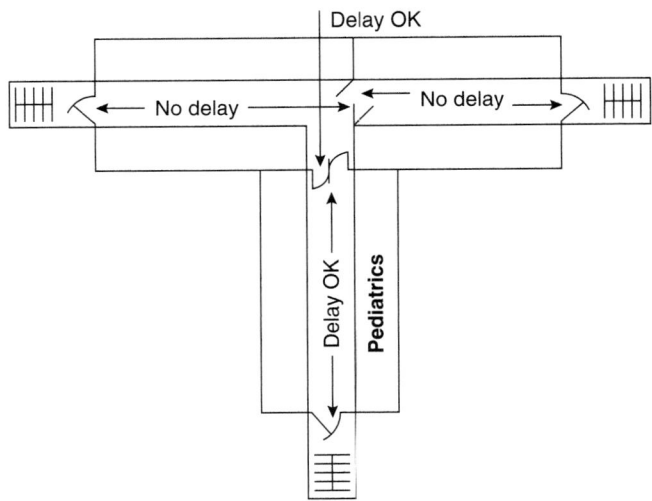

Figure 16.12 Egress arrangement using delayed egress locks—Example 1.

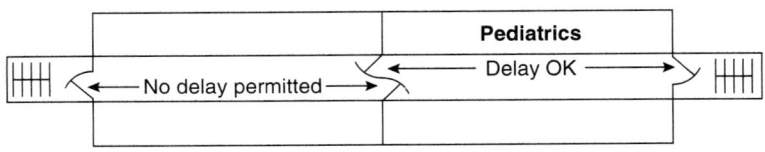

Figure 16.13 Egress arrangement using delayed egress locks—Example 2.

Access control systems are also permitted in health care facilities. These systems do not provide any egress security but can be used to increase security against entry. Access control systems may be useful for maternity units, nurseries, pediatrics, pharmacies, emergency departments, or surgical areas. The systems permit the door to be locked; however, a motion sensor unlocks the door in the direction of egress when a person approaches the door from the egress side. A touchpad, card reader, or similar device can be used on the ingress side. This lock must be failsafe and must be operable by a button near the door on the egress side in case the motion sensor fails. The fire alarm system and sprinkler system must operate this system as well, in buildings with such equipment, although the *Life Safety Code* does not mandate the presence of sprinklers or alarm systems in order to have access control systems. The most common mistake with this type of security device is failing to provide the motion sensor or manual release button or locating the release button in a place that is not obvious. As with delayed egress devices, refer to the *Life Safety Code* for specific requirements for this device, as some occupancies do have additional restrictions.

The *Life Safety Code* has no provisions for locks, such as magnetic locks, that release upon activation of the fire alarm system. If the device is not a delayed egress lock, an access control lock, or a lock as permitted for clinical needs (see the next subsection), it is not permitted by the *Life Safety Code*. To be installed, such devices need the specific approval of the authority having jurisdiction.

Special Restrictions for Health Care Facility Locks

There are some additional restrictions with regard to locks for health care occupancies. The first of these restrictions prohibits locking in either direction for patient sleeping room doors. This allows staff to have rapid access to patients in emergencies as well as allowing patients free egress. However, the *Life Safety Code* allows rooms such as unused sleeping rooms to be equipped with a lock that can be operated only from the corridor side by staff but that does not restrict egress. This type of lock does not allow a patient to lock out staff or for staff to lock in a patient.

Another provision for health care facilities is that egress doors can be locked in the direction of egress where the clinical needs of the patient require special security measures. This is a clinical need, not a security need. The original intent of this provision was to be able to deal with psychiatric patents, but over the years the rule was expanded to cover any patient whose clinical needs required it. Some facilities have used this exception to secure nurseries, although it is difficult to claim that the clinical needs of infants can be met only by locking them in. Security needs of nurseries should be addressed using other means, not by locking the patients in.

When using such locks to meet the *clinical* needs of a patient or group of patients, it is important to evaluate several items. First, the locks must be able to be released rapidly. This must be accomplished by requiring staff to carry keys at all times or by using a reliable method of remote release or other equivalent system. In addition, the lock must be arranged so that it does not affect the egress of others. For example, if a required egress path for a "normal" medical–surgical unit must egress through a psychiatric unit, then locking would not be permitted that affects the egress from the medical–surgical unit unless it was delayed egress locking, an access control system, or equivalent. Where there is a need to lock large portions of a facility, such as in a mental health hospital, a prison hospital, or a similar facility, it will be useful to review the requirements for detention and correctional occupancies. Although the *Life Safety Code* does not mandate this, the requirements for detention and correctional occupancies were originally developed based on the requirements for health care occupancies. The differences are few, but they do address security issues.

Another problem commonly found in health care facilities is with the latches on cross-corridor doors. When both doors open in the same direction, it is common to have an astragal on one door to prevent smoke from passing between the doors. This requires that one door be opened before the other. The *Life Safety Code* clearly requires that when this occurs, the leaf that must be opened second must not have any surface-mounted hardware and must be equipped with automatic flush bolts. This forces people to open the correct door first, and the opening of that door automatically unlatches the other. In most situations, it is advisable to replace the doors with a pair of opposite-swinging doors, which eliminates the problem with the astragal.

16.3.1.6 Closers

Any door required to have a fire resistance rating must be self-closing. In addition, other doors may be required to be self-closing, such as smoke barrier doors, even if they do not need to be rated. This provision applies to exit stair doors, doors to hazardous areas, horizontal exit doors, smoke barrier doors, doors to shafts, and others. One major problem with self-closing doors is that many doors are installed in locations where the self-closing feature is a nuisance. As a result, doors are blocked, wedged, or tied open. To address this breach of the requirements, the *Life Safety Code* permits the use of automatic closing doors (Section 7.2.1). These doors must close upon activation of smoke detectors that sense smoke in the area of the door, rather than close by activation of a fusible link. *NFPA 72®, National Fire Alarm Code®*, provides details for this type of smoke detection. Figure 16.14 is a diagram from *NFPA 72* that illustrates some of the requirements for these smoke detectors.

A very common problem found in health care facilities is automatic door closers with no smoke detector nearby [within 5 ft (1525 mm)]. NFPA *101* (Section 7.2.1), relies totally on *NFPA 72* for the provisions for smoke detectors for doors. *NFPA 72* has been revised to clarify its provisions in this area, as well. This provision benefits health care facilities in that a corridor smoke detection system with detectors possibly as far as 15 ft (4.5 m) from the door can be used to close cross-corridor doors. Note that for health care occupancies, water flow in the sprinkler system and any activation of the building fire alarm system must also close the doors. Doors can be closed by zone or throughout the entire building.

16.3.1.7 Special Doors (Power Operated, Revolving, Turnstiles, etc.)

The *Life Safety Code* has specific provisions that apply to special doors (Sections 7.2.1.9 through 7.2.1.14). Facility engineers should review the *Life Safety Code* if any of these doors are used in their facilities.

Chapter 16. Egress Components 265

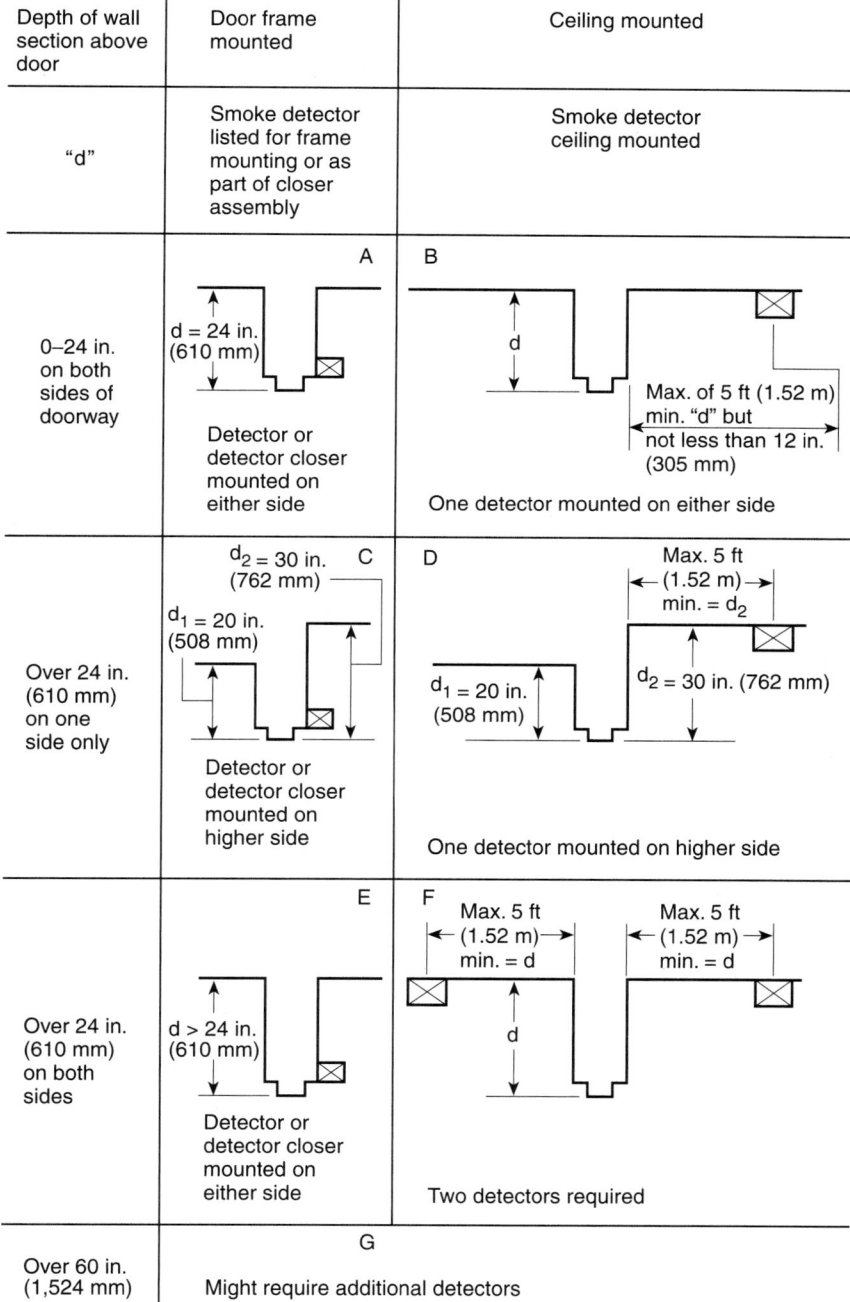

Figure 16.14 Requirements for location of smoke detectors installed for door release.
Source: NFPA 72, 2002, Figure 5.14.6.5.1.1.

Numerous health care facilities use power-operated doors. As discussed earlier in Sections 16.3.1.2 and 16.3.1.4, these doors can be in the means of egress as long as they swing from any position when a force of 50 lb (222 N) is applied in the egress direction at the latch stile. There are some exceptions to this provision, as per Section 7.2.1.9 of the *Life Safety Code*, 2006 edition. In addition, when a power-operated door must be self-closing as discussed earlier, it is very important that a smoke detector at the door disable the power-opening feature and result in the door self-latching, if required.

Very frequently, doors to surgical suites and to intensive care units and similar areas are power opening. If these doors are the doors to a corridor, which in most cases they are, they must be self-latching. In such cases, they must have smoke detection to deactivate the automatic opener and make the door self-latching. In some cases, these doors are in a smoke barrier or a horizontal exit. When this occurs, the smoke detection must also result in the door becoming self-closing.

Revolving doors are not allowed in the means of egress from a health care occupancy, but they can be used outside the egress system. They are also permitted in the egress system of other occupancies often found in health care facilities, such as business occupancies. The *Life Safety Code* has restrictions on revolving doors that are both in the egress path and not in the egress path. The code also addresses the use of turnstiles, but they are not commonly found in health care facilities. Should they occur, the *Life Safety Code* should be reviewed. Doors in folding partitions are also not common in health care facilities. However, they may be present in the folding partitions used in cafeterias or conference areas. As with turnstiles, the *Life Safety Code* should be consulted when such a door is present.

Balanced doors are sometimes used for exterior doors, especially in high-rise buildings or other locations where there may be significant pressure differentials between the outside and the inside of the building. Figure 16.15 illustrates a balanced door. As illustrated by the figure, this type of door will not open if a person pushes on the wrong end of the door. Therefore, the *Life Safety Code* restricts the panic or fire exit hardware when installed on such a door to the push-pad type, with the pad not extending more than halfway across the door from the latch stile.

The special horizontal sliding door is becoming popular in health care facilities. As discussed under door types, this is a very special door not typically thought of as a sliding door. The fact that the *Life Safety Code*, 2006 edition (Section 7.2.1.14), requires the door to slide open when a force is applied in the direction of egress clearly indicates that this is not a typical sliding door. [The typical sliding door can extend across an egress path only if it *breaks away* when a force of 50 lb (222 N) is applied to the latch side of the door.] Figure 16.16 illustrates an example of a special horizontal sliding door.

16.3.2 Stairs

Stairs are the second most common egress element in most buildings, especially multistory buildings. They are allowed as an egress element in all occupancies. The *Life Safety Code* goes into extensive detail on stairs, including both interior and exterior

Chapter 16. Egress Components **267**

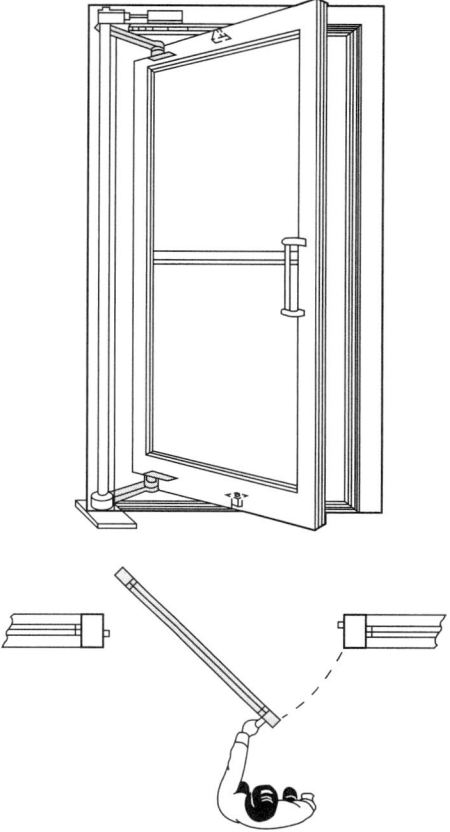

Figure 16.15 A balanced door.
Source: *Life Safety Code Handbook*, 2003, Exhibit 7.21.

Figure 16.16 A horizontal sliding door.
Source: *Life Safety Code Handbook*, 2003, Exhibit 7.24; photo courtesy of Won-Door Corp.

types of stairs (Section 7.2.2). Stairs can serve as an exit when properly protected, an exit access or exit discharge, or merely as a convenient means of movement and not part of the required means of egress for the rest of the building.

The provisions for stairs that serve as exits and the protection of stairs that are exit enclosures are discussed in Section 16.2 and in Chapter 15, Section 15.3.3.

Regardless of their use, if a stair pierces a floor between two stories, it is a vertical opening and must be protected as such. When it is both an exit and a vertical opening, the exit enclosure requirements are more stringent and must be met. In many cases, the stair is neither a vertical opening nor an exit and does not have to be protected as either one. Examples of this are stairs to stages, platforms, mezzanines, and similar locations.

Many stair details are determined when the stair is built, and it is important that the architect or designer be aware that the requirements of the *Life Safety Code* must be met. Many errors generated by the lack of this knowledge are expensive to correct, and if it is not clearly spelled out in the contract between the building owner and the architect that the provisions in the *Life Safety Code* must be met, the facility may end up paying for the corrections.

Requirements for handrails and guards are included in the stair details. These are critical safety items, for both new and existing installations. It is important to ensure that all guards and handrails comply with the *Life Safety Code*. See the discussions in Section 16.3.2.2 of this chapter and in the *Life Safety Code*, Section 7.2.2.4.

16.3.2.1 Smokeproof Enclosures

The *Life Safety Code* also has provisions for smokeproof enclosures, which are a type of exit stair enclosure. With very few exceptions, the *Life Safety Code* does not mandate the use of smokeproof enclosures for stairs, but many building codes require them for high-rise buildings.

There are many ways of accomplishing a smokeproof enclosure for stairs, including an open-air balcony, a naturally or mechanically ventilated vestibule, or stair pressurization. Figure 16.17 illustrates four methods of providing a smokeproof enclosure.

16.3.2.2 Handrails and Guards

Other common problems with stairs involve handrails and guards. Details for handrails and guards are found in the *Life Safety Code*, 2006 edition, Section 7.2.2.4. For existing stairs, handrails must be between 30 in. (76 cm) and 38 in. (96 cm) above the foremost projection of the tread. If the handrail also serves as a guard, it can be as high as 42 in. (107 cm). In many cases, especially where a residential unit has been converted to a medical office building or similar facility, the handrail is too low. Another issue that often arises with stairs is that the guards are too low. The *Life Safety Code* requires means of egress that are more than 30 in. (76 cm) above the floor or grade below to have guards. That is, if someone can fall 30 in. (76 cm) or more over the open side of the egress, a guard is needed. The guard must be 42 in. (107 cm) high, unless an existing handrail [30 to 38 in. (76 to 96 cm)] is serving as the guard. It is not at all

Chapter 16. Egress Components **269**

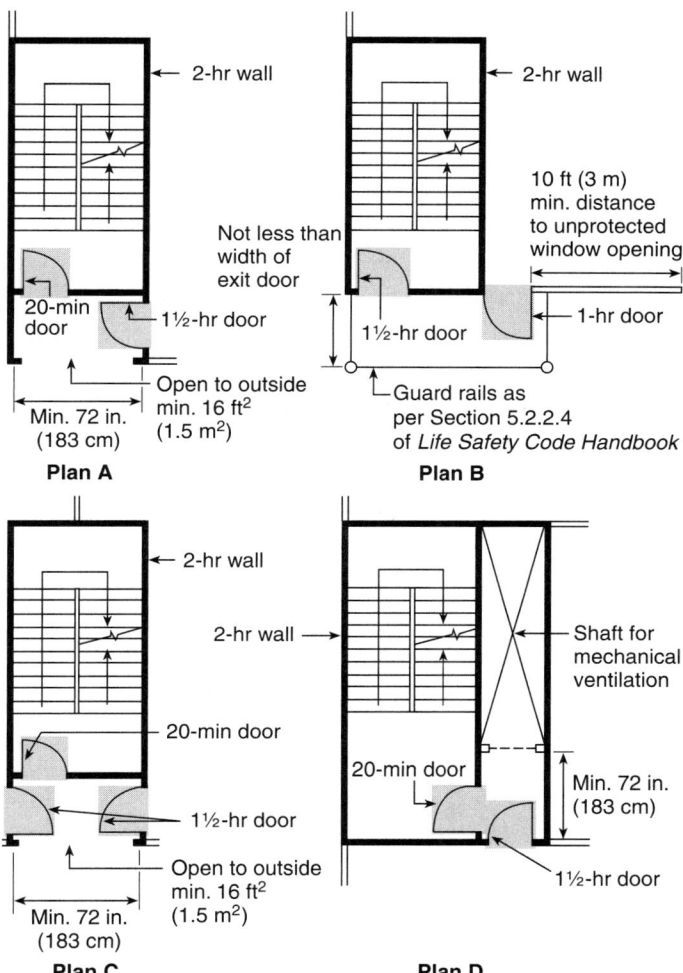

Figure 16.17 Four variations of smokeproof enclosures.
Source: *Life Safety Code Handbook*, 2003, Exhibit 7.47.

uncommon to find the guard at the top landing of a set of stairs to be below the 42-in. (107-cm) height. This presents a significant hazard that should be corrected.

The *Life Safety Code* requires, with some exceptions, that new guards be constructed such that a sphere 4 in. (10.2 cm) in diameter is not able to pass through the guard. Although this is not required for existing guards, health care facilities should evaluate all existing guards. If they do not meet this requirement, an assessment should be conducted to determine the probability of children under age 10 having access to the area, as these children may be able to pass through the guard. If it is a public area, serves a public area, or is in pediatrics or similar areas, serious consideration should be given to retrofitting the guard so that it will comply with the 4-in. (10.2-cm-) sphere rule.

16.3.2.3 Stair Identification

Two other issues that often arise with regard to stairs deal with signage within the stair enclosure. First, the *Life Safety Code* requires stair identification signs wherever a stair serves five or more stories. Starting with the 2006 *Code*, stairs need this sign in new construction when serving five or more stories. Figures 16.18 and 16.19 illustrate the requirements for these signs, which are often lacking in mid- and high-rise health care facilities.

Second, stairs that require upward movement for more than one story for egress require signage within the stair that clearly indicates the direction of egress. If one enters a stair, the natural action is to head down to reach an exit. This sign will point people in the proper direction.

Figure 16.18 Stair sign providing all required information.
Source: Life Safety Code, 2006, Figure 7.2.2.5.4.

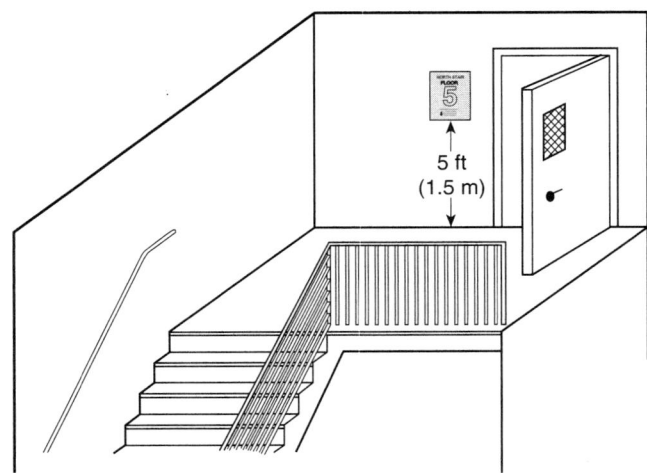

Figure 16.19 Required placement of stair sign to ensure that it is readily visible.
Source: Life Safety Code Handbook, 2003, Exhibit 7.42.

16.3.3 Horizontal Exits

Horizontal exits are very beneficial to health care facility occupants, such as bedridden patients or patients requiring movement aids, as these exits allow horizontal movement for exiting a multistory building without the use of stairs. Being able to move surgical patients, intensive care patients, emergency room patients, and others across a fire barrier into a fire compartment that is separated from the area of the fire has significant fire safety and medical care benefits.

Horizontal exits are considered to be exits. They are essentially the same as an enclosed exit stair or a door to the outside when evaluating capacity of means of egress, measuring travel distance, and evaluating the number and remoteness of exits. See Chapter 17 of this book for additional discussion of these principles. The only drawback related to horizontal exits is that no more than 50 percent of a facility's exits, in terms of total number of exits and total capacity of exits for the compartments formed, can be horizontal exits (66⅔ percent for health care occupancies). Since most health care occupancies must have smoke barriers, establishing the smoke barriers as horizontal exits results in limited extra cost (especially in new construction), with the significant added benefit of being able to reduce the number of stairs and doors to the outside. The added benefit for patient care of keeping a patient inside during a fire is a significant benefit.

As with many egress elements, the basic requirements for horizontal exits are found in the means of egress chapter in the *Life Safety Code* (Chapter 7), and the occupancy chapters generally just reference this section. However, the health care occupancy provisions provide some important modifications to these requirements. Table 16.6 illustrates some of the differences between a horizontal exit and a smoke barrier in a health care occupancy.

The fire compartment on each side of a horizontal exit must have enough room to accommodate the total number of people from both sides. The size of the compartment

Table 16.6 Requirements for Horizontal Exits and for Smoke Barriers

Item	Horizontal Exit New	Horizontal Exit Existing	Smoke Barrier New	Smoke Barrier Existing
Wall	2-hr	2-hr	1-hr	½-hr
Door	1½-hr	1½-hr	"20 min"[1]	"20 min"[1]
Door latch	Yes	Yes	NR	NR
Door auto or self-closing	Yes, cross-corridor auto-closing Req'd.	Yes, cross-corridor auto-closing Req'd.[2]	Yes	Yes
Door vision panel	Yes	NR	Yes	NR
HVAC duct	Permitted if AS	Permitted	Permitted	Permitted
Damper	Fire	Fire	Smoke[3]	Smoke

[1]Door equivalent to a 20-min fire door but no latch required.
[2]Existing self-closing doors may be continued in use where approved by the authority having jurisdiction.
[3]Required only for transfer openings, not for ducted systems.
AS = protected with an automatic sprinkler, NR = not required.

on each side of the horizontal exit must be the same size as that required for smoke compartments (see Chapter 18). Figure 16.20 illustrates many of the provisions for a new horizontal exit, while Figure 16.21 illustrates the provisions for existing horizontal exits. Both figures apply to health care occupancies.

16.3.4 Ramps

As with stairs, ramps are an acceptable means of egress component in all occupancies. The *Life Safety Code* contains details for ramps in the means of egress chapter (Section 7.2.5). The provisions for new construction are very similar to the regulations that accommodate the mobility impaired. It is important to check these requirements, especially when dealing with new construction including renovations. This is very critical in the United States, due to the Americans with Disabilities Act (ADA). The guidelines developed by the U.S. Department of Justice to comply with this act, *ADA Accessibility Guidelines for Buildings and Facilities* (ABA/ADA Guidelines), go into detail on this and many other aspects of making facilities usable and safe for people with disabilities.

Ramps are most commonly used as part of an exit access and exit discharge. To properly enclose a ramp as an exit is usually not practical, but it is done at times, especially for shorter ramps. Note that ramps are often required to have handrails that meet the requirements for stairs.

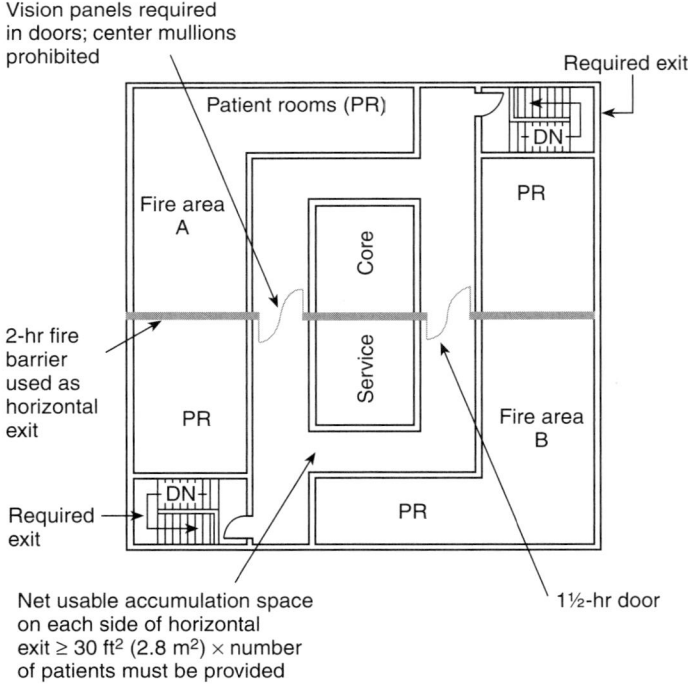

Figure 16.20 Horizontal exit in a new health care facility.
Source: Life Safety Code Handbook, 2003, Exhibit 18/19.3.

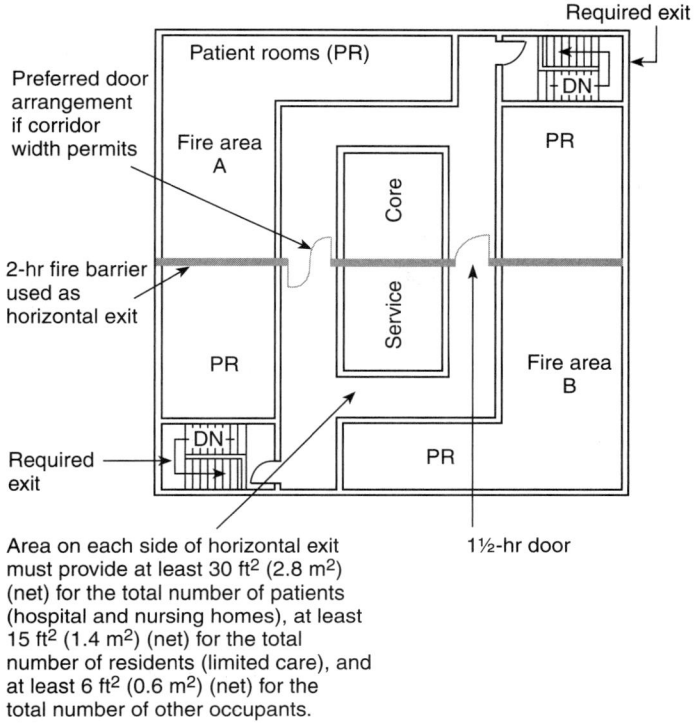

Figure 16.21 Horizontal exit in an existing health care facility.
Source: Life Safety Code Handbook, 2003, Exhibit 18/19.5.

16.3.5 Exit Passageways

The concept of an exit passageway is one that many people have a hard time understanding. An exit passageway is a horizontal means of egress element similar to a corridor or passageway, but it is enclosed in a manner that meets all the requirements for an exit. It might be thought of as an exit stair enclosure on its side. The most common use for an exit passageway is to connect an exit stair to the outside.

Two common problems can be resolved by using exit passageways. The first problem (as discussed in more detail in Chapter 17) is that only 50 percent of the total number of exits may discharge within the building, and then only under limited conditions. One way to correct this problem is to connect one or more exit stairs to the outside with an exit passageway. The second problem relates to shortening the travel distance to an exit, which is less commonly done in health care facilities but is often used in malls. The problem is that exit passageways must meet all of the requirements for an exit enclosure, including the restrictions on openings and penetrations. Unfortunately, many designers think of exit passageways as "beefed up" corridors and allow doors to open into areas that are not normally occupied or install HVAC penetrations, neither of which is permitted in an exit enclosure. Figure 16.22 illustrates the use of an exit passageway to connect an exit stair to the outside, while Figure 16.23 illustrates the use of an exit passageway to shorten travel distance to an exit.

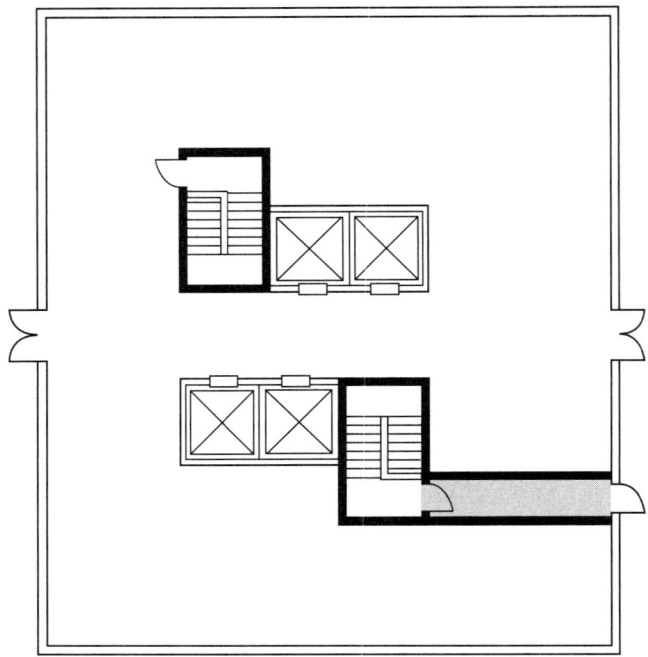

Figure 16.22 Exit passageway used to connect exit stair to the outside.
Source: Life Safety Code Handbook, 2003, Exhibit 7.55.

16.3.6 Escalators and Moving Walks

Escalators and moving walks cannot be part of the means of egress except where previously approved, and then only in certain occupancies (existing assembly, dormitories, and business would be most common in health care facilities). This does not mean that escalators and moving walks are prohibited—it means only that they cannot be in the required means of egress path.

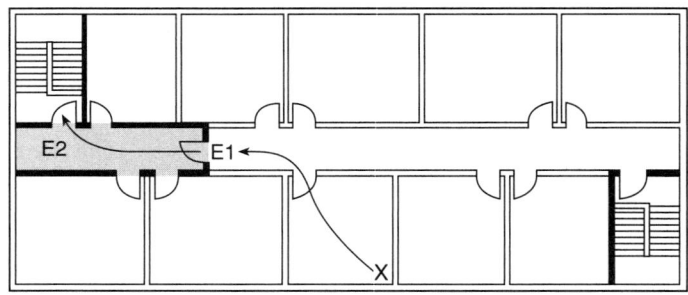

Figure 16.23 Exit passageway used to reduce travel distance to an exit.
Source: Life Safety Code Handbook, 2003, Exhibit 7.56.

16.3.7 Fire Escape Stairs and Ladders and Alternating-Tread Devices

Fire escape stairs, fire escape ladders, and alternating-tread devices are severely restricted in their use. First, it must be noted that a fire escape stair and an outside stair are two very different items. A fire escape stair is narrower and steeper than an outside stair. An outside stair must meet all the requirements of the *Life Safety Code*, 2006 edition, for stairs (Section 7.2.2), while fire escape stairs have significantly fewer restrictive requirements (Section 7.2.8). As a result, the *Life Safety Code* prohibits new fire escape stairs from being built, except under very limited conditions.

All stairs located outside a building must be evaluated to determine if they are an outside stair or a fire escape stair. All stairs determined to be fire escape stairs must be further evaluated to determine if the stair can be given any credit at all as a means of egress. Existing fire escape stairs can be given limited credit as egress stairs, but only in certain occupancies, not including health care occupancies. Table 16.7 provides a comparison of some of the differences between an outside stair and a fire escape stair. Starting with the 2006 edition of the *Life Safety Code,* there are no longer two classes of existing stairs. Class A has been eliminated and all existing stairs must comply with the former Class B requirements.

Most occupancies allow the use of fire escape ladders and alternating-tread devices but with limited conditions on where they can be located and how many people they can serve. Fire escape ladders are permitted only if they are used for the following:

- Access to unoccupied roof spaces
- A second means of egress from storage elevators
- A means of egress from towers and elevated platforms around machinery or similar spaces, subject to occupancy by no more than three persons, all capable of using the ladder

Table 16.7 Differences Between Outside Stairs and Fire Escape Stairs

Design Factor	New Outside Stair	Existing Outside Stair	Fire Escape Stair Normal	Fire Escape Stair Small Buildings
Accepted as exit	Yes	Yes	Existing buildings	Existing buildings
Width	44 in.*	44 in.*	22 in.	18 in.
Maximum rise	7 in.	8 in.	9 in.	12 in.
Minimum tread	11 in.	9 in.	9 in.	6 in.
Tread construction	Solid	Solid	Solid	Metal bars
Access by windows	No	No	Yes	Yes
Swinging stair accepted	No	No	Yes	Yes
Ladder accepted	No	No	No	Yes

For SI units, 1 in. = 2.54 cm.
The capacity of normal fire escape stairs is 45 persons if accessed by doors and 20 if accessed by windows where it is necessary to climb over a sill. On small buildings, the capacity is 10 persons; 5 if winders or a ladder from bottom landing; or 1 if both winders and a ladder from bottom landing.
*36 in. where serving occupant load of fewer than 50.
Source: Life Safety Code Handbook, 2003, Table 7.1.

- A secondary means of egress from boiler rooms or similar spaces, subject to occupancy by no more than three persons, all capable of using the ladder
- Access to the ground from the lowest balcony or landing of a fire escape stair for very small buildings

Alternating-tread devices are permitted in the same locations as listed above, except for the last item. Figure 16.24 is an example of an alternating-tread device.

Figure 16.24 An alternating-tread device.
Source: Life Safety Code Handbook, 2003, Exhibit 7.60.

16.3.8 Accessible Means of Egress and Areas of Refuge

An *accessible means of egress* is a path of travel usable by a person with a severe mobility impairment that leads to a public way, a horizontal exit, or an area of refuge. For a ground-level story or a single-story building, the accessible means of egress can be a door or ramp that leads directly to grade and the public way. To have an accessible means of egress in a multistory building, however, horizontal exits or areas of refuge must be available. The *Life Safety Code* permits basically two types of areas of refuge:

- For buildings protected by supervised automatic sprinklers (i.e., sprinklers that are electronically monitored using alarms such that, for example, if a valve closes or power to a fire pump is disconnected, an alarm goes off; refer to Chapter 21), an area of refuge can be the story one is on. Unless a specific occupancy exempts the two-room requirement, there would be no further requirements except to provide two accessible rooms that are separated from each other by smoke-resisting partitions.
- A space, in a path of travel leading to a public way, that is protected from the effects of fire, either by means of separation from other spaces in the same building or by virtue of location, that could provide an area of refuge and permit the delay of egress travel from any level.

These provisions, which first appeared in the *Life Safety Code* in 1991, are included in the means of egress chapter. The *Life Safety Code* provides extensive detail on how to provide accessible means of egress for areas that are not on a first (ground) floor and in buildings that are not protected by sprinklers, and how to arrange and construct the space and egress from the space to the public way (Section 7.2.12). Although the specific details are too extensive to be discussed here, the *Life Safety Code Handbook* should be referenced in this regard and the following points noted.

First, the area of refuge must have access to the public way. A special stair or special elevator, both of which can be costly, can provide this access. The *Life Safety Code* provides details on the provisions for stairs and elevators that lead from areas of refuge to the public way. Second, providing automatic sprinkler protection can eliminate many of the special requirements. Third, the use of a horizontal exit for the area of refuge also reduces some of these requirements.

To meet the more complex requirements for a multistory building where it is impractical to have a door or ramp that leads directly to grade, several options are available. Some of the most common options are as follows:

- Provide automatic sprinkler protection and treat the entire story as the area of refuge (two accessible rooms separated by smoke-resisting barriers are required in most occupancies)
- Subdivide the story by a horizontal exit with access to standard stairs on each side (illustrated by Figure 16.25)
- Subdivide the story with a smoke barrier having a 1-hour fire resistance rating, each resulting area of refuge requiring access to a special elevator or special stair [48 in. (122 cm) between handrails] (illustrated by Figure 16.25)
- Provide oversize landings on the exit stair landing along with the special stair restrictions (illustrated by Figure 16.26).

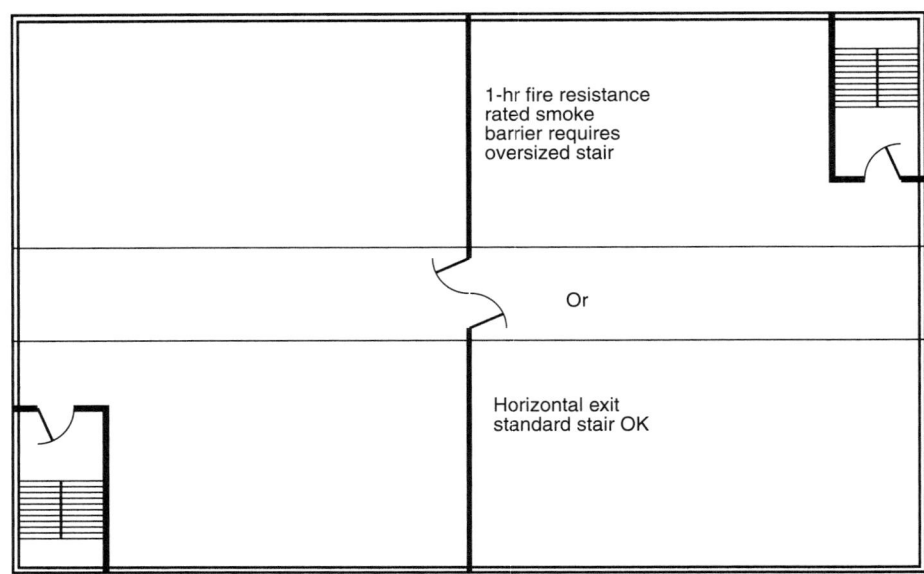

Figure 16.25 A story subdivided by a horizontal exit, with access to standard stairs provided on each side. Also a story subdivided by a 1-hr smoke barrier with access to oversized stairs.

According to the *Life Safety Code*, all new construction must have accessible means of egress (Section 7.5). For stories in a new facility that do not have doors or ramps to grade, an area of refuge is required. Health care occupancies that are protected by automatic sprinklers are not required to have accessible means of egress. The provisions for health care adequately address the needs of the mobility impaired. The provisions apply to all other new occupancies or mixed occupancies.

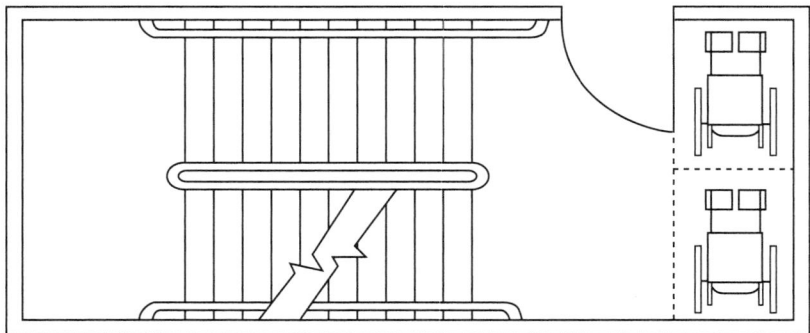

Figure 16.26 Oversized landings on exit stairway.

17

Principles of Egress

In addition to using properly rated materials in the proper configurations, building a facility with safe egress also involves applying several principles of egress. These principles include such factors as the number of exits a facility must have to be safe, the arrangement of means of egress, the distance people must travel to reach an exit, the number of people who can safely occupy an area, and door and corridor widths, to name a few.

When designing a *new* facility, all principles of egress must be applied to calculate the capacity of each egress element, the number of elements that must be installed for safety, and their arrangement. When responsible for maintaining an *existing facility*, these principles come into play to ensure, for example, that areas used for egress are not used inappropriately for equipment storage, corridors and doors are not blocked shut or inappropriately propped opened, exit discharges are cleared of snow and ice, exit signs are properly illuminated, and other key issues are dealt with.

17.1 Egress Capacity

It is important for the available egress capacity to meet or exceed the total number of people who might occupy a building or a space or story within that building at any one time (i.e., the *occupant load*). In health care occupancies, egress capacity is not often an issue. However, in assembly occupancies, such as a cafeteria or conference room, it is often a problem. There are five steps to evaluating egress capacity:

1. Determine the occupant load
2. Determine the clear width of each means of egress component
3. Determine capacity of each component based on its clear width
4. Determine the most restrictive component of each route
5. Determine if egress capacity is sufficient

Egress capacity also must consider the minimum dimensions for egress elements.

17.1.1 Determining the Occupant Load

Several pieces of information are needed to determine the total number of people who might occupy a building or a portion of the building at any one time. This information includes the area of a floor and an occupant load factor, which is based on the use (not necessarily the occupancy) of the space. To calculate the occupant load, the floor area (in either square feet or square meters) is divided by the appropriate occupant load factor. The occupant load for an area or story is the *actual* maximum number of people that the area or story would most likely hold at any given time, but not less than the number of people determined by dividing the floor area by the appropriate occupant load factor.

It should be noted that occupant load is an often-misunderstood concept. Occupant load factors are used to determine the *minimum* number of people who must be *provided with egress*, not the *maximum* number of people who can *occupy an area*. The intent of NFPA *101*®, *Life Safety Code*®, is for facilities to provide sufficient egress for the people in the building or space. It is not the intent, other than in assembly occupancies, to limit the number of people in the area. It is acceptable to have more than the calculated number of people in the area in question as long as the proper egress is provided for that number of people. (There are limits for assembly occupancies so that occupancies such as nightclubs do not become overcrowded, but this is generally not an issue with health care facilities.) Thus, the occupant load used to determine required egress capacity is the actual maximum number of people who would likely fill a space but not less than the number determined by calculation. And, egress capacity must always exceed occupant load. With only one exception for existing assembly occupancies, the egress capacity must be increased to meet occupant load; occupant load can never be reduced below that calculated.

The *Life Safety Code* provides occupant load factors to use in determining occupant loads (see Section 7.3.1.2 in the 2006 edition of the *Code*).

Table 17.1 provides the occupant load factors commonly used in health care facilities. (Note that starting with the 2006 edition of the *Code*, storage areas are calculated at 500 ft² per person.)

Table 17.1 Common Occupant Load Factors Used in Health Care Facilities

Use	ft² (m²)	Net or Gross
Cafeteria, conference room	15 (1.4)	Net
Auditorium (without fixed seats)	7 (0.65)	Net
Kitchens	100 (9.3)	Gross
Fixed seating	Seat count	NA
Classroom seating	20 (1.9)	Net
Offices	100 (9.3)	Gross
Industrial (shops, boiler rooms, etc.)	100 (9.3)	Gross
Inpatient sleeping departments	120 (11.1)	Gross
Inpatient treatment departments	240 (22.3)	Gross
Day care facilities	35 (3.3)	Net
Storage		
In storage occupancies	No factor	NA
In mercantile occupancies	300 (27.9)	Gross
In all other occupancies	500 (46.5)	Gross

NA = not applicable.

Note that in some cases the net area is used, and in other cases the gross area is used. *Gross area* is the entire area involved, with no deductions for walls, columns, corridors, or areas occupied by fixed furniture—it is that area within the exterior walls. *Net area* is the actual area that can be occupied, discounting the space occupied by walls, columns, corridors, and fixed furniture. Figure 17.1 illustrates the difference between net and gross area.

It is also important to note that the occupant load is determined based on the *use* of the space, not necessarily the occupancy of the space. For example, a room for assembly purposes is calculated based on the assembly occupant load factors, regardless of whether there are 50 people in the area. Similarly, a classroom is evaluated using the educational classroom factors, even if it is used for students beyond the twelfth grade.

Remember that the occupant load is the minimum number of people who must be provided with sufficient egress. If more people are in an area, more capacity must be provided by adding more exits or increasing the width of the exits already available. Conversely, capacity cannot be reduced if fewer people are in an area. If the calculated occupant load is 300 people and only 200 people are in the area, capacity must be provided for 300 people.

17.1.2 Determining the Clear Width of Egress Components

Once the occupant load has been determined, the next step in determining egress capacity is to determine the available clear width of the available egress paths. To do this, it is necessary to measure each element in the means of egress to determine the clear width it provides. When taking these measurements, projections up to $3\frac{1}{2}$ in. (8.9 cm) (2000 and older editions) or $4\frac{1}{2}$ in. (114 mm) (2003 and newer editions) at and below handrail height [38 in. (96 cm)] can be ignored. When measuring doors, it

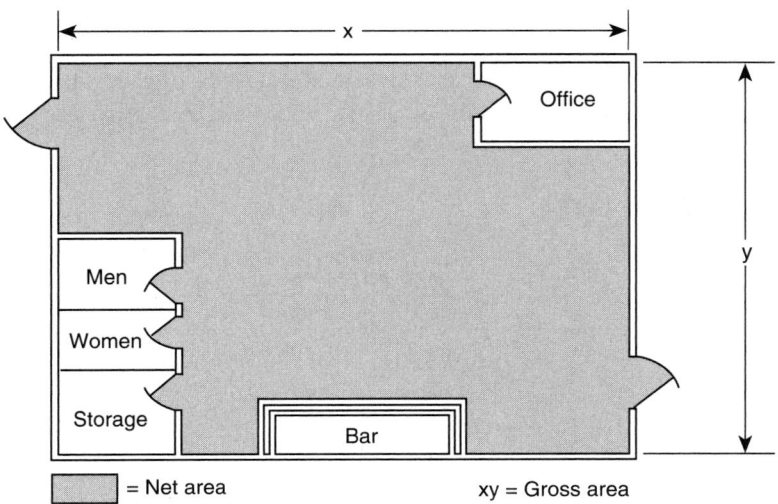

Figure 17.1 Net and gross area.

is important to check the provisions of the current *Life Safety Code*—the way to take this measurement has changed recently, and the edition of the *Life Safety Code* in use should be checked. Door measurements are covered under the door provisions in Section 7.2.1.2 of the *Code*.

17.1.3 Determining the Capacity of Each Egress Component

After determining the clear width of each element in the egress path, the next step is to use the table in the *Life Safety Code* (Section 7.3.3, 2006 edition) to determine the capacity of each element. This table is reproduced here as Table 17.2. In most cases, other than existing nonsprinklered health care occupancies, the figures of 0.3 in. (0.8 cm) per person for stairs and 0.2 in. (0.5 cm) per person for other egress elements are used. For example, given a 36-in. (91-cm) door that provides 32 in. (81 cm) of clear width, the capacity of the door is 32 in./0.2 in. (81 cm/0.5 cm) per person, or 160 people. In another example, given a 44-in. (112-cm-) wide stair with handrails that project not more than 4½ in., the capacity of the stair is 44 in./0.3 in. (approximately 112 cm/0.8 cm) per person, or 147 people. In a third example, given a corridor 44 in. (112 cm) in clear width, the capacity of the corridor is 44 in./0.2 in. (112 cm/0.5 cm) per person, or 220 people.

17.1.4 Determining the Most Restrictive Egress Component

After determining the capacity of each egress element, the capacity of the egress system must be calculated by determining the element in the system that has the least capacity (a chain is as strong as its weakest link). For example, using the example above for the egress system with a door, a corridor, and a stair, the egress capacity of the system is based on the capacity of the stair because this element provides the least capacity. Therefore, the system capacity is 147 people.

Some cautions to note when determining egress capacity are as follows:

- The least capacity is not always provided by the smallest element, since there are two different factors, 0.3 and 0.2 in. (0.8 and 0.5 cm) per person.
- When elements have the same capacity factor (doors, aisles, corridors, ramps, and passageways), only the capacity of the smallest element must be calculated.

Table 17.2 Capacity Factors

Area	Stairways (width per person) in.	Stairways (width per person) cm	Level Components and Ramps (width per person) in.	Level Components and Ramps (width per person) cm
Board and care	0.4	1.0	0.2	0.5
Health care, sprinklered	0.3	0.8	0.2	0.5
Health care, nonsprinklered	0.6	1.5	0.5	1.3
High hazard contents	0.7	1.8	0.4	1.0
All others	0.3	0.8	0.2	0.5

Source: Adapted from *Life Safety Code*, 2006, Table 7.3.3.1.

Usually doors are the smallest level component, and aisles and corridors do not need to be evaluated. In the preceding example, all that needed to be calculated was the door and the stair.
- Doors are level components and use 0.2 in. (0.5 cm) per person, regardless of the fact that they may be part of a stair enclosure.

Egress capacity is done on a floor-by-floor basis; occupant loads are not accumulated from adjacent stories unless specified by the code.

17.1.5 Determining Sufficiency of Egress Capacity

Once the capacity of each egress system is determined, the final step in evaluating egress capacity is to determine whether sufficient capacity is provided. Sufficient capacity is determined by comparing the egress capacity to the occupant load of the area served. If the egress capacity equals or exceeds the occupant load of the area served, there is compliance; if not, additional egress capacity is required. For example, a story with an occupant load of 400 people and two egress systems, each capable of handling 160 people, provides insufficient egress capacity because the total capacity of the two egress systems (i.e., 160 + 160 = 320 people) is less than the occupant load of 400 people. The egress capacity needs to be increased to handle 400 people.

17.1.6 Minimum Dimensions for Egress Elements

Based on the formula that is used to calculate the capacity of means of egress, a door to a room with an occupant load of 10 people would have to be 10 people × 0.2 in. per person, or 2 in. wide. This method of calculation obviously does not work, so the *Life Safety Code* establishes certain minimum dimensions for egress elements (Section 7.3.4). General minimums are provided in the chapter on means of egress, and specific minimums are provided in many of the occupancy chapters.

17.1.6.1 General Minimums

There are two general minimums for egress elements—one applies to all egress elements except doors, and the other applies to doors. In addition, the specifications for stairs establish the general minimum widths for stairs. Table 17.3 summarizes the general minimum egress widths. There are some cases, including health care occupancies, where different numbers are provided.

Table 17.3 General Minimum Egress Widths

Egress Element	Minimum—*New*	Minimum—*Existing*
General	36 in. (91 cm) clear[1]	28 in. (71 cm) clear[1]
Doors	32 in. (81 cm) clear	28 in. (71 cm) leaf
Stairs	44 in. (112 cm) clear[1, 2, 3]	44 in. (112 cm) clear[1, 2, 3]
Stairs serving 7	2000 occupant load total	56" clear (1420 mm)

[1] Projections up to 4½ in. (114 mm) at each side are permitted up to a height of 38 in. (96 cm).
[2] 36 in. (91 cm) is permitted where total occupant load of all stories served by the stair is fewer than 50 people.
[3] 56 in. (1420 mm) where total cumulative occupant load assigned to the stair equals or exceeds 2000 people.

Table 17.4 Minimum Corridor Widths for Health Care Facilities

Location	New	Existing
Hospital, nursing home	8 ft (244 cm)	4 ft (122 cm)
Limited care, psychiatric	6 ft (183 cm)	4 ft (122 cm)
Non-patient areas	44 in. (112 cm)	44 in. (112 cm)

See Chapter 16 for additional information on door minimum widths. Corridors are regulated by the general egress minimums. However, many occupancies establish wider minimum corridor widths. For example, business occupancies require a minimum corridor width of 44 in. (112 cm) for new and existing, and assembly occupancies require corridors to be a minimum of 44 in. (112 cm) in new construction when serving more than 50 people.

17.1.6.2 Health Care Occupancy Corridors

Because patients are often moved on beds or gurneys and equipment must be able to be moved both normally and during emergencies, the minimum widths for health care facility corridors are wider than the widths needed for general facilities. Table 17.4 summarizes minimum corridor widths for health care facilities.

Obstructions up to 4½ in. (114 mm) may project on each side of a corridor up to a height of 38 in. (96 cm). An 8-ft (244-cm) requirement has been in the *Life Safety Code* for many decades, and therefore most facilities have corridors at least 8 ft (244 cm) wide. As always, if the corridor is more than 4 ft (122 cm) wide, the additional width up to 8 ft (244 cm) must be maintained.

Furniture or equipment cannot reduce this width. However, it is recognized that certain equipment, such as patient gurneys, are moved in and out of corridors on a routine basis and may be parked in the corridor for some time. Food delivery carts, housekeeping carts, IV carts, code carts, and so forth are also routinely parked in corridors, sometimes for many hours. This short-term parking is acceptable and is part of the reason for the 8-ft (244-cm) width.

It is not acceptable, however, to store gurneys, wheelchairs, IV poles, trash and recycle containers, and similar items in the corridor. Even if this equipment does not interfere with the minimum width, it still may not be acceptable, since storage areas are normally considered hazardous areas that need to be protected appropriately. Alcoves for equipment such as clean linen carts, food carts, gurneys, and so forth are generally acceptable, but storage of more than 50 ft² is generally not. Although the line between storage and parking may be vague, it is usually not hard to determine which is which. Similar to short-term parking at the airport, leaving a piece of equipment in a corridor for more than several hours starts to push the limits on short-term parking and enters the realm of storage. The presence of cobwebs and dust and just simple knowledge of how a facility operates can easily indicate the difference between parking and storage. The Joint Commission on Accreditation of Healthcare Organization (JCAHO) requires that the item be attended. For example, a trash cart should have someone from housekeeping in the area that is using it.

Starting with the 2006 edition of the *Code*, a major change has been instituted to allow projections by alcohol-based hand rub cleaners (ABHR), flip-up charting stations, and similar items. When using the 2006 or newer editions, check the *Code* for details on these items. NFPA issued Tentative Interim Amendments (TIAs) to the 2000 and 2003 editions of the *Code* with regard to the alcohol hand cleaners. The basic concepts of the TIAs have been added to the 2006 editions of NFPA *101* and *NFPA 5000®, Building Construction and Safety Code®*. A TIA can be used only if the agency adopting the *Code* specifically allows the use of the TIA. JCAHO allows its use, but other authorities having jurisdiction may not. The Centers for Medicare and Medicaid Services (CMS) released an "Interim Final Rule" in March 2005 that formally adopts the criteria established in the TIAs for ABHR cleaners.

17.1.6.3 Health Care Occupancy Doors

Similar to corridors, minimum widths for doors in health care facilities are greater than the minimum width for doors in general facilities. Refer to the earlier discussion for general minimum door widths. Table 17.5 provides the minimum door widths for health care occupancies.

Stair enclosures and nurseries do not require the 41.5 in. (105 cm) of minimum clear width because it is not desirable to move beds or gurneys into stairs, and babies are usually moved in bassinets or by hand. Similarly, non-patient areas of a health care occupancy do not require the larger doors.

Similar to corridor widths, the 44-in. (112-cm) door leaf has been required by the *Life Safety Code* in health care occupancies for many years. It is not permissible to reduce the door width to that required for existing structures. See the earlier discussion and Chapter 16 for more information on door widths.

17.2 Number of Exits

Providing the proper number of exits is not usually an issue in health care occupancies. However, having the proper number of exits can be an issue in other occupancies associated with health care, such as a medical office building. Health care occupancies have some special requirements for the number of exits, in addition to those found in the general provisions. In general, the *Life Safety Code* requires all buildings, stories,

Table 17.5 Minimum Door Widths for Health Care Occupancies

Location	New–Clear Width	Existing–Clear Width
Hospital, nursing home	41.5 in. (105 cm)[1]	32 in. (81 cm) or 34 in. (86 cm) leaf width[2]
Limited care, psychiatric	32 in. (81 cm)	
Stair enclosures	32 in. (81 cm)	
Nurseries	32 in. (81 cm)	
Non-patient areas	32 in. (81 cm)	

[1] A 44-in. (112-cm) door leaf usually provides at least 41.5 in. (105 cm) of clear width.
[2] A 34-in. (86-cm) door leaf does not always provide 32 in. (81 cm) of clear width, so an exception for existing doors with a leaf width of 34 in. (86 cm) is provided.

mezzanines, or portions thereof to have a minimum number of means of egress in accordance with Section 7.4.1.2 of the *Code*, as shown in Table 17.6.

The *Life Safety Code* does provide several important exceptions, including when a particular occupancy can have a single means of egress or when existing buildings of a particular occupancy do not need to have a third and fourth means of egress. An additional exemption is for mezzanines that have a common path of travel compliant with the occupancy involved.

Health care occupancies must have two exits from every building, every story in a building, and any fire compartment. In other words, single exits are never allowed. In addition, every floor and every fire area must have at least one exit that is not a horizontal exit. This is illustrated by Figure 17.2.

Table 17.6 Number of Means of Egress

Occupant Load	Minimum Number of Means of Egress
Up to 500 people	2
501 to 1000 people	3[*]
More than 1000 people	4[*]

[*]Not retroactively applicable to existing buildings other than assembly.

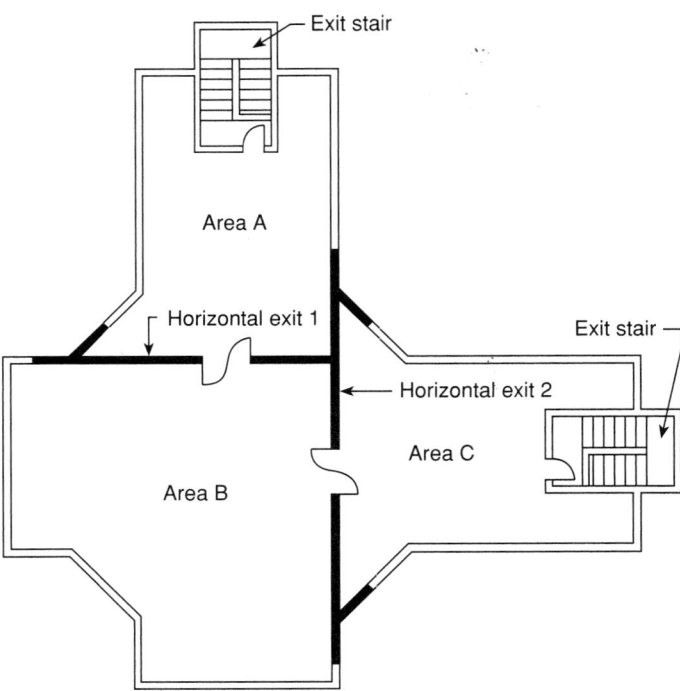

Figure 17.2 Arrangement of exits for fire compartments formed by horizontal exit fire barriers. Area B is deficient.
Source: Life Safety Code Handbook, 2003, Exhibit 18/19.9.

In addition, every smoke compartment must have access to at least two exits, not necessarily in the smoke compartment, so that travel is not required through the compartment of the fire origin. In other words, dead-end smoke compartments are prohibited. Figure 17.3 illustrates this concept.

Another occupancy that is often associated with health care facilities that does not allow a single exit is assembly occupancies. However, several business occupancy situations can have a single exit. The most common of these situations is a room or area with an occupant load of fewer than 100 people and a total travel distance to the outside of less than or equal to 100 ft (30 m). (See Section 17.3 in this text for more on travel distance.) This circumstance is most common in single-story buildings. However, single exits can be used in two-story structures as long as the vertical distance does not exceed 15 ft (4.5 m) and the stairs are either outside stairs or, if interior stairs are used, they discharge directly outside and have no doors to any other story. Figures 17.4 and 17.5 illustrate this exception. There are other exceptions as well, so the *Life Safety Code* must be referenced for other single-exit situations not mentioned here.

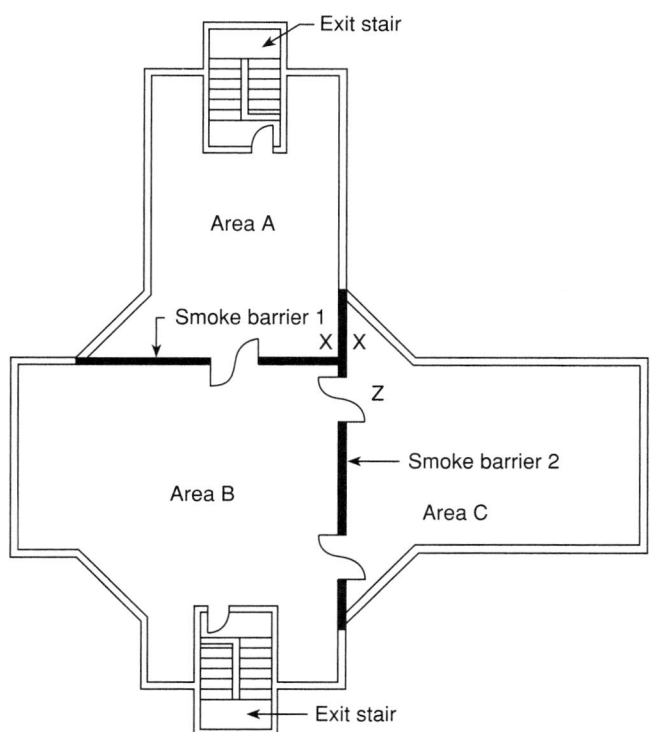

Figure 17.3 Arrangement of exits for smoke compartments formed by smoke barriers. Area C is deficient.
Source: Life Safety Code Handbook, 2003, Exhibit 18/19.10.

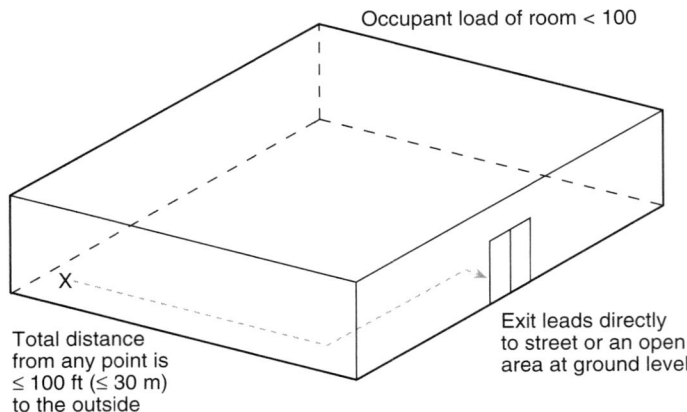

Figure 17.4 Single exit from area or room—travel to exit without stairs.
Source: Life Safety Code Handbook, 2003, Exhibit 38/39.8.

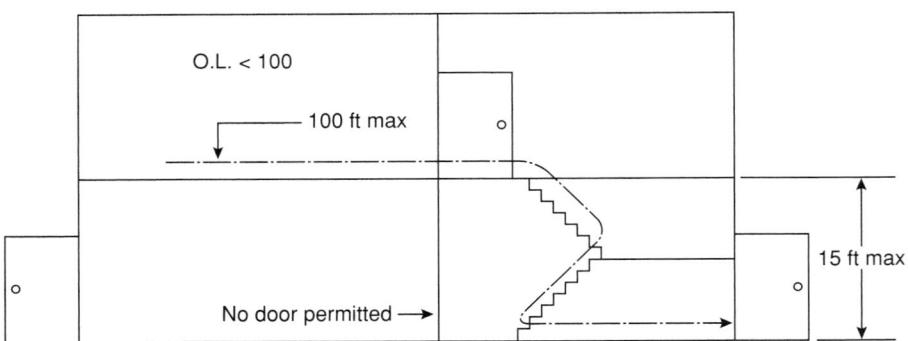

Figure 17.5 Single exit from area or room with an occupant load less than 100 ft (30 m) and total travel distance less than 100 ft (30 m).

17.3 Travel Distance

As the name implies, *travel distance* is the distance one must travel to reach an exit. The distance is measured along the natural path of travel, from the most remote occupiable point to the exit—it is not measured to both exits. Unfortunately, measuring travel distance is not easy to accomplish in health care occupancies or in any occupancy where people sleep, which includes health care occupancies. In sleeping-type occupancies, the travel distance within the room and within the corridor are treated separately and not as one measurement. Figure 17.6 illustrates how to measure travel distance in general, and Table 17.7 provides travel distance limits in occupancies typically associated with health care facilities.

Figure 17.7, in conjunction with Table 17.8, illustrates travel distances in health care occupancies and ambulatory health care facilities.

Chapter 17. Principles of Egress **289**

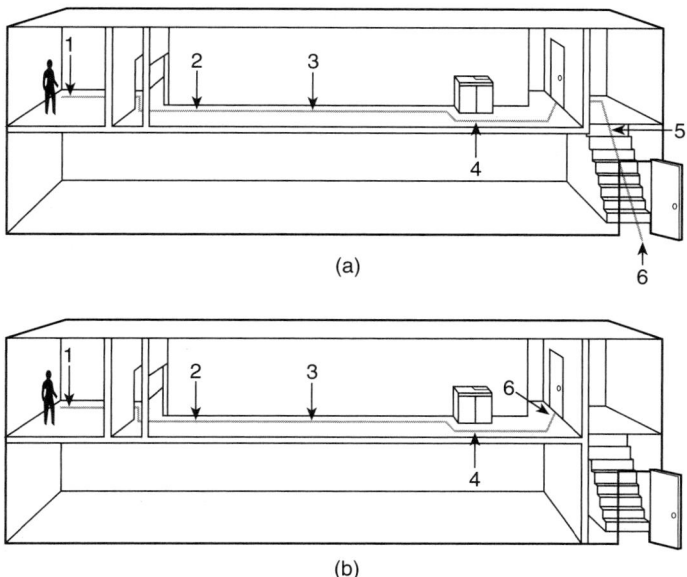

Figure 17.6 Measuring travel distance to an exit. The stair in part (a) opens to the first floor and therefore is not an exit; travel distance continues to the door to the outside.
Source: Life Safety Code Handbook, 2003, Exhibit 7.87.

Travel distance is rarely an issue for existing sprinklered buildings. Careful consideration must be given to travel distance limitations for all nonsprinklered situations and in new construction, renovations, or additions.

17.4 Arrangement of Means of Egress

Arranging the means of egress is an extremely important subject and one that is often violated in many occupancies. One of the issues that makes this difficult for health care facilities is that health care occupancies treat the arrangement of the means of egress quite differently than do other occupancies.

Table 17.7 Travel Distance Limits in Occupancies Typically Associated with Health Care Facilities

Occupancy	Travel Distance Limit–Nonsprinklered	Travel Distance Limit–Sprinklered
Assembly	150 ft (45 m)	200 ft (60 m)
Business	200 ft (60 m)	300 ft (91 m)
Industrial	200 ft (60 m)	250 ft (75 m)*
Storage (ordinary hazard)	200 ft (60 m)	400 ft (122 m)

*See the "Industrial Occupancies" chapter (Chapter 40, 2000 edition) in the *Life Safety Code* for further details and exceptions.

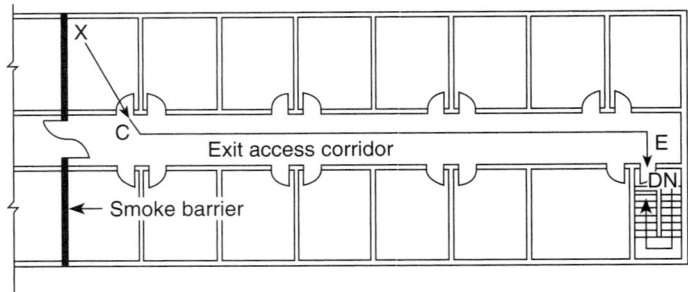

Figure 17.7 Travel distance limitations (see Table 17.8).
Source: Life Safety Code Handbook, 2003, Exhibit 18/19.19.

Arranging the means of egress involves arranging the egress systems such that a single fire event will not block more than one way out. In most occupancies, this is accomplished using performance criteria, restricting dead-end corridors, and limiting the common path of travel (i.e., that part of the exit access that must be traversed before two separate and distinct paths of travel to two separate exits are available; see Section 17.4.2 in this text). In health care occupancies, performance criteria are used and restrictions are placed on the use of dead-end corridors, but common paths of travel are not limited. Specific requirements for suites, such as for the number of ways out of any room or suite of rooms, are also established for health care occupancies (see the *Life Safety Code*, Section 18.2.5 or 19.2.5).

17.4.1 General Arrangements

Several issues regarding general arrangements for means of egress apply universally regardless of the occupancy (*Life Safety Code,* Section 7.5). One general concept is that from any point in a facility, safe and continuous passageways, aisles, or corridors must lead directly to every exit and be arranged such that there is access to two exits by separate ways of travel. There are two important exceptions to this concept. The first

Table 17.8 Travel Distances in Health Care Occupancies and Ambulatory Health Care Facilities

Location	Nonsprinklered	Sprinklered	Figure 17.7 Reference
Any room corridor door to exit	100 ft (30 m)	150 ft (45 m)	C to E
Any point to exit	150 ft (45 m)	200 ft (60 m)	X to E
Within sleeping room	50 ft (15 m)	50 ft (15 m)	X to C
Within sleeping suite	100 ft (30 m)	100 ft (30 m)	X to C
Within nonsleeping suite—1 intervening room	100 ft (30 m)	100 ft (30 m)	X to C
Within nonsleeping suite—2 intervening rooms	50 ft (15 m)	50 ft (15 m)	X to C
Ambulatory health care room corridor to exit	100 ft (30 m)	150 ft (45 m)	C to E
Ambulatory health care any point to exit	150 ft (45 m)	200 ft (60 m)	X to E

exception is that in single-exit situations, it is impossible to arrange access to two separate exits, so these situations are exempted. The other exception allows access to a single exit for a limited distance, referred to as common path of travel, when allowed by the occupancy chapter. Without this exception, every room in any occupancy would need two doors out of each room.

Another general principle is that once in a corridor, a person must have access to two exits without having to pass through any intervening room other than a corridor, lobby, or space allowed to be open to the corridor by the occupancies involved. This concept has been in the *Life Safety Code* for many years but was just recently clarified (Section 7.5.1.2). Figure 17.8 illustrates this requirement. Before this clarification, the *Life Safety Code* could have been interpreted to allow travel from a corridor to a room for accessing an exit, so the code permits this arrangement in existing situations. The *Life Safety Code* should be reviewed by those responsible for compliance for any facility where this situation is present and the exception is allowed. The exception is not permitted in health care occupancies but could potentially apply in non–health care occupancies within health care facilities.

The *Life Safety Code* has included a very powerful general performance statement for decades regarding the arrangement of means of egress (Section 7.5.1.3, 2006 edition):

> Where more than one exit is required from a building or portion thereof, such exits shall be remotely located from each other and shall be arranged and constructed to minimize the possibility that more than one has the potential to be blocked by any one fire or other emergency condition. [101–06: 7.5.1.3.1]

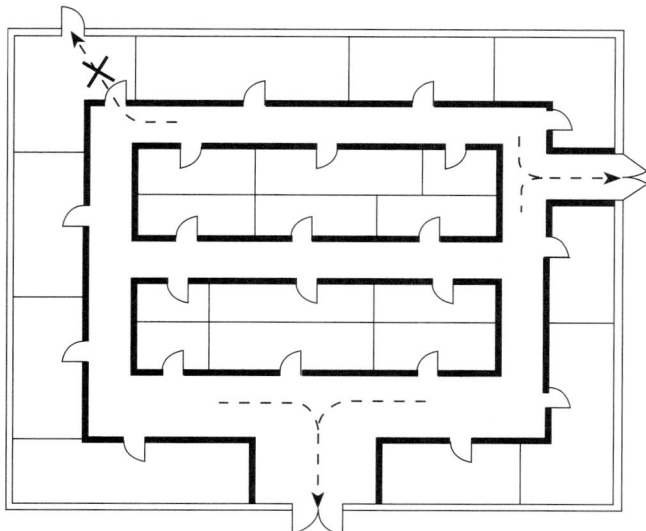

Figure 17.8 Exit access arrangement allowing two separate means of egress and travel paths from the corridor without the need to pass through intervening rooms.

If the requirement were properly met, one would always have access to an exit. Unfortunately, this is not always the case, as exits are often placed too close to each other. To combat this, a new requirement was added to the *Life Safety Code* several years ago, as follows (Sections 7.5.1.3.2 and 7.5.1.3.4):

> Where two exits or exit access doors are required, they shall be located at a distance from one another not less than one-half the length of the maximum overall diagonal dimension of the building or area to be served, measured in a straight line between the nearest edge of the exit doors. . . . Where exit enclosures are provided as the required exits specified in 7.5.1.3.2 or 7.5.1.3.3 and are interconnected by not less than a 1-hour fire resistance–rated corridor, exit separation shall be permitted to be measured along the line of travel within the corridor. [101–06: 7.5.1.3.2 and 7.5.1.3.4]

This provision is commonly referred to as the *one-half diagonal rule*. The rule does not apply to existing buildings; however, the performance statement above does. In most cases, it is easiest to apply the one-half diagonal rule to both new and existing facilities. If an existing arrangement does not comply with the provision, then the performance statement should be applied. Since the performance statement is more difficult to measure, the violation of the one-half diagonal rule often results in a violation of the performance statement. Also, if the building is protected by a supervised automatic sprinkler system, the one-half diagonal becomes one-third diagonal.

The diagonal measurement is always measured in a straight line between the doors involved, except where exits are connected by minimum 1-hour fire-rated corridors, in which case the measurement can be made within the corridor. Figures 17.9 and 17.10 illustrate this concept.

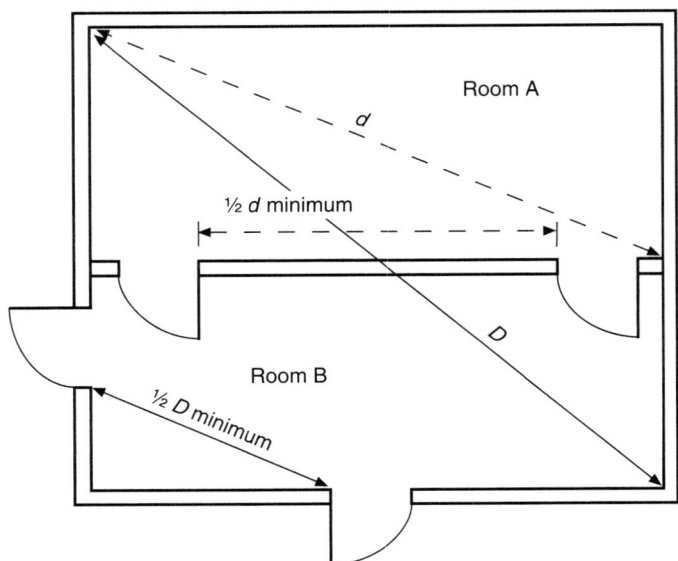

Figure 17.9 Diagonal rule for exit and exit access remoteness.
Source: Life Safety Code, 2006, Figure A.7.5.1.3.2(b).

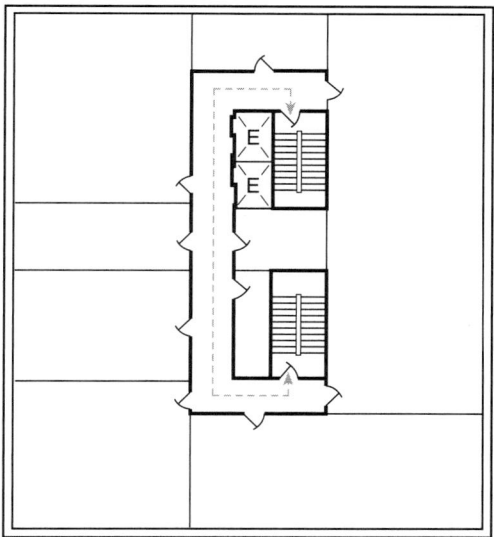

Figure 17.10 Exit remoteness measured along 1-hour rated corridor
Source: Life Safety Code Handbook, 2003, Exhibit 7.78.

17.4.2 Common Path of Travel

Common path of travel is a concept that was unique to the *Life Safety Code* but has more recently been introduced into building codes in the United States. It is an important concept in that it measures the distance one must travel before reaching two different paths to two separate exits. Consider, for example, a typical office with only one door. The distance a person must travel to get to the point where two distinct separate paths of travel to two different exits are present is the common path of travel. The basic concept of the *Life Safety Code* is that any common path of travel is prohibited unless a particular occupancy specifically allows a certain limited common path of travel. Figure 17.11 illustrates the concept of common path of travel.

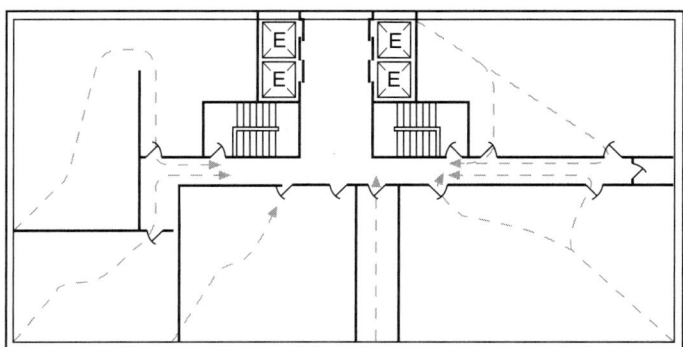

Figure 17.11 Common paths of travel.
Source: Life Safety Code Handbook, 2003, Exhibit 7.83.

Table 17.9 Common Path of Travel Limits for Occupancies Commonly Found in Health Care Facilities

Occupancy	Nonsprinklered	Sprinklered
Assembly	20 ft (6.1 m) [75 ft (23 m) serving 50 people or less]	Same
Business	75 ft (23 m) (several exceptions provided)	100 ft (30 m) (several exceptions provided)
Industrial–general	50 ft (15 m)	100 ft (30 m)
Storage–ordinary	50 ft (15 m)	100 ft (30 m)
Health care	No requirement	No requirement

Note that when a space has two doors, this does not necessarily totally eliminate a common path of travel. If the two paths allowed by the two doors merge before reaching access to two different exits, this is still a common path of travel. Table 17.9 provides common path of travel limits for occupancies typically found in health care facilities.

Since health care occupancies do not have any specific requirements for common paths of travel, the limit is technically zero. However, this rule would then require every room to have two doors. Thus, in health care occupancies, common paths of travel are in essence limited by the requirement for rooms of certain sizes to have two ways out rather than by specifically limiting the common path of travel. Rooms (or suites of rooms) used for patient sleeping that are more than 1000 ft^2 (93 m^2) and rooms (or suites of rooms) other than those used for patient sleeping that are more than 2500 ft^2 (230 m^2) must have two remote ways out. This rule is important to remember for mixed-occupancy situations as well, since other occupancies place limits on the common path of travel. Common path of travel is also a critical element to consider whenever rearranging an area, especially an office area, where this rule is commonly violated.

17.4.3 Dead-End Corridors

A dead-end corridor is one in which a person could mistakenly travel past an exit and have to retrace his or her steps to get back to the exit, or where one can mistakenly travel down a corridor only to discover that there is no exit and then have to backtrack. The rules for dead-end corridors are similar to common path of travel, but there are some differences. Dead-end corridors are measured only in corridors, whereas common paths of travel are measured anywhere in a building. Figure 17.12 illustrates the concept of dead-end corridors, and Figure 17.13 illustrates the difference between dead-end corridors and common paths of travel. Table 17.10 provides typical dead-end corridor limitations for health care facilities.

17.4.4 Impediments to Egress

Needless to say, the *Life Safety Code* prohibits any impediment to the use of a means of egress (Section 7.5.2). One impediment, the locking of doors, is discussed

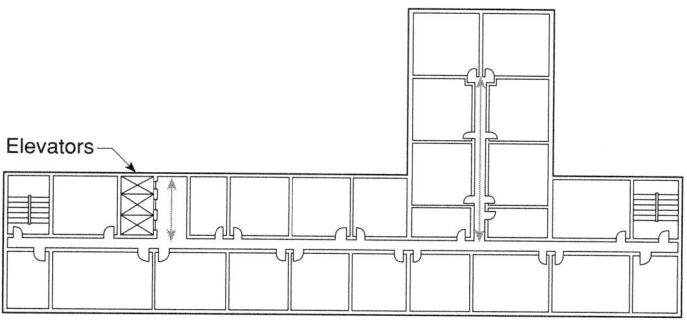

Figure 17.12 Commonly found dead-end corridors.
Source: *Life Safety Code Handbook,* 2003, Exhibit 7.81.

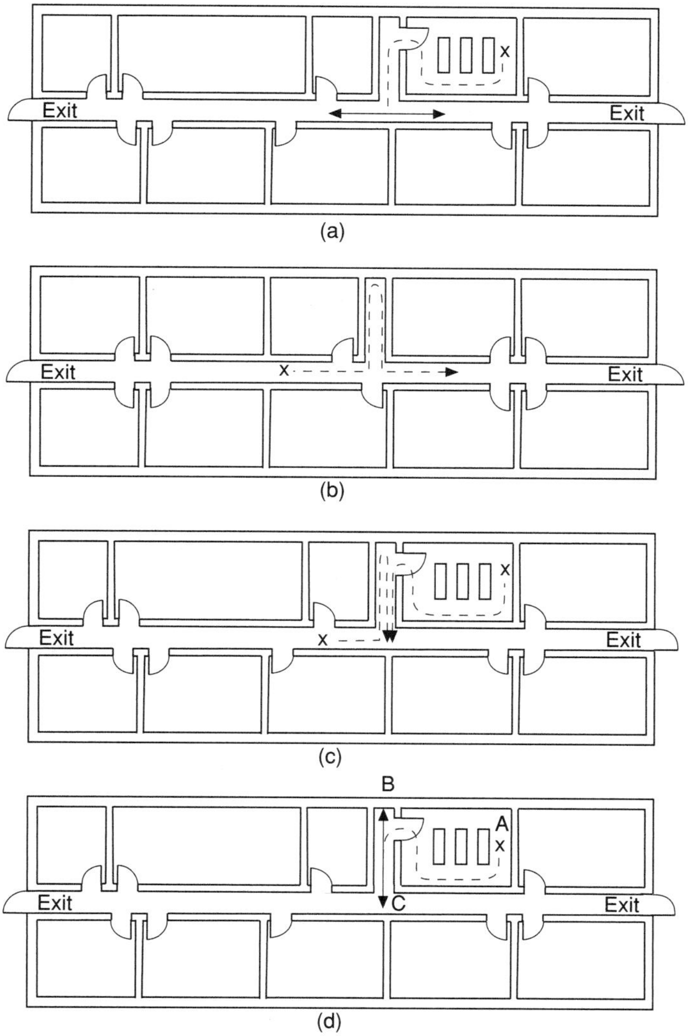

Figure 17.13 Common paths of travel and dead-end corridors. (a) Common path of travel. (b) Dead-end corridor. (c) Common path of travel and dead-end corridor (d) A to C, common path of travel, and B to C, a dead end.
Source: *Life Safety Code,* 2006, Figure A.7.5.1.5.

Table 17.10 Typical Dead-End Corridor Limitations for Occupancies Commonly Found in Health Care Facilities

Occupancy	New Sprinklered	New Nonsprinklered	Existing Sprinklered	Existing Nonsprinklered
Assembly	20 ft (6.1 m)	20 ft (6.1 m)	20 ft (6.1 m)	20 ft (6.1 m)
Health care	30 ft (9.1 m)	30 ft (9.1 m)	NR*	NR*
Ambulatory health care	50 ft (15 m)	20 ft (6.1 m)	50 ft (15 m)	50 ft (15 m)
Business	50 ft (15 m)	20 ft (6.1 m)	50 ft (15 m)	50 ft (15 m)
Industrial–general	50 ft (15 m)	50 ft (15 m)	50 ft (15 m)	50 ft (15 m)
Storage–ordinary	100 ft (30 m)	50 ft (15 m)	100 ft (30 m)	50 ft (15 m)

*NR indicates that no specific limit is established, but the *Life Safety Code* states that dead-end corridors are undesirable and should be eliminated wherever practical. Since 30 ft (9.1 m) is permitted in new, it logically should be permitted in existing also.

in Chapter 16. Snow and ice in the exit discharge are other examples of obvious impediments to the means of egress. In any area where snow or ice accumulations are possible, arrangements must be made for their removal. This is especially true for exit doors that are not commonly used, which tend to be forgotten by snow removal crews; snow or ice could prevent the use of these doors and create a significant hazard. Every facility should have a written snow removal policy that makes it a high priority to clear "back" exit paths. One way to ensure that these exit paths are cleared promptly is to make it a higher priority to clear these doors than to clear the parking lot.

Other impediments include mirrors and curtains that hang across doors, which can make the egress path confusing. For example, convex mirrors at corridor intersections often cause confusion with exit signs. The *Life Safety Code* also prohibits egress through kitchens, storerooms, restrooms, workrooms, closets, and similar spaces.

Another major impediment in a health care facility is the placement of beds, chairs, gurneys, IV poles, and so forth, in locations so as to block exit access or the exit itself. Storage space always is at a premium, but the means of egress must be kept clear.

17.4.5 Exterior Exit Access

Exterior exit access is outside the building but not yet at the exit. A classic example of exterior exit access is a two-story motel, where one must still travel to an exit stair even after leaving a second-floor room. The path from the room to the exit stair is exterior exit access. The *Life Safety Code* treats this type of access similarly to corridors but with some common exceptions (Section 7.5.3). Figure 17.14 illustrates some of the concepts for exterior exit access.

Exterior exit access is not common for health care facilities, except in warm climates. Wherever they are used, the *Life Safety Code* should be reviewed.

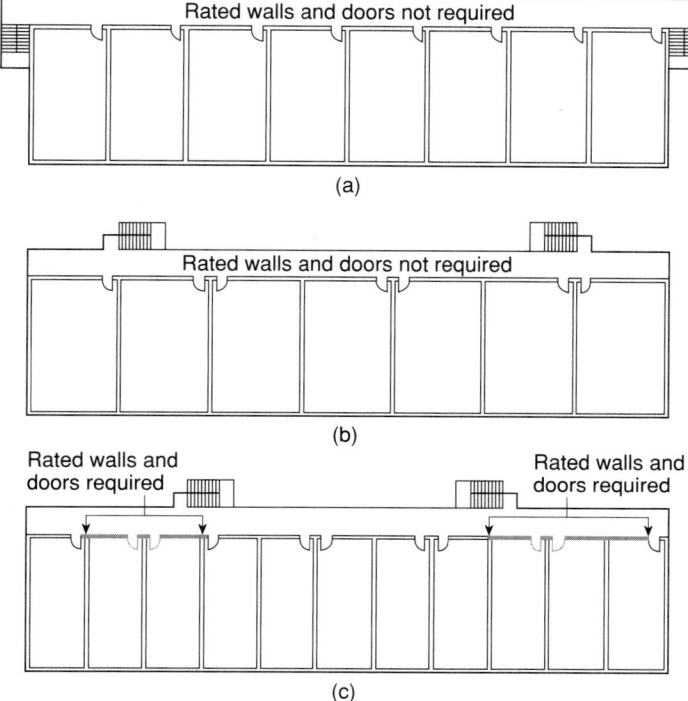

Figure 17.14 Exterior exit access.
Source: Life Safety Code Handbook, 2003, Exhibit 7.85.

17.4.6 Health Care Occupancy Suites

The general rule for health care facilities is that *every* habitable room must have a door directly to a corridor, with *no* intervening rooms. The *Life Safety Code* includes four exceptions for not having a door directly to the corridor (Sections 18.2.5 and 19.2.5), two of which involve the use of suites. These exceptions make the concept of suites very important for a health care occupancy, particularly when considering the restrictions related to separating use areas from corridors and those related to corridor walls and widths. Figure 17.15 illustrates the requirement that all rooms open directly onto a corridor, along with a simple example of each exception.

The *Life Safety Code* allows a single intervening room between a patient sleeping room and a corridor, provided the intervening room does not serve more than eight beds or bassinets (the "eight-bed exception"). These rooms are commonly used as isolation rooms, small intensive care units (ICUs), small coronary care units (CCUs), and similar spaces. This single intervening room can serve more than one room, provided the total number of patients who must go through this room to egress does not exceed eight people. This type of multiple room that adjoins a single intervening room is not uncommon in nursing homes or limited care facilities. The single intervening room serving a maximum of eight beds or bassinets is illustrated in Figure 17.16.

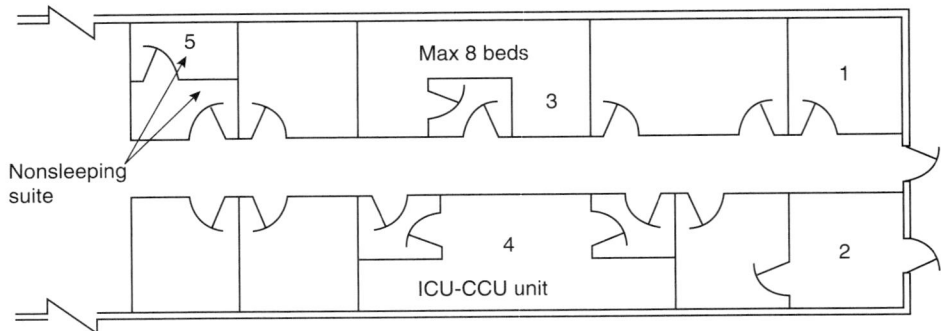

Figure 17.15 Habitable rooms with access to the corridor Room 1 meets the general requirement that rooms have a door directly to the corridor Room 2 illustrates that a door to the outside exempts this requirement. Rooms 3, 4, and 5 illustrate three different suites.

It is important to know that the suite provisions were totally rewritten for the 2006 edition of the *Code*. While the concepts remained the same, the provisions were reorganized for both clarity and ease of use. In addition, more options were added. When using the 2006 or newer editions of the *Life Safety Code*, the *Life Safety Code®Handbook* should be reviewed for more detailed discussion.

There are two different types of suites, sleeping and nonsleeping. The provisions for sleeping suites allow the use of a single intervening room, but there is no limit on the number of beds or bassinets that can be in the single intervening room, and other restrictions apply. These restrictions are that the suite cannot exceed 5000 ft² (460 m²), and there must be direct and constant visual supervision by nursing staff of the patients

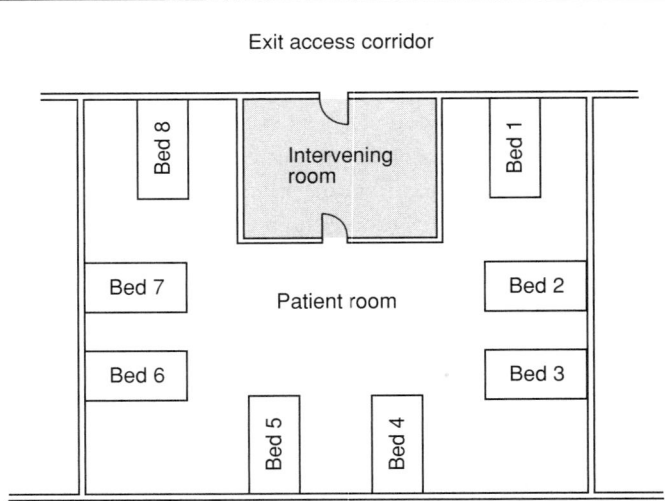

Figure 17.16 Intervening room between patient room and corridor.
Source: Life Safety Code Handbook, 2003, Exhibit 18/19.12.

within the whole suite. The 2006 edition of the *Code* allows for 7500 ft² if there is both staff supervision and smoke detection. It also allows rooms that are not visible from the staff location to be equipped with smoke detection as an option to the direct visual supervision. This exception is the most common way to create ICUs, CCUs, nurseries, neonatal ICUs, and similar spaces in that it allows significant flexibility within a limited space. Figure 17.17 illustrates the use of the special nursing suite.

The other type of suite is for nonsleeping areas, most commonly used for surgical suites, emergency departments, radiology, pharmacy, offices, and similar areas. This type of suite does not need to be supervised like the special nursing suites do and can be up to 10,000 ft² (930 m²) rather than 5000 ft² (460 m²). To meet the requirements for these suites, the *Life Safety Code* offers an option—with only one intervening room, the travel distance to the corridor can be 100 ft (30 m); with two intervening rooms, the travel distance to the corridor is limited to 50 ft (15 m). Figure 17.18 illustrates the use of this provision for a single intervening room, and Figure 17.19 illustrates the use of this provision when there are two intervening rooms.

It should be noted that the intervening rooms within the suites are not considered to be corridors and are not subject to the requirements for minimum widths or construction features or to the limitations for placing items in the area. This makes the concept of suites extremely important for surgical areas, due to equipment in the space and the door hardware often found here. The 10,000-ft² (900-m²) limitation can be a problem, especially in existing arrangements, as many nonsleeping suites are more than 10,000 ft² (900 m²) and they are difficult, if not impossible, to subdivide. Often equivalencies must be developed in order to provide a compliant situation.

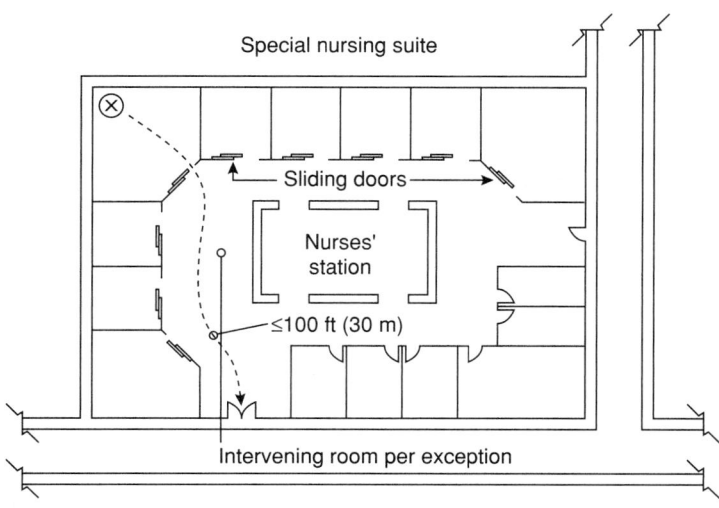

Figure 17.17 A special nursing suite.
Source: OPUS Communications and Koffel Associates, Inc., redrawn from 2000 Life Safety Code. Workbook & Study Guide for Health Care Facilities, 2003, Figure 16.6.

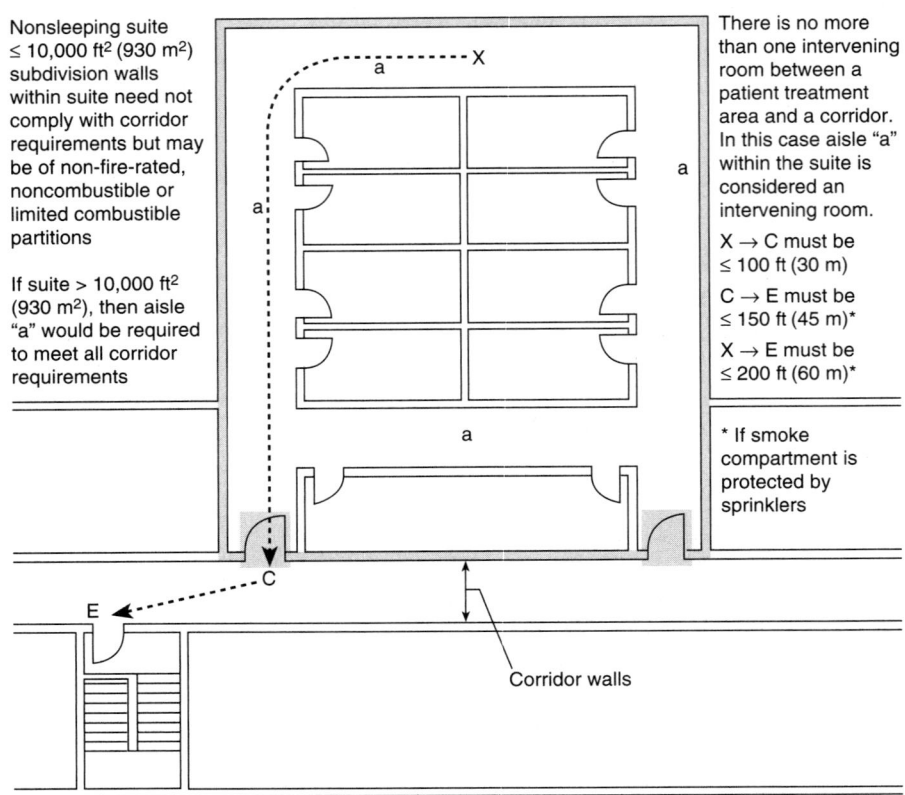

Figure 17.18 Treatment (nonsleeping) suite with one entervening room to corridor. *Source: Life Safety Code Handbook, 2003, Exhibit 18/19.16.*

One problem that often arises in complying with the requirements for suites is that most architects do not indicate them as such on the building plans. It is therefore necessary for facilities managers to work backwards to determine where they exist. That is, they must inspect and evaluate their facility to find any suites and then make the proper indications on the facility diagrams for future use. In bringing suites into compliance, remember that suites used for sleeping that exceed 1000 ft^2 (93 m^2), or those not used for sleeping that exceed 2500 ft^2 (230 m^2), must have two ways out, one of which must be a door directly to a corridor. The 2006 edition of the *Code* specifically allows the second way out of a suite to go into another suite.

17.4.7 Mechanical Equipment Rooms and Furnace and Boiler Rooms

Several new provisions were added to the *Life Safety Code* several years ago that specifically address the issues of when boiler rooms need two ways out and when mechanical penthouse rooms need two exits, two issues that were difficult to deal with in the past (Section 7.12, 2006 edition).

First, any mechanical equipment room, boiler room, or furnace room must have a maximum common path of travel of 50 ft (15 m) unless it is either existing, the building is protected by sprinklers, or no fuel-fired equipment is in the room. Under *any* of these conditions, the common path of travel can be up to 100 ft (30 m). In addition, an exception in the *Life Safety Code* states that if *all three* of these conditions are met (i.e., existing, sprinklered, and no fuel-fired equipment), the common path of travel may be as long as 150 ft (45 m). Another exception for existing health care occupancies places no limit on the common path of travel if the occupancy meets all the arrangement and travel distance limits for health care. Under that concept, since boiler rooms, furnace rooms, and mechanical equipment rooms are nonsleeping rooms in a health care occupancy, they must have two ways out when they exceed 2500 ft^2 (230 m^2).

These provisions basically say that the common path of travel for these spaces is regulated by the basic provisions of the *Life Safety Code* rather than by the occupancy chapter involved. This makes sense, since the hazard presented by these spaces is independent of the occupancy within which it is located. Figure 17.20 illustrates the intent of this provision.

Another issue relates to the number of exits provided for a story used exclusively for such purposes (e.g., elevator machine room penthouses or basement boiler rooms). For many occupancies, a minimum of two exits on each story is required. The *Life Safety Code* allows elevator penthouses and small basement boiler rooms to have a

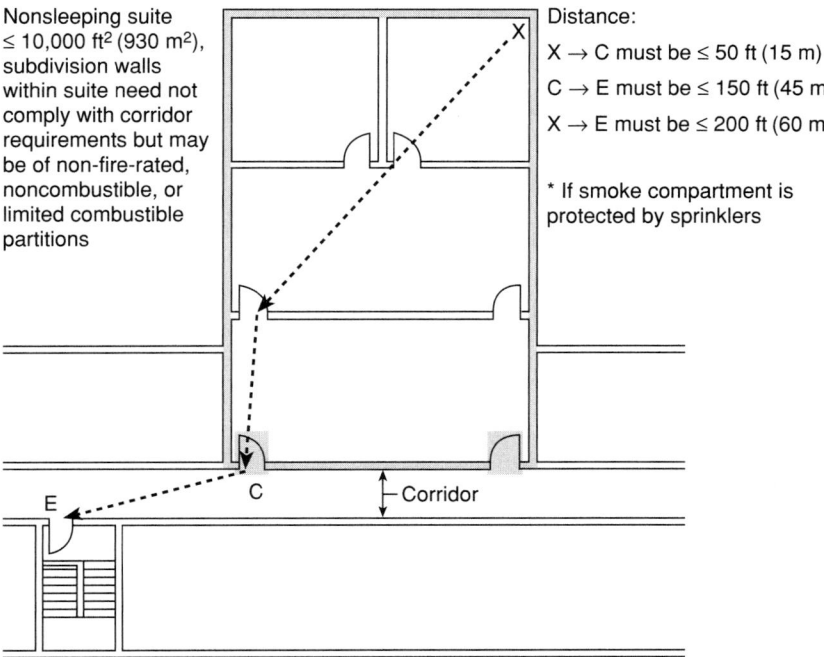

Figure 17.19 Treatment (nonsleeping) suite with two intervening rooms to corridor.
Source: Life Safety Code Handbook, 2003, Exhibit 18/19.17.

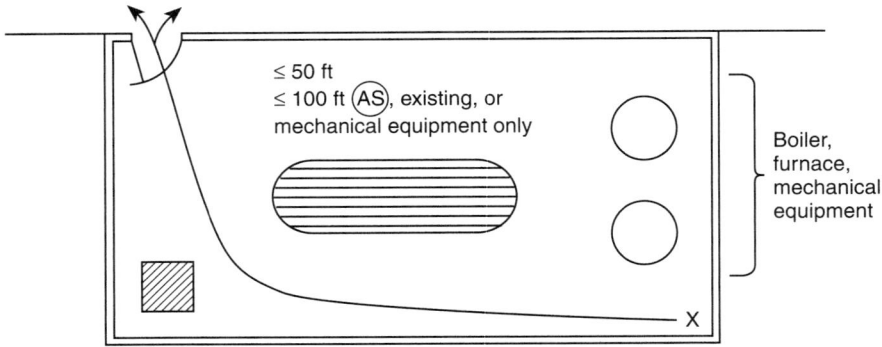

Figure 17.20 Requirement for maximum common path of travel from a mechanical room.

single exit as long as one can reach that exit within the limitations for the common path of travel, discussed previously. In other words, if a person is allowed to travel a certain distance before having the choice of two ways to go (i.e., the common path of travel), he can go the same maximum distance to reach a single exit on a mechanical story, regardless of the number of exits required by the occupancy chapter involved. Figure 17.21 illustrates the intent of this provision.

17.5 Accessible Means of Egress for the Mobility Impaired

As discussed in Chapter 16, the requirements for accessible means of egress are included in the *Life Safety Code* in the section on arrangement of means of egress. In any *new* construction, two accessible means of egress are required for any area accessible to people with severe mobility impairments. Accessible means of egress are required in the same number and travel distance limits as normal means of egress. Note that this does not apply to *existing* facilities.

Since the whole basis for the health care occupancy provisions is to provide defend-in-place protection for the mobility impaired, special protection for the mobil-

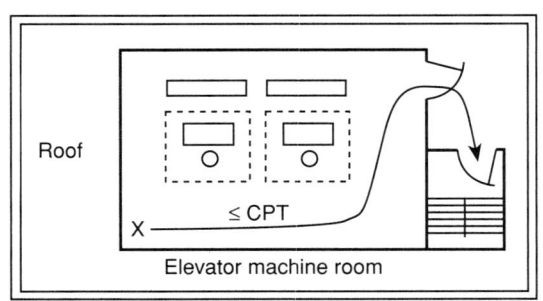

Figure 17.21 Requirements for a single exit from a mechanical room.

ity impaired is not needed. Health care occupancies protected with automatic sprinklers, which includes all new health care facilities, are exempted from having two accessible means of egress for areas accessible to people with severe mobility impairments.

This subject must be addressed for health care facilities that include occupancies other than health care occupancies. Any new business, assembly, industrial, or storage occupancy must provide the two accessible means of egress. If the facility is not protected throughout by a supervised automatic sprinkler system, this can cause a significant impact on the design of the building. The accessible means of egress must provide a path, usable by a person with severe mobility impairment, to a public way or an area of refuge. Therefore, if the egress path is other than by a level component or a ramp, an area of refuge must be provided. See the area-of-refuge discussion in Chapter 16, Section 16.3.8, of this book.

17.6 Illumination of Egress and Emergency Lighting

Illuminating the means of egress is a relatively simple subject and not a major issue. The *Life Safety Code* requires that the means of egress be illuminated to 1 ft-candle [10 lx (lux)] (Section 7.8). Note, though, that the requirement applies to the entire means of egress, including the exit discharge. However, for the purposes of exit access and exit discharge, the *Life Safety Code* requires that only those designated areas be illuminated. In other words, the lights in a room can be turned off, but not those in the corridors, designated aisles, exits, and designated portions of the exit discharge. This level of illumination must be provided by a reliable light source, typically the public or private utility service. In addition, the failure of any single lighting unit must not leave the area in darkness. This may apply to a small lobby or stair enclosure. It should be noted that this is a common problem in many facilities, especially in the exit discharge. Beginning with the 2003 edition of the *Code*, in new construction, stair illumination is required to be 10 ft-candle (108 lx).

Compliance with the provisions for emergency lighting is not as straightforward, particularly for business occupancies. The *Life Safety Code* establishes how to provide emergency lighting (Section 7.9) but then leaves it up to each occupancy to determine whether emergency lighting is required. Needless to say, most health care facilities require emergency lighting. Table 17.11 summarizes the major provisions for emergency lighting in occupancies usually associated with health care facilities.

Note that for business occupancies, any building three or more stories above ground level (two or more stories above the level of exit discharge) must have emergency lighting regardless of the occupant load. For one- or two-story buildings, the occupant load must be determined in order to evaluate the mandate for emergency lighting. Figure 17.22 helps illustrate the thresholds for emergency lighting in business occupancies.

When emergency lighting is required by the *Life Safety Code,* the performance requirements are quite simple but are often overlooked in occupancies other than

Table 17.11 Major *Life Safety Code* Provisions for Emergency Lighting in Occupancies Typically Associated with Health Care Facilities

Occupancy	Emergency Lighting	Exceptions
Assembly	Yes	*Existing*, when ≤ 300 for worship, small party tents
Health care	Yes (*new*–life safety branch, per NFPA 99)	None
Business	≥ 300 (1000 *existing*) people, or ≥ 50 (100 *existing*) people above or below level of exit discharge, or two stories above the level of exit discharge, or any windowless or underground story	
Industrial	Yes	See *Life Safety Code*—exceptions not common in health care facilities
Storage	Yes	Where not normally occupied; also see *Life Safety Code*
Ambulatory day care	Yes	None

health care. In new health care occupancies, the source of electrical power for emergency lighting must be the life safety branch, in accordance with NFPA 99, *Standard for Health Care Facilities* (refer to the definitions in Chapter 2 of NFPA 99 and Chapter 8 of this book). Health care facilities typically do not have a problem complying with the *Life Safety Code*; however, medical office buildings, administrative buildings, powerhouses, and so on often do. The *Life Safety Code* requires that the emergency lighting system provide an average of 1 ft-candle (10 lx) on any single failure of the electrical system, including failure of the public utility. Therefore, connections ahead of the main disconnect, or a second drop from the public utility, are not acceptable for emergency lighting under the code. In addition, the emergency lighting must consider any single failure within the building, such as an open breaker, switch, or fuse.

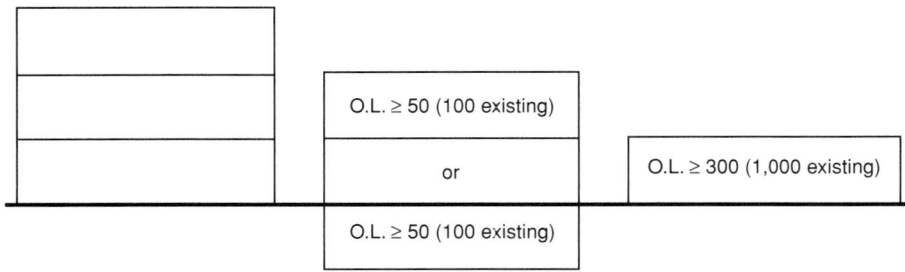

Figure 17.22 Thresholds for emergency lighting in a business occupancy (elevation view of a building).

Within health care occupancies, emergency power provided by generators operated by a transfer switch is the most common form of emergency lighting. Electrical configurations that are properly wired, tested, and maintained in accordance with NFPA 110, *Standard for Emergency and Standby Power Systems*, should be satisfactory. Other occupancies can use an emergency generator, central battery systems, or unit lighting (i.e., battery backup light packs), as long as they, too, are properly installed and maintained. Some common problems with the installation and maintenance of emergency lighting systems include

- Improper installation
  - Unit lighting is powered off the power circuits rather than the lighting circuits.
  - Emergency lighting circuits are energized only when the utility fails; internal failures are not considered.
  - Unit lighting is provided with insufficient units to illuminate the egress path adequately to an average of 1 ft-candle (10 lx).
- Improper maintenance
  - Unit lighting is not tested monthly for 30 seconds and yearly for 1½ hours.
  - Generator is not tested or maintained in accordance with NFPA 110.

The following case study illustrates several of these problems.

Problems with Emergency Lighting

In a new state-owned building, the state fire marshal was conducting a test of the emergency lighting. When the power to the building was disconnected to test the emergency lighting, the power shut down and the generator started. However, the type of lighting was such that it took many minutes for the lights to turn back on. The contractor subsequently hired to compensate for the problem installed unit lights such that the power was connected to the emergency circuits. When the test was redone, the power went out, the unit lights came on, and the generator started, but then the unit lights shut off. Thus, the new unit lights helped for less than 10 seconds, and the facility was still in darkness until the lamps relit. The first problem with this installation was that the lights on the emergency circuit should not have been the type that takes time to reilluminate. The second problem was that the unit lights should not have been tied into the emergency power circuits.

17.7 Marking the Means of Egress

Marking the means of egress deals with the placement of signage, such as exit signs, directional exit signs, "No Exit" signs, and similar items. Similar to emergency lighting, the *Life Safety Code* establishes specific requirements for marking of means of egress for all occupancies (Section 7.10).

Problems with exit signage are prevalent in health care facilities. The *Life Safety Code* does not mandate that a person be able to step into a corridor and see exit signs in both directions, but there are two simple rules that must be followed. One requirement is that all exits must be marked with signs (an exception allows for no signs at the main entrance, where it is obvious). The second requirement is that the way to reach an exit must be marked when it is not readily apparent to the occupants. Of course, it is this judgment issue that creates uncertainties. The best way to evaluate exit signs is to have a person who is not familiar with the facility try to find the exits. A person from engineering who is thoroughly familiar with the facility may not provide the most objective evaluation in this regard. Remember that the inspectors or surveyors who evaluate the facility are not very familiar with the facility, either. If *they* have a hard time finding the exits, the way to reach the exits is not obvious.

How the signs are installed is also important. It is not at all uncommon to find an exit sign placed over a door such that it is parallel to the path of egress travel. The *Life Safety Code* mandates that the sign be visible from any direction of exit access. Figure 17.23 illustrates the intent of this requirement.

Another problem with exit signs in health care facilities is their obstruction by *way-finding signs* (signs that direct a visitor to a particular destination within the building). Way finding is a problem in many health care facilities, and it is not at all uncommon to find way-finding signage obstructing the view of an exit sign. Another problem deals with the convex mirrors often installed at corridor intersections. Not only may these block the exit sign from view, but they can cause confusion with regard to the location of the sign because of reflection.

The *Life Safety Code* requires signs to be illuminated (Section 7.10.5). The signs do not have to be internally illuminated, although some local codes may require this feature. If the occupancy requires emergency lighting, the signs must be illuminated in both the normal and emergency lighting modes. Since most health care facilities require emergency lighting, internally illuminated signs are most commonly found. Alternatively, externally illuminated signs and photoluminescent signs are also acceptable.

This requirement results in another very common problem in most health care facilities, that of burned out bulbs in signs. With incandescent and in some cases fluorescent bulbs, this can be a major maintenance headache. Replacing older signs with newer signs that utilize LED or similar technology can significantly reduce if not eliminate this problem. It can also help by eliminating signs that do not meet the *Life Safety Code* requirements for newer signage. The lower power usage, in conjunction with lower maintenance costs, can rapidly pay for the changes in signage.

It should be pointed out that the *Life Safety Code* has fairly new requirements regarding directional indicators on directional exit signs. The indicator must be a

Chapter 17. Principles of Egress **307**

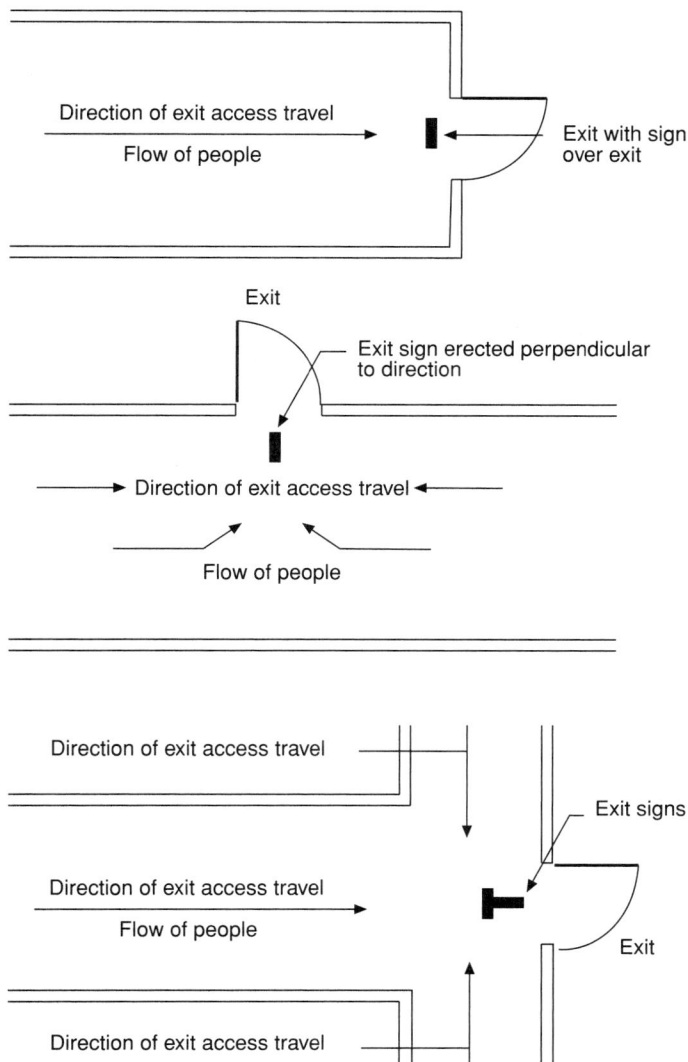

Figure 17.23 Required orientation of exit signs.
Source: Life Safety Code, 2006, Figure A.7.10.1.2.

chevron, not an arrow, and must be located outside the word "EXIT" at the end of the sign for the direction indicated. Locating the directional indicator within the legend is no longer acceptable; neither is putting a right-hand indicator on the left end of the sign and vice versa. These provisions do not apply to existing signs.

Incidentally, the *Life Safety Code* does not establish any specific color for exit signs, but it mandates that the signs contrast with the background. The intent is that they be visible, legible, and identifiable. Some local codes may require the signs to be red, whereas other codes may require them to be green. If the local code does not mandate a specific color, contrast and visibility must be considered. Also, in the international scheme, red means danger—stop; green means safety—go. Since an exit is supposed to be safe, green-colored signs are more appropriate in buildings where there may be many international travelers and visitors.

18

Fire Protection Features

A facility's fire protection features include a diversity of items, such as protecting and sealing vertical openings (i.e., floor penetrations); enclosing and/or protecting hazardous areas; addressing concealed spaces; providing smoke barriers or smoke partitions where needed; using appropriate interior finishes, contents, and furnishings; and installing the proper fire alarm and extinguishing systems. NFPA *101*®, *Life Safety Code*®, also considers fire barriers and corridors to be fire protection features, as covered in Chapters 15 and 16 of this book. Likewise, fire extinguishing systems and alarm systems are addressed to a limited degree in this chapter and in more detail in Part IV.

18.1 Vertical Opening Protection

Essentially, any hole in the floor is a *vertical opening*. More important, it is a way for fire and smoke to travel between one story and another. Examples of vertical openings include elevator hoistways, light wells, HVAC (heating, ventilating, and air-conditioning) shafts, telephone and electrical shafts, pipe shafts, stairs, atria, and seismic and expansion joints. One note of caution: Most stairs that communicate between floors are also considered to be exits and as such must meet the more stringent requirements for exit enclosures—see Chapter 16.

Figure 18.1 illustrates some examples of vertical openings that might spread the effects of fire from floor to floor if not properly enclosed.

NFPA fire records show that unprotected or improperly protected vertical openings have been responsible for multiple deaths in numerous fires. In response, the *Life Safety Code* has traditionally placed very stringent requirements on protecting vertical openings (within the main chapter on fire protection features, Chapter 8). Moreover, virtually every occupancy includes additional requirements for protecting vertical openings, and almost every occupancy includes some exceptions to the basic requirements.

In general, vertical openings must either be sealed at the floor line, as is typically done with pipes, tubes, and wiring, or the vertical opening must be enclosed in a shaft. In new construction, shafts that connect a maximum of three stories must have a fire resistance rating of 1 hour or more, while shafts that connect four or more stories must have a fire resistance rating of 2 hours or more. Table 18.1 summarizes the new and existing requirements for vertical openings.

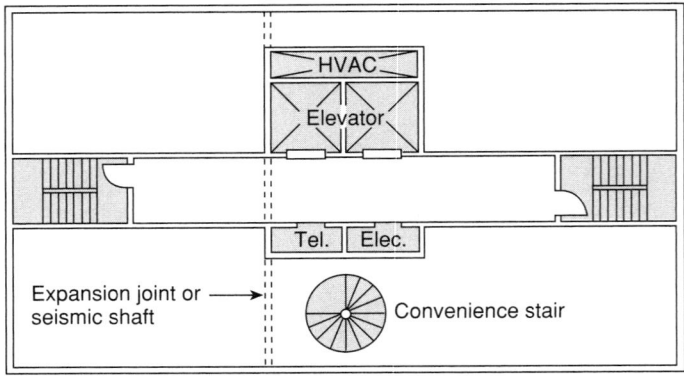

Figure 18.1 Typical vertical openings in floors.
Source: Life Safety Code Handbook, 2003, Exhibit 8.11.

As can be seen in Table 18.1, the requirements for new and existing health care facilities up to three stories are the same. The main difference in the requirements for vertical openings are between existing health care occupancies and other existing occupancies that may be associated with health care facilities. Shafts used in existing health care occupancies are required to have ratings of 1 hour, whereas shafts for other existing occupancies can have ratings as low as ½ hour.

Again, it is important to remember that although exit stairs are vertical openings, they must meet the more stringent requirements for exit enclosures. There are no limitations on the type or number of openings or penetrations that shafts can have, as long as the shafts are properly protected. This is significantly different from there for exit enclosures, for which most penetrations are prohibited (see Section 16.2 in this book).

Within the chapter on fire protection features, the *Life Safety Code* provides certain exceptions to the protection of vertical openings. Some of these exceptions include "mini-atria," atria, convenience openings, and other miscellaneous vertical openings. The *Life Safety Code* is set up so that some of these exceptions apply to all occupancies, some apply unless prohibited by the occupancy chapter, and some apply if allowed by the occupancy chapter.

Table 18.1 Fire Resistance/Fire Protection Requirements for Vertical Openings

	New		Existing	
Vertical Opening Connections	Walls	Openings	Walls	Openings
≤3 stories	1 hr	1 hr	30 min	20 min
≥4 stories	2 hr	1½ hr	30 min	20 min
Health care ≤3 stories	1 hr	1 hr	1 hr	1 hr*
Health care ≥4 stories	2 hr	1½ hr	1 hr	1 hr*

*Existing doors with a ¾-hr fire protection rating may be continued in use.

18.1.1 Mini-Atria

The term *mini-atrium* does not appear in the *Life Safety Code*. It is referred to there as "unenclosed floor openings communicating between three floors." The term *mini-atrium* is commonly used as a shorthand method of referring to this provision. Beginning with the 2003 edition of the *Code*, it is referred to as a "communicating space." In this book, we will continue to refer to this as a *mini-atrium*, as it is more descriptive than *communicating space*. Health care occupancies do *not* permit the mini-atrium, but most occupancies associated with health care facilities do permit them. The method for referencing the mini-atrium changed between the 1997 and 2000 editions of the *Life Safety Code*. In pre-2000 editions, the provision could be used only when specifically permitted in the occupancy chapter. With the 2000 edition, the provision has been changed such that it is allowed unless prohibited by the occupancy chapter.

Briefly, mini-atrium openings are limited to three floors, of which the lowest or next-to-lowest level is a street floor. There are no size limitations and no requirements for smoke control. Mini-atria do not necessarily require sprinkler protection; however, compliance without sprinkler protection is very difficult. There are stringent limitations on egress through the space and on the minimum required openness of the space. Figure 18.2 illustrates the use of this exception.

18.1.2 Atria

Virtually all occupancy chapters allow atria, although some occupancies apply additional restrictions. The method of allowing the atrium is like that described above for mini-atria, and like mini-atria, the method was changed between the 1997 and 2000

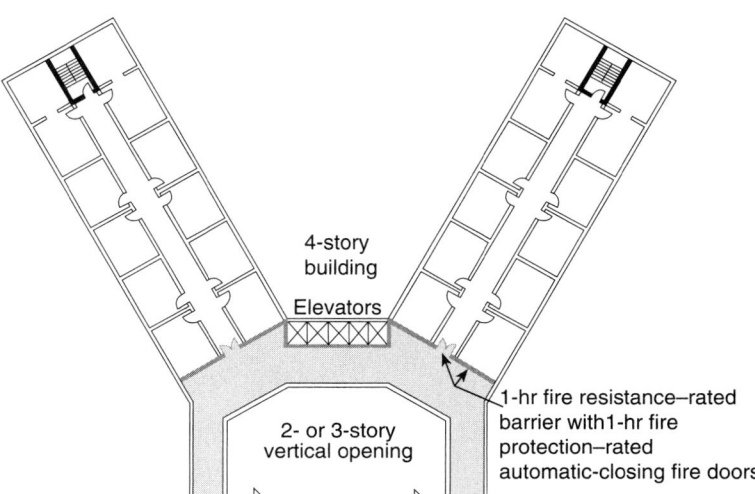

Figure 18.2 Separation of two- or three-story communicating space from the remainder of the building.
Source: Life Safety Code Handbook, 2003, Exhibit 8.16.

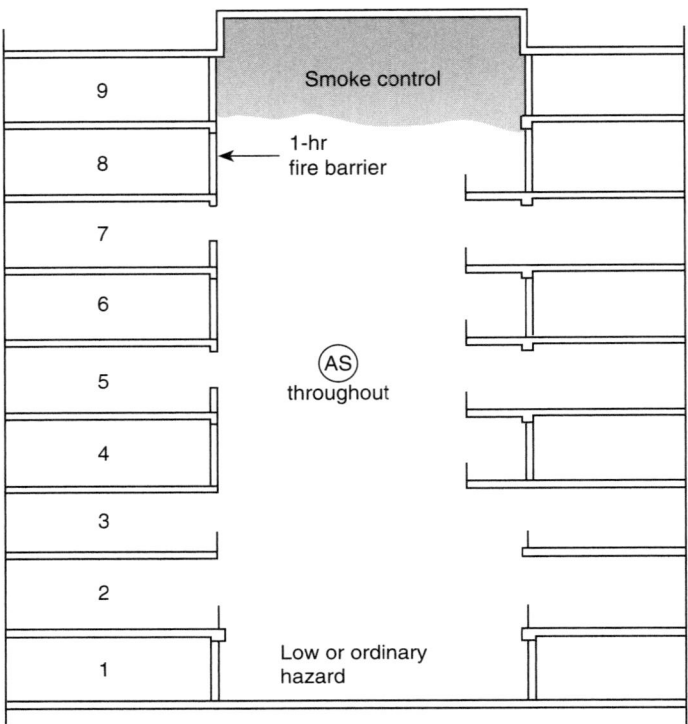

Figure 18.3 Requirements for protecting an atrium.

editions of the *Life Safety Code*. There are advantages and disadvantages of having atria over mini-atria. Atria require total sprinkler protection and some form of smoke management, but there is no restriction on size or egress through the atrium. Atria have separation requirements, but alternatives are provided. The main restriction in health care occupancies is that patient sleeping and treatment rooms are not allowed to be open to the atrium. Figure 18.3 illustrates some of the provisions for an atrium. Table 18.2 provides some comparisons between atria and mini-atria.

Table 18.2 Features of Atria and Mini-Atria

Feature	Mini-Atrium*	Atrium
Maximum number of stories	3	No limit
Location limit in building	Yes	No
Egress restrictions	Yes	No
Openness requirement	Yes	No
Separation	Usually needed	Design option
Sprinkler mandate	Usually needed	Throughout building
Smoke management	Not required	Required

*Now called a three-story convenience opening in the *Code*.

18.1.3 Convenience Openings and Other Unprotected Vertical Openings

Another exception for vertical openings permitted in most occupancies associated with health care facilities, but not assembly occupancies, is the two-story convenience opening. The *Life Safety Code* allows an opening between any two stories, provided the opening is separated from the corridors and unprotected openings to other floors and does not serve as part of the required means of egress. Examples of this feature include a "convenience" stair between a two-level office area, a two-level storage area, or a similar configuration. Figure 18.4 illustrates one use of the two-story convenience opening.

The *Life Safety Code* includes special provisions for several specific items, such as mail chutes, escalators, pneumatic tubes, and expansion and seismic joints. See the *Life Safety Code* for details. In addition, several of the occupancies often found in health care facilities have particular exceptions, such as those for

- Multilevel sleeping areas in psychiatric facilities
- Two-story townhouse-type office buildings
- Existing sprinkler-protected business, industrial, and storage occupancies

18.1.4 Mezzanines

Mezzanines are similar to vertical openings, but they are different. The *Life Safety Code* defines a *mezzanine* as an intermediate level between the floor and ceiling of any room or space. A code-compliant mezzanine is not a vertical opening, because it is considered to be a level within a story, not a separate story. A mezzanine that does not comply with the *Life Safety Code* is considered to be another story, and the opening between the mezzanine and the floor below is then an unprotected vertical opening that must be protected or meet one of the *Life Safety Code*'s exceptions for unprotected vertical openings.

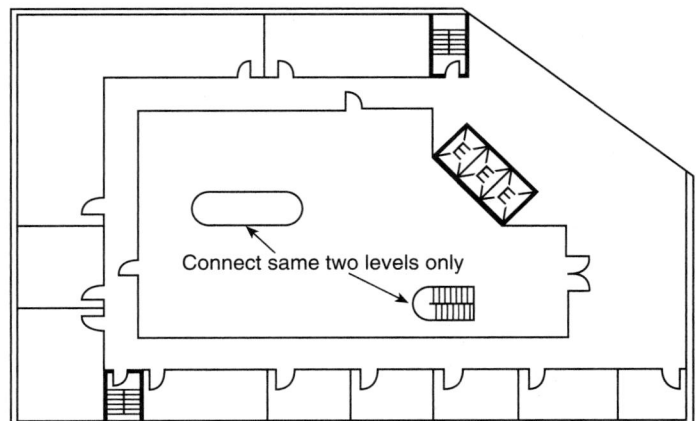

Figure 18.4 A two-story convenience opening.

There are two main limitations on mezzanines, related to size and openness. With regard to size, mezzanines are limited to one-third the open area of the room in which they are located (i.e., the room below). In calculating the area of the room, the area of the mezzanine and any enclosed areas within the room are not included. Figure 18.5 illustrates this point.

With regard to openness, the mezzanine must be open to the level below. The intent of this requirement is to provide the occupants of the mezzanine with awareness (by sight, smell, and sound) of the situation below. Since mezzanines are allowed to egress to the level below, it is important that this awareness be provided.

There are three exceptions related to the openness of mezzanines. The first exception is very logical, in that it allows the mezzanine to have walls not more than 42 in. (107 cm) high, which is guardrail height, plus columns and posts. In fact, this wall would be required due to the guard requirements in the *Life Safety Code* (Chapter 7). This exception has little if any impact on the required aspects of awareness. The next two exceptions allow some enclosure of the mezzanine level. One exception allows the presence of enclosed areas on the mezzanine, as long as the total occupant load of all enclosed areas on the mezzanine does not exceed 10 people. Thus, a mezzanine can include enclosed areas such as rest rooms, small office areas, or small locker rooms.

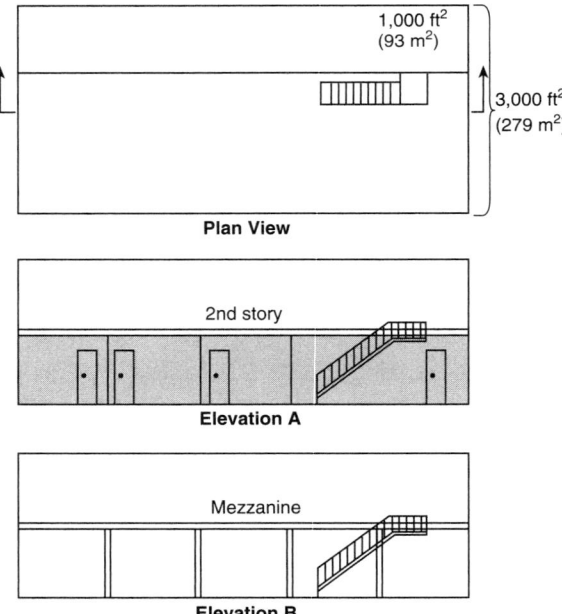

Figure 18.5 The one-third rule for determining whether a partial upper level qualifies as a mezzanine. In Elevation A, the space under the mezzanine is enclosed, and the open space of the room is 2000 ft²; the mezzanine exceeds two-thirds and is considered a second story. In Elevation B, the space is open, and the mezzanine complies.
Source: Life Safety Code Handbook, 2003, Exhibit 8.28.

The other exception allows any area to be enclosed as long as there is an exit from the mezzanine that permits egress directly to the outside without going through the level below. This exception addresses the lack of awareness caused by the enclosure of the mezzanine. By providing a direct exit from the mezzanine, a fire on the lower level is not as much a threat to the people on the mezzanine. Figure 18.6 illustrates these exceptions.

It is important to remember that if a mezzanine does not meet the above requirements, it is considered to be another story, and the opening to the story below must be treated as an unprotected vertical opening.

18.2 Protection of Concealed Spaces

Concealed spaces, especially combustible concealed spaces, can form an area for fire to spread undetected in a building. Even once a fire has been detected, it is very difficult for fire-fighting personnel to stop it in a combustible concealed space. *Concealed spaces* include but are not limited to spaces between the ceiling and the roof or floor above, hollow spaces in walls, and voids created by renovations. Although it is best to avoid building a structure that has combustible concealed spaces, it is not always

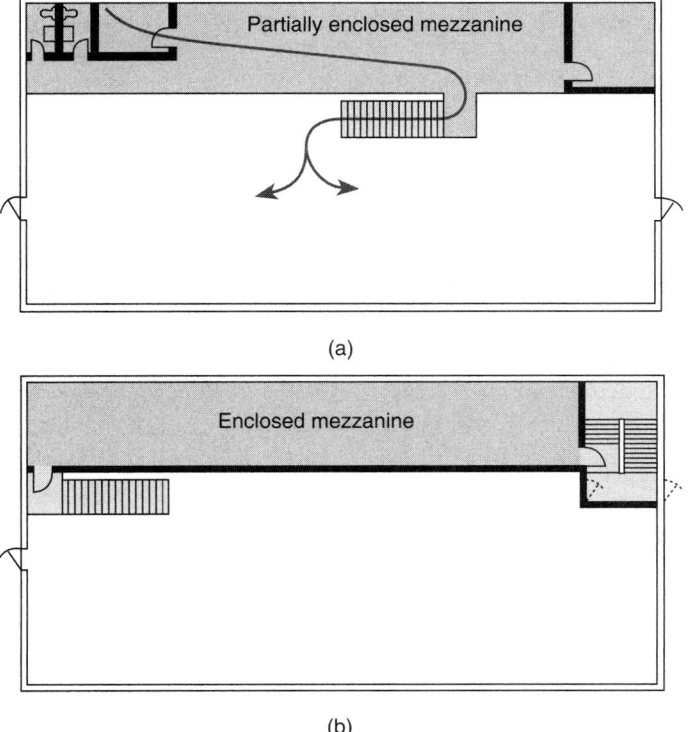

Figure 18.6 (a) Open mezzanine and (b) enclosed mezzanine.
Source: Life Safety Code Handbook, 2003, Exhibits 8.29 and 8.30.

possible to do so, especially in buildings of combustible construction. Therefore, most codes require protection of combustible concealed spaces and limit the amount of combustibles located in noncombustible concealed spaces. For example, see NFPA 90A, *Standard for the Installation of Air-Conditioning and Ventilating Systems,* for limitations on combustible items in concealed spaces used for plenums.

Protecting combustible concealed spaces is accomplished either with fire blocking, draft stopping, or automatic sprinkler protection. Fire blocking is usually accomplished by placing wood blocking in a small cross section in the concealed space, like that often placed between studs in wood stud construction. The blocking is usually done at the floor line, the ceiling line, the roof line, and similar locations. Draft stopping is accomplished by using some sort of continuous membrane across such spaces, such as those typically found between ceilings and the floor above, or between a ceiling and the roof above, to restrict the passage of smoke, heat, and flames. The *Life Safety Code* addresses concealed spaces in Chapter 8. Buildings that are properly protected by automatic sprinklers may not have to install draft stopping but usually do need to have fire blocking.

Most codes treat the protection of combustible concealed spaces similarly. The *Life Safety Code* requires roof/ceiling spaces in new construction to be subdivided into areas not exceeding 3000 ft^2 (280 m^2) and floor/ceiling spaces to be subdivided into 1000-ft^2 (93-m^2) spaces. Most codes have required fire blocking and draft stopping for many years, so many existing structures include this protection as well.

When designing, building, and operating new construction, it is essential to protect the facility's combustible concealed spaces properly. In existing buildings, it is important that draft stopping and fire blocking be maintained. It is not at all uncommon to find draft stopping that has had openings cut into it and not repaired.

It is possible to create combustible concealed spaces inadvertently, even in noncombustible construction. This can happen, for instance, when a noncombustible drop tile ceiling is installed below an old combustible tile ceiling. The resulting space is a combustible concealed space. This type of situation creates a problem for both sprinklered and nonsprinklered construction. In sprinklered buildings, this new space is now a nonsprinklered combustible area to which sprinkler protection must be added. In nonsprinklered buildings, this newly created space must be subdivided at 1000-ft^2 (93-m^2) intervals.

18.3 Smoke Barriers

A *smoke barrier* is a continuous membrane, either vertical or horizontal, such as a wall, floor, or ceiling assembly, that is designed and constructed to restrict the movement of smoke. Practically speaking, smoke barriers in health care facilities permit the horizontal movement of patients into a safe area, which may allow a delay in evacuation or even eliminate the need for evacuation. The important point to remember about smoke barriers is that they need to resist the passage of smoke.

18.3.1 General Provisions for Smoke Barriers

The *Life Safety Code* contains general provisions for smoke barriers within the chapter on fire protection features (Chapter 8). These provisions probably encompass some of the most important fire protection requirements for health care facilities, so much so that even buildings with sprinkler protection are not exempt.

18.3.1.1 Difference Between Smoke Barriers and Smoke Partitions

It should be noted that starting with the 2000 edition, the *Life Safety Code* addresses the requirements for a new type of partition called a *smoke partition*, which should not be confused with a smoke barrier. The provisions for smoke partitions were added so that the phrase "a barrier that resists the passage of smoke" can be replaced by the term *smoke partition*. The phrase has been used in numerous locations in the *Life Safety Code* for many years. It is not the intention of the code to have smoke partitions meet the full requirements for smoke barriers.

The difference between smoke barriers and smoke partitions is not always clear. Neither type of barrier is required to have a fire resistance rating unless the referencing requirement states so. For example, new health care occupancies require smoke barriers with a 1-hour fire resistance rating. (As such, smoke compartments in health care occupancies are really formed with combination fire/smoke barriers.) Most model building codes require smoke barriers to have a 1-hour fire resistance rating, primarily for new construction or renovations. The main issue behind requiring a rating for smoke barriers is to ensure that the barrier has some integrity and that someone does not use a sheet of plastic as a smoke barrier. Unprotected transfer grills, louvers, and screen or jalousie doors are not permitted.

In general, the requirements for smoke barriers are more stringent than those for smoke partitions (similar to the requirements for a fire barrier being more stringent than a fire partition in the *International Building Code*). Originally, the *Life Safety Code* required smoke barriers to have smoke dampers, with relatively few exceptions. However, now many smoke barriers do not require smoke dampers. Table 18.3

Table 18.3 Comparison of Smoke Barriers and Smoke Partitions

Feature	Smoke Barrier	Smoke Partition
Resistance to passage of smoke	Yes	Yes
Extension to above floor	Yes—rare exception	Yes—common exception
Fire rating	Optional	Not required
Smoke damper	Yes—many exceptions	Only for air-transfer openings
Door—self- or automatic closure by smoke	Yes	Yes—can be exempted
Door resistance to smoke	Resists smoke—20-min rating; if wall, 1-hr rating	Resists smoke
Sealed penetrations	Yes	Yes

compares smoke barriers and smoke partitions. As the table shows, there is not a significant difference between the two types of smoke barriers, but the differences are important.

Typical problems associated with smoke barriers in health care facilities are unsealed penetrations, improperly fitting doors, and damper issues.

18.3.1.2 Difference Between Smoke and Fire Barriers

Smoke barriers are very similar to fire barriers in that continuity is very important. (Refer to Chapter 15 for a discussion of horizontal and vertical continuity of fire barriers.) The major differences between smoke barriers and fire barriers are that smoke barriers are not required to have a fire resistance rating, unless specifically required by code. Also, smoke barriers must resist the passage of smoke, which some fire barriers do not. One word of caution is important here. Some occupancies, such as health care, require smoke barriers to have a fire resistance rating of ½ hour or 1 hour. These barriers are referred to as 1-hour smoke barriers, but in reality, they are both smoke barriers and 1-hour fire barriers and must meet the requirements of both types of barriers. Additionally, most of the U.S. model building codes require smoke barriers to have a 1-hour fire resistance rating (see Chapter 15 for a discussion of fire barriers).

18.3.1.3 Smoke Barrier Doors

Since a smoke barrier is intended to resist the passage of smoke, it is very important for facility designers and engineers to pay close attention to the doors installed in smoke barriers. Smoke barrier doors must close the opening with only the minimum clearances necessary to operate the door. There must be no undercuts, louvers, or grilles. The *Life Safety Code* does not specify the maximum gap allowed between doors or between the door and the frame, but an annex note defines ⅛ in. (3 mm) as the clearance necessary for the proper operation of a smoke door, as per NFPA 105, *Standard for the Installation of Smoke Door Assemblies*. (NFPA 80, *Standard for Fire Doors and Fire Windows*, is the original source for this gap size; see Chapter 15.)

When smoke barriers must have a fire resistance rating, the smoke barrier doors must have at least a 20-minute fire protection rating when tested in accordance with NFPA 252, *Standard Methods of Fire Tests of Door Assemblies*. However, doors are not required to pass the hose stream test required by that test method, and latches are not required when so exempted by a particular occupancy. Smoke barrier doors in health care occupancies do not require latches, unless the doors serve another purpose that requires a latch, such as a corridor door or a smoke barrier door located in a horizontal exit. Since smoke barrier doors are intended to resist the passage of smoke, they must be either self-closing or automatic closing by activation of a smoke detector (see Chapter 15). Smoke barrier doors in health care occupancies are not required to have a 20-minute fire protection rating but are required to resist the passage of fire for 20 minutes; thus, a 1¾ in.-solid bonded wood core door without label is acceptable. The use of field-applied protective plates up to 48 in. from the bottom of these doors are permitted beginning with the 2000 edition of the *Code*.

Common problems with doors in smoke barriers in health care facilities are as follows:

- Gap between doors is greater than ⅛ in. (3 mm) (astragals are required in new health care occupancies).
- Doors do not close properly due to weak, broken, or missing closers; missing or broken coordinators; or broken or misadjusted latches.
- There is an excessive gap under the door. With tile floors, gaps that are greater than ⅝ in. (12.7 mm) are excessive. Many inspectors do not think this is a problem in existing installations until the gap exceeds 1 in. (25.4 mm), especially for cross-corridor doors in sprinklered facilities.
- Doors are automatic closing and no smoke detector is nearby. [Smoke detectors generally need to be within 5 ft. (1.5 m), but with a corridor smoke detection system, detectors can be up to 15 ft. (4.5 m) away and be compliant.]

18.3.1.4 Smoke Dampers

Another important feature of smoke barriers is the protection of HVAC penetrations using smoke dampers. Smoke dampers are required at each air transfer opening or duct penetration of a smoke barrier. However, under the *Life Safety Code,* smoke dampers are not needed in the following locations:

- Where the opening or duct is part of an engineered smoke control system
- On ducts where the air continues to move and the system prevents the recirculation of air during a fire emergency (Examples may include bathroom exhaust systems or systems with 100 percent supply and exhaust that operate all the time.)
- Where openings in the duct are limited to a single smoke compartment
- On ducts that penetrate floors (All floors must be designed to resist the passage of smoke, but smoke dampers are not required.)
- Where exempted by an occupancy chapter

New health care occupancies exempt dampers in fully ducted systems, as all new health care facilities must be protected throughout with automatic sprinkler protection. Because the sprinklers installed in smoke compartments containing patient sleeping areas must be quick-response or residential sprinklers, and sprinklers in light hazard areas must be quick-response sprinklers (as per NFPA 13, *Standard for the Installation of Sprinkler Systems in Residential Occupancies up to and Including Four Stories in Height*), most new health care facilities have quick-response sprinklers. Therefore, the NFPA Committee on Health Care Occupancies felt that it was justified to eliminate the requirement for dampers in fully ducted systems.

Since taking advantage of this exception might lead to significant savings in maintenance costs, the exception is also provided for existing buildings that have similar sprinkler protection. And instead of the whole building requiring protection, only the smoke compartments on each side of the barrier being evaluated require protection. This exception provides a very realistic way for existing facilities to protect duct penetrations in smoke barriers where dampers might not have been installed previously, or where a new barrier or duct penetration is planned. If the smoke compartments are already provided with sprinkler protection, changing the sprinklers to quick-response ones may be a viable alternative. This exemption does not apply to air transfer openings or in partially ducted systems.

Since smoke dampers aim to prevent the passage of smoke, fusible links do not activate smoke dampers, because they are heat-activated devices. Therefore, smoke detection is required for smoke dampers. These detectors must be installed in accordance with *NFPA 72®*, *National Fire Alarm Code®*. There are several annex notes to *NFPA 72*, which provide assistance on locating detectors within ducts, as shown in Figures 18.7 (a)–(c) and in Figures 18.8 (a) and (b). Also see Chapter 27 for a more thorough discussion of *NFPA 72*.

There are two exceptions for compliance with smoke detectors for smoke dampers. The first exception is that approved detectors can be located within the duct in existing installations, even though the detector may not comply with *NFPA 72*. The second exception allows the damper to be activated by a smoke detector for door-releasing service when the duct is located above the doors being released.

Many common problems are associated with smoke dampers for HVAC penetrations in health care facilities, including

- Lack of smoke dampers where required
- Insufficient clearance from the wall for smoke damper actuator to work properly
- Lack of smoke detector to activate damper
- Lack of access panel in duct to work on or inspect damper

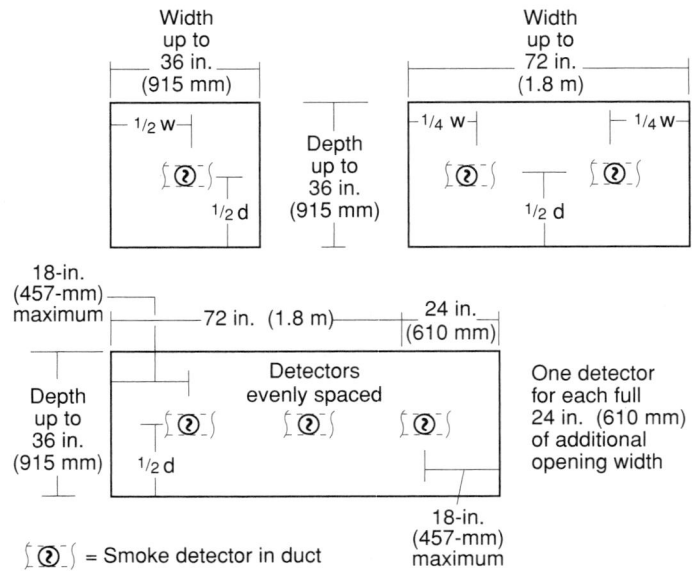

Figure 18.7 (a) Location of smoke detector(s) in return air systems for selective operation of equipment.
Source: NFPA 72, 2002, Figure A.5.14.4.2.2(a).

Chapter 18. Fire Protection Features **321**

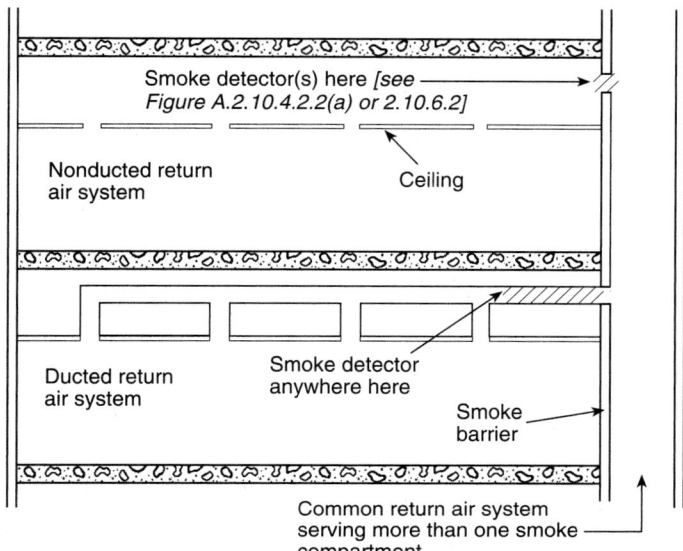

Figure 18.7 (b) Location of smoke detector(s) in return air systems for selective operation of equipment.
Source: NFPA 72, 2002, Figure A.5.14.4.2.2(b).

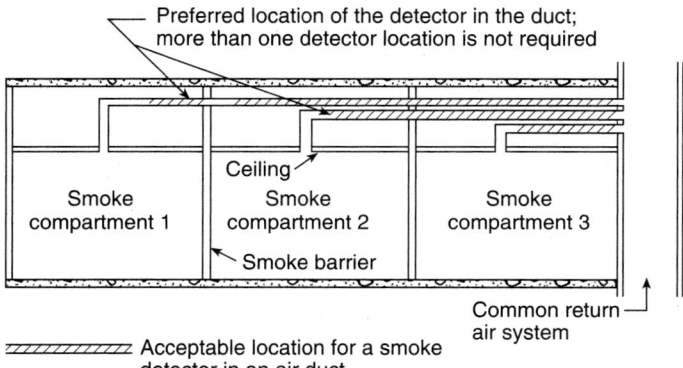

Figure 18.7 (c) Detector location in a duct that passes through smoke compartments not served by the duct.
Source: NFPA 72, 2002, Figure A.5.14.4.2.2(c).

Another common problem with HVAC ducts is that often there is no sleeve or similar device in place to prevent the smoke from leaking around a duct.

18.3.1.5 Sealing Smoke Barrier Penetrations

Similar to fire barriers, penetrations of smoke barriers must be sealed. All seals must retain the smoke resistance of the wall, and if the wall requires a fire resistance rating, the seal must maintain this rating as well. These requirements present a very common problem in health care facilities. It is not at all uncommon to find penetrations of smoke barriers above the ceiling line that have not been sealed. These include, but are not limited to, penetrations for pipes, wires, cable trays, pneumatic tube systems, and others. It is critical to inform the trades as to the location of smoke barriers and the need to seal any penetration properly. Many facilities have contracts with firms that periodically review all barriers and seal any penetrations found, while other facilities have established preventive maintenance programs in place for which in-house staff review all of a facility's barriers. Regardless of the method used, this type of inspection needs to be done periodically.

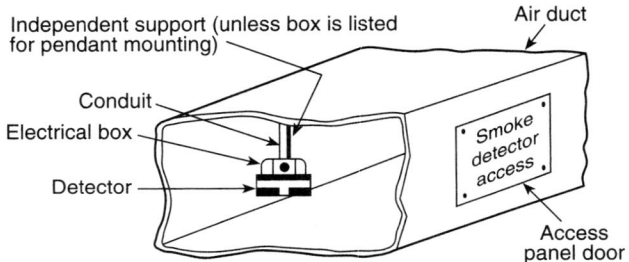

Figure 18.8 (a) Pendant-mounted air duct installation.
Source: NFPA 72, 2002, Figure A.5.14.5.2(a).

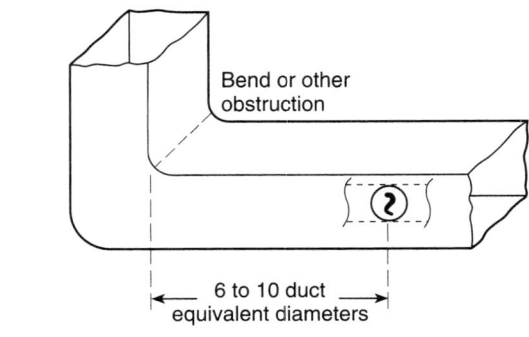

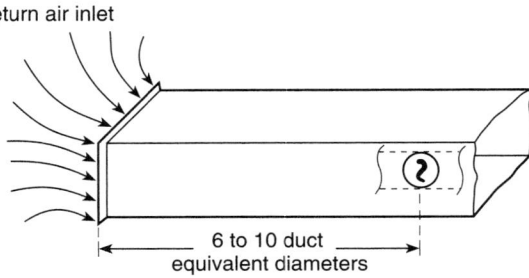

Figure 18.8 (b) Typical duct detector placement.
Source: NFPA 72, 2002, Figure A.5.14.5.2(b).

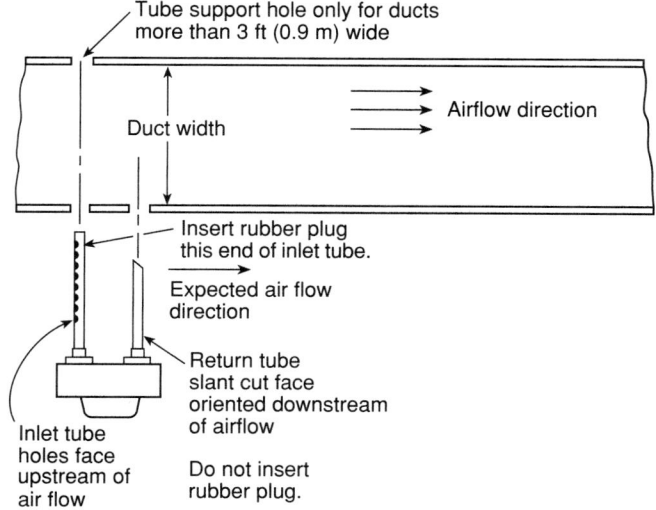

Figure 18.8 (c) Inlet tube orientation.
Source: NFPA 72, 2005, Figure A.5.14.5.2(c).

18.3.2 Requirements for Smoke Barriers in Health Care Occupancies (and Related Occupancies)

The primary reason for installing smoke barriers in new and existing health care facilities is to meet the requirement for "subdivision of building spaces." In new nonsprinklered buildings, smoke barriers can help meet the requirements for accessible means of egress. Smoke barriers are also required in existing nonsprinklered and all new educational occupancies and in some existing nonsprinklered residential occupancies.

The requirements for smoke compartmentation and barriers are significantly different between new and existing health care occupancies. However, it is *very important* to remember that smoke barriers installed in accordance with the requirements for *new* cannot be removed even though they are not required for *existing*. It is a basic *Life Safety Code* element that one cannot eliminate or reduce the level of performance of an item in an existing situation that is required for new construction.

One of the first things a facility must do when preparing a Joint Commission on Accreditation of Healthcare Organizations (JCAHO) Statement of Conditions (SOC) is to determine where the required smoke barriers are. If one reviews only the requirements for existing facilities and determines that some existing barriers are not required, a serious and potentially fatal mistake can be made if these smoke barriers are removed or not maintained. To determine if an existing barrier is not needed, the requirements for new construction must be reviewed. It is not uncommon to survey a floor in a health care facility and find the remnants of a smoke barrier. Although the smoke barrier was not required on the story according to the requirements for existing, it was required for new and should not have been eliminated. This problem is exacerbated due to the significant differences in the requirements for smoke barriers in new and existing health care occupancies. This problem cannot be emphasized enough.

18.3.2.1 Smoke Barriers in New Health Care Occupancies

In new construction, smoke barriers are required to subdivide into at least two compartments every story used for inpatient sleeping or treatment and *any* story in the building, regardless of occupancy, that has an occupant load of more than 50 people. There are several exceptions to this requirement. The following three exceptions can be used on any story that does not contain a health care occupancy and is properly separated from the health care occupancy, which bypasses the requirements for mixed occupancies (e.g., 2-hour separation, among other provisions):

- Any story above the health care occupancy
- Any areas horizontal to the health care occupancy that are separated by a horizontal exit
- Stories below the health care occupancy that are located more than one story below the health care occupancy (There is a one-story buffer zone in addition to the 2-hour separation.)

In addition, sprinkler-protected open-air parking structures do not need to be subdivided. These circumstances are illustrated in Figure 18.9.

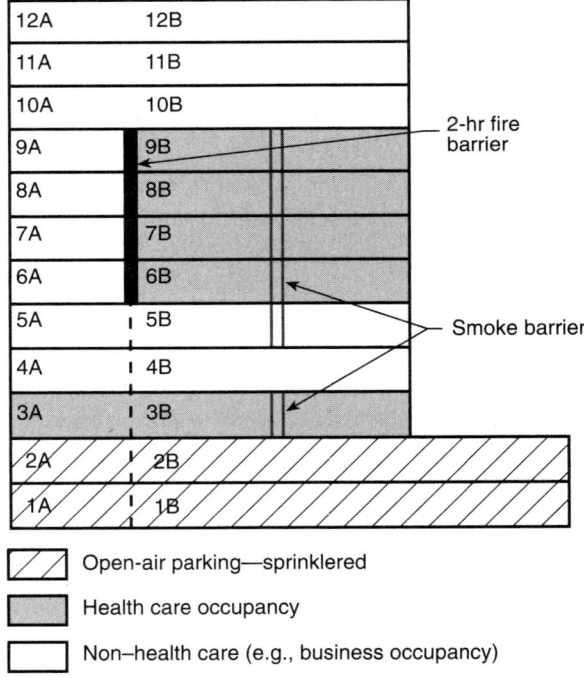

Figure 18.9 Smoke barriers for new health care occupancy buildings, with floors and supporting elements having a 2-hour fire resistance rating.
Source: Life Safety Code Handbook, 2003, Exhibit 18/19.25.

Table 18.4 Areas Needed on Both Sides of Smoke Barriers and Horizontal Exits

Type of Space	Net Area Available on Both Sides of Barrier
Hospitals, nursing homes	30 net ft² (2.8 m²) per patient
Limited care facilities	15 net ft² (1.4 m²) per resident
Health care, but no bed or litter patients	6 net ft² (0.56 m²) per person
Non-health care–horizontal exit	3 net ft² (0.28 m²) per person

For stories that must be subdivided, there are two limitations on how large the smoke compartment formed can be, related to area and travel distance from any point to a door in a smoke barrier. No smoke compartment can exceed 22,500 ft² (2100 m²), nor can the travel distance from any part of a smoke compartment to the door in a smoke barrier exceed 200 ft (60 m). In addition, the minimum size of an area must be able to accommodate a certain number of people in an emergency, based on the same calculations used to estimate horizontal exit capacity; see Chapter 16. Table 18.4 provides the area needed on each side of a smoke barrier and a horizontal exit in health care facilities. These requirements can be easily met if very small compartments are not formed and the compartments that are formed are reasonably equal in terms of space per person provided.

Additional features that are required in new health care occupancies and not in the general provisions for smoke barriers include the following:

- Smoke barriers must have a 1-hour fire resistance rating.
- Cross-corridor doors must be a pair of doors swinging in opposite directions; vision panels must be provided.
- With pairs of doors, the meeting edge must have an astragal, bevel, or rabbet; no center mullion is permitted.
- Stops are required at the head and sides of door frames.
- Positive latching hardware is not required. (If the smoke barrier also serves as a horizontal exit, the door is required to have a 1½-hour fire protection rating, and positive latching hardware is required.)

Figures 18.10 and 18.11 illustrate some of the requirements for smoke barriers in new health care facilities.

18.3.2.2 Smoke Barriers in Existing Health Care Occupancies

In *existing* facilities, subdividing a space with a smoke barrier is required only on stories with patient sleeping areas that can accommodate more than 30 patients. When subdividing is required, the requirements are similar to those for new health care occupancies with regard to maximum area. One difference is that since the 200-ft (60-m) maximum travel distance to reach a smoke barrier door is fairly new to the *Life Safety Code*, one exception for existing health care occupancies is that the travel distance to reach the smoke barrier door is not limited where neither the length nor width of the smoke compartment exceeds 150 ft (45 m). Other differences are noted in Table 18.5.

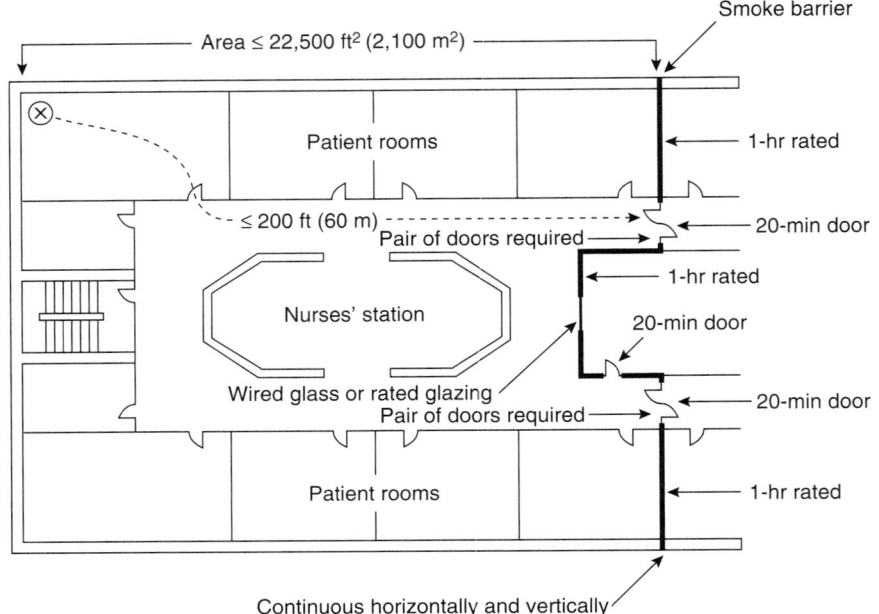

Figure 18.10 A smoke compartment in a new health care facility
Source: OPUS Communications and Koffel Associates, Inc., redrawn from *2000 Life Safety Code Workbook & Study Guide for Health Care Facilities*, 2003, Figure 26.3.

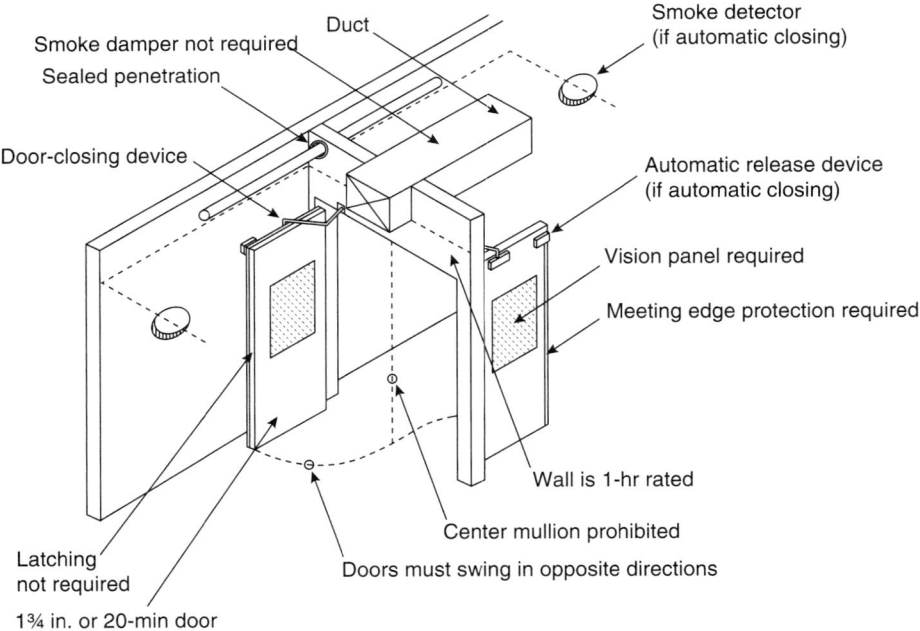

Figure 18.11 Smoke barrier opening protection in a new health care facility.
Source: OPUS Communications and Koffel Associates, Inc., redrawn from *2000 Life Safety Code Workbook & Study Guide for Health Care Facilities*, 2003, Figure 26.4.

Table 18.5 Smoke Barrier Features in New and Existing Construction

Feature	New Construction	Existing Conditions
Smoke barrier fire rating	1-hr	½-hr
Cross-corridor doors swing opposite directions	Required	Not required
Astragal, rabbet or bevel	Required	Not required
Vision panel in cross-corridor doors	Required	Not required
Center mullion	Prohibited	No restriction
Stops at head and sides of frame	Required	Not required

It must be emphasized that the doors in existing health care occupancies still must resist the passage of smoke. Therefore, either a tight tolerance must be provided at the meeting edges, top, and sides or protection such as stops and astragals must be provided. When astragals are provided on doors that must swing in the same direction, a coordinator is required, although maintenance of such devices is difficult. It may be advisable to replace the doors with a pair of opposite-swinging doors.

As mentioned earlier and cannot be overemphasized, an existing smoke barrier that is in an existing building must be maintained if it is required in new construction, even though it may not be required for existing structures.

Figures 18.12 and 18.13 illustrate some of the provisions for smoke barriers in existing health care occupancies.

18.4 Special Hazard Protection

The *Life Safety Code* defines a hazardous area as follows:

> An area of a structure or building that poses a degree of hazard greater than that normal to the general occupancy of a building or structure. Hazardous areas include areas for the storage or use of combustibles or flammables; toxic, noxious, or corrosive materials; or heat-producing appliances. [101–06: 3.3.17.4 and A.3.317.4]

Special hazard protection, or more commonly called the protection of hazardous areas, is covered in two locations in the *Life Safety Code*. General provisions that apply to all occupancies are provided in the chapter on fire protection features, and more specific provisions are provided in each occupancy chapter.

18.4.1 General Provisions for Protecting Hazardous Areas

In general, the *Life Safety Code* requires that hazardous areas be protected by one of the following:

1. Enclose the area with a fire barrier having a 1-hour fire resistance rating . . . without windows.
2. Protect the area with automatic extinguishing systems.
3. Apply the provisions in both (1) and (2) where the hazard is severe or where otherwise specified by the occupancy chapters.

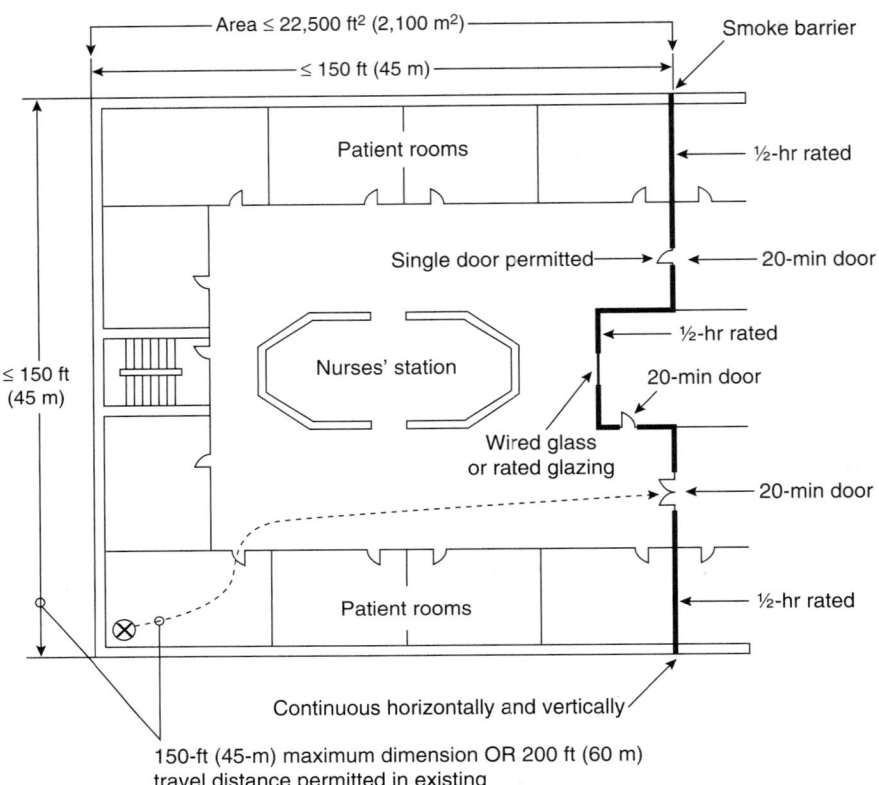

Figure 18.12 A smoke compartment in an existing health care facility
Source: OPUS Communications and Koffel Associates, Inc., redrawn from 2000 *Life Safety Code Workbook & Study Guide for Health Care Facilities,* 2003, Figure 26.1.

When the 1-hour enclosure is used, the doors must be ¾-hour fire protection–rated, self-closing, and self-latching doors. In new construction (other than mercantile and industrial occupancies) and in any health care occupancy, when a hazardous area is sprinkler protected without having a fire resistance–rated enclosure, the area must be protected with at least a smoke-resisting enclosure equipped with smoke-resisting self- or automatic-closing doors. It should be noted that Chapter 9 of *NFPA 101* contains a special provision that allows up to six sprinklers to be supplied from the domestic water piping to protect isolated hazardous areas. This allowance can be considered as a remedy for some existing health care occupancies. In general, the door in this smoke-resisting wall does not have to latch, unless specifically required to elsewhere. In health care facilities, corridor doors are normally required to latch, so that these doors in health care facility hazardous areas are typically required to be self- or automatic-closing and self-latching.

The *Life Safety Code* also includes specific provisions for protecting certain hazardous areas, such as where flammable liquids and gases are used or stored, laborato-

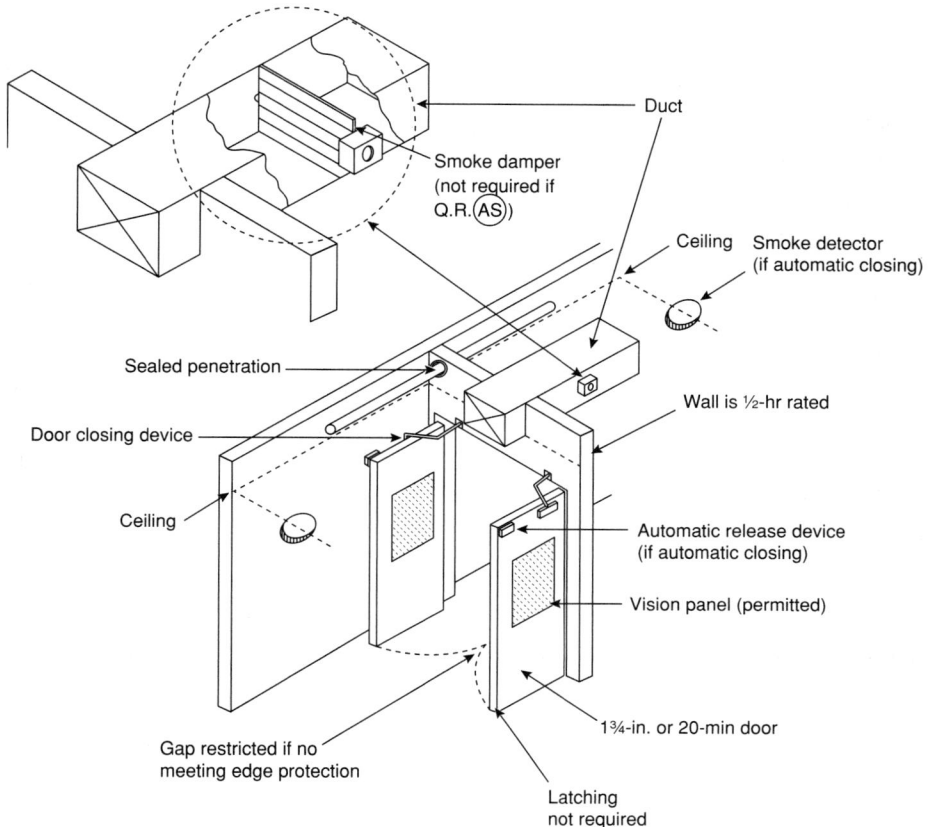

Figure 18.13 Smoke barrier opening protection in an existing health care facility
Source: OPUS Communications and Koffel Associates, Inc., redrawn from *2000 Life Safety Code Workbook & Study Guide for Health Care Facilities*, 2003, Figure 26.2.

ries, and where explosion potentials occur, although the explosion potential is not normally an issue for health care facilities, as it is more related to industrial occupancies.

One of the biggest problems with protecting hazardous areas is determining which areas require protection, and then determining if the area is merely hazardous or severely hazardous, which requires the double protection of both a fire-rated barrier and sprinkler protection. The definition is helpful but leaves a lot up to judgment. Many occupancy chapters, including health care, residential, and to some extent assembly, include lists of typically hazardous areas. The chapter on business occupancies provides a short list of hazardous areas (e.g., general storage, boiler and furnace rooms, maintenance shops, and woodworking and painting areas) and categorizes areas with severe hazards as those containing high hazard contents (i.e., contents that present the potential for a flash fire or explosion). Since most health care facility business occupancy areas do not have flash fire or explosion potential, most hazardous areas in these areas are either protected by sprinklers or enclosed in 1-hour fire resistance–rated construction, but not both.

There is one very important precaution to take in judging whether an area is hazardous or not. The lists of potentially hazardous areas included in the *Life Safety Code* are not all-inclusive. Based on the definition of hazardous area, other areas that are hazardous can exist in a facility that are not included on one of the lists.

Figure 18.14 illustrates the general provisions for protection of hazardous areas.

18.4.2 Protecting Hazardous Areas in Health Care Occupancies

Similar to the provisions for new versus existing smoke barriers, the differences between the requirements for protecting new and existing hazardous areas can cause several significant problems in health care occupancies. In addition to the significant differences in the requirements for protecting new versus existing areas (the same as in the general provisions), it must first be determined if the hazardous area is new or existing. To make this determination, it does not matter when the building was built; the date that counts is when the area became a hazardous area. For example, a hospital built in 1965 is an existing health care facility. If an existing patient room was converted to a storage room last month, it is considered to be a new hazardous area and must meet the requirements for protecting hazardous areas in a new health care facility.

It is also very important that existing facilities use caution when converting an area to a hazardous area to make sure the required retrofits can be made. Using the previous example, the room that was converted from a patient room to a storage room for combustible materials is over 100 ft^2 (4.6 m^2). As such, the room must have sprinkler protection and be enclosed in 1-hour construction with 45-minute self-closing,

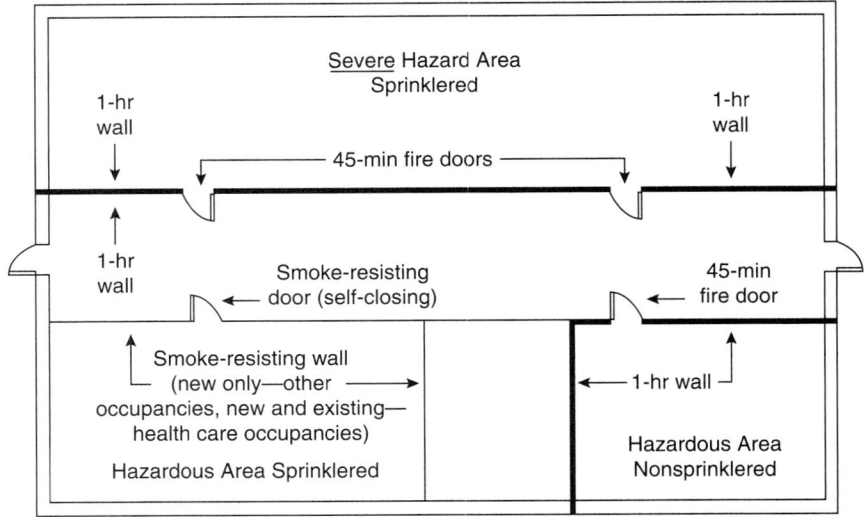

Figure 18.14 Protection of hazardous areas in business occupancies.
Source: OPUS Communications and Koffel Associates, Inc., redrawn from 2000 *Life Safety Code Workbook & Study Guide for Health Care Facilities*, 2003, Figure 31.2.

self-latching doors. If the facility is not sprinklered, adding the sprinkler protection may be difficult. It is also difficult to install 1-hour construction with 45-minute self-closing, self-latching doors, since patient rooms are rarely so equipped. Within health care occupancies, almost any room over 50 ft² (4.6 m²) containing combustible storage is a hazardous area.

The list of hazardous areas is virtually the same for new and existing health care facilities. The main difference is that many rooms listed for *new* are considered severely hazardous areas and must have the double level of protection, while all the rooms on the list for *existing* are given the option of either being a protected enclosure or having sprinkler protection. Again, it is important to remember that the lists are not all-inclusive and that the general provisions must still be met and enforced. Table 18.6 summarizes the lists for both new and existing health care occupancies.

18.4.2.1 Protecting Gift Shops, Laundry and Trash Chutes, and Other Hazardous Areas

The *Life Safety Code* requires laboratories, anesthetizing locations, and medical gas systems to comply with the provisions of NFPA 99, *Standard for Health Care Facilities*, that relate to protecting hazardous areas. It should also be noted that new heliports must comply with NFPA 418, *Standard for Heliports*. Cooking facilities need to be protected in accordance with NFPA 96, *Standard for Ventilation Control and Fire Protection of Commercial Cooking Operations*, which is discussed in Chapter 24 of this book.

Gift shops and trash and laundry chutes are two additional areas that sometimes need to be protected as hazardous areas. Gift shops must be protected as a hazardous

Table 18.6 Typical Hazardous Areas

New		Existing
1-hr Enclosure	Smoke-Resisting Enclosure with Self-Closing Doors	1-hr Enclosure or Sprinklers
Boiler rooms		Boiler rooms
Fuel-fired heater rooms		Fuel-fired heater rooms
Soiled linen rooms		Soiled linen room
Storage rooms > 100 ft² (9.3 m²)	Storage rooms > 50 ft² (4.6 m²) but ≤ 100 ft² (9.3 m²)	Rooms > 50 ft² (4.6 m²) designated by AHJ
Laundries > 100 ft² (9.3 m²)		Laundries > 100 ft² (9.3 m²)
Maintenance shops		Repair shops
Paint shops		Paint shops
Trash collection rooms		Trash collection rooms
Laboratories* (with severe hazard)	Laboratories* (without severe hazard)	Laboratories* (without severe hazard)

AHJ = Authorities having jurisdiction.
*Laboratories considered severe hazard shall comply with NFPA 99.

area when they are used for storing or displaying combustibles in quantities considered hazardous. The *Life Safety Code* provides no guidance on this, but the typical gift shop with separately protected storage will usually be allowed to comply with the following special provisions. The *Life Safety Code* allows significant flexibility for gift shops protected by sprinklers that have a separated storage area. Gift shops smaller than 500 ft^2 (46.5 m^2) are allowed to be open to a corridor or lobby. Larger shops can be separated from the corridor or lobby by a nonrated barrier such as plain glass. This allows most gift shops to be easily seen from the lobby, without creating a fire hazard to the facility.

Trash chutes and linen chutes can also present significant problems for health care occupancies. The *Life Safety Code* requires these areas to comply with NFPA 82, *Standard on Incinerators and Waste and Linen Handling Systems and Equipment*. In addition, the trash rooms and soiled linen rooms on each story, as well as the discharge room at the bottom of the chute, are hazardous areas. These rooms have traditionally been considered severely hazardous areas and have required protection by both 1-hour fire resistance–rated enclosures and sprinklers. As discussed previously, existing conditions require only enclosure protection or sprinkler protection. However, if both currently exist, they must be maintained. It is not at all uncommon to find damaged fire-rated doors, closers missing or broken, or door labels missing or painted over. Even if sprinkler protection is provided, the enclosure must be kept to code. Since the requirements for new construction must be maintained, it is important that the protection originally provided for these hazardous spaces be maintained as well.

An additional concern is that the chute is considered to be a vertical opening. As such, it must be enclosed in a shaft as required for vertical openings as discussed previously in Section 18.1. The doors into the chute on each floor must be labeled, self-closing, and self-latching. These doors are often blocked open and missing the proper labels. Also, self-closers may be disconnected and latches taped open. The doors at the bottom of the chute can be held open with a fusible link, one of the few places in the *Life Safety Code* where fusible links are acceptable on an item that is to be self-closing. Also, both the *Life Safety Code* and NFPA 82 require sprinkler protection in the chute. See NFPA 82 for details.

Hazardous areas are moving targets in many larger health care facilities. Rooms that were not hazardous areas last year may be this year. It is important to survey the facility periodically to find new storage areas. What may have been a patient room last year may be used this year for storing oxygen, transferring liquid oxygen, or storing broken beds.

18.5 Interior Finish

Interior finishes are regulated to help control the spread of fire in a facility. Improper interior finishes can lead to the rapid spread of fire, and in the past they have contributed to multiple-fatality fires. Today, due to improved control of interior finish, fire spreading rapidly on interior finish is not a common problem in health care facilities. It should be noted that the material on interior finish was moved to its own chapter

(Chapter 10) starting with the 2000 edition of the *Life Safety Code*. Before the 2000 edition, this material was in Section 6.5.

Two categories of interior finish include interior wall and ceiling finish and interior floor finish. As the names imply, the first category applies to the exposed surfaces of walls, ceilings, columns, and partitions in a building's interior, while the second applies to a building's floor surfaces.

18.5.1 Interior Wall and Ceiling Finish

Traditionally, interior wall and ceiling finish has been a significant problem during fires. Today, several tests are used to determine the hazard of interior finishes, depending on the material and where it is used in a building. The most common test used to evaluate interior finish is NFPA 255, *Standard Method of Test of Surface Burning Characteristics of Building Materials* (also known as ASTM E 84, or the *Steiner Tunnel Test*). This test results in a flame spread index and a smoke developed index. Based on these indices, the codes assign a flame spread classification. Some codes, including the *Life Safety Code,* use letter classifications, while other codes use number classifications, both of which are shown in Table 18.7. These ratings should be available from the manufacturer of the finish and are included in the directories of testing laboratories.

The biggest problem with interior finish in health care facilities tends to be determining the classification of an existing interior finish for which no files exist. To judge existing interior finish, it is helpful to understand what the flame spread and smoke developed indices are based on. The index of 0 is assigned to a noncombustible cement board, while 100 is assigned to red oak. Therefore, unless the finish is treated, it is doubtful that a wood product would be indexed lower than 100. In fact, some varnished Douglas fir plywood can be rated as high as 500. Following are some typical interior finish ratings. These ratings can be used for guidance for existing installations, but they are not absolutes and should never be used for determining the finishes for new installations.

- Redwood: 75 to 200
- Oak: 70 to 95
- Pine: 105 to 190
- Douglas fir: 100

Table 18.7 Flame Spread Classifications

Classification	Flame Spread Index	Smoke Developed Index
A (1 or I)	0 to 25	0 to 450
B (2 or II)	26 to 75	0 to 450
C (3 or III)	76 to 200	0 to 450
Not allowed*	Over 200	Over 450

*"Not allowed" is not a classification but is included to indicate that materials with such indices are not permitted to be used.

- Cedar: 110
- Spruce: 75 to 110
- Plywood
  - Douglas fir: 120
  - Latex paint: 100
  - Varnish: 160 to 500
  - Wallpaper: 65
  - Shellac: 300 to 800
- Mineral acoustical tile: 12 to 20
- Gypsum board: 10 to 20
  - Latex paint: 8 to 10
  - Wallpaper: 35
- Fiberboard: 200 to 350
- Cork: 560 to 640

Each occupancy establishes limits of where each of the A, B, and C interior finishes can be used. Table 18.8 provides the requirements for occupancies usually associated with health care facilities. Where sprinklers are present, the requirement for A becomes B, and the requirement for B becomes C. There is no reduction in Class C for sprinklers. There are numerous exceptions to the classifications given, so it is important to consult the *Life Safety Code* for specific information.

Given these limitations on interior finish, facilities should carefully consider using the following types of interior finishes:

- Wood panels or paneling that have no documented flame spread, especially in nonsprinklered facilities
- Cork
- Cellulose ceiling tiles (These tiles were commonly square and glued or stapled, not hung, and were popular in the 1950s and 1960s. This item does not refer to the modern suspended mineral ceiling tile.)

Table 18.8 Flame Spread Classification Requirements for Interior Finishes in Occupancies Typically Associated with Health Care Facilities

Occupancy	Exits	Corridors, Lobbies	Other Spaces
Health care–new	A	A*	A*
Health care–existing	A	A*	A*
Assembly ≤ 300 people	A	A or B	A, B, or C
Assembly > 300 people	A	A or B	A or B
Day care centers–new	A	A	A or B
Day care centers–existing	A or B	A or B	A or B
Business	A or B	A or B	A, B, or C
Industrial, storage	A or B	A, B, or C	A, B, or C

*Class B in certain locations.

- Burlap
- Textile wall coverings (see commentary following)
- Any interior finish that is not installed over a noncombustible base
- Foam padding that may be installed in psychiatric units

Items that are usually used safely include

- Masonry units—painted or not
- Gypsum board—painted or wallpapered
- Mineral tile ceilings—suspended or not

Caution must be exercised when multiple layers of finishes have been installed. For example, two or three layers of wallpaper may behave significantly differently than a single layer, even if the newest layer has a Class A interior finish rating. The same applies to multiple layers of paint or vinyl wall coverings.

Another issue that should be carefully investigated is the use of foamed plastics. The *Life Safety Code* basically prohibits foamed plastics from being used as interior finishes, but there are some limited exceptions. Anyone planning to use these materials or who is responsible for maintaining a facility where these materials are used should thoroughly review the relevant provisions in the *Life Safety Code*.

The *Life Safety Code* is also quite stringent with regard to the use of textiles as interior finishes. Again, reviewing the code is highly recommended for anyone planning to use this material or maintaining a facility where it is already in use. Another test used to evaluate and categorize textile wall coverings is NFPA 265, *Standard Methods of Fire Tests for Evaluating Room Fire Growth Contribution of Textile Coverings on Full Height Panels and Walls*. Fire tests have shown a potentially significant problem with textiles being used as interior finish; therefore, their use is limited. Note that expanded vinyl wall coverings are treated the same way as textiles.

A new test method was introduced with the 2000 edition of the *Life Safety Code*—NFPA 286, *Standard Methods of Fire Tests for Evaluating Contribution of Wall and Ceiling Interior Finish to Room Fire Growth*. This test is similar to NFPA 265 but can be used to evaluate both wall and ceiling finishes and for any interior finish (including expanded vinyls) except textiles. The *Life Safety Code* allows the use of NFPA 286 as a testing option in any occupancy in any location if the nontextile material passes the "pass–fail" criteria in the *Life Safety Code*. If a product passes NFPA 286, the *Life Safety Code* allows the product to be used anywhere. Beginning with the 2006 edition of the *Code*, even textile or textile-like products are permitted, but not mandated, to be tested using NFPA 286.

18.5.2 Interior Floor Finish

Interior floor finish used to be regulated the same as interior room finish. This changed during the mid-1980s as a result of significant research conducted in the United States. Most traditional noncarpet floor finishes, such as tile, linoleum, wood, and concrete, are not affected by the modern restrictions on interior floor finish. Carpet and carpet-like materials, as well as unusual finishes, are regulated, but only in certain locations.

The test used to evaluate floor finish is NFPA 253, *Standard Method of Test for Critical Radiant Flux of Floor Covering Systems Using a Radiant Heat Energy Source*. Since this test measures a resistance to ignition, the higher numbers are more favorable. The *Life Safety Code* and U.S. model building codes establish two classes of interior floor finish based on critical radiant flux, as shown in Table 18.9. As with interior wall and ceiling finish, where sprinklers exist, Class I can shift to Class II, but in this case, Class II reverts to having no requirement.

Within occupancies associated with health care facilities there are very few limitations on interior floor finish, as shown in Table 18.10. Alcohol-based hand rub (ABHR) units are not permitted to be installed in areas with carpeted floor coverings unless such areas are protected with automatic sprinklers.

There is no restriction for floor finish within rooms or in sprinklered areas. However, regardless of occupancy or location, if the authority having jurisdiction (AHJ) determines the interior floor finish to be of an unusual hazard, the AHJ can require testing.

18.6 Contents and Furnishings

Very few occupancy chapters regulate contents and furnishings. However, health care and several other occupancy chapters regulate loosely hanging furnishings, such as curtains and drapes, including privacy curtains typically used in health care facilities. Where these furnishings are regulated, NFPA 701, *Standard Methods of Fire Tests for Flame Propagation of Textiles and Films*, is used. Health care and assembly occupancies require that all curtains and drapes be flame retardant in accordance with NFPA 701. There is no exception for facilities that are protected with sprinklers. In fact, special precautions regarding sprinklers must be exercised to ensure that the curtains and drapes do not interfere with sprinkler operation.

The *Life Safety Code* also prohibits the use of furnishings and decorations of an explosive or highly flammable nature in any occupancy. This restriction might be very

Table 18.9 Classes of Interior Floor Finish

Class	Critical Radiant Flux	Class	Critical Radiant Flux
I	> 0.45 W/cm^2	II	≤ 0.45 W/cm^2 but > 0.22 W/cm^2

Table 18.10 Floor Finish Limitations

Occupancy	Restriction
Health care	Newly installed floor finish in nonsprinklered corridors and exits–Class I
Business	Newly installed floor finish in nonsprinklered corridors and exits–Class II
Day care centers	Newly installed floor finish in nonsprinklered corridors and exits–Class II

unpopular during the winter holiday season, because it severely curtails the use of natural Christmas trees and related wreaths and trimmings. Some authorities allow live trees, but most prohibit any form of live or cut tree. NFPA 1, *Uniform Fire Code*™, does provide some guidance on the use of live Christmas trees.

Another issue regarding contents and furnishings applies only to nonsprinklered facilities. In certain occupancies, including health care occupancies (but not business, industrial, storage, or assembly occupancies), newly introduced upholstered furniture and mattresses have additional limitations. The intent of these limitations is to allow only the use of furnishings that reduce the hazard caused by cigarettes or other small ignition sources and to prevent flashover by restricting the heat-release rate to a maximum of 250 kW (kilowatts) and total energy released to a maximum of 40 MJ (megajoules) in the first 5 minutes of the test. Different tests are used depending on the item and the occupancy. Managers of facilities not protected with sprinklers should review the requirements in the code and work with the purchasing department to make sure that they are aware of these restrictions. The facility also should have a written policy regarding such purchases.

18.7 Fire Alarm Systems

Most occupancies associated with a health care facility require a fire alarm system under the *Life Safety Code*. The exceptions are small business, industrial, and storage occupancies. Table 18.11 summarizes the thresholds for requiring a fire alarm system. See Chapter 27 of this book for more details. Most of the categories are fairly straightforward, except for business occupancies, which have multiple thresholds. Figure 18.15 illustrates when a fire alarm system is required in a business occupancy.

Table 18.11 Thresholds for Requiring a Fire Alarm System

Occupancy	Threshold	Exceptions
Health care	All	None
Assembly	>300 people	None
Ambulatory health care	All	None
Business—new	≥ 2 stories above LED	None
	≥ 50 people above or below LED	
	≥ 300 people	
Business—existing	≥ 2 stories above LED	None
	≥ 100 people above or below LED	
	≥ 1000 people	
Industrial	All	< 100 people and < 15 people above or below LED
Storage	Nonsprinklered only	≤ 100,000 ft² (9300 m²)

LED = level of exit discharge.

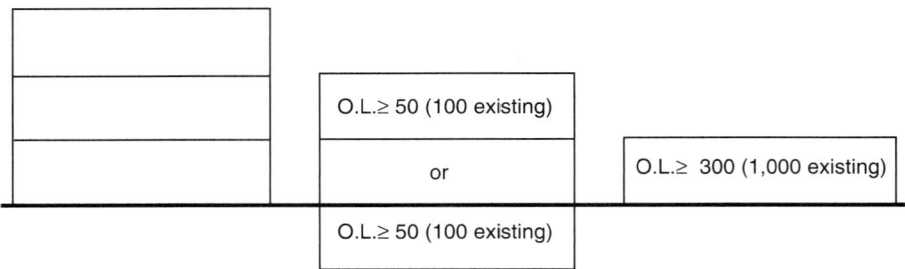

Figure 18.15 Required fire alarm system in a business occupation (building section view).

18.8 Extinguishing Equipment

The *Life Safety Code* mandates that all new health care occupancies be fully protected with an approved supervised automatic sprinkler system. In addition, many requirements address the use of sprinklers in existing health care occupancies, including construction types, egress capacities, corridor walls, spaces open to corridors, travel distance, and interior finish, to name a few. Beginning with the 2006 edition of the *Code*, *all* existing nursing homes are required to be protected throughout by a supervised automatic sprinkler system. The provisions for automatic sprinklers are discussed in detail in Chapter 21.

Although the *Life Safety Code* does not have mandatory sprinkler requirements for new business occupancies, numerous factors encourage sprinklers. Some of the most important factors include accessible means of egress, corridor protection, and travel distance (see Chapters 38 and 39 of the *Life Safety Code*, and Chapters 16 and 17 of this book). Sprinklers must protect all new high-rise business buildings. All existing high-rise business occupancies must be sprinkler protected as well, or else shown to have an adequate level of life safety based on an engineering analysis that is acceptable to the authority having jurisdiction.

One of the biggest problems regarding sprinklers in health care occupancies tends to occur with renovations. Except for minor renovations, the *Life Safety Code* requires sprinklers to be installed throughout smoke compartments in which renovations are being made. The big issue is determining just what is minor. In 2003, the *Life Safety Code* added a provision that essentially defines major as any renovation over 4500 ft^2 or more than 50 percent of the smoke compartment involved. There are exceptions to this, so refer to the *Code*. It is the intent of the *Life Safety Code* that sprinklers eventually protect all health care facilities. Therefore, if there is any doubt, it is wise to sprinkler the compartment. The *Life Safety Code* is organized so that the smoke compartment receives most of the benefits of being sprinkler protected, even though the building is not fully sprinklered.

19

Building Service Equipment

Building service equipment consists of all equipment installed in or around a building for the operation of the facility. NFPA *101*®, *Life Safety Code*®, divides building service equipment into five categories:

1. Utilities
2. Heating, ventilating, and air-conditioning equipment (HVAC)
3. Smoke control
4. Elevators, escalators, and vertical conveyors
5. Rubbish chutes, incinerators, and laundry chutes

The *Life Safety Code* refers to various standards for installing and maintaining this equipment. It is not the intent of the *Life Safety Code* to require newer standards to apply to the installation of existing equipment, unless the authority having jurisdiction (AHJ) finds an existing installation to be a hazard. However, all facilities must adhere to maintenance and testing requirements.

The *Life Safety Code* also addresses by reference the installation, use, and maintenance of cooking equipment and fire-extinguishing equipment used to protect cooking areas. However, the *Life Safety Code* treats building service equipment separately from the equipment used for fire protection, such as automatic sprinklers, fire alarm systems, standpipe systems, and similar equipment. For information on fire protection equipment, see Part IV of this book.

The following provides a brief review of the standards referenced by the *Life Safety Code* for building service equipment, cooking equipment, and equipment that protects cooking areas. Many of these standards, or those found in similar documents, are referenced by some of the U.S. and other model building codes, as well. Refer to Chapter 18 of this book for a discussion of rubbish chutes and laundry chutes.

19.1 Utilities

Five standards for utilities are referenced by the *Life Safety Code*. These are NFPA 54, *National Fuel Gas Code*; NFPA 58, *Liquefied Petroleum Gas Code*; NFPA 70, *National Electrical Code*® (*NEC*®); NFPA 110, *Standard for Emergency and Standby Power Systems*; and NFPA 111, *Standard on Stored Electrical Energy Emergency and Standby Power Systems*.

339

19.1.1 NFPA 54, *National Fuel Gas Code*

NFPA 54 is unusual for NFPA documents in that the NFPA and the American Gas Association (AGA) produce it jointly. As a result, it carries the designation ANSI Z223.1 as well as NFPA 54. Some of the requirements are extracted into the *International Fuel Gas Code (IFGC)*, The IFGC does not cover all the subjects covered in NFPA 54, especially the higher gas pressures likely to be used in larger health care facilities. Use of NFPA 54 is recommended.

The *National Fuel Gas Code* addresses the gas system from the utility connection (meter or valve depending on the system), including all piping, valves, regulators, and so on. It also covers the installation of gas equipment, including location and venting, but not the equipment itself, fuel gas utilization equipment, and related accessories as follows:

- Piping systems, beginning from the point of delivery to the connections with each gas utilization device (The point of delivery is considered to be the outlet of the service meter assembly or, where no meter is provided, the outlet of the service regulator or service shutoff valve. For liquefied petroleum gas systems, the point of delivery is the outlet of the final pressure regulator, exclusive of line gas pressure regulators in the system.)
- Maximum operating pressure covered by NFPA 54 is 125 psi (pounds per square inch). For higher gas pressures, use another code, such as ASME B31.1, *Power Piping*, or ASME B31.3, *Process Piping*.
- Piping systems, regarding design, materials, components, fabrication, assembly, installation, testing, inspection, operation, and maintenance.
- Installation, combustion, and ventilation air and venting for gas utilization equipment and related accessories.

This code does not apply to portable liquefied petroleum (LP) gas equipment that is not connected to a fixed fuel piping system, oxygen-fuel-gas cutting and welding systems, or liquefied natural gas (LNG) installations. The chapters of NFPA 54, 2006 edition, basically cover the following material:

- Chapter 5, "Gas Piping System Design, Materials, and Components," includes information on types of piping, location of meters, regulators, and pressure protection devices.
- Chapter 6, "Pipe Sizing," provides numerous tables for sizing gas piping systems.
- Chapter 7, "Gas Piping Installation," includes information on underground and above-ground piping, piping in buildings, drips and sediment traps, and outlets; it also addresses electrical bonding and grounding, electrical circuits, and electrical connections.
- Chapter 8, "Inspection, Testing, and Purging," contains information on pressure testing and inspection, system and equipment leakage testing, and purging.

- Chapter 9, "Appliance, Equipment, and Accessory Installation," addresses general installation requirements, limitations on locations, appliance approval, protection of equipment, building structural members, flammable vapors, garages, and general provisions for venting, accessibility and clearance, and air for combustion and ventilation.
- Chapter 10, "Installation of Specific Appliances," in combination with Chapter 5, covers the installation of specific equipment such as air-conditioners and heat pumps, central heating boilers and furnaces, clothes dryers, gas fireplaces, food service equipment, incinerators, and water heaters; it also addresses restrictions based on the type of equipment.
- Chapter 11, "Procedures to Be Followed to Place Appliance in Operation," covers adjusting burner input and primary air adjustment, as well as checking the safety shutoff device, automatic ignition protective devices, and draft, and providing operating instructions.
- Chapter 12, "Venting of Appliances," provides requirements on all different types of venting and lists the types of equipment that are not required to be vented.
- Chapter 13, "Sizing of Category I Venting Systems," provides numerous tables for sizing Category I venting systems.

This code also provides numerous annexes (A through L) that include examples and recommended procedures.

19.1.2 NFPA 58, *Liquefied Petroleum Gas Code*

The *Liquefied Petroleum Gas Code* covers the use of liquefied petroleum gas, otherwise known as LP gas. This code applies to containers, piping, and associated equipment when delivering LP gas to a building for all uses. This code does not apply to LP gas used with oxygen or those portions of LP gas systems covered by NFPA 54, where NFPA 54 is adopted, used, or enforced. (NFPA 51, *Standard for the Design and Installation of Oxygen–Fuel Gas Systems for Welding, Cutting, and Allied Processes*, and ANSI Z49.1, *Safety in Welding and Cutting,* apply when LP gas is used with oxygen.)

In general, this code covers a much broader area than does the *National Fuel Gas Code*. The chapter headings of the 2004 edition indicate this wider coverage.

- Chapter 5, "LP-Gas Equipment and Appliances," includes information on containers and their appurtenances; piping, including hoses, fittings, and valves; equipment, including pumps, compressors, vaporizers, strainers, meters, and regulators; and appliances.
- Chapter 6, "Installation of LP-Gas Systems," includes a vast amount of material of interest to a health care facility, such as location of containers, location of transfer operations, installation of containers and their appurtenances (including on roofs), regulator installations, piping systems, equipment installation, LP gas systems within buildings or on roofs or exterior balconies (this has specific occupancy requirements and provisions for construction and renovations as well as for temporary installations), and LP gas systems on vehicles.

- Chapter 7, "LP-Gas Liquid Transfer," covers the transfer of LP gas from one container to another wherever this transfer involves connections and disconnection, or the venting of LP gas to the atmosphere. This operation is not normally an issue, as cylinders and tanks are not usually filled by health care facilities.
- Chapter 8, "Storage of Cylinders Awaiting Use, Resale, or Exchange," covers storage of cylinders of 1000-lb (454-kg) water capacity or less. It includes location of cylinders, protection of cylinder valves, and storage within buildings and outside buildings.
- Chapter 9, "Vehicular Transportation of LP-Gas," covers various modes of transporting LP gas by vehicles, including portable containers, cargo vehicles, trailers, and the parking and garaging of vehicles used to transport LP gas.
- Chapter 10, "Buildings or Structures Housing LP-Gas Distribution Facilities," is not normally of concern to health care facilities.
- Chapter 11, "Engine Fuel Systems," addresses fuel systems that use LP gas for internal combustion engines, including engines installed on vehicles used for any purpose (automobile or forklift truck), as well as stationary and portable engines. This chapter also addresses the garaging of vehicles with LP gas engine fuel systems.
- Chapter 12, "Refrigerated Containers," contains provisions for refrigerated containers. We normally think of LP gas containers as pressurized containers containing LP gas at high pressures but basically at ambient temperature. To store a lot of LP gas, large pressure vessels must have very thick walls, which make them very difficult to fabricate and expensive. Refrigerated LP gas tanks usually have a minimum volume of 1 million gal (3785 m^3) and regularly exceed 10 million gal. They are not a concern of health care facilities.
- Chapter 13, "Marine Shipping and Receiving," is not usually a concern to most health care facilities.
- Chapter 14, "Operations and Maintenance," contains requirements related to the operations and maintenance of bulk plant, industrial plant, refrigerated, marine, and pipeline LP gas systems.
- Chapter 15, "Pipe and Tubing Sizing Tables," provides numerous tables for sizing gas piping systems.

As with NFPA 54, NFPA 58 contains several annexes that provide additional information that may be of use.

19.1.3 NFPA 70, *National Electrical Code*

The *NEC* is perhaps the most widely adopted code promulgated by the NFPA. It is used throughout the United States and in numerous other countries. It is available in several languages. As its name implies, it deals with electrical safety within buildings. Chapters 4 through 12 of Part II of this book contain details of how the *NEC* is applied to attain electrical safety in health care facilities.

19.1.4 NFPA 110, *Standard for Emergency and Standby Power Systems*

NFPA 110 covers installation, maintenance, operation, and testing of emergency power supply systems, primarily generators. This is a very important standard for health care facilities. Chapter 8 of this book contains further details on this standard.

19.1.5 NFPA 111, *Standard on Stored Electrical Energy Emergency and Standby Power Systems*

NFPA 111 covers installation, maintenance, operation, and testing for stored emergency power supply systems. This standard is similar to NFPA 110 but covers stored electrical energy systems, such as batteries, rather than generators. An uninterruptable power supply (UPS) system, as discussed in Chapter 10, is an example of such a system. See Chapter 8 of this book for further details on this standard.

19.2 Heating, Ventilating, and Air-Conditioning Equipment

Heating, ventilating, and air-conditioning (HVAC) systems are regulated by a series of standards, depending on the type of system and the fuel used. Following is a brief discussion of the various standards involved.

19.2.1 NFPA 90A, *Standard for the Installation of Air-Conditioning and Ventilating Systems*

NFPA 90A covers safety to life and property from fire within the air distribution system, including such items as air intakes, air cleaners and filters, fans, air ducts and related components, air outlets and inlets, plenums, and fire and smoke dampers. This document also addresses HVAC shafts, penetrations of fire-rated walls and partitions, and fire-rated floors and floor–ceiling assemblies. HVAC penetrations of smoke barriers are also addressed. Although it is not the intent of this book to address all of the provisions of NFPA 90A, it is necessary to point out significant issues that typically apply to health care facilities.

19.2.1.1 Air Connectors

Air connectors are limited-use flexible air ducts that are not required to conform to the provisions for air ducts. They are prohibited from exceeding 14 ft (4.27 m) in length; from passing through any wall, partition, or shaft enclosure required to have a fire resistance rating of 1 hour or more; and from passing through any floor.

19.2.1.2 Air Duct Coverings

Air duct coverings are not allowed to extend through walls or floors that are required to be fire stopped or that have a fire resistance rating unless they have been tested as part of the assembly. However, for ducts that do not require fire dampers, air duct coverings are allowed to extend through walls or floors under limited conditions.

Air duct coverings are also prohibited from concealing or preventing the use of any service opening. Service openings are required in ducts adjacent to each fire damper, smoke damper, and smoke detector. This opening must be located and sized to permit the maintenance and resetting of the device. Service openings must be properly marked to indicate the location of the device within.

19.2.1.3 Smoke Dampers

With only a few exceptions, smoke dampers must be installed in systems with a capacity greater than 15,000 ft^3/minute (7080 L/second) to isolate the air-handling equipment, including filters, from the remainder of the system to restrict the movement of smoke into occupied spaces. These dampers must close automatically whenever the system is not in operation unless they are part of an approved, engineered smoke control system. This type of installation may not be found on older systems and would not be required, per the discussion at the beginning of this chapter, unless they are determined to be a significant hazard.

Smoke dampers must be installed at or adjacent to the point where an air duct passes through a smoke barrier. The smoke damper should be between the barrier and the first outlet or inlet but not more than 2 ft (0.6 m) from the barrier. There are numerous exceptions to this requirement, including those for the following:

- Engineered smoke control systems
- Systems where isolation smoke dampers are provided
- Where the duct serves only one smoke compartment
- Systems where the air continues to move and the system prevents recirculation of exhaust or return air under fire conditions
- When specifically exempted by the *Life Safety Code*

19.2.1.4 Ceiling-Cavity Plenums

There are numerous restrictions on the material used in ceiling-cavity plenums. These restrictions must be observed any time that items including, but not limited to, wiring, cables, piping, tubing, light diffusers, and other items are put into a ceiling cavity used as part of the air distribution system. Ceiling-cavity air plenums are common for office buildings but unusual for hospitals, although they do exist. Raised-floor plenums have similar restrictions. Incidentally, when a mechanical room is used as a plenum, the room cannot be used for storage or occupancy.

19.2.1.5 Exits and Corridors

Exits, such as enclosed exit stairs and exit passageways, cannot be used as a part of supply, return, or exhaust air systems that serve other parts of the building. Also, the *Life Safety Code* prohibits the penetration of an exit enclosure by HVAC systems (see Chapter 15, Section 15.3.3, in this book). In health care and residential occupancies, corridors cannot be used as part of supply, return, and exhaust systems, and air transfer openings are specifically prohibited in these areas. However, air transfer to toilet rooms, bathrooms, shower rooms, sink closets, and similar spaces that open directly onto the corridor is not prohibited, nor is air transfer caused by pressure differentials, as long as the clearances around the doors do not exceed those specified by NFPA 80, *Standard for Fire Doors and Fire Windows*.

19.2.1.6 Fire Dampers

Unless the building code or the *Life Safety Code* specifically exempts such protectives, fire dampers must be installed in the following circumstances:

- Where air ducts penetrate or terminate at openings in walls or partitions that must have a fire resistance rating of 2 hours or more
- Where air transfer openings are in a partition or wall required to have any fire resistance rating

Appropriate smoke and/or fire dampers must be located in the wall, and the ducts must be designed so that they will not pull the damper out of the wall when the duct collapses. Fire dampers must also be installed in each duct or opening into a shaft enclosure. There are two specific exceptions for exhaust ducts, one of which is commonly used to omit dampness in bathroom exhaust systems.

19.2.1.7 HVAC Shafts

There are numerous restrictions placed on HVAC shafts. These prevent HVAC shafts from being used for the following:

- Exhaust ducts to remove smoke- and grease-laden vapors from cooking equipment
- Ducts that remove flammable or corrosive vapors or fumes
- Ducts for removing nonflammable corrosive fumes and vapors
- Refuse and linen chutes
- Piping (except noncombustible piping conveying nonhazardous or nontoxic materials)
- Combustible storage

19.2.1.8 Smoke Detectors

Smoke detectors that shut down the air distribution system are required in the following locations:

- Downstream of air filters and ahead of any branch connections in supply air systems having a capacity greater than 2000 ft^3/minute (944 L/second)
- At each story prior to the connection to a common return and prior to any recirculation or fresh air inlet connection in return air systems having a capacity greater than 15,000 ft^3/minute (7080 L/second) and serving more than one story

There are some limited exceptions to these requirements. One of the more common exceptions is that fan units whose sole function is to remove air from the inside of the building to the outside of the building do not require smoke detectors for system shutdown.

19.2.1.9 Maintenance Provisions

NFPA 90A includes several maintenance provisions. At least once every four years, fusible links must be removed from the mechanical fire dampers. Dampers must be operated to verify that they close properly (and latch properly, if so provided). Moving parts must be lubricated as necessary at that time.

All automatic shutdown equipment including related dampers must be tested at least annually.

Chapter 7 of NFPA 90A covers the very important subject of acceptance testing.

19.2.2 NFPA 91, *Standard for Exhaust Systems for Air Conveying of Vapors, Gases, Mists, and Noncombustible Particulate Solids*

As the title implies, NFPA 91 covers the minimum requirements for the design, construction, installation, operation, testing, and maintenance of exhaust systems for air conveying of vapors, gases, mists, and noncombustible solids, except as modified or amplified by other applicable NFPA standards. Numerous other NFPA standards may also be involved. Those that might affect a health care facility include

- NFPA 30, *Flammable and Combustible Liquids Code*
- NFPA 33, *Standard for Spray Application Using Flammable or Combustible Materials*
- NFPA 45, *Standard on Fire Protection for Laboratories Using Chemicals*
- NFPA 92A, *Standard for Smoke-Control Systems Utilizing Barriers and Pressure Differences*
- NFPA 92B, *Standard for Smoke Management Systems in Malls, Atria, and Large Spaces*
- NFPA 96, *Standard for Ventilation Control and Fire Protection of Commercial Cooking Operations*
- NFPA 204, *Standard for Smoke and Heat Venting*
- NFPA 211, *Standard for Chimneys, Fireplaces, Vents, and Solid Fuel–Burning Appliances*

Subjects included in NFPA 91 cover such items as

- Design and construction of the systems, including duct materials, access, hangers and supports, and clearances
- Special precautions when handling corrosive materials
- Air-moving devices
- Controls on ignition sources such as static electricity, open flames, and sparks, removal of ferrous materials, belt drives, and bearings
- Automatic extinguishing requirements
- Testing and maintenance, including a requirement that all system components be inspected monthly and that an operational maintenance log be kept

19.2.3 NFPA 211, *Standard for Chimneys, Fireplaces, Vents, and Solid Fuel–Burning Appliances*

As indicated in the title, NFPA 211 covers a wide variety of issues. It contains provisions for chimneys, fireplaces, venting systems, and solid fuel–burning appliances, including their installation, and applies to residential as well as commercial, industrial, and all other installations. This standard sets general performance requirements for draft, termination (height) of chimneys and vents, enclosure, flue linings, and caps and spark arresters.

The standard contains a rather long list of definitions and extensive requirements for chimney inspection and maintenance. An entire chapter is dedicated to the selection of chimney and vent types. Other chapters contain requirements for factory-built chimneys and chimney units and extensive details on masonry chimneys, including such items as support, corbelling, size or shape changes, cleanouts, firestopping, smoke testing, structural design, thimbles, relining, construction, and clearance from combustibles. Provisions for unlisted metal chimneys (smokestacks) in nonresidential applications address construction, termination (height), and clearance for these chimneys. Chimney and vent connectors are regulated as well regarding materials, location, length, size, installation, and clearances. Vents are similarly covered.

19.2.4 NFPA 31, *Standard for the Installation of Oil-Burning Equipment*

NFPA 31 covers the installation of oil-fired heating appliances (boilers, air heaters, water heaters), including all accessory equipment and controls, including multifueled appliances using fuel oil as one of the fuels. It also includes requirements for fuel oil storage and supply systems; air for combustion and ventilation; venting of combustion (flue) gases; and chimneys and chimney connections, including appropriate references to NFPA 211, discussed earlier. Specific requirements are included for appliances that burn used oil and appliances that burn fuel oil and gas.

NFPA 31 includes specific requirements for fuel oil storage tanks installed inside buildings and for fuel storage tanks up to 2500 L (660 gal) installed outside, aboveground. These requirements govern acceptable tank design standards, maximum storage capacities, supports and foundations, testing, and maintenance. Extensive guidance is provided for abandonment and removing tanks from service. Underground tanks and outside aboveground tanks that exceed 2500 L (660 gal) are covered by reference to NFPA 30.

Fuel oil supply piping and piping system components are also covered, including centralized oil distribution systems.

NFPA 31 also includes specific provisions for kerosene-fired room heaters and portable heaters, appliances not typically found in health care occupancies.

19.3 Smoke Control

Before 2006 there were no enforceable standards available for smoke control. The *Life Safety Code* simply stated that where smoke control systems are required or provided, they must have an approved maintenance and testing program. The *Life Safety Code* also states that the purpose of a smoke control system is to confine smoke to the general area of origin and maintain the means of egress as usable. What is important to note is the reference to the annex (or the appendix in pre-2000 editions). The information in the annex (or appendix) provides a list of documents available to assist in the design of smoke control systems. NFPA 92A and NFPA 92B were recently revised to become standards; thus, nationally recognized design documents now exist.

19.4 Elevators, Escalators, and Vertical Conveyors

A common form of movement in multistory buildings involves the use of elevators, escalators, and vertical conveyors. Not only do these systems present concerns regarding potential fire problems because they are considered vertical openings, as addressed in Chapter 18, but they also present mechanical problems. To address these issues, NFPA *101* references standards promulgated by the American Society of Mechanical Engineers (ASME), including ASME/ANSI A17.1, *Safety Code for Elevators and Escalators,* for new units, and ASME/ANSI A17.3, *Safety Code for Existing Elevators and Escalators,* for existing installations.

19.4.1 ASME/ANSI A17.1, *Safety Code for Elevators and Escalators,* and ASME/ANSI A17.3, *Safety Code for Existing Elevators and Escalators*

Both ASME/ANSI A17.1 and ASME/ANSI A17.3 contain significant information, much of which is normally addressed by elevator maintenance companies or original installers. ASME/ANSI A17.1 is widely used and recognized in the United States for these devices. However, ASME/ANSI A17.3 is not as widely used, and its application may not be as automatic and thus may cause some problems for existing installations.

ASME specifically developed ASME/ANSI A17.3 to address safety issues in existing installations. Much of the material in A17.1 does not appear in A17.3, similar to the way that many requirements of the *Life Safety Code* for new construction do not appear in the requirements for existing buildings. It should be brought to the attention of elevator maintenance contractors that health care facilities must comply with A17.3, and they should be able to document their ability to use and apply A17.3.

It is not the intent of this book to discuss all the issues covered by these two codes. There are several other useful sources of information on these issues, however. One source is a handbook on A17, published by ASME. NFPA's *Life Safety Code® Handbook* includes a supplement that excerpts information from A17.1 and the ASME handbook. The *Life Safety Code Handbook* supplement includes Rule 211.1, on Car Emergency Signaling Devices; Rule 211.2, on Standby Power; Rules 211.3 through 211.7, on Fire Fighters' Service, as well as related ASME handbook commentary. These extracts address some of the issues that are of most concern from the aspect of the *Life Safety Code* and issues that inspectors enforcing the *Life Safety Code* may likely check.

The *Life Safety Code* not only references A17.1 and A17.3 but further requires all new elevators to conform to the requirements for fire fighters' service and all existing elevators with a travel of 25 ft (7.6 m) or more above or below the level of fire fighter access to be provided with fire fighters' service. In other words, the *Life Safety Code* requires existing elevators that have a travel of 25 ft (7.6 m) or more above or below the level of fire fighter access to be retrofitted with fire fighters' service if not so equipped. This means that an existing elevator could potentially travel almost 50 ft (15 m) and not have fire fighters' service [25 ft (7.6 m) up and 25 ft (7.6 m) down]. Since the measurement is for elevator travel distance, not building height, with 10-ft (3-m) stories, an elevator going up to the second floor travels 10 ft (3 m), to the third floor 20 ft (6.1 m), and to the fourth floor more than 25 ft (7.6 m). Therefore, with floor-to-floor distances of 12 ft

(3.7 m) or less, buildings with three stories (and up to three basements) do not need fire fighters' service for existing elevators. However, if fire fighters' service is provided, it must be maintained, since the *Life Safety Code* requires such service for all new installations.

Fire fighters' service has two phases, Phase I, emergency recall operation, and Phase II, emergency in-car operation. ASTM A17.1 should be reviewed for details, but, briefly, Phase I consists of the following elements:

- A three-way key switch ("bypass," "off," and "on") in an accessible location near the elevators on the *designated level* (i.e., the main level or the level that best serves the needs of fire fighters)
- Smoke detectors in every elevator lobby, in the elevator machine room, and in non-sprinklered hoistways
- Instructions within or adjacent to the switch

In short, if the switch is turned to "on," or if the smoke detectors operate, the elevators recall to the designated level. If the smoke detector at the designated level activates, the elevators recall to an alternative level. The standard includes much more detail, but this level of information can be used to determine whether an elevator has Phase I operation. The elevator maintenance company must determine whether this phase is operating properly or not.

Phase II operation consists of the following items on the operating panel in each car:

- A three-way key switch ("off," "hold," and "on")
- A "call cancel" button
- Instructions for operating under Phase II can be adjacent to the operating panel in each car

Phase II allows the fire service to operate the elevator under specific controls. The operation of the doors is strictly controlled. See ASME A17.1 and the ASME handbook for more details.

Another area that needs to be checked periodically is the two-way communication system from each elevator. This is an important safety feature that is often overlooked. It is not uncommon to hear of a Joint Commission on Accreditation of Healthcare Organizations (JCAHO) surveyor checking the phone in the elevator and the elevator inspection certificate.

An effort has been under way since 2003 to determine the equipment needs, building design regulation, and public education needs that may permit elevators to be recognized as a means of egress component. It is possible that future editions of NFPA 101, NFPA 5000®, *Building Construction and Safety Code*®, and other model building codes might contain the requirements for such use.

19.5 Rubbish Chutes, Incinerators, and Laundry Chutes

Rubbish and laundry chutes are first of all vertical openings and must be protected as such (see Chapter 18). In addition, the *Life Safety Code* restricts the location of chute inlet openings to separately protected rooms, such as laundry rooms or trash rooms.

Existing installations can continue to have chute doors that open into a corridor. The *Life Safety Code* requires all rubbish and linen chutes, both new and existing, to be protected by automatic sprinklers. NFPA 82, *Standard on Incinerators and Waste and Linen Handling Systems and Equipment,* provides significant detail on this subject. Figure 19.1 illustrates some of the provisions in NFPA 82.

The laundry or trash room is considered a hazardous area in a health care occupancy. As such, new construction must be 1-hour enclosed and protected by sprinklers, and existing situations must be either 1-hour enclosed or protected by sprinklers. Remember that if both means of protection are provided, they must be maintained. It is very common for soiled linen and trash rooms to maintain both forms of protection, since the double level of protection has been a requirement for many years.

Some common problems with trash and linen chutes in health care occupancies include

- Chute access doors are not self-closing due to missing or broken closers or wedges or other hold-open devices.
- Chute access doors are not self-latching due to missing or broken latches or added tape.
- Chute access or discharge room doors are held open improperly or closers are missing or broken.

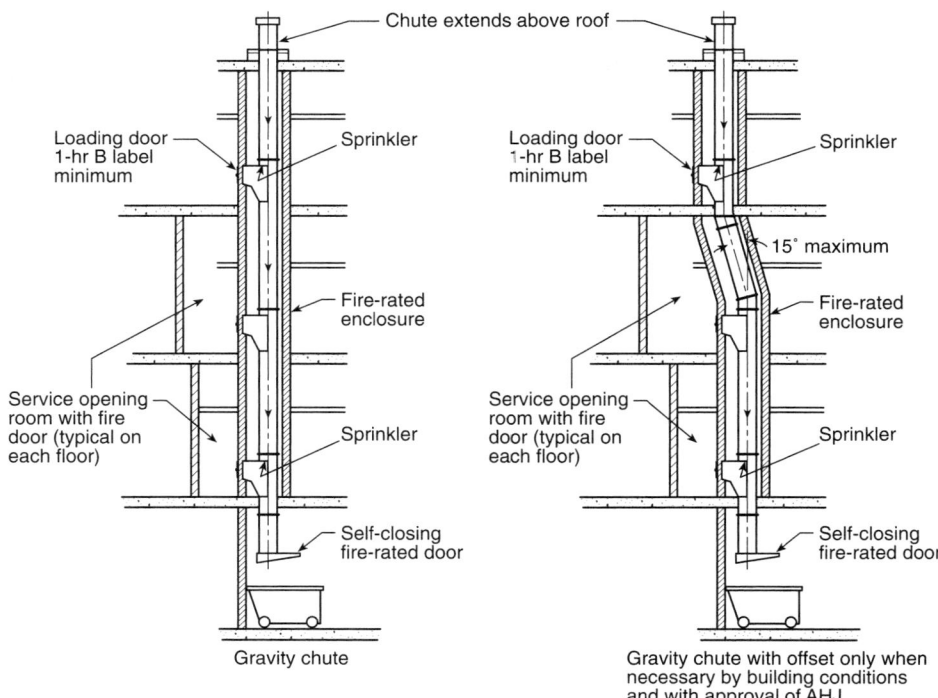

Figure 19.1 Gravity chutes protected internally by automatic sprinklers.
Source: NFPA 82, 2004, Figures A.5.2(a) and A.5.2(b).

- Chute access or discharge room door latches are broken or made nonworking.
- Chute access or discharge room door labels are missing or painted over.
- Chute access or discharge rooms are not protected by sprinklers.
- Chute discharge doors are not equipped with fusible links, or links are missing and have been replaced with nonfusible elements.
- Trash or linens are so deep in discharge rooms that chute discharge doors cannot close even if activated.
- Trash or linen chutes are not equipped with sprinkler protection.

19.6 Cooking Equipment

Numerous standards apply to cooking and cooking equipment, including those that cover cooking fuels consumed and the protection of facilities from the effects of cooking. NFPA 58 covers equipment that uses LP gas, and NFPA 70 covers the installation of electrical equipment. Additionally, NFPA 96, *Standard for Ventilation Control and Fire Protection of Commercial Cooking Operations,* must be applied to control the production of grease-laden vapors, and NFPA 10, *Standard for Portable Fire Extinguishers,* addresses the installation, use, and marking of portable fire extinguishers.

19.6.1 Scope of NFPA 96

The scope of NFPA 96 is quite important, since it applies to many health care facilities and all occupancies (except one- and two-family dwellings).

NFPA 96 provides the minimum fire safety requirements (preventive and operative) related to the design, installation, operation, inspection, and maintenance of all public and private cooking operations except for cooking equipment located in a single dwelling unit. The requirements include, but are not limited to, all manner of cooking equipment, exhaust hoods, grease removal devices, exhaust ductwork, exhaust fans, dampers, fire-extinguishing equipment, and all other auxiliary or ancillary components or systems that are involved in the capture, containment, and control of grease-laden cooking effluent NFPA 96 does apply to residential cooking equipment used for commercial cooking operations. However, it does not apply to facilities where *all* of the following are met:

1. Only residential equipment is being used.
2. Fire extinguishers are located in all kitchen areas in accordance with NFPA 10.
3. The facility is not an assembly occupancy.
4. Compliance is subject to the approval of the authority having jurisdiction.

This judgment should take into account the type of cooking being performed, items being cooked, and the frequency of cooking operations. Examples of operations that might not require compliance with this standard include the following:

- Day care centers warming bottles and lunches
- Therapy cooking facilities in health care

- Churches and meeting operations that are not cooking meals that produce grease-laden vapors
- Employee break rooms where food is warmed

NFPA *101* provides some guidance for health care facilities in that the standard states that both new and existing cooking facilities must be protected in accordance with the requirements of NFPA 96. One exception to this requirement is that protection or segregation of food preparation facilities is not required where domestic cooking equipment is used for food warming or limited cooking. An appendix note in NFPA *101* states that this exception intends to exempt small appliances used for reheating, such as microwave ovens, hot plates, toasters, and nourishment centers, from the requirements for commercial cooking equipment.

Although this exemption provides some relief in health care occupancies, authorities having jurisdiction still must judge the use of this equipment in other occupancies and also where residential-type units are used for limited cooking regardless of the occupancy. Thus, it may be best not to install such equipment unless NFPA 96 protection is provided.

It should be noted that the standard residential range hood does not comply with NFPA 96. If such equipment is installed, it implies that grease is in fact being produced or that the facility anticipates that it will be produced, and its existence generates a question as to whether it should be replaced with an NFPA 96–compliant system.

19.6.2 Exhaust Systems

NFPA 96 also requires cooking equipment used for cooking processes that produce smoke or grease-laden vapors to be equipped with an exhaust system that complies with all the equipment and performance requirements of this standard. All such equipment, including the cooking equipment per se, as well as hoods, ducts (if applicable), fans, fire suppression systems, and special effluent or energy control equipment, must be maintained in good working condition, according to proper maintenance intervals, as provided in Table 19.1. All air flows must be maintained as well.

Table 19.1 Exhaust System Inspection Schedule

Type or Volume of Cooking Frequency	Frequency
Systems serving solid fuel cooking operations	Monthly
Systems serving high-volume cooking operations such as 24-hour cooking, charbroiling, or work cooking	Quarterly
Systems serving moderate-volume cooking operations	Semiannually
Systems serving low-volume cooking operations, such as churches, day camps, seasonal businesses, or senior centers	Annually

Source: NFPA 96, 2004, Table 11.3.

19.6.2.1 Clearance

One of the major issues covered by NFPA 96 is clearance for hoods, ducts, grease-removal devices, and fans. Fires in hoods and ducts are very common, and preventing ignition of adjacent combustible material is important. An entire chapter in this standard is dedicated to the hood itself, which includes information on construction (steel, with liquid-tight external welds), size, and related issues.

19.6.2.2 Filters

Grease-removal devices are addressed in a separate chapter, as is the installation of filters, including requirements for the distance filters must be above the cooking surface, accessibility and removability, angle of installation, and the provision of drip trays. One important item to note that is often found as a violation in health care facilities is that mesh filters (common on residential hoods) are prohibited by NFPA 96.

19.6.2.3 Exhaust Ducts

Extensive material is provided in NFPA 96 on exhaust duct systems. Some specific requirements of interest are that exhaust duct systems must not pass through fire walls or partitions or be interconnected and have any other building ventilation or exhaust system. These ducts must lead as directly as possible to the exterior of the building and must be installed without dips or traps that might collect residues.

Clearances, openings, and materials are also addressed for these duct systems. As with hoods, ducts must be made of steel and have liquid-tight continuous external welds (some exceptions apply). Additional information is provided for exterior and interior duct installations. For interior installations, the enclosure must be continuous from the lowest fire-rated ceiling or floor above the hood to or through the roof, including through all concealed spaces. The requirements for the exhaust duct systems also address the termination, including rooftop terminations and wall terminations. The code includes many other requirements regarding the enclosure of ducts.

NFPA 96 also addresses air movement, including upblast fans and in-line fans, air flow, replacement air, and common duct systems. One item of specific interest is that openings provided for replacing air must not be restricted by covers, dampers, or any other means. Another requirement is that hood exhaust fans must continue to operate after the extinguishing system activates, unless a listed component of the ventilation system or the design of the extinguishing system requires the fans to shut down. If the exhaust fan and all cooking equipment served by the fan had previously shut down, restarting the hood exhaust fan is not required when the extinguishing system activates.

Other requirements are that exhaust systems must be operated whenever cooking equipment is turned on, filter-equipped exhaust systems must not be operated without filters, and cooking equipment must not be operated while its exhaust system (or fire-extinguishing system) is nonoperational or otherwise impaired.

19.6.3 Auxiliary Equipment and Other Devices

NFPA 96 prohibits the installation of dampers unless specifically listed for such use where the dampers are part of a listed or approved device or system. In addition, wiring systems of any type are prohibited from being installed in the duct and motors, and lights and other electrical devices are prohibited from being in ducts or hoods or in the path of travel of exhaust products, unless specifically approved for such use. There are several restrictions on hood lights. Needless to say, facility workers should not conduct any electrical work within ducts or hoods unless they are qualified to do such work, and the work is within the limitations of NFPA 96.

An entire chapter in NFPA 96 is dedicated to fire-extinguishing equipment. See NFPA 96 and Part IV of this book for additional details. NFPA 96 also covers minimum safety requirements for the installation of cooking equipment, recirculating systems, and solid-fuel cooking operations, not a common application in health care facilities.

19.6.4 Extinguishing Systems

See Chapter 24 of this book for information on kitchen hood and duct fire-extinguishing systems including

1. Dry chemical
2. Wet chemical
3. Water-assisted wet chemical
4. Wet-pipe sprinkler
5. Carbon dioxide

The chapter also addresses alarms, manual release stations, and portable fire extinguishers and the inspection, testing, and maintenance of these systems.

For additional information on portable fire extinguishers for kitchens, see Chapter 26 of this book.

Building Construction and Life Safety Frequently Asked Questions

1. Why should a health care facility comply with NFPA 101®, *Life Safety Code*®?

There are several reasons why a health care facility should comply with the *Life Safety Code,* as follows:

- Local law (i.e., city, county, or state) may mandate compliance. More than half the states in the United States mandate compliance with the *Life Safety Code,* and several other countries also mandate the *Life Safety Code* or use it as a guide.
- Facilities receiving Medicare or Medicaid funds must comply with the *Life Safety Code* or be Joint Commission on Accreditation of Healthcare Organizations (JCAHO) accredited; JCAHO accreditation requires compliance with the *Life Safety Code.*
- NFPA 101 is an ANSI (American National Standards Institute) document. As such, it can be used in civil litigation to establish a minimal level of safety that a prudent person would provide.
- NFPA 101 is recognized by the Occupational Safety and Health Administration (OSHA), in the Occupational Safety and Health Act, Subpart E, on means of egress. OSHA is on record as accepting the latest edition of NFPA 101 for compliance with Subpart E.

2. If a facility must comply with both a building code and the *Life Safety Code,* will it be satisfactory to design the building to the building code?

Most likely not. With the exception of *NFPA 5000®, Building Construction and Safety Code®,* which is totally compatible with *NFPA 101,* the other model building codes in the United States are generally compatible with the *Life Safety Code,* but there are differences. Other building codes could be significantly different.

Amendments that are added to building codes when they are adopted can add even more differences. It is important to make all architects contracted to design health care facilities aware that compliance with the local building code will not be sufficient and that facilities must also comply with the *Life Safety Code.*

3. **An existing facility has smoke barriers on a story used only for treatment, not patient sleeping. Chapter 19 of the *Life Safety Code* does not require smoke barriers on nonsleeping floors. Is it acceptable to remove or stop maintaining the smoke barriers on this story?**

No. To evaluate whether a smoke barrier can be removed or not, it is necessary to use the *Life Safety Code*'s criteria for new construction (Chapter 18). Smoke barriers can be eliminated only if they are not required for new construction. This is an important concept, due to the significant differences between new and existing.

4. **An existing hospital or nursing home is having trouble maintaining smoke dampers at the smoke barriers (or none exist). Chapter 18 of the *Life Safety Code* does not require smoke dampers in new construction. Can an existing facility delete smoke dampers, since they are not required in new construction?**

Most likely not. This question raises several issues. First, the reason that new health care facilities no longer require smoke dampers at smoke barriers is due to the requirement that not only are automatic sprinklers required, but smoke compartments containing patient sleeping rooms must be protected throughout with quick-response or residential sprinklers. If the smoke compartments on both sides of the barrier have sprinkler protection in accordance with the requirements for new construction, the dampers could be eliminated. Refer to the *Life Safety Code* (Sections 19.1.1.4.3.3 and 19.1.1.4.3.4 in the 2006 edition). Second, it should be noted that as long as sprinklers are provided, dampers can be eliminated in smoke compartments that do not have quick-response or residential sprinklers if the compartments do not contain patient sleeping areas. Another point to remember is that the local building code should be checked before removing dampers, since it may not allow the deletion of the dampers in question.

5. **What can be done to secure health care facility nurseries that must prevent unauthorized persons from removing any infants?**

There are several *Life Safety Code*–compliant methods to provide this type of security. The best way to prevent unauthorized people from taking infants is to prevent them from having access to the babies in the first place. The *Life Safety Code* does not restrict locking or otherwise securing doors against ingress. Various methods are available to secure nurseries from unwanted entries without restricting egress.

Another very viable option allowed by the *Life Safety Code* to improve security is the use of delayed egress devices, provided the building is either fully sprinklered or fully fire detected (i.e., has a complete heat or smoke detection system). The *Life Safety Code* also allows the use of alarm devices that do not restrict egress, but these devices do not slow down a perpetrator. Some professionals in the field have advocated using the *Life Safety Code*'s provision that allows locking an area when the clinical needs of the patients require that they be locked in. However, it is very difficult to justify that the clinical needs of infants require that they be locked in, as babies do not wander off—they are taken. This issue is a matter of security, not meeting the criteria related to the clinical needs of patients.

Although they are not addressed by the *Life Safety Code,* other devices that provide security are available and can be used. However, if a device restricts egress, it must be approved by the authority having jurisdiction. One such device locks the door when a baby approaches and then requires the use of a code or card to bypass the security. However, the use of this type of device raises the question of how to remove the babies in a nursery during an emergency when the card is lost or the code forgotten, which has proved to be a problem in real emergencies. Items such as tying disablement of the device into smoke detector operation in the unit may help compensate for this problem.

6. According to the *Life Safety Code*, a *major* rehabilitation to a smoke compartment must include providing sprinkler protection throughout the compartment. What is a *major* rehabilitation?

Beginning with the 2003 edition of the *Life Safety Code*, this question is now answered in the *Code*. A major rehabilitation involves the modification of more than 50 percent, or more than 4500 ft² (420 m²), of the area of the smoke compartment. A minor rehabilitation shall involve the modification of not more than 50 percent, and not more than 4500 ft² (420 m²), of the area of the smoke compartment. Work that is exclusively plumbing, mechanical, fire protection system, electrical, medical gas, or medical equipment is not included in the computation of the modification area within the smoke compartment.

The original intent of the *Code* was to require all renovations to be sprinklered, with an exception provided for minor renovations. An annex note (Annex A.19.1.1.4.5) helped somewhat by saying that it is not the intent of the *Life Safety Code* to exempt significant renovations or modernization projects. A prior attempt to allow sequential or staged renovations to be exempt from having sprinkler protection was rejected by the NFPA Health Care Committee overseeing this code. It should be noted that since 1991, when sprinklers were first mandated for new construction, the overall intent of the *Life Safety Code* has been to get existing buildings protected by sprinklers at time of renovation without having to mandate an across-the-board requirement for existing buildings to be sprinklered.

7. Is a hemodialysis unit a business or an ambulatory health care occupancy?

There are several possible answers to this question. If the hemodialysis unit is within a hospital, it could be treated as part of the health care occupancy. If it is properly separated from the health care occupancy, or if it is freestanding, it could be either a business or an ambulatory health care occupancy. To make this determination, the following question must be answered: Are there four or more patients in the hemodialysis unit at one time who are not capable of self-preservation? The best way to obtain this information is to ask the staff working in the hemodialysis unit how long it would take to evacuate all of the patients. (Note that this is not a question for management or other agencies—it is for the people who have to do the evacuation.)

It must be remembered that if patients need assistance getting out, then they are not capable of self-preservation. Under ambulatory health care, if they came into the facility by themselves, they must be able to get out by themselves. If they cannot, ithe unit cannot be considered a business occupancy, but must be at least ambulatory health care (four or more such patients).

When in doubt about estimated evacuation times, it may be helpful to apply the requirements for ambulatory health care occupancies. This is the conservative route that would err on the side of safety. It should be noted that the differences in requirements between business and ambulatory health care are not that onerous. Most features incorporated in ambulatory health care occupancies are needed for everyday operations in business occupancies (e.g., 32-in. clear width doors), or it is common sense to have such features for these facilities (i.e., emergency lighting and a fire alarm system).

8. Do elevators have to be recalled when the fire alarm system is activated?

NFPA *101* basically relies on ANSI A17.1 for elevator provisions. Neither of these codes requires elevators to be recalled when the fire alarm system activates, although some local codes do include this requirement. NFPA *101* and ANSI A17.1 require elevators to be recalled (i.e., Phase I fire fighters' service) upon the activation of smoke detectors in the elevator lobby or elevator machine room. This action allows the elevator cabs to operate until elevators and their related controls are exposed to smoke. This may be an important feature, especially in hospitals.

9. Why doesn't the *Life Safety Code* require closers on corridor doors in a health care occupancy?

In most cases, the patient or patients in health care facility rooms need assistance to leave, which requires staff to enter the room. If the door to the room is self-closing, when the staff go into the room, the door closes behind them. If the fire is in the room and if the door is closed, the conditions in the room will deteriorate more rapidly. When staff attempt to drag or carry the patient out of the room, they will be hampered by the closed door or, worse, will most likely prop the door open to help remove the patient. This type of situation will be even less safe when removing the second patient from the typical semiprivate room.

During day-to-day operations, the ability to see into the room is important, which might make propping open the door a normal condition. However, propping the door will likely result either in a less safe situation (i.e., a propped open, self-closing door is harder to close than a normal door that is not propped open) or in the door being smoke-activated automatic closing. Using smoke-activated automatic closing doors eliminates the problem with propping open the door under normal conditions but does not address the problem associated with propping the door open to rescue the patients.

10. Why are employee locker rooms no longer considered a hazardous area?

First, it must be clarified that the removal of locker rooms from the list of areas deemed to be hazardous does not mean that locker rooms are not hazardous areas. Rather, it means only that they are not *automatically* considered to be hazardous areas.

The *Life Safety Code* defines a hazardous area as an area that poses a degree of hazard greater than normal to the general occupancy of the building. Chapter 8 of the *Life Safety Code*, 2006 edition, contains requirements for the protection of hazardous areas. These requirements do not require occupancy chapter references. Most occupancy chapters, including new and existing health care occupancies, provide additional guidance (Sections 18.3.2 and 19.3.2), including referencing Chapter 8 and providing a list of areas deemed hazardous. These lists are not all-inclusive. In other words, areas on the list are hazardous, and those not on the list must still be evaluated for a determination.

For many years, employee locker rooms were included on the occupancy lists as needing sprinkler protection or rated enclosure, based on items typically found in locker rooms, including large quantities of soiled clothing, trash, and racks of clean clothing. Unfortunately, some enforcers considered *any* lockers to be hazardous areas and inappropriately cited purse lockers in nurses' stations and civilian lockers in waiting rooms. Note that the code did not consider the *lockers* to be a hazard, but did consider *employee locker rooms* to be hazardous areas. By removing employee locker rooms from the list, the NFPA felt that the problem of inappropriate citations would be prevented. However, a true employee locker room could still be judged to be a hazardous area (and possibly a severe hazardous area requiring both sprinklers and rated separation), based on the amount of trash, soiled linen, and clean storage in the room.

11. A typical two-story single-family dwelling is being converted into a medical office building. What are some of the typical problems that might be encountered?

The biggest problem is to deal with the usual single open stairway serving the second floor. Since this will be a business occupancy, and due to the fact that it is a change of occupancy, the requirements for new business occupancies need to be followed. [Starting with the 2006 edition of the *Code,* a conversion from a single-family dwelling to a business occupancy no longer requires compliance with the "new" business requirements. The requirements for existing businesses (except new for sprinkler and fire alarms) would apply. The issues discussed here remain the same.] There are two exceptions that, used together, allow a single open stair to serve the second story. These exceptions deal with number of exits and protection of vertical openings. If the building meets all of the following conditions, the open arrangement of the stairway would be acceptable:

- Maximum two story
- Single-tenant occupancy
- Protected throughout by sprinklers
- Total travel distance to the outside less than 100 ft (30 m)

If any of these conditions cannot be met, alternative arrangements must be developed. Adding an outside stair will help, but the unprotected vertical opening and the second means of egress being an open stair will still be issues. Another possibility is to develop an equivalency. Use of NFPA 101A, *Guide on Alternative Approaches to Life Safety,* could provide assistance in finding acceptable alternatives.

12. In existing buildings, corridors in hospitals and nursing homes must be a minimum of 4 ft (1.2 m) wide. If the corridor is 7 ft (2 m) wide, can this excess width be used for other purposes, or can the corridor be narrowed to 4 ft (1.2 m)?

No. The *Life Safety Code* (Chapter 4) clearly prohibits any alteration that would reduce the level of protection below the requirements for new construction. Since new construction requires 8-ft- (2.4-m-) wide corridors, the current width cannot be reduced further. There is no requirement to increase the corridor width to 8 ft (2.4 m), but it cannot be reduced to 4 ft (1.2 m). (Also, see Question 13.)

13. If the corridor in a hospital or nursing home is over 8 ft (2.4 m) wide, is it permissible to use the excess corridor width for chairs for waiting?

Further analysis is needed. The *Life Safety Code* does not require that width in excess of *new* be maintained. However, two issues must be evaluated. The first issue is to determine if the excess width is needed for some other reason, such as egress capacity. If not, the next issue to be evaluated is the prohibition in health care occupancies about use space being open to the corridor, unless it meets one of the requirements based on the use of the space and the protection provided. Waiting is a use that must be separated from the corridor (by a corridor wall), unless the corridor has, as a minimum, smoke detection, and possibly sprinkler protection. The use (waiting area) also could not interfere with egress. As such, a very wide corridor would be required, since the chair and the person in the chair would need to be clear of the 8-ft (2.4-m) width.

14. What is the occupancy of a physician "on-call" that provides sleeping accommodations for physicians or other staff?

The answer to this question depends on how many sleeping accommodations are provided. The *Life Safety Code* defines occupancies with up to 16 sleeping accommodations as lodging or rooming houses and those with more than 16 as dormitories. If the rooms are spread all over the facility, each one must be evaluated as a lodging and rooming house. Chapter 6 of the *Life Safety Code* has clarified that residential occupancies cannot be considered incidental and must be addressed. This was also true before 1994, but the 1994 and 1997 editions allowed any occupancy to be considered incidental. Either way, smoke alarms are required in the sleeping room, and a door closer will probably be required (exempted for sprinklered lodging and rooming houses). In nonsprinklered facilities that qualify as a dormitory, there may be issues with the corridor wall that will need to be addressed.

15. The *Life Safety Code* limits nonsleeping suites to 10,000 ft^2 (930 m^2). If a suite of operating rooms exceeds this limit, what can be done?

Several options are available to bring the area into compliance. First, the suite could be subdivided such that each resulting suite is under the limit. In most cases, however, this solution is not going to be feasible. The next solution would be to make the circulation areas within the suite comply with corridor requirements. This might be possible, but again most likely not. Lastly, an equivalency could be developed using NFPA 101A, *Guide on Alternative Approaches to Life Safety*. This solution provides a practical use of this code, for which the entire suite can be considered to be a zone and the violation of corridors charged in Safety Parameters 5 and 6 as listed in Table 3.4 of NFPA 101A, 2004 edition.

Building Construction and Life Safety Checklist

This checklist can be used for evaluating an existing facility. This checklist is for the building construction and life safety section of this book. The reader is cautioned that this is not a complete list, but one that touches on many of the major items. To ensure compliance, the reader should be familiar with all of the codes and standards that are applicable.

General Information

✓	Item	Inspection Activity	Comments
	1.	Determine whether the building is protected throughout with a supervised automatic sprinkler system.	
	2.	Determine whether the building construction type is acceptable based on the occupancy, *new* versus *existing*, sprinkler protection, and building height.	
	3.	Ensure that plans are available indicating all *required* exits, smoke barriers, hazardous areas, horizontal exits, building separation walls, occupancy separations, shafts, suites, and related fire protection features.	

Means of Egress
Exit Enclosures (enclosed exit stairs and exit passageways)

✓	Item	Inspection Activity	Comments
	1.	Ensure that all required exits are properly enclosed.	
	2.	Ensure that fire resistance ratings of enclosing wall are adequate.	
	3.	Verify that penetrations are limited and properly protected.	
	4.	Verify that door labels are legible and indicate the proper rating.	
	5.	Ensure that doors self-close and self-latch properly.	
	6.	Verify that door protective plates, if provided, are compliant with NFPA 80, *Standard for Fire Doors and Fire Windows*.	

✓	Item	Inspection Activity	Comments
	7.	Verify that the distance between the bottom of the door and the floor is compliant with NFPA 80.	
	8.	Ensure that informational signs are provided and visible in exit enclosures serving five or more stories.	
	9.	Ensure that exit signs are properly placed.	
	10.	Verify that the enclosure is properly illuminated.	
	11.	Ensure that exit enclosures are kept free of storage or any material that could affect the use of the exit.	

Stairs

✓	Item	Inspection Activity	Comments
	1.	Ensure that stair treads and risers are the proper dimensions.	
	2.	Ensure that differences in riser and tread sizes is within tolerance.	
	3.	Ensure that handrails are provided and are of proper design and height.	
	4.	Ensure that guards are provided and are of proper design and height.	
	5.	Ensure that outside stairs are properly protected.	

Doors

✓	Item	Inspection Activity	Comments
	1.	Determine whether doors are unlocked in the direction of egress except where specifically permitted by the *Life Safety Code*.	
	2.	Ensure that latches and locks are arranged so that one action releases the door.	
	3.	Verify that the forces required to open doors are within acceptable limits.	
	4.	Ensure that doors swing in the proper direction.	
	5.	Ensure that when required to be fire protection rated, the door is of the proper rating and the labels are legible.	
	6.	Ensure that doors required to be self-closing properly self-close.	
	7.	Ensure that doors required to be self-latching properly self-latch.	

Horizontal Exits (If the horizontal exit also serves as a smoke barrier, see requirements for smoke barriers also.)

✓	Item	Inspection Activity	Comments
	1.	Verify that the walls have a 2-hour fire resistance rating.	
	2.	Verify that walls extend from outside wall to outside wall.	

✓	Item	Inspection Activity	Comments
	3.	Verify that walls extend from the floor to the floor above.	
	4.	Ensure that all penetrations of the wall are properly sealed or protected.	
	5.	Ensure that HVAC penetrations are properly protected.	
	6.	Ensure that doors are 1½-hour rated.	
	7.	Ensure that door labels are legible.	
	8.	Verify that the doors swing in the direction of egress.	
	9.	Ensure that the gap between pairs of doors is within limits.	
	10.	Ensure that if a coordinator is used, it is functioning properly.	
	11.	Ensure that the doors self-close and self-latch.	
	12.	Verify that if automatic-closing doors are used, the activating smoke detector is properly located.	
	13.	Ensure that exit signs are properly placed.	
	14.	Ensure that fire alarm manual stations are nearby.	

Ramps

✓	Item	Inspection Activity	Comments
	1.	Ensure that ramps are of the proper slope.	
	2.	Ensure that handrails are provided where required.	
	3.	Ensure that handrails are of the proper design and height.	
	4.	Ensure that guards are provided where required.	
	5.	Ensure that guards are of the proper design and height.	

Escalators and Moving Walks

✓	Item	Inspection Activity	Comments
	1.	Determine whether escalators and moving walks are used for egress. (Note: Not permitted in health care occupancies and limited in other occupancies.)	

Fire Escape Stairs

✓	Item	Inspection Activity	Comments
	1.	Determine whether fire escape stairs are present.	
	2.	Determine whether fire escape stairs are permitted for the occupancies involved.	
	3.	Ensure that the fire escape stairs meet the dimensional requirements of the *Life Safety Code*.	
	4.	Ensure that the fire escape stairs are properly protected.	

Fire Escape Ladders and Alternating Tread Devices

✓	Item	Inspection Activity	Comments
	1.	Determine whether fire escape ladders or alternating tread devices are present.	
	2.	Ensure that they are used in the proper locations for occupancies that allow them.	
	3.	Ensure that they are arranged properly.	

Capacity

✓	Item	Inspection Activity	Comments
	1.	Verify that adequate egress capacity is provided for each story.	
	2.	Ensure that doors comply with the minimum widths.	
	3.	Ensure that corridors comply with minimum widths.	

Arrangement

✓	Item	Inspection Activity	Comments
	1.	Ensure that all corridors lead to exits without passing through intervening rooms.	
	2.	Ensure that dead ends are limited.	
	3.	Ensure that patient sleeping rooms over 1000 ft^2 (93 m^2) are provided with two remote doors.	
	4.	Ensure that rooms, other than patient sleeping rooms over 2500 ft^2 (230 m^2), are provided with two remote doors.	
	5.	Ensure that patient sleeping suites are limited to 5000 ft^2 (460 m^2); 7500 ft^2 (690 m^2) permitted if smoke detection and visual observation are provided	
	6.	Ensure that suites of rooms other than patient sleeping rooms are limited to 10,000 ft^2 (930 m^2).	
	7.	Verify that travel distances within suites are within limits.	

Travel Distance

✓	Item	Inspection Activity	Comments
	1.	Verify that travel distances within patient sleeping rooms are less than 50 ft (15 m).	
	2.	Ensure that travel distances from any corridor door to an exit are limited to 100 ft (30 m) [150 ft (45 m) if sprinklered].	
	3.	Ensure that the total travel distance from *any* point to an exit is limited to 150 ft (45 m) [200 ft (60 m) if sprinklered].	

Exit Discharge

✓	Item	Inspection Activity	Comments
	1.	Ensure that all exits discharge directly to the outside.	
	2.	Ensure that if exits do not discharge directly to the outside, the exits that discharge within the building meet the 50 percent rules and limitations.	
	3.	Ensure that exit stairs that continue past the level of exit discharge are provided with gates, doors, or other effective means to prevent accidental passage past the exit discharge.	
	4.	Ensure that illumination (and emergency lighting if required) is adequate in the exit discharge.	
	5.	Ensure that exit discharges are kept free of accumulations of snow, ice, and similar obstructions.	

Illumination

✓	Item	Inspection Activity	Comments
	1.	Verify that corridors, aisles, ramps, stairs, exits, and exit discharges are illuminated to a minimum of 1 ft-candle (10 lx).	
	2.	Ensure that illumination is such that a single failure of a bulb will not leave an area in total darkness.	

Emergency Lighting

✓	Item	Inspection Activity	Comments
	1.	Verify that emergency lighting is provided.	
	2.	Ensure that emergency lighting is tested and maintained in accordance with NFPA 101®, *Life Safety Code*® (for unit lighting), or NFPA 110, *Standard for Emergency and Standby Power Systems* (for generators) (also see electrical section).	

Exit Signage

✓	Item	Inspection Activity	Comments
	1.	Ensure that all exits marked with exit signs are visible from the direction of exit access.	
	2.	Ensure that where the direction to exit is not obvious, directional exit signs are provided.	
	3.	Verify that convex mirrors are arranged so as not to block visibility of exit signs and such that confusion with exit location is not created by the mirror.	
	4.	Verify that exit signs will be illuminated in both the normal and emergency lighting modes.	

	5.	Verify that there is a program to ensure that exit sign illumination is maintained.	
	6.	Ensure that direction indicators on exit signs are clear and understandable and pointing in the proper direction.	

Protection

Vertical Openings

✓	Item	Inspection Activity	Comments
	1.	Verify that vertical openings (other than exit stair enclosures) are properly enclosed or protected.	
	2.	Verify that penetrations of shafts are properly sealed or protected.	
	3.	Ensure that HVAC penetrations are properly protected.	
	4.	Ensure that doors are the proper rating.	
	5.	Ensure that doors are self-closing and self-latching.	

Linen and Trash Chutes

✓	Item	Inspection Activity	Comments
	1.	Verify that chutes are properly enclosed.	
	2.	Ensure that charging doors to chutes are properly rated with legible labels.	
	3.	Ensure that charging doors to chutes are self-closing and self-latching.	
	4.	Ensure that charging and discharge rooms are protected as hazardous areas.	
	5.	Verify that sprinklers are properly provided within the chute.	

Protection from Hazards

✓	Item	Inspection Activity	Comments
	1.	Verify that all hazardous areas have been evaluated.	
	2.	Determine whether there are any new hazardous areas since the last survey.	
	3.	Verify that new hazardous areas meet the requirements for *new* rather than *existing*.	
	4.	Ensure that all doors are self-closing.	
	5.	Ensure that doors to corridors are self-latching.	
	6.	Ensure that if the area is a severe hazard or if there are no sprinklers, the doors are at least 45-minute rated with legible labels.	
	7.	Ensure that if the area is a severe hazard or if rated enclosure is not provided, sprinklers are provided.	
	8.	Verify that sprinklers are clear and unobstructed.	
	9.	Verify that laboratories are protected in accordance with NFPA 99, *Standard for Health Care Facilities*.	

✓	Item	Inspection Activity	Comments
	10.	Verify that anesthetizing locations are protected in accordance with NFPA 99.	
	11.	Verify that medical gases are stored in accordance with NFPA 99.	
	12.	Verify that cooking facilities are protected in accordance with NFPA 96, *Standard for Ventilation Control and Fire Protection of Commercial Cooking Operations*.	

Interior Finish

✓	Item	Inspection Activity	Comments
	1.	Ensure that interior finish within exits is adequate.	
	2.	Ensure that interior finish in corridors is adequate.	
	3.	Ensure that interior finish within rooms is adequate.	
	4.	Verify that textile interior finishes have been properly addressed.	
	5.	Verify that expanded vinyl wall or ceiling coverings have been properly addressed.	
	6.	Determine whether any exposed foam plastics have been used for interior finish.	
	7.	Ensure that interior floor finishes are adequate.	

Corridor Walls

✓	Item	Inspection Activity	Comments
	1.	Ensure that corridor walls and doors resist the passage of smoke.	
	2.	Ensure that the walls are at least ½-hour fire protection rated in nonsprinklered facilities.	
	3.	Ensure that the walls extend to the floor or roof deck above (unless sprinklered and the ceiling and wall together resist the passage of smoke—no above-ceiling plenums).	
	4.	Verify that all penetrations of the wall are properly sealed.	
	5.	Verify that doors latch or provide equivalent protection.	
	6.	Ensure that the clearances under the doors are limited to not more than 1 in. (2.5 cm).	
	7.	Ensure that Dutch doors are properly arranged.	

Subdivision of Building Spaces

✓	Item	Inspection Activity	Comments
	1.	Determine whether required smoke barriers have been identified. (Note: To evaluate whether an existing barrier can be deleted, the evaluation must be done using the requirements for new construction.)	

✓	Item	Inspection Activity	Comments
	2.	Ensure that all smoke barriers extend to the floor or roof deck above.	
	3.	Ensure that smoke barriers are at least ½-hour-rated construction (1 hour in *new*).	
	4.	Verify that all penetrations are properly sealed.	
	5.	Verify that all HVAC openings are properly protected.	
	6.	Verify that doors are self-closing or automatic closing.	
	7.	Verify that if automatic opening, the smoke detectors are properly located.	
	8.	Ensure that doors close properly with a maximum of ⅛-in. (0.3-cm) clearance between the doors under all humidity conditions.	
	9.	Ensure that if a coordinator is used, it functions properly.	
	10.	Verify that any vision panels in the door are of wired glass or other fire-rated glazing.	

Operating Features

✓	Item	Inspection Activity	Comments
	1.	Verify that fire evacuation and relocation plans are provided, up to date, and available to all supervisory personnel.	
	2.	Verify that all personnel are properly instructed and drilled on the plans.	
	3.	Verify that fire drills are held on all shifts at least quarterly.	
	4.	Ensure that all draperies, curtains, cubicle curtains, and similar hanging fabrics are flame retardant in accordance with NFPA 701, *Standard Methods of Fire Tests for Flame Propagation of Textiles and Films.*.	
	5.	Verify that polices are in place to properly regulate the purchase of upholstered furniture and mattresses in nonsprinklered facilities.	
	6.	Ensure that furnishings and decorations of an explosive or highly flammable character are prohibited.	
	7.	Determine if the facility has in place rules regarding holiday decorations.	
	8.	Verify that smoke control systems are properly tested and maintained.	
	9.	Verify that portable space heaters are prohibited except in nonsleeping staff locations.	
	10.	Ensure that adequate precautions are exercised during construction repair and improvement operations.	

Part IV
Fire Protection Systems

This part of the book covers the active systems that protect health care facilities from fire. These systems are examined in terms of their background in health care facilities, where they are required within a facility, the various system types, considerations for system design and modification, and inspection, testing, and maintenance requirements to keep the equipment in good operating order. (Refer to Chapter 35 on acceptance and performance testing procedures that apply to systems overall.)

This information is intended to assist the health care facility engineer in understanding these systems, so that he can select the system with the most appropriate and desired features for the protected hazard with due regard for the facility and its functions. The information also should assist in better managing existing systems and project managing contracts with both design engineers and installers of these systems when new systems are added or existing systems are modified.

The chapters of this part specifically cover the following topics:

- The importance of fire protection systems in health care facilities
- Automatic fire sprinkler systems, including sprinkler facts and myths; history of automatic sprinklers; system activation and controls; system types; design considerations for new construction and retrofits; and requirements for inspection, testing, and maintenance
- Standpipe system history, requirements, types, classes, design considerations, and inspection, maintenance, testing, and record keeping requirements
- Fire pumps, including information on location requirements, pump types, pump drivers, design considerations, and inspection, testing, and maintenance
- Kitchen hood fire-extinguishing system methods of control, UL 300–compliant suppression systems, system types, considerations for system design and modification, inspection and maintenance, and testing
- Specialized suppression systems, including clean fire suppression agents and water mist extinguishing systems

- Portable fire extinguishers, including information on where they are required, fire types, extinguisher ratings, extinguisher selection and placement, and inspection, testing, maintenance, and recharging requirements
- Fire detection and alarm systems, in terms of system types; control panels; alarm devices; specialized features for fire alarm systems in health care facilities; and inspection, testing, and maintenance

The intent of this part is to provide health care facility engineers with enough information about fire protection equipment and systems to select the right equipment and systems with the desired features. The information should also assist facility personnel in managing fire protection installations and contracts with installers and maintenance personnel for a protected hazard with due regard for the facility, function, and aesthetics. At a minimum, this part should allow the facility engineer to "speak the same language" as the engineers, designers, and vendors of these systems. General testing and maintenance procedures that should be followed to keep equipment in good operating condition are also presented.

20

Importance of Fire Protection Systems

As discussed in Part III, a building's features, such as its height, area, construction type, and compartmentation, are built-in passive fire protection features. Active fire protection systems also play a vital role in determining a building's design and are imperative for limiting the growth of a fire, protecting the people situated in the smoke compartment that contains the fire, and limiting the spread of fire products to adjacent compartments. The value of the combination of passive protection measures with active suppression systems has been proved over and over again in actual health care facility fires.

Each health care facility may contain a number of active fire protection systems and components, including the following:

- Automatic fire sprinkler systems—integrated systems of underground and overhead piping that include one or more water supplies
- Standpipe systems—arrangements of piping, valves, hose connections, and other equipment installed in tall or otherwise very large buildings or structures to aid in the manual delivery of water to a fire
- Fire pumps—equipment that enhances water supply pressure available from municipal water mains, gravity tanks, and other sources, usually to supplement the available water pressure for sprinkler and standpipe systems
- Kitchen hood fire protection systems—equipment that detects and suppresses grease fires on cooking appliances or in kitchen hood ventilation systems
- Portable fire extinguishers—portable containers for a number of different types of fire extinguishing agents
- Fire detection and alarm systems—integrated systems of manual devices and automatic sensors that, when activated, cause the systems to warn building occupants and emergency forces as to the location of the fire signal and that usually interface with several other building systems
- Specialized suppression systems—specific suppression equipment used where an item being protected may be damaged by the application of a particular suppression agent or destroyed if a fire is not suppressed quickly enough

371

Based on the date of construction, the applicable codes and standards, the prevalent technology of the time, user needs, budget, and other factors, each building contains various fire protection systems to various levels of protection. Fire alarm systems, sprinkler and standpipe systems, and portable fire extinguishers are the most commonly installed fire protection in health care facilities, but these facilities are complex and ever changing, which requires some unique fire protection systems and configurations.

Buildings and their occupants are best protected from fires by a coordinated system of active fire protection systems, passive fire protection features, and proper maintenance. The facilities engineers must know enough about all of the fire protection within their facilities to provide some level of inspection and to call upon specialty engineers/contractors to conduct detailed inspection, regular maintenance, and testing services. *Additionally, the facility's interface with the local fire department should be structured.* The fire department should be invited into the facility at least yearly, to increase the department's knowledge of the facility and its fire protection. While on site, fire department personnel can update their preplanned response to the facility, and the facility personnel can learn how to best work with the fire department in case of an emergency. It is incumbent on the facility staff to explain the unique "protect-in-place" concept of a health care facility.

20.1 Codes and Standards for Active Fire Protection Systems

As with most equipment and systems in health care facilities, codes and standards govern the manufacture, design, installation, use, testing, inspection, and maintenance of active fire protection systems. The major codes and standards include the following:

- *Building codes*—detail where and when fire protection systems are required inside or outside structures.
- NFPA 1, *Uniform Fire Code™*—contains criteria on inspection; investigation; review of construction plans, drawings, and specifications; storage, use, processing, handling, and on-site transportation of hazardous materials; access requirements for fire department operations; and interior finish, decorations, furnishings, and other combustibles that contribute to fire hazards.
- NFPA 10, *Standard for Portable Fire Extinguishers*—provides criteria for portable fire extinguisher use, including distribution and placement, required maintenance, proper operation, timely inspection and testing, and safe recharging.
- NFPA 12, *Standard on Carbon Dioxide Extinguishing Systems*—provides minimum requirements for installing and maintaining carbon dioxide extinguishing systems, and covers total flooding systems, local application systems, hand hose line system, standpipe systems, and mobile supplies.
- NFPA 13, *Standard for the Installation of Sprinkler Systems*—details how sprinkler systems must be designed and installed for all types of fire hazards.

- NFPA 14, *Standard for the Installation of Standpipe and Hose Systems*—includes information on installing private fire hydrants and hose houses, guidance for the design and application of standpipe systems, and information about how to conduct water supply testing.
- NFPA 17, *Standard for Dry Chemical Extinguishing Systems*—provides minimum requirements for dry chemical extinguishing systems and addresses total flooding, local application, hand hose lines, and engineered and pre-engineered extinguishing systems.
- NFPA 17A, *Standard for Wet Chemical Extinguishing Systems*—applies to the design, installation, operation, testing, and maintenance of wet chemical extinguishing systems and includes minimum requirements for restaurant and institutional hoods, plenums, ducts, and associated cooking appliances.
- NFPA 20, *Standard for the Installation of Stationary Pumps for Fire Protection*—includes information on proper fire pump design, electrical and mechanical construction, acceptance testing, and operation of stationary pumps.
- NFPA 25, *Standard for the Inspection, Testing, and Maintenance of Water-Based Fire Protection Systems*—provides details on fire pumps; foam–water sprinklers, standpipe and hose systems; water-spray fixed systems; valves and connections; private fire service mains; water storage tanks; and proper impairment handling for all types of water-based fire protection systems.
- *NFPA 72®*, *National Fire Alarm Code®*—covers the application, installation, location, performance, and maintenance of fire alarm systems and their components.
- NFPA 96, *Standard for Ventilation Control and Fire Protection of Commercial Cooking Operations*—covers all aspects of commercial cooking operations, including requirements for cooking equipment; hoods; grease removal devices; exhaust duct systems; fans; fire suppression systems; and clearance to combustibles. It also includes requirements and information on fire-extinguishing equipment, special effluent equipment, recirculating systems, solid fuel cooking operations, and clearance and enclosure requirements.
- *NFPA 101®*, *Life Safety Code®*—provides minimum requirements, with due regard to function, for the design, operation, and maintenance of buildings and structures for safety to life from fire. Its provisions will also aid life safety in similar emergencies.
- NFPA 2001, *Standard on Clean Agent Fire Extinguishing Systems*—covers minimum requirements for total flooding clean agent fire-extinguishing systems.

As discussed in Chapter 13, the International Code Council publishes the *International Fire Prevention Code*. Similar to NFPA 1, this fire prevention code deals with such issues as fire reporting, fire lanes, hazardous materials (flammable liquids, flammable gases, dusts, etc.), housekeeping, control of ignition sources, and sometimes education.

There is usually little question as to what requirements must be followed for systems, such as sprinkler systems, which are usually required to comply with just one edition of NFPA 13 (although legislative bodies can amend the standard, and authorities having jurisdiction can develop their own interpretations of it). Other systems may require compliance with more than one code. For example, the hydraulic design requirements for standpipe systems in the *International Building Code* are different from those in NFPA 14, and some facilities are required to comply with both of these codes. In such cases, it is best to follow the most stringent requirements to ensure compliance; unfortunately, what is considered "most stringent" is not always easily defined.

21

Automatic Fire Sprinkler Systems

According to NFPA 13, *Standard for the Installation of Sprinkler Systems,* an automatic sprinkler system is an integrated system of underground and overhead piping that includes one or more automatic water supplies. The system is usually activated by heat from a fire, after which it discharges water over the fire. The aboveground portion of the system is a network of specially sized and hydraulically designed piping installed in a building or area to which the sprinklers are attached in a systematic pattern. A valve controls each system riser, which includes a device for actuating an alarm when the system is in operation.

As already suggested in previous parts of this book, new buildings containing health care facilities must be protected throughout by an approved, supervised automatic sprinkler system (NFPA *101*®, *Life Safety Code*®, 2006 edition, Section 12.3.5). There are limited exceptions to this provision where the authority having jurisdiction (AHJ) may permit alternative protection measures, which will not cause a building to be classified as nonsprinklered. Starting with the 1997 edition of the *Life Safety Code* (Section 12.3.5.2), smoke compartments that contain patient sleeping rooms are required to be equipped throughout with listed quick-response sprinklers (QRSs). Residential sprinklers are one type of QRS that operates by spraying water up higher on adjacent walls than other quick-response types and can be used in patient sleeping rooms.

It was not always the case that health care facilities needed sprinkler protection. Compartmentation, or passive fire protection features, was the prevalent means of protection for many years, and the *Life Safety Code* contained many exceptions for facilities that were sprinkler protected. The 1991 edition of the *Life Safety Code* was a landmark edition because it contained numerous new requirements for mandatory sprinklers in new health care facilities. Now that sprinklers are mandatory, the exceptions related to sprinkler protection have been deleted. The *International Building Code* also contains sprinkler system requirements and influences where and when sprinklers are required in new health care facilities. (*NFPA 5000*®, *Building Construction and Safety Code*®, also contains sprinkler requirements, but they are the same as those contained in the *Life Safety Code* for new construction.)

This chapter details the application of automatic fire sprinkler systems in health care facilities. Automatic sprinkler systems are examined in terms of their history; system activation and controls; system types; system design and modification considerations; and inspection, testing, and maintenance concerns. The intent of this information is to assist health care facility engineers in managing existing fire sprinkler systems and (project) manage contracts with design engineers/system installers when new systems are added or existing systems are modified. The information also is intended to assist facilities engineers in contracting for and maintaining the right system with the desired features for the protected hazard with due regard for the facility, function, and aesthetics.

21.1 Sprinkler Facts and Myths

Some interesting sprinkler facts that may be considered when contemplating the installation of a sprinkler system are as follows:

- In fully sprinklered buildings housing health care occupancies, the corridor walls are not required to extend above a ceiling that is designed to resist the passage of smoke.
- The costs for installing fire sprinkler systems in buildings six to eight stories high range from less than $1.00 to about $3.00 per square foot in most new construction and from about $1.50 to $6.00 per square foot to retrofit sprinklers in existing buildings (costs may vary from these estimates depending upon many project-specific conditions).
- Installation of fire sprinklers can provide discounts on insurance premiums.
- The cost of a sprinkler system is a deductible item on federal income tax returns.
- The interest on a loan secured to pay for a sprinkler system installation is tax deductible.
- In downtown Fresno, California, there has been residential fire damage of only $42,000 during a 10-year period in which a residential sprinkler law has been in effect.
- According to the National Fire Protection Association (NFPA), property damage in health care fires was 63 percent less in structures with sprinklers than it was in structures without sprinklers from 1987 to 1996. (Average loss per fire was $1600 in sprinklered buildings and $4300 in nonsprinklered buildings.)

Although automatic fire sprinklers are widely recognized as a very effective method for stopping the spread of fires in their early stages—before they cause severe injury to people and damage to property—there are many misconceptions about these systems. Some of the more popular myths are as follows:

- **Myth:** "Water damage from a sprinkler system is more extensive than fire damage."
  Fact: Water damage from a sprinkler system is much less severe than the damage caused by water from fire-fighting hose lines or smoke and fire damage if the fire goes unabated. Sprinklers release 8 to 24 gal (30.3 to 90.8 L) of water per minute, compared to 50 to 200 gal (189.3 to 757.0 L) per minute released by a fire hose nozzle. Additionally, fire-fighting hose lines are deployed much later in a fire, so the fire is much larger and more dangerous before it is controlled or suppressed.
- **Myth:** "When a fire occurs, every sprinkler goes off."
  Fact: Most sprinklers are individually activated by fire. Most fires in sprinklered buildings are controlled with only a few sprinklers, and residential fires are usually controlled with fewer than four sprinklers.

- **Myth:** "Sprinklers are designed to protect property but are not effective for life safety."
  Fact: Sprinklers provide a high level of life safety. Statistics demonstrate that there has never been any multiple loss of life in a fully sprinklered building where the system was not shut off or otherwise impaired. NFPA statistics also indicate that in health care properties, civilian deaths due to fire are reduced by 73 percent when comparing sprinklered facilities to nonsprinklered facilities. There are case histories in health care facilities where quick-response sprinklers extinguished fires involving bedding material while a patient was in the bed, and the patient survived.

- **Myth:** "Sprinklers spray water that has been superheated by the fire, resulting in the scalding of building occupants."
  Fact: Water discharges from sprinklers early in a fire's development such that only a small portion of the water is converted to steam, but the volume of steam compared to the volume of the building's interior does not result in scalding of occupants. Actually, the discharging water quickly lowers the interior temperature of affected area(s).

- **Myth:** "Sprinkler system operation causes drowning or electrocution of building occupants."
  Fact: For a multitude of reasons, fire history in occupied structures indicates that neither drowning nor electrocution occurs.

- **Myth:** "Sprinkler system operation increases the amount of smoke generated by the fire."
  Fact: Actual fires and fire testing have confirmed that the operation of sprinklers quickly and significantly decreases the amount of smoke generated by a fire.

21.2 History of Automatic Sprinklers

21.2.1 Early Sprinkler Systems

The first sprinkler systems, which were not automatic, were invented in the early 1800s. These systems were simply a series of pipes with holes in them (called perforated pipes) that wetted an entire area (i.e., the *protected space*) after a water supply valve was opened. Although these systems were somewhat effective, they would "save" the building at the expense of water damage to areas not involved in the fire.

The first practical automatic sprinkler was invented in the late 1800s. All early systems as well as all that followed operate using a simple heat-sensing element that "fuses" at a particular temperature, releasing water from an orifice. The water then sprays against a deflector and spreads out in a pattern. The fusible element results in sprinklers that discharge water only over the fire area, and the deflector breaks the water into smaller droplets for better distribution and to better absorb the heat from the fire.

The fusible element of many early sprinklers consisted of a link-and-lever system held together by solder. The solder could be applied such that the manufacturers could determine the activation temperature of the sprinkler very accurately. An operating element that has enjoyed much popularity of late is the glass bulb (also called a frangible bulb). The bulbs are filled with a liquid that expands due to the heat of the fire, breaking the glass, which releases a cap over the orifice, allowing water to flow from the piping.

21.2.2 Quick-Response Sprinkler Technology

Quick-response sprinkler technology is a by-product of the fire protection community's effort in the late 1970s and early 1980s to design automatic sprinkler protection for one- and two-family dwellings and mobile homes (now called manufactured housing). During this effort, engineers determined through fire testing that simply installing standard sprinklers in dwellings did not result in tenable conditions, and therefore occupants near or in the room of fire origin would not survive. The goal of the residential sprinkler tests was to maintain tenability in the structure during the early stages of a fire so that occupants could escape. Although property protection was not a goal of the system design, it was a natural result because the fire was suppressed and often it was extinguished.

The test engineers concluded that to maintain tenable conditions, sprinklers needed to operate much earlier in a fire. Of course, a sprinkler with a low activation temperature would accomplish this, but problems of unwanted activation were definitely undesirable. The standard sprinkler activation temperatures were not too high, but their response time was too slow because of thermal lag—they just took too long to heat up. What was needed was a sprinkler that activated at the same temperatures as standard sprinklers but with an operating element that responded more quickly.

After a series of prototypes and tests, the quick-response sprinkler was born. These sprinklers provide increased life safety because they respond so much more quickly to a fire. Since that time, QRS technology has been used and is required extensively in many occupancies and especially in health care facilities, where patients usually have a lower tolerance for untenable conditions than the general public. Quick-response sprinklers start to fight a fire before too much smoke and heat have been introduced into the structure. This of course makes it easier to control or suppress a fire, as compared to fires that have more time to grow.

21.3 System Activation and Controls

Through one means or another, an automatic sprinkler system detects a fire and discharges water to control or suppress it. Water flow may be the result of any of the following conditions:

- A sprinkler that has fused due to a fire
- Damage to the system
- An open drain or fire hose valve (on the standpipe system)

Typically these systems are divided into zones by floor level and further by smoke zone or other prominent building area. System zoning allows smaller portions of the system to be shut down during maintenance and more clearly defines the area of the building where the water flow alarm indicates a system flow. The more sprinkler zones a system has, the smaller the area that must be investigated to find the source of a water flow. *Of course, the key to limiting water damage, whether the water flow is desired or not, is keeping the health care facility's staff knowledgeable and trained as to the configuration of the sprinkler zones and the location of the control valves for each.*

21.3.1 Sprinkler Activation

A common misconception is that when one sprinkler activates, all the sprinklers in the system activate (see Section 21.1). Except for deluge systems (see Section 21.4.7), only those sprinklers that individually react to the heat of the fire (fuse) will discharge water on a fire. Usually the sprinkler closest to the fire activates first. Sprinklers farther from the fire operate only if the fire continues to produce enough heat to raise the temperature of these sprinklers to their activation temperature.

This activation of subsequent sprinklers is anticipated by NFPA 13, in the hydraulic calculations that are used to determine the sprinkler pipe sizing. The calculation procedure requires designers to design the system with all the sprinklers within a "design area" operating at the same time. This calculation takes a conservative approach, which is a major reason why sprinkler systems have such an excellent safety record. It is conservative because the typical design area contains 10 to 20 sprinklers, depending on the building's configuration, but most fires are controlled with fewer sprinklers. Since fewer sprinklers are actually flowing as compared to the number designed to be flowing, those actually flowing have the advantage of higher pressures and therefore flow more water than calculated.

Also, for many years fire incident reports have shown that in most cases, fires have been controlled by automatic sprinkler systems with 10 or fewer sprinklers. Table 21.1 provides sprinkler performance data from "U.S. Experience with Sprinklers," Kimberly D. Rohr, Fire Analysis and Research Division, NFPA, November 2003.

21.3.2 Control Valves

Valves that control the water flow to sprinkler systems must be *indicating* valves (i.e., they must have some means to visually indicate their position as either open or closed).

Table 21.1 Percentage of Fires Controlled by Number of Sprinklers by System Type

Number of Sprinklers Operating	Wet Pipe	Dry Pipe
1 or fewer	62.3%	21.5%
2 or fewer	79.9%	42.9%
3 or fewer	86.3%	61.0%
4 or fewer	90.0%	78.7%
5 or fewer	90.9%	78.7%
6 or fewer	92.7%	82.3%
7 or fewer	93.6%	82.3%
9 or fewer	95.4%	82.3%
10 or fewer	96.3%	92.9%

This indication allows maintenance and inspection personnel to see the valve's orientation from several feet away. Also, the valves must be identified with signage as to the area or zone of the system that they control.

At one time, codes allowed valves to be locked or chained in the open position. In most jurisdictions today, the fire alarm system must supervise the position of the sprinkler control valve using electronic supervisory switches (slang: tamper switches), because locking a valve in the open position is not effective when someone maliciously cuts the lock and closes the valve. Conversely, tamper switches connected to the fire alarm system provide an audible and visual alarm at the fire alarm control panel, which is usually located in, or reports to, a constantly attended location. The aim of electronic supervision is for the person on duty to respond to the impairment at hand, even when the impaired portion of the sprinkler system is located in a remote (often locked) mechanical space that is not usually occupied.

Another disadvantage of locked valves is that they take longer to close when shutting the system off, such as after a controlled or suppressed fire. A delay in closing the control valve(s) can contribute to excessive water damage after the system has performed its intended function.

The best way to limit the amount of water damage from operating sprinklers is to know the location of all system control valves. Building engineering staff should close system control valves only to conduct system repairs, when there has been accidental water flow (leaks or system damage), or after the fire department orders such closure during an actual fire incident. Any time a control valve is closed, the local fire department or fire marshal must be notified of the system's "out of service" status.

Typical valves approved or listed as indicating control valves are outside stem and yoke (OS&Y) valves and butterfly valves with a position indicator. See Figure 21.1 for examples of indicating valves that are listed for fire protection service.

21.4 System Types

There are several different types of sprinkler systems, each of which has been developed to address a specific need of the protected area's environment or hazard. The different types of systems are basically the same but have discretely different configurations or control equipment. The different system types include the following:

- Wet pipe
- Dry pipe
- Antifreeze
- Preaction
- Double interlock preaction
- Combined dry-pipe preaction
- Deluge

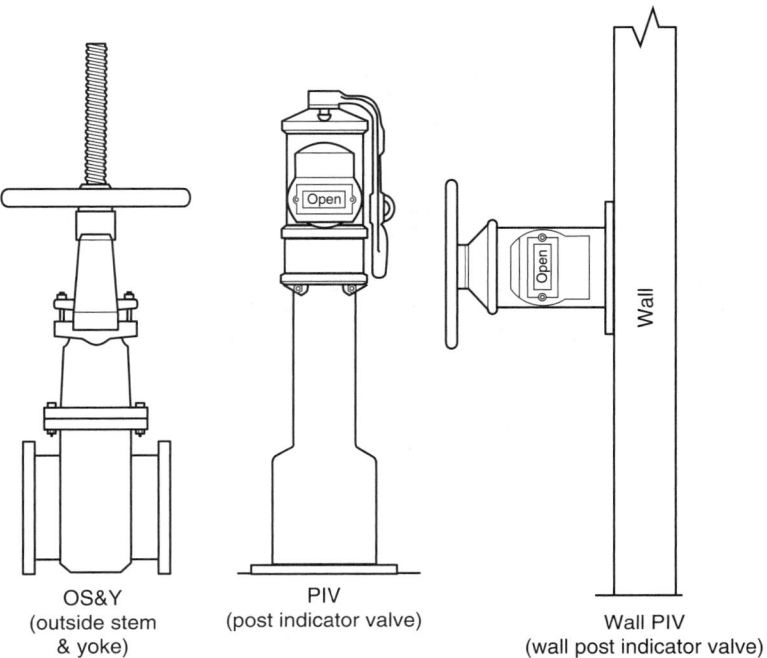

Figure 21.1 Sprinkler system indicating valves.
Source: *Fire Protection Handbook,* NFPA, 2003, Figure 10.11.16.

21.4.1 Wet-Pipe Sprinkler Systems

The wet-pipe sprinkler system is not only the most common sprinkler system, it is also the least expensive and most reliable. Health care facilities have more wet-pipe sprinkler systems than any other type. A wet-pipe sprinkler system consists of a series of interconnected pipes and sprinklers that is constantly charged with water. The sprinklers contain a fusible element that releases water upon being heated to its designed activation temperature. Standard spray sprinklers are designed to discharge water in a semihemispherical pattern such that all portions of a space are protected where the sprinklers are located in accordance with NFPA 13. See Figure 21.2 for an example of a wet-pipe sprinkler system.

Wet-pipe sprinkler systems can be used in any space that is maintained at a temperature of 40°F (4°C) or higher. These systems are used in most buildings and are the preferred type of system except in the following circumstances:

- The occupant or owner prefers an even higher degree of protection from accidental water flow that could result from a sprinkler or pipe that is damaged (e.g., an MRI equipment room).
- The hazard is so severe that a deluge system is required to discharge the system's full flow upon detection (e.g., power transformers).

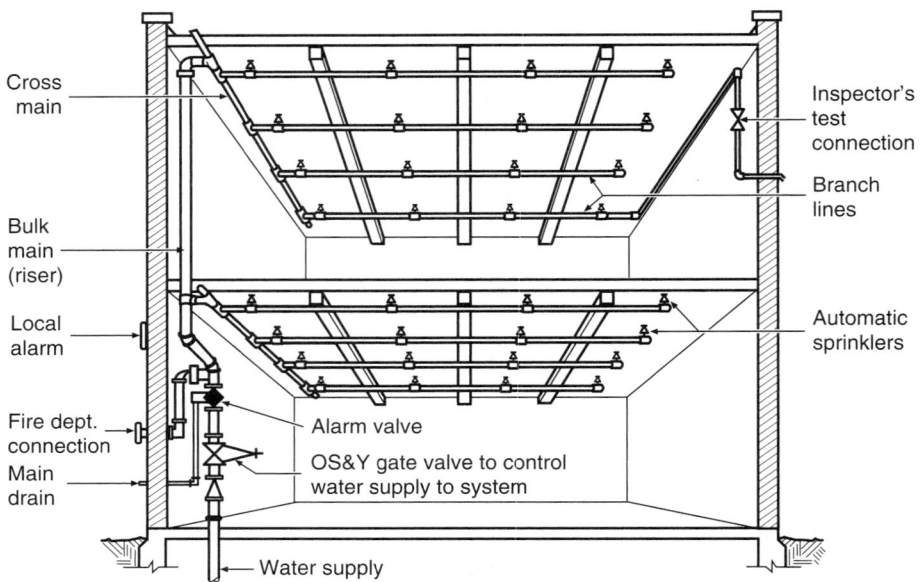

Figure 21.2 Wet-pipe sprinkler system. The system is under water pressure at all times so that water will discharge immediately when an automatic sprinkler operates. The automatic alarm valve causes a warning sound (bell) when water flows through the sprinkler piping.
Source: Fire Protection Handbook, NFPA, 2003, Figure 10.11.8.

For many years, *alarm check valves* (see Figure 21.3) were used to detect water flow from wet-pipe sprinkler systems. As their name indicates, these valves not only detected water flow in the system, they also acted as check valves to restrict flow "backwards" toward the supply. The check valve function allowed the fire department to pump water into the sprinkler system through a fire department connection (FDC) without pressurizing the municipal water supply mains. Because the water supply mains supply the fire department pumpers, without a check valve the fire department would effectively be pumping the water in a loop. The check valves also kept the piping on the system side of the valve at the highest pressure over time as the municipal system pressure fluctuated, thereby decreasing the number of false alarms.

Before modern fire alarm systems, alarm check valves were connected to water motor gongs (a water powered bell) mounted on the exterior of buildings, which alerted people outside a building that the sprinkler system was flowing. Figure 21.4 illustrates a water motor gong. Fire alarm systems can also be connected electrically to an alarm check valve through a pressure switch that annunciates a flow condition. Due to the cost, size, and maintenance requirements of alarm check valves, their use in new installations has dwindled in favor of newer premanufactured riser-type control valves combined with vane-type flow switches connected to a fire alarm system. This arrangement is typically lower in cost and smaller, requires less maintenance, and is more reliable (Figure 21.5).

Chapter 21. Automatic Fire Sprinkler Systems **383**

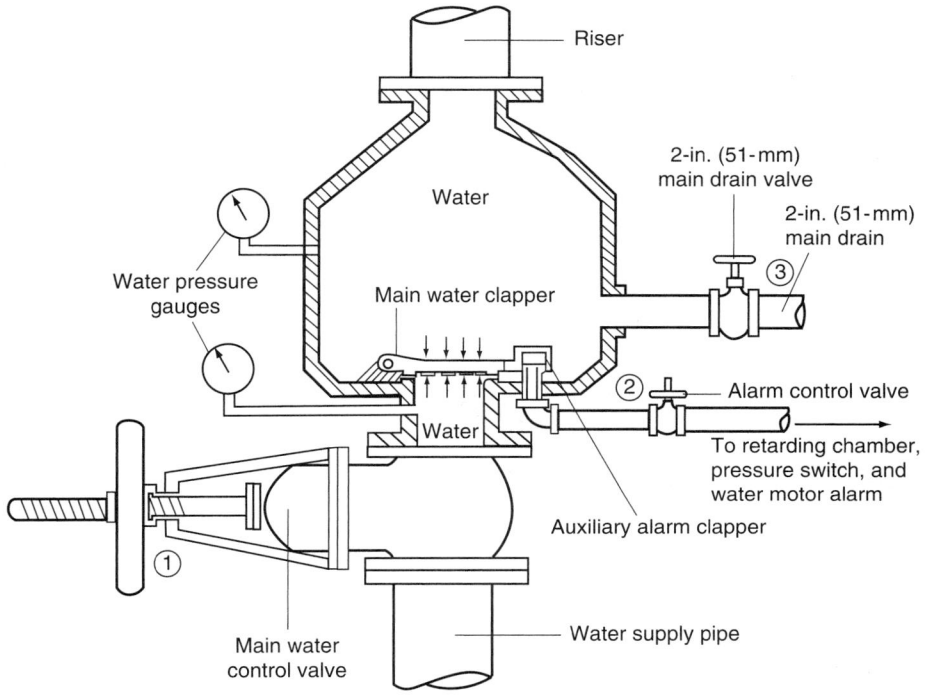

Note: Valves 1 and 2 are normally open.   Close valve 1 to shut off water.
Valve 3 is normally closed.   Close valve 2 to silence alarm.
Open valve 3 to drain system.

Figure 21.3 Section view of a typical alarm check valve for a wet-pipe sprinkler system.
Source: Fire Protection Systems: Inspection, Test & Maintenance Manual, by Wayne G. Carson and Richard L. Klinker, NFPA, 2000, Figure 2.2.

21.4.2 Dry-Pipe Sprinkler Systems

The second most popular type of sprinkler system found in health care facilities is the dry-pipe type of system. Dry-pipe sprinkler systems are much like the wet-pipe type of sprinkler system, except that the piping is filled with air (or nitrogen) instead of water. The air in the piping is compressed, and this air pressure holds a special dry-pipe valve closed against the pressure of water supplied to the bottom of the valve (usually from the municipal water supply). Dry-pipe sprinkler systems are used in areas subject to freezing, such as open parking structures, loading docks, and large industrial freezers. NFPA 13 requires that sprinkler system piping be protected from freezing wherever temperatures cannot be maintained at 40°F (4°C) or higher.

Dry-pipe valves are designed with a differential clapper valve that gives the system air pressure the mechanical advantage to hold the clapper closed against higher water supply pressures. Figure 21.6 is a section view of a typical dry-pipe valve. When one or more sprinklers in a dry-pipe system fuse due to a fire, the air in the piping discharges through the sprinkler(s) orifice until the air pressure in the piping decreases. At some

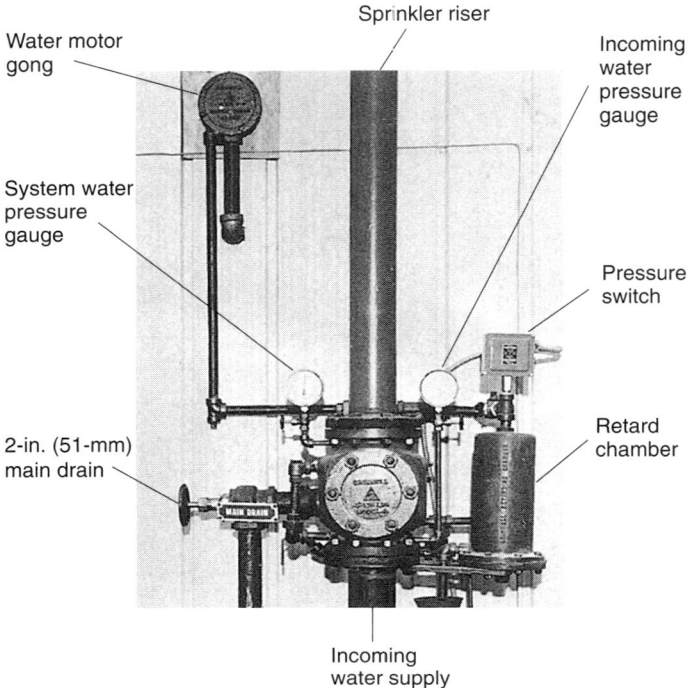

Figure 21.4 Fire alarm system using a water motor gong.
Source: Fire Protection Systems: Inspection, Test & Maintenance Manual, by Wayne G. Carson and Richard L. Klinker, NFPA, 2000, Figure 2.3.

Figure 21.5 Ready-Riser® riser check valve.
Source: Courtesy of Grinnell Fire Protection Systems Co., Inc.

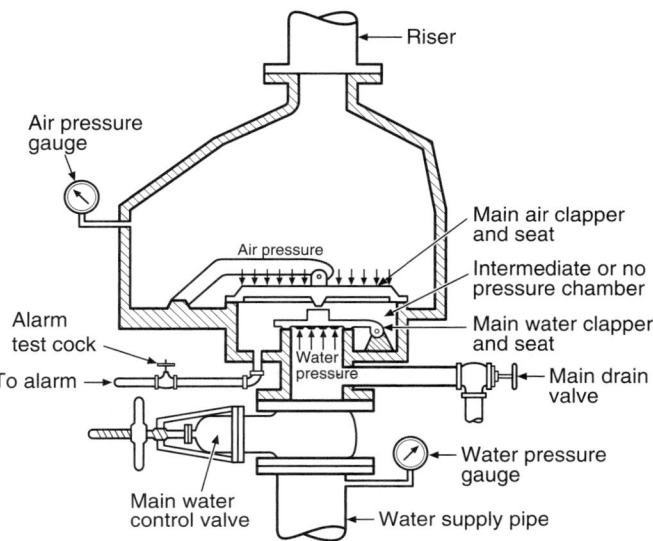

Figure 21.6 Typical dry-pipe valve.
Source: *Fire Protection Handbook,* NFPA, 2003, Figure 10.11.9.

point, the water pressure overcomes the air pressure and opens the clapper valve, flooding the piping with water. Because of the time required to lower (or bleed off) the air pressure in the piping, dry-pipe systems are slower than wet-pipe systems in delivering water to a fire.

Because of this slower response, hydraulic calculations for dry-pipe systems are required to account specifically for the expected larger number of sprinklers that may operate due to a fire. As a result, dry-pipe systems are generally not used where the protected space is not subject to freezing temperatures. Like the alarm check valve, the dry-pipe valve provides a means to connect water motor gongs and pressure switches to initiate a water flow alarm. Additional pressure switches are provided to supervise the air pressure for both high and low (nonflow) conditions in case of an air compressor malfunction or failure of the piping/fittings.

Many dry-pipe systems in buildings with fire pumps are supplied water from the municipal water supply and not the pump because the dry-pipe systems are usually in lower portions of the building (e.g., loading docks), where the additional water pressure is not necessary. Also, such an arrangement avoids the possibility of the fire pump's start-up pressure causing the dry-pipe valve(s) to falsely trip.

21.4.3 Antifreeze Sprinkler Systems

An antifreeze system is a wet-pipe system in which the piping system attached to the automatic sprinklers contains an antifreeze solution and is connected to a water supply. Upon operation of those sprinklers that have opened due to heat from a fire, the antifreeze solution discharges immediately, followed by water. Like dry-pipe systems, antifreeze systems are intended for areas that are subject to freezing, but they are not economically feasible for systems that contain more than 20 to 30 sprinklers, due to the high level of maintenance required (e.g., mixing of antifreeze solution and checking freeze point with a hydrometer).

Antifreeze systems can be part of larger standard wet-pipe systems because the *antifreeze loop,* which is a specific piping arrangement located at the point where the antifreeze solution is added to the system, is designed to keep the antifreeze from permeating the wet-pipe portion of the system. When contemplating the installation of an antifreeze system, three major considerations are as follows:

- The local water purveyor may require special backflow prevention because of the antifreeze solution.
- Where the connection between the antifreeze solution and the wet-pipe system incorporates a backflow prevention device, a listed expansion chamber of appropriate size and precharged air pressure shall be provided to compensate for thermal expansion of the antifreeze solution.
- The required antifreeze solution is dependent on whether the water supply connection is potable or nonpotable.

Never use commercially available automobile antifreeze because it will clog the sprinkler system. Refer to NFPA 13 for specific guidance on acceptable antifreeze solutions.

21.4.4 Preaction Sprinkler Systems

Preaction sprinkler systems are a lot like their dry-pipe counterparts because they are filled with air as opposed to water. The major difference between the two types of systems is that in preaction sprinkler systems, the valve that controls the flow of water to the piping system opens upon a signal from a detection system located within the protected space. The air in the preaction system can be at atmospheric pressure, or it may be maintained at a slightly higher pressure. This "slight" pressurization provides a means to supervise the integrity of the piping network. If a leak develops in a pressurized system, a pressure switch configured to sense low pressure can send a signal to the fire alarm control panel indicating a problem.

After the detection system signals the control valve to operate, the valve opens and floods the piping network with water. Water discharges from the sprinklers if a fire in the protected space is large enough to fuse one or more sprinklers. Preaction systems usually do not respond as slowly as dry-pipe systems because early detection, usually from smoke detectors, often results in a flooded piping network before a fire fuses a sprinkler.

21.4.5 Double-Interlock Preaction Sprinkler Systems

Double-interlock preaction systems are the same as standard preaction systems except that two separate and distinct events must occur before the piping floods with water. The detection system must achieve an alarm condition and a sprinkler must fuse, bleeding off the air pressure in the sprinkler piping. Only after both of these events occur will the control mechanism open the valve that allows water to fill the piping and discharge from the fused sprinkler(s). Double-interlock sprinkler systems are usually installed where there is an extremely low tolerance to accidental water discharge (e.g., clean rooms or rooms holding high-value art collections). This protection from accidental water damage comes at the expense of the sprinkler system's slower response time due to the two actions required to open the supply valve and at a higher installed cost. Double-interlock preaction systems are not common in health care settings.

21.4.6 Combined Dry-Pipe–Preaction Sprinkler Systems

In a combined dry-pipe–preaction system, the piping system attached to automatic sprinklers contains air under pressure and has a supplemental detection system installed in the same areas as the sprinklers. Activation of this detection system actuates tripping devices that open dry-pipe valves simultaneously and without loss of air pressure in the system. Operation of the detection system also opens listed air exhaust valves at the end of the feed main, which usually precedes the opening of the sprinklers. The detection system also serves as an automatic fire alarm system. Like double-lock preaction systems, combined dry-pipe–preaction systems are not common in health care settings.

21.4.7 Deluge Sprinkler Systems

A deluge sprinkler system is similar to a preaction sprinkler system except that all the sprinklers are *open* (i.e., they have no fusible elements or other operating mechanisms that stop water from discharging from the sprinkler). It is the only system in which all the sprinklers operate at once. When a detection system located in an area protected by a deluge system automatically signals the control valve to open, water floods the piping network and discharges from all of the sprinklers simultaneously. Because water flows from all deluge system sprinklers at once, the detection mechanism is usually robust and provided with some means to verify the detection of a fire before opening the deluge valve. Activation by a manual pull station, however, is not typical except at the deluge valve location.

These systems are used to protect high-challenge hazards such as power transformers, aircraft hangars, fuel loading platforms, and conveyor belt openings in fire walls. Their use in health care facilities is unlikely, with the exception of protecting externally located power transformers.

21.5 Design Considerations for New Construction and Retrofits

Although the responsibility for designing a sprinkler system rests mostly with the fire protection engineer and contractor, a high level of involvement by the health care facility's engineering or maintenance staff can greatly enhance the final installation of the system. The spaces being protected, their associated hazards, and their functions have a profound effect on a system's design. In many cases, one system configuration may be acceptable, but another may be more suitable for the facility personnel because of the way the space is to be used. If the system designer does not know about facility functions that are specific to a space, then the system most likely will not accommodate those functions in the best manner possible. This is especially true in a health care facility, where there are a large number of specialized spaces, equipment, and procedures programmed throughout the building.

In a retrofit situation, the designer needs building use information from the facility personnel who work in areas where the use is expected to change. If you want a particular style of sprinkler or one with a particular finish, or if you want the sprinklers located in the center of ceiling tiles, this information must be communicated to the designer. The designer should ask questions regarding these types of owner requirements, but if he does not ask, the facility most likely will not receive such specific treatment. Facility personnel who carefully interface with system designers can create a more suitable system before it is installed. Such exchange of information may be accomplished by completion of the Owner's Information Certificate (Figure 21.7).

Sprinkler systems are often modified after the initial installation because of changes either outside or within the health care facility. The system should be evaluated and modified in the following circumstances:

- When the hazard to be protected has changed to a more hazardous or challenging type
- When the space configuration has changed, such as in a renovation, where walls, ceilings, and doors are moved, added, deleted, or modified
- When the water supply has degraded or changed in some fashion
- When fire alarm or smoke zones have changed

Owner's Information Certificate

Name/Address of property to be protected with sprinkler protection:

Name of Owner:

Existing or planned construction is:

❑ Fire resistive or noncombustible

❑ Wood frame or ordinary (masonry walls with wood beams)

❑ Unknown

Is the system installation intended for one of the following special occupancies:

Aircraft hangar	❑ Yes	❑ No
Fixed guideway transit system	❑ Yes	❑ No
Race track stable	❑ Yes	❑ No
Marine terminal, pier, or wharf	❑ Yes	❑ No
Airport terminal	❑ Yes	❑ No
Aircraft engine test facility	❑ Yes	❑ No
Power plant	❑ Yes	❑ No
Water-cooling tower	❑ Yes	❑ No

If the answer to any of the above is "yes," the appropriate NFPA standard should be referenced for sprinkler density/area criteria.

Indicate whether any of the following special materials are intended to be present:

Flammable or combustible liquids	❑ Yes	❑ No
Aerosol products	❑ Yes	❑ No
Nitrate film	❑ Yes	❑ No
Pyroxylin plastic	❑ Yes	❑ No
Compressed or liquefied gas cylinders	❑ Yes	❑ No
Liquid or solid oxidizers	❑ Yes	❑ No
Organic peroxide formulations	❑ Yes	❑ No
Idle pallets	❑ Yes	❑ No

If the answer to any of the above is "yes," describe type, location, arrangement, and intended maximum quantities.

(NFPA 13, 1 of 2)

Figure 21.7 Owner's Information Certificate.

```
Indicate whether the protection is intended for one of the following specialized occupancies or areas:
    Spray area or mixing room                              ❏ Yes    ❏ No
    Solvent extraction                                     ❏ Yes    ❏ No
    Laboratory using chemicals                             ❏ Yes    ❏ No
    Oxygen-fuel gas system for welding or cutting          ❏ Yes    ❏ No
    Acetylene cylinder charging                            ❏ Yes    ❏ No
    Production or use of compressed or liquefied gases     ❏ Yes    ❏ No
    Commercial cooking operation                           ❏ Yes    ❏ No
    Class A hyperbaric chamber                             ❏ Yes    ❏ No
    Cleanroom                                              ❏ Yes    ❏ No
    Incinerator or waste handling system                   ❏ Yes    ❏ No
    Linen handling system                                  ❏ Yes    ❏ No
    Industrial furnace                                     ❏ Yes    ❏ No
    Water-cooling tower                                    ❏ Yes    ❏ No

If the answer to any of the above is "yes," describe type, location, arrangement, and intended maximum quantities.
_____

Will there be any storage of products over 12 ft (3.6 m) in height?
                                       ❏ Yes    ❏ No
If the answer is "yes," describe product, intended storage arrangement, and height.  _____
_____

Will there be any storage of plastic, rubber, or similar products over 5 ft (1.5 m) high except as described above?
                                       ❏ Yes    ❏ No
If the answer is "yes," describe product, intended storage arrangement, and height.  _____
_____

I certify that I have knowledge of the intended use of the property and that the above information is correct.

Signature of owner's representative or agent:   _____
Date:  _____

Name of owner's representative or agent completing certificate (print):  _____
Relationship and firm of agent (print):  _____

                                                                    (NFPA 13, 2 of 2)
```

Figure 21.7 Cont'd

Depending on the local jurisdiction, modifications to sprinkler systems can be categorized as minor or major and may be defined by the cost or number of sprinklers affected by the associated work. Some jurisdictions consider minor modifications as those involving five or fewer sprinklers. All such work usually requires the local building or fire department to issue a permit, but not with the same level of plan review and permit costs as larger modifications. Some jurisdictions keep track of the total number of sprinklers that have been modified in minor projects to prevent a facility from conducting a large project as a series of many smaller modifications. The facility should check with local officials to determine specific requirements before proceeding with modification projects.

Other important considerations when designing a new system or modifying an existing one include aesthetics of the sprinklers and piping, ceiling type/configuration, the supply of water, interface with the fire alarm systems, and location of the control valves. In health care facilities in particular, another design consideration is possible interference between the sprinkler system and patient-room privacy curtains.

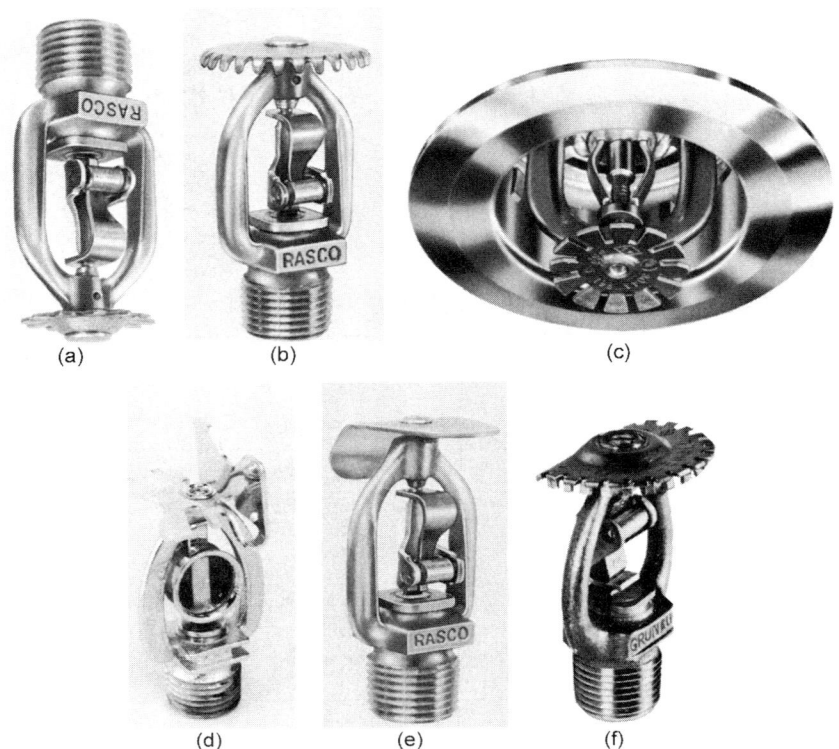

Figure 21.8 (a) Pendent and (b) upright sprinklers, two models of the same issue; (c) a Central model H recessed sprinkler; (d), (e), (f) a representative selection of listed sidewall sprinklers showing various deflector shapes.
Source: Fire Protection Handbook, NFPA, 2003, Figures 10.10.8, 10.10.12, and 10.10.20.

21.5.1 Aesthetics

The aesthetics of the sprinklers and piping are usually very important to the management and staff of the health care facility and should be discussed with the system designers.

21.5.1.1 Sprinklers

The style, configuration, finish, and location of the sprinklers (slang: sprinkler heads) are very important. Sprinklers are available in many styles, such as flush, semi-recessed, and concealed. Sprinkler configurations include pendent, upright, and sidewall (Figure 21.8). The number of available sprinkler finishes has increased greatly in recent years, in response to owners' requests. In addition to the many different colors and textures available, certain sprinklers may be factory painted to match adjacent surfaces. Note that sprinklers must never be painted except at the factory, by the manufacturer. Sprinklers with field-applied paint (purposely applied or by accident) must be replaced with new sprinklers.

The general location of sprinklers is dictated by the spacing requirements of NFPA 13, but specific locations within these requirements are the choice of the system designer. Locating pendent sprinklers in the center of ceiling tiles is desirable in many

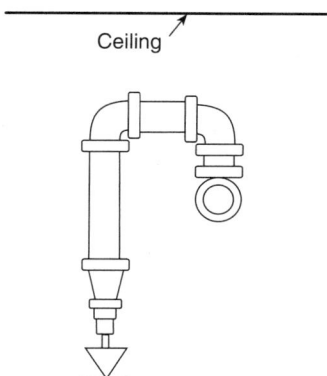

Figure 21.9 Return bend (slang: gooseneck) arrangement for sprinkler system.
Source: NFPA 13, 2002, Figure 8.14.18.2.

spaces where aesthetics are of great importance, such as main lobbies, conference rooms, and other assembly spaces. It is important to note that a system installation project where the sprinklers are "center of tile" has major cost and time implications. Sprinkler contractors typically charge considerably more for center-of-tile installations because of several additional tasks they must conduct:

- Spend more time coordinating with the ceiling contractor (new building)
- Document the existing ceiling location (existing building) including lights; heating, ventilating, and air-conditioning (HVAC) diffusers; speakers; convex mirrors; ceiling fans; and so on. Facilities that do not have reflected ceiling drawings of their existing ceilings must procure such documents.
- Provide sprinklers on return bends (slang: goosenecks) or some other method of fine adjustment (Figure 21.9)
- Install more sprinklers (based on standard ceiling tile sizes; maximum allowable NFPA 13 sprinkler spacing cannot be utilized)
- Revisit each sprinkler location to check and make a final adjustment to the sprinkler's position

The point here is not to discourage the specification of center-of-tile installations but to make facility personnel aware that this design requires more time, coordination, and money. Without careful planning, many projects have resulted in more expensive center-of-tile installations in back-of-house spaces where aesthetics are not important, which wastes valuable health care dollars. In the early stages of a design project, the facility should meet with the design engineer to identify the style, configuration, and finish and those spaces where a center-of-tile design is desirable.

21.5.1.2 Piping

The aesthetics of a sprinkler system installation depend greatly on the concealment of the piping network. In health care facilities, there are always portions of the system that are exposed, such as riser piping and control valve piping in the stairways, piping in mechanical rooms and attic spaces, and so on. Where the piping is exposed, facility managers should consider whether it is desirable to have the contractor paint the piping, and if so, they must select the color and number of coats and the primer to be applied. It is usually desirable to conceal system piping in most occupied spaces.

In new construction, most system piping is concealed above ceilings, which requires coordination between the sprinkler contractor and other trades (e.g., HVAC, electrical, plumbing) such that all these systems fit in the space above the ceiling. In existing construction where the system is to be retrofitted, the sprinkler contractor must identify space above the ceilings where the piping can be installed. The difficulty of this effort varies throughout a facility, depending on existing conditions. Facility personnel should make sure that the installation of the sprinkler piping does not significantly obstruct the future access to equipment that requires maintenance, such as HVAC dampers and lighting fixtures.

In areas that have plaster or gypsum ceilings, access to the ceiling space may be difficult and usually requires some partial demolition. The design engineer usually has no idea how piping should be routed in these spaces and must rely on the installation contractor to access the space before these decisions can be made. Once plaster or gypsum ceilings are breached, the facility should consider the installation of access doors or panels in the ceiling before it is repaired and patched. Older concealed-spline and friction-fit metal tile ceilings are difficult to remove and replace to their existing condition. With these ceilings, the facility may want to consider a ceiling upgrade to a modern lay-in tile ceiling as part of the sprinkler upgrade.

Where there is no ceiling or no access above the ceiling, or where a ceiling space cannot be disturbed (i.e., it contains asbestos or lead), the piping may be installed exposed and then painted to match the adjacent surfaces, or it may be installed within a soffit. Soffits may be custom built on the job site out of studs and gypsum, or they may be premanufactured (Figure 21.10).

Figure 21.10 Sprinkler system installation concealed within a Soffi-Steel premanufactured enclosure.
Source: Courtesy of Grice Engineering and Schirmer Engineering Corporation.

21.5.2 Water Supply

The water supply available to a health care facility has a major impact on the cost of a sprinkler system and should be one of the first things a fire protection engineer evaluates. Most health care facility water supplies come from a municipal water purveyor, with the water delivered under pressure through underground piping. Large campus facilities may even supplement the municipal supply with on-site elevated water storage tanks. The question the system designer must answer for the facility is whether the existing water source is adequate to supply the new (or modified) sprinkler system.

One of the first concerns of a fire protection engineer is to determine the hazards that must be protected within the health care facility. Second, the engineer must determine which hazard requires the most hydraulically demanding system, which may or may not be the most physically remote system. Such determinations are usually made by reviewing building plans, surveying the facility, and preparing an estimate of the new system's water demand. Hydraulic calculations (estimates or actual) are prepared starting from the most hydraulically remote area of the system, back through the piping network to the water source. This demand is usually expressed as the flow and pressure required at the *base of the riser*. The actual location of the base of the riser must be identified by the fire protection engineer. Many times, the base of the riser for the system is the point where a fire pump is installed (if required), the point where the water supply enters the building, or the point where the underground street main has been tapped to supply the building. The base of the riser may be located anywhere in the system, however, as long as the fire protection engineer identifies its position and provides documentation of the piping network and any assumptions.

Next, the demand for water must be compared to the available water supply, which can be determined by contacting the local water purveyor, which may have recent test results on file. Since the water supply can change over time, tests that are more than 1 year old should not be used. If more recent data are necessary, a flow test must be conducted. The test may be conducted on the municipal supply system, an on-site system of underground mains, or at a fire pump if available. Water purveyors usually conduct tests of municipal water supply systems for a fee (most purveyors do not allow anyone to open fire hydrants except themselves and the fire department). With the assistance of a contractor, the engineer can test an on-site underground system of water mains or a fire pump, if the facility has such equipment. In any water supply test, a large volume of water is discharged, so coordination with all affected parties both on and off site is essential. Where appropriate, a 24-hour test of the water system's pressure should be conducted.

At this point in the design, the engineer is aware of a system demand that can be compared to an available water supply. The results of this comparison can significantly affect the cost of the sprinkler system if the existing supply is not adequate, because most of the solutions are costly. Solutions include increasing the size of the supply main to the building, adding a second supply main to the building, adding a fire pump to the building (see Chapter 23), and relocating the hydraulically most demanding hazard to a level or area closer to the incoming water supply. Another part of the "cost" of these solutions is the time it takes to implement them. These solutions all involve materials that usually require long lead times in terms of ordering and delivery. If underground piping modifications are required, the excavation may be affected by seasonal temperatures and must be coordinated with other facility operations. Fire pumps involve costs for the pump itself, space for the pump installation, electrical power supply or diesel fuel, emergency power supply, connection to the fire alarm system, a test header location, and future maintenance.

21.5.3 Interface with the Fire Alarm System

All sprinkler systems containing 20 or more sprinklers must have a means to alarm water flow conditions, and in most cases the fire alarm system must supervise the position of control valves electrically. The sprinkler and fire alarm systems should be coordinated in terms of their zones. Although these two systems are not required to have zones that coincide, coordination does streamline the delivery of critical information to the facility staff and responding fire department. Fire protection zones should coincide with smoke zones in patient or resident areas; in non–smoke zone portions of a facility, they should subdivide floors (if necessary) along some break in building areas or function (e.g., along a wall).

In new construction, the design engineer(s) should coordinate the sprinkler and fire alarm systems. As conditions change during the construction process, the installation contractors should continue with this coordination. Proper coordination results in fire alarm systems that contain the appropriate number of alarm points to accurately annunciate all required sprinkler water flow and supervisory alarms. If the design is not accomplished by a single engineer, the facility should make sure that the various designers communicate on a regular basis.

In a retrofit situation, the number of additional sprinkler system water flow and valve supervisory switches that are needed may be greater than the available spare zones on the existing fire alarm control panel. Even if there are enough spare fire alarm zones, the addition of the new sprinkler alarm devices may use all spare fire alarm system capacity, such that it cannot be expanded further to accommodate future building modifications. In such cases, the fire alarm system must be evaluated as to its expansion capability, age, and availability of parts and service. Often, when fire alarm systems are not capable of needed expansion, a small alarm panel can be installed to monitor the sprinkler system, which can be interfaced with the existing alarm system to annunciate any outputs. If the existing fire alarm system is old or problematic, the facility may want to consider a fire alarm system upgrade as part of the sprinkler upgrade.

21.5.4 Location and Marking of Control Valves

The location of system valves is vital to the expeditious control of the system, because control of these systems is usually performed during a water flow condition or during maintenance. In either case, time must not be wasted searching for the location of the appropriate valve for that portion of the system. With concurrence of the facility personnel, the designer should determine control valve locations so that the valves are easily accessed and the locations make sense in relation to the area protected by the zone. Facility personnel are responsible for directing others to the locations of these valves, including the fire department, maintenance contractors, and facility inspectors [e.g., for the Joint Commission on Accreditation of Healthcare Organizations (JCAHO)].

A typical control valve location is within a fire-rated stair enclosure (see Figure 21.11).

Chapter 21. Automatic Fire Sprinkler Systems **395**

Figure 21.11 Control valve located in a fire-rated enclosure.
Source: Courtesy of Thomas W. Gardner and Schirmer Engineering Corporation.

Stair enclosures provide a protected location for the fire department to access the valves during a fire. Valves located within the same stair or stairs for every floor level of the structure also help to eliminate confusion as to their location. Since valves are not required to be within stair enclosures, they are sometimes located above ceiling tiles, in mechanical rooms, and in closets. No matter where they are located, the following are extremely important:

- Drawings on file with the building facility personnel that indicate the locations of all control valves
- Signs permanently affixed to the valves that indicate the zone controlled by that valve
- Signs on ceiling tiles or doors where control valves are located above a ceiling or within a room
- Special signs on the control valves where the associated zone is supplied from more than one valve. These special signs should indicate the zone controlled by that valve and the location of the *other* control valve that also supplies that zone

21.5.5 Privacy or Cubical Curtains

Health care facilities utilize curtains in non–private patient or resident rooms to provide privacy during examinations and other procedures. The issue of these curtains blocking sprinkler spray patterns has been evaluated and successfully addressed for years. NFPA 13 provides guidance on the allowable distances between sprinklers and obstructions such as curtains. Where the allowable criteria are not provided, additional sprinklers are required. The distances given in NFPA 13 were determined through tests in which privacy curtains with either a solid fabric or close mesh [¼ in. (6.4 mm)] top panel were installed. For broader-mesh top panels [e.g., ½ in. (13 mm)], the obstruction of the sprinkler spray is not likely to be severe and the authority having jurisdiction may not need to apply these spacing requirements.

The Centers for Medicare and Medicaid Services (CMS) [formerly the Health Care Financing Administration] has a written policy that older privacy curtains that do not provide the broader-mesh ½-in. (13-mm) top panels may remain in use and be replaced with the approved type as part of the normal life-cycle replacement program for the curtain. This policy is based on fire risk analysis and the fact that in fires that are large enough to activate a sprinkler [commonly rated for 165°F (74°C)], the curtain or its curtain support system might sometimes drop to the floor, thereby no longer obstructing the sprinkler spray.

21.6 General Requirements for Inspection, Testing, and Maintenance

Since the invention of the automatic sprinkler, system detractors have argued that accidental sprinkler operation would cause so much damage that system installation is not prudent. Both laboratory and actual fires have proved the detractors wrong: The chances of a fire sprinkler going off accidentally are extremely remote. Sprinkler components and systems are among the most tested systems in an average building.

Although sprinkler system failures are very rare, when failures do occur, they are usually the result of improper design, installation, or maintenance. Therefore, to avoid problems, health care facilities should carefully select a qualified fire protection engineer to design the sprinkler system and a qualified contractor to perform the installation, and they should be committed to proper system maintenance.

Several codes and standards detail the requirements for inspection, testing, and maintenance in terms of system components that require this work, testing frequency, and personnel responsible for conducting these tasks.

21.6.1 Applicable Codes

Owners of health care facilities are legally responsible for maintaining sprinkler systems and keeping the systems in good operating condition. Sprinkler systems installed in accordance with NFPA 13 must be properly inspected, tested, and maintained in accordance with NFPA 25, *Standard for the Inspection, Testing, and Maintenance of Water-Based Fire Protection Systems*, to provide at least the same level of performance and protection as originally designed. NFPA 25 also governs related systems, including underground piping, fire pumps, storage tanks, water spray systems, and foam–water sprinkler systems. NFPA 25 provides instruction on how to conduct inspection, testing, and maintenance activities. It also stipulates how often such activities must be completed.

Requirements are provided for impairment procedures, notification processes, and system restoration. Incorporating this type of information into a building maintenance program can enhance the demonstrated success of the water-based fire protection system.

21.6.2 Frequency of Inspection, Testing, and Maintenance

The required frequency of any inspection, testing, and maintenance program is a function of the system component in question. Table 21.2 is a summary of all inspection,

Table 21.2 Summary of NFPA 25 References for Sprinkler System Inspection, Testing, and Maintenance

Item	Activity	Frequency	NFPA 25 Reference	Performed by
Gauges (dry, preaction, and deluge systems)	Inspection	Weekly/monthly	5.2.4.2, 5.2.4.3	Facility staff
Control valves	Inspection	Weekly/monthly	Table 12.1	Facility staff
Alarm devices	Inspection	Quarterly	5.2.6	Facility staff
Gauges (wet-pipe systems)	Inspection	Monthly	5.2.4.1	Facility staff
Hydraulic nameplate	Inspection	Quarterly	5.2.7	Facility staff
Buildings	Inspection	Annually (prior to freezing weather)	5.2.5	Facility staff
Hanger/seismic bracing	Inspection	Annually	5.2.3	Facility staff
Pipe and fittings	Inspection	Annually	5.2.2	Facility staff
Sprinklers	Inspection	Annually	5.2.1	Facility staff
Spare sprinklers	Inspection	Annually	5.2.1.3	Facility staff
Fire department connections	Inspection	Quarterly	Table 12.1	Facility staff
Valves (all types)	Inspection		Table 12.1	Facility staff
Alarm devices	Test	Quarterly/semiannually	5.3.3	Facility staff
Main drain	Test	Annually	Table 12.1	Sprinkler contractor
Antifreeze solution	Test	Annually	5.3.4	Sprinkler contractor
Gauges	Test	5 years	5.3.2	Sprinkler contractor
Sprinklers—extra-high temperature	Test	5 years	5.3.1.1.1.3	Sprinkler contractor
Sprinklers—fast response	Test	At 20 years and every 10 years thereafter	5.3.1.1.1.2	Sprinkler contractor
Sprinklers	Test	At 50 years and every 10 years thereafter	5.3.1.1.1	Sprinkler contractor
Valves (all types)	Maintenance	Annually or as needed	Table 12.1	Sprinkler contractor
Obstruction investigation	Maintenance	5 years or as needed	13.2.1, 13.2.2	Sprinkler contractor
Low-point drains (dry-pipe system)	Maintenance	Annually prior to freezing and as needed	12.4.4.3.3	Sprinkler contractor

Source: Adapted from NFPA 25, 2002, Table 5.1.

testing, and maintenance requirements, with an indication of the section of NFPA 25 that details the requirements and who can perform such functions.

21.6.3 Inspection, Testing, and Maintenance Personnel

Sprinkler system inspections are performed by various persons; the health care facilities engineering staff, sprinkler contractors, and authorities having jurisdiction. Qualified health care facility engineering staff typically performs all of the monthly and weekly inspections indicated in Table 21.2, including inspection of the fire department connection and all valves. The purpose of these inspections is to check visually that a particular device is indeed still in place and not damaged or otherwise impaired and that control valves are open and drain valves are closed.

The activities that require testing or maintenance of the system are best performed by a qualified sprinkler contractor. All health care facilities should have a testing and maintenance agreement with a sprinkler contractor requiring compliance with NFPA 25. In negotiating such an agreement, the health care facility should review the details of the contractor's scope of work to determine which activities are to be performed by the facility staff and which activities will be contracted. The contract should require that a current copy of NFPA 25 (not just the latest edition, but the edition that is current for the local jurisdiction) be provided to the facility at the commencement of the contract and whenever a newer edition is locally adopted. See the applicable edition of NFPA 25 for specific information on the required inspections, testing, and maintenance including frequency and procedures.

21.7 Inspection of Sprinkler Systems

Various system components, including sprinklers, pipes and fittings, gauges, building alarm devices, hydraulic nameplate, and hose connections, must be inspected periodically.

21.7.1 Inspection of Sprinklers

Sprinklers, sprinkler pipes and fittings, pipe hangers, and seismic braces must be visually inspected annually. Sprinklers must be installed in the proper orientation (e.g., upright, pendent, or sidewall) and free of corrosion, foreign materials, paint, and physical damage. Any sprinkler that is painted, corroded, damaged, loaded (covered with dirt, dust, or lint), or in an improper orientation must be replaced. Any obstructions to the sprinkler spray patterns that have been identified must be corrected. These inspections are important because the conditions described can have a detrimental effect on the performance of sprinklers by affecting water distribution patterns, insulating thermal elements, delaying operation, or otherwise rendering the sprinkler inoperable or ineffectual.

Obstructions to spray patterns include horizontal obstructions near the ceiling, vertical obstructions, suspended or floor-mounted obstructions, and low clearances between sprinklers and storage below. The clearance requirement between sprinkler deflectors and the top of storage is typically 18 in. (457 mm). Specific guidance for clearance and obstructions is found in NFPA 13 and other standards and specific sprinkler listings (see the sprinkler manufacturer's technical data sheets or "catalog cuts"). Additional information on system obstructions is provided later.

Health care facilities must keep a spare supply of sprinklers on site, which must be inspected annually for the proper number and type of sprinklers and for the presence of one sprinkler wrench for each type of sprinkler installed in the facility.

21.7.2 Inspection of Pipes and Fittings

Qualified health care facility staff should look to see if pipes and fittings are in good condition and free of mechanical damage, leakage, corrosion, and misalignment. Sprinkler piping must not be subjected to external loads by materials either resting on the pipe or hung from the pipe. Hangers and seismic braces (in localities requiring such devices) that are damaged or loose must be replaced or refastened.

Sprinkler system components installed in concealed spaces do not require inspection. Examples of such spaces include areas above suspended ceilings, some floor/ceiling and roof/ceiling assemblies, areas under theater stages, pipe chases, and other inaccessible areas. Sprinkler components installed in areas that are inaccessible for safety considerations due to operations in that area must be inspected during scheduled shutdowns. In a health care facility, conducting this inspection may require coordination with staff to access sensitive areas while they are not in use, such as operating rooms, imaging laboratories, and darkrooms.

Over time, all pipes are subject to corrosion. Where inspection of pipes reveals evidence of potential internal or external corrosion that could lead to internal obstructions or weakening of the pipe, the affected sections of the pipe should be replaced.

One particular type of corrosion called microbiologically influenced corrosion (MIC) has been a serious problem in certain areas, depending upon the local water supply. The following is an excerpt from NFPA 25 on MIC:

> Microbiologically Influenced Corrosion (MIC). The most common biological growths in sprinkler system piping are those formed by microorganisms, including bacteria and fungi. These microbes produce colonies (also called biofilms, slimes) containing a variety of types of microbes. Colonies form on the surface of wetted pipe in both wet and dry systems. Microbes also deposit iron, manganese, and various salts onto the pipe surface, forming discrete deposits (also termed nodules, tubercles, and carbuncles). These deposits can cause obstruction to flow and dislodge, causing plugging of fire sprinkler components. Subsequent under-deposit pitting can also result in pinhole leaks.
>
> Microbiologically influenced corrosion (MIC) is corrosion influenced by the presence and activities of microorganisms. MIC almost always occurs with other forms of corrosion (oxygen corrosion, crevice corrosion, and under-deposit corrosion). MIC starts as microbial communities (also called biofilms, slimes) growing on the interior surface of the wetted sprinkler piping components in both wet and dry systems. The microbial communities contain many types of microbes, including slime formers, acid-producing bacteria, iron-depositing bacteria, and sulfate-reducing bacteria, and are most often introduced into the sprinkler system from the water source. The microbes deposit iron, manganese, and various salts onto the pipe surface, forming discrete deposits (also termed nodules, tubercles, and carbuncles). These deposits can cause obstruction to flow and dislodge, causing plugging of fire sprinkler components. MIC is most often seen as severe pitting corrosion occurring under

deposits. Pitting is due to microbial activities such as acid production, oxygen consumption, and accumulation of salts. Oxygen and salts, especially chloride, can greatly increase the severity of MIC and other forms of corrosion.

In steel pipe, MIC is most often seen as deposits on the interior surface of the pipes. The deposits may be orange, red, brown, black, and white (or combinations thereof), depending on local conditions and water chemistry. The brown, orange, and red forms are most common in oxygenated portions of the system and often contain oxidized forms of iron and other materials on the outside, with reduced (blacker) corrosion products on the inside. Black deposits are most often in smaller diameter piping farther from the water source and contain reduced forms (those with less oxygen) of corrosion products. White deposits often contain carbonate scales.

MIC of copper and copper alloys occurs as discrete deposits of smaller size, which are green to blue in color. Blue slimes may also be produced in copper piping or copper components (e.g., brass heads).

MIC is often first noticed as a result of pinhole leaks after only months to a few years of service. Initial tests for the presence of MIC should involve on-site testing for microbes and chemical species (iron, pH, oxygen) important in MIC. This information is also very important in choosing treatment methods. These tests can be done on water samples from source waters and various locations in the sprinkler system (e.g., main drain, inspector's test valve). Confirmation of MIC can be made by examination of interior of pipes for deposits and under-deposit corrosion with pit morphology consistent with MIC (cup-like pits within pits and striations).

The occurrence and severity of MIC is enhanced by the following:

(1) Using untreated water to test and fill sprinkler piping. This is made worse by leaving the water in the system for long periods of time.
(2) Introduction of new and untreated water containing oxygen, microbes, salts, and nutrients into the system on a frequent basis (during repair, renovation, and/or frequent flow tests).
(3) Leaving dirt, debris, and especially oils, pipe joint compound, and so forth in the piping. These provide nutrients and protection for the microbes, often preventing biocides and corrosion inhibitors from reaching the microbes and corrosion sites.

Once the presence of MIC has been confirmed, the system should be assessed to determine the extent and severity of MIC. Severely affected portions should be replaced or cleaned to remove obstructions and pipe not meeting minimal mechanical specifications.(25–02: D.2.6)

21.7.3 Inspection of Gauges

Gauges on wet-pipe sprinkler systems must be inspected monthly to ensure that they are in good condition and that normal water supply pressure is being maintained. On dry, preaction, and deluge systems, the gauges must be inspected weekly to ensure that normal air and water pressures are being maintained. If the air pressure is supervised at a constantly attended location (e.g., security station or central service station), then the inspection of the gauges can be relaxed to a monthly interval. The fire alarm system often provides such supervision at these attended locations.

21.7.4 Inspection of Buildings

Annually, before the onset of freezing weather, buildings with wet-pipe systems should be inspected to verify that windows, skylights, doors, ventilators, other openings and closures, blind spaces, unused attics, stair towers, roof houses, and low spaces under buildings do not expose water-filled sprinkler piping to freezing and to verify that adequate heat [minimum 40°F (4.4°C)] is available.

21.7.5 Inspection of Alarm Devices

Alarm devices must be inspected quarterly to verify that they are free of physical damage.

21.7.6 Inspection of the Hydraulic Nameplate

The hydraulic nameplate, if provided, must be inspected quarterly to verify that it is attached securely to the sprinkler riser and is legible (see Figure 21.12 for an example of a hydraulic nameplate). The hydraulic nameplate should be secured to the riser with durable wire, chain, or some other equivalent method.

```
This system as shown on . . . . . . . . . . . . . . . . . . . . . . company

print no . . . . . . . . . . . . . . . . . . . . . . . . . dated . . . . . . . . . . . . . .

for . . . . . . . . . . . . . . . . . . . . . . . . . . . . . . . . . . . . . . . . . . . . . . . . . . .

at . . . . . . . . . . . . . . . . . . . . . . . . . . . . . . . contract no . . . . . . . . .

is designed to discharge at a rate of . . . . . . . . . . . . . . gpm/ft²

(L/min/m²) of floor area over a maximum area of . . . . . . . . . .

ft² (m²) when supplied with water at a rate of . . . . . . . . . . . . .

gpm (L/min) at . . . . . . . . . . . psi (bar) at the base of the riser.

Hose stream allowance of  . . . . . . . . . . . . . . . . . . gpm (L/min)

is included in the above.

Occupancy classification . . . . . . . . . . . . . . . . . . . . . . . . . . . . . . . .

Commodity classification . . . . . . . . . . . . . . . . . . . . . . . . . . . . . . . .

Maximum storage height . . . . . . . . . . . . . . . . . . . . . . . . . . . . . . . .
```

Figure 21.12 A hydraulic nameplate.
Source: NFPA 13, 2002, Figure A.16.5.

21.7.7 Inspection of Hose Connections

Where they are provided, hose, hose couplings, and nozzles that are connected to the sprinkler system must be inspected annually in accordance with NFPA 1962, *Standard for the Inspection, Care, and Use of Fire Hose, Couplings, and Nozzles and the Service Testing of Fire Hose*.

21.8 Testing of Sprinkler Systems

Sprinkler systems are tested for a variety of reasons at various periods. Installation contractors normally conduct performance tests on small portions of sprinkler systems as they install these systems before moving on to other areas of the building. AHJs normally conduct acceptance tests of sprinkler systems before accepting them as complete and operational. Owners (themselves or through maintenance contractors) conduct routine sprinkler system tests throughout the life of the system to ensure that the systems are in proper operating order. Refer to Chapter 35 for more information on system testing.

21.8.1 Performance Testing of Sprinkler Systems

Performance testing of a sprinkler system by definition is a test to determine if the installation is ready for further testing (acceptance testing) under supervision of an AHJ. Performance testing includes hydrostatic testing of the piping network. The purpose of the hydrostatic test is to demonstrate the integrity of the piping network (i.e., to ensure that there are no leaks). The hydrostatic test is accomplished by pressurizing water in the system to 200 psi (13.8 bar) [or 50 psi (3.4 bar) above the highest anticipated system pressure, whichever is higher] for 2 hours while inspectors observe the entire system, checking for leaks at all joints. Of course, the system piping must be exposed during this test, so ceiling tiles, wallboard, or soffits must be open. Incidentally, these enclosures should not be replaced until successfully completing a hydrostatic *acceptance* test for the AHJ (performance testing is not acceptance testing).

In addition to hydrostatic testing, pneumatic testing of the piping network would be required for dry-pipe systems and certain preaction systems. These systems must be pressure-tested using both air and water because they are subjected to pressures from both. Differences in the physical properties between air and water make it impossible to check for system leakage based on only one of these tests. The pneumatic test is specified in NFPA 13.

Additional performance testing efforts should include full-flow testing of each inspector's test connection to demonstrate that each water flow switch works and that the fire alarm system reports the switch as being in the correct zone. A secondary benefit of this test is that it demonstrates the appropriateness of the drain discharge, which must accept the full water flow under pressure. Valve supervisory switches (slang: tamper switches) must also be tested to ensure that their zoning is correct and, most important, that they activate a supervisory alarm when a valve is "off-normal." The off-normal signal must be obtained during the first two revolutions of the valve's hand wheel or during one-fifth of the travel distance of the valve control apparatus from its normal position.

It is also beneficial to activate the fire alarm–indicating devices (e.g., bells, horns, chimes, speakers) upon activation of the first water flow switch. However, the alarm-indicating devices should be silenced for subsequent flow tests so as not to desensitize building occupants to the importance of the alarm signal (i.e., in an attempt to avoid the "boy who cried wolf" syndrome). Once testing has established that the alarm-indicating devices actually work when the alarm system initiates them, they can be silenced at the fire alarm control panel, and subsequent flow tests may be considered successful if the alarm system "calls" for them to sound. This will also reduce the number of complaints about the testing that will be lodged with the facilities engineering staff from other staff, patients, and occupants.

A system's main drain [often 2 in. (50 mm) in diameter] should be tested for sufficient pressure. To give the inspector an indication of whether the underground water supply valves are fully open, the system's main drain can be opened. A dramatic drop in gauge pressure during full flow of a main drain test may indicate a weak water supply or a partially closed supply valve. After the main drain valve is closed, the gauge pressure should quickly return to the normal static pressure; if it does not, this may also indicate a partially closed valve. Many times partially closed valves are located underground (curb-box valves) in the main water supply piping. Be aware that some older valves are left-hand turn valves and may actually be closed by persons who think that they are opening the valve.

21.8.2 Acceptance Testing of Sprinkler Systems

Acceptance tests by definition are performed to demonstrate that a system is properly installed so that it can be approved by an AJH. Prior performance testing increases the chance of a successful acceptance test. Facility personnel should secure the required Contractor's Material and Test Certificate(s) from the installer (see Figures 21.13 and 21.14). It is important to coordinate hydrostatic testing with the AHJ, sprinkler contractor, and facility personnel to minimize problems and delays. If the sprinkler system is retrofitted into an occupied facility, coordination with the facility staff or occupants is essential. Again, the piping must be exposed during this test.

Acceptance testing of all water flow and valve supervisory switches requires coordination with the fire alarm contractor. Most AHJs measure the time between the start of water flow and the initiation of the alarm, and many require a water flow switch to activate the fire alarm system between 30 and 60 seconds after the initiation of water flow. Water flow switches have adjustment mechanisms to delay the activation of alarm signals to avoid their initiation as a result of pressure surges in the water supply.

Supervisory switches attached to OS&Y-type control valves often must be adjusted to indicate the valve's status correctly. Sometimes just operating the valve to test the supervisory switch can result in a switch that is out of adjustment. Facility personnel must make sure that all supervisory switches are adjusted correctly before the sprinkler contractor leaves the site. Butterfly control valves approved for fire protection service have integral supervisory switches and are not susceptible to such adjustment problems.

Contractor's Material and Test Certificate for Aboveground Piping

PROCEDURE

Upon completion of work, inspection and tests shall be made by the contractor's representative and witnessed by an owner's representative. All defects shall be corrected and system left in service before contractor's personnel finally leave the job.

A certificate shall be filled out and signed by both representatives. Copies shall be prepared for approving authorities, owners, and contractor. It is understood the owner's representative's signature in no way prejudices any claim against contractor for faulty material, poor workmanship, or failure to comply with approving authority's requirements or local ordinances.

Property name		Date	
Property address			

Plans	Accepted by approving authorities (names)			
	Address			
	Installation conforms to accepted plans	☐ Yes	☐ No	
	Equipment used is approved	☐ Yes	☐ No	
	If no, explain deviations			

Instruction	Has person in charge of fire equipment been instructed as to location of control valves and care and maintenance of this new equipment? If no, explain	☐ Yes	☐ No
	Have copies of the following been left on the premises?	☐ Yes	☐ No
	1. System components instructions	☐ Yes	☐ No
	2. Care and maintenance instructions	☐ Yes	☐ No
	3. NFPA 25	☐ Yes	☐ No

Location of system	Supplies buildings

Sprinklers	Make	Model	Year of manufacture	Orifice size	Quantity	Temperature rating

Pipe and fittings	Type of pipe _____ Type of fittings _____

Alarm valve or flow indicator	Alarm device			Maximum time to operate through test connection	
	Type	Make	Model	Minutes	Seconds

Dry pipe operating test	Dry valve			Q. O. D.		
	Make	Model	Serial no.	Make	Model	Serial no.

	Time to trip through test connection[1,2]		Water pressure	Air pressure	Trip point air pressure	Time water reached test outlet[1,2]		Alarm operated properly	
	Minutes	Seconds	psi	psi	psi	Minutes	Seconds	Yes	No
Without Q.O.D.									
With Q.O.D.									
If no, explain									

[1] Measured from time inspector's test connection is opened
[2] NFPA 13 only requires the 60-second limitation in specific sections

Figure 21.13 Contractor's Material and Test Certificate for Aboveground Piping.
Source: NFPA 13, 2002, Figure 16.1.

Deluge and preaction valves	**Operation:** ☐ Pneumatic ☐ Electric ☐ Hydraulics **Piping supervised:** ☐ Yes ☐ No **Detecting media supervised:** ☐ Yes ☐ No **Does valve operate from the manual trip, remote, or both control stations?** ☐ Yes ☐ No **Is there an accessible facility in each circuit for testing?** ☐ Yes ☐ No **If no, explain** Make	Model	Does each circuit operate supervision loss alarm? Yes/No	Does each circuit operate valve release? Yes/No	Maximum time to operate release: Minutes / Seconds
Pressure reducing valve test	Location and floor | Make and model | Setting | Static pressure — Inlet (psi) / Outlet (psi) | Residual pressure (flowing) — Inlet (psi) / Outlet (psi) | Flow rate — Flow (gpm)
Test description	**Hydrostatic:** Hydrostatic tests shall be made at not less than 200 psi (13.6 bar) for 2 hours or 50 psi (3.4 bar) above static pressure in excess of 150 psi (10.2 bar) for 2 hours. Differential dry-pipe valve clappers shall be left open during the test to prevent damage. All aboveground piping leakage shall be stopped. **Pneumatic:** Establish 40 psi (2.7 bar) air pressure and measure drop, which shall not exceed 1½ psi (0.1 bar) in 24 hours. Test pressure tanks at normal water level and air pressure and measure air pressure drop, which shall not exceed 1½ psi (0.1 bar) in 24 hours.
Tests	All piping hydrostatically tested at ____ psi (____ bar) for ____ hours If no, state reason Dry piping pneumatically tested ☐ Yes ☐ No Equipment operates properly ☐ Yes ☐ No Do you certify as the sprinkler contractor that additives and corrosive chemicals, sodium silicate or derivatives of sodium silicate, brine, or other corrosive chemicals were not used for testing systems or stopping leaks? ☐ Yes ☐ No Drain test | Reading of gauge located near water supply test connection: ____ psi (____ bar) | Residual pressure with valve in test connection open wide: ____ psi (____ bar) Underground mains and lead-in connections to system risers flushed before connection made to sprinkler piping Verified by copy of the (U Form No.85B) ☐ Yes ☐ No Other Explain Flushed by installer of underground sprinkler piping Sprinkler piping ☐ Yes ☐ No If powder-driven fasteners are used in concrete, has representative sample testing been satisfactorily completed? ☐ Yes ☐ No If no, explain
Blank testing gaskets	Number used | Locations | Number removed
Welding	Welding piping ☐ Yes ☐ No If yes. . . Do you certify as the sprinkler contractor that welding procedures comply with the requirements of at least AWS B2.1? ☐ Yes ☐ No Do you certify that the welding was performed by welders qualified in compliance with the requirements of at least AWS B2.1? ☐ Yes ☐ No Do you certify that the welding was carried out in compliance with a documented quality control procedure to ensure that all discs are retrieved, that openings in piping are smooth, that slag and other welding residue are removed, and that the internal diameters of piping are not penetrated? ☐ Yes ☐ No
Cutouts (discs)	Do you certify that you have a control feature to ensure that all cutouts (discs) are retrieved? ☐ Yes ☐ No

Figure 21.13 Cont'd

Hydraulic data nameplate	Nameplate provided ☐ Yes ☐ No	If no, explain
Remarks	Date left in service with all control valves open	
Signatures	Name of sprinkler contractor		
	Tests witnessed by		
	For property owner (signed)	Title	Date
	For sprinkler contractor (signed)	Title	Date
Additional explanations and notes			

Figure 21.13 Cont'd

21.8.3 Routine Testing

In accordance with NFPA 25, several components of automatic sprinkler systems are tested on a routine basis.

21.8.3.1 Routine Testing of Sprinklers

The requirements for the routine testing of sprinklers vary depending on the age of the sprinkler:

- Sprinklers manufactured before 1920 must be replaced.
- Sprinklers that have been in service for 75 years must be replaced, or representative samples from one or more sample areas must be submitted to a recognized testing laboratory acceptable to the authority having jurisdiction for field service testing. Test procedures shall be repeated at 5-year intervals.
- Sprinklers that have been in service for 50 years must be replaced, or representative samples from one or more sample areas must be submitted to a recognized testing laboratory acceptable to the authority having jurisdiction for field service testing. Test procedures must be repeated at 10-year intervals.
- Sprinklers manufactured using fast-response elements that have been in service for 20 years must be tested and retested at 10-year intervals.
- Representative samples of solder-type sprinklers with a temperature classification of extra high [325°F (163°C)] or greater and exposed to semicontinuous or continuous maximum allowable ambient temperature conditions must be tested at 5-year intervals.

A representative sample of sprinklers consists of a minimum of not less than four sprinklers or 1 percent of the number of sprinklers per individual sprinkler sample, whichever is greater. Where one sprinkler within a representative sample fails to meet the test requirement, all sprinklers represented by that sample must be replaced.

Contractor's Material and Test Certificate for Underground Piping

PROCEDURE
Upon completion of work, inspection and tests shall be made by the contractor's representative and witnessed by an owner's representative. All defects shall be corrected and system left in service before contractor's personnel finally leave the job.

A certificate shall be filled out and signed by both representatives. Copies shall be prepared for approving authorities, owners, and contractor. It is understood the owner's representative's signature in no way prejudices any claim against contractor for faulty material, poor workmanship, or failure to comply with approving authority's requirements or local ordinances.

Property name		Date	
Property address			

Plans	Accepted by approving authorities (names)		
	Address		
	Installation conforms to accepted plans	☐ Yes	☐ No
	Equipment used is approved	☐ Yes	☐ No
	If no, state deviations		
Instructions	Has person in charge of fire equipment been instructed as to location of control valves and care and maintenance of this new equipment? If no, explain	☐ Yes	☐ No
	Have copies of appropriate instructions and care and maintenance charts been left on premises? If no, explain	☐ Yes	☐ No
Location	Supplies buildings		
Underground pipes and joints	Pipe types and class	Type joint	
	Pipe conforms to _____ standard	☐ Yes	☐ No
	Fittings conform to _____ standard	☐ Yes	☐ No
	If no, explain		
	Joints needed anchorage clamped, strapped, or blocked in accordance with _____ standard	☐ Yes	☐ No
	If no, explain		
Test description	Flushing: Flow the required rate until water is clear as indicated by no collection of foreign material in burlap bags at outlets such as hydrants and blow-offs. Flush at flows not less than 390 gpm (1,476 L/min) for 4-in. pipe, 880 gpm (3,331 L/min) for 6-in. pipe, 1,560 gpm (5,905 L/min) for 8-in. pipe, 2,440 gpm (9,235 L/min) for 10-in. pipe, and 3520 gpm (13,323 L/min) for 12-in. pipe. When supply cannot produce stipulated flow rates, obtain maximum available. Hydrostatic: Hydrostatic tests shall be made at not less than 200 psi (13.8 bar) for 2 hours or 50 psi (3.4 bar) above static pressure in excess of 150 psi (10.3 bar) for 2 hours. Leakage: New pipe laid with rubber gasketed joints shall, if the workmanship is satisfactory, have little or no leakage at the joints. The amount of leakage at the joints shall not exceed 2 quarts per hour (1.89 L/hr) per 100 joints irrespective of pipe diameter. The leakage shall be distributed over all joints. If such leakage occurs at a few joints, the installation shall be considered unsatisfactory and necessary repairs made. The amount of allowable leakage specified above can be increased by 1 fluid OZ/in valve diameter per hr. (30 mL/25 mm/hr) for each metal seated valve isolating the test section. If dry barrel hydrants are tested with the main valve open so the hydrants are under pressure, an additional 5 OZ/min (150 mL/min) leakage is permitted for each hydrant.		
Flushing tests	New underground piping flushed according to _____ standard by (company)	☐ Yes	☐ No
	If no, explain		
	How flushing flow was obtained: ☐ Public water ☐ Tank or reservoir ☐ Fire pump	Through what type opening: ☐ Hydrant butt ☐ Open pipe	
	Lead-ins flushed according to _____ standard by (company)	☐ Yes	☐ No
	If no, explain		
	How flushing flow was obtained: ☐ Public water ☐ Tank or reservoir ☐ Fire pump	Through what type opening: ☐ Y connection to flange and spigot ☐ Open pipe	

Figure 21.14 Contractor's Material and Test Certificate for Underground Piping.
Source: NFPA 13, 2002, Figure 10.10.1.

Hydrostatic test	All new underground piping hydrostatically tested at _____ psi for _____ hours		Joints covered ☐ Yes ☐ No
Leakage test	Total amount of leakage measured _____ gallons _____ hours		
	Allowable leakage _____ gallons _____ hours		
Hydrants	Number installed	Type and make	All operate satisfactorily ☐ Yes ☐ No
Control valves	Water control valves left wide open If no, state reason		☐ Yes ☐ No
	Hose threads of fire department connections and hydrants interchangeable with those of fire department answering alarm		☐ Yes ☐ No
Remarks	Date left in service		
Signatures	Name of installing contractor		
	Tests witnessed by		
	For property owner (signed)	Title	Date
	For installing contractor (signed)	Title	Date
Additional explanation and notes			

Figure 21.14 Cont'd

21.8.3.2 Routine Testing of Gauges

Gauges must be replaced every 5 years or tested every 5 years by comparison with a calibrated gauge. Gauges that are not accurate to within 3 percent of full scale must be recalibrated or replaced.

21.8.3.3 Routine Testing of Alarm Devices

Water flow alarm devices including, but not limited to, mechanical water motor gongs, vane-type waterflow devices, and pressure switches that provide audible or visual signals must be tested quarterly. Testing these devices should not consist of merely opening the cover over the water flow switch and manually moving it to the alarm position. The testing of water flow alarms on wet-pipe systems must be accomplished by opening the inspector's test connection and actually causing a water flow to initiate the alarm. NFPA 25 allows an exception to using the inspector's test connection where freezing weather conditions or other circumstances prohibit its use; in such cases, the bypass connection on alarm check valves may be used.

NFPA 25 does not recommend that fire pumps be turned off during alarm-device testing unless all impairment procedures contained in Chapter 11 of NFPA 25 are followed. If fire pumps are active during testing, it is advisable to disable any automatic shut-off timers to avoid a situation in which the pump could cycle on and off at the beginning and end of each test. Also, a small bypass connection in the pump room that discharges to a proper drain should be left open. This connection keeps water moving past the pump so that it will not overheat during periods of the tests when no other water is flowing.

21.8.3.4 Routine Testing of Antifreeze Systems

The freezing point of solutions in antifreeze systems must be tested annually by measuring the specific gravity with a hydrometer or refractometer and adjusting the solutions if necessary. The solutions used must be in accordance with NFPA 25, 2002 edition [Table 5.3.4.1(a), "Antifreeze Solutions to Be Used if Nonpotable Water Is Connected to Sprinklers"; and Table 5.3.4.1(b), "Antifreeze Solutions to Be Used if Potable Water Is Connected to Sprinklers"]. The health care facility engineer should make sure the maintenance or testing contractor uses the proper type of antifreeze solutions in accordance with all state or local health regulations.

21.8.3.5 Routine Testing of Hose Connections

Hose connections to sprinkler systems should be service tested in accordance with NFPA 1962 at least 5 years after initial installation and then every 3 years thereafter. After each service test, each hose connection must be flow tested to ensure that water discharges from the hose and a water flow alarm operates.

Having said this, it is important to note that hose connections to a sprinkler system may be intended for fire department use, fire brigade use, or occupant use. Since fire brigades are employed more often in industrial situations, the fire department use and occupant use is the most likely scenario in a health care setting. Occupant-use hose has been out of favor for quite some time with many members of the fire protection community. There is a concern that untrained occupants may not have the ability to safely use a 100-ft- (30.5-m-) long hose flowing up to 100 gpm (378 L/min). Their ability to use the hose may be further hampered if the drop in pressure caused by flowing the hose causes the fire pump to start, thereby significantly increasing the pressure delivered to the nozzle.

Additionally, others in this occupant-use hose debate have questioned the wisdom of encouraging occupants to fight a fire without protective clothing or breathing apparatus instead of evacuating immediately upon discovering a fire. Fire service objections have stemmed from the fact that they have no control over the condition of the preconnected hose and would rather depend on "standpipe packs" of their own hose that they carry into buildings. Preconnected hoses that are not properly maintained may burst under pressure during fire-fighting operations. Often the nozzles provided for these hoses are vandalized or missing. These concerns have led to a trend toward reducing the requirements for and eliminating the installation of systems with preconnected hoses. Some jurisdictions have gone so far as to ban the installation of occupant-use preconnected hose.

21.9 Sprinkler Maintenance

As with any building system, sprinkler systems require some level of maintenance. Maintenance must be performed to keep the system equipment operable or to make repairs. NFPA 25 requires facilities to retain as-built system installation (record) drawings, original acceptance test records, and device manufacturer's maintenance bulletins to assist in the proper care of the system and its components. There are three types of maintenance:

- *Preventive maintenance* includes, but is not limited to, lubricating control valve stems; bleeding moisture and condensation from air compressors, air lines, and dry-pipe system auxiliary drains; and cleaning strainers.
- *Corrective maintenance* includes, but is not limited to, replacing loaded, corroded, or painted sprinklers; replacing missing or loose pipe hangers; replacing valve seats and gaskets; and restoring heat in areas subject to freezing temperatures where water-filled piping is installed.
- *Emergency maintenance* includes, but is not limited to, repairs due to piping failures caused by freezing or impact damage; repairs to broken underground fire mains; and replacement of frozen or fused sprinklers, defective electric power, or alarm and detection system wiring.

21.9.1 Obstruction Investigation and Prevention

A major sprinkler system maintenance issue is preventing obstructions to the system's water flow. Many conditions require facilities to investigate a sprinkler system for obstructions, but often the onset of these conditions is not known to the sprinkler contractor who is providing the system maintenance. Since the owner is legally responsible for maintaining the system and keeping the system in good operating condition, it is the health care facility's obligation to bring these conditions to the attention of the maintenance contractor.

To ensure that piping remains clear of all obstructive foreign matter, an obstruction investigation must be conducted for system or yard main piping wherever any of the following conditions exists:

- Defective intake for fire pumps takes suction from open bodies of water.
- Routine water tests show the discharge of obstructive material.
- Foreign material is present in fire pumps, dry-pipe valves, or check valves.
- Foreign material is present in the water during drain tests or plugs inspector's test connection(s).
- Sprinklers are plugged.
- Plugged piping in sprinkler systems is dismantled during building alterations.
- Yard piping or surrounding public mains were not flushed following new installations or repairs.
- There is a record of broken public mains in the vicinity.
- One or more dry-pipe valves abnormally and frequently experiences false tripping.
- A system is returned to service after an extended shutdown (more than 1 year).
- There is reason to believe that the sprinkler system contains sodium silicate or highly corrosive fluxes in copper systems.
- A system has been supplied with raw water via the fire department connection.
- Pin hole leaks
- A 50 percent increase in the time it takes water to travel to the inspector's test connection from the time the valve trips during a full-flow trip test of a dry pipe system when compared to the original system acceptance test

Systems should always be examined internally for obstructions where conditions could cause obstructed piping. Systems where these conditions have not been corrected, or where a condition could result in obstructing piping despite any previous flushing procedures that have been performed, must be examined internally for obstructions every 5 years. Such investigations are accomplished by examining the interior of a dry valve or a preaction valve and by removing two cross-main flushing connections (see Figure 21.15).

If the obstruction investigation indicates the presence of sufficient material to obstruct sprinklers, a complete flushing program is necessary. In such cases, the health care facility should have the work conducted by a qualified sprinkler contractor. During the flushing program, if pipe conditions indicate internal or external corrosion that could lead to serious weakening of the pipe, the affected section of the pipe should be replaced. Pendent sprinklers should be removed and inspected until it is reasonably certain that all are free of obstructive material. Painting the ends of branch lines and cross mains is a convenient method for keeping a record of those pipes that have been flushed.

Dry-pipe or preaction sprinkler system piping that protects or passes through freezers or cold storage rooms should be visually inspected internally on an annual basis for ice obstructions at the point where the piping enters the refrigerated area. All penetrations into the cold storage areas should be inspected, and if an ice obstruction is found, additional pipe must be examined to ensure that no ice blockage exists.

21.9.2 In-House Maintenance

In-house maintenance of sprinkler systems should be performed only when the maintenance staff have been specifically trained to conduct such work. Qualified personnel can perform basic repairs, such as reconnecting hangers or bleeding condensation from air compressor lines, but in-house personnel usually do not have the qualifications or experience for more technical sprinkler system maintenance.

21.9.3 Contracted Maintenance

Outsourcing sprinkler system maintenance to a qualified sprinkler contractor provides a health care facility with several benefits. The benefits include the following:

- The maintenance personnel who work on a system are specialists in the field of sprinkler systems.
- Sprinkler contractors have the specialized tools to maintain the sprinkler system.

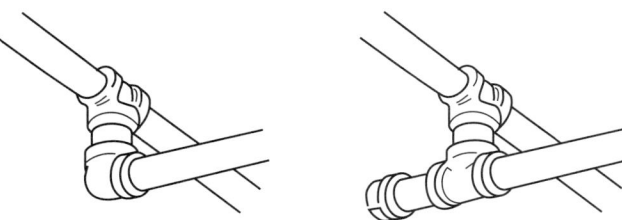

Figure 21.15 Replacement of elbow at end of cross main with a flushing connection consisting of a 2-in. (50-mm) nipple and cap.
Source: NFPA 25, 2002 Figure D.3.2(a).

- Sprinkler contractors have working relationships with the manufacturers and suppliers of the components or spare parts for valves, fittings, and attachments.
- The maintenance liability is shared by the sprinkler professional and the facility.
- The workload is transferred away from the health care facility's maintenance personnel.

When contracting sprinkler system maintenance, the health care facility should make sure that the contractor conducts this type of work regularly on a day-to-day basis. Many companies work on piping systems, and it is important to make sure that the selected company is a true sprinkler company. True sprinkler companies usually are members of one of the two sprinkler contracting associations in the United States: the American Fire Sprinkler Association and the National Fire Sprinkler Association.

22

Standpipe Systems

Standpipe systems are arrangements of piping, valves, hose connections, and allied equipment installed in tall or otherwise very large buildings or structures. The hose connections are located in such a manner that water can be discharged in streams or spray patterns through attached hoses and nozzles, which aids in the manual delivery of water to a fire. Without standpipe systems, it is extremely difficult and time consuming for fire fighters to extend hoses from fire engines into remote areas of these larger buildings. Standpipe systems allow fire fighters, industrial fire brigades, and in some cases occupants to attack a fire quickly with a manageable amount of hose attached to the standpipe system.

This chapter examines standpipe systems in health care facilities in terms of their history; where they are required; system classes and types; considerations for system design and modification; and inspection, testing, and maintenance requirements. Although this book details standpipe and sprinkler systems in separate chapters (similar to the separate NFPA standards for each system), most modern health care facilities have combined sprinkler/standpipe systems. It is more economical for a single riser in an enclosed stair to supply both the standpipe hose valves and the sprinkler system. Figure 22.1 shows a combination sprinkler/standpipe riser with a 2½-in. (65-mm) hose outlet and a sprinkler "floor control valve station."

22.1 History of Standpipes in Health Care Facilities

Standpipe systems have been installed in health care facilities for many years to provide hose valves in enclosed stairs and throughout selected occupied rooms and corridors. The valves in the rooms and corridors are usually recessed into a wall cavity and often enclosed in a cabinet with or without hose and a nozzle. Figure 22.2 shows a familiar fire hose assembly.

Over the years, members of the fire protection community have debated the value of standpipe systems that provide occupant-use hose. Many professionals in the field are concerned that untrained occupants cannot safely use a 100-ft- (30.5-m-) long hose that flows up to 100 gpm (378 L/min) of water at high pressures. Also, the fire protection community generally does not want to encourage occupants to fight a fire in lieu of evacuating a building, as illustrated by the following case study.

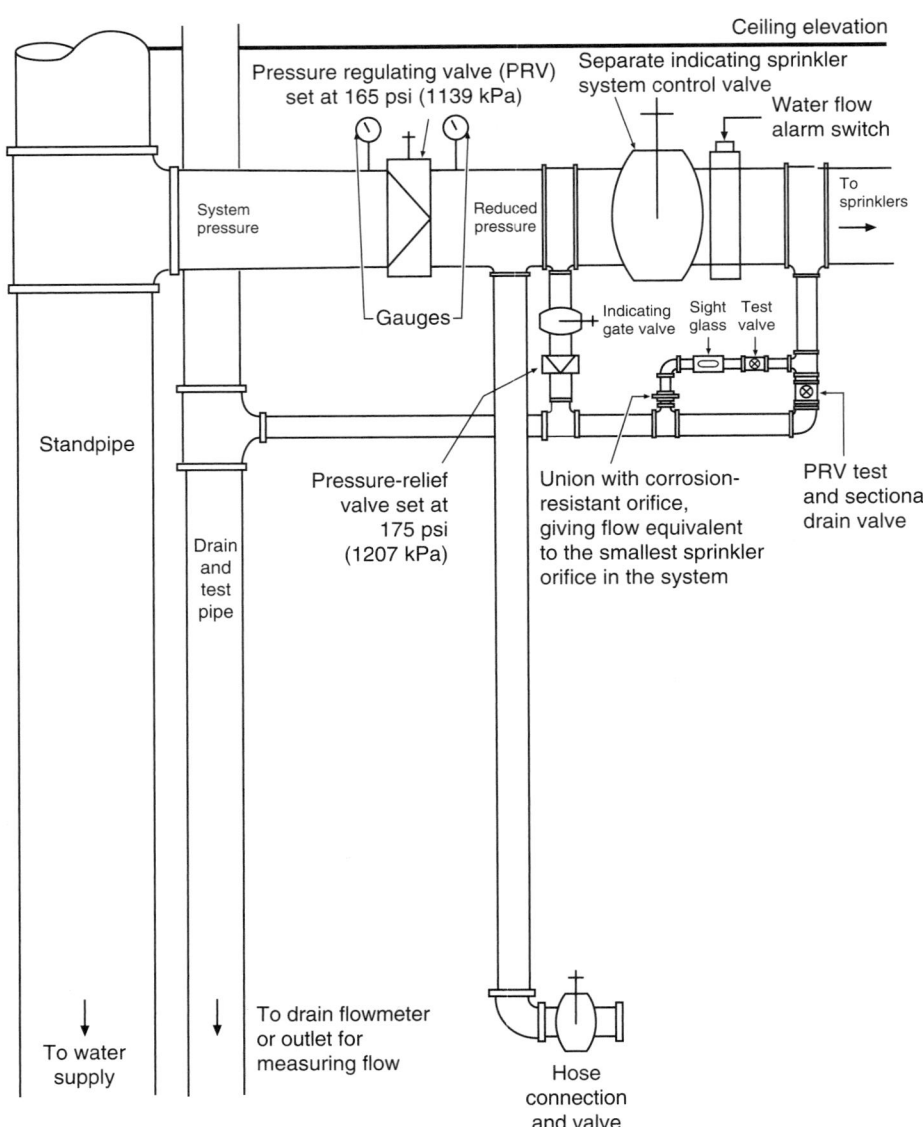

Figure 22.1 Valving at a floor connection for a combined standpipe sprinkler system. *Source: Fire Protection Handbook,* NFPA, 2003, Figure 10.18.9.

Difficulty with Occupant-Use Hose

During a fire in a hospital, an on-duty orderly, who was also a volunteer fire fighter, responded to a fire in a patient room. The orderly advanced an occupant-use hose to the fire and began an aggressive attack on the fire. The drop in water pressure resulting from the operation of the hose caused the fire pump to start automatically. The resulting increase in water pressure almost caused the orderly to lose control of

(Continued)

Figure 22.2 Fire hose assembly and valves.
Source: Health Care Facilities Handbook, NFPA, 1999, Exhibit 11.8.

the hose and nozzle. He remained pinned against the corridor wall holding the nozzle but was unable to continue to apply water to the fire.

Although the orderly was trained in handling such fire streams (as part of being a volunteer fire fighter), additional personnel were not present to assist him with handling the hose and nozzle. Also, he had no radio contact with others to communicate the need to control the pressure supplied to the standpipe system. Someone with less training than this volunteer fire fighter might have lost all control of the nozzle and might well have been injured or killed by it. In addition, there was extensive damage to the hospital as a result of water being applied to areas that were not involved in the fire.

As with fire extinguishers, it is always important to receive training on how to use standpipes and hoses that are capable of flowing high volumes of water.

For years, fire departments have not had faith that occupant-use hoses were adequately maintained. They would routinely disconnect such hose and connect their own "standpipe pack" hose to the hose outlet to fight a fire. Even in health care facilities that maintain their occupant-use hose properly, the possibility of vandalism could render it unusable. These concerns have led to a reduction in the requirements for occupant-use hose and the elimination altogether of some occupant-use hose installations on standpipe systems.

22.2 Standpipe System Requirements

Since 1915, NFPA has provided a standard (NFPA 14, *Standard for the Installation of Standpipe and Hose Systems*) for the installation of standpipe and hose systems. Substantive changes to the 1993 edition of this standard were made as a result of experience with standpipe systems under fire conditions. Flow rates, pressures, and the specific locations of the hose connections were studied for the 1993 edition to determine optimum combinations for each feature. The 1996 edition of NFPA 14 contained several additional changes, including those related to system testing. In the 2003 edition hydraulic calculation requirements have been rewritten for clarification, and requirements for horizontal standpipes have been added.

NFPA 101®, *Life Safety Code*®, requires all high-rise health care facilities to have standpipes. *NFPA 5000*®, *Building Construction and Safety Code*®, requires standpipe systems in buildings four or more stories in height, or having four or more basement levels. Other model building codes require standpipe systems to be installed in buildings that have a story located more than 30 ft (9 m) above the lowest level of fire department vehicle access [or a story located more than 30 ft (9 m) below the highest level of fire department vehicle access]. Requirements for standpipe systems are also dependent on whether a building is fully sprinklered.

For buildings under construction or demolition, most building codes require standpipes to be provided to the level below the highest level under construction or demolition. Even noncombustible structures present extreme fire hazards during construction because of the large amount of combustible concrete forms and shoring that are used and the various ignition sources that are present. The standpipes provided during construction may be temporary or permanent and do not necessarily require an automatic water supply. Fire department connections that supply standpipes must be clearly marked and constantly accessible, even during construction or demolition operations.

As always, health care facility engineers must check with local authorities for specific standpipe requirements (e.g., thread sizes of the connection to fire department hoses) and the applicable editions of required codes, standards, and amendments that must be followed.

22.3 Standpipe System Types

NFPA 14, 2000 edition, Chapter 5, categorizes standpipe systems according to the system's water supply arrangement. Standpipe systems include the following types:

- *Automatic-dry*—a dry standpipe system normally filled with pressurized air that, through the use of a device such as a dry-pipe valve, is arranged to admit water into the system piping automatically upon the opening of a hose valve. The water supply for an automatic-dry standpipe system is capable of supplying the system demand.

- *Automatic-wet*—a wet standpipe system that has a water supply capable of supplying the system demand automatically.
- *Semiautomatic-dry*—a dry standpipe system that, through the use of a device such as a deluge valve, is arranged to admit water into the system piping upon activation of a remote control device (similar to a fire alarm pull station) located at a hose connection. A remote-control activation device is provided at each hose connection. The water supply for a semiautomatic-dry standpipe system is capable of supplying the system demand.
- *Manual-dry*—a dry standpipe system that does not have a permanent water supply attached to the system. To supply the system demand, manual-dry standpipe systems need a fire department pumper (or the like) to pump water into the system through the fire department connection (i.e., a connection through which the fire department can pump water into the sprinkler system, standpipe, or other system, furnishing supplemental water for fire extinguishment in addition to existing water supplies).
- *Manual-wet*—a wet standpipe system connected to a small water supply that can maintain water within the system but does not have the capability to deliver the system demand attached to the system. To supply the system demand, manual-wet standpipe systems require a fire department pumper (or the like) to pump water into the system.
- *Combined standpipe and sprinkler system*—a system in which the water piping services both 2½-in. (63.5-mm) outlets for fire department use and outlets for automatic sprinklers.

22.4 Standpipe System Classes

According to NFPA 14, Chapter 5, standpipe systems are also divided into *classes* depending on the intended user of the system. The classes of standpipe systems are as follows:

- *Class I standpipe system*—provides 2½-in.- (65-mm-) diameter hose connections to supply water for use by fire departments and those trained in handling heavy fire streams (no hose is provided).
- *Class II standpipe system*—provides 1½-in.- (38.1-mm-) diameter hose stations to supply water for use primarily by trained personnel or by the fire department during the initial response.
- *Class III standpipe system*—provides 1½-in. (38.1-mm) hose stations to supply water for use by trained personnel and 2½-in. (65-mm) hose connections to supply a larger volume of water for use by fire departments and those trained in handling heavy fire streams.

22.5 Design Considerations

Although the responsibility for designing a standpipe system rests mostly with the fire protection engineer and contractor, a high level of involvement by the health care

418 Part IV. Fire Protection Systems

facility's engineering or maintenance staff can greatly enhance the final installation. If the system designer does not know about facility functions that are specific to a space, then the system most likely will not accommodate those functions in the best manner possible. Facility personnel who carefully interface with the system designers can improve the system before it is even installed.

22.5.1 Hose Valve Locations

Design considerations for hose valve locations include that they must be properly located and accessible at all times. Standpipes and hose valves in enclosed stairs must be located such that they are in the corner of the stair landing and a distance from the stair newel post that is greater than the required landing radius (see Figure 22.3). Installing standpipes and hose valves closer to the newel post than the required landing radius results in an impediment to egress.

In Class I systems, hose valves must be located as follows:

- At each intermediate landing between floor levels in every required exit stairway. [*Note:* This requirement is a major change that has been required since the 1996 edition of NFPA 14. Previously, hose valves needed to be installed at the main floor landings, but for fire department operational reasons, the location requirement has been changed to the intermediate landing. There is an exception that allows the hose connections to be located at the main floor landings in exit stairways where approved by the authority having jurisdiction (AHJ).]
- On each side of the wall adjacent to the exit openings of horizontal exits
- In each exit passageway at the entrance that leads from the building areas into the passageway
- At the highest landing of stairways in buildings with stairway access to a roof, and on the roof where stairways do not access the roof
- Where the most remote portion of a nonsprinklered floor or story is located in excess of 150 ft (45.7 m) of travel distance from a required exit containing or adjacent to a hose connection

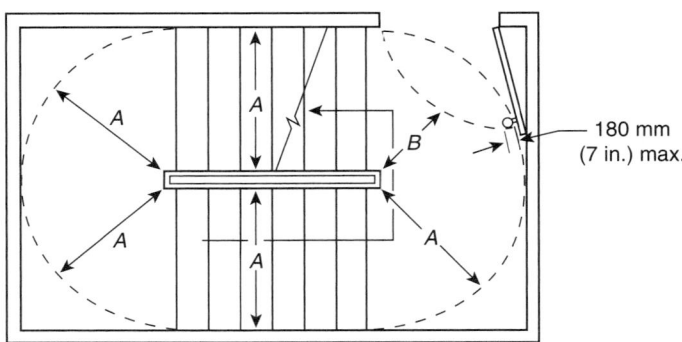

A — Required width
B — At least A/2

Figure 22.3 Enclosed exit stair housing hose valves, with minimum unobstructed clearance. *Source: Life Safety Code Handbook,* NFPA, 2003, Exhibit 7.17.

- Where the most remote portion of a sprinklered floor or story is located in excess of 200 ft (61 m) of travel distance from a required exit containing or adjacent to a hose connection
- Additional hose connections shall be provided in approved locations, where required by the local fire department or the authority having jurisdiction

In Class II systems, hose valves must be located so that all portions of each floor level of the building are within 130 ft (39.7 m) of a hose connection provided with 1½-in. (38.1-mm) hose or within 120 ft (36.6 m) of a hose connection provided with less than 1½-in. (38.1-mm) hose. Distances must be measured along a path of travel originating at the hose connection. In Class III systems, hose valves must be located as required for both Class I and Class II systems.

22.5.2 Aesthetics

The aesthetics of a standpipe system installation usually are not an issue in health care buildings because most of the piping network is installed in enclosed stairs and mechanical spaces. Where the piping is exposed, the facility must decide if the installation contractor should paint the piping and if so, what color the piping should be and the number of coats and primer that should be applied. In accordance with NFPA 14, Chapter 6, piping must be protected from mechanical damage.

22.6 Inspection, Testing, and Maintenance

A standpipe system installed in accordance with NFPA 14 must be properly inspected, tested, and maintained in accordance with NFPA 25, *Standard for the Inspection, Testing, and Maintenance of Water-Based Fire Protection Systems,* 2002 edition, so that it provides at least the same level of performance and protection as when the system was originally designed. NFPA 25 is the reference standard for inspection, testing, and maintenance of water-based fire protection systems, including standpipe systems (as well as sprinkler systems, underground piping, fire pumps, storage tanks, water spray systems, and foam–water sprinkler systems). The standard provides instructions on how to conduct inspection, testing, and maintenance activities and stipulates how often such activities must be completed. Requirements are provided for impairment procedures, notification processes, and system restoration. When incorporated into a building maintenance program, these activities enhance the favorable operation of all water-based fire protection systems.

The required frequency of any inspection, testing, and maintenance activity is a function of the system component in question. Table 22.1 is a summary of all inspection, testing, and maintenance requirements, with references to NFPA 25 sections that detail the requirements and who should perform such activities.

Health care facility owners are legally responsible for maintaining standpipe systems and keeping them in good operating condition. Health care facility engineering staff, sprinkler contractors, and AHJs all may be involved in inspecting the systems. The health care facility engineering staff usually conducts all of the monthly and weekly and most of the quarterly inspections shown in Table 22.1, including the inspection of the fire department connection and all valves. The activities that require testing or maintenance of the system are best performed by a qualified sprinkler contractor.

Table 22.1 Summary of Standpipe and Hose System Inspection, Testing, and Maintenance

Item	Activity	Frequency	NFPA 25 Reference	Performed by
Control valves	Inspection	Weekly/monthly	Table 6.1	Facility staff
Pressure regulating devices	Inspection	Quarterly	Table 6.1	Facility staff
Piping	Inspection	Quarterly	6.2.1	Facility staff
Hose connections	Inspection	Quarterly	Table 6.1	Facility staff
Cabinet	Inspection	Annually	NFPA 1962	Facility staff
Hose	Inspection	Annually	NFPA 1962	Sprinkler contractor
Hose storage device	Inspection	Annually	NFPA 1962	Sprinkler contractor
Alarm device	Test	Quarterly	Table 6.1	Sprinkler contractor
Hose nozzle	Test	Annually	NFPA 1962	Sprinkler contractor
Hose storage device	Test	Annually	NFPA 1962	Sprinkler contractor
Hose	Test	5 years/3 years	NFPA 1962	Sprinkler contractor
Pressure control valve	Test	5 years	Table 6.1	Sprinkler contractor
Pressure reducing valve	Test	5 years	Table 6.1	Sprinkler contractor
Hydrostatic test	Test	5 years	6.3.2	Sprinkler contractor
Flow test	Test	5 years	6.3.1	Sprinkler contractor
Main drain test	Test	Annually	Table 6.1	Sprinkler contractor
Hose connections	Maintenance	Annually		Sprinkler contractor
Valves (all types)	Maintenance	Annually/as needed	Table 6.1	Sprinkler contractor

Source: Adapted from NFPA 25, 2002, Table 6-1.

For systems that require compliance with NFPA 25, the facility should have a testing and maintenance agreement with a sprinkler contractor. In negotiating such an agreement, the health care facility should review the details of the contractor's scope of work to determine which activities the facility staff will perform and which activities will be contracted. The contract should require that a current copy of NFPA 25 (not just the latest edition of the code, but the edition that is current for the local jurisdiction) be provided to the facility at the beginning of the contract and whenever a newer edition is locally adopted. See the applicable edition of NFPA 25 for specific information on the required inspections, testing, and maintenance activities including frequency and procedures.

When the results of tests, inspections, and preventive maintenance activities reveal that a standpipe system is out of service, those responsible for making the repairs must follow procedures contained in Chapter 14 of NFPA 25. These procedures cover how to handle preplanned and emergency impairments for any water-based fire protection system including standpipe systems. Details of the equipment that must be used for restoring systems into service are also provided.

NFPA 25 (Table 6.2.2) contains all of the inspection, testing, and maintenance requirements for standpipe systems along with corrective actions for each component or checkpoint. The table is reproduced here as Table 22.2 for reference.

Table 22.2 Inspection, Testing, and Maintenance Requirements for Standpipe Systems, with Corrective Action

Component/Checkpoint	Corrective Action
Hose Connections	
Cap missing	Replace
Fire hose connection damaged	Repair
Valve handles missing	Replace
Cap gaskets missing or deteriorated	Replace
Valve leaking	Close or repair
Visible obstructions	Remove
Restricting device missing	Replace
Manual, semiautomatic, or dry standpipe—valve does not operate smoothly	Lubricate or repair
Piping	
Damaged piping	Repair
Control valves damaged	Repair or replace
Missing or damaged pipe support device	Repair or replace
Damaged supervisory devices	Repair or replace
Hose	
Inspect	Remove and inspect the hose, including gaskets, and rerack or rereel at intervals in accordance with NFPA 1962, *Standard for the Care, Use, and Service Testing of Fire Hose Including Couplings and Nozzles*
Mildew, cuts, abrasions, and deterioration evident	Replace with listed, lined, jacketed hose
Coupling damaged	Replace or repair
Gaskets missing or deteriorated	Replace
Incompatible threads on coupling	Replace or provide thread adapter
Hose not connected to hose rack nipple or valve	Connect
Hose test outdated	Retest or replace in accordance with NFPA 1962, *Standard for the Care, Use, and Service Testing of Fire Hose Including Couplings and Nozzles*
Hose Nozzle	
Hose nozzle missing	Replace with listed nozzle
Gasket missing or deteriorated	Replace
Obstructions	Remove
Nozzle does not operate smoothly	Repair or replace
Hose Storage Device	
Difficult to operate	Repair or replace
Damaged	Repair or replace
Obstruction	Remove
Hose improperly racked or rolled	Remove
Nozzle clip in place and nozzle correctly contained?	Replace if necessary

(Continued)

Table 22.2 Cont'd

Component/Checkpoint	Corrective Action
If enclosed in cabinet, will hose rack swing out at least 90 degrees?	Repair or remove any obstructions
Cabinet	
Check overall condition for corroded or damaged parts	Repair or replace parts; replace entire cabinet if necessary
Difficult to open	Repair
Cabinet door will not open fully	Repair or move obstructions
Door glazing cracked or broken	Replace
If cabinet is break-glass type, is lock functioning properly?	Repair or replace
Glass break device missing or not attached	Replace or attach
Not properly identified as containing fire equipment	Provide identification
Visible obstructions	Remove
All valves, hose, nozzles, fire extinguishers, etc. easily accessible	Remove any material not related

Source: NFPA 25, 2002, Table 6.2.2.

22.6.1 Inspecting Standpipe Systems

Per NFPA 25, components of standpipe systems must be visually inspected quarterly or as specified in Table 22.1. Standpipe systems must be free of corrosion, foreign materials, and physical damage and must be installed so that they are accessible at all times. Devices must be in place, free of foreign material and corrosion, and not damaged, otherwise impaired, or experiencing other conditions that could prevent operation. These inspections are important because the conditions described can have a detrimental effect on the performance of or the ability to use the standpipe system.

Health care facility staff should look to see if pipe and fittings are in good condition and free of mechanical damage, leakage, corrosion, and misalignment. The system piping must not be subjected to external loads by materials either resting on the pipe or hung from the pipe. Hangers and seismic braces (in localities requiring such devices) that are damaged or loose must be replaced or refastened.

22.6.2 Testing Standpipe Systems

Standpipe systems are tested for a variety of reasons at various periods. Installation contractors test standpipe systems during installation and again in the presence of the AHJ, before the systems go on line. Owners (themselves or through maintenance contractors) conduct routine standpipe system tests throughout the life of the system to ensure that the system remains in proper operating order. Testing is characterized in two groups, performance testing and acceptance testing.

22.6.2.1 Installation (Performance) and Acceptance Testing

Installation contractors normally hydrostatically test small portions of standpipe systems (usually one riser at a time) as they install these systems before moving on to other areas of the building. Facility personnel should secure the required Contractor's Material and Test Certificate(s) from the installer as evidence that these performance tests were successful (Figures 22.4 and 22.5). Authorities having jurisdiction normally witness the testing of standpipe systems to make sure that they are properly installed

Contractor's Material and Test Certificate for Aboveground Piping
Standpipe System NFPA 14

PROCEDURE
Upon completion of work, inspection and tests shall be made by the contractor's representative and witnessed by an owner's representative. All defects shall be corrected and system left in service before contractor's personnel finally leave the job.

A certificate shall be filled out and signed by both representatives. Copies shall be prepared for approving authorities, owners, and contractor. It is understood the owner's representative's signature in no way prejudices any claim against contractor for faulty material, poor workmanship, or failure to comply with approving authority's requirements or local ordinances.

PROPERTY NAME		DATE
PROPERTY ADDRESS		

PLANS	ACCEPTED BY APPROVING AUTHORITIES (NAMES)		
	ADDRESS		
	INSTALLATION CONFORMS TO ACCEPTED PLANS	☐ YES	☐ NO
	EQUIPMENT USED IS APPROVED OR LISTED IF NO, EXPLAIN DEVIATIONS	☐ YES	☐ NO
TYPE OF SYSTEM	AUTOMATIC-DRY	☐ YES	
	AUTOMATIC-WET	☐ YES	
	SEMIAUTOMATIC-DRY	☐ YES	
	MANUAL-DRY	☐ YES	
	MANUAL-WET	☐ YES	
	COMBINATION STANDPIPE/SPRINKLER	☐ YES	
	OTHER, IF YES EXPLAIN	☐ YES	
WATER SUPPLY DATA USED FOR DESIGN AND AS SHOWN ON PLANS	FIRE PUMP DATA MANUFACTURER _____ MODEL _____ TYPE: ☐ ELECTRIC ☐ DIESEL ☐ OTHER, EXPLAIN _____ RATED GPM _____ RATED PSI _____ SHUT-OFF PSI _____		
WATER SUPPLY SOURCE CAPACITY, GALLONS	PUBLIC WATERWORKS SYSTEM ☐ STORAGE TANK ☐ GRAVITY TANK ☐ OPEN RESERVOIR ☐ OTHER ☐ EXPLAIN		
IF PUBLIC WATERWORKS SYSTEM:	STATIC PSI _____ RESIDUAL PSI _____ FLOW IN _____ GPM		
HAVE COPIES OF THE FOLLOWING BEEN LEFT ON THE PREMISES?	☐ SYSTEM COMPONENTS INSTRUCTIONS ☐ CARE AND MAINTENANCE OF SYSTEM ☐ NFPA 25 ☐ COPY OF ACCEPTED PLANS ☐ HYDRAULIC DATA/CALCULATIONS		
SUPPLIES BUILDING(S)	MAIN WATERFLOW SHUT-OFF LOCATION _____ NUMBER OF STANDPIPE RISERS _____ DO ALL STANDPIPE RISERS HAVE BASE OF RISER SHUT-OFF VALVES?	☐ YES	☐ NO
VALVE SUPERVISION	LOCKED OPEN ☐ SEALED AND TAGGED ☐ TAMPERPROOF SWITCH ☐ OTHER ☐ IF OTHER, _____		
PIPE AND FITTINGS	TYPE OF PIPE _____ TYPE OF FITTINGS _____		
BACKFLOW PREVENTOR	A) DOUBLE CHECK ASSEMBLY ☐ SIZE _____ MAKE AND MODEL _____ B) REDUCED-PRESSURE DEVICE ☐		

(NFPA 14, 1 of 3)

Figure 22.4 Contractor's Material and Test Certificate for Aboveground Piping.
Source: NFPA 14, 2000, Figure 9.1.2(a).

CONTROL VALVE DEVICE			
TYPE	SIZE	MAKE	MODEL

TIME TO TRIP THROUGH REMOTE HOSE VALVE _____ MIN _____ SEC WATER PRESSURE _____ AIR PRESSURE _____
TIME WATER REACHED REMOTE HOSE VALVE OUTLET _____ MIN _____ SEC TRIP POINT AIR PRESSURE _____ PSI
ALARM OPERATED PROPERLY ☐ YES ☐ NO IF NO, EXPLAIN _____

TIME WATER REACHED REMOTE HOSE VALVE OUTLET _____ MIN _____ SEC
HYDRAULIC ACTIVATION ☐ YES
ELECTRIC ACTIVATION ☐ YES
PNEUMATIC ACTIVATION ☐ YES
MAKE AND MODEL OF ACTIVATION DEVICE _____
EACH ACTIVATION DEVICE TESTED ☐ YES ☐ NO IF NO, EXPLAIN _____

EACH ACTIVATION DEVICE OPERATED PROPERLY ☐ YES ☐ NO IF NO, EXPLAIN _____

PRESSURE-REGULATING DEVICE

LOCATION & FLOOR	MODEL	NONFLOWING (PSI) INLET	NONFLOWING (PSI) OUTLET	FLOWING (PSI) INLET	FLOWING (PSI) OUTLET	GPM

ALL HOSE VALVES ON SYSTEM OPERATED PROPERLY ☐ YES ☐ NO IF NO, EXPLAIN _____

(NFPA 14, 2 of 3)

Figure 22.4 Cont'd

before accepting the systems as complete and operational. Successful performance testing conducted by the installation contractor before the acceptance testing also increases the chances that acceptance tests will be successful.

For conducting acceptance tests for standpipe systems, it is important to coordinate hydrostatic testing with the AHJ, sprinkler contractor, and facility personnel to minimize problems and delays. If the sprinkler system is retrofitted into an occupied facility, coordination with the facility staff and occupants is essential. Again, the piping must be exposed during this test.

TEST DESCRIPTION	<u>HYDROSTATIC</u>: HYDROSTATIC TESTS SHALL BE MADE AT NOT LESS THAN 200 PSI (13.6 BAR) FOR 2 HOURS OR 50 PSI (3.4 BAR) ABOVE STATIC PRESSURE IN EXCESS OF 150 PSI (10.2 BAR) FOR 2 HOURS. DIFFERENTIAL DRY PIPE VALVE CLAPPERS SHALL BE LEFT OPEN DURING TEST TO PREVENT DAMAGE. ALL ABOVEGROUND PIPING LEAKAGE SHALL BE STOPPED. <u>PNEUMATIC</u>: ESTABLISH 40 PSI (2.7 BAR) AIR PRESSURE AND MEASURE DROP, WHICH SHALL NOT EXCEED 1½ PSI (0.1 BAR) IN 24 HOURS. TEST PRESSURE TANKS AT NORMAL WATER LEVEL AND AIR PRESSURE AND MEASURE AIR PRESSURE DROP, WHICH SHALL NOT EXCEED 1½ PSI (0.1 BAR) IN 24 HOURS.	
TESTS	ALL PIPING HYDROSTATICALLY TESTED AT_____ PSI FOR _____ HRS DRY PIPING PNEUMATICALLY TESTED ☐ YES ☐ NO EQUIPMENT OPERATES PROPERLY ☐ YES ☐ NO	IF NO, STATE REASON
	DO YOU CERTIFY AS THE STANDPIPE CONTRACTOR THAT ADDITIVES AND CORROSIVE CHEMICALS, SODIUM SILICATE, OR DERIVATIVES OF SODIUM SILICATE, BRINE, OR OTHER CORROSIVE CHEMICALS WHERE NOT USED FOR TESTING SYSTEMS OR STOPPING LEAKS? ☐ YES ☐ NO	
	DRAIN TEST	READING OF GAUGE LOCATED NEAR WATER SUPPLY TEST CONNECTION _____ PSI	RESIDUAL PRESSURE WITH VALVE IN TEST CONNECTION OPEN WIDE _____ PSI
	UNDERGROUND MAINS AND LEAD-IN CONNECTIONS TO SYSTEM RISERS FLUSHED BEFORE CONNECTION MADE TO STANDPIPE PIPING. VERIFIED BY COPY OF THE U FORM NO. 85B ☐ YES ☐ NO OTHER EXPLAIN FLUSHED BY INSTALLER OF UNDER- GROUND STANDPIPE PIPING ☐ YES ☐ NO	
BLANK TESTING	NUMBER USED	LOCATIONS	NUMBER REMOVED
WELDING	WELDED PIPING ☐ YES ☐ NO	
	IF YES . . .	
	DO YOU CERTIFY AS THE STANDPIPE CONTRACTOR THAT WELDING PROCEDURES COMPLY WITH THE REQUIREMENTS OF AT LEAST AWS D10.9, LEVEL AR-3 ☐ YES ☐ NO DO YOU CERTIFY THAT THE WELDING WAS PERFORMED BY WELDERS QUALIFIED IN COMPLIANCE WITH THE REQUIREMENTS OF AT LEAST AWS D10.9, LEVEL AR-3 ☐ YES ☐ NO DO YOU CERTIFY THAT WELDING WAS CARRIED OUT IN COMPLIANCE WITH A DOCUMENTED QUALITY CONTROL PROCEDURE TO ENSURE THAT ALL DISCS ARE RETRIEVED, THAT OPENINGS IN PIPING ARE SMOOTH, THAT SLAG AND OTHER WELDING RESIDUE ARE REMOVED, AND THAT THE INTERNAL DIAMETERS OF PIPING ARE NOT PENETRATED ☐ YES ☐ NO	
CUTOUTS (DISCS)	DO YOU CERTIFY THAT YOU HAVE A CONTROL FEATURE TO ENSURE THAT ALL CUTOUTS (DISCS) ARE RETRIEVED? ☐ YES ☐ NO	
HYDRAULIC DATA NAMEPLATE	NAMEPLATE PROVIDED IF NO, EXPLAIN ☐ YES ☐ NO	
REMARKS	DATE LEFT IN SERVICE WITH ALL CONTROL VALVES OPEN:	
NAME OF SPRINKLER/ STANDPIPE CONTRACTOR	NAME OF CONTRACTOR _____ ADDRESS _____ STATE LICENSE NUMBER (IF APPLICABLE) _____	
SYSTEM OPERATING TEST WITNESSED BY	PROPERTY OWNER _____ TITLE _____ DATE _____ SPRINKLER/STANDPIPE CONTRACTOR _____ TITLE _____ DATE _____ APPROVING AUTHORITIES _____ TITLE _____ DATE _____	
ADDITIONAL EXPLANATION AND NOTES		

(NFPA 14. 3 of 3)

Figure 22.4 Cont'd

22.6.2.2 Flow Testing

Flow testing of a standpipe system must be conducted every 5 years at the hydraulically most remote hose connection (usually near the roof) of each zone of a standpipe to verify that the water supply provides the required flow at the required pressure. The required flow and required pressure are dependent on the applicable code that was in

Contractor's Material and Test Certificate for Underground Piping

PROCEDURE
Upon completion of work, inspection and tests shall be made by the contractor's representative and witnessed by an owner's representative. All defects shall be corrected and system left in service before contractor's personnel finally leave the job.

A certificate shall be filled out and signed by both representatives. Copies shall be prepared for approving authorities, owners, and contractor. It is understood the owner's representative's signature in no way prejudices any claim against contractor for faulty material, poor workmanship, or failure to comply with approving authority's requirements or local ordinances.

PROPERTY NAME	DATE
PROPERTY ADDRESS	

PLANS	ACCEPTED BY APPROVING AUTHORITIES (NAMES)		
	ADDRESS		
	INSTALLATION CONFORMS TO ACCEPTED PLANS	☐ YES	☐ NO
	EQUIPMENT USED IS APPROVED IF NO, STATE DEVIATIONS	☐ YES	☐ NO
INSTRUCTIONS	HAS PERSON IN CHARGE OF FIRE EQUIPMENT BEEN INSTRUCTED AS TO LOCATION OF CONTROL VALVES AND CARE AND MAINTENANCE OF THIS NEW EQUIPMENT? IF NO, EXPLAIN	☐ YES	☐ NO
	HAVE COPIES OF APPROPRIATE INSTRUCTIONS AND CARE AND MAINTENANCE CHARTS BEEN LEFT ON PREMISES? IF NO, EXPLAIN	☐ YES	☐ NO
LOCATION	SUPPLIES BUILDINGS		
UNDERGROUND PIPES AND JOINTS	PIPE TYPES AND CLASS	TYPE JOINT	
	PIPE CONFORMS TO _____ STANDARD	☐ YES	☐ NO
	FITTINGS CONFORM TO _____ STANDARD IF NO, EXPLAIN	☐ YES	☐ NO
	JOINTS NEEDING ANCHORAGE CLAMPED, STRAPPED, OR BLOCKED IN ACCORDANCE WITH _____ STANDARD IF NO, EXPLAIN	☐ YES	☐ NO
TEST DESCRIPTION	FLUSHING: Flow the required rate until water is clear as indicated by no collection of foreign material in burlap bags at outlets such as hydrants and blow-offs. Flush at flows not less than 390 gpm (1,476 L/min) for 4-in. pipe, 880 gpm (3,331 L/min) for 6-in. pipe, 1,560 gpm (5,905 L/min) for 8-in. pipe, 2,440 gpm (9,235 L/min) for 10-in. pipe, and 3,520 gpm (13,323 L/min) for 12-in. pipe. When supply cannot produce stipulated flow rates, obtain maximum available. HYDROSTATIC: Hydrostatic tests shall be made at not less than 200 psi (13.8 bar) for 2 hours or 50 psi (3.4 bar) above static pressure in excess of 150 psi (10.3 bar) for 2 hours. LEAKAGE: New pipe laid with rubber gasketed joints shall, if the workmanship is satisfactory, have little or no leakage at the joints. The amount of leakage at the joints shall not exceed 2 qt/hr (1.89 L/hr) per 100 joints irrespective of pipe diameter. The leakage shall be distributed over all joints. If such leakage occurs at a few joints the installation shall be considered unsatisfactory and necessary repairs made. The amount of allowable leakage specified above can be increased by 1 fl oz/in. valve diameter per hr (30 mL/25 mm/hr) for each metal seated valve isolating the test section. If dry barrel hydrants are tested with the main valve open, so the hydrants are under pressure, an additional 5 oz/min (150 mL/min) leakage is permitted for each hydrant.		
FLUSHING TESTS	NEW UNDERGROUND PIPING FLUSHED ACCORDING TO _____ STANDARD BY (COMPANY) IF NO, EXPLAIN	☐ YES	☐ NO
	HOW FLUSHING FLOW WAS OBTAINED ☐ PUBLIC WATER ☐ TANK OR RESERVOIR ☐ FIRE PUMP	THROUGH WHAT TYPE OPENING ☐ HYDRANT BUTT ☐ OPEN PIPE	
	LEAD-INS FLUSHED ACCORDING TO _____ STANDARD BY (COMPANY) IF NO, EXPLAIN	☐ YES	☐ NO
	HOW FLUSHING FLOW WAS OBTAINED ☐ PUBLIC WATER ☐ TANK OR RESERVOIR ☐ FIRE PUMP	THROUGH WHAT TYPE OPENING ☐ Y CONN. TO FLANGE ☐ OPEN PIPE & SPIGOT	

(NFPA 14, 1 of 2)

Figure 22.5 Contractor's Material and Test Certificate for Underground Piping.
Source: NFPA 14, 2000, Figure 9.1.2(b).

effect when the system was designed. The facility should check with the original installation contractor for the correct flow and pressure values.

Most systems provide two 2½-in. (63.5-mm) hose valves at the top of the most remote standpipe or at the roof level, which must be protected from freezing, to facilitate flow testing of 500 gpm (1892 L/min) through the standpipe. Often, hoses can be

HYDROSTATIC TEST	ALL NEW UNDERGROUND PIPING HYDROSTATICALLY TESTED AT _____ PSI FOR _____ HOURS		JOINTS COVERED ☐ YES ☐ NO
LEAKAGE TEST	TOTAL AMOUNT OF LEAKAGE MEASURED _____ GAL _____ HOURS		
	ALLOWABLE LEAKAGE _____ GAL _____ HOURS		
HYDRANTS	NUMBER INSTALLED	TYPE AND MAKE	ALL OPERATE SATISFACTORILY ☐ YES ☐ NO
CONTROL VALVES	WATER CONTROL VALVES LEFT WIDE OPEN IF NO, STATE REASON		☐ YES ☐ NO
	HOSE THREADS OF FIRE DEPARTMENT CONNECTIONS AND HYDRANTS INTERCHANGEABLE WITH THOSE OF FIRE DEPARTMENT ANSWERING ALARM		☐ YES ☐ NO
REMARKS	DATE LEFT IN SERVICE		
SIGNATURE	NAME OF INSTALLING CONTRACTOR		
	TESTS WITNESSED BY		
	FOR PROPERTY OWNER (SIGNED)	TITLE	DATE
	FOR INSTALLING CONTRACTOR (SIGNED)	TITLE	DATE
ADDITIONAL EXPLANATION AND NOTES			

(NFPA 14. 2 of 2)

Figure 22.5 Cont'd

connected to the valves at the top of the riser and routed to the roof through a nearby access door.

Discharging a large volume of water from the hydraulically most remote outlet may not be practical, especially when tall buildings are involved. In such cases, the AHJ should be consulted for an appropriate alternative location for the test. Where such testing must be conducted at the roof level, there are testing products that minimize the force of the flowing water, thereby avoiding associated damage. Even when these testing products are used, the total volume of water (and associated weight) must be safely directed to drain. Figure 22.6 is an example of a flow-testing device.

If the standpipe system contains pressure-reducing valves or pressure-regulating valves, such valves should be inspected, tested, and maintained by a qualified contractor. These tests must be conducted every 5 years.

22.6.2.3 Main Drain Test

Another type of flow test is a main drain test. Main drains are usually 2 in. (50 mm) in diameter, located either near the low point to drain the system or at the connection where the supply main enters the building. A main drain test must be performed annually on all classes of standpipe systems. Where systems have 2-in. (50-mm) main drains, the drain can be opened to give the inspector an idea of whether the water supply valves are fully open.

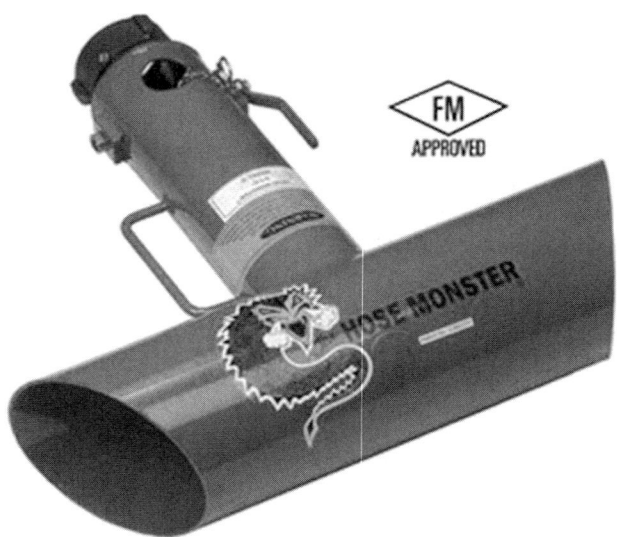

Figure 22.6 Example of a flow-testing device that disperses pressure.
Source: Hydro Flow Products, Inc.

In a 2-in. (50-mm) main drain test, if the gauge pressure drops dramatically during full flow, it may indicate a weak water supply or partially closed supply valve. After the 2-in. (50-mm) main drain valve is closed, the gauge pressure should return quickly to the normal static pressure; if it does not, this may also indicate a partially closed valve. Many times, partially closed valves are located underground (curb-box valves) in the main water supply piping.

22.6.2.4 Hydrostatic Testing

The purpose of hydrostatic testing is to demonstrate the integrity of the piping network (i.e., show that the system has no leaks). The hydrostatic test is accomplished by pressurizing water in the system to 200 psi (13.8 bar) for 2 hours or 50 psi (3.4 bar) in excess of the maximum pressure, where the maximum pressure is greater than 150 psi (10.3 bar). The hydrostatic test pressure must be measured at the low elevation point of the individual system or zone being tested.

Authorities having jurisdiction require hydrostatic testing before a system is operational. During the hydrostatic test, inspectors observe the entire system, checking for leaks at all joints. Of course, the system piping must be exposed during this test, so ceiling tiles, wallboard, or soffits must be open, and they should not be replaced until the hydrostatic acceptance tests for the AHJ have been completed successfully.

Hydrostatic testing should be conducted every 5 years on dry standpipe systems and on dry portions of wet standpipe systems. Hydrostatic tests should also be conducted on any system that has been modified or repaired or where the results of an inspection indicate that the system could fail to operate properly in an emergency. Of course, the 5-year hydrostatic testing does not require all ceilings and soffits to be open;

this test is subsequent to the original acceptance test and is intended to check for pressure drop. If the system piping does leak, water damage is likely to result.

Pneumatic testing is recommended where water damage is possible, the piping network is in occupied areas, or the area in which the standpipe is installed is very sensitive to any amount of water leakage. Since air is a compressible fluid, NFPA 25 recommends pneumatic testing at 25 psi (1.7 bar).

22.6.2.5 Alarm Device Testing

Not all standpipe systems are provided with water flow alarms, but for systems that are so equipped, the alarm devices and supervisory devices should be tested on a quarterly basis. For health care facilities located in climates that experience freezing temperatures, quarterly testing usually results in conducting tests during freezing conditions. Because testing standpipe system alarm devices may result in a substantial amount of water being discharged to the grade surface around the facility, freezing conditions may delay testing, but the tests should be performed as soon as weather conditions allow.

22.6.3 Standpipe System Maintenance

NFPA 25 (2002) requires maintenance and repairs to be conducted in accordance with Table 6.2.2 of that standard (see Table 22.2). In-house maintenance of standpipe systems should be performed only where the maintenance staff has been specifically trained to conduct such work. With the exception of some basic repairs, such as reconnecting hangers or bleeding condensation from air compressor lines, in-house personnel usually do not have the qualifications or experience to perform these functions.

Outsourcing standpipe system maintenance tasks to a qualified sprinkler contractor provides a health care facility with several benefits:

- The maintenance personnel are specialists in the field of standpipe and sprinkler systems.
- Sprinkler contractors have the specialized tools to maintain the standpipe system.
- Sprinkler contractors have working relationships with the manufacturers and suppliers of the components or spare parts for valves, fittings, and attachments.
- The maintenance liability is "passed" to the sprinkler professional.
- The workload is transferred away from the health care facility's maintenance personnel.

When contracting standpipe system maintenance, the health care facility should make sure that the contractor regularly conducts this type of work on a day-to-day basis. Many companies work on piping systems, but it is essential to make sure that the selected company is well qualified (such as by being a member of one of the two sprinkler contracting associations in the United States: the American Fire Sprinkler Association and the National Fire Sprinkler Association).

22.6.4 Record Keeping for Standpipe Systems

NFPA 14 requires that records of inspections, tests, and maintenance of the system and its components be made available to the authority having jurisdiction upon request. Typical records include, but are not limited to, reports of valve inspections; flow, drain, and pump tests; and trip tests of dry-pipe, deluge, and preaction valves. Records should indicate the procedure performed (e.g., inspection, test, or maintenance), the organization that performed the work, the results of the activities, and the date the work was performed. Facility owners must maintain original records for the life of the system and subsequent records for a period of 1 year after the next inspection, test, or maintenance required by the standard.

23

Fire Pumps

Fire pumps provide or enhance the water supply pressure available from municipal water mains, gravity tanks, or other sources. Fire pumps have been used for many years in health care facilities where it has been necessary to supplement the available water supply so as to provide water to sprinkler and standpipe systems.

Today's fire pumps are designed to provide maximum reliability and specific discharge pressure characteristics. Unlike commercial or "house" pumps that remain in operation much of the time, fire pumps are usually idle. The commercial pumps that supply domestic water and other services are designed for maximum efficiency and economy because of the amount of work they perform. With fire pumps, economy is sacrificed for reliability and performance so that they perform in the rare instances when they are needed.

The first modern type of fire pump was a wheel-and-crank reciprocating type, belt driven by mill machinery. These pumps gave way to rotary-type pumps that were driven by horizontal water wheels. With the popularity of steam power, the reciprocating steam pump was universally accepted for many years. In the early 1900s, fire pumps were driven by gasoline engines, but this method proved to be relatively unreliable.

Today, the standard fire pump is the centrifugal fire pump, which comes with a choice of drivers—electric motors, steam turbines, or diesel engines. Another type of modern pump is the positive-displacement pump, used for both water mist and foam fire pump systems. The NFPA Committee on Fire Pumps, organized in 1899, is responsible for NFPA 20, *Standard for the Installation of Stationary Pumps for Fire Protection*, 2003 edition, which covers centrifugal pumps as well as positive displacement pumps.

This chapter focuses on the centrifugal-type fire pump, because positive-displacement pumps are not commonly found in health care facilities. Details on stationary fire pumps are provided in terms of location requirements, pump types, considerations for system design and modification, and requirements for inspection, testing, and maintenance.

23.1 Location Requirements

Unlike with other fire protection systems discussed in this book, building or fire codes do not specifically require fire pumps to be installed in health care facility buildings based on a building's use, construction type, configuration, or height and area. The need for a fire pump is a function of the pressure of the available water supply and the pressure

431

demand of the fire protection system being supplied. Although rare, some relatively tall buildings (as high as 10 stories) have sprinkler and standpipe systems but no fire pump, because their municipal water supplies have static pressures of up to 120 psi (828 kPa).

Early in the design stage of a building or a sprinkler/standpipe system, it is essential for the fire protection engineer to calculate whether the building requires a fire pump, because of the impact that this equipment has on a construction project. If a fire pump is required, then the fire protection engineer must determine space requirements for the pump, account for fuel or electric power supplies, and detail it on the contract documents such that the installing contractor can order the pump, potentially well in advance, to allow time for it to be manufactured and shipped to the project site.

23.2 Fire Pump Types

As mentioned, the centrifugal-type pump is the standard for pumps supplying sprinkler and standpipe systems. Four types of centrifugal pumps are common in the fire protection industry:

- Horizontal split-case
- Horizontal end suction
- Vertical in-line
- Vertical turbine

The horizontal split-case fire pump is probably the most common type of fire pump in use today in health care facilities (see Figure 23.1). The vertical in-line pump is basically a horizontal split-case pump turned on its end, such that the electric motor is situated on top of the pump instead of next to it (see Figure 23.2). Vertical in-line pumps require less floor space than horizontal split-case pumps but can cost more to maintain, if the motor must be lifted off the pump to service the pump casing or interior parts.

Vertical turbine pumps (see Figure 23.3) were originally designed to pump water from wells. These pumps are excellent for pumping water from static sources, where the pump must operate with suction lift (i.e., the water has no pressure forcing it into the pump). Vertical turbine pumps operate without priming. They are not extremely common in health care settings because most of these facilities are supplied with water from pressurized municipal sources. Vertical turbine pumps are more likely to be found in

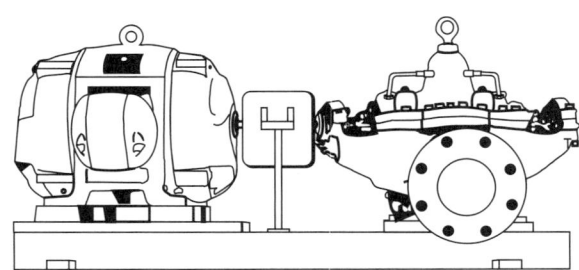

Figure 23.1　A horizontal-shaft, single-stage centrifugal pump with cutaway view of the pump.

Chapter 23. Fire Pumps   **433**

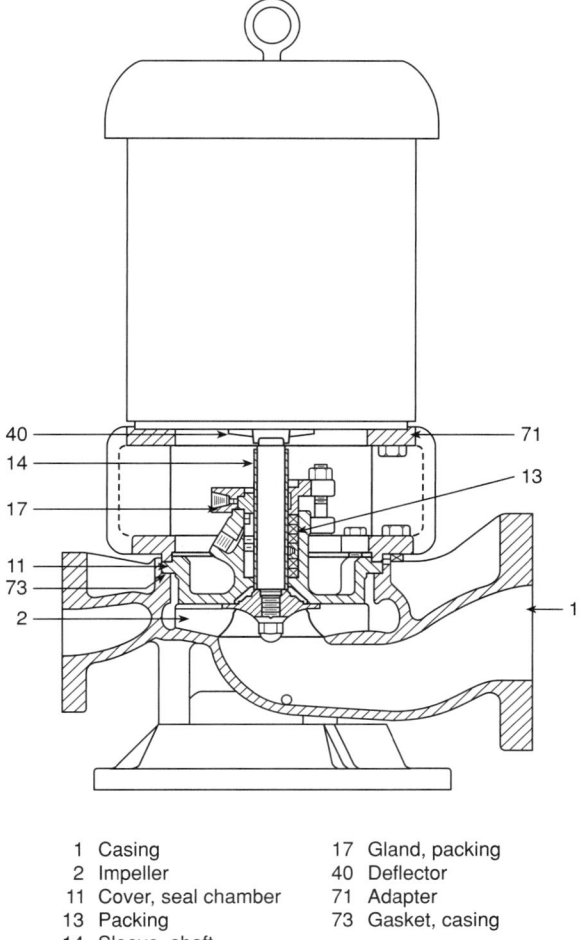

1	Casing	17	Gland, packing
2	Impeller	40	Deflector
11	Cover, seal chamber	71	Adapter
13	Packing	73	Gasket, casing
14	Sleeve, shaft		

Figure 23.2 Overhung impeller—close coupled single stage—in-line pump (showing seal and packaging).

industrial settings, where they draw water from streams, ponds, wells, or underground storage tanks.

23.3 Pump Drivers

The approved devices that power pumps (i.e., pump drivers) consist of electric motors, steam turbines, or diesel engines. Although still used, steam-powered pump drivers are rarely found in use today. Electric motors are the most commonly used pump drivers in health care facilities, although diesel engines are also excellent drivers for these purposes. The pump driver should be selected based on reliability, adequacy, safety, and economy. An excellent technical discussion on this subject is "Making the Right Choice for a Fire

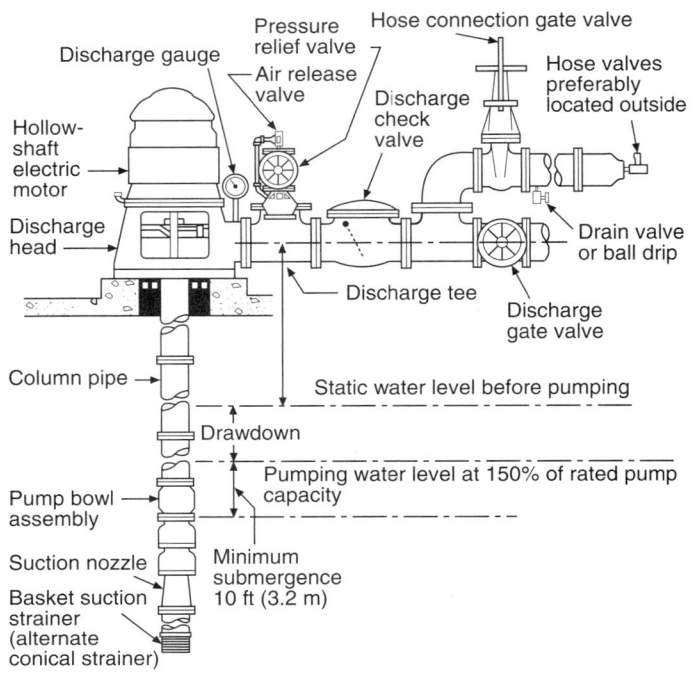

Figure 23.3 Vertical shaft turbine-type pump installation.

Pump Driver and Power Source, Diesel Engine, or Electric Motor," by Schirmer Engineering Corporation, September 20, 2003, available at www.clarkesystem.com.

23.3.1 Electrically Driven Fire Pumps

Electric motors are extremely reliable devices for powering fire pumps. The question of reliability for electrically driven fire pumps must focus on the reliability of the electrical power delivery system, which may be judged by reviewing the supplier's record of power outages as well as the electrical infrastructure of the facility. According to NFPA 20, electric motors must be powered either by a reliable source of power or by two or more approved independent sources. The two or more approved independent sources may consist of a separate electrical service (i.e., separate connections to separate power grids), an on-site standby generator, a redundant diesel engine–driven fire pump, or a redundant steam turbine–driven fire pump. Of course, where more than one electrical supply is provided, a listed transfer switch must be provided to connect the electric motor to the emergency source of power upon failure of the normal electrical supply.

When a fire pump is called upon to perform in a fire situation, there is no question as to the importance of its mission. Accordingly, electric fire pump motor drivers are designed to withstand severe ranges of operating conditions, and the electrical service must be designed to accommodate large electrical loads. In the name of reliability, it has been a common practice for many years for electrically driven fire pumps to be designed and configured to run to the destruction of the motor. In contrast, commercial pumps are designed to shut down under such severe conditions, to save the pump package.

If the power supply to an electric pump is not reliable, or if the height of the structure is beyond the pumping capacity of the fire department apparatus, a second electrical source is required, and a connection to an emergency generator is usually provided. Emergency generators must be installed in accordance with NFPA 20; NFPA 37, *Standard for the Installation and Use of Stationary Combustion Engines and Gas Turbines*, 2002 edition; and NFPA 110, *Standard for Emergency and Standby Power Systems*, 2005 edition. Emergency transfer switches may be an integral part of the fire pump controller or be included as a separate piece of equipment (see Figures 23.4 and 23.5). To enhance reliability and to maintain the listing of the devices, however, the transfer switch should be an integral part supplied by the controller manufacturer.

Figure 23.4 Across-the-line electric fire pump controller.
Source: Fire Pump Handbook, NFPA, 1998, Exhibit 7-2; courtesy of Firetrol, Inc.

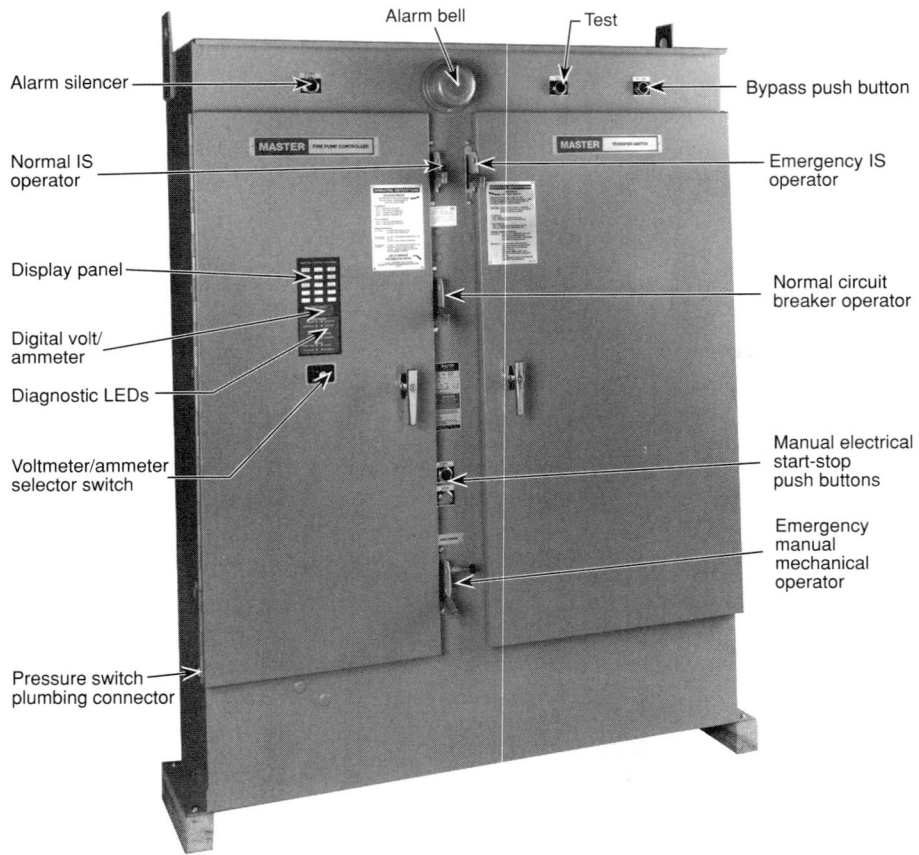

Figure 23.5 Combination fire pump controller and power transfer switch.
Source: Fire Pump Handbook, NFPA, 1998, Exhibit 7-8; courtesy of Master Controls, Inc.

The electrical wiring that supplies the fire pumps must be carefully designed by an electrical engineer in concert with a fire protection engineer. NFPA 20 and NFPA 70, *National Electrical Code*® (*NEC*®), 2005 edition, require fire pumps to be supplied by a separate service or by a tap located ahead of the service disconnection means. The connection must be located and arranged so as to minimize the possibility of damage by fire from within the premises and from exposing hazards. Such service taps ensure that the fire pump continues to be provided with electrical power even after the fire department secures (i.e., shuts off) all utilities, which is a common and good practice during fire-fighting operations.

NFPA 20 and the *NEC* also require the supply conductors to be physically routed outside a building and to be installed as service entrance conductors (see NFPA 20, Chapter 9). Where supply conductors cannot be physically routed outside a building, they can be routed through a building when encased in 2 in. (50.8 mm) of concrete or within enclosed construction that is dedicated to the fire pump circuits and has a minimum of a 1-hour fire resistance rating. The conductors can also be installed within listed electrical circuit–protective systems with a minimum of 1-hour fire resistance. An example of a listed electrical circuit-protective system is mineral insulated (MI) cable. The code does not require the supply conductors that are located in the electrical equipment room where they originate or the fire pump room to have the minimum 1-hour fire separation or fire resistance rating.

Refer to Chapters 5, 6, and 8 of this book for more information regarding the electrical supply at the entrance to the facility, electrical cables, and emergency power supplies in health care facilities, and Chapters 14 and 15 for more information on fire resistivity and fire resistance ratings.

23.3.2 Diesel-Driven Pumps

Engine-driven fire pumps have been available for many years (see Figure 23.6). The compression-ignition diesel engine has proved to be the most dependable of the internal combustion engines for driving fire pumps. Health care facility engineering personnel should be fairly familiar with diesel engines, since such facilities require emergency power supplies (refer to Chapter 8). Selecting diesel engine–driven fire pump equipment for each situation should be based on careful consideration of the following factors:

- Reliability of control
- Fuel supply

Figure 23.6 Diesel engine drive.
Source: Fire Pump Handbook, NFPA, 1998, Exhibit 8-1; courtesy of Patterson Pump.

- Installation requirements
- Starting and running operation of the diesel engine

With diesel-driven pumps, the question of reliability switches from the source of electrical power to the need for the following: (1) constantly charged batteries for starting the equipment, unless hydraulic or pneumatic starting is provided; (2) engine cooling; (3) maintenance of the pump room temperature; (4) ventilation; and (5) a fuel supply.

Although diesel engines require a fuel supply, lubricating oil, engine-speed governors, overspeed shutdown devices, combustion air, and an exhaust discharge, their application is quite beneficial because they are very reliable and do not require an emergency generator as a backup power supply. These features can be extremely advantageous in a retrofit situation, where it would likely be very expensive to increase the size of an emergency generator or add another generator set. Installing a residential-grade muffler on all diesel-driven fire pumps is suggested to avoid disturbing facility patients and staff with an excessive amount of noise.

Ventilation of the fire pump room is very important because it performs the following functions (see Figure 23.7):

- Controls the maximum temperature to 120°F (49°C) at the combustion air cleaner inlet with the engine running at the rated load
- Supplies air for engine combustion
- Removes any hazardous vapors
- Supplies and exhausts air as necessary for radiator cooling of the engine when required

NFPA 20 requires the components of the ventilation system to coordinate with the operation of the engine.

23.4 Design Considerations

To ensure that the fire pump installation is acceptable, health care facility managers and engineering personnel should be involved in several aspects of designing the fire pump system, as follows:

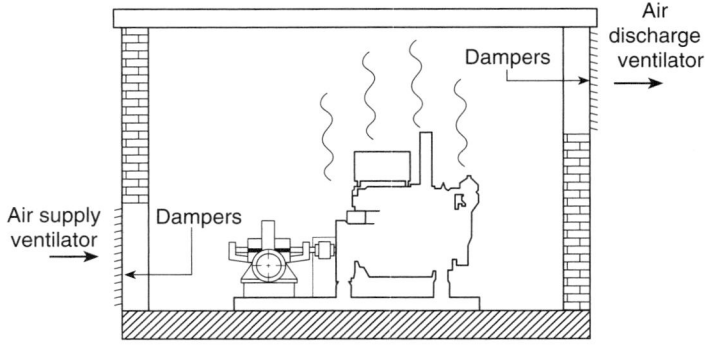

Figure 23.7 Typical ventilation system for a heat exchange–cooled diesel-driven pump. *Source*: NPFA 20, 2003, Figure A.11.3.2(a).

- When trying to decide between using electrically driven and diesel-driven pumps, the facility should require the designer to justify the type of pump chosen for the facility's application. Often a good source of advice can be the local authority having jurisdiction (AHJ).
- For electrically driven pumps, the facility should make sure that the fire protection engineer and the electrical engineer fully coordinate the horsepower for the motor, equipment location, emergency power, lighting (on the emergency power), and the means of starting the motor provided by the selected controller (see Figure 23.8). Also, the fire protection engineer must coordinate with the mechanical engineer in terms of heating, ventilation, and air conditioning (controllers will fail at elevated temperatures).

Figure 23.8 Diesel drive controller.
Source: Fire Pump Handbook, NFPA, 1998, Exhibit 9-1.

- For diesel-driven pumps, the fire protection engineer and the mechanical engineer must coordinate fire pump room heating, ventilation, air conditioning, exhaust location, combustion air, and room drainage.

- The fire protection engineer and the architect (of a new building) or facility personnel (in a retrofit installation) must work together to design a pump room that is adequate in size and provides access for future maintenance or replacement of the pump. The fire pump, driver, and controller must be protected against possible interruption of service through damage caused by explosion, fire, flood, earthquake, rodents, insects, windstorm, freezing, vandalism, and other adverse conditions. Indoor pump rooms must be separated from all other areas by 2-hour fire-rated construction (1-hour fire-rated construction if the building is fully sprinklered). Outdoor pump houses must be at least 50 ft (15.3 m) from the protected building. The location of the test header and its distance from the pump can have a profound effect on the ability and cost to perform future flow testing.

- In systems equipped with relief valves, carefully planning the location where such devices discharge water is of prime importance. During system tests, these devices may constantly discharge thousands of gallons of water per minute, which can affect the adjacent landscape and drainage and cause problems with ice in freezing weather. Where the storm water drainage facilities can be designed and coordinated to accept the flow from the relief valve(s), flooding and ice problems may be avoided.

- The design of the fire pump and the fire alarm system must also be coordinated. For electrically and diesel engine–driven pumps where the pump room is not constantly attended, audible or visible alarms must be provided at a point of constant attendance (usually through connection to the fire alarm system). For electrically driven pumps, these alarms must indicate (1) if the pump or motor is running, (2) if there is a loss of any phase, (3) phase reversal, and (4) if the controller is connected to an alternative power source. For diesel engine–driven pumps, the alarms must separately indicate (1) if the engine is running, (2) if the controller main switch has been turned to the off or manual position, or (3) if there is any "trouble" on the controller or engine, which includes the following 11 items monitored by the engine controller:
    1. Critically low oil pressure in the lubrication system
    2. High engine jacket coolant temperature
    3. Failure of engine to start automatically
    4. Shutdown from overspeed
    5. Battery failure or missing battery. Each controller is provided with a separate visible indicator for each battery
    6. Battery charger failure
    7. Low air or hydraulic pressure (where air or hydraulic starting is provided)
    8. System overpressure (for engines equipped with pressure-limiting controls)
    9. ECM selector switch in alternate ECM position (for engines with ECM controls only)
    10. Fuel injection malfunction (for engines with ECM only)
    11. Low fuel level

23.5 Inspection, Testing, and Maintenance

A fire pump installed in accordance with NFPA 20 must be properly inspected, tested, and maintained in accordance with NFPA 25, *Standard for the Inspection, Testing, and Maintenance of Water-Based Fire Protection Systems*, 2002 edition. NFPA 25 is the reference standard for inspection, testing, and maintenance of water-based fire protection systems and governs fire pumps as well as related systems, including sprinkler systems, underground piping, standpipe systems, storage tanks, water spray systems, and foam–water sprinkler systems. NFPA 25 provides instructions on how to conduct inspection, testing, and maintenance activities and stipulates how often such activities must be completed. Requirements are provided for impairment procedures, notification processes, and system restoration. This type of information, where incorporated into a building maintenance program, enhances the favorable performance of all water-based fire protection systems.

The required frequency of any inspection, testing, and maintenance activity is a function of the system component in question. Table 23.1 is a summary of all inspection, testing, and maintenance requirements for fire pumps with references to the parts of NFPA 25 that detail the requirements and who must perform such activities.

Table 8.1 of NFPA 25 contains all of the inspection, testing, and maintenance requirements for fire pumps along with specific frequencies for each component/checkpoint. The table is reproduced here as Table 23.2 for reference.

Health care facility owners are legally responsible for maintaining fire pumps and keeping them in good operating condition. Health care facility engineering staff, the sprinkler contractor, and the AHJ may all be involved in inspecting fire pump systems.

Table 23.1 Summary of NFPA 25 References for Fire Pump Inspection, Testing, and Maintenance

Item	Activity	Frequency	NFPA 25 Reference	Performed by
Pump house, heating and ventilating louvers	Inspection	Weekly	8.2.2(1)	Facility staff
Fire pump system	Inspection	Weekly	8.2.2(2)	Facility staff
Pump operation				
No-flow condition	Test	Weekly	8.3.1	Facility staff
Flow condition	Test	Annually	8.3.3.1	Sprinkler contractor
Hydraulic	Maintenance	Annually	8.5	Sprinkler contractor
Mechanical transmission	Maintenance	Annually	8.5	Sprinkler contractor
Electrical system	Maintenance	Varies	8.5	Sprinkler contractor
Controller, various components	Maintenance	Varies	8.5	Sprinkler contractor
Motor	Maintenance	Annually	8.5	Sprinkler contractor
Diesel engine system, various components	Maintenance	Varies	8.5	Sprinkler contractor

Source: Adapted from NFPA 25, 2002, Table 8.1.

Table 23.2 Summary of Fire Pump Inspection, Testing, and Maintenance Requirements

Complete as Applicable	Visual Inspection	Check	Change	Clean	Test	Frequency
A. Pump System						
1. Lubricate pump bearings			X			Annually
2. Check pump shaft end play		X				Annually
3. Check accuracy of pressure gauges and sensors		X	X			Annually (change or recalibrate when 5% out of calibration)
4. Check pump coupling alignment		X				Annually
5. Wet pit suction screens		X		X		After each pump operation
B. Mechanical Transmission						
1. Lubricate coupling			X			Annually
2. Lubricate right-angle gear drive			X			Annually
C. Electrical System						
1. Exercise isolating switch and circuit breaker					X	Monthly
2. Trip circuit breaker (if mechanism provided)					X	Annually
3. Operate manual starting means (electrical)					X	Semiannually
4. Inspect and operate emergency manual starting means (without power)	X				X	Annually
5. Tighten electrical connections as necessary		X				Annually
6. Lubricate mechanical moving parts (excluding starters and relays)		X				Annually
7. Calibrate pressure switch settings		X				Annually
8. Grease motor bearings			X			Annually

Table 23.2 Summary of Fire Pump Inspection, Testing, and Maintenance Requirements—cont'd

Complete as Applicable	Visual Inspection	Check	Change	Clean	Test	Frequency
D. Diesel Engine System						
1. Fuel						
(a) Tank level	X	X				Weekly
(b) Tank float switch	X				X	Weekly
(c) Solenoids valve operation	X				X	Weekly
(d) Strainer, filter, or dirt leg, or combination thereof				X		Quarterly
(e) Water and foreign material in tank				X		Annually
(f) Water in system		X		X		Weekly
(g) Flexible hoses and connectors	X					Weekly
(h) Tank vents and overflow piping unobstructed		X			X	Annually
(i) Piping	X					Annually
2. Lubrication System						
(a) Oil level	X	X				Weekly
(b) Oil change			X			50 hours or annually
(c) Oil filter(s)			X			50 hours or annually
(d) Lube oil heater		X				Weekly
(e) Crankcase breather	X		X	X		Quarterly
3. Cooling System						
(a) Level	X	X				Weekly
(b) Antifreeze protection level					X	Semiannually
(c) Antifreeze			X			Annually
(d) Adequate cooling water to heat exchanger		X				Weekly
(e) Rod out heat exchanger					X	Annually
(f) Water pump(s)	X					Weekly
(g) Condition of flexible hoses and connections	X	X				Weekly
(h) Jacket water heater		X				Weekly

Table 23.2 Summary of Fire Pump Inspection, Testing, and Maintenance Requirements—cont'd

Complete as Applicable	Visual Inspection	Check	Change	Clean	Test	Frequency
(i) Inspect duct work, clean louvers (combustion air)	X	X	X			Annually
(j) Water strainer				X		Quarterly
4. Exhaust System						
(a) Leakage	X	X				Weekly
(b) Drain condensate trap		X				Weekly
(c) Insulation and fire hazards	X					Quarterly
(d) Excessive back pressure					X	Annually
(e) Exhaust system hangers and supports	X					Annually
(f) Flexible exhaust section	X					Semiannually
5. Battery System						
(a) Electrolyte level		X				Weekly
(b) Terminals clean and tight	X	X				Quarterly
(c) Remove corrosion, case exterior clean and dry	X		X			Monthly
(d) Specific gravity or state of charge					X	Monthly
(e) Charger and charge rate	X					Monthly
(f) Equalize charge		X				Monthly
6. Electrical System						
(a) General inspection	X					Weekly
(b) Tighten control and power wiring connections		X				Annually
(c) Wire chafing where subject to movement	X	X				Quarterly
(d) Operation of safeties and alarms		X			X	Semiannually
(e) Boxes, panels, and cabinets				X		Semiannually
(f) Circuit breakers or fuses	X	X				Monthly
(g) Circuit breakers or fuses			X			Biennially

Source: NFPA 25, 2002, Table 8.1.

The health care facility engineering staff typically conducts all of the weekly inspections shown in Table 23.1, including the inspection of the water supply and all valves. It is essential for qualified operating personnel to be in attendance during weekly pump operations.

The activities that require testing or maintenance of the system are best performed by a qualified sprinkler contractor. The testing and maintenance agreement the facility has with a sprinkler contractor should stipulate that the system must comply with the requirements of NFPA 25. In negotiating such an agreement, the health care facility should review the details of the contractor's scope of work to determine which activities the facility staff will perform and which activities will be contracted. The contract should require that a current copy of NFPA 25 (not just the latest edition but the edition that is current for the local jurisdiction) be provided to the facility at the beginning of the contract and whenever a newer edition is adopted locally. See the applicable edition of NFPA 25 for specific requirements for inspection, testing, and maintenance, including the required frequency of the activities and required procedures.

When the results of tests, inspections, and preventive maintenance activities reveal that fire pumps are out of service, those responsible for repairing the equipment must follow procedures listed in Chapter 14 of NFPA 25. These procedures cover how to handle preplanned and emergency impairments for any water-based fire protection system, including fire pumps. Details on the equipment that must be used for restoring systems to service are also provided.

23.5.1 Inspecting Fire Pumps

Health care facility engineering personnel should inspect the fire pump. The purpose of an inspection is to check visually that a particular device is indeed still in place; in operating condition; free from physical damage, foreign material, and corrosion; and not otherwise damaged or impaired. NFPA 25 requires the following visual observations to be performed weekly:

- Pump house/room conditions
  - Heat is adequate, not less than 40°F (4.4°C) [70°F (21°C) for pump room with diesel pumps without engine heaters].
  - Ventilating louvers are free to operate.
- Pump system conditions
  - Pump suction and discharge and bypass valves are fully open.
  - Piping is free of leaks.
  - Suction line pressure gauge reading is normal.
  - System line pressure gauge reading is normal.
  - Suction reservoir is full.
  - Wet pit suction screens are unobstructed and in place.
- Electrical system conditions
  - Controller pilot light (power on) is illuminated.
  - Transfer switch normal pilot light is illuminated.
  - Isolating switch is closed—standby (emergency) source.

- Reverse phase alarm pilot light is off or normal phase rotation pilot light is on.
- Oil level in vertical motor sight glass is normal.
- Diesel engine system conditions
  - Fuel tank is at least two-thirds full.
  - Controller selector switch is in AUTO position.
  - Voltage readings of the batteries (2) are normal.
  - Charging currents of the batteries (2) are normal.
  - Battery pilot lights (2) are on or battery failure pilot lights (2) are off.
  - All alarm pilot lights are off.
  - Engine running-time meter is reading.
  - Oil level in right-angle gear drive is normal.
  - Crankcase oil level is normal.
  - Cooling water level is normal.
  - Electrolyte level in batteries is normal.
  - Battery terminals are free from corrosion.
  - Water-jacket heater is operating.
- Steam system conditions
  - Steam pressure-gauge reading is normal.

23.5.2 Testing Fire Pumps

Fire pumps are tested for a variety of reasons at various periods—during installation, before the equipment is approved for going on line, and throughout the life of a piece of equipment. Testing can be characterized in two groups, performance testing and acceptance testing. See Chapters 12 and 35 of this book for a discussion of general system testing.

23.5.2.1 Installation (Performance) and Acceptance Testing

Installation contractors normally conduct start-and-stop tests after they have completely installed the pump to make sure it operates properly. Authorities having jurisdiction normally test the installation of the fire pumps before accepting them as complete and operational. Conducting successful performance tests before the AHJ conducts the acceptance tests increases the chance that the acceptance tests will be successful as well.

Testing the pump assembly ensures that the pump operates automatically or manually on demand and continuously delivers the required system output. The test also detects deficiencies of the pump assembly that are not always evident by inspection. It is important to coordinate fire pump testing with the AHJ, sprinkler contractor, and facility personnel to minimize problems, delays, and damage to the surrounding landscape and objects.

23.5.2.2 Weekly Fire Pump Testing

Owners (themselves or through maintenance contractors) conduct routine fire pump tests throughout the life of the device to ensure that the equipment remains in proper operating order. NFPA 25 requires facilities to conduct weekly testing of fire pumps without flowing water. This test must be conducted by starting the pump automatically and running electrically driven pumps for a minimum of 10 minutes and diesel-driven pumps for 30 minutes. Steam turbine–driven pump assemblies must also be tested weekly, but the standard does not specify a minimum run time for this test. Before 1992, the codes did not require weekly testing of electric motor–driven pump assemblies.

The following is a checklist of pertinent visual observations or adjustments that must be made while the pump is running:

- Pump system procedure
  - Record the system suction and discharge pressure gauge readings.
  - Check the pump packing glands for slight discharge.
  - Adjust gland nuts if necessary.
  - Check for unusual noise or vibration.
  - Check packing boxes, bearings, or pump casing for overheating.
  - Record the pump starting pressure.
- Electrical system procedure
  - Observe the time for the motor to accelerate to full speed.
  - Record the time the controller is on the first step (for reduced-voltage or reduced-current starting).
  - Record the time the pump runs after starting (for automatic-stop controllers).
- Diesel engine system procedure
  - Observe the time for the engine to crank.
  - Observe the time for the engine to reach running speed.
  - Observe the engine oil pressure gauge, speed indicator, water, and oil temperature indicators periodically while the engine is running.
  - Record any abnormalities.
  - Check the heat exchanger for cooling water flow.
- Steam system procedure
  - Record the steam pressure gauge reading.
  - Observe the time for the turbine to reach running speed.

23.5.2.3 Annual Testing

Annual tests of each pump assembly must be conducted under minimum, rated, and peak flows of the fire pump by controlling the quantity of water discharged through approved test devices. The contractor who normally services the health care facility's fire protection systems should conduct these tests. Health care facility personnel should make sure that the testing contractor takes precautions to prevent water damage by verifying that there is adequate drainage for the high-pressure water discharge from the hoses. Even where there are adequate drainage facilities, the discharge water can cause damage to nearby landscaping, automobiles, and anything else in the discharge area.

The following is a checklist of pertinent visual observations, measurements, and adjustments that should be conducted annually while the pump is running and flowing water under the specified output condition. This information is not intended to be a complete list of all of the requirements of NFPA 25. Rather, it is a list to help health care facility personnel prepare for the test and assist them in checking the performance of the testing contractor, who is responsible for thoroughly conducting the tests. The contractor should test the following:

- At no-flow condition/churn. (Conduct this test first.)
  - Check the circulation relief valve for operation to discharge water.
  - Check the pressure relief valve (if installed) for proper operation.
  - Continue the test for ½ hour.
- At each flow condition
  - Record the electric motor voltage and current (all lines).
  - Record the pump speed in revolutions per minute (rpm).
  - Record the simultaneous (approximate) readings of pump suction and discharge pressures and pump discharge flow.
  - Observe the operation of any alarm indicators or any visible abnormalities.

Where a fire pump installation has an automatic transfer switch, the overcurrent protective devices (i.e., fuses or circuit breakers) must be tested to make sure that they do not open. A normal power failure must be simulated while the pump is delivering peak power output to cause the pump motor to connect to the alternate power source. Although the sprinkler contractor or pump company representative should conduct this test, the test requires close coordination with the health care facility engineering staff to *simulate* a power failure to the fire pump while not actually causing a power failure to the balance of the facility.

Other items that must be tested annually include the following:

- All local and remote alarm indicating devices (visual and audible), by simulating alarm conditions
- Emergency generators supplying emergency or standby power to fire pump assemblies (see NFPA 110)
- Appropriate environmental pump room space conditions (e.g., heating, ventilation, illumination) to ensure proper manual or automatic operation of the associated equipment

23.5.2.4 Test Results and Evaluation

The contractor (or others skilled in such matters) should interpret the test results. Health care facilities should require the contractor to compare the pump test curve to the unadjusted field acceptance test curve and the previous annual test curve(s). The pump must be capable of supplying the maximum sprinkler or standpipe system demand.

Any abnormality observed during inspection or testing should be reported promptly to the facility's chief engineer, and the abnormality should be corrected as soon as possible. The health care facility should record and retain the contractor's test report for future comparison purposes. It cannot be assumed that the contractor keeps complete records of any facility's fire pump installation.

23.5.3 Maintaining Fire Pumps

Fire pump system maintenance and repairs must be conducted in accordance with Table 8.1 of NFPA 25 (see Table 23.2). In-house maintenance of certain fire pump items should be performed only when the maintenance staff has been specifically trained to conduct such work. Except for conducting some basic repairs, such as to the pump room's heating, ventilation, and illumination, in-house personnel usually do not have the qualifications or experience to maintain fire pump equipment.

Outsourcing fire pump system maintenance to a qualified sprinkler contractor provides a health care facility with several benefits, such as:

- Maintenance personnel are specialists in the field of fire pumps.
- Contractors have the specialized tools to maintain the system.
- Contractors have working relationships with the manufacturers and suppliers of the components or spare parts for valves, fittings, and attachments.
- The maintenance liability is "passed" to the contracted professional.
- The workload is transferred away from the health care facility's maintenance personnel.

The health care facility should require the testing contractor to enlist the services of a manufacturer's representative for each brand of pump and pump controller being tested and maintained. Note that the pump and pump controller are usually manufactured by separate companies.

NFPA 25 requires facilities with fire pumps to establish a preventive maintenance program for all components of the pump assembly in accordance with the manufacturer's recommendations. Records should be maintained on all work performed on the pump, driver, controller, and auxiliary equipment. In the absence of manufacturer's recommendations for preventive maintenance, Table 8.1 of NFPA 25 (see Table 23.2) provides alternative requirements. This preventive maintenance program must be initiated immediately after the pump assembly has passed acceptance tests.

24

Kitchen Hood Fire-Extinguishing Systems

Note: *The engineering staff should advise all kitchen employees that if a fire starts on the cooking equipment, they should pull the manual release station for the extinguishing system first. This could be faster than an automatic activation, and it should shut off the cooking energy source, activate any attached alarm devices, and alert the local fire department. Fire extinguishers should be used only as a supplement to the fixed system and must be compatible with its extinguishing agent.*

Kitchen hood extinguishing systems are designed to detect and suppress a special class of fire: a grease fire on the cooking appliances or in the kitchen hood ventilation system. The grease may be rendered animal fat, vegetable shortening, and other such oily matter used for or resulting from cooking or preparing foods. Grease might be liberated and entrained with exhaust air or might be visible as a liquid or solid. A significant fire hazard is created when cooking grease condenses inside ducts and on exhaust equipment. These accumulations of grease can be ignited by sparks or by small fires when cooking oil or fat gets overheated. When cooking oil reaches its auto-ignition temperature, it ignites spontaneously.

Cooking fires occur frequently when frying, because cooking oils and fats are heated to their flash points during this process and may reach their auto-ignition temperatures when accidentally overheated or spilled on hot kitchen equipment. Such cooking appliances are equipped with a thermostat that is supposed to keep the cooking oil from reaching its auto-ignition temperature, but if the thermostat malfunctions, a fire can occur.

NFPA 96, *Standard for Ventilation Control and Fire Protection of Commercial Cooking Operations*, 2004 edition (Chapter 10), requires that fire-extinguishing equipment be provided for the protection of grease removal devices, hood exhaust plenums, and exhaust duct systems. The cooking equipment that produces grease-laden vapors and that might be a source of ignition of grease in the hood, grease removal device, or duct must be protected by fire-extinguishing equipment.

The following types of equipment must be protected:

- Grease removal devices
- Hood exhaust plenums
- Exhaust duct systems

- Deep-fat fryers
- Ranges
- Griddles
- Broilers
- Woks
- Tilting skillets
- Braising pans
- Any equipment that may ignite grease in the ventilation hood

Fire-extinguishing equipment in cooking areas must include both automatic and portable devices. NFPA 10, *Standard for Portable Fire Extinguishers*, addresses requirements for portable fire extinguishers in cooking areas.

This chapter provides information on codes and standards in place to protect fixed and portable cooking equipment installed in public and private cooking operations (except single-family residences) from fire hazards and the types of equipment approved for use in this regard. Knowledge about the methods used to protect cooking equipment and the design, installation, operation, inspection, testing, and maintenance of this equipment should assist facility engineers in contributing to safe health care facilities.

24.1 Methods of Control

Most existing commercial kitchen equipment in health care facilities is protected by a listed dry or wet chemical type of fire-extinguishing system. These systems suppress fires by several means:

- *Saponification*—chemically converting the fatty acid in cooking oil to a soap or foam that smothers the fire
- *Cooling*—lowering the temperature of the grease below its auto-ignition temperature
- *Radiation shielding*—using the cloud of extinguishing agent to shield the cooking oil from heat that is radiating from the flames
- *Chemical interaction*—breaking down the chemical chain reaction of the fire

In saponification, the foam blanket created by the interaction of the extinguishing agent with the oil must be maintained long enough for the oil to cool or the fire will reflash. Higher temperatures cause this foam blanket to break down even faster.

Gradual changes in commercial cooking equipment have necessitated rethinking how to control these fires. Specifically, two changes have affected the design of fryers, griddles, ranges, charbroilers, and woks and the associated fire protection systems. The first change relates to a shift in the type of cooking oil used. With the move toward preparing more low-cholesterol, low-fat foods, health care organizations have been changing the oils they use to fry foods. Vegetable oils are becoming more popular than the traditional animal fats used for many years. The difference between vegetable oils and animal fats is that vegetable oils have higher auto-ignition temperatures. Therefore, if a vegetable oil ignites spontaneously, it results in a much hotter fire that is more difficult to suppress.

The second change is that today's cooking equipment is more energy efficient and is designed a bit differently than equipment of just a few years ago. Energy efficiency may be achieved by providing more insulation, changing the shape of the vat, or relocating the heating elements. These enhancements result in equipment that takes more time to cool down after a fire has been suppressed. Equipment that stays hotter longer means that the foam blanket must stay intact that much longer. Also, some new shapes of vats with heating elements installed along the front of the equipment cause the cooking oil to circulate or "roll." This circulation of the oil is beneficial for cooking items at a uniform temperature but can more readily break up a foam blanket and allow a fire to reflash.

24.2 UL 300–Compliant Suppression Systems

To respond to changes in cooking equipment and the types of oils used, Underwriters Laboratories adopted UL 300, *Standard for Fire Testing of Fire Extinguishing Systems for Protection of Commercial Cooking Equipment,* an updated fire test standard effective as of November 21, 1994. To pass this standard, newer extinguishing system must satisfy several criteria, all of which are more stringent than previous requirements, and the equipment requires greater amounts of extinguishing agent and different placement of the nozzle than with previous systems. Today, the older and newer systems are being referred to as pre–UL 300 and new–UL 300 systems, respectively.

Determining the need to change an existing system to meet the new standards is not a straightforward issue. As such, the authorities having jurisdiction (AHJs) who govern the acceptability of fire protection installations should be consulted as to the interpretation of the applicable codes or standards and whether they require a system to be upgraded. AHJ interpretation aside, the following are generally accepted:

- Existing pre–UL 300 systems may remain in service as long as the system is installed in its original location.
- When new cooking appliances are added to a kitchen protected by a pre–UL 300 extinguishing system, the entire system does not need to be upgraded to the new standard *if the existing system is listed to protect the new equipment*. In most cases, the use of new cooking appliances or oils requires an extinguishing system upgrade.
- If new–UL 300 extinguishing equipment is added (such as a new extinguishing-agent cylinder), then the system may need to be upgraded. Checking with the AHJ is recommended.
- When upgrading an existing kitchen extinguishing system with new–UL 300–compliant equipment, the kitchen hood duct and plenum protection do not need to be replaced if they are protected by a separate system. Only the cooking appliance protection needs to be upgraded.

- All new cooking appliance protection systems manufactured on or after November 21, 1994, must be in compliance with UL 300. Existing systems that do not need to be upgraded may be serviced with replacement parts manufactured before November 21, 1994. An exception to this requirement is that automatic fire-extinguishing equipment provided as part of listed recirculating systems that comply with standard UL 197, *Standard for Commercial Electric Cooking Appliances,* do not need to comply with UL 300. Thus, fire protection codes and standards do not necessarily require an extinguishing system to be retrofitted to meet new standards. Required or not, the new–UL 300 systems provide a higher level of protection to facility occupants, equipment, and buildings.

24.3 System Types

A complete kitchen extinguishing system includes a detection–automatic release mechanism, a suppression-agent supply, a piping distribution system, discharge nozzle(s), manual release station(s), and a means to shut off the fuel or electric supply to the cooking appliances upon system operation (see Figure 24.1). All kitchen hood

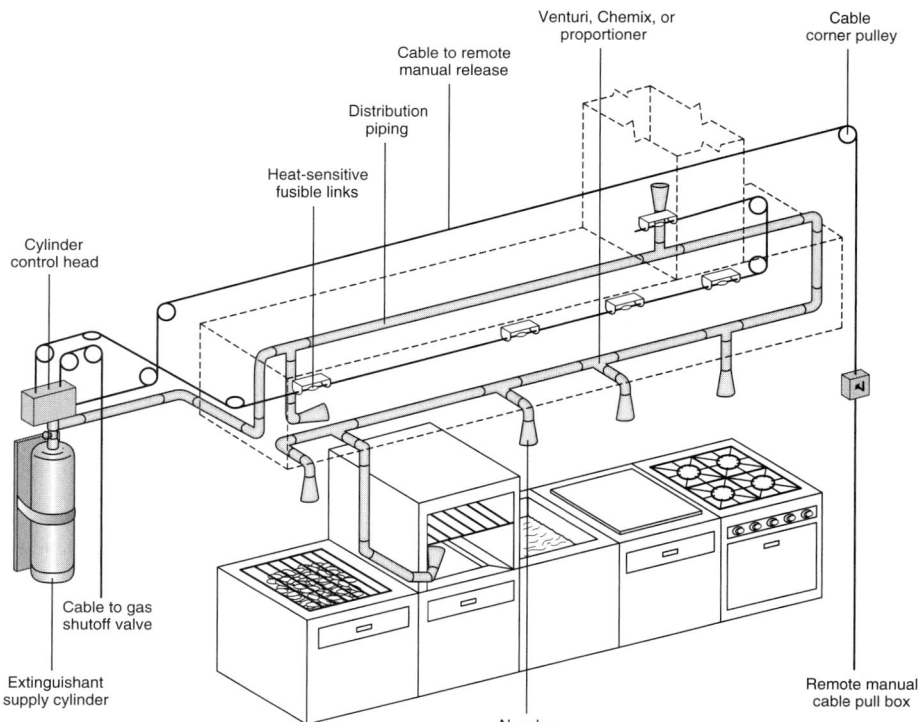

Figure 24.1 Schematic of typical kitchen hood fire suppression system.
Source: Courtesy of Chemetron Fire Systems.

extinguishing systems must be connected to the building's fire alarm system to annunciate an alarm if the system is activated.

Many kitchen extinguishing systems are pre-engineered. Pre-engineered systems have predetermined flow rates, nozzle placements, and quantities of extinguishing agent. They also incorporate specific nozzles and methods of application that can differ from those detailed elsewhere in installation standards and those that are listed by a testing laboratory. The hazards protected by these systems are specifically limited as to type and size. Limitations on hazards that can be protected by these systems are included in the manufacturer's installation manual, which is referenced as part of the listing.

Several specific types of extinguishing systems are available and used in health care facilities, although NFPA 96 does not dictate what type shall be utilized. These extinguishing systems are often referred to as *restaurant systems* and are classified as *special-hazard systems*. Each automatic fire-extinguishing system type must be installed in accordance with the terms of its listing, the manufacturer's instructions, and the standards listed in Table 24.1.

24.3.1 Dry Chemical Kitchen Hood Systems

The first special-hazard systems designed to protect kitchen exhaust hoods and ducts were dry chemical types. These systems usually include a pressure vessel similar to having a fire extinguisher permanently mounted adjacent to the hazard, with fixed piping and nozzles in the hood and ductwork. Fusible links connected to the actuating mechanism by a cable-and-pulley system melt or fuse at predetermined temperatures to discharge the chemical dry powder. Only alkaline dry chemicals can be used to protect deep-fat fryers (sodium bicarbonate, potassium bicarbonate, etc.) and saponify fats and oils. Multipurpose dry chemical agents must never be used for these hazards.

Although these systems are not being provided in new installations today because no dry chemical extinguishing system has yet passed the new UL 300 test, many existing systems are still in service. When dry chemical kitchen hood systems were manufactured, they were pre-engineered, which kept their design and installation inexpensive and quick. This in turn made them very popular, except when they discharged and soiled the entire kitchen area.

Table 24.1 Standards Governing Requirements for Installation of Automatic Extinguishing Systems

Extinguishing System (Agents)	Applicable NFPA Standard
Dry chemical	NFPA 17, *Standard for Dry Chemical Extinguishing Systems*, 2002 edition
Wet chemical	NFPA 17A, *Standard for Wet Chemical Extinguishing Systems*, 2002 edition
Wet-pipe sprinkler	NFPA 13, *Standard for the Installation of Sprinkler Systems*, 2002 edition
Carbon dioxide	NFPA 12, *Standard on Carbon Dioxide Extinguishing Systems*, 2005 edition

24.3.2 Wet Chemical Kitchen Hood Systems

Eventually, manufacturers of dry chemical systems tested and brought to market wet chemical versions of their dry chemical systems. These systems tend to do a better job of suppression, as the wet agent absorbs more heat than its dry chemical predecessor. Additionally, because the wet fire-fighting agent does not become airborne and reach all areas of the kitchen, cleanup is less extensive than when dry chemical systems discharge. Similar to dry chemical systems, these systems are pre-engineered for application to various kitchen equipment configurations. Wet chemical extinguishing systems have been listed in accordance with the UL 300 standard.

24.3.3 Water-Assisted Wet Chemical Kitchen Hood Systems

The water-assisted wet chemical system is essentially a wet chemical system with a water supply that continues to flow after the entire supply of the wet chemical agent has been released. As with the wet chemical system, the agent reacts with the hot grease to form a smothering foam blanket (saponification), and then the water continues to cool the grease and the affected cooking equipment. Pre-engineered water-assisted wet chemical extinguishing systems have been listed in accordance with the UL 300 standard.

24.3.4 Wet-Pipe Sprinkler Kitchen Hood Systems

NFPA 13, 2002 edition (Section 7.9), details the requirements for protecting commercial-type cooking and ventilation equipment by wet-pipe sprinkler systems. These systems were pioneered by the research and development efforts of Gaylord Industries, Inc. The greatest benefit of these systems is the use of water as a suppression agent. In Gaylord's Quencher system, a control panel is connected to a water supply and provides all alarms, supervision, and associated control functions, such as signals to shut off the energy source to the cooking equipment. These systems do not have a local manual release, because to release water, the sprinklers must fuse due to the heat of the fire.

Wet-pipe sprinkler systems extinguish grease fires through a massive amount of cooling of the fuel and cooking surfaces. The advantages of using this type of system are that water is a much cleaner suppression agent and is less expensive than dry or wet chemicals. Additionally, when connected to a sprinkler system, the supply of the agent is essentially endless. This is opposed to the other kitchen suppression agents, which are designed for one-time application and then must be refilled. Wet-pipe automatic sprinkler systems are not pre-engineered and must be designed within the limitations of NFPA 13 and the manufacturers' recommendations.

24.3.5 Carbon Dioxide Kitchen Hood Systems

Although protecting kitchen range hoods and ducts with carbon dioxide is not very common, it is an approved system for such applications. Carbon dioxide kitchen hood protection is accomplished by a combination of total flooding and using local application systems. Carbon dioxide extinguishes fires by displacing oxygen and providing a tremendous amount of cooling.

The installation of dampers at either the top or the bottom of the duct is considered essential, with provisions for automatic closing of the dampers when carbon dioxide begins to discharge. For long ducts, the gas is introduced at intermediate points to ensure proper distribution. The quantities of carbon dioxide required for protecting deep-fat fryers and other specific fire hazards below the hood must be in addition to the quantities required in the exhaust duct. Automatic fire detection and actuation of the system is required for concealed spaces above the filter, in the duct system, and below filters over any deep-fat cookers.

Special attention should be given to the choice of heat detectors, considering the normal operating temperature level and temperature-rise conditions of the range equipment. Actuation of the system must automatically close dampers, shut off forced ventilating fans, and shut off the master fuel valve or power switch to all cooking equipment associated with the hood. In addition to normal system maintenance, particular care should be given to keeping heat detectors and discharge nozzles clean of grease accumulation. Generally, nozzle seals or caps are required to keep nozzle orifices free of obstruction.

The major advantage of using carbon dioxide kitchen hood systems is the agent. Carbon dioxide is clean, which is important when considering that the activation of a "dirty" agent can have health department implications throughout the cooking and serving areas, even when the discharge occurs in only one part of the kitchen.

24.4 System Design and Modification

A number of factors should be considered when designing or renovating health care facility cooking areas. How the systems must be configured, installed, operated, and monitored and where to locate manual release stations, storage cylinders, and control panels are a few main concerns when designing a new system. Selecting a portable fire extinguisher that is compatible with the fixed kitchen hood fire-extinguishing system is another important concern. As with any fire protection system, modifying the protected space, such as moving equipment around or experiencing an increase in the hazard, usually necessitates modifying the system.

24.4.1 System Configuration and Operation

Some of the main requirements for designing and configuring kitchen fire-extinguishing systems are as follows:

- In some cases, fixed-pipe extinguishing systems in single-hazard areas (i.e., areas that must be protected by an automatic sprinkler system *or* with a fire barrier having a 1-hour fire resistance rating; see Chapter 15) must be arranged for simultaneous automatic operation on actuation of any one of the extinguishing systems. Simultaneous automatic operation is not required where the fixed-pipe extinguishing system is an automatic sprinkler system or where a dry or wet chemical system is used to protect common exhaust ductwork by one of the methods specified in NFPA 17 or NFPA 17A.

- All heat-producing sources of fuel and electric power for cooking equipment that requires protection must shut off automatically when any of the equipment's fire-extinguishing systems activate, except for equipment that is supplied with steam from an external source.
- Any gas appliance not requiring protection but located under the same ventilating equipment as protected equipment must also shut off automatically upon activation of any extinguishing system. (Many existing systems, and some relatively new systems, may not have this feature. Due to its importance, retrofitting with automatic shutdown should be done.)
- The automatic and manual means of activating the system external to the control head or releasing device must be separate and independent of each other, so that failure of one will not impair the operation of the other. The manual means of system activation can be common with the automatic means, if the manual activation device is located between the control head or releasing device and the first fusible link. (See NFPA 96, 2004 edition, Section 10.5, for more on manual releasing.)
- Hoods containing automatic fire-extinguishing systems are protected areas. Therefore, these hoods are not considered obstructions to overhead sprinkler systems and do not require floor coverage underneath.
- Fixed-baffle hood assemblies containing a constant or fire-actuated water-wash system listed to extinguish a fire in these devices can be used as protective equipment. The cooking equipment extinguishing system must activate the water wash in the hood.
- Cooking equipment must not be operated while its fire-extinguishing system or exhaust system is nonoperational or otherwise impaired.
- Instructions for manually operating the fire-extinguishing system must be posted conspicuously and reviewed periodically with employees.

24.4.2 Alarms

Requirements regarding alarms for kitchen fire-extinguishing systems include the following:

- An audible alarm or visual indicator must be provided to show that an automatic fire-extinguishing system has activated.
- The automatic fire-extinguishing system must activate the fire alarm signaling system that serves the occupancy within which the kitchen extinguishing system resides.
- The means for the manual actuators of fire-extinguishing equipment must be mechanical and not rely on electrical power for actuation. Electrical power can be used to operate the automatic fire-extinguishing system, if the system is monitored by a supervisory alarm and a standby power supply is provided, except in two situations. One situation is where the automatic fire-extinguishing system includes automatic mechanical detection and actuation as a backup detection system. The second situation is where the fire-extinguishing system is interconnected or

interlocked with the cooking equipment power sources so that if the fire-extinguishing system becomes inoperable due to power failure, all sources of fuel or electric power that produce heat to all cooking equipment serviced by that hood shut off automatically.

24.4.3 Manual Release Stations

During a fire, personnel cannot be expected to travel to a dead-end area of the kitchen (where they could be trapped by the fire) to activate the manual release station. The manual activation should not be too close to the cooking appliances, either, so that employees do not have to get close to the fire to activate the device. Therefore, manual release stations should be readily accessible and located between 42 and 60 in. (1067 and 1524 mm) above the floor along the means of egress from the portion of the kitchen where cooking is conducted. The manual means of activation should also clearly identify the hazard protected.

Health care facilities should be aware of existing manual release station locations during any effort to reconfigure a kitchen space and make sure that the stations are not blocked by carts, items being stored, or similar items during normal kitchen operations. Facility involvement with the designer and installer can result in selecting the most suitable locations for manual release stations. Kitchen personnel should be consulted during the design phase of a project to determine where kitchen apparatus can be placed to avoid blocking or physically damaging the extinguishing system or the manual release stations.

Shutoff devices for shutting off the fuel supply (gas or electric) to the cooking equipment must be reset manually.

24.4.4 Storage Cylinders and Control Panels

The location of suppression-agent storage cylinders or sprinkler control panels is another important feature of kitchen hood fire-extinguishing systems. These devices must be located so that they are readily accessible for inspection and maintenance. Storage cylinders that are located at heights above 4 ft (1219 mm) may be "out of the way" of the kitchen staff but also very difficult to inspect or remove to recharge.

24.4.5 Portable Fire Extinguishers

Portable fire extinguishers are required throughout the health care facility, and the kitchen is no exception. Specifically, per NFPA 10, portable fire extinguishers must be provided in areas where there is a potential for fires that involve combustible cooking media. The most important facet of portable fire extinguishers in kitchen areas is the type of agent they contain. If the portable fire extinguisher's agent is not compatible with the fixed kitchen hood system's agent and the two agents mix, they can become completely ineffective in suppressing a fire.

Fires in cooking appliances that involve combustible cooking media (vegetable or animal oils and fats) are considered Class K fires in accordance with NFPA 10 and must use extinguishers intended specifically for these hazards. Portable fire extinguishers and extinguishing agents for protecting Class K hazards must be either of wet chemical or dry chemical type in that they must saponify fires. Class A extinguishers, Class B gas-type portables such as CO_2 and halon, and Class C–listed extinguishers are not permitted in kitchen cooking areas. Manufacturers' recommendations must be followed. Extinguishers installed prior to June 30, 1998, do not have to be Class K–listed extinguishers, but due to the nature of the hazard, health care facilities would be wise to replace all non–K-listed extinguishers.

Portable fire extinguishers must be installed in kitchen cooking areas in accordance with NFPA 10. Similar to the manual control for kitchen fire-extinguishing systems, portable extinguishers for Class K fires must be accessible during a fire in the cooking equipment. As such, they must be located so that the maximum travel distance from the hazard to the extinguishers does not exceed 30 ft (9.15 m).

A placard must be placed conspicuously near the extinguisher, stating that the fixed fire protection system must be activated before using the fire extinguisher (i.e., the portable extinguisher is secondary to the automatic fire suppression system).

24.4.6 Equipment Relocation

Many kitchen extinguishing systems have nozzles that are located within specific limits to protect the type of cooking appliance being protected. However, many cooking appliances are mounted on wheels, allowing them to be easily relocated to configure the kitchen to produce various meals with greater efficiency. The health care facility must be aware of the sensitivity of the extinguishing system and the impact of relocating cooking equipment. Before any equipment is moved, consult the fire protection installation or maintenance contractor.

24.5 Inspection and Maintenance

As with all types of systems installed in health care facilities, kitchen hood fire-extinguishing systems must be inspected and maintained. Types of inspections include those that owners must conduct as well as those conducted by service contractors.

24.5.1 Owner's Inspection and Maintenance

Facility owners must conduct monthly inspections of kitchen hood fire protection systems in accordance with the manufacturer's listed installation and maintenance manual. As a minimum, this "quick check" or inspection shall include verification of the following:

- The extinguishing system is in its proper location.
- The manual actuators are unobstructed.
- The tamper indicators and seals are intact.
- The maintenance tag or certificate is in place.
- The system shows no physical damage or condition that might prevent operation.
- The pressure gauge(s), if provided, is in operable range.

- The nozzle blowoff caps, where provided, are intact and undamaged.
- Neither the protected equipment nor the hazard has been replaced, modified, or relocated.

24.5.2 Contractor's Inspection and Servicing

NFPA 96 requires fire-extinguishing systems and listed exhaust systems to be inspected and serviced at least every 6 months (semiannually) either by properly trained and qualified persons or by companies acceptable to the AHJ. The entire exhaust system is required to be inspected for grease buildup by a properly trained, qualified, and certified company or person(s) acceptable to the authority having jurisdiction in accordance with Table 24.2.

24.5.2.1 Cleaning

One of the most basic aspects of maintenance with regard to preventing fires in cooking areas is to make sure that exhaust ducts are cleaned often. If a stovetop fire occurs and the hood and duct are free of grease accumulations, the fire can be extinguished or allowed to burn out without causing appreciable damage. Of course, fire prevention is the best protection.

Upon inspection, if exhaust systems are found to be contaminated with deposits from grease-laden vapors, a properly trained, qualified, and certified company or person acceptable to the AHJ must clean the entire system. Hoods, grease-removal devices, fans, ducts, and other appurtenances must be cleaned to bare metal at frequent intervals, before the surfaces become heavily contaminated with grease or oily sludge. After the exhaust system is cleaned to bare metal, it must not be coated with powder or other substance.

At the start of the cleaning process, electrical switches that could be activated accidentally must be locked out. Components of the fire suppression system must not be rendered inoperable during the cleaning process, except during servicing by properly trained and qualified persons. Flammable solvents or other flammable cleaning aids cannot be used.

When a vent-cleaning service is used, the facility must keep on the premise a certificate that shows the date of inspection or cleaning. After cleaning is completed, the vent-cleaning contractor must place or display within the kitchen area a label that indi-

Table 24.2 Schedule of Inspection for Grease Buildup

Type or Volume of Cooking	Frequency
Systems serving solid-fuel cooking operations	Monthly
Systems serving high-volume cooking operations such as 24-hour cooking, charbroiling, or wok cooking	Quarterly
Systems serving moderate-volume cooking operations	Semiannually
Systems serving low-volume cooking operations, such as churches, day camps, seasonal businesses, or senior centers	Annually

Source: NFPA 96, 2004, Table 11.3.

cates the date the vent was cleaned and the name of the servicing company. The label must also indicate the areas not cleaned.

24.5.2.2 Actuators

All actuation components, including remote manual pull stations, mechanical or electrical devices, detectors, actuators, and fire-actuated dampers, must be checked for proper operation during the inspection in accordance with the manufacturers' listed procedures. A discharge of the agent is not normally part of this test. The contractor should also check to see that the hazard has not changed.

24.5.2.3 Sprinklers and Links

Fusible links (including fusible links on fire-actuated damper assemblies) and automatic sprinkler heads must be replaced at least semiannually, or more frequently if necessary where required by the equipment manufacturer. Other detection devices should be serviced or replaced in accordance with the manufacturers' recommendations. One exception is that automatic bulb-type sprinklers or spray nozzles do not need to be changed annually if annual examination shows no buildup of grease or other material on the sprinkler or spray nozzles.

24.5.2.4 Agent Distribution

The installation or maintenance contractor must verify that the agent distribution piping is not obstructed and that dry chemical agents are not caked. Where semiannual maintenance of any dry chemical containers or system components reveals corrosion or pitting in excess of the manufacturer's limits; structural damage or fire damage; repairs by soldering, welding, or brazing; or other types of damage, the affected part(s) must be replaced or tested hydrostatically in accordance with the recommendations of the manufacturer or the listing agency.

All kitchen hood extinguishing systems (except wet-pipe systems) must be recharged after use or as indicated by an inspection or a maintenance check. Systems must be recharged in accordance with the manufacturers' listed installation and maintenance manuals. The building owner is responsible for hiring a qualified contractor to recharge the system.

24.5.2.5 System Maintenance

When the maintenance of a system reveals that a system contains defective parts that could cause the system to work improperly or fail, the contractor must replace or repair the affected parts in accordance with the manufacturer's recommendations. The contractor is required to file a maintenance report, including any recommendations, with the health care facility.

Each system must have a tag or label that indicates the month and year that maintenance was performed and identifies the person who performed the service. Only the current tag or label must remain in place. It is good practice for the health care facility engineer to forward certificates of inspection and maintenance to the AHJ.

The health care facility personnel who conduct these inspections must keep records for those extinguishing systems that were found to require corrective actions. At least monthly, the date the inspection is performed and the initials of the person performing the inspection must be recorded on inspection tags attached to the suppression-agent cylinder. The records should be retained until the next maintenance.

24.6 Testing

Hydrostatic testing should be performed only by persons trained in pressure-testing procedures and safeguards. These persons should have available suitable testing equipment, facilities, and an appropriate service manual(s). The suppression-agents containers of kitchen hood extinguishing systems must be subjected to a hydrostatic pressure test at intervals not exceeding 12 years. Dry and wet chemical agent removed from the containers prior to hydrostatic testing should be properly discarded by the contractor.

Care must be taken to ensure that all tested equipment is thoroughly dried prior to reuse. To protect the hazard (cooking equipment) during hydrostatic testing, if there is no connected reserve, alternative protection acceptable to the AHJ is required, such as a fire watch with Class K fire extinguishers.

25

Specialized Suppression Systems

Suppression systems are used where the item or hazard being protected, such as sensitive electronic equipment, may be damaged by the application of a different suppression medium or destroyed if the fire is not suppressed quickly enough. Many suppression systems do not have a particular history of being used in health care facilities because they are not particular to health care—they are particular to the hazard being protected. Suppression systems are also not necessarily required in health care occupancies. Building codes and NFPA *101*®, *Life Safety Code*®, require suppression systems in health care facilities, but they do not specify what type of system must be installed. Some examples of where these systems are used include a computer room or a magnetic resonance imaging (MRI) installation.

This chapter details some of the more popular specialized fire suppression systems that may be used in health care facilities. Not every type of suppression system is detailed, because of the great number of systems that are available and the quick pace at which research is being conducted on new suppression agents. Information on system testing, inspection, and maintenance is also provided. Facilities that are considering the installation of such systems or those that already have such systems should employ the services of a qualified fire protection engineer or contractor who has specific expertise in these systems.

25.1 Clean Fire Suppression Agents

The use of specialized fire suppression systems depends on the need to protect a special piece of equipment or process or the environmental need for the suppression agent to be *environmentally acceptable*. NFPA 2001, *Standard on Clean Agent Fire Extinguishing Systems,* 2004 edition, states that clean agent fire-extinguishing systems are useful (within certain limits) in extinguishing fires in specific hazards or equipment, in occupancies where an electrically nonconductive medium is essential or desirable, or where cleanup of other media presents a problem. Total-flooding clean agent fire-extinguishing systems are used primarily to protect hazards that are in enclosures or equipment that, in itself, includes an enclosure to contain the agent. Some typical hazards that could be suitable for using clean agent fire-extinguishing systems include, but are not limited to, the following:

- Electrical and electronic hazards
- Subfloors and other concealed spaces with combustible loads such as electrical cabling

- Flammable and combustible liquids and gases
- High-value assets
- Telecommunications facilities

25.2 Gaseous Fire Suppression Agents

With the exception of kitchen hood fire suppression systems, the earliest "gaseous" system was carbon dioxide, which was followed by halon systems. These systems are gaseous systems that achieve suppression by "flooding" the protected space with the extinguishing agent (gas) to a particular concentration. NFPA 12, *Standard on Carbon Dioxide Extinguishing Systems,* was first published in 1929, and NFPA 12A, *Standard on Halon 1301 Fire Extinguishing Systems,* was first adopted in 1970.

The following agents are presented (in no particular order) for comparison purposes. This book is not intended to cover all total-flooding clean agent fire-extinguishing systems. Technology in this area is under constant development, and it is anticipated that this list of agents will be expanded in future revisions of this book.

25.2.1 Halon 1301

Halocarbon name: BCF-1301
Chemical name: Bromotrifluoromethane
Chemical formula: $CBrF_3$

Halon 1301 was the agent of choice for many years because of its excellent fire suppression ability. However, according to NFPA, because of the deleterious effect on the environment (ozone depletion), "global production of fire protection halons ceased in many developing countries on January 1, 1994" ("Halon and Beyond," 1994). Thus, the availability of Halon 1301 today and in the future is extremely restricted.

The environmental reason why halon is not a good choice for a suppression agent has also resulted in a several-fold increase in the price of existing halon supplies from *halon banks*. These banks store the agent and sell it to those who, for a variety of reasons, choose or need to maintain Halon 1301 systems. Some Halon 1301 systems still in use today are those installed in the past that have not yet actuated or those protecting mission-critical facilities. Many owners of equipment containing Halon 1301 have decided to keep these systems until the next time the systems discharge and then provide suitable replacements. One caution with such an approach is that those responsible for this equipment must be aware of the amount of time required to design, procure, and install a replacement system (i.e., how long can the facility risk not having a suppression system in place?). If a health care facility has existing Halon 1301 systems that are going to be decommissioned, the facility should investigate the benefit of selling the halon supply to a halon bank to offset the cost of a new replacement system.

25.2.2 FE-13™

Halocarbon name: HFC-23
Chemical name: Trifluoromethane
Chemical formula: HCF_3
Manufacturer: DuPont

25.2.2.1 Overview

FE-13 is a clean, environmentally acceptable, and human-compatible fire suppression agent. It has been developed as an alternative fire suppression agent to Halon 1301 as a result of the phaseout of the latter agent. (Since FE-13 contains no chlorine or bromine, its ozone depletion potential is zero.) FE-13 is an existing DuPont Chemical product that currently has several uses, which lowers the cost of the product through the economy of scale.

FE-13 is a UL-listed gaseous fire-extinguishing agent and has obtained Factory Mutual approval. HFC-23 is also listed as an acceptable extinguishing agent in NFPA 2001.

25.2.2.2 Physical and Chemical Properties

FE-13 extinguishes fires by both physical and chemical means. Primarily, it raises the total heat capacity of the environment to the point that the atmosphere will not support combustion. Extinguishment is also achieved because the agent removes the free radicals that serve to maintain the combustion process.

The FE-13 extinguishing concentration for normal heptane is approximately 12 percent by volume, versus 3.5 percent for Halon 1301. However, on a weight basis, when compared to Halon 1301, only 1.7 times by weight of FE-13 is needed for extinguishment. The manufacturer's recommended minimum design concentration for total-flooding application is 16 to 18 percent.

FE-13 has a much higher vapor pressure than Halon 1301 and therefore requires storage and a piping system designed for these higher pressures. When exposed to high temperatures (above 850°F, 454°C), FE-13 decomposes to form hydrogen fluoride (HF), which can potentially corrode sensitive metals and materials. However, when an FE-13 extinguishing system is properly designed and installed, temperatures within the enclosure should not rise significantly.

Short-term and extended inhalation studies on animals, including histological (tissue) examination, indicates that FE-13 is chemically and biologically unreactive. Although FE-13 is considered safe for use in occupied areas, occupants should still evacuate the room before a system discharge.

25.2.3 FE-227™ or FM200

Halocarbon name: HFC-227ea
Chemical name: 1,1,1,2,3,3,3 Heptafluoropropane
Chemical formula: CF_3CHFCF_3
Manufacturer: DuPont or Great Lakes Chemical

25.2.3.1 Overview

FE-227 or FM200 is a liquefied compressed gaseous fire suppression agent also developed as a result of the phaseout of Halon 1301. It is stored as a liquid and is discharged into the protected space as a clear, colorless vapor that does not obscure vision. FE-227 or FM200 extinguishes fire by a combination of physical and chemical mechanisms.

FE-227 or FM200 is a UL-listed gaseous fire-extinguishing agent that has obtained Factory Mutual approval. FE-227 or FM200 is also listed as an acceptable extinguishing agent in NFPA 2001.

25.2.3.2 Safety of FE-227 or FM200 and Design Considerations

FE-227 or FM200 is safe to use in occupied spaces because it does not displace oxygen, leaves no residue when discharged, and has acceptable toxicity for use in occupied spaces at design concentrations. Based on a review of the toxicity data, the U.S. Environmental Protection Agency determined that FE-227 or FM200 is safe for exposure to concentrations as great as 9.0 percent by volume. Higher exposures up to 10.5 percent in concentration are allowed for up to 1 min. Although FE-227 or FM200 is considered safe for use in occupied areas, occupants should still evacuate the room before a system discharge.

The design concentration for FE-227 or FM200 for extinguishing fires involving electrical equipment is determined by testing as part of the listing program (7.4 to 8.7 percent). The minimum design concentration includes a 20 percent safety factor.

FE-227 or FM200 is electrically nonconductive and is safe for use on delicate electronic equipment. Therefore, it is typically used to protect electronic, telecommunications, and data processing equipment. Consideration should be given to the possibility of FE-227 or FM200 migrating to adjacent areas outside the protected space.

When exposed to extremely high temperatures (above 850°F, 454°C), FE-227 or FM200 decomposes to form hydrogen fluoride (HF), which can potentially corrode sensitive metals and materials. However, when FE-227 or FM200 extinguishing systems are properly designed and installed, temperatures within the enclosure should not rise significantly.

25.2.4 Inergen

Chemical name: IG-541
Chemical makup: 52% N, 40% Ar, 8% CO_2
Manufacturer: Ansul

25.2.4.1 Overview

Inergen (inert gases and nitrogen) is another alternative fire suppression agent developed as a result of the phaseout of Halon 1301. Inergen consists of 52 percent nitrogen, 40 percent argon, and 8 percent carbon dioxide. These are all elements that occur naturally in the atmosphere.

Inergen is a UL-listed gaseous fire-extinguishing agent that has obtained Factory Mutual approval. NFPA 2001 lists Inergen as an acceptable extinguishing agent for the protection of occupied spaces. Although Inergen is considered safe for use in occupied areas, occupants should still evacuate the room before system discharge.

Inergen extinguishes fires in Class A (ordinary combustibles), Class B (flammable liquids), and Class C (electrical hazards). This agent is typically used to protect MRI rooms, computer rooms, and tape storage rooms, as well as areas containing electronic equipment that is either very sensitive or irreplaceable.

25.2.4.2 Properties of Inergen

Inergen is a noncorrosive, electrically nonconductive gas that does not support combustion and does not react with most substances. Inergen uses carbon dioxide to promote breathing characteristics that are intended to sustain life in the oxygen-deficient environment for the protection of personnel.

Inergen extinguishes fire by lowering the oxygen content below the level that supports combustion. When Inergen is discharged into a room, it still allows a person to breathe in a reduced-oxygen atmosphere. The normal atmosphere in a room contains approximately 21 percent oxygen and less than 1 percent carbon dioxide. When the oxygen content is reduced below 15 percent, most ordinary combustibles will not burn. Inergen reduces the oxygen content to approximately 12.5 percent while increasing the carbon dioxide content to approximately 4 percent. The increase in the carbon dioxide content increases a person's respiration rate and the body's ability to absorb oxygen. In other words, the human body is stimulated by the carbon dioxide to breathe more deeply and rapidly to compensate for the lower oxygen content of the atmosphere.

Inergen is stored as a dry compound gas. Inergen can be refilled at any medical gas refill station that has a filling license from Ansul. Therefore, the availability of places to refill these systems might be high in some geographical locations, depending on the number of facilities taking advantage of this program.

Inergen does not produce decomposition products such as hydrogen fluoride (HF) or bromides (HBr, Br_2), which could potentially corrode sensitive materials.

25.2.5 Novec™ 1230 (or Sapphire™)

Halocarbon name: C6-fluoroketone
Chemical formula: $CF_3CF_2C(O)CF(CF_3)_2$
Manufacturer: 3M

25.2.5.1 Overview

Novec 1230 is a long-term halon alternative and is an environment-friendly extinguishing medium. It is stored as a colorless and almost odorless liquid at room temperature. When Novec 1230 fluid is discharged under pressure by a special extinguishing nozzle, it evaporates instantly and is distributed as a gaseous extinguishing medium in the protected area. Novec 1230 is also marketed by Ansul under the name "Sapphire."

25.2.5.2 Properties of Novec 1230

Novec 1230 offers the lowest global warming potential for halocarbon alternatives and the lowest atmospheric life for halocarbons (five days). As a liquid agent it has an advantage in that it can be shipped in drums and totes rather than pressurized cylinders. That means that Novec 1230 fluid can be air freighted in bulk quantities if needed for refills, instead of the very limited quantities of gases that can be air shipped. Additionally, if a leak occurs in the extinguisher or system after superpressurization, the nitrogen can easily be vented and the agent retained while repairing the cylinder seal or gasket. With gases, the agent would be lost. The liquid is pourable, low in viscosity, and easy to handle. It can easily be pumped with hand or electric pumps. Novec 1230 fluid can be used both as a streaming agent (e.g., hand-held extinguishers) or as a total-flooding agent in fixed systems. The liquid is compatible with a wide range of materials of construction and is stable in storage.

Novec 1230 is a UL- and ULC-listed gaseous fire-extinguishing agent and is also listed as an acceptable extinguishing agent in NFPA 2001.

25.2.6 Agent Cost Comparison

Table 25.1 provides a cost comparison of several clean agents. These data were developed in 2004 by Michael J. Rzeznik, P.E., and Jeffery M. Amato of Schirmer Engineering Corporation and should not be considered applicable to a particular situation, but the data are useful when comparing relative costs between clean agents. The data consider the agent only and not the detection or agent delivery system. This table also provides data on two clean agents not previously discussed in this chapter.

25.2.7 Inspection, Testing, and Maintenance of Gaseous Systems

Inspecting, testing, and maintaining of specialized gaseous fire suppression systems is similar to conducting these tasks for other fire protection systems. Health care facility personnel can perform some of these functions, but they are usually restricted to conducting inspections. Only specialists who are factory trained for the specific system in question should perform detailed testing and maintenance of gaseous fire protection systems. The following information provides guidance as to the inspections that the health care facility can perform and what is expected of a vendor who is contracted to test or maintain these systems.

NFPA 2001 requires that, at least annually, competent personnel thoroughly inspect and test all systems for proper operation. Discharge tests are not required. At least semiannually, the agent quantity and pressure of refillable containers must be checked. The inspection report, with recommendations, must be filed with the owner. It is important to make sure that the contractor supplies the health care facility with all such reports.

All halocarbon clean agent removed from refillable containers during service or maintenance procedures must be collected and recycled or disposed of in an environmentally sound manner and in accordance with existing laws and regulations. Cylinders continuously in service without discharging shall be given a complete external visual inspection every 5 years, or more frequently if required. Only competent personnel should inspect these systems and record the results, both on

Table 25.1 Cost of Agent to Flood a 10,000 ft³ Room at 70°F

Commercial Name	Design Concentration	Amount of Agent	Unit Cost	Total Cost
FM200	7%	341 lb.	$12/lb.	$4092
FE-13	18%	450 lb.	$11/lb.	$4950
FE-25	1%	274 lb.	$13/lb.	$3562
Novec 1230	7%	372 lb.	$12/lb.	$4464
Inergen	34%	4270 ft³	$1.65/ft³	$7046
Argonite	37.9%	4760 ft³	$1.13/ft³	$5379
Halon 1301	5%	184 lb.	$19/lb.	$3496

record tags permanently attached to each cylinder and in suitable inspection reports. These records must be retained by the health care facility for the life of the system.

All system hose must be examined annually for damage. Health care facility staff can conduct these visual examinations in addition to the examinations conducted by the maintenance contractor. If a visual examination shows any deficiency, the hose should be immediately replaced or tested.

At least every 12 months, the health care facility must also inspect the enclosure (e.g., room) protected by the clean agent. The enclosure should be thoroughly inspected (including above ceilings and below raised floors) to determine if penetrations or other changes have occurred that could adversely affect agent leakage, change the volume of hazard, or both. Where an inspection indicates conditions that could result in the inability to maintain the concentration of the clean agent, the conditions must be corrected. If uncertainty regarding the conditions still exists, NFPA 2001 requires the enclosures to be retested for integrity. There is an exception to the enclosure inspection requirement if a documented administrative control program exists that addresses barrier integrity.

When any actuation, impairment, and restoration of these systems results in an impairment of the protection, the health care facility must report the situation to the authority having jurisdiction. All persons who could be expected to inspect, test, maintain, or operate fire-extinguishing systems must be thoroughly trained and remain knowledgeable in these functions. If the health care facility personnel are not comfortable with inspecting gaseous suppression systems, it is recommended that all such work be explicitly contracted with the system vendor. The health care facility personnel working in an enclosure protected by a clean agent should receive training regarding agent safety issues.

25.3 Water Mist Extinguishing Systems

The following excerpt from Section 10, Chapter 1, by John A. Frank, in NFPA's *Fire Protection Handbook,* 19th edition (2003), clearly explains the use of water as a fire suppression agent:

> Water is the most widely used and available fire-extinguishing agent. Water is inexpensive, abundant, and effective in fire suppression. Water is transportable and can be pumped from a source to the fire. Water is available from domestic water distribution systems (fire hydrants), streams, wells, ponds, lakes, and swimming pools.
>
> Water is a very effective agent for controlling and extinguishing combustion. Water is the most abundant and available substance on the earth's surface. Water comes in three states:
>
> (1) liquid (water), (2) vapor (steam), and (3) solid (ice).
>
> Life safety considerations must be weighed when choosing an effective fire-extinguishing agent. Water as an agent is safe, nontoxic, relatively noncorrosive, and stable. Water (H_2O) remains stable when applied to a fire and does not, except in very

special circumstances, break down into its basic elements of hydrogen (H) and oxygen (O), both of which would encourage fire growth. Water can be applied as an extinguishing agent with building occupants in compartments, unlike some gaseous extinguishing agents, which could cause asphyxiation or adverse side effects.

The physical properties that allow water to be an effective extinguishing agent are as follows.

1. At ordinary temperatures, water exists as a stable liquid. Water's viscosity in the temperature range of 34 to 210°F (1 to 99°C) remains consistent, which allows it to be transported and pumped.
2. Water has a high density, which allows it to be discharged from and projected from nozzles, etc. Water's surface tension allows it to exist from small droplets to a solid stream.
3. The latent heat of fusion is the amount of energy required to change the state of water from solid (ice) at 32°F (0°C) to a liquid. Water absorbs 143.4 Btu per lb (333.2 kJ/kg) in this process.
4. The specific heat of water is 1.0 Btu per lb (4.186 kJ/kg K). For example, raising the temperature of 1 lb (0.45 kg) of water 180°F from 32°F (0°C) to 212°F (100°C) requires 180 Btu.
5. Water is effective as a cooling agent because of its high latent heat of evaporation (changing water from a liquid to vapor), which is 970.3 Btu per lb (2260 kJ/kg) as described in Section 2, Chapter 3, "Chemistry and Physics of Fire."
6. Water expands in its conversion from a liquid state to a vapor state to 1600 to 1700 times the liquid volume. One gallon (3.8 L) of liquid [which occupies 0.1337 cu ft (0.004 m³)] produces over 223 cu ft (6.3 m³) of steam.

Therefore, it follows that 1 gal (3.8 L) of water at room temperature applied to a fire and converted to steam (complete conversion) will absorb heat both in rising to the temperature at which it becomes a vapor and in the phase change from liquid to vapor. (For SI units: F° − (°C × 9/5) = 32; 1 Btu = 1.055 kJ; 1 lb = 0.45 kg; 1 gal = 3.785 L; 1 cu ft = 0.0283 m³.)

Heat required to raise temperature of water to boiling:

- 212°F − 68°F (room temperature) = 144 Δ°F
- 144 Δ°F × 1 Btu/lb × 8.33 lb (weight of 1 gal) = 1200 Btu

Heat required to change water from a liquid to a vapor:

- 970.3 Btu/lb × 8.33 lb (weight of 1 gal) = 8083 Btu
- Total heat absorbed is 1200 + 8083 + 9283 Btu/gal of water

Therefore, a fire department hose stream discharging at 100 gpm will absorb 928,300 Btu per minute in a complete conversion. The same fire department hose stream will create 22,300 cu ft per min of steam in a complete conversion.

25.3.1 Water Mist Fire System Applications

NFPA 750, *Standard on Water Mist Fire Protection Systems*, 2003 edition, states that water mist systems were introduced in the 1940s and were utilized for specific applications, such as on passenger ferries. The renewed interest in water systems is due partially to the phasing out of halon and their potential as a fire safety system for spaces where the amount of water that can be stored or that can be discharged is limited. In addition, their application and effectiveness for residential occupancies, flammable liquids storage facilities, and electrical equipment spaces continue to be investigated with encouraging results. NFPA began developing NFPA 750 in 1993 and released the first approved standard in 1996.

There are three application systems for water mist systems:

1. Local application—direct application and distribution of the mist around the hazard or object being protected
2. Total compartment application—total flooding of the compartment or enclosure with water mist
3. Zoned application—protection of a predetermined portion of a compartment with the mist

There are four primary components for water mist systems:

1. Water supply
2. Water distribution network
3. Nozzles
4. Fire detection and releasing system

The International Marine Organization (IMO) is the organization that sets safety standards including fire protection standards for commercial marine applications. (Note that the United States, Canada, and most Western European countries are members of the IMO.) The IMO has approved water mist systems for shipboard use, including passenger and crew compartments in which the fire hazard consists of Class A–type combustibles. Standard fire tests have been established as a basis to evaluate these systems. Although the results of the tests are intended for evaluation of water mist systems for use on ships, the information is useful for health care facilities, since the test fires involved Class A, ordinary combustibles.

A published report by Hughes and Associates, Inc. (1995), provides the results of a feasibility study of water mist systems against residential fires. This information is useful for health care facilities because the fires represented Class A types of fires with relatively low average fuel loads.

25.3.2 System Design Characteristics

Four characteristics of water mist have been found to impact its effectiveness as an extinguishing agent: (1) distribution, (2) flux density, (3) spray momentum, and (4) additives. Past research and development work have demonstrated that mild saline (i.e., saltwater) solutions have been of some benefit in extinguishing flammable liquid pan fires.

Various manufacturers are taking different approaches in the active development of water mist hardware. The variety of approaches is evident with the span of nozzle operating pressures, which range from less than 150 psi (1034 kPa) up to approximately 3000 psi (20,684 kPa). The nozzle types fall into three general categories: impingement nozzles, pressure jet nozzles, and air-atomizing nozzles.

Impingement nozzles are hydraulic nozzles in which a water jet strikes a deflector that breaks the water up into fine droplets. These nozzles can operate at generally lower pressure, less than 150 psi (1034 kPa), to develop a mist. These types also have a benefit in that the orifices are not easily blocked. However, since the stream hits a deflector and changes direction, the spray momentum is reduced. Also, it is difficult to produce a high percentage of fine droplets with this type of nozzle.

The second type, the *pressure jet nozzle,* produces fine droplets with one or more jets, which flow at high velocity through small orifices. These nozzles can operate from very low to very high pressures. These nozzles can produce sprays with fine droplet sizes and higher momentum than impingement-type nozzles. However, these systems require more energy to obtain fine sprays, and special measures must be taken to prevent blockage of the orifices of small nozzles.

The third type of mist-generating nozzle is the *air-atomizing nozzle.* This nozzle combines low-pressure compressed air and nitrogen or some other gas with water and also operates at a relatively low pressure. The water jet is broken up into finer droplets by air injected in the water stream. The water jet then is projected out of larger orifices than the pressure jet types. This nozzle is less prone to blockage, but the system requires a separate gas cylinder and possibly a separate water storage container, and separate water and air pressure piping must be installed.

25.3.3 Inspection, Testing, and Maintenance of Water Mist Systems

Water mist systems must be inspected, tested, and maintained in accordance with NFPA 750. It is the responsibility of the health care facility to ensure that inspection, testing, and maintenance are conducted at the required intervals. Tables 25.2, 25.3, and 25.4, reproduced from NFPA 750, contain this information.

Table 25.2 Inspection Frequencies

Item	Activity	Frequency
Water tank (unsupervised)	Check water level	Weekly
Air receiver (unsupervised)	Check air pressure	Weekly
Dedicated air compressor (unsupervised)	Check air pressure	Weekly
Water tank (supervised)	Check water level	Monthly
Air receiver (supervised)	Check air pressure	Monthly
Dedicated air compressor (supervised)	Check air pressure	Monthly
Air pressure cylinders (unsupervised)	Check pressure and indicator disk	Monthly
System operating components, including control valves (locked/unsupervised)	Inspect	Monthly
Air pressure cylinders (supervised)	Check pressure and indicator disk	Quarterly
System operating components, including control valves	Inspect	Quarterly
Water flow alarm and supervisory devices	Inspect	Quarterly
Initiating devices and detectors	Inspect	Semiannually
Batteries, control panel, interface equipment	Inspect	Semiannually
System strainers and filters	Inspect	Annually
Control equipment, fiber optic cable connections	Inspect	Annually
Piping, fittings, hangers, nozzles, flexible tubing	Inspect	Annually

Source: NFPA 750, 2003, Table 13.2.2.

Table 25.3 Testing Frequencies

Item	Activity	Frequency
Pumps	Operation test (no flow)	Weekly
Compressor (dedicated)	Start	Monthly
Control equipment (functions, fuses, interfaces, primary power, remote alarm) (unsupervised)	Test	Quarterly
System main drain	Drain test	Quarterly
Remote alarm annunciation	Test	Annually
Pumps	Function test (full flow)	Annually
Batteries	Test	Semiannually
Pressure relief valve	Manually operate	Semiannually
Control equipment (functions, fuses, interfaces, primary power, remote alarm) (supervised)	Test	Annually
Water level switch	Test	Annually
Detectors (other than single use or self-testing)	Test	Annually
Release mechanisms (manual and automatic)	Test	Annually
Control unit/programmable logic control	Test	Annually
Section valve	Function test	Annually
Water	Analysis of contents	Annually
Pressure cylinders (normally at atmospheric pressure)	Pressurize cylinder (discharge if possible)	Annually
System	Flow test	Annually
Pressure cylinders	Hydrostatic test	5–12 years
Automatic nozzles	Test (random sample)	20 years

Source: NFPA 750, 2003, Table 13.2.2.

Table 25.4 Maintenance Frequencies

Item	Activity	Frequency
Water tank	Drain and refill	Annually
System	Flushing	Annually
Strainers and filters	Clean or replace as required	After system operation

Source: NFPA 750, 2003, Table 13.2.2.

As with all fire protection system maintenance, a qualified contractor should be retained to assist the health care facility with keeping the systems in proper working order.

26

Portable Fire Extinguishers

An extremely important point that must be communicated to all facility staff is that the fire department should be notified *before or at the same time as* any staff member attempts to fight a fire. History has shown that delaying the notification of the fire department and/or the annunciation of a fire alarm allows a fire to grow to a more dangerous size before the fire department can arrive, thereby threatening patients, visitors, and staff.

Portable fire extinguishers have been an effective means to control or suppress fires for more than 100 years. In 1918, the NFPA Committee of Field Practice was active in developing a standard on first aid protection. The earliest official NFPA standard on portable fire extinguishers was adopted in 1921. Early portable fire extinguishers such as the "soda and acid extinguisher" and the cartridge-operated water extinguisher provided years of good service but today are obsolete because of hazards they created to the user and their limited effectiveness. Modern portable fire extinguishers provide a safe extinguishing agent, relatively uniform operation, and a reliable (safe) container for the agent.

This chapter details the application of portable fire extinguishers in health care facilities. Portable fire extinguishers are examined in terms of the buildings where they are required, their layout (i.e., locations within a building), various types, and how they must be inspected, tested, and maintained. Staff training is also discussed.

26.1 Buildings Where Required

NFPA *101*®, *Life Safety Code*®, requires all new and existing health care occupancies to have portable fire extinguishers. Similarly, the *International Building Code* requires health care occupancies, laboratory spaces, commercial kitchens, shop areas, and completed floors of buildings under construction to have portable fire extinguishers. As always, facility personnel must check with the local authority having jurisdiction (AHJ) for specific requirements for portable fire extinguisher requirements.

26.2 Classification of Fire Types and Extinguisher Ratings

The effectiveness of various extinguishing agents is not uniform on different types of fires. NFPA 10, *Standard for Portable Fire Extinguishers*, 2002 edition, discusses the size and placement of fire extinguishers for the following five classes of fires:

1. **Class A:** Fires involving ordinary combustible materials such as wood, paper, cloth, rubber, and many plastics. These materials usually form ashes when they burn. This class of fire can be remembered by thinking of "A" for *ashes*.
2. **Class B:** Fires involving flammable or combustible liquids, flammable gases, oils, and greases. These materials usually form heavy black smoke when they burn. Thinking of "B" for *black* smoke can assist in remembering this class of fire.
3. **Class C:** Fires involving energized electrical equipment. Note that if the equipment is de-energized, then the fire is no longer a Class C fire and usually becomes a Class A fire. It is easy to remember "C" for *current* (electricity).
4. **Class D:** Fires in certain combustible metals (e.g., magnesium, sodium). This is a very special type of fire that may be found in an industrial or laboratory situation but would be highly unlikely in other buildings.
5. **Class K:** Fires in cooking appliances that involve combustible cooking media (i.e., vegetable or animal oils and fats). It is easy to remember "K" for *kitchen* (refer to Chapter 24).

Some portable fire extinguishers are appropriate for use on only one class of fire, and some are suitable for two or three classes, but no portable fire extinguisher type is suitable for use on all five classifications of fires. Most portable fire extinguishers are labeled so that users can quickly identify the class of fire for which the extinguisher is suitable. NFPA 10 details typical pictorial extinguisher marking labels. See Figures 26.1 and 26.2

Portable fire extinguishers are also rated for relative extinguishing effectiveness. They are assigned a numeral that indicates the amount of fire the portable fire extinguisher should extinguish, as compared to another portable fire extinguisher. For example, a portable fire extinguisher rated at *4-A: 20-B:C* indicates the following:

- The extinguisher should extinguish approximately twice as much Class A fire as a 2-A- (2½ gal water) rated portable fire extinguisher.
- It should extinguish approximately 20 times as much Class B fire as a 1-B-rated portable fire extinguisher.
- It is suitable for use on energized electrical equipment.

See Table 26.1 for a summary of portable fire extinguisher types, sizes, and classes.

26.3 Portable Fire Extinguisher Selection and Placement

The classification of fire types and portable fire extinguisher ratings is key to matching an extinguisher to the hazard to be protected. The nature of the area to be protected and the relative hazard in a building vary according to its fire load. NFPA 10 defines three categories of hazards:

1. Light-(or low-) hazard occupancy
2. Ordinary-(or moderate-) hazard occupancy
3. Extra-(or high-) hazard occupancy

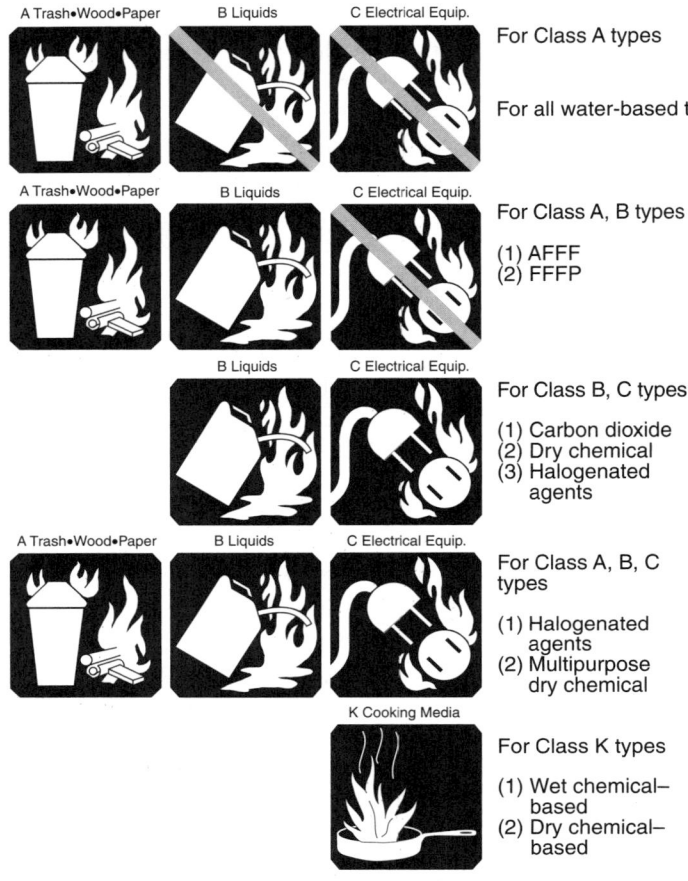

Figure 26.1 Pictorial marking system for recommended extinguishers.
Source: NFPA 10, 2002, Figure B1.1.

Most areas in health care occupancies are classified as light- or ordinary-hazard occupancies for portable fire extinguisher placement. Certain laboratory areas may be considered as extra-hazard occupancies.

In areas of a health care facility where the building is a similar shape or configuration from one floor to the next, it is suggested that portable fire extinguishers be placed in typical locations. Typical locations increase the chance that occupants will know the location or be able to find the nearest extinguisher quickly. Portable fire extinguishers are often found in cabinets (with and without fire department standpipe valves) or suspended from hooks mounted on walls or columns.

Chapter 26. Portable Fire Extinguishers **477**

Ordinary

Combustibles

Extinguishers suitable for Class A fires should be identified by a triangle containing the letter "A." If colored, the triangle is colored green.*

Flammable

Liquids

Extinguishers suitable for Class B fires should be identified by a square containing the letter "B." If colored, the square is colored red.*

Electrical

Equipment

Extinguishers suitable for Class C fires should be identified by a circle containing the letter "C." If colored, the circle is colored blue.*

Combustible

Metals

Extinguishers suitable for fires involving metals should be identified by a five-pointed star containing the letter "D." If colored, the star is colored yellow.*

*Recommended colors, per PMS (Pantone Matching System) include the following:
GREEN — Basic Green
RED — 192 Red
BLUE — Process Blue
YELLOW — Basic Yellow

Figure 26.2 Letter-shaped symbol markings for extinguisher classifications. *Source:* NFPA 10, 2002, Figure B2.2.

The code includes several requirements for the placement of portable fire extinguishers in health care occupancies. Portable fire extinguishers that weigh 40 lb (18.14 kg) or less must be installed so that the top of the extinguisher is not more than 5 ft (1.53 m) above the floor. Extinguishers that weigh more than 40 lb (18.14 kg) should be installed so that the top of the extinguisher is not more than 3½ ft (1.07 m) above the floor. All extinguishers should be mounted so that the clearance between the bottom of the fire extinguisher and the floor is not less than 4 in. (10.2 cm). The minimum 4-in. (10.2-cm) clearance prevents the extinguisher cylinder from getting wet or otherwise damaged from floor cleaning operations.

Table 26.1 Characteristics of Extinguishers

Extinguishing Agent	Method of Operation	Capacity	Horizontal Range of Stream	Approximate Time of Discharge	Protection Required below 40°F (4°C)	UL or ULC Classifications[a]
Water	Stored-pressure	6 L	30 to 40 ft	40 sec	Yes	1-A
	Stored-pressure or pump	2½ gal	30 to 40 ft	1 min	Yes	2-A
	Pump	4 gal	30 to 40 ft	2 min	Yes	3-A
	Pump	5 gal	30 to 40 ft	2 to 3 min	Yes	4-A
Water (wetting agent)	Stored-pressure	1½ gal	20 ft	30 sec	Yes	2-A
Loaded stream	Stored-pressure or cartridge	2½ gal	30 to 40 ft	1 min	No	2 to 3-A:1-B
Water mist	Stored-pressure	1¾ gal	10 to 15 ft	60 to 70 sec	Yes	2-A:C
	Stored-pressure	2½ gal	10 to 15 ft	70 to 80 sec	Yes	2-A:C
AFFF, FFFP	Stored-pressure	2½ gal	20 to 25 ft	50 sec	Yes	3-A:20 to 40-B
	Stored-pressure	6 L	20 to 25 ft	50 sec	Yes	20-A:10-B
Carbon dioxide[b]	Self-expelling	2½ to 5 lb	3 to 8 ft	8 to 30 sec	No	1 to 5-B:C
	Self-expelling	10 to 15 lb	3 to 8 ft	8 to 30 sec	No	2 to 10-B:C
	Self-expelling	20 lb	3 to 8 ft	10 to 30 sec	No	10-B:C
	Self-expelling	50 to 100 lb (wheeled)	3 to 10 ft	10 to 30 sec	No	10 to 20-B:C
Dry chemical (sodium bicarbonate)	Stored-pressure	1 to 2½ lb	5 to 8 ft	8 to 12 sec	No	2 to 10-B:C
	Cartridge or stored-pressure	2¾ to 5 lb	5 to 20 ft	8 to 25 sec	No	5 to 20-B:C
	Cartridge or stored-pressure	6 to 30 lb	5 to 20 ft	10 to 25 sec	No	10 to 160-B:C
	Stored-pressure	50 lb (wheeled)	20 ft	35 sec	No	160-B:C
	Nitrogen cylinder or stored-pressure	75 to 350 lb (wheeled)	15 to 45 ft	20 to 105 sec	No	40 to 320-B:C

Dry chemical (potassium bicarbonate)	Cartridge or stored-pressure	2 to 5 lb	5 to 12 ft	8 to 10 sec	No	5 to 60-B:C
	Cartridge or stored-pressure	5½ to 10 lb	5 to 20 ft	8 to 20 sec	No	10 to 80-B:C
	Cartridge or stored-pressure	16 to 30 lb	10 to 20 ft	8 to 25 sec	No	40 to 160-B:C
	Cartridge or stored-pressure	48 to 50 lb (wheeled)	20 ft	30 to 35 sec	No	120 to 160-B:C
	Nitrogen cylinder or stored-pressure	125 to 315 lb (wheeled)	15 to 45 ft	30 to 80 sec	No	80 to 640-B:C
Dry chemical (potassium chloride)	Cartridge or stored-pressure	2 to 5 lb	5 to 8 ft	8 to 10 sec	No	5 to 10-B:C
	Cartridge or stored-pressure	5 to 9 lb	8 to 12 ft	10 to 15 sec	No	20 to 40-B:C
	Cartridge or stored-pressure	9½ to 20 lb	10 to 15 ft	15 to 20 sec	No	40 to 60-B:C
	Cartridge or stored-pressure	19½ to 30 lb	5 to 20 ft	10 to 25 sec	No	60 to 80-B:C
	Cartridge or stored-pressure	125 to 200 lb (wheeled)	15 to 45 ft	30 to 40 sec	No	160-B:C
Dry chemical (ammonium phosphate)	Stored-pressure	1 to 5 lb	5 to 12 ft	8 to 10 sec	No	1 to 3-A[c] and 2 to 10-B:C
	Stored-pressure or cartridge	2½ to 9 lb	5 to 12 ft	8 to 15 sec	No	1 to 4-A and 10 to 40-B:C
	Stored-pressure or cartridge	9 to 17 lb	5 to 20 ft	10 to 25 sec	No	2 to 20-A and 10 to 80-B:C
	Stored-pressure or cartridge	17 to 30 lb	5 to 20 ft	10 to 25 sec	No	3 to 20-A and 30 to 120-B:C
	Stored-pressure or cartridge Nitrogen cylinder or stored-pressure	45 to 50 lb (wheeled)	20 ft	25 to 35 sec	No	20 to 30-A and 80 to 160-B:C
		110 to 315 lb (wheeled)	15 to 45 ft	30 to 60 sec	No	20 to 40-A and 60 to 320-B:C

(Continued)

Table 26.1 Characteristics of Extinguishers—Cont'd

Extinguishing Agent	Method of Operation	Capacity	Horizontal Range of Stream	Approximate Time of Discharge	Protection Required below 40°F (4°C)	UL or ULC Classifications[a]
Dry chemical (foam compatible)	Cartridge or stored-pressure	4¾ to 9 lb	5 to 20 ft	8 to 10 sec	No	10 to 20-B:C
					No	20 to 30-B:C
	Cartridge or stored-pressure	9 to 27 lb	5 to 20 ft	10 to 25 sec	No	40 to 60-B:C
	Cartridge or stored-pressure	18 to 30 lb	5 to 20 ft	10 to 25 sec	No	80 to 240-B:C
	Nitrogen cylinder or stored-pressure	150 to 350 lb (wheeled)	15 to 45 ft	20 to 150 sec	No	40 to 80-B:C
Dry chemical (potassium bicarbonate urea based)	Stored-pressure	5 to 11 lb	11 to 22 ft	18 sec	No	60 to 160-B:C
	Stored-pressure	9 to 23 lb	15 to 30 ft	17 to 33 sec	No	480-B:C
	Nitrogen cylinder or stored-pressure	175 lb (wheeled)	70 ft	62 sec		
Wet chemical	Stored-pressure	3 L	8 to 12 ft	30 sec	Yes	K
	Stored-pressure	6 L	8 to 12 ft	35 to 45 sec	Yes	2-A:1-B:C:K
	Stored-pressure	2½ gal	8 to 12 ft	75 to 85 sec	Yes	2-A:1-B:C
Halon 1211 (bromo-chlorodifluoromethane)	Stored-pressure	0.9 to 2 lb	6 to 10 ft	8 to 10 sec	No	1 to 2-B:C
	Stored-pressure	2 to 3 lb	6 to 10 ft	8 to 10 sec	No	5-B:C
	Stored-pressure	5½ to 9 lb	9 to 15 ft	8 to 15 sec	No	1-A:10-B:C
	Stored-pressure	13 to 22 lb	14 to 16 ft	10 to 18 sec	No	2 to 4-A and 20 to 80-B:C
	Stored-pressure	50 lb	35 ft	30 sec	No	10-A:120-B:C
	Stored-pressure	150 lb (wheeled)	20 to 35 ft	30 to 44 sec	No	30-A:160 to 240-B:C

Halon 1211/1301 (bromochloro-difluoromethane–bromotrifluoromethane) mixtures	Stored-pressure or self-expelling	0.9 to 5 lb	3 to 12 ft	8 to 10 sec	No	1 to 10-B:C
	Stored-pressure	9 to 20 lb	10 to 18 ft	10 to 22 sec	No	1-A:10-B:C to 4-A:80-B:C
Halocarbon-type	Stored-pressure	1.4 to 150 lb	6 to 35 ft	9 to 23 sec	No	1-B:C to 10-A:90-B:C

Note: Halon should be used only where its unique properties are deemed necessary.

[a] UL and ULC ratings checked as of July 24, 1987. Readers concerned with subsequent ratings should review the pertinent lists and supplements issued by these laboratories: Underwriters Laboratories Inc., 333 Pfingsten Road, Northbrook, IL 60062, or Underwriters Laboratories of Canada, 7 Crouse Road, Scarborough, Ontario, Canada M1R 3A9.

[b] Carbon dioxide extinguishers with metal horns do not carry a C classification.

[c] Some small extinguishers containing ammonium phosphate–based dry chemical do not carry an A classification.

Source: Fire Protection Handbook, NFPA, 2003, Table 11.6.1.

26.3.1 Placement for Class A Hazards

Portable fire extinguishers provided to protect Class A hazards must be located throughout the facility in accordance with Table 26.2. Where the area of the floor of a building is less than that specified in Table 26.2, at least one portable fire extinguisher of the minimum size recommended should be provided. NFPA 10 allows the protection requirements to be fulfilled with portable fire extinguishers of higher ratings, provided the travel distance to the extinguishers does not exceed 75 ft (22.7 m).

26.3.2 Placement for Class B Hazards

Portable fire extinguishers provided to protect Class B hazards must be located throughout the facility in accordance with Table 26.3. Portable fire extinguishers of lesser rating provided for small special hazards may be used but are not considered as fulfilling any part of the requirements presented in Table 26.3. NFPA 10 allows the protection requirements to be fulfilled with portable fire extinguishers of higher ratings, provided the travel distance to the extinguishers does not exceed 50 ft (15.25 m).

If the hazard to be protected consists of flammable liquids of appreciable depth, the placement of portable fire extinguishers must be modified from those shown in Table 26.3. Flammable liquids are considered of appreciable depth where the depth is greater than ¼ in. (0.64 cm). Health care facilities that have flammable liquids of appreciable depth should contact a qualified portable fire extinguisher contractor for selection and placement.

26.3.3 Placement for Class C Hazards

Fire extinguishers with Class C ratings must be placed where energized electrical equipment can be encountered that would require a nonconducting extinguishing medium. Obviously, using a conducting extinguishing medium (e.g., water) to put out a fire caused by this type of equipment could result in electrocution of the extinguisher operator. Class C extinguishers are also required where a fire either directly involves or surrounds electrical equipment. Since the fire involves a Class A or Class B hazard

Table 26.2 Fire Extinguisher Size and Placement for Class A Hazards

	Light- (Low-) Hazard Occupancy	Ordinary- (Moderate-) Hazard Occupancy	Extra- (High-) Hazard Occupancy
Minimum rated single extinguisher	2-A[a]	2-A[a]	4-A[b]
Maximum floor area per unit of A	3000 ft²	1500 ft²	1000 ft²
Maximum floor area for extinguisher	11,250 ft²[c]	11,250 ft²[c]	11,250 ft²[c]
Maximum travel distance to extinguisher	75 ft	75 ft	75 ft

For SI units: 1 ft = 0.305 m; 1 ft² = 0.0929 m².
[a]Up to two water-type extinguishers, each with 1-A rating, can be used to fulfill the requirements of one 2-A-rated extinguisher.
Source: NFPA 10, 2002, Table 5.2.1.
[b]Two 2½-gal (9.46-L) water-type extinguishers can be used to fulfill the requirements of one 4-A-rated extinguisher.
[c]See NFPA 10, Section E.3.3.

Table 26.3 Fire Extinguisher Size and Placement for Class B Hazards

		Maximum Travel Distance to Extinguishers	
Type of Hazard	Basic Minimum Extinguisher Rating	(ft)	(m)
Light (low)	5-B	30	9.15
	10-B	50	15.25
Ordinary (moderate)	10-B	30	9.15
	20-B	50	19.25
Extra (high)	40-B	30	9.15
	80-B	50	15.25

Notes:
1. The specified ratings do not imply that fires of the magnitudes indicated by these ratings will occur, but rather they are provided to give the operators more time and agent to handle difficult spill fires that could occur.
2. For fires involving water-soluble flammable liquids, see NFPA 10, Section 2.3.4.
3. For specific hazard applications, see Section 2.3.
Source: NFPA 10, 2002, Table 5.3.1.

once the electrical circuit is de-energized, the fire extinguishers must be sized and located on the basis of the anticipated Class A or Class B hazard.

26.3.4 Placement for Class D Hazards

Portable fire extinguishers with Class D ratings must be placed in locations where fires involving combustible metals might be expected. Class D portable fire extinguishers must be located not more than 75 ft (23 m) of travel distance from the Class D hazard. Portable fire extinguishers for Class D hazards are required in work areas where combustible metal powders, flakes, shavings, chips, or similarly sized products are generated. Such fires are not common in health care facilities.

26.3.5 Placement for Class K Hazards

Portable fire extinguishers are required in locations where there is a potential for fires involving combustible cooking media (i.e., vegetable or animal oils and fats). The maximum travel distance from the hazard to the extinguishers is limited to 30 ft (9.15 m).

26.4 Extinguisher Inspection, Maintenance, and Recharging

NFPA 10 contains requirements for the inspection, maintenance, and recharging of portable fire extinguishers, as summarized here.

26.4.1 Inspection

Health care facility personnel and outside contractors are allowed to inspect a facility's portable fire extinguishers. Extinguishers must be inspected when initially placed in service and thereafter at approximately 30-day intervals, but at more frequent intervals when circumstances require. Periodic inspection of extinguishers must include verifying at least the following conditions:

- The extinguisher is located in its designated place.
- There are no obstructions to access or visibility.
- Operating instructions are on the nameplate, legible, and facing outward.
- Condition of tires, wheels, carriage, hose, and nozzle are checked (for wheeled units).
- HMIS label is in place.
- Seals and tamper indicators are not broken or missing.
- Fullness is determined by weighing or "hefting."
- The extinguisher is examined for obvious physical damage.
- The pressure gauge indicator is in the operable range or position.

When an inspection of any portable fire extinguisher reveals a deficiency in any of the conditions listed above, the facility should correct the deficiency immediately. Personnel making inspections are required to keep records of those extinguishers that were found to require corrective actions. At least monthly, the date the inspection was performed and the initials of the person performing the inspection should be recorded on the tag affixed to the extinguisher.

26.4.2 Maintenance

Health care facilities should hire qualified portable fire extinguisher contractors to maintain portable fire extinguishers. Maintenance procedures must include a thorough examination of the three basic elements of an extinguisher—mechanical parts, extinguishing agent, and expelling means.

Extinguishers should be subjected to maintenance no more than 1 year apart or when specifically indicated by an inspection. *Stored-pressure types* of the loaded stream agent type must be disassembled on an annual basis and subjected to complete maintenance. Before disassembly, the extinguisher must be fully discharged to check the operation of the discharge valve and pressure gauge. A conductivity test must be conducted annually on all carbon dioxide hose assemblies. Hose assemblies found to be nonconductive must be replaced.

Stored-pressure extinguishers that require a 12-year hydrostatic test must be emptied every 6 years and subjected to applicable maintenance procedures. The removal of the agent from *halogenated agent extinguishers* must be done using a halon closed-recovery system. When the applicable maintenance procedures are performed during periodic recharging and hydrostatic testing, the 6-year requirement begins from that date. An exception to this requirement is that *nonrechargeable extinguishers* must not be hydrostatically tested but removed from service at a maximum of 12 years from the date of manufacture. *Nonrechargeable halogenated agent extinguishers* must be disposed of properly at the time when the 12-year hydrostatic test is required. Portable fire extinguishers that are taken out of service for maintenance or recharge must be replaced by spare extinguishers of the same type and at least equal rating.

During annual maintenance, it is not necessary to internally examine nonrechargeable extinguishers, *carbon dioxide extinguishers,* or *stored-pressure extinguishers* except for those types containing a loaded stream agent. However, such extinguishers must be thoroughly examined externally.

26.4.3 Recharging

Health care facilities should hire qualified portable fire extinguisher contractors to recharge the extinguishers, as well. All rechargeable-type extinguishers must be recharged after any use, as indicated by an inspection, or during maintenance. The recommendations of the manufacturer must be followed when performing the recharging. The amount of recharge agent used must be verified by weighing. The recharged gross weight must be the same as the gross weight that is marked on the label. After recharging, a leak test must be performed on stored-pressure and self-expelling types.

Extinguishers are recharged after any use or during maintenance operations necessitating agent removal. Only those agents specified on the nameplate, or agents proven to have equal chemical composition and physical characteristics, can be used. Multipurpose dry chemicals must not be mixed with alkaline-based dry chemicals. For all nonwater types of extinguishers, any moisture must be removed before recharging.

Halogenated agent extinguishers must be charged only with the proper type and weight of agent as specified on the nameplate. The removal of the agent from halogenated agent extinguishers must be done using a closed-recovery system. The extinguisher cylinder must be examined internally for contamination and corrosion. The agent retained in the system recovery cylinder can be reused only if there is no evidence of internal contamination in the extinguisher cylinder. Halon removed from extinguishers that exhibit evidence of internal contamination or corrosion must be processed in accordance with the extinguisher manufacturer's instructions.

The vapor phase of carbon dioxide must not be less than 99.5 percent carbon dioxide. The water content of the liquid phase must not be more than 0.01 percent by weight [−30°F (−34.4°C) dew point]. Oil content of the carbon dioxide shall not exceed 10 ppm by weight.

When recharging stored-pressure extinguishers, care must be taken not to overcharge. The proper amount of agent is determined by using one of the following methods:

- Exact measurement by weight
- Exact measurement by volume
- Use of an antioverfill tube when provided
- Use of a fill mark on extinguisher shell, if provided

26.4.4 Hydrostatic Testing

Hydrostatic testing must be performed by persons who are trained in conducting pressure testing procedures and their safeguards and who own suitable testing equipment, facilities, and an appropriate servicing manual (outside contractors). If, at any time, an extinguisher shows evidence of corrosion or mechanical injury, it must be tested hydrostatically, except for nonrechargeable fire extinguishers. At intervals not exceeding those specified in Table 26.4, extinguishers shall be tested hydrostatically.

For high-pressure gas cylinders and cartridges passing a hydrostatic test, the month, year, and the U.S. Department of Transportation identification number must be stamped into the cylinder in accordance with the requirements set forth by the

Table 26.4 Hydrostatic Test Intervals for Extinguishers

Extinguisher Type	Test Interval (years)
Stored-pressure water, loaded stream, or antifreeze	5
Wetting agent	5
AFFF (aqueous film-forming foam)	5
FFFP (film-forming fluoroprotein foam)	5
Dry chemical with stainless steel shell	5
Carbon dioxide	5
Wet chemical	5
Dry chemical, stored-pressure, with mild steel shell, brazed brass shell, or aluminum shell	12
Dry chemical, cartridge- or cylinder-operated, with mild steel shell	12
Halogenated agents	12
Dry powder, stored-pressure, cartridge- or cylinder-operated, with mild steel shell	12

Note: Stored-pressure water extinguishers with fiberglass shells (pre-1976) are prohibited from hydrostatic testing due to manufacturer's recall.
Source: NFPA 10, 2002, Table 7.2.

Department of Transportation. Extinguisher shells of the noncompressed gas type that pass a hydrostatic test must have the test information recorded on a suitable metallic label or equally durable material. The label must be affixed to the shell by means of a heatless process. These labels shall be self-destructive when removal from the extinguisher shell is attempted. The label must include the following information:

- Month and year the test was performed, indicated by a perforation, such as by a hand punch
- Test pressure used
- Name or initials of person performing the test, or name of agency performing the test

Hose assemblies passing a hydrostatic test do not require recording.

26.4.5 Labeling

Each portable fire extinguisher must have a tag or label securely attached that indicates the months and years the inspections, maintenance, and recharging were performed and identifies the person performing the service. Information regarding the 6-year maintenance requirement must be included on a separate label. Expired labels must be removed. Labels indicating inspection, maintenance, hydrostatic retests, and 6-year maintenance should not be placed on the front of the extinguisher.

26.5 Employee Training

Virtually all fires are small at first and can be extinguished easily if the proper type and amount of extinguishing agent are applied promptly. Portable fire extinguishers are designed specifically for this purpose, but their successful use depends on the following conditions:

- The extinguisher must be located properly and be in good working order.
- The selected extinguisher must be the proper type for the fire.
- The fire must be discovered while still small enough for the extinguisher to be effective.
- The fire must be discovered by a person who is ready, willing, and able to use the extinguisher.

Of the four conditions above, three of them depend on the person who will be using the portable fire extinguisher. Extinguishers are not effective if the users are not knowledgeable in their selection and operation. NFPA has training materials on the use of fire extinguishers, including a workbook—*Portable Fire Extinguishers*—and two videos—*Fire Extinguishers at Work* and *Fire Extinguishers: Fight or Flight*. The use of this material could be added to a health care facility's in-service training program to improve employees' knowledge about the proper selection and use of portable fire extinguishers. The next level of training would be actual hands-on drills using portable fire extinguishers on real fires in a controlled environment, in conjunction with the local fire department.

27

Fire Detection and Alarm Systems

In a health care facility, the purpose of any fire alarm system is to identify a fire early in its development and to indicate and warn of abnormal conditions, to allow the summoning of appropriate aid, and to allow the control of occupancy facilities to enhance protection of life. For example, in addition to activating chimes or strobe lights, fire alarm systems may also return elevators to a main floor (so that occupants do not continue to use them and the elevators are available to the fire department), shut down electrical systems and air-handling equipment, and initiate some automatic suppression systems.

Fire alarm systems have been required in health care facilities for many years, but the codes did not always require them to be complex and their area of coverage to be extensive. For example, the 1967 edition of NFPA *101®*, *Life Safety Code®*, required every building to have an electrically supervised, manually operated fire alarm system; no automatic detection was required, with the exception of sprinkler water flow alarms. The 1967 *Life Safety Code* also prohibited *presignal* systems (i.e., fire alarm systems that alert only persons in certain constantly attended locations who are responsible to respond, and then after a timed delay alert persons throughout the facility; see the definitions section). Additionally, the 1967 *Life Safety Code* required audible alarms in all nonpatient areas but permitted visible alarm devices only in patient sleeping room spaces.

Today's fire alarm system requirements are more extensive and specific than those of the past. There are two reasons for this. First, the level of understanding of fire protection in health care settings has increased, which is reflected in the codes and standards. Second, the increase in technology has resulted in very sophisticated alarm systems that are microprocessor based and can detect, alarm, and control to a level not even imagined just 10 years ago. Today, automatic smoke detection is required in specific locations and situations, with a variety of output functions initiated by the alarm system. Presignal systems are still prohibited by the *Life Safety Code*, 2006 edition. Where occupants are incapable of self-preservation, only the attendants and other personnel required to evacuate occupants from a zone, area, floor, or building are required to be notified of the actuation of a fire alarm.

The *Life Safety Code,* the *Building Construction and Safety Code,* and the model building codes require fire alarm systems in hospitals, limited care facilities, nursing homes, and ambulatory health care occupancies. These codes require such systems because of a building's height, area, and occupancy (also called "use group classification"). These codes include general requirements for fire alarm systems and then enhance those requirements in special sections specific to health care facilities. Installation and performance requirements are covered by *NFPA 72®*, *National Fire Alarm Code®*. Before any system is designed or modified, a fire protection engineer who regularly practices in the field of fire alarm system design and specification should be retained to determine the specific requirements of the applicable codes and standards required by the authorities having jurisdiction (AHJ).

This chapter details the application of fire alarm systems to health care facilities. Fire alarm systems are examined in terms of system types, fire alarm control panels, alarm devices, specialized features of fire alarm systems in health care facilities, and considerations for inspection, testing, and maintenance. Procedures to follow in case of a fire are also discussed.

27.1 System Types

Fire alarm systems can have devices that must be activated manually or those that detect fires automatically.

27.1.1 Manual Fire Alarm Systems

Manual fire alarms systems consist of input devices that require some action from the building occupants to initiate the alarm signal. These systems are usually relatively simple and consist of manual fire alarm boxes (often referred to as *manual pull stations*—see Figure 27.1), control equipment, and alarm notification appliances. Manual fire alarm boxes must be approved for the particular application and must be used only for fire-protective signaling purposes, except where the boxes are used as a combination fire alarm and guard's tour station. The actuation of a manual fire alarm box usually results in the transmission of an alarm signal within the protected area or to a constantly attended location (e.g., a telephone operator's office, private central station service, or the local fire department).

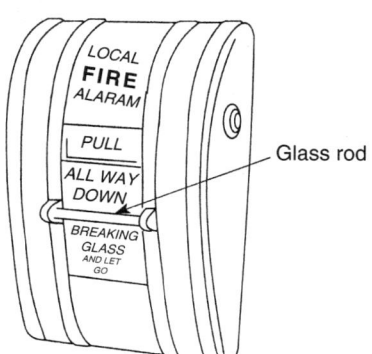

Figure 27.1 Manual break-glass station.
Source: Fire Alarm Signaling Systems, third edition, NFPA, 2003, Figure 5.1.

27.1.2 Automatic Fire Alarm Systems

Automatic fire alarm systems, as their name implies, use automatic means to detect a fire. Automatic devices include smoke detectors, heat detectors, flame detectors, and sprinkler water flow switches (see Figure 27.2). Where fire alarm systems utilize automatic means to initiate alarms, at least one manual fire alarm box shall be provided to initiate a fire alarm signal. This manual fire alarm box must be located where required by the AHJ. This requirement for at least one manual fire alarm box may seem strange to persons familiar with health care settings, because all such facilities have manual pull boxes throughout the facility. However, it is required because there can be an isolated detection system (e.g., a detection system that actuates a deluge sprinkler system protecting a hyperbaric chamber—see Chapter 21 of this book) that may not otherwise provide a manual pull box to actuate the system if a fire was first discovered by an occupant.

The actuation of an automatic fire alarm device may result in several different outputs from the system, depending on the type of device and its location within the structure. The input and output functions of every fire alarm system are very important to a successful installation. Facility engineers should insist that the system designers detail the functions in an "input/output matrix" and carefully review this with the facility and the AHJ early in the design phase. These reviews ensure that the system is programmed with functions that not only are required by code but that also satisfy the particular needs of the facility.

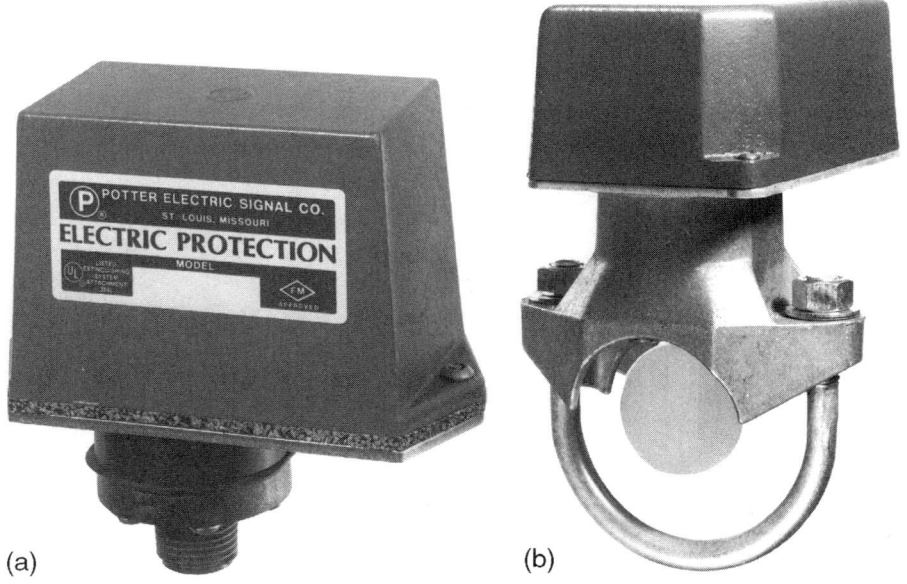

Figure 27.2 Water flow alarm switches: (a) pressure operated; (b) vane type.
Source: *Fire Protection Handbook*, NFPA, 2003, Figures 9.4.5 and 9.4.6.

27.2 Fire Alarm Control Panels

The fire alarm control panel (FACP) is the "brain" of the fire detection and alarm system (see Figure 27.3). It monitors the various alarm input devices, such as manual and automatic detection devices, and when required, it actuates alarm *output* devices (called *notification appliances*) such as horns, bells, speakers, chimes, strobe lights, door holders, digital dialers, and building systems controls. FACPs may range from simple units with a few input and output zones to complex computer-driven systems that monitor several buildings over an entire campus.

27.2.1 Initiating Device (Conventional) Circuit–Type Fire Alarm Systems

The first electronic fire detection and alarm systems consisted of initiating device circuits, or *zone-type* arrangements. These systems are connected to their devices much like an electrical lighting system. In a conventional fire alarm system, all devices are connected by circuits, and the configuration of these circuits determines which devices can be monitored and controlled individually or as a group (e.g., third floor west). Upon actuation of an alarm initiating device, the initiating device (conventional) circuit is completed and the fire control panel recognizes this as an alarm condition. Upon an alarm condition, the panel actuates one or more notification appliance circuits to sound building alarms and summon emergency forces. The panel that is "in alarm" may also send a signal to another panel that is monitoring the associated building from a remote location.

All fire alarm systems must have circuits that are monitored for integrity. This is accomplished by providing a resistor at the end of the circuit (known as the *end-of-line resistor,* or EOL) to complete the circuit, such that the FACP can energize that circuit with a small amount of current. The total amount of resistance through the circuit and the EOL is a known quantity, and if the circuit is broken, the FACP senses that the circuit is no longer complete and sounds a trouble signal. A trouble light on the FACP also illuminates to indicate which circuit requires service.

With an initiating device arrangement, all alarm initiating and signaling is accomplished by circuits that are installed to correspond to a particular zone of the building. Because of this arrangement, these systems are actually monitoring and controlling circuits, not individual devices. Typical initiating device systems provide responding personnel with an indication of the location of the alarm by type of device and the zone within the facility. The size of the zone varies depending on the system design. Where a floor area exceeds 20,000 ft² (1860 m²), additional zoning should be provided. The length of any zone should not exceed 300 ft (91 m) in any direction. Where the building is provided with automatic sprinklers throughout, the area of the alarm zone should be permitted to coincide with the allowable area of the sprinkler zone.

Emergency personnel responding may need to search the entire zone to determine which device is reporting an alarm condition (slang: "which device is in alarm"). Where zones have several rooms or locked spaces, this response can be time consuming because responding personnel must look for manual pull boxes with their handles in the actuated position and smoke detectors with an illuminated red light. (This red light

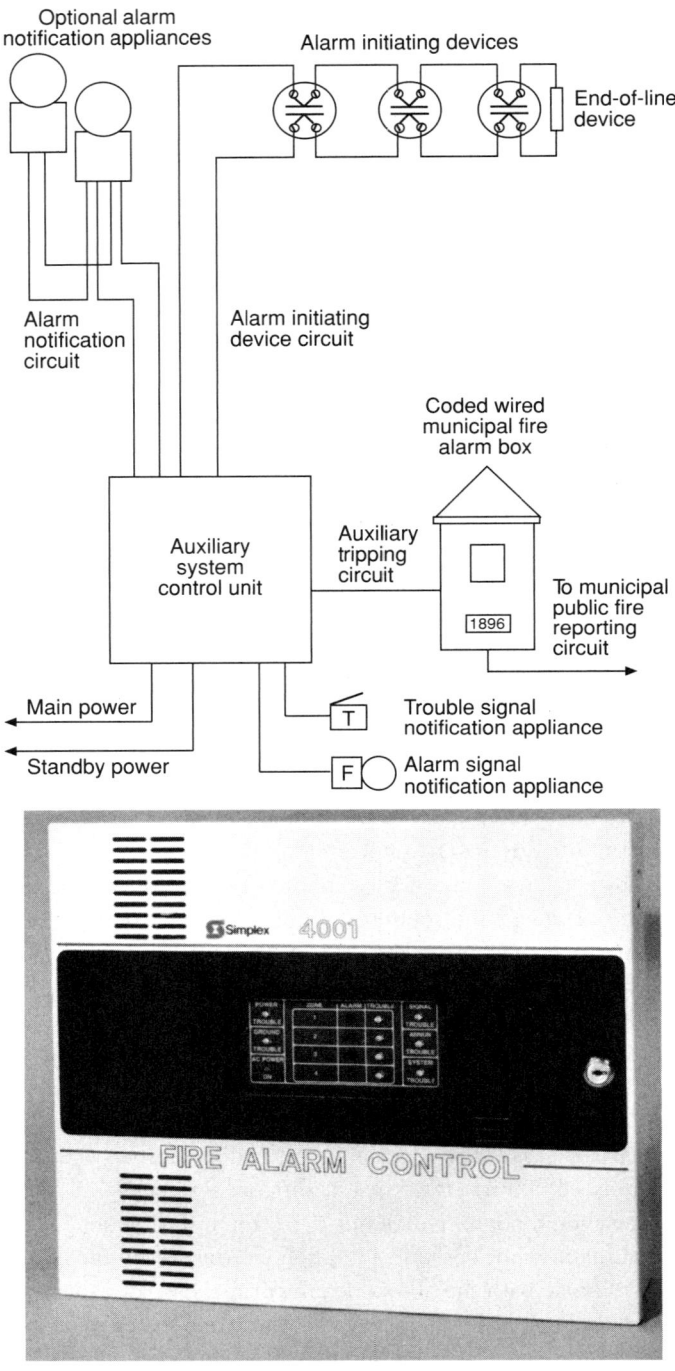

Figure 27.3 A typical arrangement of a fire alarm system (top) and a fire alarm control panel (bottom).
Source: Fire Protection Handbook, NFPA, 2003, Figure 9.1.7; courtesy of Simplex Time Recorder Company.

normally flashes to indicate that the detector has power, but the light usually remains lighted if the device has actuated.) For initiating device– (zone-) type circuits to indicate each device individually, circuits would have to be wired from each device to the FACP, resulting in a massive amount of wire at great expense. System outputs are also circuit dependent and usually must be rewired to accomplish different responses.

Initiating device circuit technology is older and delivers reliable service, but it is difficult to reconfigure. Such systems are perfectly acceptable for smaller buildings, such as an ambulatory surgical center located in a stand-alone one- or two-story building. A disadvantage to initiating device circuits is that they can be expensive to install in large buildings because of the extensive amount of wire that is necessary to monitor initiating devices accurately.

Initiating device circuit– (zone-) type fire alarm systems may also be inherently labor intensive and expensive to maintain. Each detection device may require some form of operational test to verify that it is in working condition. Smoke detectors must be periodically removed, cleaned, and recalibrated to prevent improper operation. With a conventional system, there is no way of determining which detectors are in need of servicing. Therefore, each detector must be removed and serviced at periodic intervals, or must be tested in place to determine its position.

27.2.2 Addressable Fire Alarm Systems

Addressable and analog systems represent the current state of the art in fire detection and alarm technology. Unlike conventional fire alarm systems, these systems monitor and control the capabilities of each alarm initiating and signaling device through microprocessors and system software. Like a conventional system, an addressable system consists of circuits that are installed throughout the space or building. Also like a conventional system, one or more alarm initiating devices are located along these circuits.

The major difference between system types involves the way in which each device is monitored. In an addressable and analog system, each initiating device is given a specific identification, or address. This address is programmed into the FACP computer with information such as the type of device and its location. The desired output as a result of an alarm signal is also programmed. To change the system response, a new program can be loaded at any time without the need to rewire signaling line or indicating circuits.

The FACP sends a constant interrogation signal over each circuit, in which each initiating device is contacted to inquire its status. These systems also monitor the condition of each circuit, identifying any faults that may occur. One of the advancements offered by some of these systems is their ability to identify specifically where a fault has developed. Therefore, instead of merely showing a fault along a wire, they indicate the location of the problem.

Advantages provided by these alarm systems include more detailed alarm information, stability, enhanced maintenance, and ease of modification. The alarm information may be very detailed because each device is uniquely identifiable, and such information is usually presented on an alpha-numeric display. As a result, alarm conditions can be indicated as, for example, "Smoke detector, Surgery, Room 3452."

The stability of these systems is achieved by the system software. If a detector senses a condition that could be indicative of a fire, the FACP can attempt a system reset. For most spurious situations (e.g., dust), the atmosphere around the detector returns to normal after the system is reset, thereby reducing the probability of nuisance alarms. If a true smoke or fire condition exists, the detector will again sense this condition after the reset attempt, and the FACP will process it as a confirmed alarm condition.

Maintenance of some analog fire alarm systems is enhanced because the system is able to monitor the status of each detector. As a detector becomes dirty, the FACP is able to compensate and adjust the detector's sensitivity to maintain the same sensitivity as if it were clean. This ability is known as *drift compensation*. The system is able to make such compensation until the detector becomes so dirty that the FACP provides a maintenance alert indicating exactly which detector is affected. Functional testing of the detectors may be accomplished from the FACP, as compared to conventional systems, in which each device must be actuated with smoke or a magnet. Testing of addressable and analog systems obviously saves time and money.

Modifying these systems involves connecting or removing devices from the addressable and analog circuit and changing the FACP programming. This memory change is accomplished either at the FACP or on a personal computer, with the information downloaded into the FACP microprocessor. Input and output functions of the system can be simply reprogrammed.

A disadvantage of addressable and analog systems is that each manufacturer's system has its own unique operating system. Therefore, service technicians must be trained for the particular system by the manufacturer. Periodic update training may be necessary as new operating software is developed to upgrade the system.

27.3 Alarm Devices and Appliances

There are numerous types of alarm devices, including manual pull boxes; heat, smoke, and flame detection devices; and alarm output devices that indicate an emergency is under way.

27.3.1 Manual Fire Alarm Boxes

Manual fire alarm boxes provide occupants a readily identifiable means to actuate the building fire alarm system. Manual fire alarm boxes usually must be located in the natural exit access path near each required exit from an area. "Near" each exit is defined by *NFPA 72* as not more than 5 ft (1.52 m) from the entrance to each exit. *NFPA 72* requires each manual fire alarm box to be securely mounted with the operable part (the handle) of each manual fire alarm box not less than 42 in. (1068 mm) and not more than 54 in. (1370 mm) above floor level.

The Americans with Disabilities Act (ADA) requires similar mounting heights that are determined by a minimum clear floor space that allows a forward or a parallel approach by a person using a wheelchair. If the clear floor space allows only forward approach, the maximum high-forward reach allowed is 48 in. (1220 mm). If the clear floor space allows parallel approach by a person in a wheelchair, the maximum ADA high-side reach allowed is 54 in. (1370 mm) and the low-side reach is no less than 9 in. (230 mm) above the floor. If the high-forward or high-side reach is over an obstruction, reach and clearances are further modified.

Many design professionals specify manual fire alarm pull boxes to be mounted such that the handle is located typically 48 in. (1220 mm) above the floor surface. This height satisfies multiple codes regardless of the direction of approach. Each individual location must still be evaluated for reach over an obstruction. Additional manual fire alarm boxes must also be located so that, from any part of the building, not more than 200 ft (60 m) horizontal distance on the same floor shall be traversed in order to reach a manual fire alarm box. All manual fire alarm boxes must be accessible, unobstructed, visible, and of contrasting color to the background on which they are mounted.

27.3.2 Heat Detectors

Thermal detectors are the oldest type of automatic detection device, having originated in the mid-1800s, with several styles still in production today. The most common units are *fixed-temperature devices* that operate when the room reaches a predetermined temperature [usually in the range of 135° to 165°F (57° to 74°C)]. The second most common type of thermal sensor is the *rate-of-rise detector,* which identifies an abnormally fast temperature climb over a short time period. Both of these units are *spot-type detectors,* which means that they are spaced periodically along a ceiling or on a wall [wall-mounted units must be mounted with the top edge between 4 and 12 in. (100 and 305 mm) below the ceiling]. See Figure 27.4. The third detector type is the *fixed-temperature line-type detector,* which consists of two cables and an insulated sheathing that is designed to break down when exposed to heat. The advantage of line-type over spot detection is that thermal-sensing density can be increased at lower cost.

Thermal detectors are highly reliable and have good resistance to operation from nonhostile sources. They are also very easy and inexpensive to maintain. On the down side, they do not function until room temperatures have reached a substantial temperature, at which point the fire is well under way and damage is growing exponentially. Consequently, thermal detectors are usually not permitted in life safety applications. They are also not recommended in locations where there is a desire to identify a fire

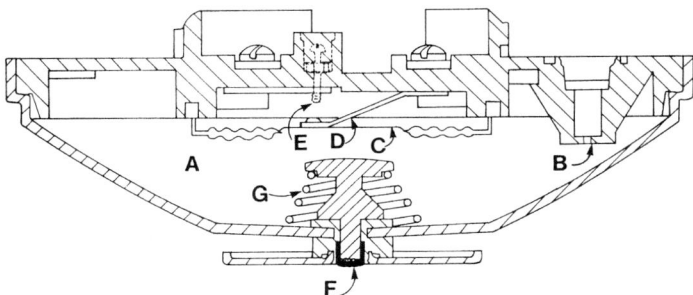

Figure 27.4 A spot-type combination rate-of-rise, fixed-temperature device.
Source: Fire Alarm Signaling Systems, third edition, 2003, Figure 5.5.

before substantial flames occur, such as spaces where high-value thermal-sensitive contents are housed.

27.3.3 Smoke Detectors

Smoke detectors are a much newer technology than heat detectors, having gained wide usage during the 1970s and 1980s in residential and life safety applications. As the name implies, these devices are designed to identify a fire while in its smoldering or early flame stages, replicating the human sense of smell. The most common smoke detectors are *spot-type units,* which are placed along ceilings or high on walls in a manner similar to spot-type thermal units (see Figure 27.5). Spot-type units operate on either an ionization or photoelectric principle, with each type having advantages in different applications. For large open spaces, such as galleries and atria, a frequently used type of smoke detector is the *projected beam unit.* This detector consists of two components, a light transmitter and a receiver, that are mounted at some distance [up to 300 ft (100 m)] apart. As smoke migrates between the two components, the transmitted light beam becomes obstructed and the receiver is no longer able to see the full beam intensity. This is interpreted as a smoke condition, and the alarm actuation signal is transmitted to the fire alarm panel.

A third type of smoke detector, which has become widely used in extremely sensitive applications, is the *air-sampling system.* This device consists of a control unit that houses the detection chamber, an aspiration fan, and operation circuitry and a network

Figure 27.5 A typical smoke detector installation. The smoke detector, wiring, and sprinkler system are surface mounted, which is typical of a retrofit installation.
Source: Thomas W. Gardner and Schirmer Engineering Corporation.

of sampling tubes or pipes. Along the pipes is a series of ports designed to permit air to enter the tubes and be transported to the detector. Under normal conditions, the fan/pump constantly draws an air sample into the detection chamber, via the pipe network. The sample is analyzed for the existence of smoke and then returned to the atmosphere. If smoke is present in the sample, it is detected and an alarm signal is transmitted to the main fire alarm control panel. Air-sampling detectors are extremely sensitive and are typically the fastest responding automatic detection method. Many high-technology organizations, such as telephone companies, have standardized on air-sampling systems. In cultural properties, these systems are used for areas such as collection storage vaults and rooms housing highly valuable art. These detectors may become popular in surgical suites, where early detection is a must since surgery patients may have the body cavity open. These are also frequently used in aesthetically sensitive applications, since components are often easier to conceal compared to other detection methods.

The key advantage of smoke detectors is their ability to identify a fire while it is still in its incipient stage. As such, they provide added opportunity for emergency personnel to respond and control the developing fire before severe damage occurs. They are usually the preferred detection method in life safety and high-content-value applications. The disadvantages of air-sampling smoke detectors are that they are usually more expensive to install, compared to thermal sensors, and are more resistant to inadvertent alarms. However, when properly selected and designed, they can be highly reliable, with a very low probability of false alarm.

27.3.4 Radiant Energy (Flame) Detectors

Radiant energy (flame) detectors represent the third major type of automatic detection method and imitate the human sense of sight. They are line-of-sight devices that operate on either an infrared or an ultraviolet principle (or a combination of the two). As radiant energy in the range of approximately 4000 to 7700 angstroms occurs, indicative of a flaming condition, their sensing equipment recognizes the fire signature and sends a signal to the fire alarm panel.

The advantage of flame detection is that it is extremely reliable in a hostile environment. Flame detectors are not often used in the health care field. They are usually used in high-value energy and transportation applications where other detectors would be subject to spurious actuation. Common uses include locomotive and aircraft maintenance facilities, refineries and fuel-loading platforms, and mines. A disadvantage is that they can be very expensive and labor intensive to maintain.

27.3.5 Notification Appliances

Upon receiving an alarm notification, the fire alarm control panel must now tell someone that an emergency is under way. This is the primary function of the alarm output aspect of a system. Occupant signaling components include various audible and visual alerting components and are the primary alarm output appliances. Bells are the most common and familiar alarm sounding appliance and are appropriate for most building applications. Horns are another option and are especially well suited to areas where a loud signal is needed, such as library stacks and architecturally sensitive buildings where appliances need partial concealment. Chimes may be used where a soft alarm tone is preferred, such as health care facilities and theaters.

Speakers, the fourth alarm sounding option, sound a reproducible signal such as a recorded voice message. They are often ideally suited for large, multistory, or other similar buildings where phased evacuation is preferred. Speakers also offer the added flexibility of emergency public address announcements. Many in the fire protection community feel that the voice alarm type of system is the only type that should be used, because studies show that persons react to the voice alarm and are quite suspect of the validity of chimes, horns, or bells.

With respect to visual alert, there are a number of strobe and flashing light appliances. Visual alerting is required in spaces where ambient noise levels are high enough to preclude hearing the sounding equipment or where hearing-impaired occupants may be found. Standards such as the ADA mandate visual notification appliances in numerous museum, library, and historic building applications.

Another key function of the fire alarm system is emergency response notification. The most common arrangement is an automatic telephone or radio signal that is communicated to a constantly staffed monitoring center such as a hospital telephone operator's office. Upon receiving the alert, the center then contacts the appropriate fire department, providing information about the location of the alarm. In some instances, the monitoring station may be the police or fire department or a 911 center. In other instances, it will be a private monitoring company ("central station") that is under contract to the facility.

Other output functions include shutting down electrical equipment such as computers, shutting off air-handling fans to prevent smoke migration, and shutting down operations such as chemical movement through piping in the alarmed area. Output devices may also activate fans to extract smoke, which is a common function in large atria spaces. These systems can also actuate discharge of gaseous fire-extinguishing systems or preaction sprinkler systems.

27.4 Specialized Features for Health Care Fire Alarm Systems

In health care facilities, initiation of the required fire alarm systems is by manual means and by means of any required sprinkler system water flow alarms, detection devices, or detection systems. Occupant notification must be accomplished automatically. As previously stated, presignal systems are prohibited. In lieu of audible alarm signals, the *Life Safety Code* (Section 18.3.4.3.1) allows visible notification appliances to be used in critical care areas.

Health care facilities are specifically required to have automatic notification to the local fire department. One exception to this is where smoke detection devices or smoke detection systems equipped with reconfirmation features are not required to notify the fire department automatically unless the alarm condition is reconfirmed after a maximum 180-second time period in new construction (a 120-second time period is still permitted in existing health care fire alarm systems). Although the alarm annunciation must be provided in accordance with the *Life Safety Code* (Section 9.6.7.3), there is an exception where the alarm zone is permitted to coincide with the allowable area of smoke compartments.

In new nursing homes, an approved automatic smoke detection system must be installed in corridors throughout smoke compartments containing patient sleeping rooms and in spaces permitted to be open to the corridors. There are two exceptions to this requirement:

- Where each patient sleeping room is protected by an approved smoke detection system
- Where patient room doors are equipped with automatic door-closing devices with integral smoke detectors on the room side installed in accordance with their listing, provided that the integral detectors provide occupant notification

The *Life Safety Code* allows an exception to the requirement for manual fire alarm boxes to be located throughout the facility. Manual fire alarm boxes in patient sleeping areas are not required at exits if they are located at all nurses' control stations or other continuously attended staff locations, provided such boxes are visible and continuously accessible and that travel distances required by the code are not exceeded. Also, in psychiatric hospitals and psychiatric areas in general hospitals, the authority having jurisdiction may approve an alternative method for locking and unlocking manual fire alarm boxes, such as by having all staff members carry a distinctively marked key. This measure is for security reasons and to prevent patients from taking action to cause a false alarm.

Any door in an exit passageway, horizontal exit, smoke barrier, stairway enclosure, or hazardous area enclosure can be held open only by an approved automatic release device. The required manual fire alarm system must be arranged to initiate the closing action of all such doors by zone or throughout the entire facility. Where doors in a stair enclosure are held open by an automatic device, initiation of a door-closing action on any level must cause all doors at all levels in the stair enclosure to close.

27.5 Inspection, Testing, and Maintenance

NFPA 72 requires service personnel to be qualified and experienced in the inspection, testing, and maintenance of fire alarm systems. According to this code, qualified personnel include, but are not limited to, individuals with the following qualifications:

- Factory trained and certified
- Fire alarm certified by the National Institute for Certification in Engineering Technologies
- Fire alarm certified by the International Municipal Signal Association
- Certified by a state or local authority
- Trained and qualified personnel employed by an organization listed by a national testing laboratory for the servicing of fire alarm systems

27.5.1 Testing

Testing should include verifying that the releasing circuits and components energized or actuated by the fire alarm system are electrically monitored and operate as intended on alarm. The facility engineer should make sure that the testing contractor has returned suppression systems and releasing components to their functional operating condition upon completion of system testing.

Of course, all new systems must be acceptance tested (see Chapter 10 of *NFPA 72*, 2002 edition). Also, these systems must be reacceptance tested after any of the following takes place:

- Adding or deleting system components
- Conducing any modification, repair, or adjustment to system hardware or wiring
- Making any change to site-specific software

All components, circuits, system operations, or site-specific software functions known to be affected by the change or identified by a means that indicates the system operational changes shall be 100 percent tested. In addition, 10 percent of initiating devices that are not directly affected by the change, up to a maximum of 50 devices, also shall be tested, and correct system operation shall be verified. A revised record of completion should be presented to the facility engineer reflecting any changes. Where the FACP is changed, all control units connected to or controlled by the system executive software must have a 10 percent functional test of the system. This test includes testing at least one device on each input and output circuit to verify critical system functions, such as notification of appliances, control functions, and off-premises reporting.

Before proceeding with any testing, all persons and facilities receiving alarm, supervisory, or trouble signals, and all building occupants must be notified of the testing to prevent unnecessary response (and the "boy who cried wolf" syndrome). At the conclusion of testing, those previously notified (and others, as necessary) must be notified that testing has been concluded. It is extremely important that the facility engineer coordinate system testing to prevent interruption of critical building systems or equipment.

Where a fire alarm system is designed to discharge a suppression system, a discharge of the suppression system is not required. It is just necessary to prove that the detection system will send the appropriate "discharge signal." The suppression systems can be secured from inadvertent actuation by disconnecting the releasing solenoids or electric actuators, closing valves, or taking other actions or combinations thereof for the specific system for the duration of the fire alarm system testing.

27.5.1.1 Smoke Detector Sensitivity

Detector sensitivity shall be checked within 1 year after the system is installed and every alternate year thereafter. After the second required calibration test, if sensitivity tests indicate that the detector has remained within its listed and marked sensitivity range (or 4 percent obscuration of light gray smoke, if not marked), the length of time between calibration tests can be extended to a maximum of 5 years. If the frequency is extended, records of detector-caused nuisance alarms and subsequent trends of these alarms must be maintained. In zones or in areas where nuisance alarms show any increase over the previous year, calibration tests must be performed.

To ensure that each smoke detector is within its listed and marked sensitivity range, it must be tested using any of the following methods:

- Calibrated test method
- Manufacturer's calibrated sensitivity test instrument
- Listed control equipment arranged for the purpose
- Smoke detector or control unit arrangement whereby the detector causes a signal at the control unit when its sensitivity is outside its listed sensitivity range
- Other calibrated sensitivity test methods approved by the authorities having jurisdiction

Detectors found to have a sensitivity outside the listed and marked sensitivity range must be cleaned and recalibrated or replaced. If the detector is listed as field adjustable, it can either be adjusted within the listed and marked sensitivity range and cleaned and recalibrated, or it must be replaced. The detector sensitivity shall not be tested or measured using any device that administers an unmeasured concentration of smoke or other aerosol into the detector. This precludes aerosol smoke from being used as a sensitivity testing method.

27.5.1.2 Heat Detector Testing

For restorable fixed-temperature, spot-type heat detectors, two or more detectors must be tested on each initiating circuit annually. Different detectors must be tested each year, and records must be kept by the facility engineer specifying which detectors have been tested. Within 5 years, each detector must have been tested.

27.5.2 Inspections

Fire alarm systems must undergo visual inspection to ensure that no changes have been made that affect equipment performance. Where devices or equipment are inaccessible for safety considerations (for example, continuous process operations, energized electrical equipment, radiation, and excessive height), they must be inspected during scheduled shutdowns, if approved by the AHJ. Extended intervals between inspections must not exceed 18 months. Also, a fire alarm system can be visually inspected annually if a remotely monitored fire alarm control unit specifically listed for conducting automatic system inspections conducts such inspections at a frequency of not less than weekly. Tables 27.1 and 27.2 show the visual inspection and testing frequencies allowed by *NFPA 72*.

27.6 Procedure in Case of a Fire

Because this chapter details fire alarm systems, and because their main purpose is to notify the facility staff of a fire condition, the appendix material from the *Life Safety Code* (Section A.18.7.2.1), which discusses procedures to follow in case of a fire, is reproduced here. Facility engineering staff must make sure that all other staff know what to do in case of a fire.

> Each facility has specific characteristics that vary sufficiently from other facilities to prevent the specification of a universal emergency procedure. The following recommendations, however, contain many of the elements that should be considered and adapted as appropriate to the individual facility.

Table 27.1 Visual Inspection Frequencies

Component	Initial/ Reacceptance	Monthly	Quarterly	Semiannually	Annually
1. Control Equipment—Fire Alarm Systems Monitored for Alarm, Supervisory, and Trouble Signals					
(a) Fuses	X	—	—	—	X
(b) Interfaced equipment	X	—	—	—	X
(c) Lamps and LEDs	X	—	—	—	X
(d) Primary (main) power supply	X	—	—	—	X
2. Control Equipment—Fire Alarm Systems Unmonitored for Alarm, Supervisory, and Trouble Signals					
(a) Fuses	X (weekly)	—	—	—	—
(b) Interfaced equipment	X (weekly)	—	—	—	—
(c) Lamps and LEDs	X (weekly)	—	—	—	—
(d) Primary (main) power supply	X (weekly)	—	—	—	—
3. Batteries					
(a) Lead-acid	X	X	—	—	—
(b) Nickel-cadmium	X	—	—	X	—
(c) Primary (dry cell)	X	X	—	—	—
(d) Sealed lead-acid	X	—	—	X	—
4. Transient Suppressors	X	—	—	X	—
5. Control Panel Trouble Signals	X (weekly)	—	—	X	—
6. Fiber-Optic Cable Connections	X	—	—	—	X
7. Emergency Voice/Alarm Communications Equipment	X	—	—	X	—
8. Remote Annunciators	X	—	—	X	—
9. Initiating Devices					
(a) Air sampling	X	—	—	X	—
(b) Duct detectors	X	—	—	X	—
(c) Electromechanical releasing devices	X	—	—	X	—
(d) Fire-extinguishing system(s) or suppression system(s) switches	X	—	—	X	—
(e) Fire alarm boxes	X	—	—	X	—
(f) Heat detectors	X	—	—	X	—
(g) Radiant energy fire detectors	X	—	X	—	—
(h) Smoke detectors	X	—	—	X	—
(i) Supervisory signal devices	X	—	X	—	—
(j) Waterflow devices	X	—	X	—	—
10. Guard's Tour Equipment	X	—	—	X	—
11. Interface Equipment	X	—	—	X	—
12. Alarm Notification Appliances — Supervised	X	—	—	X	—

Table 27.1 Visual Inspection Frequencies—cont'd

Component	Initial/Reacceptance	Monthly	Quarterly	Semiannually	Annually
13. Supervising Station Fire Alarm Systems – Transmitters					
(a) DACT	X	–	–	X	–
(b) DART	X	–	–	X	–
(c) McCulloh	X	–	–	X	–
(d) RAT	X	–	–	X	–
14. Special Procedures	X	–	–	X	–
15. Supervising Station Fire Alarm Systems – Receivers					
(a) DACR*	X	X	–	–	–
(b) DARR*	X	–	–	X	–
(c) McCulloh systems*	X	–	–	X	–
(d) Two-way RF multiplex*	X	–	–	X	–
(e) RASSR*	X	–	–	X	–
(f) RARS*	X	–	–	X	–
(g) Private microwave*	X	–	–	X	–

*Reports of automatic signal receipt shall be verified daily.
Source: NFPA 72, 2002, Table 10.3.1.

Table 27.2 Testing Frequencies

Component	Initial/Reacceptance	Monthly	Quarterly	Semiannually	Annually	Table 10.4.2.2 Reference
1. Control Equipment – Building Systems Connected to a Supervising Station						1, 7, 16, 17
(a) Functions	X	–	–	–	X	–
(b) Fuses	X	–	–	–	X	–
(c) Interfaced equipment	X	–	–	–	X	–
(d) Lamps and LEDs	X	–	–	–	X	–
(e) Primary (main) power supply	X	–	–	–	X	–
(f) Transponders	X	–	–	–	X	–
2. Control Equipment – Building Systems Not Connected to a Supervising Station	–	–	–	–	–	1
(a) Functions	X	–	X	–	–	–
(b) Fuses	X	–	X	–	–	–
(c) Interfaced equipment	X	–	X	–	–	–
(d) Lamps and LEDs	X	–	X	–	–	–
(e) Primary (main) power supply	X	–	X	–	–	–
(f) Transponders	X	–	X	–	–	–

(Continued)

Table 27.2 Testing Frequencies—cont'd

Component	Initial/ Reacceptance	Monthly	Quarterly	Semiannually	Annually	Table 10.4.2.2 Reference
3. Engine-Driven Generator – Central Station Facilities and Fire Alarm Systems	X	X	–	–	–	–
4. Engine-Driven Generator – Public Fire Alarm Reporting Systems	X (weekly)	–	–	–	–	–
5. Batteries – Central Station Facilities						
(a) Lead-acid type	–	–	–	–	–	6b
1. Charger test (replace battery as needed)	X	–	–	–	X	–
2. Discharge test (30 minutes)	X	X	–	–	–	–
3. Load voltage test	X	X	–	–	–	–
4. Specific gravity	X	–	–	X	–	–
(b) Nickel-cadmium type	–	–	–	–	–	6c
1. Charger test (replace battery as needed)	X	–	X	–	–	–
2. Discharge test (30 minutes)	X	–	–	–	X	–
3. Load voltage test	X	–	–	–	X	–
(c) Sealed lead-acid type	X	X	–	–	–	6d
1. Charger test (replace battery within 5 years after manufacture or more frequently as needed)	–	X	X	–	–	–
2. Discharge test (30 minutes)	X	X	–	–	–	–
3. Load voltage test	X	X	–	–	–	–
6. Batteries – Fire Alarm Systems						
(a) Lead-acid type	–	–	–	–	–	6b
1. Charger test (replace battery as needed)	X	–	–	–	X	–
2. Discharge test (30 minutes)	X	–	–	X	–	–
3. Load voltage test	X	–	–	X	–	–
4. Specific gravity	X	–	–	X	–	–
(b) Nickel-cadmium type	–	–	–	–	–	6c
1. Charger test (replace battery as needed)	X	–	–	–	X	–

Table 27.2 Testing Frequencies—cont'd

Component	Initial/ Reacceptance	Monthly	Quarterly	Semiannually	Annually	Table 10.4.2.2 Reference
2. Discharge test (30 minutes)	X	–	–	–	X	–
3. Load voltage test	X	–	–	X	–	–
(c) Primary type (dry cell)	–	–	–	–	–	6a
1. Load voltage test	X	X	–	–	–	–
(d) Sealed lead-acid type	–	–	–	–	–	6d
1. Charger test (replace battery within 5 years after manufacture or more frequently as needed)	X	–	–	–	X	–
2. Discharge test (30 minutes)	X	–	–	–	X	–
3. Load voltage test	X	–	–	X	–	–
7. Batteries – Public Fire Alarm Reporting Systems Voltage tests in accordance with Table 10.4.2.2, items 7(1) – (6)	X (daily)	–	–	–	–	–
(a) Lead-acid type	–	–	–	–	–	6b
1. Charger test (replace battery as needed)	X	–	–	–	X	–
2. Discharge test (2 hours)	X	–	X	–	–	–
3. Load voltage test	X	–	X	–	–	–
4. Specific gravity	X	–	–	X	–	–
(b) Nickel-cadmium type	–	–	–	–	–	6c
1. Charger test (replace battery as needed)	X	–	–	–	X	–
2. Discharge test (2 hours)	X	–	–	–	X	–
3. Load voltage test	X	–	X	–	–	–
(c) Sealed lead-acid type	–	–	–	–	–	6d
1. Charger test (replace battery within 5 years after manufacture or more frequently as needed)	X	–	–	–	X	–
2. Discharge test (2 hours)	X	–	–	–	X	–
3. Load voltage test	X	–	X	–	–	–
8. Fiber-Optic Cable Power	X	–	–	–	X	12b
9. Control Unit Trouble Signals	X	–	–	–	X	9
10. Conductors – Metallic	X	–	–	–	–	11
11. Conductors – Nonmetallic	X	–	–	–	–	12

(Continued)

Table 27.2 Testing Frequencies—cont'd

Component	Initial/ Reacceptance	Monthly	Quarterly	Semiannually	Annually	Table 10.4.2.2 Reference
12. Emergency Voice/Alarm Communications Equipment	X	–	–	–	X	18
13. Retransmission Equipment (The requirements of 10.4.7 shall apply.)	X	–	–	–	–	–
14. Remote Annunciators	X	–	–	–	X	10
15. Initiating Devices	–	–	–	–	–	13
(a) Duct detectors	X	–	–	–	X	–
(b) Electromechanical releasing devices	X	–	–	–	X	–
(c) Fire-extinguishing system(s) or suppression system(s) switches	X	–	–	–	X	–
(d) Fire–gas and other detectors	X	–	–	–	X	–
(e) Heat detectors (The requirements of 10.4.3.4 shall apply.)	X	–	–	–	X	–
(f) Fire alarm boxes	X	–	–	–	X	–
(g) Radiant energy fire detectors	X	–	–	X	–	–
(h) System smoke detectors – functional	X	–	–	–	X	–
(i) Smoke detectors – sensitivity (The requirements of 10.4.3.2 shall apply.)	–	–	–	–	–	–
(j) Single- and multiple-station smoke alarms (The requirements for monthly testing in accordance with 10.4.4 shall also apply.)	X	–	–	–	X	–
(k) Single- and multiple-station heat alarms	X	–	–	–	X	–
(l) Supervisory signal devices (except valve tamper switches)	X	–	X	–	–	–
(m) Water flow devices	X	–	–	X	–	–
(n) Valve tamper switches	X	–	–	X	–	–

Table 27.2 Testing Frequencies—cont'd

Component	Initial/ Reacceptance	Monthly	Quarterly	Semiannually	Annually	Table 10.4.2.2 Reference
16. Guard's Tour Equipment	X	–	–	–	X	–
17. Interface Equipment	X	–	–	–	X	19
18. Special Hazard Equipment	X	–	–	–	X	15
19. Alarm Notification Appliances	–	–	–	–	–	14
(a) Audible devices	X	–	–	–	X	–
(b) Audible textual notification appliances	X	–	–	–	X	–
(c) Visible devices	X	–	–	–	X	–
20. Off-Premises Transmission Equipment	X	–	X	–	–	–
21. Supervising Station Fire Alarm Systems – Transmitters	–	–	–	–	–	16
(a) DACT	X	–	–	–	X	–
(b) DART	X	–	–	–	X	–
(c) McCulloh	X	–	–	–	X	–
(d) RAT	X	–	–	–	X	–
22. Special Procedures	X	–	–	–	X	21
23. Supervising Station Fire Alarm Systems – Receivers	–	–	–	–	–	17
(a) DACR	X	X	–	–	–	–
(b) DARR	X	X	–	–	–	–
(c) McCulloh systems	X	X	–	–	–	–
(d) Two-way RF multiplex	X	X	–	–	–	–
(e) RASSR	X	X	–	–	–	–
(f) RARSR	X	X	–	–	–	–
(g) Private microwave	X	X	–	–	–	–

Source: NFPA 72, 2002, Table 10.4.3.

Upon discovery of fire, personnel should immediately take the following action:

(1) If any person is involved in the fire, the discoverer should go to the aid of that person, calling aloud an established code phrase. The use of a code provides for both the immediate aid of any endangered person and the transmission of an alarm. Any person in the area, upon hearing the code called aloud, should actuate the building fire alarm using the nearest manual fire alarm box.

(2) If a person is not involved in the fire, the discoverer should actuate the building fire alarm using the nearest manual fire alarm box.

(3) Personnel, upon hearing the alarm signal, should immediately execute their duties as outlined in the facility fire safety plan.

(4) The telephone operator should determine the location of the fire as indicated by the audible signal. In a building equipped with an uncoded alarm system, a person on the floor of fire origin should be responsible for promptly notifying the facility telephone operator of the fire location.

(5) If the telephone operator receives a telephone alarm reporting a fire from a floor, the operator should regard that alarm in the same fashion as an alarm received over the fire alarm system. The operator should immediately notify the fire department and alert all facility personnel of the place of fire and its origin.

(6) If the building fire alarm system is out of order, any person discovering a fire should immediately notify the telephone operator by telephone. The operator should then transmit this information to the fire department and alert the building occupants.

[101-03: A.18.7.2.1]

Fire Protection Systems Frequently Asked Questions

1. Do I have to provide sprinkler protection in a newly constructed health care facility?

Yes. NFPA *101*®, *Life Safety Code*® (2006 edition), Section 12.3.5, requires buildings containing health care facilities to be protected throughout by an approved, supervised automatic sprinkler system in accordance with Section 9.7. There is one exception: In Type I and Type II construction, where approved by the authority having jurisdiction, alternative protection measures are permitted to be substituted for sprinkler protection in specified areas where the authority having jurisdiction has prohibited sprinklers without causing a building to be classified as nonsprinklered.

2. If I renovate my health care facility, are sprinklers required?

Where major renovations, alterations, or modernizations are made in a nonsprinklered facility, the automatic sprinkler requirements of NFPA *101*, Chapter 12, apply to a smoke compartment undergoing the renovation, alteration, or modernization.

3. The fire department responded to my facility for a fire and they wanted to evacuate the building. Why?

In most occupancies, fire departments will evacuate the structure when there is a fire inside. Health care facilities should work together with their local fire department to explain the protect-in-place philosophy of a health care occupancy, where removing the fragile occupants of the building could be as hazardous to them as the fire. Unlike most occupancies, health care facilities have smoke compartments to allow horizontal relocation on each patient care or resident floor and a staff to move the patients or residents. Health care facilities should be evacuated to the exterior *only* when no other option is available to protect the occupants. Unless you coordinate with your local fire department, they probably won't understand the special needs of your facility.

4. I need a contractor to do some fire protection work on my facility. How do I select one?

It is advisable to involve a fire protection engineer on such projects in health care facilities. These engineers specialize in fire protection and life safety requirements, systems, and procedures and can develop a set of contract documents that can be used for bidding purposes by installation contractors. If you are working with a fire protection engineer, ask that person to supply you with a contractor bidders' list. If you are developing this list yourself, ask for references and check with each one of them.

509

5. How many bids should I get from contractors?

You should always get a minimum of three bids for all fire protection contracting work. Three bids will result in a high bid, a low bid, and something in between. With fewer than three bids, no good comparison can be made among the contractors' prices. Remember, depending on the business climate at the time, you might have to solicit a larger number of contractors to obtain the minimum three bids.

6. How much does it cost to retrofit automatic sprinkler protection or a fire alarm system in a health care building?

This is almost impossible to answer with any degree of accuracy. System costs vary greatly depending on the available water supply, the possible need for a fire pump, the business climate at the time, existing installed systems (which may need to be modified or removed), the condition and configuration of the building, access to all spaces, and the time available to complete the project. There is no way to predict such a cost without knowing the specific situation. Having said that, sprinkler installations typically cost in excess of $5 / ft^2 and fire alarm installations typically cost in excess of $2 / ft^2 in existing, active health care areas. Contact a fire protection engineer, sprinkler contractor, or fire alarm contractor on a case-by-case basis for more accurate estimates.

7. I had a contractor do some work in my facility. The contractor penetrated many of the smoke barrier walls and did not seal those penetrations. How do I prevent this from happening?

All health care facilities should keep a set of fire safety drawings on file. These drawings should show the smoke compartments, smoke barrier walls, smoke-resistive assemblies and any other fire-rated assemblies, exits, and the areas protected by fire protection systems. Provide each contractor with a set of these drawings upon commencement of any project in the facility. Require the contractor, by contract, to properly firestop any and all penetrations made in any and all smoke-resistive or fire-rated assemblies shown on the drawings. Penetration seals should be noted on the drawings, including the location, date, and type of material used to protect the penetration. This will be a useful management tool to prepare for Joint Commission on Accreditation of Healthcare Organizations (JCAHO) inspections.

8. I have fire hose cabinets in my facility and my insurance company wants me to replace the old hose with new. This will be an expensive project. Is this required?

Check with your local fire and building official, because many jurisdictions no longer require occupant-use hose for standpipe systems. For years, fire departments have not had faith in the maintenance of the occupant-use hose and routinely would disconnect such hose and connect their own "standpipe pack" hose to the hose outlet to fight a fire. Even where a health care facility provides the proper maintenance to occupant-use hose, the possibility of vandalism could render it unusable. These concerns have led to a trend toward reducing the requirements for, or eliminating the installation of, occupant-use hose on standpipe systems. Check with your local authorities and present the resulting information to your insurance company.

Fire Protection Systems Checklist

This checklist can be used to evaluate an existing facility before a survey by JCAHO or an inspection by any authority having jurisdiction. This checklist is for the fire protection systems section of this book. The reader is cautioned that this is not a complete list, but one that touches on many of the major items. To ensure compliance, the reader should be familiar with all of the codes and standards that are applicable.

Automatic Sprinkler Systems

✓	Item	Inspection Activity	Comments
	1.	Verify that fire department connections are unobstructed and in good condition.	
	2.	Ensure that valves are locked. If not locked, relock and make note.	
	3.	Inspect alarm valves to ensure that there is no leakage from retard chamber or alarm drains.	
	4.	Ensure that there are the proper numbers and types of sprinklers and a sprinkler wrench.	
	5.	Check for physical damage and that electrical connections are secure.	
	6.	Check pressure readings (psi) monthly (a loss of more than 10% should be investigated).	

Standpipe and Hose Systems

✓	Item	Inspection Activity	Comments
	1.	Inspect hose visually for damage and ensure that it is properly racked and that the nozzle is attached.	
	2.	Verify that signs are posted at each hose station.	
	3.	Ensure that fire department connections are unobstructed and in good condition.	
	4.	Confirm that signs are provided at the fire department connections.	
	5.	Verify that all valves are open, locked, and in good condition.	

Fire Pumps

✓	Item	Inspection Activity	Comments
	1.	Check that heat in pump room is 40°F or higher.	
	2.	Verify that operating louvers in pump room appear operational.	
	3.	Verify that pump suction, discharge, and bypass valves are open.	
	4.	Ensure that there are no leaks in piping or hoses.	
	5.	Ensure that suction line and system line pressures are normal.	
	6.	Confirm that suction reservoir is full.	
	7.	Verify that controller pilot light (power on) is illuminated.	
	8.	Verify that transfer switch is normal and that power light is illuminated.	
	9.	Ensure that isolating switch for standby power is closed.	
	10.	Check that reverse-phase alarm light is not illuminated.	
	11.	Check that normal-phase rotation light is illuminated.	
	12.	Determine that oil level in vertical motor sight glass is normal.	
	13.	Confirm that diesel fuel tank is at least two-thirds full.	
	14.	Confirm that controller selector switch is in "AUTO" position.	
	15.	Confirm that voltage readings for batteries (2) are normal.	
	16.	Ensure that charging current readings are normal for batteries.	
	17.	Confirm that pilot lights for batteries are "ON" or battery failure pilot lights are "OFF."	
	18.	Ensure that all alarm pilot lights are "OFF."	
	19.	Ensure that oil level is normal in right-angle-gear-drive pumps.	
	20.	Confirm that crankcase oil level is normal.	
	21.	Confirm that cooling-water level is normal.	
	22.	Confirm that electrolyte level in batteries is normal.	
	23.	Verify that battery terminals are free from corrosion.	
	24.	Ensure that water jacket heater is operational.	
	25.	Confirm for steam-driven pumps that steam pressure is normal.	
	26.	Examine exhaust system for leaks.	
	27.	Check lube oil heater for operation (diesel pumps).	
	28.	Check for water in diesel fuel tank.	

Kitchen Hood Fire-Extinguishing Systems

✓	Item	Inspection Activity	Comments
	1.	Ensure that nozzle caps are in place.	
	2.	Ensure that there is no significant grease accumulation on nozzles and link.	

✓	Item	Inspection Activity	Comments
	3.	Ensure that corrosive cleaners have not been used on cable or link.	
	4.	Verify that no cooking equipment has been added or existing equipment moved since last inspection.	
	5.	Ensure that nozzles aim at cooking surfaces they protect.	
	6.	Verify that manual actuators are not obstructed.	
	7.	Verify that tamper indicators and seals are intact.	
	8.	Verify that maintenance tag or certificate is in place.	
	9.	Confirm that pressure gauges, if provided, are in operable range.	

Specialized Suppression Systems

✓	Item	Inspection Activity	Comments
	1.	Ensure that nozzle caps are in place.	
	2.	Ensure that system is free from physical damage.	
	3.	Verify that no alterations have been made to the room being protected since last inspection.	

Portable Fire Extinguishers

✓	Item	Inspection Activity	Comments
	1.	Verify that the extinguisher is clean and well cared for.	
	2.	Confirm that the extinguisher has been charged and tested hydrostatically within the prescribed periods and tagged to show dates.	
	3.	Verify that if a seal is provided, it is intact.	
	4.	Ensure that the discharge orifice is unobstructed.	
	5.	Verify that the shell of the extinguisher is not corroded, damaged, or dented.	
	6.	Ensure that connections between the hose and the shell and nozzle are secure.	
	7.	Confirm that, if the extinguisher is a pump-operated type, the pump shaft operates freely.	
	8.	Confirm that the extinguisher is readily accessible, plainly indicated, and visible from a distance.	
	9.	Ensure that if the extinguisher is a type that is subject to freezing, it is protected.	
	10.	Check that the hanger is fastened solidly so that the extinguisher is well supported.	
	11.	Verify that the extinguisher is not located too close to the hazard that it is intended to protect, so that it cannot be reached in case of fire.	

Fire Detection and Alarm Systems

✓	Item	Inspection Activity	Comments
	1.	Verify that the fire alarm panel is operational.	
	2.	Verify that lights and LEDs on fire alarm and annunciator panels are operational.	
	3.	Confirm that battery electrolyte level is satisfactory.	
	4.	Verify that all heat detectors are operational.	
	5.	Verify that all smoke detectors are operational.	
	6.	Verify that all flame detectors are operational.	
	7.	Verify that all manual stations are operational.	
	8.	Verify that all bells or horns are operational.	
	9.	Verify that all speakers are operational.	
	10.	Verify that preamplifier is operational.	
	11.	Verify that amplifier is operational.	
	12.	Verify that voice-recording equipment is operational.	
	13.	Verify that power supplies are operational.	
	14.	Verify that all radio fire alarm transmitting equipment is operational.	
	15.	Verify that all radio fire alarm receiving equipment is operational.	
	16.	Verify that all telegraphic fire alarm transmitting equipment is operational.	
	17.	Verify that all telegraphic fire alarm receiving equipment is operational.	

Part V
Gas and Vacuum Systems and Equipment

Health care facilities, from large medical centers to one-patient dental offices, require the distribution of pressurized gases or the creation of a vacuum to accomplish a variety of tasks. These tasks provide the following services:

- Medical gases as part of the treatment of patients
- A mechanism to remove excess fluid or matter during surgery, dental procedures, etc. (via a vacuum system)
- Gases and vacuum for conducting laboratory experiments or tests
- An energy source for heaters to control the temperature of air within a building
- Air movement to move canisters between points within a building
- Vacuum to clean rugs or floors in an area

The volume of gases required and used has dramatically increased since the technology was first applied for nonmedical purposes in the 1800s and for medical purposes in the 1930s, as the complexity and length of medical procedures has increased. Today, medical gas and vacuum systems are used in almost every health care facility. While patient systems are prevalent in hospitals, surgical facilities, and dental facilities, they are also beginning to be used in some nursing homes, as these facilities begin to provide medical care for residents, in addition to the traditional nursing care.

Because the installation and use of gas and vacuum systems and associated equipment present fire hazards of varying degrees, NFPA and other organizations have developed performance, installation, and maintenance standards to help mitigate these hazards. This part of *Fire and Life Safety in Health Care Facilities* is intended to show that gas and vacuum systems in all health care facilities must be designed, installed, used, maintained, repaired, and renovated very carefully and according to codes and standards to create an adequate level of safety for all patients and staff. The chapters of this part cover the following information:

Chapter 28: History and hazards of piped gas and vacuum systems and the disposal of waste anesthetic gas

Chapter 29: Patient piped gas system levels, design considerations, sources, distribution, station outlets, monitors and alarms, testing and verification, and staff training

Chapter 30: Patient piped vacuum system pumps, distribution, station inlets, monitors and alarms, system performance criteria, renovations, system testing and verification, and staff training

Chapter 31: Waste anesthetic gas disposal methods and performance and testing criteria; non-patient piped gas systems used in laboratories and for heating, cooking, cleaning, pneumatic switching, and maintenance; and non-patient piped central vacuum cleaning systems and pneumatic tube systems

Chapter 32: Patient gas and vacuum equipment, including oxygen cylinders and containers, respirators and ventilators, and devices connected to vacuum systems and self-contained vacuum devices

28

History and Hazards of Piped Gas and Vacuum Systems

A major area of health care treatment involves the use of piped gas and vacuum systems that serve such essential patient care functions as providing oxygen for breathing, anesthesia for sedation, and suction for removing fluids during patient operations. Another important activity affecting patients, staff, and the environment is the collection of waste anesthetic gases from the operating arena, which can in some configurations take advantage of piped vacuum system technology. Gas and vacuum systems also are used for non-patient activities, such as heating, cleaning, and maintenance.

While gas systems may seem similar to vacuum systems (e.g., the two systems use networks of piping fitted together to move gas, albeit in the opposite direction), their design criteria are technically very different. With a gas system, after the gas has been connected at the source end, the whole system is pressurized with the gas until it reaches and stabilizes at a narrow range of pressures. In a vacuum system, a pump is trying to evacuate the space within the piping and provide a specified degree of vacuum [measured in inches of Hg (negative) and volume displacement (flow)] at each inlet. In a gas system, the gas within the system provides the positive pressure and flow, while in a vacuum system, a pump is required to create the subatmospheric pressure and flow.

Patient piped gas systems came into use as the practice of medicine changed during the 1930s, when the volume of gases used increased dramatically and electrical devices began to appear in operating rooms. Hazards inherent in the early use of these technologies led health care facilities, equipment manufacturers, and other interested parties in the industry to develop codes and standards to mitigate these hazards. In the 1950s, the use of vacuum had increased to a level that piped vacuum systems became viable. Still later, waste anesthetic gas disposal was considered necessary for biological or environmental reasons.

Patient piped gas and vacuum systems are often viewed much like electrical systems are viewed—as reliable, fail-safe, and inexpensive systems. They should also be viewed as important providers of vital services, and used and maintained very seriously, as the very life and breath of patients are often at stake.

28.1 Piped Gas Systems

The piping of natural gas for non-patient use, that is, *non-patient piped gas systems*, started in the early 1800s for cooking and heating. By the 1870s, iron pipelines made long-distance distribution of gas relatively safe, and in the 1920s, newly introduced seamless, electrically welded steel pipes allowed gas to be profitably distributed under higher pressure over much longer distances.

The precursor to the use of medical gas piping systems to induce anesthesia was to place a sponge soaked in ether over a patient's nose and mouth for a short period of time (1846). In 1868, the use of oxygen mixed with nitrous oxide was introduced as an adjunct to inhalation anesthesia. In 1887, the first gas anesthesia machine that used compressed gas in cylinders was introduced. During the next 40 years, the practice of medicine experienced significant changes in terms of the dramatic increase in the volume of gases used and the appearance of electrical devices in operating rooms.

All of these advancements allowed a greater number and variety of surgical procedures to be performed, but they also resulted in a growing number of cylinders that had to be wheeled around a facility—cylinders that contained pressurized gases. Full cylinders had to be unloaded from a delivery truck, then stored, then taken to operating rooms as needed. Empty (though actually not completely empty) cylinders had to be returned and stored, then transferred back to a delivery truck. Unfortunately, accidents happened—not very pleasant ones.

The first *patient piped gas system* that used high-pressure cylinders of oxygen connected to pipes and then outlets located in nearby operating rooms was installed in the 1920s. This method was believed to be an improved way to provide gases to operating rooms, based on the following distinct advantages:

- The cost of operating the system was less than that of using cylinders. Instead of many small cylinders, fewer larger cylinders were utilized, with a concurrent reduction in the unit cost per cubic foot of gas. (It has been reported that the amount saved at one hospital was sufficient to pay the salaries of the anesthesiology departmental staff.) Fewer cylinders also meant less loss of the residual gas that remained in "empty" cylinders. When individual cylinders were used, they would be replaced when down to about 500 psig (lb/sq in.) or 3448 kPa. However, when two or more cylinders were connected by a manifold to form a single source, individual cylinders could be allowed to be used to about 40 psig (276 kPa), since other cylinders in the system were available for use.

- Gases were immediately accessible. Operating room staff needed only to connect hoses to gas outlets. One person (instead of each anesthesiologist handling his own individual small cylinders) could monitor the large supply of cylinders at a central dispersion point. Since several large cylinders were grouped together, when one became empty, or nearly empty, others could be switched on line and the empty one replaced. Thus, operating room staff were assured of a constant supply of gas.

- Safety was improved. No longer were cylinders, with their inherent hazards, inside the operating room. In addition, cylinder movement around the hospital was dramatically reduced.

28.1.1 Codifying Requirements for Patient Piped Gas Systems

When piped medical gas systems were first installed, system users generally followed the standards and practices then in use for the piping of nonmedical gases. In 1932,

Chapter 28. History and Hazards of Piped Gas and Vacuum Systems

however, the National Fire Protection Association (NFPA) Committee on Gases noted that the installation of these systems posed several hazards for hospitals. With the exception of the hazards associated with the use of flammable anesthetics, which are no longer used in the United States, these hazards are still present today, as follows:

- Pipes that run through a building into operating rooms and patient rooms carry gases that are oxidizing (i.e., they support and intensify the burning of combustibles that have been ignited).
- A large quantity of gas in cylinders is concentrated and stored in one area. Storing large volumes of oxygen in a liquid state is a fire hazard. (The problem of storing a large volume of oxygen on site in a liquid state became a concern in the late 1940s. While this replaced the use of many cylinders containing pressurized oxygen, it vastly increased the amount of oxygen in one location and introduced the cryogenic aspects associated with gas in this state.)
- Concentration of potentially hazardous gas could build up should the pipes leak; gas buildup within a building is dangerous whether the gas is flammable or not.
- The possibility of explosion exists in an operating room if a hose on an anesthesia machine is inadvertently connected to the wrong gas.
- A cross-connection or mixing of gases, if more than one gas is piped, could be injurious or even fatal to a patient.
- Uncontrolled static electricity could be a source of ignition in the presence of flammable vapors.

Other problems associated with patient piped gas systems involve preventing an incident that occurs in one location from becoming an incident in another location or preventing a greater incident from occurring at the site of an initial problem. In a large facility, an incident could spread to hundreds of rooms.

As a result of a request by the National Board of Fire Underwriters, in 1934 NFPA developed and adopted a *Recommended Good Practice Requirements for the Construction and Installation of Piping Systems for the Distribution of Anesthetic Gases and Oxygen in Hospitals and Similar Occupancies, and for the Construction and Operation of Oxygen Chambers*. The guidelines, referred to as the *Recommended Good Practices*, contained recommendations for acceptable types of piping, length of pipe runs, pipe identification, manifolds, and shutoff valves.

Periodic revisions to the *Recommended Good Practices* prohibited the distribution of flammable anesthetic gases via a piped system; added safety requirements for the storage of gases, shutoff valve locations, check valves, line pressure gauges, pressure switches, and alarm panels; and included installation and testing criteria. Performance criteria, such as operating pressure limits for different gases, were also added because no other organization had addressed them in their documents, and uniformity of systems operations is helpful to medical staff, designers, and industry.

NFPA has since developed other documents that affect medical piped gas systems, including those that cover electric power [e.g., NFPA 70, *National Electrical Code® (NEC®)*], bulk oxygen supplies [NFPA 50, incorporated into NFPA 55, *Standard for the Storage, Use, and Handling of Compressed Gases and Cryogenic*

Fluids in Portable and Stationary Containers, Cylinders, and Tanks)], and building construction (NFPA *101*®, *Life Safety Code*®). In addition, other organizations have prepared standards addressing other aspects of piped gas systems. For example, the Compressed Gas Association (CGA), whose members include manufacturers of gases and gas equipment, publishes many documents on the subject of gases. Some of these documents apply directly to medical gas piping systems; others are generic and affect any closed gas system. Topics addressed by CGA include gas cylinder criteria; noninterchangeable connectors for cylinders and terminal outlets; liquefied gas transfer connections; compressed gas transfer connections; and commodity specifications for nitrogen, air, nitrous oxide, and oxygen. Refer to Chapter 1 and the References section for specific citations for other NFPA and CGA documents.

28.1.2 Codifying Requirements for Non-Patient Piped Gas Systems

Along with the increased use of flammable natural gas came increased hazards of distributing it via pipelines. Although discussion of a single national standard covering piped fuel gas systems began in 1967 among members of the American Gas Association (AGA), the American Society of Mechanical Engineers (ASME), and NFPA, it was not until 1972 that NFPA issued NFPA 54, *National Fuel Gas Code*. This code covered fuel gas piping systems from the point of delivery at the building to the connections for each gas utilization device. The scope of the document was expanded in 1988 to include piping systems up to 125 psig.

Today, the use of piped gas (and vacuum systems) is quite common. For the safety of patients and staff, these systems, large and small, must be designed, installed, used, maintained, and renovated with extreme care.

28.2 Piped Vacuum Systems

28.2.1 Patient Piped Vacuum Systems

The precursors to *patient piped vacuum systems* were portable suction machines, which removed fluids, blood, tissue, particles, and so on from within patients' bodies. These machines created a vacuum much the same way vacuum cleaners create suction. While vacuum cleaners typically use semiporous bags to collect dirt, medical vacuum machines use nonporous *trapping systems* to collect body fluids and semiliquid bulk material. One issue with using portable suction machines has always been that vacuum pumps can potentially move airborne infectious bacteria from one place to another. At one time, another problem with individual suction machines was their need to be safe for use in the presence of flammable anesthetics; this concern ceased to be a problem when nonflammable anesthetics replaced flammable anesthetics in the 1960s and 1970s.

Patient piped vacuum systems began to be installed in hospitals in the early 1950s, because studies showed that the system could be used economically as well as eliminate the problems associated with collection systems and their transportation of infectious air. Hazards associated with using flammable anesthetics were also reduced, since piped vacuum systems exhausted contaminated air outdoors, and no electric motor was needed in patient areas to provide the vacuum.

Initially, central piped vacuum systems served only operating rooms and specialty areas, such as postanesthesia recovery rooms and emergency rooms. Piped vacuum systems began to be used in general patient care areas as the demand for suction increased and the installation of systems became more economically viable. The reduction in the spread of airborne bacteria provided by central vacuum systems also contributed to their installation in general patient care areas, as hospitals became more aware and concerned about this particular hazard. (Note that portable suction pumps are still used in health care facilities, as are individual gas cylinders. Bacteria filters are placed over the suction pumps to make these devices quite suitable for many medical applications.) Also, traps placed between the patient and the vacuum control regulator and station inlet ensure that nothing but air is drawn into the piping system.

Pediatric and neonatal areas were the last areas to install central patient vacuum systems, because of uncertainty regarding the degree of vacuum and air flow that would be safe for very delicate tissues and newly functioning lungs (i.e., that would not collapse lungs). With improvements in regulating the degree of vacuum and greater education on the subject, these concerns abated and vacuum systems were installed in these areas as well.

28.2.2 Codifying Requirements for Patient Piped Vacuum Systems

Ineffective performance plagued many early piped vacuum systems, which were installed based on prevailing engineering expertise related to piped gas systems. Inadequate staff training, the lack of appropriate check valves, and widely divergent pump sizing also contributed to system problems. In 1961, after vacuum pump manufacturers failed to agree on vacuum system standards, CGA, whose members supplied vacuum pumps and inlet connectors to hospitals, developed and released a document (now withdrawn) that included recommendations on pumps and pump sizing, warning systems, piping, installation, and labeling.

In addition to these recommendations, staff practices were improving or becoming standardized on their own. For example, collection bottles began to be located below patient level to prevent back feeding, and regulator bypasses began to be used so that the system or a portion of the system did not have to be shut down if a problem developed with a regulator. These helped improve system performance as well.

Because engineers continued to differ regarding vacuum system design, NFPA subsequently tested various pumps and suction therapy equipment, surveyed actual systems installed in hospitals, and in 1980 adopted a recommended practice, NFPA 56K, *Recommended Practice on Medical Surgical Vacuum Systems*. It contained the same topics as the 1961 CGA document on vacuum systems. After several years, NFPA 56K was incorporated into the 1984 edition of NFPA 99, *Standard for Health Care Facilities*, with major portions revised from recommendations to requirements.

Over the years, other criteria that affect vacuum systems have been developed and incorporated into NFPA 99, covering such subjects as cleaning and purging, pressure testing, and connection for emergency electrical power. These criteria are the same as those listed earlier for piped gas systems (see Section 28.1.1).

Note that although the criteria for centrally piped patient vacuum systems differ from those for gas systems, if a piped vacuum system is installed at the same time as a piped gas system, use of installation standards of gas systems should be considered to avoid confusion and possible mistakes.

The requirements for patient vacuum systems have been revised over the years, with growing emphasis on performance criteria and less inclusion of design criteria. In 2002, all design guidance was deleted from the appendix since it was argued that the "designing" of piped vacuum systems was not within the scope of the technical committee responsible for vacuum system safety criteria. It was also noted that this guidance had not been restudied or updated since it was originally developed in the 1970s.

28.2.3 Non-Patient Piped Vacuum Systems

Non-patient piped vacuum systems have seen limited use in health care facilities. The two piped vacuum systems most often used are those for cleaning rugged-floors (i.e., through a central vacuum cleaner system) and those for moving small objects, such as tubes of blood, from one area of a building to another area in the building or another building. These are essentially commercial systems in a health care setting.

28.3 Disposal of Waste Anesthetic Gas

The idea of capturing waste anesthesia gases from anesthesia machines and disposing of them to the outside environment, that is, *waste anesthetic gas disposal* (WAGD), began after a study was conducted in the Soviet Union in the 1960s that attempted to determine whether trace anesthetic gases were harmful to operating room personnel. It appeared from the results of a limited survey that some operating room personnel experienced deterioration in health, and trace anesthetic gases increased the incidence of miscarriages and spontaneous abortions among some female staff. Although the quality of the data was questioned and subsequent studies could not duplicate the results of the initial study, manufacturers in the United States nonetheless began promoting the idea of capturing and disposing of trace anesthetic gases. Additionally, a U.S. government agency, the National Institute of Occupational Safety and Health (NIOSH), developed recommendations relative to controlling the atmosphere in the working environment, including areas where nitrous oxide and halogenated agents are used (see Chapter 31, Section 31.1).

It is important to note that a study conducted from 1977 to 1984 of female anesthesiologists in the United Kingdom showed no correlation between (1) spontaneous abortions or development of congenital abnormalities in live-born children and (2) the occupation of the mother, the hours exposed to the operating room environment, or the use of scavenging equipment.

Capturing waste gases began when the administration of anesthesia involved supplying several gases and circulating ether in a closed system so that it would not get too cold. As a result, *pop-off* valves and safety valves were needed to prevent overpressurization of anesthetic tubes leading to and from the patients. This practice has continued even with the switchover to nonflammable anesthetics and the resulting ways these gases were released into the air.

There are several methods of collecting and disposing of waste anesthetic gases, as discussed in Chapter 31. One method involves connecting waste gases to the medical-surgical vacuum system. As such, a vacuum system would have more than just air traveling through it. A concern arose as to what effect this addition might have on medical-surgical vacuum systems and, if there were a failure in the mechanism separating the low-pressure WAGD portion from the much higher medical-surgical vacuum system, what harm might be done to patient lungs.

Today, NFPA 99 contains criteria for the installation, performance, and safe use of waste anesthetic gas equipment.

29

Patient Piped Gas Systems

Patient piped gas systems, like most other piped gas systems, consist of many components that work together to provide uncontaminated gas at a constant pressure to specific outlets. The gas can be delivered to a facility in cylinders or containers, or the gas can be generated and compressed on site. The installation and use of each system component require a high degree of care for delivering the gas without leaking or becoming contaminated above a level that would make the use of the gas questionable or unsafe. Contaminants can harm patients and even some equipment. For those gases that are oxidizing in character, leaks can cause explosions and accelerate fire and fire spread.

The four major components of patient piped gas systems are as follows:

- *Source of the gas* (i.e., its point of origin and storage in the facility)
- *Distribution mechanism* (i.e., the transportation of the gas from its source to its destination)
- *Outlets or interface points* (i.e., the locations that represent the end of the piped system)
- *Monitors and alarms* (i.e., the mechanisms placed at various points throughout the system that warn of real or potential problems)

Another part of the system, not as easily categorized since it is present throughout the system, is the use of *electricity* to indicate the status of the gas (e.g., its pressure) as it traverses to its destination. Figure 29.1 is a diagram of a basic patient piped gas system.

Many codes and standards cover the design, installation, use, maintenance, and testing of nonflammable patient piped gas systems. NFPA 99, *Standard for Health Care Facilities,* 2005 edition (Chapters 5 and 13 through 18), and NFPA 50, incorporated into NFPA 55, *Standard for the Storage, Use, and Handling of Compressed Gases and Cryogenic Fluids in Portable and Stationary Containers, Cylinders, and Tanks)* include many installation, performance, and testing criteria for these systems. The U.S. Department of Transportation has issued regulations governing the transport of gas containers and cylinders, and the Compressed Gas Association and the U.S. Pharmacopoeia have issued standards for the purity of medical gas, to name a few of the major groups.

Chapter 29. Patient Piped Gas Systems

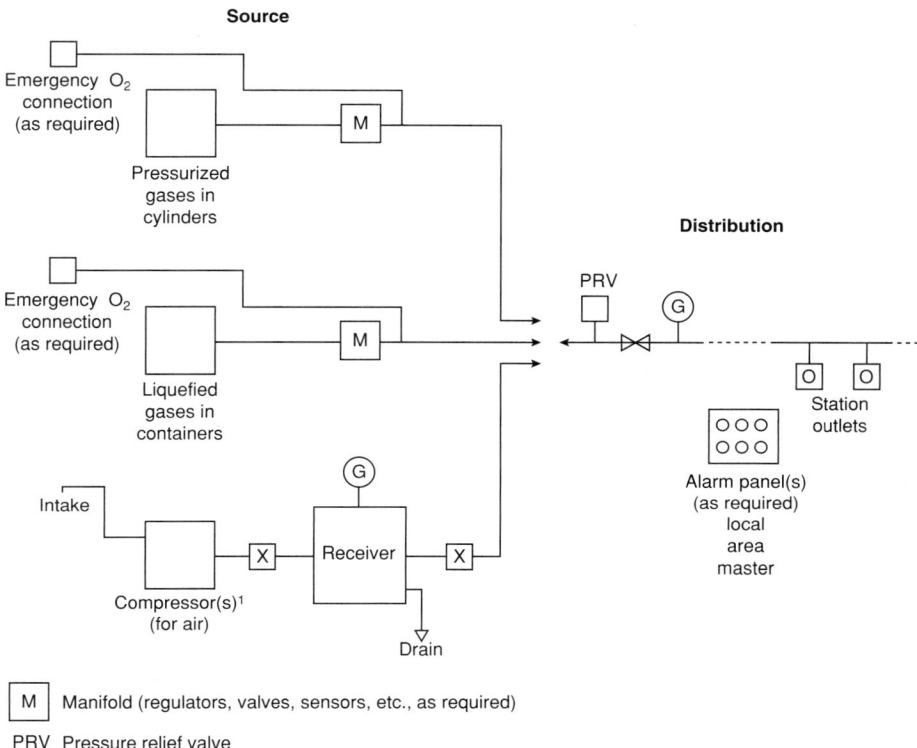

Figure 29.1 Components of a patient piped gas system (simplified).

29.1 System Levels

With the 1996 edition of NFPA 99, and retained in subsequent editions, patient piped gas systems were designated Level 1, Level 2, Level 3, and Level 4. Each level reflects the needs and conditions of the situation in which it will be used and represents a different level of risk to patients should the system fail. Note that this nomenclature ("Level") was revised from that in the 1993 edition of NFPA 99, which referred to "Types" of piped gas systems and did not address the element of *risk to patient*.

All patient piped gas systems provide nonflammable medical gases at operating pressures below 300 psig (2068 kPa), the most prevalent one in the United States being 50 to 55 psig (350 to 385 kPa). Level 1 gas systems are intended to be the most reliable and redundant systems because they serve the most critical patients (e.g., those on mechanical ventilation or assisted mechanical ventilation). Level 2 gas systems are intended for use by patients who are not as dependent on the system as patients using Level 1 systems but who still need supplemental gas(es) (e.g., patients who have had medical, surgical, or diagnostic intervention and are still dependent on a piped gas system). Level 3 gas systems are intended for use by patients for whom the interruption of the supply of gas would not be life threatening, such as for mild sedation. Level 4 gas systems are intended for use in situations that require greater regulation than that needed for commercial piped gas systems but are not intended or safe for patient applications. For the 2002 edition of NFPA 99, Level 4 system requirements were transferred to Chapter 11 (Laboratory Requirements), since it was seen that these systems were used only in these locations. (See Chapter 31 in this book for more information.) NFPA 99, 2005 edition (Section 5.3.3), divides Level 3 piped gas systems into two general categories based on the use of the compressed gas. The first type (Section 5.3.3.4) supplies medical gases that may be ingested by patients, such as compressed air systems that clear away debris after a cavity is drilled or medical gas systems that sedate patients but do not anesthetize them. The second category of Level 3 piped gas systems (Section 5.3.3.5) supplies a compressed gas (such as nitrogen) to power devices. If nitrogen is used, it must be oil free and dry (as defined in Chapter 3 of NFPA 99, 2005 edition).

NFPA 99 (Chapters 13 through 18) contains requirements regarding the levels of gas systems that are to be installed in hospitals, nursing homes, limited care facilities, and other health care facilities to meet various conditions and thus address certain levels of risk. Thus, health care facility designers must know which procedures involving medical gas are anticipated within a facility so that an appropriate system is installed for the conditions that will occur. For example, if an ambulatory surgical facility installs a medial piped gas system, designers must know what kinds of surgery will be performed.

This approach to determining which particular system will be installed places some burdens on all decision makers involved. If a facility expands its range of services, such as to treat people who require more intensive types of care, the system initially installed may expose these patients to more risks than necessary. If this change should occur, it is essential for facility personnel to review the system to determine whether a higher level of patient piped gas system is needed. However, it is not necessary to re-evaluate a system in which the treatment of patients requires less intensive levels of care.

29.2 Design Considerations

In addition to knowing which system level to install, designers of piped gas systems must also be aware of the anticipated system size and expected utilization rates. They must also try to determine how long the system will or can remain static in terms of size; how much of the system should or can be oversized; and what and where expansion might be considered and thus planned for. These are not easy questions to answer, since designing for optimal future use always involves added costs, although spending dollars now compared to many more dollars later often leads to fewer expenses when a system is expanded. Thus, to facilitate the design of piped gas systems, system users should discuss the potential capacity of the source (i.e., potential size of the compressors or manifolds—see later sections) and ways and means of adding future shutoff valves in strategic locations and pre-piping to an area.

Designers of piped gas systems must also know that most safety documents include a caveat that allows the use of equivalent methods, materials, designs, and the like, as long as the equivalents are acceptable to the authority having jurisdiction (AHJ) and meet the intent or requirements of the standard. While this fosters creativity, it also places additional burdens on AHJs in terms of determining whether another method, design, or material is indeed equivalent to the requirements listed in a code or standard.

For all types of projects, whether large or small, complete or partial, simple or complex, all parties involved in designing piped gas systems must review the physical considerations surrounding the installation of the systems before the first piece of tubing is cut. Such physical considerations include weather in the area, seismic activity, the extent of possible unwanted human intervention, potential for nearby excavation work, and the like.

After a system has been in use for either a few or many years, users sometimes begin to question the continued viability and reliability of the system and whether it should be replaced or renovated. If this occurs, some of the factors that will need to be reviewed include (1) whether any stress fractures have been found, (2) whether replacement parts are still available for older equipment, and (3) whether portions of the existing system must be brought up to comply with more recent codes and requirements that apply to new systems.

Note, finally, that piped gas systems are *not* to be designed with the intent of being used as an electrical grounding connection.

29.3 System Sources

The gases most piped into patient care areas are oxygen, medical air, nitrogen, and nitrous oxide. The only restrictions regarding which patient gases can be piped either to the next room or throughout a building are (1) that they be nonflammable and (2) that they be used only for patients or the testing of patient gas equipment (e.g., an anesthesia service area used to test anesthesia equipment). This prohibition has been in effect in NFPA documents on piped patient gas systems since the 1930s.

The gases used for piped gas systems can be purchased in cylinders and containers from outside sources or produced (compressed) on site. (See Section 32.1.1 in Chapter 32 for the difference between a cylinder and a container.) The only patient gas that facilities typically generate on site is air for medical, surgical, or dental purposes. This is justifiable given the amount of compressed air used by health care facilities and the fact that normal air outside a facility is generally clean enough (or can be made clean enough) for use in a medical compressed air system.

Factors that can affect the performance or the quality of the source gas in piped gas systems include the accumulation of particulates, hydrocarbons, water, and other contaminants and internal failures of system components. NFPA and other organizations have developed standards for gas source safety, whether purchased or generated on site.

All systems supplying a gas (except a compressed air system that uses outside air) require a policy and procedures to resupply the gas before reserve supplies are depleted. This is generally in the form of audible and visual alarms at constantly attended location(s) that activate in sufficient time (i.e., when reserve supplies begin to be used) for the facility to order a refill.

It should be noted that oxygen for medical purposes is classified as a drug by the U.S. Food and Drug Administration (FDA), and thus its distribution and use are regulated as a drug. Suppliers of the gas must meet FDA current good manufacturing practices (GMPs). The installation of a manifold to connect gas cylinders or containers for liquid oxygen must also meet FDA policies. These policies address system design and material, technician training, brazing procedures for connecting joints, testing by qualified third-party inspectors, and documentation. NFPA 99 covers many of the topics listed in the FDA policy.

29.3.1 Purchased Gases

Gases in either cylinders or containers can be purchased and delivered to a facility by a supplier. Major considerations when using purchased gases include in what state they are purchased (gaseous or liquid); how safely and cleanly they have been stored, transported, and delivered; and how resupply is accomplished when cylinders or containers become empty. [For example, bulk systems require refilling a large tank (or tanks) from tanker trucks that connect directly to the large tank(s), rather than swapping empty tanks for filled tanks.]

29.3.1.1 Cylinders

Cylinders are tanks that hold gas under high pressure. This pressure can be as high as 3000 psig (20,680 kPa). Oxidizing gases at this pressure can create significant heat when flowing through small openings; temperatures, in fact, can reach several thousand degrees. This is high enough to ignite combustibles, such as plastic O-rings, grease, and the like. The proper use of cylinders should be reviewed and practiced.

When more than one cylinder is used to supply gas to a system, a device called a *manifold* links the cylinders together in such a way as to allow the valves of each cylinder to be opened (i.e., have its valve opened) at the same time (see Figure 29.2). Several manifolds can be installed for one system, along with a mechanism that automatically switches to the other manifold when one bank of cylinders reaches a preset low level of gas. Several gas manifold configurations are described in NFPA 99 [Sections 5.1.3.4.10 (manifolds for gas cylinders without a reserve supply), 5.1.3.4.11 (manifolds for gas cylinders with a reserve supply), and 5.1.3.4.12 (manifolds for cryogenic liquid containers)] for Levels 1 and 2 piped gas systems.

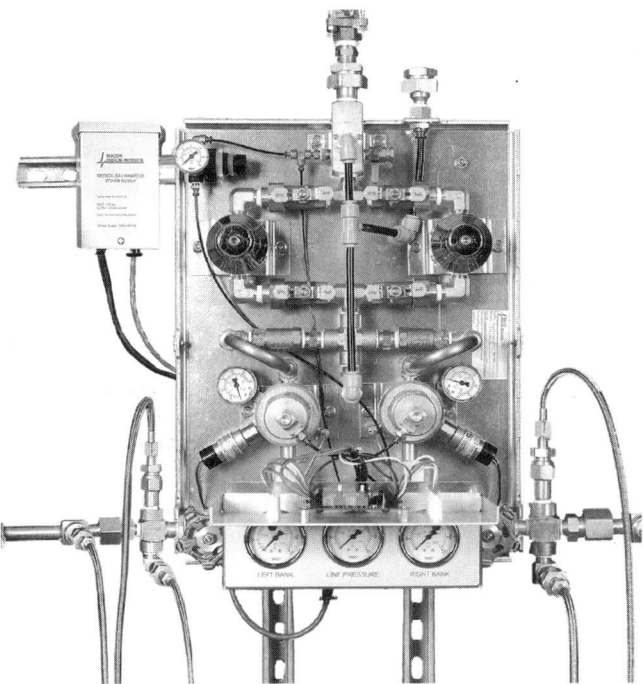

Figure 29.2 Details of a typical manifold that is supplying gas from cylinders to the piping system.
Source: Health Care Facilities Handbook, NFPA, 2005, Exhibit 5.4.

The following case study is an example of a group of cylinders that was not used properly.

Bulk Oxygen System and Reserve Supply

A verifier testing a bulk oxygen system in a government hospital discovered that 100 high-pressure cylinders located inside the facility were being used as the "reserve supply" for the system. All the cylinders were connected to a hand-made header bar fabricated from regular type L tubing instead of schedule 80, high-pressure brass, which should be used because of the pressure of the cylinders [i.e., the cylinders are filled to 2000 psig (13,787 kPa)].

Some time later, the main underground line from the bulk tank to the building was accidentally severed while other work was being done. It was then discovered that all 100 cylinders were empty. It was also found that the system had been operating at 58 psig (406 kPa), but that the reserve supply was set to open when pressure fell to 55 psig (385 kPa). Over time, gas in all the cylinders was drained out. No "reserve in use—low-pressure alarm" had been installed (see later sections in this chapter for more on alarms).

Figure 29.3 Cylinders of nitrogen connected to the manifold of a Level 1 piped gas system.

For Level 3 systems that power devices, the gas generally used (oil-free, dry nitrogen) is produced by manufacturers and supplied in cylinders. The cylinders providing the gas for these devices must meet the same criteria as cylinders used for Level 1 systems, as listed in NFPA 99 (Section 5.3.3.1.1). Figure 29.3 shows cylinders of nitrogen connected to the manifold of a Level 1 piped gas system. Note that the primary supply (left) has six cylinders, while the reserve supply (right) has three cylinders. Figure 29.4 shows a cylinder of nitrous oxide connected to the manifold of a Level 1 piped gas system. Note that the primary supply is on the right in this installation.

29.3.1.2 Containers

The use of *containers* holding liquefied medical gases, originally just oxygen, from 50 up to 200 psig (350 to 1400 kPa) began around 1980. One of the main features that attracted their use by health care facilities was the amount of gas that could be stored in the same size of tank as that used for gas under high pressure. This reduced the number of tanks that needed to be changed, which in turn reduced staff burden. There is about 20 times more oxygen in a liquid state in a container than oxygen in a gaseous state in a cylinder of the same volume. This benefit must be tempered against the fact that liquid oxygen presents 20 times the fire hazard for the same volume of gas, although this hazard is somewhat mitigated in that smaller containers are generally used. Another feature that has to be considered is the temperatures involved. For gas to be in the liquid state, it must be supercooled to about −300°F (−184.4°C). This handling issue presents the hazard of skin burns of the *cryogenic* (i.e., super-cold) variety.

Containers for liquefied gases range considerably in size from mobile "walkers" light enough to be carried with one hand to large stationary supplies that need to be mounted on a concrete pad outside the facility.

Figure 29.4 Cylinders of nitrous oxide connected to the manifold of a Level 1 piped gas system.

29.3.1.3 Storage Limitations

When source tanks (cylinders and/or containers) exceed 20,000 ft^3 (566 m^3) of gas, tanks must be located outdoors and comply with NFPA 55. The source portion of the system must then meet the requirements of both NFPA 55 and NFPA 99 (Sections 5.1.3.3.1.8 and 5.2.3.3) for Level 1 and 2 systems. (See Section 29.3.3. in this chapter for more information on tank storage and management.) Figure 29.5 shows an outdoor oxygen container for a Level 1 system.

29.3.1.4 Resupplying Source

Piped gas systems that use cylinders or containers have a *primary* and a *secondary* supply or a primary and a *reserve* supply (much like the old Volkswagen cars in the 1950s and 1960s), instead of just an indicator to warn when a supply is getting low (much like today's cars, which have a single gasoline supply tank with a needle or number to indicate the fuel level and a light that illuminates when the fuel is below a certain level). The former method (i.e., a separate secondary reserve supply) was chosen and continues to be used for the following reasons:

- Having a separate supply and identifying it as the *reserve* has more impact on persons responsible for managing the supply. The *reserve* cannot be lumped into the *primary* and *secondary* supplies and thus not be seen.
- An indicator or sensor might fail, which could result in the facility not knowing about the dwindling supply of gas until a very hazardous situation developed (i.e., the pressure and flow of gas begins to drop).
- Maintenance of the supply system can be accomplished without having to shut down the piped gas system.
- Failure of the *primary* supply will not result in the shutdown of the piped gas system.

Figure 29.5 Container of oxygen for a Level 1 piped gas system located outdoors, since capacity exceeds 20,000 ft³ (566 m³).

29.3.1.5 Performance Criteria and Standards for Purchased Gases

Because gases are transported to consumers, the U.S. Department of Transportation (DOT) has developed regulations for the tanks in which gases are shipped and used (i.e., cylinders for gases under high pressure; containers for gases under low pressure). The regulations cover tank materials, construction, strength, useful life span, and other parameters. Because DOT enforcement agents cannot constantly check all gas tanks presently in use, it is prudent for individual facilities to (1) check that cylinder and container end-of-use dates have not passed, (2) inspect cylinders for any new cracks, (3) ensure that only oxygen has been transfilled into a cylinder marked "oxygen," and (4) transport cylinders in the proper manner. Having facility staff conduct these checks adds another level of safety in the management of cylinders and containers during their usage within the health care facility.

Purchased gases intended for ingestion by patients must be *clean*. The U.S. Pharmacopoeia and the Compressed Gas Association have developed standards for the purity of purchased gases used for medical purposes. Suppliers of these gases in either the gaseous or liquid state must comply with the following standards in this regard:

- Oxygen: CGA G-4.3, *Commodity Specification for Oxygen*, and U.S. Pharmacopoeia Standards
- Nitrous oxide: CGA G-8.2, *Commodity Specification for Nitrous Oxide*, and U.S. Pharmacopoeia Standards
- Medical air in cylinders: CGA G-7.1, *Commodity Specification for Air*
- Medical air from a compressor: CGA G-7, *Compressed Air for Human Respiration*
- Carbon dioxide: CGA G-6.2, *Commodity Specification for Carbon Dioxide*
- Nitrogen: CGA G-10.1, *Commodity Specification for Nitrogen*

To make sure that the purchased gases meet standards, facilities should obtain documentation that certifies that the gases delivered meet the specifications of the referenced standard(s). In addition, periodic sampling of gases might be prudent to make sure the gas on the label is actually the gas coming out of the cylinder or container.

All the requirements for the operation and management of cylinders (e.g., training, handling, special precautions) have been moved to Section 9.7 in Chapter 9 of NFPA 99, 2005 edition. This guidance on cylinders is applicable whether cylinders are used in conjunction with a piped gas system or individually on patients. Refer to Chapter 32 in this book for more information on the use of cylinders and containers as "patient gas equipment."

29.3.2 Compressors

The most common method used to produce medical air within health care facilities is through the use of air compressors. These devices eliminate the need to establish a delivery service of cylinders of medical air, since "local air" is used. However, this means that care must be exercised in locating the intake pipe to the compressors, because the quality of the air that compressors use as source air influences the quality of compressed air produced. Since it is required that compressed air so produced not degrade air (this air will be used only for patients), it needs to be as clean as possible. These intake pipes, which represent the beginning of the air compressor system, can be located at one of two places. They can either be mounted outside the facility, turned down, screened, and connected to the compressor, or be connected within the facility to an air source equal to or better than outside air and then connected to the compressor. For intakes located outdoors, major concerns are the weather that can occur, the exhaust from vehicular traffic, and any facility exhausts that might drift into the intake. The following case study shows where not to install a compressor.

Improperly Located Air Compressor

In an urban surgicenter, a contractor installed the air compressor for the piped medical air system in the same room as manifolds. Due to lack of space, the contractor mounted the compressor approximately 8 ft (2.5 m) above the floor and directly over the manifolds. This arrangement required the use of a portable ladder for viewing,
(Continued)

servicing, and testing the compressor. The affiliated medical center planned to provide facility staff for managing the compressor. As indicated in his report, however, the verifier did not approve the system because of the location of the compressor. NFPA 99 (Section 5.1.3.3.13) requires compressors (and vacuum pumps) to be located separately from cylinder patient gas systems or cylinder storage enclosures.

29.3.2.1 Compressor Technology

Compressors use a variety of technologies to compress air. NFPA 99 (Section 5.1.3.5.4.1) now permits three types of compressor technology. One method operates using pumps with no oil anywhere within the compression chambers. Another type uses reciprocating compressors, provided there is a separation of the oil-containing section from the compression chamber by at least two seals, as well other safeguards to prevent the introduction of oil into the air stream. A third type uses rotating element compressors that are provided with a compression chamber free of oil, as well as other safeguards.

Compressors have been referred to by various names, some of which have been confusing (e.g., oil-less and oil-free compressors). NFPA 99 never used the term *oil-less compressor*. The term *oil-free compressor* was deleted from use in NFPA 99 after the 1990 edition to eliminate further confusion. As will be discussed further, the greater the possibility of oil getting into the air stream, the greater the number of safety features that are required to be included on the compressor.

To increase the total reliability of the compressors used for Level 1 piped gas systems, NFPA 99 (Section 5.1.3.5.11.2) requires facilities to install at least two compressors. This duplex arrangement allows for the alternation of compressors, and thus the repair of one compressor without shutting down the system, and provides an overall increased reliability of the system. Level 2 and Level 3 piped gas systems are permitted to operate with just one compressor because of lower patient risk if the source fails. The extent of system redundancy and ability to supply air is based on the facility's need to repair the system without interrupting service and the extent to which patients rely on the air supply.

While cylinders of compressed gas can be connected almost directly to a system's distribution pipes, mainly using pressure regulators and valves, systems that use air compressors must use these devices plus additional ones before connecting the air stream to the distribution portion of the system. These devices include, as needed or specified, *receivers, dryers, aftercoolers, filters,* and *air quality monitors*. As with compressors, all of these devices (except the receiver) must be duplexed in Level 1 systems to ensure that the system does not have to be shut down to repair or replace one of these devices. Level 2 and Level 3 systems do not require as much duplexing of these devices, as stated in NFPA 99 (Sections 5.2.3.5 and 5.3.3.5, respectively).

Receivers are airtight reservoirs that can hold air. They are necessary for storing significant quantities of compressed air. The quantity of air stored, and thus the size of the receiver, depends on the amount of compressed air used by a facility. By storing a sufficient amount of air (depending on the size of the system, the type of patients treated, the number of patients who are expected to be using medical air at any one time, and the medical services provided), the pressure of the air throughout the piped system can remain relatively constant, even as usage varies. The size of the receiver and associated piping is calculated by the engineering firm that designs the system.

When compressed, intake air can condense in receivers, piping, and other system components. Because moisture in a system is a mechanical hazard to the piping distribution system and to equipment connected to the piping system, dryers are necessary to eliminate any water before the air is distributed. Moisture in air also represents a health hazard to patients because, in the extreme, moisture can cause a patient essentially to drown. Moisture can also support the growth of bacteria, which can be carried within the piping system and beyond. System designers must carefully consider the size of dryers for the effective removal of moisture. Figure 29.6 shows a medical compressed air source for a Level 1 piped gas system with all components duplexed except the receiver. In the figure, a person is standing next to the dryers. Notice that spacing between items allows room to maintain and service the equipment properly.

The entire compressor package can create considerable vibration (and thus noise) when fully operating. Therefore, appropriate *antivibration mountings* are required to dampen the effects of the compressor. Manufacturers generally provide recommendations on the types of mountings necessary to reduce vibration noise from their units.

29.3.2.2 Compressor Air Quality

The most important reason why source gas must not be contaminated or degraded is that the tolerance of many patients is lower than that of healthy people, and in some cases it is so low that they may not be able to fend off any ill effects caused by air

Figure 29.6 Medical compressed-air source, with components duplexed (except receiver), for a Level 1 piped gas system.

contaminants. For example, patients' lungs, circulatory systems, or immune systems may not be able to tolerate any undesirable components in the piped air or oxygen. Furthermore, degraded or contaminated gas, especially medical air that contains particulates or water vapor, can negatively affect the operating characteristics of anesthesia equipment.

In addition, when considering weather conditions and the location of the intake point for a system, the intake site must be some distance from exhausts, vehicular traffic, and other devices that generate toxic atmospheres. A minimum of 10 ft (3 m) is required by NFPA 99 (Section 5.1.3.5.13.1), but greater distances could be necessary, depending on factors such as wind and laboratory exhausts in the vicinity. In all cases, the intake has to be at least 20 ft (6.1 m) above street level. It might even be necessary to install air filters within the intake to ensure that the air collected is relatively clean. Facilities must also be vigilant to note *any changes in distances* from pollution sources that could increase the chances of foul or contaminated air entering the air intake pipe and subsequently pass through compressor(s) and filters and be distributed to patients. Changes might include new parking spaces near the intake pipe, the installation of an exhaust port, or the erection of a fence, all of which could block airflow or concentrate air with undesirable components.

In addition to contaminants entering the system via the air intake, contaminants can also find their way into the distribution system by other means, including oil used in the compressors, metal particles, and internally generated gases [e.g., chlorine from liquid rings or polytetrafluoro-ethylene (PTFE) by-products]. (See several subsequent paragraphs for Level 3 requirements.)

The compressor itself can be a main source of contamination in compressed air systems. For this reason, special attention must be paid to the following factors that can influence compressors and their performance:

- Compressor construction
- Air treatment (i.e., filters, dryers, regulators)
- Compressor maintenance
- System operating parameters

For Levels 1 and 2 piped gas systems, NFPA 99 requires on-site generated air, *before* it is connected to the piping distribution system, to meet minimum criteria for dew point [39°F (3.9°C) at 50 psig (350 kPa)], carbon monoxide (10 ppm maximum), carbon dioxide (500 ppm maximum), gaseous hydrocarbons (25 ppm maximum), and halogenated hydrocarbons (2 ppm maximum). It is not necessary to identify the various hydrocarbons (there are many); rather, the sum total of hydrocarbons must not exceed these values (see NFPA 99, Section 5.1.12.3.12.3).

The criteria for the compressor package for Level 3 gases that will be ingested by patients are less stringent than those for Level 1 or 2 systems, since the patients for which this type of system is intended are in much better health than those who use Level 1 or 2 systems. For Level 3 systems, a *moisture indicator* is considered sufficient for indicating that the compressed air has not exceeded 40 percent at line pressure and temperature, and a oil indicator is considered sufficient for indicating that oil concentration downstream of the receiver has not exceeded 0.05 ppm maximum (see NFPA 99, Sections 5.3.3.5.5 and 5.5.3.3.5.6, respectively).

The NFPA technical committee responsible for piped gas systems has noted the need to keep air generated for "patient use" separate from air generated for all other purposes (NFPA 99, Section 5.1.3.5.3.1). Patient use includes that air used to test anesthesia equipment for proper operation. This restriction aims to prevent contamination of medical or dental air from equipment backflow (which can occur when certain devices are used or fail) as well as reduce excess wear on machinery. This restriction also serves to reduce maintenance and other costs incurred in operating the supply system. Refer to Chapter 31 for information on air generated for purposes other than patient use.

NFPA 99 also states that medical air produced on site must not add any more contaminants to the gas than are already present in the air surrounding patients. Making such air cleaner than surrounding air, however, is encouraged so as to reduce the amount of undesired components in the atmosphere that reaches patients. Improving the quality of the gas that is ingested by patients can only be beneficial for them. This issue of contamination applies to the distribution of piped gas, irrespective of how the source gas is provided (see later sections in this chapter).

If there is still concern about the level of particulate matter in the air or oxygen beyond the piped system, *end-stage filters* can be used. These filters are installed ahead of flow meters or ventilators to reduce the buildup of static electricity, which can occur when particulates in a narrow gas line pass by plastic. Static discharge can ignite plastic, and some flow meters use plastic extensively. (Static buildup occurs more easily in oxygen lines than in air lines.) By properly grounding the filter, static buildup can be mitigated. Operating the system at lower pressure or flow can also ease the problem of static buildup. Refer to Chapters 6, 7, and 11 for information on grounding equipment.

Based on the increased desire to prevent atmospheric pollution and to prevent contaminated air generated by patients from being distributed to other patients, NFPA 99 (Sections 5.1.3.5.15 and 5.2.3.5 for Level 1 and 2 systems, respectively) has included requirements since 1993 for monitoring the quality of medical air created by compressors. The parameters that must be monitored include dew point, carbon monoxide, and, for some compressors, liquid hydrocarbons. Monitoring can be conducted on either a continuous or a periodic basis. It is important for facilities managers to determine which edition of NFPA 99 was in effect when the facility's piped gas system was approved or which edition of the code was being enforced at the time of initial construction/installation; this determination will reveal what requirements were in effect when the system was installed.

29.3.3 Physical Considerations for Source Rooms and Piping

All gas source rooms must be designed and installed to handle extremes in weather and undesired human actions. (Source rooms can be inside a facility, stand-alone *sheds,* or an area enclosed by a chain-link fence.) When facilities are located in seismic activity–prone areas, local building codes need to be reviewed to properly brace source equipment (and piping) as needed. Piping and fittings of the distribution system (see Section 29.4) must also be sturdy enough to maintain integrity under extreme conditions such as fire and physical stress.

29.3.3.1 Indoor Source Rooms and Storage Areas

Gas source and storage rooms for piped gas systems must meet the requirements of NFPA 99 (Section 5.1.3.3 for Level 1 and 2 systems and Section 5.3.3.3 for Level 3 systems). This applies to rooms in which just the manifolds are housed, as well as rooms used to store extra cylinders. The provisions in these sections cover protection from heat sources and dislocation, the fire resistance of the enclosures, compliance with electrical codes, the exclusion of flammable and combustible materials, the use of valve protection caps, and the location of the areas.

NFPA 99 (Section 5.1.3.3.1.4) requires compressors to be located in rooms separate from those having manifolds or those containing stored cylinders or containers, except instrument air reserve headers that meet certain requirements. Vacuum system pumps can be in the same room as a compressor, since the intake for the compressor and the exhaust for the pump must both be outside the building *and* sufficiently separated so that the vacuum exhaust cannot contaminate the air intake. (The actual distance between portals is subject to many factors, including weather conditions, height differentials, items between the two portals, Environmental Protection Agency requirements, etc.)

NFPA 99 does not require compressed air supplies to be located in a room or enclosure separate from other utilities because such supplies do not store thousands of cubic feet of compressed gas. The system compresses only enough air to maintain pressure at approximately 50 to 55 psig (350 to 355 kPa) of air (the most common pressure) or other relatively low pressures as needed. However, the system can take up considerable space (with a receiver, at least two pumps, dryers, etc.). While it may save floor space to mount compressors, dryers, and other equipment atop each other, it may prove very difficult to service them. Conversely, mounting items horizontally very close to each other to save floor space can also make it very difficult to service the equipment. Providing adequate spacing to install, maintain, *and service* items of a compressor system should be done early in the design stage to preclude extra expenses (and probably much frustration) after installation.

29.3.3.2 Outdoor Source Areas

Although the source supply of any piped gas system can be located outdoors, whenever the total amount of gas in use or stored for use, connected and not connected, exceeds 20,000 ft^3 (566 m^3), the gas source must be located outdoors (per NFPA 99, Section 5.1.3.3.1.9). For freestanding outdoor supplies, the basic considerations are to protect the supplies from environmental conditions (e.g., sun, wind, and snow) and undesired human actions (e.g., vandalism, vehicle crashes, and the firing of weapons into the site) that could cause tanks or cylinders to rupture or explode. Some facilities have had to construct a wall around a supply area to protect it adequately.

If local codes allow cylinders or containers to be placed against the side of a building, such installations must meet appropriate requirements from NFPA 50 (now part of NFPA 55), to protect both the facility and the equipment. Other hazards are addressed in NFPA 55 as well. Other considerations when designing an outdoor storage area include providing accessibility for safely refilling or exchanging tanks or cylinders, preventing unauthorized entry, and providing the types of alarms necessary to prevent or warn of unauthorized entry, as illustrated by the following case study.

Outdoor Storage Areas Near Occupied Buildings

In designing the outdoor storage area for liquid oxygen tanks (a main tank and a reserve tank), an inner-city hospital had two major concerns: (1) the size and location of the site available and (2) the exposure to a street where shooting at such objects was well within the realm of possibilities. To reduce the effect of fire or explosion and at the same time eliminate exposure of tanks to the street, a three-sided brick wall was built around the tanks. The wall was high enough so that the tanks could not be seen from the street. No cover was placed on the top, to allow any forces (from fire or explosion) to be exerted upward rather than against the brick walls. Steel rods were placed across the top (from the corners to the middle of the opposite sides) as an added measure of strength to help keep the walls intact during an incident. The facility did have to make sure there was plenty of drainage from rain and snow and any water that might have to be applied to the tanks in an incident. The remainder of the design was normal for an outdoor installation of its size and nature.

29.4 Gas Distribution

Once the gas or gases to be piped have been selected, they must be safely distributed. The distribution of piped gas can range from simple (one pipe with several outlets) to complex (many parallel pipes serving outlets on many floors of a building), and the gas can flow just one room away or throughout an entire facility.

The goal of any piped patient gas distribution systems is similar to any closed-loop gas or liquid distribution system in that it aims to distribute gases without losing any gas and without adding any undesired components along the way to outlets. The reasons why it is not desirable to degrade gases within piping systems are mentioned in Section 29.3. There also are other reasons why gases must not be lost. First, leaky systems waste finite resources (i.e., gas and money). Second, the loss of gas creates a hazard. Oxidizing gases, such as oxygen, air, and nitrous oxide, support combustion. Since the gas within piping is under pressure, a leak means the gas escapes from the system at about 55 psig (385 kPa) and increases the percentage of oxygen in the area around the leak. The air becomes oxygen enriched, and items in such atmospheres burn more easily and more intensely and ignite at a lower temperature than in normal atmospheres. It should be pointed out that the characteristics of the place where a leak occurs, such as the space within a wall or the entire space above a dropped ceiling, could also influence how hazardous the problem can be.

The main components of the distribution portion of a piped gas system are the piping, valves, and alarms. Piping materials, the methods of joining piping together and keeping it clean, and the paths taken by the gas are some of the key aspects of gas distribution. NFPA 99 specifies a number of parameters for these components, some of which follow, to minimize the hazards of distributing gases and make these systems safe:

- Minimum acceptable grade of material for piping
- Cleaning methodology to be used for components (piping, fittings, and valves)
- Location of shutoff valves
- Minimum qualification of brazers
- Quality of brazing
- Location of alarms
- Minimum operating pressure throughout the system
- Methods of testing for compliance

These and the rest of the specifications listed in NFPA 99 help ensure that staff are provided with the gas they expect from each outlet at needed pressure, patients are provided gas of acceptable quality, and building integrity is not compromised because of the installation of the piping system to distribute the gas. Should some incident occur, it should also be possible to shut down or isolate the portion of the system involved or affected, thereby interrupting gas to the fewest patients. Per NFPA 99 (Section 5.2.10), the distribution systems for Level 1 and Level 2 systems are to be installed to the same requirements.

29.4.1 Piping

There are a number of considerations with respect to the piping that distributes medical gas. The type of material used cannot be affected by the gas that will contact it, pipe joints must meet a number of standards to prevent what could be a fatal accident, and all pipes must be marked in a manner that is legible and permanent. There are also considerations regarding proper installation of piping in order to maintain pipe pressure. The piping system also has to meet a number of performance criteria.

29.4.1.1 Piping Materials

As new types of tubing have been made available over the years, the type of tubing found to be acceptable for distributing patient gases has changed. Currently, Section 5.1.10.1 of NFPA 99 specifies ASTM B 819, *Standard Specification for Seamless Copper Tube for Medical Gas System*. For the *usual and normal* operating pressure of 50 to 55 psig (350 to 385 kPa), type K or L tubing is acceptable. If operating pressures of 200 to 300 psig (1400 to 2100 kPa) are used (and some new equipment operates best at these pressures), then only type K tubing can be used. In addition, tubing that meets these criteria is to be marked "OXY," "MED," "OXY/MED," "OXY/ACR," or "ACR/MED" by the manufacturer.

The ASTM B 819 standard was written specifically for medical gas applications and includes all the requirements found in three previously separate standards on this subject. Other types of tubing exist and are acceptable, however. For example, stainless steel is acceptable, although the cost of stainless steel is significantly higher than that of ASTM B 819 tubing. Note that a wider range of tubing is permitted for piped vacuum systems, as discussed in Chapter 30.

If another type of piping is desired to be used, it has to be demonstrated to the authority having jurisdiction that it is of such quality, durabilty, strength, etc., as to be "equivalent" to that listed in NFPA 99. (See criteria for "equivalency"' in Section 1.4 of Chapter 1 in NFPA 99.)

29.4.1.2 Joining of Pipes

To join lengths of metal piping together to create a closed-loop system, each length must be joined to another length of pipe using some type of coupling mechanism that adequately holds the two pieces of tubing together (i.e., a chain is only as strong as its weakest link). All fittings for piped gas systems must be able to withstand temperatures of up to 1000°F (537°C) and still maintain piping integrity. Because solder melts below 1000°F, all fittings that require a heating process to join and seal a fitting to piping must be brazed to meet the temperature criterion. This was the way the original document on piped gas systems was worded. The 1000°F temperature was considered an acceptable value in light of temperatures measured in fires and the need for piping to stay intact at least until these temperatures were reached. Reference to a "brazed fitting" replaced "temperature value," since brazing cannot be achieved at temperatures below 1000°F. It was believed that referencing this method was a better way of ensuring that the criterion that fittings withstand 1000°F temperatures was met—that is, a soldered joint might look like a brazed joint, but the temperature for soldering needs to be only about 400°F (204°C). Brazed joint criteria are listed in Section 5.1.10.5 in NFPA 99.

All methods of joining tubes must meet this temperature performance criterion. *Memory-metal fittings* [Section 5.1.10.7(1) in NFPA 99] do not have to be heated to create a seal but rather rely on being cooled in liquid nitrogen before being used. (Upon removal from the liquid nitrogen, this type of fitting contracts as it warms up to room temperature.) However, this method has to meet the temperature performance criterion. Other methods of joining that are now acceptable include the welding of joints if a gas tungsten arc welding (GTAW) autogenous orbital procedure is used (Section 5.1.10.6 in NFPA 99), listed or approved metallic gas tube fittings [Section 5.1.10.7(2) in NFPA 99,], dielectric fittings required by manufacturers of special medical equipment [Section 5.1.10.7(3) in NFPA 99], and axially swaged, elastic strain preload fittings if they meet certain criteria [Section 5.1.10.7(4) in NFPA 99].

NFPA 99 (Section 5.1.10.3.1) states that any change in the direction of piping is to be accomplished using an acceptable fitting, since both stretching and compressing the walls of a tube, as a result of bending, weakens it. The joint is required to meet certain parameters (see later sections) and possess an integrity comparable to that of the tubing. If stainless steel pipes are used, the tubes can be welded together, but the performance of such joints must be equivalent to that afforded by fittings. "Bending" of piping is still not permitted by NFPA 99.

While other joining methods can be used if documented to be technically equivalent to the above criteria (i.e., meet the thermal, mechanical, and sealing integrity of a brazed joint), the technical committee responsible for this chapter in NFPA 99 has listed joints that are *not* to be allowed for piped gas (and vacuum) systems. These are flared or compression-type connections, other straight-threaded connections (including unions), and "crimping" of piping in lieu of a cap or shutoff valve. (See NFPA 99, Section 5.1.10.8.)

29.4.1.3 Pipe Identification

Criteria covering tubing identification are listed in NFPA 99 (Sections 5.1.11.1 and 5.2.11) for Level 1 and 2 gas distribution systems. Markings on piped gas system tubing indicate the type of gas in each pipeline and its direction of flow. This helps reduce the possibility of a mix-up of gases; where such gases flow directly into patients, this mix-up might have fatal consequences. Knowing *which* gas is flowing in a pipe helps preclude cross-connections if and when a system is breached; knowing *which way* a gas is flowing informs personnel, in part, of the areas controlled by each shutoff valve.

With the 2005 edition of NFPA 99, marking for new piping is now to be accomplished by labeling, stenciling, or tagging the tubing with (1) the name or chemical symbol of the gas therein, (2) the color coding as shown in Table 5.1.11 in NFPA 99, and (3) if not the standard gauge pressure, the operating pressure of the gas. (See NFPA 99, Section 5.1.11.1.1.) Direction of gas flow must also be indicated, using arrows applied directly on tubing. Whatever marking method is used, it must be *permanent* (i.e., not easily removed or altered) and *repeated* at specified maximum intervals. Multiple markings eliminate the problem of spending excess time just trying to figure out which gas is flowing. Figure 29.7 shows gas distribution system tubing identified with the name of the gas flowing within the tube. Note the arrows to indicate direction of flow. NFPA 99 requires at least one identification and directional arrow in a room through which piping passes, even if this would be less than the minimum

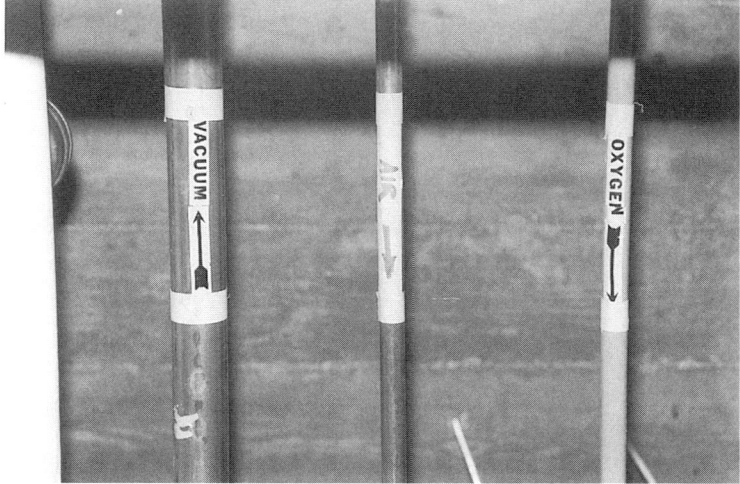

Figure 29.7 Tubing identified with the name of gas flowing within the tube.

interval distance. This requirement ensures that each room through which piping traverses will indicate what is flowing within piping traveling through the room and the direction of that flow.

29.4.1.4 Gas Distribution Pressure

Of late, some facilities have begun distributing gases at different pressures (e.g., one pressure for direct patient ingestion, another pressure for patient equipment, another pressure for dental equipment, and another pressure for hyperbaric facilities). Rather than installing completely separate systems, pressure-reducing regulators can be installed at the distribution source (per Section 5.1.3.4.7 in NFPA 99). Note that this allowance of differing pressures is to be made at the source and not within the piping distribution system.

Distributing gas at more than one pressure led the NFPA technical committee responsible for medical piped gas systems to review the issue of whether a marking is necessary to indicate the pressure of the gas within the tubing. Since most gas is distributed at a standard pressure of 50 to 55 psig (350 to 385 kPa), NFPA 99 (Section 5.1.11.1.1) currently requires pressure to be indicated only if the pressure within the tubing is different from this pressure. This requirement has a considerable number of opponents, however, who believe that pressure should be indicated whenever a gas is distributed at more than one pressure within a facility. This issue is reviewed at every revision cycle of NFPA 99.

29.4.1.5 Pipe Installation

Since the 1993 edition of NFPA 99, the standard has included good manufacturing and installation practices to increase the probability that *clean* piped gas systems are built initially and that they remain clean for the life of the system. Requirements now exist for all tubing handlers (including manufacturers, installers, and brazers) who are in a position to affect the cleanliness of the system. The standard also now addresses installation procedures for end caps, brazing, purges, and the like to minimize contamination. However, these new criteria do *not* mean that all piping installed prior to 1993 is no longer acceptable.

Another parameter that influences the reliability of a piped gas installation is the ability of the piping to remain intact and not break or leak under fire or physical stress (i.e., it must withstand physical abuse that can reasonably be expected to occur). NFPA 99 (Sections 5.1.10.10.4, 5.2.10, and 5.3.10.10.6) includes requirements for piping support and for protection from the elements (above and below ground) and indoor routing restrictions to keep tubing intact.

The NFPA technical committees responsible for piped gas system requirements over the years have made changes to requirements based on reviews of actual incidents (e.g., earthquakes in California in the 1990s). They have added references to national standards, placed added restrictions on where tubing can be run, and strengthened the requirements for tubing run underground, both outdoors and indoors.

29.4.1.6 Pipe Contamination

It is not possible to prevent all contaminants from entering a gas distribution system, because it is difficult to isolate fully the conditions under which the piping equipment is manufactured and installed. NFPA 99 (Sections 5.1.10.1, 5.1.10.5.5.1, 5.1.10.5.5.6, 5.1.10.10.2, and 5.1.12.2.2) provides much guidance on how to reduce the concentration of contaminants within the piping. The intent of the standard is to keep the tubing as clean as possible, using techniques for assembling the equipment that minimize contamination. Some of the precautions that need to be taken include the following:

- Manufacturers must meet CGA G-4.1, *Cleaning Equipment for Oxygen Service*.
- Piping must be shipped with ends capped (sealed).
- Tubes must be stored on site in a reasonably clean area with end caps in place.
- Installers must avoid handling tubes with dirty hands or tools.
- When the system is being assembled, there must be blow-downs (i.e., the lines must be blown clear), and during brazing of fittings there must be a piping purge, using oil-free dry nitrogen.
- After joint ends are cleaned, piping is to be joined within 8 hours if a brazing technique is to be used.
- After assembly, another piping purge and a piping purity test must be conducted.

All these and other activities listed in Sections 5.1.10 to 5.1.12 aim to ensure that the piping system does not add contamination to the gases beyond the following limits: For both Level 1 and 2 systems (NFPA 99, Sections 5.1.12.3.6 and 5.1.12.3.7), a heavy intermittent purging of the pipeline is not to discolor a white cloth held loosely over an adapter during the purge. Following this, a piping particulate test of 25 percent of the zones (at the outlet most remote from the source) is to be conducted, with the filter accruing no more than 0.001 g (1 mg) of matter from any outlet tested. Additional medical air purity tests for compressor systems are to be conducted, per Section 5.1.12.3.12.1 in NFPA 99. For both tests, oil-free, dry nitrogen is to be used. For Level 3 systems (NFPA 99, Sections 5.3.12.3.6 and 5.3.12.3.7), the purging and particulate testing requirements are now very much the same.

29.4.2 Valves

Valves are important—and not just because they can seal off one section of piping from another section. Valves control the flow of gases, which can mean life or death for those who rely on them. Valves are also important for fire safety purposes in that they can eliminate one source of fuel in the case of flammable gases, or in the case of gases that support combustion, they can isolate tubing of the medical gas system that has ruptured or outlets that have been damaged, either of which allows nonflammable but oxidizing gases to flow unimpeded. (Although oxygen, for example, may not start a fire, its presence increases the burning rate of combustibles that have ignited.)

Under normal conditions, the flow of gas is unimpeded by open valves. When a valve is closed, gas downstream of the closed valve flows for a short period if equipment connected to the system is still being used, although the gas will flow at a decreasing pressure and flow rate. If a valve is inadvertently closed, patients in need of the gas will be affected. Depending on the distance between a closed valve and a downstream station outlet(s) in use, it might not take very long for the decreased pressure of gas in this section of the system to fall below a level that activates a required low-pressure alarm (see the following sections).

Under emergency conditions, the closing of a valve or valves can have an opposite result in that it cuts off the supply of gas that may be contributing to the emergency. Closing any valve must be done very judiciously, however, when gas(es) going to patients dependent on the gas(es) for life-sustaining purposes are flowing through these valves. A facility should develop a policy, in conjunction with the local fire department, regarding staff responsibilities concerning the closing of valves in emergencies.

29.4.2.1 Valve Locations

Shutoff valves must be strategically installed so that their closing produces the least disruption to patient care while allowing the facility to isolate portions of a system as needed. (NFPA 99, Sections 5.1.4 and 5.2.4, lists the minimum number of valves for Level 1 and Level 2 systems, respectively, and their relative locations for minimum patient safety. However, *knowing* their specific locations is just as important so that they can be found in an emergency. In addition, installing more than the minimum number of valves in an area with a large number of outlets can help reduce the amount of patient disruption and dislocation in an emergency.) The requirements for shutoff valve placement vary in different patient care areas (i.e., anesthetizing locations, critical care units, general patient areas), based on the criticality of patient conditions. Each operating room must have a shutoff valve(s) for each gas piped into it, so that the valves to each room can be closed without affecting the gas distribution to other operating rooms. These valves must be located on a wall immediately outside each operating room.

The fact that some shutoff valves in patient care areas need to be readily accessible in an emergency is an operational problem for which facilities must take special precautions. These valves are generally located in corridors near nurses' stations, where they can be observed by staff in the area. The valves also need to be protected from normal abuse such as transport carts, intravenous (IV) poles, doors being opened, and the like. (NFPA 99, Section 5.1.4, lists these conditions.) As such, shutoff valves are typically recessed and/or have some type of cover over them, usually made of clear plastic so that the position of the valve (open or closed) can be readily seen. However, this can also mean that curious children or persons with evil intent can do considerable harm if they should turn these valves. As a precaution, these covers are sometimes fitted with alarms that indicate when they are removed. (Section 29.6 in this book discusses monitors and alarms in more detail.)

NFPA 99 lists the minimum number of valves in a piped gas system. These include a *source shutoff valve* that can isolate the entire distribution portion of a system from a source, a *main line valve* if the source shutoff valve is not in the same building with the distribution piping, *riser valves* located at the bottom of any pipeline going up through a building, and a *service valve* at the beginning of each branch line that connects tubing on a floor to a riser passing through the floor (so that servicing of a branch line can be performed without shutting down other floors in the building). Other valves that are required as necessary are *zone valves* on a floor to subdivide a floor and *in-line valves* for being able to shut down an individual room or area on a floor. NFPA 99 also requires shutoff valves to be labeled with information on the buildings, floors, zones, areas, or rooms they control. NFPA 99 (Sections 5.1.4, 5.2.4, and 5.3.4) contains a complete list of the minimum number of valves required for Levels 1, 2, and 3 piped gas systems, respectively. NFPA 99 (Section 5.1.11.2) lists labeling requirements for shutoff valves.

The placement of some shutoff valves is done in coordination with NFPA *101*®, *Life Safety Code*®, as health care facilities need to meet a *defend-in-place* concept (see Chapter 14). These valves must be installed with a wall between them and outlets downstream from them, so that staff do not have to go into an area where there is a fire in order to shut off the gas that is feeding the area.

Figure 29.8 shows zone shutoff valves mounted in a recessed box. Note the gauges mounted on the downstream side of the valves. If a patient care area is fitted with alarms, then sensors (also on the downstream side) will activate when pressure drops below a preset alarm value.

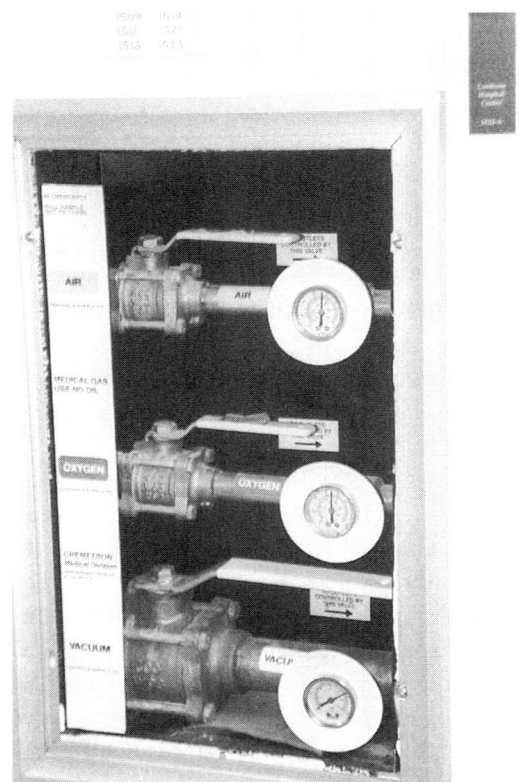

Figure 29.8 Zone shutoff valves mounted in a recessed box.

Additional shutoff valves can be installed as desired for enhanced facility operation. However, if they are readily accessible to the public, these optional valves need to be locked or fitted with alarms. Installing more valves than minimally required should be considered, since the cost of installing them during initial construction is far less than doing so after the system has been completed. It will also mean that a smaller area will be affected if an expansion is made at some time in the future. It can also mean that fewer patients may need to be moved or inconvenienced if a portion of the system must be shut down. Less movement of patients generally means a lower opportunity for medical complications to occur. Also, even if not required to by local codes, designers of facilities located in seismic activity–prone areas should consider installing more shutoff valves than normally required, to allow for shutting down areas more readily and quickly as a situation dictates.

29.4.3 Path of Least Deviation

The path that piping takes from the source to each outlet must be reviewed for its potential impact on the gas traveling inside the piping. It is particularly important to know where condensation may be created (e.g., within compressed air systems or where there are temperature swings of sufficient magnitude to produce condensation). This can occur when piping goes outside a building or is placed underground. Further, access to piping for service must also be considered, not only for new systems but also when renovations are being done.

29.5 End of the Line

The end of a piped gas system is any point in the distribution portion of the piping system where tubing is terminated, permitting something else to be connected to it. NFPA 99 permits three methods for terminating piping systems: (1) A station outlet can be installed with or without a manufactured assembly following it (see Section 29.5.1), (2) a sensor or alarm can be installed (see Section 29.5.2), or (3) a cap can be installed with or without a valve just ahead of it (see Section 29.5.3). Tubing is never to be bent and folded over, nor is it to be crimped, as this will weaken the tubing and does not ensure a seal that will remain intact even under normal conditions. The integrity of an end point is as important as the fittings used to join sections of tubing.

A valve can be installed at the end of a system to make it easier to add additional tubing if necessary. Valves installed in such locations are required to have locking mechanisms to prevent them from being opened by anyone except authorized personnel.

29.5.1 Station Outlets

Station outlets act as bridges between piping and the connectors at the end of equipment hoses (much the way electrical receptacles and plugs at the ends of equipment power cords provide a path for electricity in building wiring to be connected to electrical equipment). Figure 29.9 shows a typical gas-outlet/vacuum-inlet setup in an operating room. Station outlets enable gas to flow out of piping when the appropriate connector is inserted into the station outlet.

There are a number of points along the route of a piped gas system where a mix-up of gases can take place, resulting in a gas *other than* the one desired or expected to flow out of the station outlet. These points are

- *At the gas source*—Has the oxygen cylinder been replaced with another cylinder that does not contain oxygen?
- *In the piping system*—Has an oxygen line been crossed with a medical air line?
- *At station outlets*—Has an outlet been replaced with an outlet designed for another gas?

To prevent the wrong gas from being accessed through an outlet in systems that distribute more than one type of gas and to ensure that the end product is the gas desired or expected, several standards have been developed for outlet configurations. In this way, only matching configurations between outlets and connecting equipment will permit gas flow. Professionals in the field determined early on that just relying on the labeling of station outlets to indicate the type of gas within the pipe is not sufficient to prevent unintentional (and unfortunately, intentional) interconnections.

The major standards for connection configurations are DISS (Diameter Index Safety System) fittings (i.e., a threaded medical gas connector that complies with CGA V-5, *Diameter Index Safety System—Non-Interchangeable Low Pressure Connections for Medical Gas Applications*), and CGA V-11, *Compressed Gas Cylinder Valve Outlet and Inlet Connections,* a standard using a pin index system to prevent interchangeability. As a result of having two widely used standards, adapters have been created to make it possible to connect hoses with one type of connector to an outlet with another type of connector (though for the same gas). This practice is discouraged, however, because of problems it creates related to needing the correct adapter.

Additional safety criteria for station outlets can be found in NFPA 99, Section 5.1.5.

Station outlets, like electrical receptacles, are designed to tolerate abuse and remain useful for quite some time. However, if tolerances (i.e., the specifications) are exceeded,

Figure 29.9 A typical utility wall arrangement for gas, vacuum, WAGD, and electricity in an operating room. There will generally be one of these setups on an opposite wall as well since the orientation of a patient may be different.
Source: Health Care Facilities Handbook, NFPA, 2005, Exhibit 5.18.

the results can create a hazardous situation, the level of which varies depending on many factors. For example, if the pressure in the pipeline exceeds the limit of rings that provide a seal in the station outlet, the rings could be permanently damaged and permit gas to leak out. See the following case study regarding outlet pressure.

Nitrogen System Pressure Setting

A doctor in a new government hospital requested that the piped nitrogen system be designed to operate at 300 psig (2068 kPa) instead of the standard 160 psig (1103 kPa). The verifier involved with the project questioned this request, since DISS outlets are typically approved only up to 200 psig (1400 kPa). In addition, manufacturers do not make a gas-specific outlet for high-pressure nitrogen systems, and alarms are not available to function at other than 20 percent over the standard 160 psig. The system was redesigned to operate at 160 psig.

Excessive temperatures and mechanical damage to components can also hinder the proper functioning and safety of station outlets, leading to hazardous situations and compromising an outlet's noninterchangeability features. In hospitals, major culprits in damaging outlets are patients' beds, since they tend to be moved often.

Like some shutoff valves, one way to protect station outlets is by recessing them in the wall a sufficient depth so that connectors do not protrude beyond the wall line. Another way to protect outlets is to place them high on a wall. While there is no standard height for outlets, they are generally installed at least 5 ft (1.5 m) from the floor. This reduces the number of items that can come in contact with them, including small children's hands. This placement means that longer hoses leading from the equipment to the outlets may be required in some instances, but this requirement is offset by the reduction in the number of items that can impact the outlets.

During facility design, the protection of gas outlets may need to be reviewed further for any special features. For example, gas outlets in pediatric wards or in prison health care facilities may require the installation of some protective features to prevent children from knocking or activating them and prisoners from abusing them, respectively.

The number of gas outlets that must be installed in patient bed locations varies with the needs of the patient. The most referenced guideline is the American Institute of Architects (AIA) *Guidelines for Design and Construction of Hospital and Health Care Facilities*. This document should be viewed as a minimum guideline, since circumstances may warrant more outlets than recommended.

The following case studies show how outlets should not be used.

Oxygen for a Doctor's Fish Tank

A verifier was conducting tests on a piped oxygen system in a hospital and discovered that a doctor had connected a hose from an oxygen outlet in his office to a fish tank also located in his office. The verifier also noted that the outlet was below the
(Continued)

level of the tank. A failure of the oxygen system could backfeed water from the fish tank into the pipeline. NFPA 99 (Section 5.1.3.4.2) states that patient oxygen systems shall be used only for patient care applications. The verifier disconnected the hosing, returned later and disconnected the hosing again. Upon returning a third time and seeing the hosing connected once again, he disabled the outlet in the doctor's office.

Medical Air for Drying Instruments

In testing a piped medical air system in a hospital, a verifier found that an outlet from the medical air system was being used in the endoscopy suite for drying instruments after sterilization. However, NFPA 99, Section 5.1.3.4.2, states that piped medical air is to be used only for patient care applications, and Section 5.1.3.5.2 requires the medical air compressor to be connected only to the medical air piping distribution system. The verifier informed the facility staff that this practice was unacceptable per NFPA 99 and did not certify that portion of the system.

Nitrogen can be used as an alternative gas source. A small compressor system dedicated to supplying air just for drying instruments can also be used.

29.5.1.1 Surface-Mounted Gas Rail Systems

Traditionally, outlets for gas systems are set in place in the wall at the head of the bed. This can be a problem as the amount of equipment surrounding a patient increases and impinges on the necessary space required around an outlet. In today's hospital, the head end of an ICU bed is a very crowded place.

One space-saving solution in operating rooms and some catheterization laboratories has been to install gas outlets in the ceiling and then connect flexible hoses to them that terminate about 7 ft (2 m) from the floor with a connector that is in essence an outlet. This configuration provides more flexibility, since the hoses are flexible, and makes it easier for an anesthesiologist to connect equipment to outlets.

A method to vary the location and increase the number of outlets in patient care rooms is the installation of *surface-mounted gas rail systems*. For this system, gas outlets are mounted in what is essentially a multitube rail-mounted piping system. Outlets can be moved as desired, and each outlet is connected to a flexible cord that in turn attaches to a DISS fitting. This fitting is then connected to the regular rigid piping system.

These gas rail systems are helpful, though it should be noted that the movability feature is temporary in that once the outlet is moved to a new position, it must be secured so that it does not move while in use.

There is industry concern about the proper installation of a rail system, since this configuration could allow gas to accumulate behind a wall if not installed correctly. Further, facility engineers must check local codes to learn if this configuration is permitted. Some localities do not allow this type of installation because of the amount of exposed piping created in the patient area. (For details on minimum safety criteria, see Section 5.1.7 in NFPA 99.)

29.5.1.2 Manufactured Assemblies

As a way of making installations easier, manufacturers have designed units that contain gas outlets, vacuum inlets, electrical receptacles, brackets for mounting equipment, lighting, and so forth. The intent of including all these items in one assembly is to reduce the amount of work on the job site when finishing a new patient room or renovating an existing one. Because manufactured assemblies are considered to be permanently installed, the piping within them is considered part of the piping system, even though the connection between the assembly and the piping system may be semipermanent and tubing within the assembly may be rigid or flexible. Examples of these assemblies include headwall units, ceiling columns, and boom pendants.

It is necessary for designers of piped gas systems to know the specifications of the manufactured assembly selected, so that they know the total number of station outlets that will be available. Manufacturers of manufactured assemblies follow requirements for piping systems in general and NFPA 99 (Section 5.1.6) in particular. Installers must follow piping system installation and performance criteria, since these assemblies will be interfacing with the rigid tubing of the distribution portion of the piped gas system. Figure 29.10 shows a manufactured assembly.

29.5.2 Sensors

Other items considered to be termination points of piped gas systems (although, like station outlets, the tubing may continue on to other areas) are *sensors* that monitor what is taking place within the piping system. Sensors monitor such parameters as pressure, chemical composition, foreign gases and elements, and dew point.

As with station outlets, sensors must be installed gas tight, since they take their measurements directly from within a pipeline. Because they are connected in this

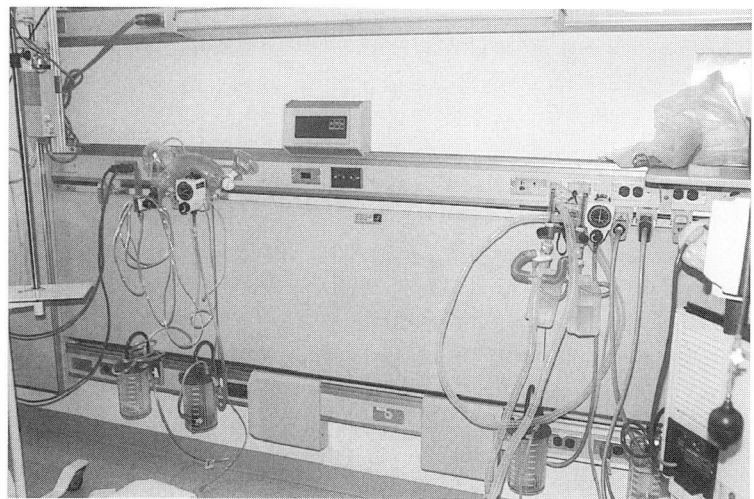

Figure 29.10 A manufactured assembly with external devices mounted (e.g., regulators, collection bottles).

fashion, demand-check valves are placed between the sensor and the pipeline to protect the pipeline in case of sensor failure, as well as to facilitate sensor servicing. Demand-check valves allow the removal of sensors, gauges, and the like without having to shut down the piping system or a portion of it. Demand-check valves, along with sensors and similar equipment, must be gas specific so that they are not interconnected to gases they are not designed for. See NFPA 99 (Section 5.1.8.2.3). See Section 29.6 for more on monitoring.

29.5.3 Temporary Endings

At one or more points in some piped gas systems, it may be necessary to terminate a line for future expansion. In these situations, either an end cap or a shutoff valve must be installed. An end cap is a fitting that has not been bored through at one end. It seals piping so that the system cannot be accessed at that point until the cap is replaced with more piping. A shutoff valve, while more expensive than a cap, allows the system to remain active while new piping is being added. The valve remains closed until the expansion is completed. An *end shutoff valve* can become an *in-line shutoff valve* after expansion takes place. This can be a useful capability. As with sensors and outlets, the installation of end caps and shutoff valves needs to be gas tight to prevent leaks.

29.6 Monitors and Alarms

Throughout a piped gas system, both source equipment and the gas flowing in the pipeline are monitored. The extent of monitoring and the specific types of alarms used depends on the level of the system; the size, extent, and complexity of the distribution system; and the system's application. A system containing its source in one room and one outlet in the next room requires relatively fewer monitors and alarms compared to a system that supplies gas to a 500-bed medical center. A system that is used only during the day (i.e., for 8 to 12 hours) has less monitoring than a system that runs 24 hours a day, 7 days a week. [To see this difference, refer to alarm requirements for Level 1 systems (Section 5.1.9) and those for Level 3 systems (Section 5.3.9) in NFPA 99.]

Piped gas monitors consist of sensors, indicators, and alarms. *Sensors* monitor such parameters as gas pressure, oil temperature, and water level. *Indicators* are calibrated scales or digital readouts that show the status of a specific parameter, such as the pressure or the temperature at a specific location or within a specific zone. The *alarm* is itself a sensor that activates when the parameter being monitored exceeds a preset range that is considered normal. The alarm must be of such brightness and volume that it can attract the attention of persons in the area in which it is located.

The entire monitoring package can be installed in one location or the individual components can be separated, with the sensor in one location, the indicator in the same or another location, and the alarm in the same or even an additional location. NFPA 99 (Chapters 4 and 5) requires *local alarms* to be located at points where items are being monitored, *area alarms* to be located at some central point for monitoring the pressure of gas serving an area being monitored, and a *master alarm* panel(s) to be located at a central continuously attended location(s) within a facility. These three types of alarms are discussed below.

29.6.1 Master Alarms

Master alarms for piped gas systems are general monitoring systems that indicate abnormalities of essential parameters of the system, including the proper performance of the source of supply and changes in pipeline pressure. Of major interest to a facility is the amount of gas remaining within a system. Monitoring the amount of gas remaining in cylinders prevents the entire supply of gas from being completely consumed without the staff first being notified that the reserve supply has begun to be used. Alarms for this parameter are set to activate at, or just before, changeover from one set of tanks to another takes place. Figure 29.11 shows a master alarm panel for various piped gas systems. Note the test button at the left.

NFPA 99 (Section 5.1.9.2) requires Level 1 piped gas systems to have master alarms so that any major problem that develops is immediately known to facility staff [this is the reason at least one master alarm panel is to be located where it can be continuously supervised by person(s) who can respond or who can call someone to immediately respond; for a hospital, this would mean 24 hours a day, 7 days a week]. NFPA 99 in fact requires Level 1 piped gas systems to have at least two master alarm panels remotely located from each other in at least two places, with both panels wired directly to sensors. A facility might want or need more panels for more efficient coverage. A centralized computer monitoring system can now (beginning with the 2005 edition of

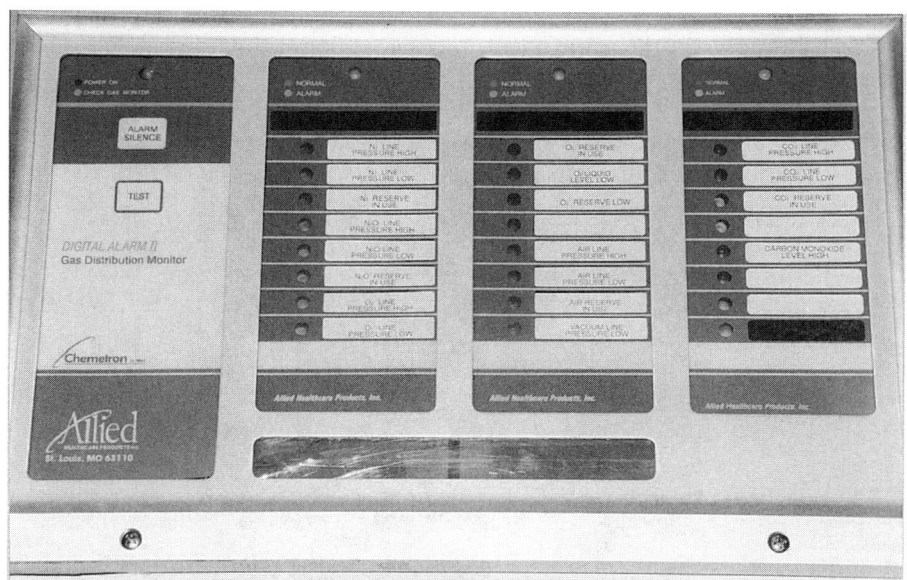

Figure 29.11 A master alarm panel for various piped gas systems within a hospital.

NFPA 99) constitute one of the two master alarm panels if it complies with Section 5.1.9.2.2 in NFPA 99. Some of these requirements include that power for the computer be of the "uninterruptable" category, that the computer be under constant supervision either directly or by a paging system, and that software ensures that alarms of higher priority supercede alarms of lower priority. Level 2 piped gas systems have the same requirements for master alarm panels as Level 1 systems, but with a few exceptions. One exception is that, unlike Level 1 systems, which must have at least two master alarm panels, Level 2 systems can have just one panel (NFPA 99, Section 5.2.9). Level 3 piped gas systems do not require a master alarm panel because of their limited size (extent) and the relatively low level of risk to patients with regard to reliance on these systems. (In general, Level 3 systems require fewer alarms, and the alarms are less extensive than those required for Level 1 and 2 systems. Alarm requirements for Level 3 systems are listed in NFPA 99, Section 5.3.9.)

29.6.2 Area Alarms

Area alarms monitor specific sections of pipeline to notify personnel in an area of any unacceptable change in gas pressure serving the area (Figure 29.12). Audible and visual alarms on a panel in the area are to be set to activate when pressure increases or decreases 20 percent from a pre-established norm. Area alarm sensors are to be located on the downstream side of a shutoff valve for the area. If the sensor were on the upstream side of the valve and someone closed the valve, the sensor would continue to detect the normal pressure and not activate the alarm. An *area alarm panel* must be located where area staff can see and hear it. The panel should not, however, be located where a door would block it when opened.

For Level 1 piped gas systems, the requirements for locating area alarms are listed in NFPA 99 (Section 5.1.9.3) and include specific areas that require area alarms, the location for the alarm panel, and the portion within the pipeline a sensor

Figure 29.12 An area alarm panel for an operating room.
Source: Health Care Facilities Handbook, NFPA, 2005, Exhibit 5.22.

is to monitor (i.e., either upstream from room shutoff valves or downstream from the zone shutoff valve). Area alarms for Level 2 piped gas systems have almost the same requirements as Level 1 systems, with a few exceptions (see NFPA 99, Section 5.2.9).

Level 3 piped gas systems for medical gases do not require the installation of formal area alarm panels, since these systems are much less complex than Level 1 and Level 2 systems and generally provide gas to just one area. However, pipelines must still be monitored for pressure swings, with an alarm panel located in an area that is continuously attended while the area is open and caring for patients (see NFPA 99, Section 5.3.9.1). The panel's location may not necessarily be in the area of gas use. NFPA 99 does not require Level 3 piped gas systems that are used to power devices to have any alarm system, since they are under constant staff control when used and present a relatively low level of risk to patients (see NFPA 99, Section 5.3.9.2.).

29.6.3 Local Monitors/Alarms

Local alarms have the sensors, indicators, and alarms installed all in one location. Several types of situations warrant the placement of local alarms, such as the position of a shutoff valve serving a patient care area. Facilities that produce gas on site do not have to monitor the supply itself, but they do need to know whether the gas is being produced and whether it is flowing in the piping system. For systems that produce compressed gas, a local monitor is installed to indicate when the system is not maintaining the preset level of pressure.

Any locally monitored item can be wired for remote observation. In the 2005 edition, NFPA 99 (Section 5.1.9.5.2) requires at least one signal from the local alarm of an equipment to activate the master alarm panel whenever a required local alarm signal activates. Depending on the parameter and the event that takes place to cause the alarm to activate, the resultant staff action may be to call the fire department, the maintenance department, or the supplier of gas or gas equipment.

29.6.4 Electrical System Reliability for Monitoring Systems

The reliability of the electrical system that powers a piped gas system's alarms and sensors varies in the same proportion as the risks associated with the reliability of the gas supplied to patients. That is, the higher the risk of loss, the more reliable the supply must be. Patients who are being provided with mechanical ventilation or assisted mechanical ventilation or who are dependent on the piped gas system due to medical or surgical intervention use the electrical equipment associated with the piped gas system for life support. Therefore, the electrical power to this equipment must be extremely reliable, and the electrical equipment must have an alternate electrical supply source in the event of an interruption of regular power.

NFPA 99 [Section 5.1.9.1(9)] thus requires Level 1 and Level 2 piped gas system alarms to be connected to the essential electrical system of the facility. As discussed in Chapter 8, essential electrical systems are designed to restore power to essential functions when there is a disruption of the regular source of power. While a maximum of a 10-second outage of power is permitted as the electrical distribution system switches to its alternate source of power, this interval may not be acceptable in some situations. Note that where this 10-second interval could jeopardize patient safety, uninterrupted power supplies are utilized. (Chapter 9 provides additional information on the wiring of alarms and electrical systems.)

Because patients in a facility with a Level 3 piped gas system are not left unattended while gas is being administered, nor are patients expected to be on any type of assisted breathing system, the loss of electrical power to Level 3 piped gas systems is not anticipated to expose patients to a level of risk that warrants the regular source of power to be supplemented by an alternate power source for piped gas systems. [Note that an alternate power source is required for other purposes, such as egress lighting (see Chapter 17) and sufficient power and lighting to complete a procedure (see Chapter 8 in this book).]

29.7 System Installation, Testing, and Verification

Of late, more and more regulatory agencies and voluntary organizations such as insurance companies are monitoring the installation of patient piped gas systems more closely—both the system itself and the persons installing and testing the system. Some of this oversight stems from the greater detail included in NFPA 99 about installation practices. This greater detail in NFPA 99 was the result of prior editions stating only that systems "shall be installed by a qualified installer." Some of this oversight can also be attributed to the legal climate of the day.

NFPA 99 (Sections 5.1.12, 5.2.12, and 5.3.12) stipulates numerous performance criteria to ensure safe delivery of gases via a piped gas system, as well as many general installation requirements for the system itself. The standard also includes performance criteria for system testing and verification. A range of tests must be conducted to check system pressure; cross connections; the operation of source equipment, valves, and alarms; system purity; and labeling of piping. There are also specific requirements for brazer and installer competence and for testing and verifying the work that brazers and installers have done.

29.7.1 System Installer Testing

One of the most critical aspects of any installation is the joining together of tubing sections through the use of fittings so that the resultant joints are structurally strong and gas does not leak out at these points. NFPA 99 (Sections 5.1.10.10.12 and 5.2.10 for Level 1 and Level 2 piped gas systems, respectively) has included requirements for brazing procedures and brazer performance since 1996. In 2002, the technical committee added performance requirements for installers of piped gas systems in the form of referencing ASSE 6010, *Professional Qualifications Standard for Medical Gas Systems Installers*. Note that although NFPA 99 (Chapter 5) lists performance criteria and qualification requirements for brazers and installers, NFPA does not certify that an installation meets these performance criteria or that a person meets these qualification requirements. Some states, however, do use these criteria as a basis for conducting training courses and for certifying "medical gas" brazers and installers as part of a licensing program.

The requirements for installer testing of Level 1 and Level 2 piped gas systems are given in NFPA 99 (Sections 5.1.12.2 and 5.2.12, respectively) and cover an initial blow-down of piping, initial pressure and cross-connection tests, and piping purges. Many of the installation requirements and testing for Level 1 and Level 2 systems are repeated for Level 3 systems (Section 5.3.10.10). Considering the risks associated with patients for whom Level 3 systems are intended, the NFPA technical committee responsible for this material believes that the safeguards selected are sufficient to address the risks to which patients are exposed.

29.7.2 System Verification

Since the 1999 edition of NFPA 99, the technical committee has required a facility to have new piped gas systems *verified* for proper system operation following installer testing of the piped system (Section 5.1.12.3 for Level 1 and 2 systems; Section 5.3.12.3 for Level 3 systems). A verifier provides a means of *independent* assurance that a system has been installed and operates according to the approved design. This verification includes ensuring that the gas at each outlet is the correct gas and at the required flow rate and pressure, that the gas is within specified contamination limits, and that all sensors and alarms are working as specified.

The verifier and installing contractor are *not* to be financially or administratively connected in any way. In 2002, the technical committee added that verifiers must also be qualified (i.e., technically competent and experienced in the field of piped gas systems used to deliver patient gases) in accordance with ASSE 6030, *Professional Qualifications Standard for Medical Gas Systems Verifiers*. (For the 2005 edition of NFPA 99, this requirement is in Section 5.1.12.3.1.3.) As with brazers and installers, formal credentialing is not included in the standard. Facility managers thus need to check the professional credentials of verifiers being considered, such as by contacting facilities where a potential verifier has done work and local authorities for any licensing requirements or any prior complaints.

Some of the tests conducted by verifiers are similar to those made by the installer (e.g., a standing pressure test, a cross-connection test, a test of alarms). However, the verifier also conducts operational tests, when the system is completely connected and operating with the gas of system designation, to ensure that the system is operating as designed. A verifier will also conduct a medical gas concentration test (Section 5.1.12.3.11 in NFPA 99), and a medical air purity test if a compressor is used (Section 5.1.12.3.12 in NFPA 99). These tests are less extensive for Level 3 systems than for Level 1 and Level 2 systems, since Level 3 systems are much less complex than Level 1 or Level 2 systems.

Figure 29.13 shows the process used to verify the cleanliness of pipelines installed in a hospital. Figure 29.14 shows the process of verifying the dew point of a compressed air system in the same area. Figure 29.15 shows the process of verifying the oxygen concentration of a compressed air system in the same area.

558 Part V. Gas and Vacuum Systems and Equipment

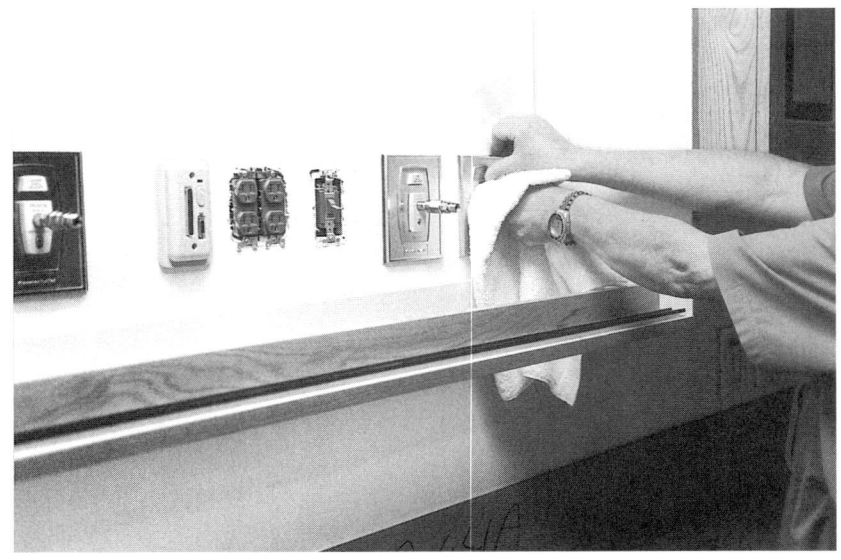

Figure 29.13 Verifying cleanliness of pipelines installed in a hospital.

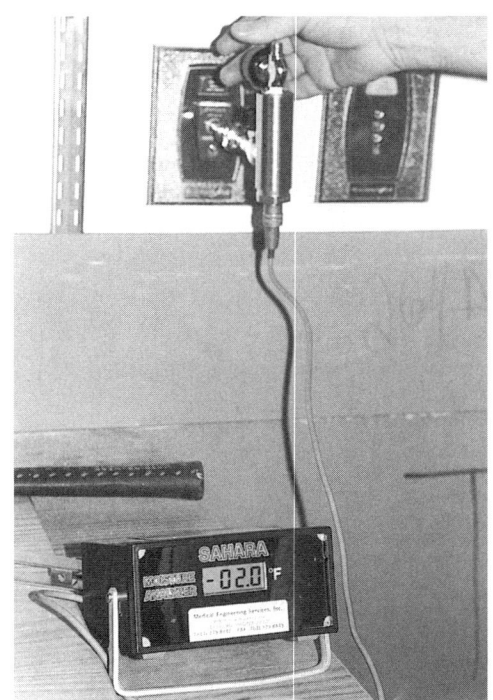

Figure 29.14 Verifying dewpoint of a compressed air system in the same area.

Figure 29.15 Verifying the oxygen concentration of a compressed air system in the same area (for cross-connection purposes, since an oxygen line was installed at the same time).

29.8 Staff Training

It is intended that piped gas systems provide a specific gas at a specific pressure at an outlet. However, a gas leak anywhere in the piped system can have serious repercussions if it is allowed to continue for an extended period of time. Also, an interchange or contamination of gases can be injurious or even fatal to the patient(s) being administered the gas. Therefore, staff must be instructed as to how the piped gas system functions overall, and they must be trained as to what they must do in various situations (such as a significant drop in gas pressure, a complete loss of gas, or a fire in an area). They must be educated about the many peripheral features and operational changes of piped gas systems that can take place both during normal use and when an emergency occurs. Staff training needs to be conducted as often as necessary to maintain a safe operating system for patients and to reduce the risk of injury to patients and staff. This education applies equally to piped vacuum systems and waste anesthetic gas disposal systems.

Information regarding how all equipment should be used must be conveyed to all nursing and medical staff who will be using the equipment. Anyone who is authorized to connect any component to an outlet of the piped gas system (or an inlet of a piped vacuum system) must know how the system is intended to be used. To help facilitate this communication, general information on system design should be provided in an understandable, nontechnical format. They should also tour the supply source and know the normal range of operating parameters and what alarms will activate when they are exceeded. Staff also should be trained on the various noninterchangeable outlet configurations that are used to prevent the wrong gas from going into equipment and thence into a patient. Staff should be clearly advised to *never* force an equipment connector into an outlet or inlet since connectors are made to easily insert in their correct outlet.

It is essential that staff recognize the sound of the alarm that activates when the pressure within the piping system changes beyond set limits (it may be over or under normal pressure). Staff should know the areas controlled by shutoff valves. Staff should also know when gas outlets and vacuum inlets look damaged or worn and how to report these problems promptly so that appropriately trained personnel can service the system.

There are several emergency situations for which staff knowledge of the system is vital. The first situation is that of a sudden loss of gas pressure and flow. Staff must know the hierarchy of steps that must be taken when this event occurs: what the alarms mean, who should be notified, and how staff should deal with those patients on life support equipment and those who are receiving air or oxygen for supplemental purposes. A procedure for addressing this type of emergency needs to be part of the disaster plan of all facilities (see Chapter 36).

The other type of emergency is one in which the piped gas system is operating normally but the facility has incurred an emergency, such as a fire or broken gas pipe, for which the system (or a portion of it) must be shut down. Staff must know if the system needs to be shut down and, if so, who is responsible for making this decision. They must know what to do when the fire department orders that the piped gas system (or a portion of it) be shut down and where the shutoff valves that need to be closed are located. Of utmost importance is knowing what provisions must be made for patients affected by such a shutdown and what steps are to be taken *before* a shutdown is made. Again, this situation must be addressed in a facility's disaster plan, and staff must be trained accordingly. The closing of an inappropriate valve may cause an even greater disaster to develop.

30

Patient Piped Vacuum Systems

Unlike a pressurized piped gas system, a piped vacuum system is one in which a *vacuum* is created within piping, with the direction of flow opposite to that of the pressurized gas system.

As with piped gas systems, NFPA 99, *Standard for Health Care Facilities*, delineates three levels of piped vacuum systems, based on assessing the same levels of risk to patients. To determine the level of risk, the following questions need to be asked: (1) Is the patient in an immediate life-threatening situation if the piped vacuum system fails or if electric power to it is interrupted for more than 10 seconds? (2) Are there alternative methods for providing vacuum? (3) If so, are they adequate to avoid compromising patient treatment and safety?

Level 1 and Level 2 piped vacuum systems consist of central vacuum-producing equipment with operating controls, shutoff valves, alarms, warning systems, gauges, and a network of piping that terminates at station inlets where appropriate vacuum equipment can be connected. Level 3 vacuum systems are either wet or dry piping systems that remove liquid, air–gas, and solids from a treatment area. Figure 30.1 shows the basic components of a patient piped vacuum system.

Overall, the physical considerations for Level 1 and Level 2 piped *vacuum* systems are different from those for Level 1 and Level 2 piped *gas* systems. For example, because there is not a constant pressure being exerted within the piping system, as is the case in piped gas systems, there is much less physical stress on the tubing of a piped vacuum system than on that of a piped gas system. Patient vacuum systems also present lesser fire-related risks than patient gas systems, because there is no buildup of gas within tubing. Physical considerations are less complex for Level 3 vacuum systems, but the concerns still prevail for patient and staff safety, both inside and outside the facility.

Another difference between piped gas and piped vacuum systems is that a piped gas system (except for a medical compressed air system) does not always need electricity to provide gas at station outlets, whereas a piped vacuum system always relies on electricity to create vacuum at station inlets (since vacuum systems require electrically operated pumps to create vacuum). (Note that sensors, alarms, and electronic meters require electricity to function but are not needed to make gas flow or create a vacuum.)

Many other differences and similarities between patient piped gas systems and patient piped vacuum systems are indicated throughout this chapter.

561

Part V. Gas and Vacuum Systems and Equipment

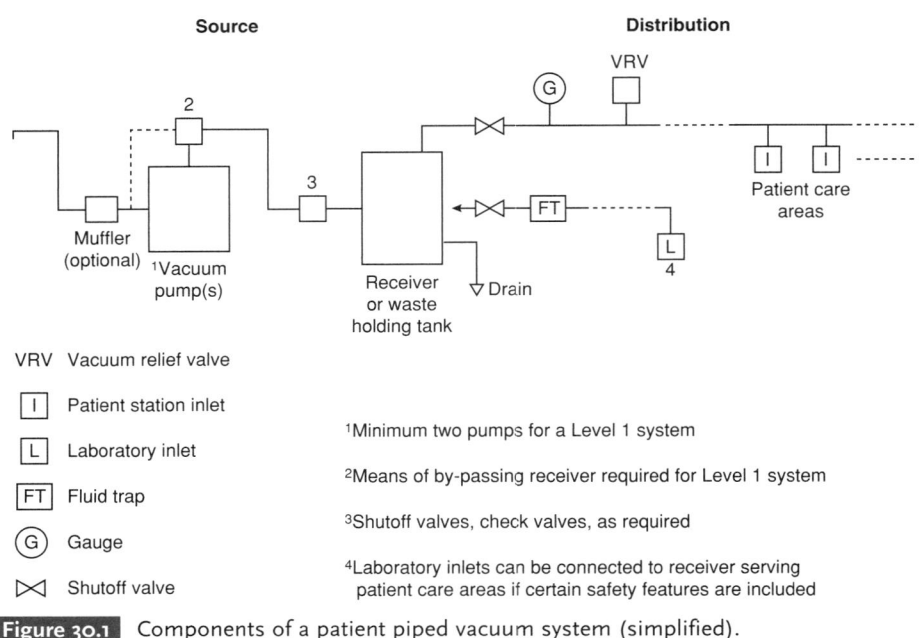

Figure 30.1 Components of a patient piped vacuum system (simplified).

30.1 Piped Vacuum Sources (Pumps)

The equipment that creates the vacuum in a patient vacuum system is a pump (or, as some like to say, a compressor hooked up backwards). Simplistically, this pump can be viewed as a vacuum cleaner (see Figure 30.1). Figures 30.2 and 30.3 show two different arrangements of the vacuum source for Level 1 piped vacuum systems. In Figure 30.2, the two vacuum pumps are mounted above and below the control box, and the receiver is in the background. This arrangement can save floor space but increase time and difficulty to maintain and service the equipment properly. In Figure 30.3, pumps are mounted on either side of the receiver. Notice the amount of spacing around the pumps.

Level 1 vacuum systems require a minimum of two pumps; Level 2 and Level 3 systems can be installed with just one pump (see Section 30.1.2 below).

Like piped patient gas systems, piped patient vacuum systems operate at various levels of vacuum, depending on the source equipment and the needs of medical, dental, or surgical staff for treating patients. The minimum vacuum that must be maintained throughout the system is 12 in. of mercury (Hg). The types of pumps used to create the vacuum pressures vary. The choice of pump depends on the following factors:

1. The amount of vacuum required based on the total number of station inlets (i.e., 2 station inlets or 200 station inlets)
2. The amount of flow required. (If the vacuum level is adequate but the flow is not, the system would not be acceptable, particularly in operating rooms.)
3. The usage rate [whether pump(s) must be capable of operating 24 hours a day, 7 days a week; 8 hours a day, 5 days a week; or intermittently]

Chapter 30. Patient Piped Vacuum Systems **563**

Figure 30.2 Vacuum source for a Level 1 patient piped vacuum system.

Figure 30.3 Another vacuum source for a Level 1 patient piped vacuum system.

4. The level of maintenance required for pumps and the ability of users to perform required maintenance. (Will in-house or local service personnel be able to maintain the equipment properly, or will only manufacturer-authorized companies be allowed or be able to maintain it?)
5. Whether one pump is used or several pumps are required. (Level 2 and Level 3 vacuum systems can have just one pump, while Level 1 vacuum systems are required to have at least two pumps.)
6. The operating economy of the pumps in terms of their operating and maintenance costs
7. The level of noise that can be tolerated, and whether the noise is dependent on the location of the pump(s).

30.1.1 Receivers

Level 1 or Level 2 piped vacuum system pumps utilize receivers, much the same way that compressed air systems use them (i.e., they act as holding tanks to reduce the swings in pressure and flow as the number of outlets or inlets in use varies). Having a receiver also allows pumps to cycle on and off and not run continuously. Receivers are equipped with drains at the bottom to prevent fluids and undesirable particles from entering the pumps if such products have been drawn into the system's piping.

In the *dry* mode, Level 3 systems are similar to Level 1 and Level 2 systems in that the inside of the piping is intended to remain dry. These systems use such devices as external traps to collect fluids and other materials and prevent them from entering the piping system. In the *wet* mode, fluids, solids, and the like are drawn into the piping and are collected in a waste-holding tank that is situated outside the patient care area. In some systems, the collected materials actually pass through the pumps and then into a waste-holding collection tank or into a drain and sewer. Specific requirements regarding disposal of collected materials into sewers, either directly or after treatment, depend on local and state codes on hazardous waste disposal.

Note that for a given size and number of pumps used and size of receiver used, there is a limit to the number of inlets that can be used simultaneously before the vacuum level at the inlets begins to fall. This sizing factor must be considered during system design.

For repairing a receiver, valves and bypass piping need to be arranged so that valves can direct the vacuum path around the receiver, thereby isolating the receiver without shutting down the entire system.

30.1.2 Total Number of Pumps

Level 1 systems must have at least two pumps. These pumps must be able to handle the peak calculated demand for vacuum with the largest pump out of service. For systems with two pumps, either pump must be able to handle the peak demand. For a system with three pumps, the two smaller pumps must be able to handle the demand (Figure 30.4). Due to the concern for maintaining required vacuum level at all times, NFPA 99, 2005, Section 5.1.9.5.4(4) requires backup or lag vacuum pumps to operate as necessary and to activate a local alarm when this condition arises. In addition, a signal to the master alarm panel is required when this condition occurs, so that the problem can be investigated [NFPA 99, Section 5.1.9.2.4(9)].

Pump alternation is not an issue for Level 2 piped vacuum systems because these systems are permitted to operate with only one pump (NFPA 99, Section 5.2.3.6). However, in all other respects, the pump(s) and other source requirements of a Level 1 piped vacuum system are applicable.

Requirements for pumps in a Level 3 piped vacuum system are different because of the different methods used (i.e., wet and dry). Criteria are listed in NFPA 99, Section 5.3.3.6.

30.1.3 Pump Room

Similar to that for compressors, the room housing vacuum pumps for Levels 1, 2, and 3 systems can also house compressors, but it cannot house any gas cylinders or

Chapter 30. Patient Piped Vacuum Systems

Figure 30.4 A vacuum source with three vacuum pumps.
Source: *Health Care Facilities Handbook*, NFPA, 2005, Exhibit 5.12.

containers either in use or in storage (NFPA 99, Sections 5.1.3.3.1.4 and 5.3.3.3.2.2). Also like compressed air supplies, vacuum pumps do not have to be placed in a room or enclosure separate from other utilities. The same space and mounting considerations apply to vacuum pumps and their accessories as apply to compressed air supplies (see Chapter 29).

For Levels 1, 2, and 3 systems, the discharge from the vacuum pumps must be located outdoors, remote from compressor intakes, windows, and the like, and sized to minimize backpressure.

30.2 Vacuum Distribution

Like the distribution of piped patient gas systems, the distribution of piped patient vacuum systems consists of tubing connected together by fittings. Although the first NFPA standard on piped vacuum systems in 1984 specified only that tubing had to be metallic and corrosion resistant, over the years the type of tubing permitted has become more restricted and more specific (as noted in Section 30.2.1 of this book).

30.2.1 Vacuum System Piping

Because of the nature of the operation of vacuum systems and because a crack along the distribution route of a vacuum system will not result in pressurized gas escaping into the surrounding atmosphere, the ranges of acceptable piping materials and methods of joining piping are wider than are permitted for piped gas systems.

30.2.1.1 Piping Materials

Tubing for new Level 1 and Level 2 vacuum systems must now (per the 2005 edition of NFPA 99) be made of either hard-drawn seamless copper tube [ASTM B 88 Type K, L, or M), ASTM B 280 (copper ACR tube), or ASTM B 819 (Type K or L)] or stainless steel (Section 5.1.10.2.1 in NFPA 99, 2005). Joining of tubing is now to be accomplished in the same way as tubing of Level 1 and Level 2 piped gas systems (see Section 29.4.1.2 in Chapter 29 of this book). All piping is to be blown clear during installation using oil-free dry nitrogen.

Because it is intended that patients using Level 3 systems will not be on life support equipment (i.e., their risks are lower than for patients using Level 1 and Level 2 systems), a wider range of materials can be used for Level 3 piped vacuum systems tubing. In addition to the copper tubing listed in the previous paragraph, PVC schedule 40 is acceptable, as listed in Section 5.3.10.2 in NFPA 99. Also listed are criteria for tubing installed underground or within floor slabs (something that is not done in Level 1 or Level 2 systems).

Whatever tubing is used, it must meet a pressure test (as specified in NFPA 99, Section 5.1.10.2 for Level 1 and Level 2 systems and Section 5.3.10.2 for Level 3 systems). The performance of tubing that will also be used for waste anesthetic gas disposal is not to be affected by such gases (see Chapter 31 in this book).

30.2.1.2 Piping Methods

It is common practice to install all patient distribution piping at the same time. When only gas systems are being installed, there is no chance for the different types of tubes to be mixed up (except with respect to cross-connections between gas systems, as covered in Chapter 29 of this book). Since piped vacuum systems are permitted a wider variation in the type of tubing that can be used, it is of concern that vacuum tubing *not* find its way into the runs of gas tubing, and vice versa. In addition, since some of the tubing requirements for piped vacuum systems are less stringent than those for piped gas systems (see later sections), there is also a concern that gas system tubing might be joined together using tubing not acceptable for piped gas systems.

One way to avoid these potential problems, though not required by NFPA 99, is to use the same type of tubing, fittings, and installation method for the vacuum system as for the gas system. This can add some expense initially but may be well worth the investment should some mix-up occur if incorrect tubing is installed and discovered *after* pressure testing, or worse, *after* some incident occurs. As a consequence, NFPA 99 (Section 5.1.10.2.2) now requires Level 1 and Level 2 vacuum system copper tubing that does *not* meet criteria for medical gas tubing and its installation to be prominently marked as such.

Like tubing differences between Level 1 and Level 3 piped gas systems, there are differences between Level 1 and Level 3 piped vacuum systems with respect to tubing permitted (see Sections 5.3.10.1 and 5.3.10.2 in NFPA 99). Distribution requirements for Level 2 piped vacuum systems, which include tubing requirements, are required to meet the same requirements as Level 1 piped vacuum systems (see Section 5.2.10 in NFPA 99).

30.2.1.3 Grounding

As with piped gas systems, piped vacuum systems are *not* designed to be reliable electrical grounding connections and should not be used as such.

30.2.2 Vacuum System Valves

Like in piped gas systems, shutoff valves must be installed throughout the distribution portion of a piped vacuum system; this is to isolate various portions of the system as needed without having to shut down the entire system. These valves help prevent the spread of smoke during a fire or other emergency and assist during maintenance or repair activities. The number and placement of such valves depends on the complexity of the system. NFPA 99 (Section 5.1.4) lists minimum valve placement criteria. Additional valves can be installed if desired, but they must not lower the safety level provided by the required group of valves (e.g., the valves should not be accessible to the public, and they should be locked in the open position so that they cannot be inadvertently closed).

A change in the 1999 edition of NFPA 99 [Section 4.2.2.2.9(d)] (Section 5.1.9 in the 2005 edition of NFPA 99) brought valve requirements for Level 1 vacuum systems in line with those for Level 1 gas systems with respect to the location of shutoff valves and to the placement of sensors for alarms. Level 3 piped vacuum systems, because of their limited sizes, have requirements for shutoff valves only at supply source (see NFPA 99, Section 5.3.4).

30.3 End of the Line

As with piped gas systems, the end of the vacuum line (actually the beginning of the line with respect to flow) has station inlets, sensors, and temporary endings, such as caps or shutoff valves. The primary concerns with this portion of piped vacuum systems are (1) the materials that can enter the system, (2) any space constraints, and (3) the maintenance of the system.

30.3.1 Station Inlets

It is at this point (at inlets) in a piped vacuum system that vacuum equipment is connected. Like station outlets for gas systems, station inlets separate the inside of the vacuum system from normal atmospheric pressure. Once an inlet is opened, however, it will draw in whatever is small enough to fit through the opening. In the past, this could have been flammable inhalation waste gases. Today, this could mean air, liquids (both flammable and nonflammable), solids, infectious material, or almost anything else that comes near the vacuum opening. It is of great concern to those who design, use, and maintain piped vacuum systems that vacuum station inlets not allow a liquid or anything small to be drawn into the piping system. As a result, measures are taken (e.g., external traps) to prevent anything but gases from entering the piping system for Level 1 and Level 2 systems and thus maintain dryness inside the tubing.

30.3.1.1 Surface-Mounted Rail Systems

Vacuum station inlets can be installed in surface-mounted rail systems. As with gas outlets, locating inlets near patients makes these systems attractive. The extensions for vacuum inlets are made the same way as extensions for station outlets (see Chapter 29), and the same restrictions apply.

30.3.1.2 Manufactured Assemblies/Headwalls

Manufactured assemblies include vacuum station inlets in the same way they include gas station outlets. One design problem in this regard is pipe sizing, which can restrict or slow down vacuum flow if not calculated correctly. Careful attention to this detail is necessary to minimize this problem.

30.3.1.3 Inlet Space Considerations

One of the challenges in designing health care facilities today involves the amount of equipment needed to treat some patients and the space required by this equipment. While the space needed to install a few vacuum station inlets in the wall at the head of the bed is generally not a problem [the dimension of an inlet is approximately 3 in. by 4 in., or 12 in.^{2} (7.6 cm by 10.2 cm, or 77.4 cm^2)], the distance *between* each inlet is sometimes overlooked by designers. Designers must allocate sufficient space for vacuum equipment, such as collection bottles, to be placed between these inlets. In addition, where there is a vacuum inlet, there almost always is more than one gas station outlet, and there must be sufficient room to attach equipment, such as regulators, to each of these outlets as well.

As noted for piped gas station outlets, the number of vacuum station inlets varies with the needs of the patients in the area. The most referenced document for guidance on the number of inlets at a patient bed location is the American Institute of Architects (AIA) *Guidelines for Design and Construction of Hospital and Health Care Facilities*. Those designing vacuum systems should also check the codes adopted by their state and local jurisdictions for any other spatial requirements.

30.3.2 Temporary Endings

Like piped gas systems, piped vacuum systems can have temporary termination points at various places in the system. These endings can be in the form of either a cap or a shutoff valve. The latter is preferable so that it will not be necessary to shut down the system or a portion of it when the tubing is extended at that point. Refer to Chapter 29 for more discussion on temporary endings.

30.3.3 Sensors

Like sensors in piped gas systems, sensors for piped vacuum systems are placed either in the tubing to measure the vacuum inside the pipe or outside the tubing to measure a particular activity (e.g., pump operation). Given the nature of piped vacuum systems, system performance and impending system failures or abnormalities can be determined with fewer sensors than are needed for piped gas systems. The following are minimum monitoring criteria required by NFPA 99 for Levels 1 and 2 systems (Sections 5.1.9 and 5.2.9, respectively).

For Level 1 and Level 2 vacuum systems, one sensor must be placed in the main line leading to the pumps, but ahead of the distribution portion of the system, to measure vacuum. Another sensor must be placed on each pump to signal when a pump is turning on or off and thus help determine pump alternation. If several areas are connected to the vacuum source(s) (e.g., if there are inlets on different floors or there are zones within a floor), sensors must be placed in each area to detect if vacuum drops below a preset level in that area or zone.

Alarms and thus alarm sensors are not required for Level 3 vacuum systems (see Section 30.4). However, facilities staff should consider monitoring vacuum levels, since it provides an indication of potential problems. If a significant change occurs, staff would be notified and thus be able to take steps to maintain adequate vacuum.

30.4 Monitors and Alarms

Like piped gas systems, Level 1 and Level 2 piped vacuum systems require local, area, and master alarms, and much of the material on alarms presented in Chapter 29 in this regard is applicable to vacuum systems. Again, because of the nature of piped vacuum systems, the amount of monitoring required is less than that for piped gas systems (see NFPA 99, Sections 5.1.9 and 5.2.9). The wiring of monitors and alarms of piped vacuum systems follows practices as used for piped gas systems (see Chapter 29 for more information on monitors and alarms for patient piped gas systems and Chapters 6 and 9 for guidance on wiring alarms and electrical systems in general).

The only local alarm (audible and visual) required by NFPA 99 for Level 1 and Level 2 vacuum systems is one that indicates when a reserve pump comes on as a result of the main pump or pumps not being able to maintain vacuum [NFPA 99, Section 5.1.9.5.4(4)]. The problem could be the result of a large leak somewhere in the system or a failing pump. Normally, reserve pumps come on only during a pump alternation scheme. This alarm must be connected to the master alarm panel. Level 2 piped vacuum systems that use more than one pump must meet this same requirement.

NFPA 99 does not require Level 3 piped vacuum systems to have any alarms, since requirements are based on the following: (1) The procedures performed with these systems are not envisioned to be complex or life threatening, (2) the facilities in which these systems are installed serve out-patients with no overnight stays, and (3) the systems themselves are relatively small, with fewer inlets. Some designers include one alarm (vacuum level), simply to have greater knowledge about system operations.

30.5 System Performance Criteria

The bottom line in meeting performance criteria for Level 1 and Level 2 piped vacuum systems is determining what is taking place at station inlets. Taking measurements at inlets is the most accurate way to learn if design of the system meets the performance criteria listed or specified.

The factors that affect vacuum and flow at the station inlets are all the features of the components that lead up to them, including the size of the pump(s) and receiver, the length and diameter of the tubing in the distribution system, the quality of the installation in terms of the amount—if any—of leaks, the number of branches, and the number of inlets. Inlet performance is also dependent on how frequently and long each inlet is used (i.e., are only 25 percent of the inlets used at any one time, or can 75 percent of the inlets be in use at one time?). The design objective for system performance is to keep the vacuum level and flow relatively constant even as the number of inlets in use varies. This objective is applicable whether the system has as few as 2 inlets or as many as 1000 inlets.

Performance at inlets also has to take into account vacuum degradation as the distance from the source to inlets increases. Here again, it is vacuum flow—which is affected by the number of inlets on the system, among other things—that is critical.

Thus, the vacuum level and flow *at each inlet* in new Level 1 and Level 2 piped vacuum systems is measured to ascertain that the installation meets design criteria. As listed in NFPA 99, a vacuum level of 12 in. Hg must be maintained at the *farthest inlet* from the source (pump). In addition, a flow of 3 SCFM (standard cubic feet per minute) must be maintained at *any inlet* without reducing the vacuum pressure at an adjacent station inlet below this 12 in. Hg (NFPA 99, Section 5.1.12.3.10.4). This second criterion is included to determine that pipe sizing is adequate throughout the system (e.g., a value below 12 in. Hg at the adjacent inlet would indicate that tubing in this portion of the system is not large enough).

Another performance criterion for Level 1 and Level 2 piped vacuum systems involves conducting an initial pressure test that determines if tubing has been installed properly (NFPA 99, Section 5.1.12.2.3). Tubing is pressurized to at least 60 psi (415 kPa), and then soapy water or another method is used to determine if joints are leaking. If any leaks are found, they need to be corrected and the test repeated. When this test is successfully completed, a 24-hour standing vacuum test is conducted (NFPA 99, Section 5.1.12.2.7), with any leaks corrected and the test conducted again.

Because vacuum systems suck in atmosphere plus any other gases or liquids that may be near inlets (some of which can be quite unhealthy), it is impractical (often impossible) and very cost ineffective to clean used vacuum tubing to levels that meet the cleanliness requirements of patient gas systems. As a result, NFPA 99 (Section 5.1.10.10.10.2) prohibits patient vacuum systems from being converted to patient gas systems. But while new tubing for vacuum systems does not have to be cleaned and capped to gas tubing criteria, vacuum tubing must be kept clear of debris, insects, etc. Even with these measures, incidents, accidents, or poor operating practices can degrade performance criteria over time. Examples of things than can degrade performance include the following:

- Debris that is small enough to be sucked into a system through inlets but large enough to get stuck in the inlets or within tubing
- Vacuum seal rings that begin to leak and make the system operate as if more inlets are in use than actually are in use
- Body fluids and material that are drawn into the system and solidify within the inlet or tubing
- External vacuum regulators that have not been maintained properly
- Suction tubing that is too small to handle increases in flow volume
- Equipment hoses that are longer than necessary
- External vacuum regulators that are not shut off completely or at all when not needed

Taking appropriate corrective action in each of the preceding instances should correct much of the degradation that reduces performance of the vacuum system. Some actions are the responsibility of system maintenance personnel (in-house technical staff or outside contractors). Others actions are the responsibility of those operating the vacuum system (e.g., medical/surgical/nursing staff, operating room technicians). The responsibilities of each of these parties should be clearly delineated.

The only performance criterion for Level 3 piped vacuum systems is that of determining if the piping system shows any decrease in holding a vacuum for 10 minutes (see NFPA 99, Section 5.3.12.4.3).

While cleanliness of Level 3 tubing is not as critical as that for Level 1 and 2 vacuum systems, tubing still needs to be handled properly to avoid damaging it.

For any system (whether Level 1, 2, or 3) to continue providing service according to design and performance specifications, it must be operated and maintained in accordance with manufacturer guidelines.

30.6 System Renovations

Much of the discussion in Chapter 29 regarding whether to renovate or replace an existing piped gas system and how to account for future systems through the current design applies to piped vacuum systems as well. When a system is enlarged, technical as well as budgetary problems must be included in the equation.

A somewhat growing phenomenon is facilities creating *swing units*, in which a patient care area is converted to office space, only to be converted again in a few years to another type of occupancy. When a system is being made smaller, the major problem is what to do with any portion of piping that will no longer be used but also not be removed (in case another change warrants reactivation).

If unused piping is not removed, the remaining ends of piping (such as at former inlets) must be terminated (capped) in a safe manner. A facility must determine whether a removable cover plate should be installed over these ends (as is done with the ends of electrical wiring that is no longer used but left for possible future reuse). Such ends should not be plastered over, but facility staff should determine if there are local regulations on the matter.

All abandoned piping (whether removed or not) must be documented, with a complete file kept by the department(s) that may be called upon at some future date to advise on the status of such piping. Accurate documentation could save much time and money in the long run.

30.7 System Installation, Testing, and Verification

Installing patient piped vacuum systems requires many of the same skills as those used to install patient piped gas systems. Installers must have knowledge about the pumps, alarms, and other equipment used; how to install such equipment; and how to connect the piping. For Level 1 vacuum systems, the same criteria as those for braziers, installers, and verifiers of piped gas systems now apply. For Level 3 piped vacuum systems, if brazing of joints is required, the brazing procedures listed in Level 1 piped systems are generally followed (see NFPA 99, Section 5.3.10.7); if soldered joints are acceptable, the procedures listed in NFPA 99, Section 5.3.10.5, are to be followed; and if solvent-cemented joints are acceptable, the procedures in Section 5.3.10.6 are to be followed. However, braziers of Level 3 piped vacuum systems now have to be qualified like those for Level 1 systems (see NFPA 99, Section 5.3.10.10.15.1), as do installers (see NFPA 99, Sections 5.3.10.10.1.2 and 5.3.12.2.1.1), and verifiers (see NFPA 99, Section 5.3.12.3.1.4).

Since Level 1 and 2 piped vacuum systems are used in situations where patients' lives could be in jeopardy if a system failed, the tests required of the installer to ensure a sound installation are very similar to those required for piped gas systems (i.e., system blow-down, an initial pressure test, cross-connection, and initial standing vacuum tests must be conducted) (NFPA 99, Section 5.3.12.2). However, since piped vacuum systems are intended to ingest air and, in some systems, other matter throughout their operational lives, piping purge tests are not required.

Like verifiers of Level 1 and Level 2 piped gas systems, verifiers of Level 1 and Level 2 piped vacuum systems must be knowledgeable about and experienced in the testing of piped patient vacuum systems. They must also be independent of the installer per NFPA 99, Section 5.3.12.3.1.6. System verifiers must conduct an operational test of the entire system in its normal operating mode per NFPA 99, Section 5.3.12.3. If the vacuum system has been installed at the same time as a piped gas system, then a complete cross-connection check must also be conducted.

30.8 Staff Training

The same guidance noted in Chapter 29 regarding staff training on piped gas systems under both normal conditions and emergency conditions applies to piped vacuum systems.

31

Other Gas/Vacuum Systems in Health Care Facilities

Several other types of gas and vacuum systems may be used in health care facilities, some having more fire and life safety concerns than others. One system is waste anesthetic gas disposal (WAGD), otherwise referred to as *scavenging*. WAGD systems, which can be either a dedicated system or one connected to a medical-surgical vacuum system, are regulated by several federal agencies and are addressed by several non-government entities, such as the Joint Commission on Accreditation of Healthcare Organizations (JCAHO), National Fire Protection Association (NFPA), American National Standards Institute (ANSI), and American Institute of Architects (AIA). Other non-patient piped gas systems are those used to distribute nonflammable and flammable gases to laboratories and those used for heating, cooking, cleaning, and pneumatic switching (i.e., changing the setting of a switch through the use of gas under pressure), and some maintenance tasks. Non-patient piped vacuum systems also include pneumatic tube systems and central vacuum cleaning systems.

Preventing contamination of patient piped gas systems from non-patient piped gas systems is another key issue related to safety.

31.1 Waste Anesthetic Gas Disposal

As noted in Chapter 28, waste anesthetic gases are routinely collected and disposed of when general inhalation anesthetics are administered. There is still debate as to the necessity of conducting waste anesthetic gas disposal, but because of the litigious climate that prevails today, it is very unlikely that this practice will be eliminated or even reduced. The National Institute of Occupational Safety and Health (NIOSH), a federal government agency, has developed and published recommendations on collecting waste gas if nitrous oxide or halogenated anesthetics are administered in a medical procedure. While these are just recommendations, they are generally followed as if they were regulations. The American Society of Anesthesiologists published a monograph in 1999 on managing anesthetizing areas and postanesthetic care units (PACU) that covers the subject of waste anesthetic gases and their disposal.

In 1991, the Occupational Safety and Health Administration (OSHA) of the U.S. Department of Labor issued a fact sheet (OSHA 91-38, *Waste Anesthetic Gas*, 1991) that describes some of the health hazards that might be experienced by hospital staff in an operating room that did not have a waste anesthetic gas disposal system. The fact sheet includes some recommendations on maximum exposure levels to these agents, as well as sampling methods, control methods, and training. However, as noted in the fact sheet, this information does not carry the force of a legal opinion. OSHA can cite hospitals and other health care facilities under a "General Duty Clause" of the Occupational Safety and Health Act (Section 5) if waste anesthetics gases are not disposed of through a WAGD system. (The Act states that employers must furnish to each worker a place of employment that is free from recognized hazards.) This quasi-requirement for a WAGD system is independent of the facility in which the general anesthesia is administered, such as a hospital, ambulatory care facility, or surgeon's private office.

Currently, NFPA 99, *Standard for Health Care Facilities,* 2005 edition (Sections 5.1.3.7 and 5.2.3.7), includes requirements for Level 1 and Level 2 WAGDs, respectively, with Level 2 system requirements the same as those for Level 1 systems except that Level 2 systems are permitted to use one medical WAGD pump and staff are to develop an emergency plan for loss of WAGD. NFPA 99 does not presently include criteria for Level 3 WAGD systems because such systems are not intended for use in situations where gas disposal would be necessary (i.e., it is not intended that inhalation anesthetics will be administered in these locations). As with patient piped gas and vacuum systems, system levels are based on the level of risk to patients should a system fail.

31.1.1 WAGD Methods

There are two generally accepted methods in use today to collect and dispose of waste anesthetic gases from operating rooms. One method is a dedicated system much like a piped vacuum system, with vacuum pumps and piping. The second method is connecting the WAGD system to a medical-surgical vacuum system. (A third WAGD method, which is termed *venturi,* utilizes water to create a vacuum but is no longer used in clinical areas.) The advantages and disadvantages of each WAGD method are described in the remainder of this section.

The major difference between WAGD systems and piped vacuum systems is the level of vacuum and flow. While piped vacuum systems operate at 12 in. Hg and a flow of 3 SCFM (standard cubic feet per minute), WAGD systems operate at about 0.5 cm H_2O (or about 0.01447 in. Hg). This is about 800 times lower in value. The low value of vacuum is necessary since a WAGD system is connected directly to a patient's breathing circuits, which could be sucked clear of any gases, or worse, collapse a person's lungs if the vacuum were much higher than this level. The alarms and monitors of a piped vacuum system cannot be used in a WAGD system because they cannot differentiate pressure changes in this low range.

31.1.1.1 Dedicated WAGD Systems

A dedicated WAGD system is most similar to a piped vacuum system in that it consists of vacuum pump(s) and piping. In fact, NFPA 99 (Section 5.1.3.7) references criteria from piped vacuum systems for portions of dedicated WAGD system requirements.

However, these WAGD systems also involve the use of blowers and fans to capture waste gases and direct them to exhaust vents for disposal outdoors.

One of the major concerns for dedicated systems is cross-connection with piped gas and vacuum systems. A test for cross-connection must be conducted during installation or after any breach of the WAGD system.

It is important to note that captured waste gases cannot be *dumped* into normal air-handling systems. If the captured waste gases are connected to the room exhaust system (though this is rarely done today), the air-handling system must be configured for 100 percent outside exhaust to prevent the recirculation of nitrous oxide. This is in line with JCAHO's "Environment of Care Standard," EC.1.4 in *Comprehensive Accreditation Manual for Hospitals,* which requires a management plan to control hazardous material and waste (nitrous oxide is considered a hazardous vapor by JCAHO). The AIA *Guidelines for Design and Construction of Hospital and Health Care Facilities* (Sections 7.31.06 for hospitals and 9.31.06 for outpatient facilities) also requires 100 percent exhaust of waste gases (whether or not the evacuation system is combined with a room exhaust system) for hospitals and outpatient facilities. If a facility is using energy conservation measures, such as recirculating already cooled or heated air, then the room air-handling system definitely cannot accept waste anesthetic gases.

31.1.1.2 Direct Connection to a Medical-Surgical Vacuum System

When a direct connection to a medical-surgical vacuum system is made, the WAGD inlet is connected to the piped patient vacuum system as described in Chapter 30. This configuration is permitted in NFPA 99 (Section 5.1.3.7.1.2) but should be done with caution because of the effect on a patient's lungs if the vacuum reduction fails and lungs are exposed to the vacuum and flow levels of the piped vacuum system. The vacuum and flow level is hundreds of times greater than that which the lungs can tolerate.

To prevent breathing circuits from being connected directly to these levels of vacuum, anesthesia machines have what is called a *pop-off device* that limits the level of waste gas vacuum to 2 in. (5 cm) of water. When this level of vacuum is sensed, a valve opens (pops off) to prevent the vacuum from increasing any more.

31.1.2 Performance Criteria

Whatever method is used to provide WAGD, the performance criteria are similar for all methods. Each system must have an appropriate level of vacuum and flow, which is dependent on the operating characteristics of the anesthesia machines being used in the facility. Also, the vacuum and flow at one inlet must not be affected when an adjacent inlet (wherever it is located) is opened.

As might be expected, the WAGD inlet configuration must not be interchangeable with gas outlets or vacuum inlet configurations. ANSI has issued ANSI Z79.11, *Standard for Anesthetic Equipment—Scavenging Systems for Excess Anesthetic Gases,* in this regard. Also, the American Society for Testing and Materials (ASTM) has developed standards for anesthesia machines, which are integral parts of WAGD systems. These standards are ASTM F 1343, *Standard Specification for Anesthetic Gas Scavenging*

Systems—Transfer and Receiving Systems, and ASTM F 1850, *Standard Specification for Particular Requirements for Anesthesia Workstations and Their Components*.

31.1.3 WAGD Testing Criteria

Testing procedures for WAGDs are similar to those for vacuum systems. The level of vacuum and flow at each inlet must be tested while an adjacent inlet (which could be in another room) is opened. The required value of vacuum and flow is dependent on the specifications for the anesthesia machines being used. [See NFPA 99, Section 5.1.12.1.12.2(3), and ANSI Z79.11.] While quantitative tests are not listed, they should be done to verify that there is sufficient flow without having too much vacuum on the circuits. As part of ongoing maintenance, facilities should also monitor personnel exposure to waste gases to determine if the system is operating as designed.

31.2 Instrument Air Piped Gas Systems

Introduced in the 2002 edition of NFPA 99, and expanded in the 2005 edition, is a piped gas system intended "for the powering of medical devices unrelated to human respiration (e.g., surgical tools, ceiling arms)" but used in patient care areas. These systems are not to be connected in any way to piped medical air systems, but they are considered "medical support gas" as defined in NFPA 99, Chapter 3.

Section 5.1.3.8 in NFPA 99 lists the quality of "instrument air," and the requirements for sources, filters, accessories, piping, monitoring, alarms, and electric power of these systems.

31.3 Non-Patient Piped Gas Systems

Non-patient piped gas systems are used in health care facilities to conduct laboratory activities, as well as for heating, cooking, pneumatic switching, and other tasks. Nonflammable gases, such as oxygen, and flammable gases, such as hydrogen and petroleum, are used.

Although the components of non-patient piped gas systems are similar to those used in patient piped gas systems, it is crucial to make sure that the two systems do not become interconnected. Figure 31.1 shows the source for a non-patient piped air system. Note the presence of only one compressor.

NFPA 99 (Chapter 11, "Laboratories") contains requirements for non-patient piped gas systems. NFPA 45, *Standard on Fire Protection for Laboratories Using Chemicals*, and NFPA 50 (incorporated into NFPA 55, 2005 edition, *Standard for the Storage, Use, and Handling of Compressed Gases and Cryogenic Fluids in Portable and Stationary Containers, Cylinders, and Tanks*) also contain requirements that apply to non-patient piped gas systems used in health care facilities.

31.3.1 Non-Patient Piped Gas Systems in Laboratories

Gases in laboratories can be used to conduct experiments, heat chemicals, liquids, and other substances, react with other gases, and so forth. Some of these gases are provided to laboratories via piped systems.

Chapter 31. Other Gas/Vacuum Systems in Health Care Facilities

Figure 31.1 The source of a piped air system that is *not* to be used for patients (typical).

The requirements for preventing fires and explosions and ensuring electrical safety in health care facility laboratories are covered mainly by NFPA 99 and NFPA 45. NFPA 45 covers construction, ventilation systems, and related fire protection for all laboratories in all facilities. NFPA 99 (Chapter 11) covers health care facility laboratories and includes many references to the requirements in NFPA 45 rather than extracting text from this standard. NFPA 99 also includes more stringent requirements than NFPA 45, since NFPA 99 covers health care facilities. These stringent requirements include *fire protection features* based on a laboratory's proximity to inpatients, and *exiting features* based on the types of patients that may be present in the facility that has the laboratory.

As a guide, Table A.11.1 in NFPA 99 states which document (NFPA 99 or NFPA 45) should be used, depending on the ability of patients, if present, to leave an area in an emergency.

31.3.1.1 Criteria for Separation from Patient Systems

The technical committee responsible for piped gas system requirements in NFPA 99 has always been concerned about patient gas systems becoming contaminated (either inadvertently or deliberately) by non-patient gas systems. This is particularly the case with Level 1 compressed air systems, where backflow is a real possibility. For this reason, NFPA 99 (Section 5.1.3.5.2) allows connection of medical air compressors in Level 1 piped systems only to *medical* air piping distribution systems. NFPA 99 (Section 5.1.3.4.2) restricts the use and piping of oxygen and medical air of Level 1 piped systems to patient care applications. This wording is used because some laboratories contain patient care areas in which it is necessary to have medical-quality air. These patient care areas can be connected to the patient piped medical air system;

however, the rest of the laboratory must be served by a completely separate compressed air or oxygen system.

Requirements for all other piped gas systems installed in laboratories are listed in Chapter 11 of NFPA 99. (See also Sections 31.2.1.2 and 31.2.1.3.)

There is another, more operational, safety reason for *not* having laboratory piped gas systems connected to patient piped gas systems. Station outlets for patient piped gas systems have both a primary and secondary check valve built into them so that the outlet is automatically closed when something connected to the outlet (such as a hose) is removed. This prevents gas within the system from continuing to flow. The secondary check valve is for safety. Because of the wide variety of ways that devices may be connected to the gas, laboratories use a simple petcock and a short, open, tapered tube for an outlet. No check valves are used, and the petcock has to be closed manually. This is acceptable in a laboratory but not in patient care areas, where an automatic shutoff is a safer method, given patient needs and staff responsibilities.

31.3.1.2 Nonflammable Piped Gas Systems

Nonflammable piped gas systems are installed in many laboratories if the volume used justifies their being installed or if the design or space makes it practical to do so. For those laboratories located in health care facilities that house inpatients or outpatients incapable of self-preservation, the requirements in Chapter 11 of NFPA 99 are applicable. NFPA 99 (Sections 11.11.1.2 and 11.11.2.3) covers the installation of nonflammable piped gas systems in laboratories.

The quality (purity) of source gases used in health care facility laboratories can vary from *ordinary* to *ultra pure*. The facility designer must know the gas-quality needs of a laboratory *before* designing a particular laboratory or group of laboratories.

If the source of the nonflammable piped oxygen system exceeds 20,000 ft^3 (566 m^3) of gas in total (source plus the amount in the piping system), the source must be located outdoors and comply with NFPA 55. Amounts this large are located outdoors because of the fire hazard this amount of oxygen would present if it were inside a facility.

Per Chapter 11 of NFPA 99, the piping distribution portion for non-patient nonflammable piped gas systems is required to conform to those requirements set forth in NFPA 99 for Level 1 systems. Because of this reference to Level 1 systems, these systems are required to have an alarm system. Additionally, the piping of these systems must be tested in accordance with NFPA 99 (Section 11.11.5.1, which references Section 5.1.12). Note that it should not be inferred from these two references to a Level 1 system that the piping portions of a Level 1 system for patients and a nonflammable piped gas system for laboratories can be interconnected, or that a Level 1 system for patients can be extended to include those outlets that are a part of a nonflammable piped gas system for laboratories. It only means that the installation and testing of the piping portion of a nonflammable piped gas system in a laboratory must meet the same requirements as those for a Level 1 system.

In laboratories located in health care facilities that do *not* house inpatients or outpatients incapable of self-preservation, nonflammable piped gas systems follow the requirements in NFPA 45, 2004 edition (Section 8.2).

31.3.1.3 Flammable Piped Gas Systems

The distribution of flammable gases via a piping distribution system is permitted in laboratories located within health care facilities. However, because of the nature of activities in some health care facilities, the protection of the areas outside the laboratory is of considerable concern for those health care facilities housing or treating patients with various degrees of mobility, infections, weaknesses, etc.

For laboratories covered by NFPA 99, there is much concern relative to the location of the manifolds for these piped systems, particularly in relation to where patients are housed or treated. NFPA 99 (Section 11.11.1.2) provides guidance in this regard. Distribution requirements, including marking of piping and terminals, is covered in NFPA 99 (Section 11.11.2.3). NFPA 45, 2004 edition (Section 8.2), contains other requirements for flammable piped gas systems in laboratories, including storage and piping criteria.

The standards for the installation of a piped system distributing a flammable gas varies with the type of gas involved. The following list of NFPA standards address the hazards of these gases, as well as the installation procedures to mitigate these hazards.

- NFPA 50A, incorporated into NFPA 55, *Standard for the Storage, Use, and Handling of Compressed Gases and Cryogenic Fluids in Portable and Stationary Containers, Cylinders, and Tanks* (for hydrogen gas)
- NFPA 51, *Standard for the Design and Installation of Oxygen–Fuel Gas Systems for Welding, Cutting, and Allied Processes* (for gases such as acetylene)
- NFPA 54, *National Fuel Gas Code* (for fuel gases in the *gaseous* stage, such as natural gas and LP gas)
- NFPA 58, *Liquefied Petroleum Gas Code* (for fuel gases in the liquid state)

For Compressed Gas Association (CGA) documents, refer to the following documents:

- CGA G-1, *Acetylene*
- CGA G-4, *Oxygen*
- CGA G-5.3, *Commodity Specification for Hydrogen*
- CGA G-6, *Carbon Dioxide*
- CGA G-7.1, *Commodity Specification for Air*
- CGA G-8.1, *Standard for the Installation of Nitrous Oxide Systems at Consumer Sites*
- CGA G-9.1, *Commodity Specification for Helium*
- CGA G-10.1, *Commodity Specification for Nitrogen*
- CGA G-11.1, *Commodity Specification for Argon*
- CGA P-1, *Safe Handling of Compressed Gases in Containers*
- CGA P-6, *Standard Density Data, Atmospheric Gases and Hydrogen*
- CGA P-9, *The Inert Gases: Argon, Nitrogen and Helium*
- CGA P-23, *Standard for Categorizing Gas Mixtures Containing Flammable and Nonflammable Components*
- CGA P-24, *Guide to the Preparation of Material Safety Data Sheets (MSDS)*
- CGA P-24D, *Material Safety Data Sheets on Disk*

31.3.2 Other Piped Gas Systems Used for Heating, Cooking, Cleaning, Switching, and Maintenance

Piped gas systems (both flammable and nonflammable) in health care facilities can also be used for heating, cooking, cleaning, pneumatic switching, and some maintenance tasks. These systems must comply with the same requirements and regulations as those installed in commercial buildings.

31.3.2.1 Nonflammable Gas Systems

The most common non-patient nonflammable gas system used within health care facilities is a compressed air system. The compressed air typically is used in woodworking or metal fabricating areas to blow down an item after it has been cut or drilled. These systems may also provide compressed air to operate pneumatic switches that control dampers on heating, ventilating, and air conditioning (HVAC) systems. Regulators on branch lines provide the level of air pressure needed for a particular application.

The air within these systems is generally not filtered except for large particles that could clog tubing or an outlet or that could affect the devices connected to it. The compressors creating the compressed air are not monitored for the presence of contaminants that might pass into the air stream. Any tubing that might be installed also is not monitored for cleanliness the way it is for patient systems.

Of major concern within health care facilities is that this nonmedical air must be clearly identified throughout the facility so that it is not confused with medical or dental air. As noted in Chapter 29 of this book (and as specified in NFPA 99), piped medical and dental air systems are to be used *only* for medical applications. *Thus, nonmedical piped gases must never be connected in any manner with piped medical or dental gas systems.*

31.3.2.2 Flammable Gas Systems

Some of the flammable gases that are used in laboratories may be and are piped into other areas within the facility, such as kitchens and the maintenance department. It is important for personnel responsible for these systems to review the fire protection requirements for the areas in which these systems are installed. (Refer to Parts III and IV of this book for information on building codes and fire safety codes, respectively.)

31.4 Non-Patient Piped Vacuum Systems

Several types of non-patient piped vacuum systems are used in health care facilities. One type of system is a central vacuum cleaning system. Another type used is a pneumatic tube system. These systems are marketed as commercial systems and used in health care facilities the same way they are in commercial facilities. It appears that there are no national standards for non-patient piped vacuum systems.

NFPA 99 (Section 5.1.3.6.5) permits analysis, research, or teaching laboratories in health care facilities to be connected to the medical-surgical vacuum system if tubing is connected from the laboratory directly to the receiver through its own isolation valve and fluid trap that is to be located at the receiver. This is the only allowance in NFPA 99 for a piped patient care system to be connected to a piped non–patient care system.

31.4.1 Pneumatic Tube System

Pneumatic tube systems are used to move paper, specimens, and other small items around a facility when timeliness is of concern and staffing is not always available to meet time constraints. An item is transported from one point to another point by placing it in a container (or *transporter*) and then inserting the container into a pipeline, which uses vacuum to draw the container into the pipeline and move it along to its destination. The containers used to carry the items vary in size, depending on the types and sizes of items being sent by this method. Figure 31.2 shows a pneumatic tube system.

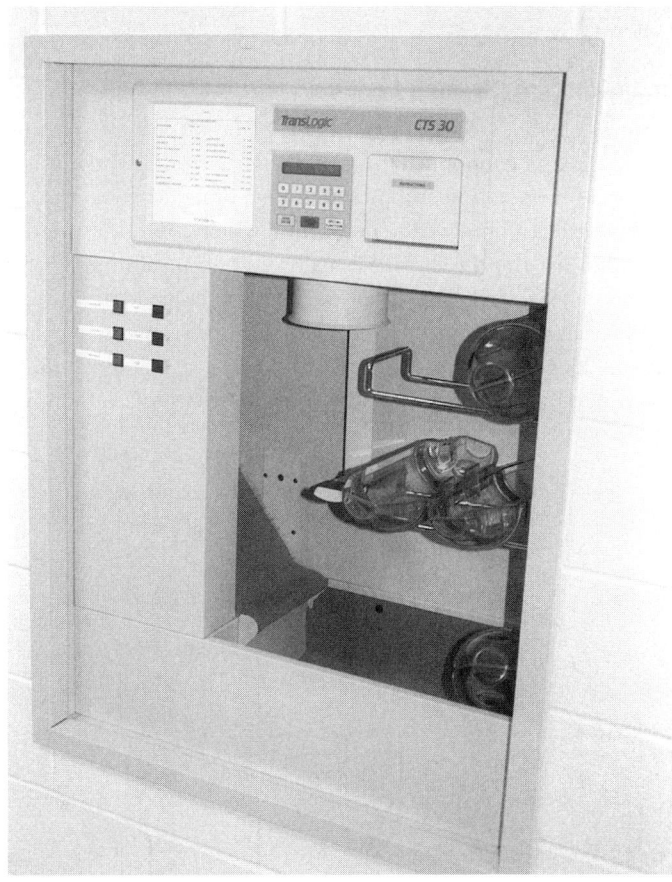

Figure 31.2 A pneumatic tube system.

One concern in health care facilities that use pneumatic tube systems is spillage of any items (liquid or otherwise) that could occur en route. Spills could cause short circuits or create increased fire threats. Spills may or may not stop the movement of one or more containers (transporters).

Because of accountability, it may be necessary to monitor each container to ensure that it does not get lost or delayed to the extent that the items being transported become spoiled or otherwise damaged. For a dedicated system in which a container can travel only from one point to one other point, as opposed to an interconnected system in which a container can travel from one point to any number of points, the major issue is just delay.

Connected to the issue of accountability is the issue of accessibility in the event of a failure of the system or the jamming of a transporter. Such an occurrence would make it necessary to retrieve a container manually. Thus, designers of these systems should make sure that retrieval of a container does not require the destruction of walls or the like (refer to Chapter 15) and that the conveyor mechanism does not make retrieving a container or transporter that has become stuck or has stopped functioning a major undertaking.

31.4.2 Central Vacuum Cleaning Systems

Central vacuum cleaner systems are used in place of portable vacuum cleaners when it is more efficient and economical to install such systems. The initial capital costs plus the annual costs for operating and maintaining a piped system versus a number of vacuum cleaners must be compared to determine which method is more advantageous for a facility. In addition, the time saved in using a central vacuum system versus portable vacuum cleaners is another factor that needs to be considered.

One design consideration for central vacuum systems is to make sure that inlets are strategically located (generally in corridors) so that the only equipment that cleaning personnel need to carry with them is a hose and some attachments (e.g., one for rugs, one for floors).

32

Patient Gas and Vacuum Equipment

There are important concerns regarding patient gas and vacuum equipment, both those used in conjunction with piped gas and vacuum systems and those used independent of such systems. For patient gas equipment, concerns include the fire hazards of and safe practices for the equipment. For patient vacuum equipment, they are the safe practices used to keep vacuum pump(s) from becoming clogged and bacteria from being spread.

32.1 Patient Gas Equipment

The most important concern for patient gas equipment that is to be connected to a piped gas system is that *at no time* should the equipment be able to put anything back into the system. This restriction is necessary to avoid compromising the cleanliness of the piping distribution system (i.e., by the backfeeding of any gases or particles into the system). If a different gas were to mix with the gas in the piped system, an explosive gaseous mixture might be created. Backfeeding can also weaken or break internal components of station outlets that are designed for gas to travel in one direction only—outward. Damage can also occur to station outlets if the pressure of the external gas is higher than the design limits of station outlets, monitoring devices, or sensors. If the external pressure of the gas being backfed into the system is high enough, leaks might develop at fittings (i.e., joints). The possibility of backfeeding is, in part, why patient station outlets have primary and secondary check valves.

The other major concern with patient gas equipment connecting to the patient piped gas system is whether the correct connector is used at the end of the hose that will allow connection to the piped system. Forcing an incorrect connector can produce disastrous results.

Patient gas equipment requirements are covered in NFPA 99, *Standard for Health Care Facilities,* 2005 edition, Chapter 9. Subjects covered include hazards associated with gas equipment, gas equipment storage, signage, construction, use, transfilling and transferring, record keeping, and personnel qualifications. Some of these subjects are discussed in this chapter.

584   Part V. Gas and Vacuum Systems and Equipment

32.1.1 Oxygen Cylinders and Containers

Figures 32.1 and 32.2 are photographs of an oxygen cylinder and an oxygen container. Figure 32.3 shows a cylinder labeled with the gas contained therein and a flame to indicate that there is a hazard associated with handling the cylinder. (See the Selected NFPA Definitions section for definitions of *cylinder* and *container*.) There are several reasons or scenarios why oxygen cylinders or containers are used or need to be available for use in health care facilities:

- A complete failure or interruption of a patient piped oxygen system has taken place as a result of an earthquake, rupture of a mainline, or the like, but there are patients who must remain on supplemental oxygen.
- A shutdown of an existing system (or portion of it) must be done for a repair, extension, or the like, but patients cannot be moved to another area.
- A patient on oxygen has to be moved to another area but must remain on oxygen during the transfer.
- Medical staff want a patient who is being administered oxygen to move around or be able to move around but remain on oxygen at all times.

Having freestanding (i.e., portable) oxygen cylinders or containers available for use at all times requires them to be stored properly so as to minimize exposure of occupants and the building to their potential dangers. As noted in previous chapters, oxygen does not burn, but its presence enables ignition to occur at a lower temperature and enhances burning of items already ignited.

According to NFPA 99 (Section 9.4), the *storage* of freestanding oxygen cylinders and containers, as well as containers for other nonflammable gases, either oxidizing or

Figure 32.1   An oxygen cylinder.

Chapter 32. Patient Gas and Vacuum Equipment **585**

Figure 32.2 An oxygen container.

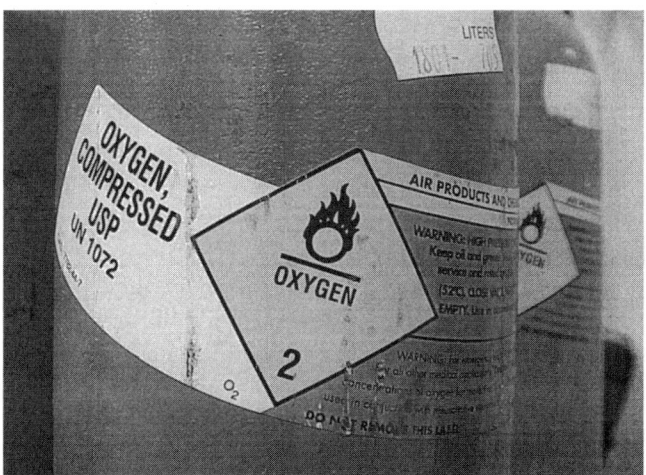

Figure 32.3 Identification and warning labeling on an oxygen cylinder.
Source: *Health Care Facilities Handbook*, NFPA, 2005, Exhibit 9.5; courtesy of Rob Swift and Prince William Health System.

inert, must meet several criteria. First, they must be stored in a separate room or enclosure with adequate fire protection features. Next, if the volume of gas of all cylinders and containers in a storage room or enclosure equals or is greater than 3000 ft^3 (85 m^3), the room or enclosure must meet the same requirements as those for cylinders or containers for piped systems in Chapter 5 in NFPA 99 (as noted in Section 9.4.1). No combustibles are permitted in such rooms. If the volume of gas in the cylinders and containers is less than 3000 ft³ (85 m³) but more than 300 ft³ (8.5 m³), requirements are not as stringent (as noted in Section 9.4.2 in NFPA 99). The most notable difference is that combustibles or incompatible materials are allowed in these areas, provided certain distances and fire suppression features are maintained within the room. If a total of less than 300 ft³ (8.5 m³) of gas is present in a fire zone [22,500 ft² maximum (2100 m²)], storage in an enclosure is not required (per Section 9.4.3 in NFPA 99) *if* certain handling procedures are followed.

Of concern to many users and authorities having jurisdiction is the question of when a cylinder or container is *in use* and when it is *in storage*. The cylinder or container that is stationed securely and positioned next to a bed with a patient who might need it at any moment is considered to be in use. A cylinder placed securely on an emergency cart for use at a moment's notice is also considered to be in use. However, a second cylinder or container next to a patient's bed would be considered an option and as such would be subject to storage requirements, because a second cylinder, in general, is redundant.

Items in storage must be in an enclosed area. If the enclosed area is located *outside* a building, then a substantive fence of some form is acceptable to restrict access. The size of the storage room for keeping items in storage *indoors* is not specified or restricted—it is the fire protection rating of the six surfaces making up the enclosure that is of concern (i.e., the floor, ceiling, and walls of a room, or the bottom, top, and sides of a cabinet). These surfaces must be of 1-hour fire-protected construction for volumes of gas 3000 ft³ (85 m³) or greater or of noncombustible or limited combustible construction if the volume of gas is less than 3000 ft³ (85 m³) in total. (Refer to Chapter 14 for more information on fire-protected construction and combustibility.) These requirements are in addition to other requirements applicable for such rooms or enclosures, such as the location of any electric switches and the absence of any combustibles [unless the total volume of gas involved is less than 3000 ft³ (85 m³)]. Refer to Chapter 9 of NFPA 99 for more information on the requirements for storing cylinders or containers.

Although mechanical safety measures differ between cylinders and containers, the precautions that must be followed when dispensing an oxidizing gas from either one are the same as those when dispensing oxidizing gas from a station outlet. That is, oxygen is oxygen is oxygen—or in this instance, an oxygen-enriched atmosphere is an oxygen-enriched atmosphere is an oxygen-enriched atmosphere. These safety measures are listed in Chapter 9 of NFPA 99.

A safety feature now permitted for cylinders is the use of valves that include some means of slowing the initial amount of gas that flows out of the cylinder when the valve is first opened. This reduction in velocity is for preventing high temperatures from developing and causing ignition of any plastic of metal surrounding the gas stream before it reaches the outside of the cylinder.

A major difference in handling gaseous oxygen in a cylinder as opposed to liquefied oxygen in a container is in the movement of oxygen from one cylinder to another cylinder (termed *transfilling*) and from one container to another container (termed *transferring*). NFPA 99 (Section 9.6.2.2) prohibits transfilling gaseous oxygen from one cylinder to another within a patient area of a health care facility. Transferring liquefied oxygen from one container to another is permitted (Section 9.6.2.3), provided certain procedures are followed. NFPA 99 (Chapter 9) refers to Compressed Gas Association (CGA) pamphlets P-2.5, *Transfilling of High Pressure Gaseous Oxygen to Be Used for Respiration*; P-2.6, *Transfilling of Liquid Oxygen Used for Respiration*; and P-2.7, *Guide for the Safe Storage, Handling, and Use of Portable Liquid Oxygen Systems in Healthcare Facilities*, for compliance requirements.

32.1.2 Respirators and Ventilators

The use of respirators and ventilators presents several issues that must be addressed before devices are placed into service. These issues relate to the supply of adequate electric power and gas (either oxygen or air or both), the sanitary storage of this equipment, and the elimination of any chance of cross-connections of gases.

With regard to electric power, the circuits to which these devices will be connected must be able to handle the power load from the respirator/ventilator along with all the other electrical devices that may be connected to these circuits. Facility staff or an electrical load specialist must conduct a load analysis to make this determination, as discussed in Chapter 12.

Another electrical issue that needs to be addressed is how these devices (in fact, any patient device that uses electricity) will function if there is a power outage followed by a restoration of power (generally within 10 seconds from an emergency power source). Staff must know whether a device will restart and function exactly as it was functioning before the power failure, or will turn on but remain in a neutral state, or will remain off. Staff also need to know whether the receptacle into which the equipment is plugged is one that will be connected to the essential electrical system source when normal power is interrupted. This issue is also particularly critical for devices that have built-in computers that control the functioning of the device. Refer to Chapters 8 and 9 for more information on power interruption to medical equipment.

Under no circumstances is a device to come back on in a mode that will injure a patient connected to it.

Newer models of many respirators and ventilators have their own uninterrupted power supplies (UPSs) that allow the device to continue to function for a limited time when there is a power interruption. This of course assumes that the UPS has been maintained properly and kept charged according to manufacturer instructions.

In terms of the piped gases that serve respirators and ventilators, *flow requirements* are critical. The gas source supply is supposed to be sized to meet the volumetric needs of a facility (i.e., the equipment that is intended to be connected to the piped gas system). The diameter of the distribution piping is supposed to be designed to provide a sufficient flow of gas to patient gas equipment such as respirators. This can become an issue, for example, in the pipelines serving operating rooms, when, on any day, all procedures might start at about the same time in the morning. Low-pressure alarms on gas-dependent equipment will be a sign that the piped gas system is becoming overloaded.

Another issue relates to the proper storage of respirators and ventilators. In addition to their size, which can create a space problem, these devices must be stored in rooms that are as dust-free as possible. These devices may be used in locations where sterile procedures are taking place (e.g., operating rooms) or on patients who depend on them for assisted breathing. Keeping these devices clean is not just good policy but is necessary for use in these particular areas and for these particular patients.

Finally, similar to the concern to prevent cross-connections in piped gas systems is the concern to prevent cross-connections within respiratory equipment that uses several different gases. The correct gas connector must be installed on the equipment, to match the gas inlet connector of the piped gas system. This concern also relates to a need to maintain the same pressure in the various pipelines serving these devices. Since the current quality of check valves installed in these devices cannot handle significant pressure gradients, the occurrence of one gas being backfed into another gas line is a real possibility. Even small particles in a line can be sufficient to keep these check valves open enough for gas to leak through them. A pressure differential would exacerbate this situation.

32.2 Patient Vacuum Equipment

There are two categories of patient vacuum equipment: devices that connect to the patient piped vacuum system to function and devices that produce their own vacuum to function.

32.2.1 Devices That Connect to Piped Vacuum Systems

Devices connected to vacuum systems use the vacuum produced by the piped system to function (i.e., to provide vacuum for use on patients). Controls on the device allow staff to vary the level of vacuum produced at the input of the device. These devices typically are equipped with *traps* (or water reservoirs) to prevent fluids, solids, and the like from being sucked into the hosing connected to these devices or to the piping system.

It is important to use these traps to keep pipelines clear. If hosing is connected directly to an inlet, then pipelines can become clogged with unwanted matter that had been sucked into the hosing. This will result in a reduced flow and lower vacuum pressure. In some cases, pipelines have become completely blocked because so much unwanted matter had been sucked into hosing and thence into pipelines.

32.2.2 Devices with Their Own Vacuum Producer

Portable suction devices are used when no central vacuum system is available, when a portion of a central vacuum system must be shut down for repairs or expansion, or as a backup to the central vacuum system (such as on an emergency crash cart for a patient who may not be located close enough to a central vacuum inlet). These devices are relatively simple in nature, consisting of a pump, trap, vacuum pressure gauge, regulator valve, and hosing. A power cord and on/off switch complete the assembly.

These devices require periodic testing and maintenance to ensure that they will function when needed. This is particularly important for machines that are placed on emergency crash carts. Cleanliness is also a major concern: Equipment must be cleaned with a disinfectant after each use, since bacteria from the products of suction can easily grow if they are not removed thoroughly and promptly.

Gas and Vacuum Systems and Equipment Frequently Asked Questions

(Answers based on 2005 edition of NFPA 99, *Standard for Health Care Facilities*)

1. Why are walls required between zone valves for a Level 1 or Level 2 piped gas systems and the gas outlets they control?

For the safety of the person closing a zone valve in an emergency, a wall separating the zone valve from the gas outlets it controls enables a person to close the valve without being exposed to heat and smoke in the area where the fire is occurring. It should *not* be necessary for that person to go into the zone of fire origin to close the valve controlling that zone. (Refer to NFPA 99, Section 5.1.4.8.7.)

2. Why is an area gas/vacuum alarm panel needed in an emergency room?

An emergency room meets the definition of a critical care area, and critical care areas are required to have an area alarm panel. (Refer to NFPA 99, Section 5.1.9.3.)

3. What can be done to prevent water from collecting in a piped gas system?

A desiccant or equivalent dryer can be installed in the system.

4. Why should fire fighters be knowledgeable about some aspects of piped medical gas systems?

Fires can affect outlets in rooms and joints in the piping distribution system and allow gas to escape into the atmosphere. Oxidizing gases, such as oxygen, will help accelerate a fire. The fire department can direct a facility to shut down its piped medical gas system if it deems that the system is contributing to the spread of fire. Knowing that patients on medical gas will be affected by a shutdown, fire fighters will understand the need to be selective in shutting down as small a portion of the system as possible.

5. Does NFPA 99 cover the cleaning of medical gas system piping on site?

Chapter 5 of NFPA 99 no longer permits cleaning of *new* piping on site. Section 5.1.10.1.2 requires that new piping be delivered to a site cleaned (i.e., suitable for oxygen service) as well as capped, plugged, or sealed in some way. On-site recleaning is limited to the surface area in the immediate vicinity of the ends of piping that are about to be joined but may have become contaminated in some way (e.g., end of pipe is dropped after cap has been removed). See NFPA 99, Section 5.1.10.5.3.

For *existing* piping that has become contaminated, a facility should have the system evaluated by experts to determine whether affected piping can be cleaned of the contamination. Factors involved in deciding whether piping can or should be cleaned include (1) the nature of the contaminant (e.g., is it water, oil, bacteria, rust?), (2) the extent of the contamination, and (3) the cost of decontaminating the piping (versus the cost of removing the existing piping and replacing it with new piping).

6. If flammable vapors are present in an operating room, can they be disposed of through the medical-surgical vacuum system?

While flammable anesthetics are no longer used in the United States, other flammable vapors, such as from isopropyl alcohol, can be present. However, they are to be diluted below the minimum concentration level for ignition by the time they reach the vacuum pump of the medical-surgical vacuum system and are exhausted. NFPA 99, Section 5.1.3.7.1.2(2), permits this practice.

7. Can "regular" room ventilation exhaust be used for scavenging if air is recirculated?

This practice is not currently permitted, per NFPA 99, Section 6.4.1.6.

8. Why doesn't NFPA 99 include a requirement for the total number of cylinders that can be stored within a facility?

Cylinders come in a variety of sizes, so specifying the number of cylinders that can be stored in a facility does not address the issue of the amount that can be stored.

NFPA 99 has developed criteria for the storage of gases within a facility, based on the volume of gas:

1. NFPA 99, Section 5.1.3.3.1.9, specifies the maximum indoor capacity for oxygen supply systems, including those cylinders connected and those in storage. It is 20,000 ft^3 (566 m^3), in storage and connected. When a system exceeds this volume of gas, the requirements of NFPA 50 (incorporated into NFPA 55, *Standard for the Storage, Use, and Handling of Compressed Gases and Cryogenic Fluids in Portable and Stationary Containers, Cylinders, and Tanks*, 2005 edition) have to be followed.

2. For nitrous oxide supply systems, including those cylinders connected and those in storage, the maximum total system capacity indoors is 3200 lb (1452 kg) [or 28,000 ft^3 (793 m^3)], per NFPA 99, Section 5.1.3.3.1.10. When a system exceeds this volume of gas, the requirements of Compressed Gas Association (CGA) Pamphlet G-8.1, *Standard for the Installation of Nitrous Oxide Systems at Consumer Sites*, are to be followed.

9. Is measuring the vacuum level of a vacuum system adequate to determine if the system is operating properly?

No, measurements made under a *no flow* or an occluded state will not provide data as to whether a system is actually able to create a vacuum. *Flow* is the test used to determine whether a system is functioning adequately as a vacuum system. This is why NFPA 99, Section 5.1.12.3.10.4, requires station inlets to meet a flow criteria of 3 SCFM (standard cubic feet per minute) "without reducing the vacuum pressure below 12 in. Hg at an adjacent inlet."

Gas and Vacuum Systems and Equipment Checklist

This checklist is an abbreviated listing of items to be checked or reviewed before, during, and after the installation of a piped gas or vacuum system. It should be used in conjunction with local and state codes and authorities. The order of completing the checklist will vary with each project (i.e., whether the project is a new system, an extension of an existing system, or a partial or complete renovation of an existing system).

Design Stage

✓	Item	Inspection Activity	Comments
	1.	Ensure that the total number of outlets and inlets that the system will support has been determined.	
	2.	Confirm paths of all pipelines.	
	3.	Determine that locations for local, area, and master alarm panels have been selected.	
	4.	Determine location of any exterior gas supplies.	
	5.	Ensure that if an emergency oxygen connection is required, its location has been determined.	
	6.	Confirm location of all storage rooms and manifolds.	
	7.	Verify that a decision has been made on whether piping for the vacuum system is to be the same as that used for the gas system.	
	8.	Ensure that specification sheets for all equipment, piping, sensors, alarms, outlets, inlets, components, etc. have been obtained and developed.	
	9.	Confirm that all drawings have been approved and signed off by all required authorities.	

Construction Stage

✓	Item	Inspection Activity	Comments
	1.	Verify that the systems have been installed in accordance with drawings and appropriate sections in NFPA 99, *Standard for Health Care Facilities,* Chapters 4 and 5; that pipelines were labeled when installed with correct name of gas or vacuum and direction of flow; and that the intended pressure levels have been indicated on piping as required.	
	2.	Ensure that *installer* tests have been conducted and that the results of testing and a certificate listing test results have been submitted to the party that contracted for the installation and testing.	
	3.	Confirm that *verifier* tests have been conducted and that results of testing and a certificate listing test results have been submitted to the party that contracted for the verification and testing.	

Postconstruction Stage

✓	Item	Inspection Activity	Comments
	1.	Ensure that a complete set of as-built drawings has been created and provided to the facility owner.	
	2.	Ensure that procedures for periodic testing of gas and related alarms have been developed and implemented.	
	3.	Confirm that a preventive maintenance program has been developed and implemented for each piped system.	
	4.	Verify that a list of spare parts has been developed in coordination with the manufacturer, distributor, and installer and that a stock of parts has been procured and inventoried.	

Part VI
Communication Among Health Care Facility Departments

This part of *Fire and Life Safety in Health Care Facilities* involves communication—communication among departments, between administrators and staff, between the facility and outside services, and between the facility and authorities having jurisdiction and inspectors. In addition to vital medical information, numerous other types of information must be communicated within health care facilities from the design stage forward. Submitting requests for equipment repairs, coordinating equipment maintenance and testing, conducting training programs, maintaining all required logs and then some, and letting all involved know when the inspectors are coming are just a few examples of typical information exchanged.

Today, there are more ways to communicate than ever. Electronic mail and cellular phone communications add significantly to the speed at which we can send and receive information, whether it is necessary or not. In addition to establishing proper lines of communications, aiming to communicate effectively can avoid the utter chaos that a facility often faces during times of in-house emergencies, when a disaster occurs near a facility, or when it receives notice from the Joint Commission on Accreditation of Healthcare Organizations (JCAHO) or another inspection agency that inspectors will be on site in 30 days.

Properly training staff on how to handle equipment on a routine basis as well as during emergencies is an essential part of communications between departments of a health care facility. Coordinating site work with departments affected by the work will ensure that the project will proceed smoothly without creating more disturbance than necessary, particularly to patients. Providing documentation of all essential activities and communications provides inspectors and surveyors with the information they require for determining whether a facility will pass inspection. Accurate and thorough record keeping—documents for new equipment, logs of all activities, tests, and repairs conducted, emergency plans, and the like—also help the facilities engineer provide a safe facility and reliable services for patients, visitors, and staff.

The chapters in this part include information as follows:

Chapter 33: Communicating with administrators, clinical and other departments, support services, and outside agencies and groups, and meeting JCAHO communication standards

Chapter 34: Record keeping, in terms of equipment manuals, equipment maintenance and testing records, and keeping and storing records

Chapter 35: Planning construction projects, conducting site work, writing progress reports, conducting performance and acceptance testing, holding walk-throughs, and handling shakedown periods

Chapter 36: Training all health care facility staff in general and the engineering staff in particular

33

Fire and Life Safety Communications in Health Care Facilities

Although in the broad spectrum of health care facility management, the facility engineer is often considered to play a supporting role to those who provide primary health care, the facility engineer actually plays a pivotal role in securing the safety of the facility and minimizing electrical, fire, explosive, physical, and other hazards. From the planning and design of a facility to its operation, maintenance, emergency preparedness, and eventual renovation, facilities engineers must communicate with many people regarding the ways and means of providing a safe infrastructure with safe medical and electrical equipment and systems that meet or exceed all codes and standards. For example, facilities engineers must communicate with architects and planners, representatives from all of a facility's clinical and support departments (i.e., medical, dental, nursing), many outside authorities, such as fire and police officers, and equipment manufacturers. Additionally, facilities engineers interface with surveyors from the Joint Commission on Accreditation of Healthcare Organizations (JCAHO) and inspectors from state and local authorities having jurisdiction before, during, and after site inspections take place.

Expanding on traditional written communications, telephone calls, and planned meetings, today's communications include electronic mail, video and telephone conferencing, and cellular phone connections. These newer versions of communication can speed up the information exchange. But no matter what form the communications take, the basic rules of communication still apply for establishing solid and productive working relationships based on trust, honesty, and respect between those superior, subordinate, and on equal footing to the facilities engineer. These good working relationships will not only make for a more pleasant working environment, they can also lead to clearer communications regarding technical and complex situations. This, in turn, can lead to the proper treatment for the patient and safer conditions overall as well as avoid the chaos that typically ensues when an inspector announces an impending site visit.

33.1 Communicating with Administrators

Ultimately, the chief operating officer (COO) and the chief executive officer (CEO) are responsible for the entire health care facility. Many forms of communication must take place between these executives (as well as the chief financial officer, CFO) and the facilities engineer, and it is important for the facilities engineer to partake in proper communications. Information regarding the physical and environmental safety of the facility, the status of major repairs that must take place, and equipment purchases are a few examples of the types of information that must be communicated, much of which involves justifying added expenditures.

A facility's organizational structure and size determine the formal lines of communication between the facilities engineer and his superiors. One problem with a long chain of command from the director of engineering to the CEO or COO is that often the administrators in between are not technically oriented, and the information that is passed on to a higher plateau can sometimes be watered down or changed completely. (It seems that the greater the complexity of the original message, the more the message changes.) In some larger institutions, one administrator with a technical background is responsible for all of the so-called technical departments, such as the engineering, clinical engineering, communications, safety, and environmental engineering departments. It is ideal for the director of engineering to report to this type of administrator, but if this is not the case, the next best order is to have the director of engineering report directly to the CEO or COO. This structure at least diminishes the number of people through whom the information passes on the way to the top decision maker.

No matter how the information flows, for the most part, the primary role the director of engineering has in relation to the administrators is that of educator—about projects, needed equipment, costs and benefits, and the like. It is essential to remember that even administrators with a technical background may appreciate having information presented in a less technical and more succinct manner. When meeting to discuss a technical item, the administrator may have very little prior or detailed knowledge (technical or otherwise) of what is being discussed. Keeping the message at the appropriate level of detail but without excess details that waste highly valued time should provide enough information for decision makers to make their decisions. Bringing along photographs, charts, or sketches may prove useful in explaining the issue at hand.

33.1.1 Necessary Facility Upgrades and Major Repairs

One of the main responsibilities of the facilities engineer is to communicate clearly with administrators regarding a facility's physical and environmental safety. It can sometimes be difficult to explain to an administrator that a code has changed and work is necessary to reach compliance. Additionally, when conditions arise that require large expenditures to ensure physical or environmental safety, the facilities engineer must provide valid explanations and justifications as to why less costly alternatives are less appealing. Therefore, the facilities engineer should always be prepared to present at least two or three alternative suggestions and explain the pros and cons of each one and the relative costs. It then becomes easier to explain why the suggested choice is the preferred option.

When a major repair must be made, it is important to present the request for action with the proper backup information and data, and to present it fairly. Having solid working relationships with administrators often makes it easier for a facilities engineer to, for example, request 10 hours of disruption, which will include contingencies when only about 5 hours of actual work are needed. It may also mean that facilities engineers do not have to justify every action after presenting all the necessary backup information, and administrators, CEOs, COOs, and CFOs are confident that the facilities engineer will accomplish the job at hand within the allotted time and budget.

At times, a facilities engineer may wish to coordinate a site visit to allow the administrator to examine an existing condition. If this becomes necessary, it is important to provide ample notification so that the administrator can set aside enough time for the visit and come prepared, such as with proper outerwear for visiting outdoor sites. If the site is located away from the main facility, the facilities engineer must make transportation plans.

33.1.2 Requests for New Equipment

In an era when many health care facilities face financial constraints, it is sometimes more convenient to keep repairing equipment rather than replacing it. For equipment that absolutely must be replaced, and to fill immediate needs for new equipment (and personnel), the facility engineer must gather and present all of the necessary information to the appropriate administrators and financial officers. Cost is an important factor. Therefore, it is useful to present the history of repair costs and lost time and an accurate analysis of the total cost of the new equipment compared to the total cost of repairing and eventually replacing the older equipment. If two pieces of equipment accomplish the same function, the facilities engineer must present a valid reason (such as better service and improved reliability) for ordering the more expensive device. When the information is presented properly and understandably and a relationship of trust, honesty, and respect has already been established among the relevant parties, the task of convincing the appropriate people to approve replacement rather than repair will be much easier.

Many facilities set up an administrative committee to discuss the purchase of new equipment. Facilities engineers should make it a point to be a member of this committee or, if no such committee exists, at least to be consulted before the purchase of any equipment. In this way, administrators can become informed of the environmental requirements for the equipment (space, electrical, plumbing, ventilation, etc.) before it is purchased and not after, when problems can arise. For example, facilities engineers often are not consulted before the purchase of a piece of equipment that is too large for its allocated space or that will require large expenditures to bring sufficient electrical power at the proper voltage, adequate cooling and ventilation, or water and drainage to the equipment area. Since administrators have final approval for all equipment purchases, facilities engineers are responsible for tactfully convincing administrators of the importance of being consulted on these matters, as the following case study illustrates.

Requirements for New Equipment

When the Lithotriptor was first introduced into this country many years ago, it required a 480/277-V (volt) system. Since the equipment is considered to be a wet location, the codes required the installation of ground-fault circuit interrupter protection (see Chapters 6 and 7). Unfortunately, at that time, no accepted ground-fault circuit interrupter was made for a 480/277-V application that was approved for that service in this country. After the equipment was delivered, expensive additional electrical protection had to be installed. Had the facilities engineer been consulted before the equipment was purchased, he could have informed the decision makers beforehand regarding the added requirements.

33.2 Communicating with Various Departments

Each department in a health care facility has a key responsibility. Only when all departments function together properly does the facility provide for the entire well-being of the patient. Once all departments accept this concept, communication among them flows more smoothly. Having a well-established protocol for informing the engineering department when repairs must be done also enhances communications, as well as the efficient conduct of the work at hand. This often takes place via work order requests, as discussed in Chapters 11 and 12.

33.2.1 Planned Meetings

Larger issues are often addressed in planned meetings between the engineering department and other departments, as well as with administrators. These issues include the progress of key projects, equipment not being used properly, the time frame necessary to make repairs and adjustments, how the departments can help to expedite repairs, and other general problems and items that affect the functioning of the engineering department and the facility overall. The facilities engineer should make every effort to attend all of these meetings, as it is very valuable for the head of the engineering department to establish and maintain excellent face-to-face relationships with the top-level personnel in each of the other departments in the facility in noncrisis situations. If good working relationships have not already been established when a crisis occurs, each department head may act more independently to solve the problem. These actions may lead to the crisis not being handled as efficiently as possible and to greater adversity among the department heads and the facilities engineer at a later time.

It is also advantageous for the engineering department to initially meet with one department at a time. These meetings provide a means for introducing the personnel from both departments to each other. Eventually, they allow for establishing a solid relationship between the departments based on mutual respect and understanding about the function and operations of each department and about typical or ongoing problems. They also aid the engineering department in helping to solve problems and address future concerns. Likewise, the other departments should cooperate with the engineering department in providing needed services.

To aid this understanding, the engineering department must disseminate information in a manner that is understood by the other departments. Here again, highly technical information must be presented in a fashion that nontechnical people can understand. All communication should be tailored to the department to which it is presented.

33.2.2 Meeting Agendas and Preparation

Every planned meeting should have an agenda, and the person in charge of the meeting usually has one, even if it is not formalized. It is ideal for all parties involved to agree on the agenda in advance, but this can be difficult when planning a meeting with senior administrators, and compromises must usually be made.

When an agenda is received in advance, there is no excuse for not being prepared for the meeting. Although there may not be time for questions and answers, it is important for the facilities engineer to prepare properly for all meetings and to be able to respond to questions regarding ongoing projects and outstanding problems. It is also important to be prepared to explain diplomatically why a problem remains unresolved. In general, preparing for meetings in advance allows for more thorough discussion of a problem and for decisions to be made about it during the meeting and not at some later date.

Meeting preparation may involve communicating in advance with the other department concerning particular problems or topics that are scheduled to be discussed. For certain issues, the facilities engineer may be able to anticipate questions and come prepared with the answers. The reverse is also true. Meetings will be more productive when the other department knows what questions the engineering department is likely to raise. The purpose of these meetings is to resolve issues, not create surprises.

In addition to being prepared for meetings, another way to keep meetings productive is to make sure that the people with authority to make decisions, and not their substitutes, attend the meetings. It may be necessary to appeal directly to the missing person to please attend, or else report to a higher authority to resolve this issue.

33.2.3 Meetings Regarding Special Projects

Some meetings are held regularly to discuss a single, ongoing, or difficult project, such as a repair or renovation that requires work to be accomplished in an active area occupied by patients or used daily for treatment. This type of project may require the temporary relocation of the patients or activity, temporary emergency procedures, and the use of additional staff. Meetings to discuss these types of major issues are usually attended by all departments involved, and depending on the size and type of the facility and the type of project, follow-up meetings may be required, as illustrated by the following case study.

Special Project Meetings

A renovation had to be made to the main electrical bus duct that fed the floor of a large hospital. To make the necessary changes, the normal electrical supply to the floor had to be turned off for 1 hour. It was a large patient floor, and the patients could not be relocated. Many meetings were held to plan the power shutdown. The nurse in charge of the floor was in charge of the meetings and planning. During the meetings, each of the departments involved, both clinical and service, stated their concerns and requirements. When all of the contingencies had been covered and all parties agreed to the plan, the renovation was scheduled.

The work was conducted just after sunrise—to ensure that patients who were awake had some light—and before medication and feeding time. Additional staff members from all departments were brought in to cover any emergencies. Communication was maintained among all staff members during the work. With this type of meeting and planning, the project was a success and patient care was not compromised.

Most meetings that involve more than two departments typically have one person in charge who is capable of running the meeting, handing out assignments to accomplish the work, and seeing the project through to completion. An administrator responsible for that area or function of the institution usually designates the person who will be in charge of the meetings for work that involves many departments (if the administrator does not take the lead).

33.2.4 Meeting Frequency

Depending on the size of the facility, the engineering department, and the other departments involved, the meetings can be as frequent as once per week or as far apart as quarterly. If an emergency occurs and a meeting must be cancelled, another meeting should be scheduled as soon as possible. Conversely, some meetings are called to attend to specific problems that cannot wait for the next regularly scheduled meeting. The frequency of the meetings can always be changed if required, but it is essential to hold regularly scheduled meetings. (Note that with short advance notice of a meeting, all parties should come prepared to discuss the items at hand.)

33.2.5 Meeting Location and Atmosphere

The location and atmosphere of a meeting are also important. The location must be convenient for all personnel attending and be held in an area that is private, where physical interruptions are at a minimum. In a large medical institution, finding such a location can be a problem. An even bigger problem today is the many interruptions that occur when people have their pagers and cellular phones turned on. There is no one solution to these problems. For relatively short meetings, a staff member who is not in attendance can take all calls. In a severe emergency, someone can hand-deliver a message to an attendee. When there is no other solution, interruptions must be tolerated.

Something as simple as having refreshments, coffee, and the like at meetings often helps to ease the tension of the day and encourage participants to relax. However, it is generally not a good policy to use these meetings as a forum to air problems regarding the various departments or administrators or the attitude and relationships of specific personnel. These problems should be attended to on a personal basis unless they affect the entire facility.

33.3 Communicating with Clinical Departments

It is clear that the primary function of the medical, dental, nursing, and other clinical departments is to attend to the needs of the patients, while the facilities engineering department ensures that all equipment within the clinical area functions properly to allow the clinical staff to provide the best patient care possible. More than just having proper lighting and functioning medical equipment, the clinical staff must have comfortable air conditioning and heating, adequate work and storage space, and a flow of materials that facilitates their work. When work must be done in a clinical area that will affect its functioning, it is important to keep the department informed of the planned physical change and to bring the department into the planning process at the beginning stages of a project.

To achieve the desired work within the physical and budgetary limitations of the various departments, compromises must usually be made, based on give and take between the departments. For example, if extra space is needed in one department, another department may be able to release it. Relationships based on mutual respect for each department's functions and personnel will assist in making any compromises.

At some point, the line of communications may include the administrator responsible for that area. If at all possible, the facilities engineer should not allow himself to be placed in the middle of a discussion between administrators and the clinical department concerned. Having good working relationships with these departments makes it easier for all to accept decisions that are less than favorable.

As with administration, the facilities engineer should establish a relationship with the clinical departments such that he is consulted when a new piece of equipment is being purchased. Sales personnel can be very convincing about a product but not necessarily knowledgeable about problems the equipment may cause in a facility or code compliance, whereas facilities engineers are more able to determine these needs.

33.4 Communicating with Support Services

Support service departments with which the facilities engineer must interact to ensure facility safety include building services, central supply, dietary, mail, and the like. The most frequent support service requests made to the engineering department are for adequate storage, repair of special equipment, proper environmental conditions, and help with transporting goods. Requests for repairs are relatively straightforward communications. If new equipment or additional space is required or a major renovation is necessary, the appropriate people must be consulted for approval. Each facility has its own protocol for these channels of communication.

One of the major problems a facilities engineer faces when dealing with support services is setting priorities. Although every department naturally feels that its needs are of utmost importance, highest priority is most often given to requests made by clinical services. Therefore, the engineering department usually cannot attend to each support service request immediately unless it is an emergency. To accomplish the work in the assigned priority, the facilities engineer often must explain (tactfully) to the various department heads involved why there is a delay in responding to a request. This form of communication will allow the department in need to make satisfactory alternative arrangements. The important thing is to keep all lines of communication open and all parties informed of the reasons why progress may be impeded.

33.5 Communicating with Outside Agencies and Groups

Facilities engineers must also communicate with representatives from agencies outside the health care facility, including authorities having jurisdiction and local police and fire departments. Here, too, it is always beneficial to develop relationships with mutual trust and respect. The results of a good relationship can be seen from the following case study.

Good Relationships Pay Off

In a large urban hospital, a small fire that appeared to be fed by oxygen developed in the basement of one of the buildings. There was no shutoff valve between the fire and the main oxygen storage tank for the facility, since the fire was located at the same entrance to the building as the oxygen line. The commander of the fire department at the scene determined that the main oxygen supply to the medical center needed to be shut off. Since this event occurred at 10:00 A.M. during a weekday, all of the operating rooms, delivery rooms, and ICUs were in operation and patients were on ventilators. The relationship between the fire department commander and the person in charge of the facilities was such that while there could not be a delay in shutting off the oxygen valve, the commander did allow a few necessary seconds for communicating this information to the entire staff. Likewise, staff were prepared when the event took place and there was no loss of life.

Facilities engineers also interact with members of construction teams, such as architects, consulting engineers, designers, equipment planners, and other professionals. These interactions take place from the planning of a project through when a new area is successfully put to use. It is not uncommon for members of a planning team to disagree regarding the safety of features and equipment that some members of the construction team wish to install or regarding the physical limitations of the existing facility. The facilities engineer must thus be able to contribute information on safety in a way that does not make it seem like he is sabotaging the project. Refer to Chapter 35 for more information on construction projects.

33.6 Meeting JCAHO Communications Standards

One of the most obvious reasons to establish proper communications and proper lines of communication is to facilitate compliance with regulatory requirements. JCAHO devotes an entire chapter to managing health care facilities ("Management of the Environment of Care") in the *Comprehensive Accreditation Manual for Hospitals*. To be accredited by JCAHO, institutions must meet many standards for creating plans of action, keeping records, and writing incident reports. JCAHO is constantly changing its approach to accreditation, the manner of the survey, and the standards. It is important for the facilities engineer to stay abreast of the latest "Management of the Environment of Care" standards as well as what will be expected during the survey.

JCAHO plans of action must include information on how the facility will correct nonconforming conditions, what procedures it will follow during a change in the normal service conditions, and how it will provide services during construction and disruptions of service during a disaster (see Chapter 36). All of these actions require exchanging information between departments. Record keeping includes passing information between departments and maintenance personnel so that service and maintenance records and records of equipment failure are kept accurately. Incident reporting includes logging information about accidents where an injury occurred and where there was the loss of the function of a necessary piece of equipment. This requires the affected departments to obtain the pertinent information and make corrections if necessary. Many federal, state, and local government agencies also require the submission of reports with tabulations of operational data, which can be accomplished only by communication among departments.

Health care facility administrators often establish various committees, such as a safety committee, to comply with JCAHO's management plan and address the many facets of facility operation. For these committees to function effectively, communications must be set up among departments, and procedures must be established for reporting problems and following up on corrective measures. Many of the committees include a representative from the engineering department, who often is responsible for coordinating the work to address the problems at hand.

When JCAHO notifies a health care facility that there will be an inspection, preparations are hectic and utter chaos usually sets in. It is at these times when the relationships that are already established between the engineering department and the other departments are put to the test. If these relationships and lines of communication are not based on mutual trust and respect, then this period can be unbearable. Typically, administrative, clinical, and support departments all are making demands to ensure that their areas of concern are prepared. The most common need is to repair the facility to ensure that it meets all codes and looks good. Additionally, there is usually a request for storage space for all items that have never been stored but must now be removed from the floors. Preparing for inspections is a time when a facilities engineer must wear many hats but be a diplomat throughout the entire process.

34

Record Keeping

Health care facility record keeping other than that which is required for the clinical services for patient care and treatment takes many forms. Minutes of meetings, letters to manufacturers, purchase orders, repair requests, equipment manuals, equipment testing and maintenance logs, preventive maintenance schedules, and lists of maintenance procedures are all considered to be records. Record keeping is required not only for compliance with codes but also to enhance operations throughout a facility and to support legal matters, as they provide evidence of maintenance and testing that has been performed.

For records to be useful, they must be true and accurate recordings of the technical maintenance, preventive maintenance, and testing activities that took place, as well as any communications that were exchanged, including those in meetings. The records should reflect what the recorder observed or heard, not what was told to him, as exemplified in the following case study.

Accurate Record Keeping

In one case, a woman who claimed to have fallen on the sidewalk in front of a facility's emergency room was suing the facility for not maintaining the sidewalk properly. Before trial, it was questioned as to whether anyone actually observed the woman fall. Although the patient's chart indicated that she fell in front of the emergency room, the attending physician who filled in the chart said that he did not observe the fall but that the patient told him that that is where she fell. He also said that he did not think to write that the woman had told him this and that this was not his actual observation. In this case, it was easy for the physician to explain what actually happened in a signed affidavit, but it is generally more difficult to get this piece of documentation from physicians who may no longer be working in the hospital and cannot be reached for testimony. In this case the patient's chart was her best witness, but it did not accurately reflect the reports of witnesses to the event.

34.1 Record-Keeping Requirements

As mentioned throughout this book, NFPA 99, *Standard for Health Care Facilities*, NFPA 70, *National Electrical Code® (NEC®)*, and most other codes require health care facilities to keep and maintain records to enhance electrical, life, fire, and gas/vacuum safety. The Joint Commission on Accreditation of Healthcare Organizations (JCAHO) lists requirements for record keeping in its *Comprehensive Accreditation Manual for Hospitals* ("Management of the Environment of Care"). To comply with JCAHO requirements, a facility must develop and maintain management plans, minutes of meetings, maintenance records, test logs, and the like. Inspection records must be kept for many items as well, such as medical equipment and utilities. These records must log system and equipment usage as well as map the layout of the systems.

34.2 Equipment Manuals

Health care facilities must keep manufacturers' manuals for most equipment, medical, electrical, electronic, and otherwise, as well as record all tests, preventive maintenance activities, and repair work conducted on that equipment. NFPA 99, 2005 edition (Section 10.2.8), includes requirements for manufacturers concerning the content of the instruction manuals and labels that must be furnished with new equipment when delivered. These include operator's, maintenance, and repair manuals. NFPA 99, 2005 edition (Section 8.5.3), lists requirements for record keeping for the electrical equipment. The first aspect of these requirements involves keeping a permanent file of the instruction and maintenance manuals for all pieces of equipment and having these documents readily accessible at all times. It is recommended that this document file be kept by the engineering group responsible for the equipment, in a place that is accessible to the maintenance personnel at all times—not in a locked filing cabinet or an office that is locked evenings, nights, weekends, and holidays.

A duplicate copy of these manuals should be kept with or near the equipment. This could save maintenance personnel valuable time not only in locating the manual but also in performing the maintenance to get the equipment back on line as soon as possible. There may be times during an equipment malfunction when the initial contact between the user of the equipment and the maintenance personnel is by telephone. If corrective action is a simple procedure, then the operator of the equipment can refer to the manual for instructions on restoring the equipment or be led through the correction by the maintenance person on the telephone with the help of the manual at hand. Again, this option could save valuable time, especially if the breakdown occurs during the night, on a weekend, or on a holiday, when a maintenance worker would otherwise need to return to the facility to make the repair.

Keeping a manual with a piece of equipment that is frequently moved can pose a problem. If possible, the manual can be attached to the equipment or the cart on which the equipment is attached. A more serious problem arises when the manual cannot be located or in some cases was never delivered with the equipment. If a duplicate copy of the manual cannot be obtained from the manufacturer, then the staff will have to do their best to write one.

34.3 Equipment Records

Many codes and authorities having jurisdiction require documentation of all equipment, including historical records of equipment and logs of all tests, test results, preventive maintenance procedures, and repairs. Much of the data for the equipment logs comes from equipment users and maintenance personnel.

These records can be most useful when a facility must decide whether to replace or repair a particular piece of equipment, particularly when the engineering department must make a recommendation to other departments concerning the repair or replacement of the equipment. If prepared properly, the records will indicate the number of times a repair is needed and the type and frequency of repair, which can indicate the potential for trouble.

At other times, records may be used in lawsuits to prove that the institution performed the proper tests or maintenance on the piece of equipment or part of the structure in question. For example, simple items such as the record of preventive maintenance performed on a bed can be helpful if a patient falls out of the bed and there is a lawsuit contending that the bed was not maintained properly.

34.3.1 Historical Records

The minimum historical information that should be included for each piece of equipment is the name and type of equipment, manufacturer's name, serial number, model number, specific individual data such as voltage or wattage, date of purchase, and cost at the time of purchase. This list can be expanded or contracted to suit the individual facility's needs.

The information included in these records should be presented in a manner that is understandable to other departments and the engineering department. Many computer programs are available to record data for all types of equipment in an understandable way. These programs typically come with or are part of a facility's computerized maintenance and preventive maintenance program.

34.3.2 Maintenance and Preventive Maintenance Records

As preventive maintenance tasks are performed, the records should indicate the frequency and type of preventive maintenance work that has taken place, items used for preventive maintenance, and amount of time spent calibrating each piece of equipment. Preventive maintenance logs should also include information such as the voltage and refrigerant used. This information should be recorded on a tag that is attached to the piece of equipment. These tags should be maintained and kept up to date, as indicated in the following case study.

Not Keeping Up-to-Date Records

In one facility, the type of refrigerant used in a piece of equipment was changed by the day shift while making a repair to the unit, but the change was never recorded on the tag. When the night shift was performing routine maintenance, someone noticed that the unit needed additional refrigerant and checked the tag to determine which refrigerant to use. As a result of the day shift not making the change on the tag, the wrong refrigerant was added. The equipment needed to be shut down to drain out the wrong refrigerant, clean the lines, and install the correct refrigerant. This was a costly operation in terms of time, labor, and dollars. Since the equipment served a patient care area, there was also an inconvenience to the patients.

Preventive maintenance work orders can be generated automatically by a computer. All data that are necessary to identify a piece of equipment or an area in the facility should be included in the order. Instructions on what procedures must be performed should also be noted to provide maintenance personnel with specific directions. There should be a place in the repair log for maintenance personnel performing the work to sign their names to ensure that the work has been completed and to provide for accountability.

Maintenance or repair logs should track the same information as preventive maintenance records. Records of the work date, type of malfunction, materials used, time spent, number of people who performed the work, and total cost of the work in time and materials, if possible, should be kept for both maintenance and preventive maintenance projects. This information should be added to the historical data to show a complete record of the performance of a piece of equipment or the usage of an area in the facility.

34.3.3 Equipment Testing Records

Records of the results of required periodic equipment tests must also be kept. In the truest sense, these records are preventive maintenance records in that they indicate what, if any, work should be done on the equipment, such as repairs or calibrations. These records also are important to track not only the performance of the equipment but also the maintenance tasks required and performed, which may be useful in the event of a lawsuit concerning the equipment in question. Comments should be made in the record concerning any recommendations that are made as a result of the test, such as a recommendation to conduct a particular repair or to replace the equipment. These comments should be discussed with the facility engineer's immediate supervisor, since they become a permanent part of the record. Refer to Chapters 11 and 12 for more information on testing and maintenance of equipment.

34.4 Keeping and Storing Records

All records should be stored in a safe place and be readily accessible. It is recommended that all equipment records be kept for at least the life of the piece of equipment if not longer. Computers, microfilm, and microfiche can be used as a means to store or archive records when storage space becomes limited, for both text and drawings.

It is important to impress on the staff that all records, whether they are minutes of meetings or operating and maintenance manuals, are not to be used as convenient notepaper for emergency note taking. The question of maintaining duplicate sets of records should be discussed. With the latest electronic equipment, the area for storage of duplicate records can be set at a minimum. With a duplicate set of records, there is obviously no concern if the original set is destroyed or compromised in any way.

35

Construction Projects

Health care facilities undergo numerous types of construction projects, from those that bring a facility up to code to those that enhance the facility beyond the requirements to improve the services provided to patients. With the advent of managed care, many facilities are constructing ambulatory care facilities as well as renovating existing facilities to accommodate new ways of providing medical care. Some facilities are constructing separate imaging centers and dialysis centers. For all types of projects, the engineering department is involved in conveying information regarding acceptable design, acceptable building materials, and the safety of the project and its effects on the facility. Final decisions are often a matter of economics, but decision makers must be informed about all safety issues.

The first order of business when planning a new project is for the chief executive officer and the affected department to determine what specific needs must be filled. Once these needs have been established, all construction projects must be planned, and all involve temporary relocation of a few or many operations and the maintenance of safe conditions while work is being done. All of these steps also involve the facilities engineering department for maintaining a safe facility during a project and after the area is put into use.

Many institutions on large projects hire a *clerk of the works,* who will supervise the work on behalf of the institution. This person is paid by and reports to the institution. The clerk of the works does not replace the general contractor's project engineer or superintendent, but rather watches the construction work for the institution. In the event that there is no clerk of the works, the facilities engineer or another person on the staff of the institution with construction experience may be designated to perform this function. In the following discussions, the title *facilities engineer* is used to denote any of the previously noted positions.

35.1 Project Planning

All projects must be planned to make sure all requirements are met. Larger facilities may have a separate planning department to oversee construction projects. Other facilities may establish a construction committee for large projects that consists of administrators, representatives from the affected department, the engineering department, and perhaps members of the outside design team, including an architect, engineer, interior designer, equipment planner, or other professionals. Still other facilities may ask a few in-house staff to coordinate the work to be done.

The facilities engineer is typically consulted at the onset of project planning for information on the effects the project might have on the existing facility and physical plant, any limitations that might exist, and safety concerns as well as to provide input into the design and construction specifications. As discussed in Chapter 33, it is important in these preliminary meetings to establish a cooperative atmosphere rather than an adversarial one, despite whatever types of previous relationships had existed. Although the staff in the engineering department may feel slighted when someone on the design team insists on installing a specific item or design that will create additional work for the engineering department, they must at times concede the point (graciously, it is hoped). The manner in which members of the engineering department conduct themselves at these planning meetings and the attitudes that they convey often are carried throughout the project.

There is always the question of how far down in a facility's organizational structure the planning team must go to consult with staff who will be using the new or renovated facility. It is up to the administrators or the planning team to decide how many viewpoints to consider. Incorporating many viewpoints may lead to a more well-received project and lessen the possibility that the administrators will be blamed for a project that ultimately does not meet the intended needs. On the other hand, speaking to the entire staff may greatly extend the time frame for project completion.

35.2 Construction

Construction work begins after the design process is completed. If the work consists of renovating an area that is presently being used, arrangements must be made to relocate the existing service to temporary quarters before the start of work. Again, constant communication with the departments affected is necessary.

Many codes, such as the Joint Commission on Accreditation of Healthcare Organizations (JCAHO) accreditation manual, the American Institute of Architects (AIA) *Guidelines for Design and Construction of Hospital and Health Care Facilities,* and many state and local codes, require that areas in use, even temporary ones, meet all safety regulations, such as for means of egress and fire detection, throughout the course of a project. If this means that emergency access must pass through a construction area, the passage of people will have to be done in stages so as not to create additional hazards. JCAHO inspectors pay particular attention to construction areas during routine surveys. NFPA 70, *National Electrical Code*® (NEC®), 2005 edition (Article 590), covers the requirements for temporary wiring. The Occupational Safety and Health Administration (OSHA) also has safety regulations for construction areas. While outside contractors are responsible for the safety conditions at the work site, OSHA can cite both the contractor and facility owners for safety violations.

35.3 Progress Reports

Periodic progress reports should be issued to the staff while work is in progress. In addition to satisfying the natural curiosity that people have toward construction projects, they also keep staff informed so that they can make their own plans for when the work is completed. These progress reports are disseminated the same way as other staff information—during department head meetings, through progress reports posted on bulletin boards throughout the facility, and through reports sent via interoffice mail.

When work is in progress, the staff may make suggestions regarding the design and the time frame. While a facilities engineer usually incorporates the useful suggestions, he must often reject the unacceptable ones, albeit without offending anyone. A prior good working relationship may help in this regard.

35.4 System Testing

All electrical, gas/vacuum, and fire protection systems are tested for a variety of reasons. Testing that is done during various stages of construction (as well as over the life of the system) to determine if the installation is proceeding correctly is referred to as *performance testing*. Testing that is conducted by an authority having jurisdiction (AHJ) to make sure the systems have been properly installed is called *acceptance testing*.

35.4.1 Performance Testing

Performance testing is usually done on smaller discrete portions of the system before the entire system is complete or before a portion is connected (slang: "tied in") to an existing system component. On completion of a project, all systems should be fully performance tested to rectify any final problems and make any last-minute adjustments before the system is inspected and acceptance tested by the AHJ. Performance testing is the only way to keep system projects on schedule so that they don't fail the acceptance testing, which subsequently affects facility use and occupancy.

The scope of a facility's performance testing should be the exact tests and procedures the AHJs use for their acceptance testing. Thus, before the performance tests are conducted, the AHJs should be contacted to determine what tests and procedures they plan to use. It is possible that the manufacturer, facilities engineer, and contractor may require other performance testing. In that case, all the performance testing must be done—that required by the AHJ for acceptance as well as the other required tests. All performance testing should be completed and witnessed by an independent third party (not the installation contractor) and also the same facility personnel who will be on site and assist the AHJ with the acceptance testing. If different personnel are on hand during the different testing processes, there can be confusion concerning the specifics of what was tested, how the testing was conducted, and what the results were. It is advisable to schedule the performance tests well in advance of the acceptance tests in order to allow enough time to make necessary repairs (which may involve ordering material) and adjustments to the system whose necessity may not be apparent until the testing commences.

35.4.2 Acceptance Testing by AHJs

Installation contractors usually conduct the acceptance testing, under the control of the AHJ. The purpose of acceptance testing is to prove to the AHJ that all aspects of the system(s) have been properly designed and installed such that the AHJ can approve the system and issue the associated certificate(s). Facilities should be prepared to prove all aspects of system installation through acceptance testing and should have all proper documentation. All approved shop drawings must be on site and available. If the system installation is properly documented (including field notes) and performance testing has been successful, acceptance testing has the greatest chance of being successful the first time. Such success saves the facility time and money; many AHJs charge for their time on site through permit and inspection fees. Also, AHJs, especially local fire marshals, often have well-defined (i.e., busy) schedules, and if the system fails the acceptance tests, the AHJ may not be able to return to the health care facility for quite some time. Facilities that have multiple AHJs involved in acceptance testing (e.g., the local fire marshal and the insurance carrier) should try to get all the AHJs to witness the acceptance testing at the same time. When all AHJs have approved the installation and systems and all paperwork is received by the facilities engineer, he can then accept the project for the facility.

35.5 Walk-Throughs

After the work has been completed, at least one meeting takes place before the opening of the area. At these meetings, the facilities people ensure that the work has been completed and staff become informed and trained on how to occupy and use the area. It usually helps to have a walk-through with all parties concerned. Although it is easy to include the clinical staff in walk-throughs, the service staff also will be using the area and may have useful comments. It is more advantageous to incorporate any necessary corrective actions while the area is still unoccupied.

35.6 Shakedown Period

After the staff occupies the area, there is a shakedown period when the staff who use the area, as well as the facilities engineer and/or facility representative, create a *punch list* of work that still needs to be done. During this time, the facilities engineering department receives requests to repair items that are not working properly or suggestions for even further changes. In most projects, the punch list will be started before occupancy. Minor punch list items can be completed without inconveniencing the staff and patients. If an outside contractor is involved with the punch list, then the facilities engineer must coordinate the work, since in all probability the area in question will be occupied. More involved changes may also be made if they are acceptable to the administration and do not hurt the budget. If these larger changes are not done, the facilities engineer again must don the diplomat's hat.

There are times when the staff is surprised by the results of a new project, both positively and negatively. When the facilities engineer receives criticisms of the new area, he may not have been responsible for the bad decision, but it is never good policy to put the blame on another department's decision.

As can be seen, good communications are necessary for the successful completion of a construction project. In most cases the facilities engineer continues to work with the staff involved with the project. Relationships established or developed during the project often carry over to other projects. One of the most important things to remember is that everyone is happy when a project is completed on time and within budget.

36

Health Care Facility Training

Health care facilities are dynamic and constantly changing. New equipment is installed, old equipment is updated, rooms are renovated, and the use of various areas changes from year to year. Yet underlying these dynamics is the fact that there are safety precautions that must be taken at all times in all conditions to reduce hazards. All personnel must know these precautions as well as keep up to date on how to use and maintain the ever-changing facility equipment and infrastructure.

Training health care facility personnel has two facets. One type of training occurs when the engineering department trains staff members from other departments on general facility safety and the proper use of some equipment. The second type of training involves the continuing education and training of the engineering staff regarding basic equipment, as well as complex electrical and gas/vacuum systems. Both types of training are essential for operating a safe health care facility. Putting to use even such straightforward information as how to use electrical plugs and outlets properly may eliminate a hazard or even save a life.

Many institutions have a separate department responsible for training, with the primary function of ensuring that training is completed for all relevant staff and is in accordance with all applicable codes and standards and the goals of the facility. These departments typically draw on in-house staff to teach the actual training sessions. Other facilities have less formal arrangements for staff training or use outside training consultants. No matter what type of program is in place, staff training and continuing education are constantly needed as hospital personnel changes, new equipment is installed, and old equipment is updated.

36.1 Training for All Personnel

All new staff need a facility overview and orientation. Some staff will also receive emergency training and become members of the disaster plan team.

Health care facility orientation programs for new employees provide one of the few opportunities for discussing general safety measures related to electrical, building construction, life safety, fire prevention and protection, and gas/vacuum systems. The engineering department and clinical engineering department should be involved in presenting information in all orientation programs. The following is a suggested list of topics that can be covered. The facilities engineer can tailor the topics to the staff in attendance and the time allowed for the presentation.

- The safe use of electricity
- Description of ground fault circuit interrupters (GFCIs) (see Chapter 7)
- When it is permitted to use a circuit breaker as a switch
- What to do when a circuit breaker, fuse, or GFCI opens
- The proper way to insert and remove a plug from a receptacle
- Proper use of medical gas outlets and vacuum inlets
- What to do and what not to do with thermostats
- Information on other items needed to protect the facility
- How to report defective equipment, such as heaters and air conditioners, leaky faucets, exit lights that are out, cracked receptacles, lightbulbs that need to be replaced, frayed electrical cords, snap switches that don't work, and other pieces of equipment that are malfunctioning
- What is an emergency condition and how to report it
- General facility safety
- Fire safety including the proper use of fire extinguishers
- The proper use of medical electronic equipment

The Joint Commission on Accreditation of Healthcare Organizations (JCAHO) and many local health departments require many types of health care facilities to have a disaster plan. To meet these standards, the plan must have a disaster team made up of staff from many departments, who are trained to know what actions to take during emergencies or disasters. Even if not required by law, almost every health care facility, even the smallest, should have some type of plan in effect, with trained personnel. In a small private physician's office, the plan can be as simple as knowing where flashlights and battery-powered radios are stored, how to keep the patients calm, and what telephone calls to make. One of the administrators is usually responsible for the entire disaster plan as well as for holding the required drills. The same administrator will be responsible for activating the plan and will be in charge while the plan is active. JCAHO requires a disaster drill semiannually. The administrator will coordinate all of the departmental plans into one cohesive document. Each department head is required to write the disaster plan for that department and to ensure that the department staff is knowledgeable about the plan as well as trained in what to do should a disaster occur. Disasters can be either internal or external. Typical items that may be included in the engineering department internal disaster plan include loss of utilities such as electricity, water, and gas; loss of heat; fire or explosion; bomb threats; and destruction of part of the facility. An external disaster is called when there has been a major accident and many patients are being brought into the emergency room and the hospital. The engineering department usually has a support role here to ensure that the facility is functioning and necessary utilities are available and to add manpower support where necessary.

It is important to keep records of all personnel attending training sessions, such as through the use of attendance or sign-in sheets. It is a way for department heads to ensure that all personnel have received timely training. These records may be reviewed by JCAHO when they visit the facility to ensure that proper training has been given to all personnel.

When delivering a training session, it is important to talk to people on their level and make them part of the solution, not treat them as part of the problem. Make sure the people in the audience understand what is being said and answer all questions. Some participants may be knowledgeable about some of the topics covered in training sessions, such as those who have attended similar sessions in other health care facilities where they may have been employed. However, each facility is unique in what it must communicate regarding the safety of the facility, and there most likely will be many important topics that participants must become informed about to enhance facility safety.

36.2 In-Service Training and Continuing Education for the Engineering Staff

In-service training is one of the best forms of preventive maintenance. Staff who are taught how to use and maintain equipment properly are less likely to misuse equipment, which should lead to fewer repairs. A large majority of service calls on medical equipment can be attributed to either misuse of equipment or lack of knowledge on how to use it properly.

All members of the engineering department must receive complete in-service training on the operation and maintenance of each new piece of equipment that is put into use in a health care facility. This includes nonmedical equipment such as air-handling units, pumps, and the like. If there is a separate biomedical engineering department, then its staff will receive the training on the medical equipment. Some manufacturers offer training schools at convenient locations around the country, while other manufacturers send their staff to the facility to do the training. The requirement for an equipment manufacturer to provide in-service training should be part of all equipment purchase orders (see Chapter 11).

When new equipment is placed into operation, there is an on-the-job instructional period for equipment operators (often the clinical staff) and maintenance personnel (i.e., from the engineering department). In most cases the equipment manufacturer will provide the first in-service training. Properly performed, in-service training means constant give-and-take of information among all staff members involved, which facilitates thorough training for all.

JCAHO requires the staff responsible for the use and maintenance of medical equipment to be trained. This in-service training should include the following topics, which again can be tailored to the facility and piece of equipment:

- The purpose of the equipment
- What the equipment can and cannot do
- How to operate it
- Basic safety in its use

JCAHO also requires health care facilities to provide personnel in every department with continuing education on the use and maintenance of equipment. The manner in which the facility fulfills this requirement is left to the facility. Most engineering departments provide this training via periodic department meetings, portions of which are devoted to continuing education regarding equipment operations and repairs. Relationships with other departments and patients as well as safety should also be topics for continuing education. Additionally, representatives from equipment manufacturers can be invited to these meetings to instruct the staff on the use of the existing equipment and talk about equipment updates and new equipment that is available.

36.2.1 Trained Staff Availability

Many codes and standards require that qualified personnel be responsible for the operation, maintenance, and testing of the equipment. As an example, NFPA 99, *Standard for Health Care Facilities*, 2005 edition, includes a requirement (Sections 4.4.4.1.1.2(C), 4.5.4.1.1.2, and 4.6.4.1.1.2) that *competent personnel* perform scheduled tests of the essential electrical system (see Chapter 8 of this book). These tests can then be used to train personnel on operating procedures. Unfortunately, there is no definition or set of requirements for "competent personnel." Thus, it is suggested that all personnel assigned to test equipment be thoroughly knowledgeable in the operation and maintenance of the equipment under normal conditions and, equally as important, know how to correct a malfunction.

The preferred staffing arrangement for health care facilities is to have someone who is thoroughly familiar with the operation and maintenance of the equipment on duty at all times. If this is not possible, then a qualified person should be available by telephone at all times and geographically located so as to be able to return quickly to the facility.

Unfortunately, many facilities have only one person on staff who is thoroughly familiar with the operation of the essential electrical system, including the generators, transfer switches, and other equipment. If that person is not on duty or immediately available when the malfunction occurs, there can be a delay in restoring emergency power until that person is contacted by telephone. He will have to give instructions to the staff on duty over the telephone or if necessary come into the facility and perform the necessary work personally. In many cases, emergency repairs could have been made immediately to restore a minimal amount of power if the personnel on duty had been properly trained.

It is unfortunate that training and continuing education are functions that are among the first to be eliminated when time and money are scarce. Similar to preventive maintenance, the value of training and education is often misunderstood. If done correctly and in a timely fashion, both can save money and provide added safety in the long run, as shown by the following case study.

Proper and Improper Staff Training

One facility installed 12 new monitors in the 12 operating rooms of an operating suite. One evening, the director of the clinical engineering department received a frantic call at home from the anesthesiologist on duty, stating that none of the 12 monitors was functioning properly. Luckily, it did not take long for the clinical engineering director to determine that the anesthesiologist was turning the gain control on the front face of the monitor to the point where the trace was enlarged and off the screen. The amount of time lost was minimal, since all members of the clinical engineering department had been properly trained. Training the anesthesiologists in using the equipment was another issue, however.

Selected NFPA Definitions

Alternate Power Source One or more generator sets, or battery systems where permitted, intended to provide power during the interruption of the normal electrical service, or the public utility electrical service intended to provide power during interruption of service normally provided by the generating facilities on the premises. [NFPA 99, 2005 edition]

Ampacity The current, in amperes, that a conductor can carry continuously under the conditions of use without exceeding its temperature rating. [NFPA 70, 2005 edition]

Anesthetizing Location Any area of a facility that has been designated to be used for the administration of nonflammable inhalation anesthetic agents in the course of examination or treatment, including the use of such agents for relative analgesia. [NFPA 99, 2005 edition]

Apartment Building A building or portion thereof containing three or more dwelling units with independent cooking and bathroom facilities. [*NFPA 5000*, 2006 edition]

Approved Acceptable to the authority having jurisdiction. [Note: The National Fire Protection Association does not approve, inspect, or certify any installations, procedures, equipment, or materials, nor does it approve or evaluate testing laboratories. In determining the acceptability of installations, procedures, equipment, or materials, the authority having jurisdiction may base acceptance on compliance with NFPA or other appropriate standards. In the absence of such standards, said authority may require evidence of proper installation, procedure, or use. The authority having jurisdiction may also refer to the listings or labeling practices of an organization that is concerned with product evaluations and is thus in a position to determine compliance with appropriate standards for the current production of listed items.] [NFPA Official Definition]

Assembly Occupancy An occupancy (1) used for a gathering of 50 or more persons for deliberation, worship, entertainment, eating, drinking, amusement, awaiting transportation, or similar uses or (2) used as a special amusement building, regardless of occupant load. [NFPA *101*, 2006 edition]

Authority Having Jurisdiction (AHJ) An organization, office, or individual responsible for enforcing the requirements of a code or standard or for approving equipment, materials, an installation, or a procedure. [Note: The phrase "authority having jurisdiction," or its acronym AHJ, is used in NFPA documents in a broad manner, since jurisdictions and approval agencies vary, as do their responsibilities. Where public safety is primary, the authority having jurisdiction may be a federal, state, local, or other regional department or individual such as a fire chief; fire marshal; chief of a fire prevention bureau, labor department, or health department; building official; electrical inspector; or others having statutory authority. For insurance purposes, an insurance inspection department, rating bureau, or other insurance company representative may be the authority having jurisdiction. In many circumstances, the property owner or his designated agent assumes the role of the authority having jurisdiction; at government installations, the commanding officer or departmental official may be the authority having jurisdiction.] [NFPA Official Definition]

Automatic That which provides a function without the necessity of human intervention. [NFPA *101,* 2006 edition]

Birth Center (Freestanding) A facility in which low-risk births are expected following normal, uncomplicated pregnancies and in which professional midwifery care is provided to women during pregnancy, birth, and postpartum. [NFPA *101,* 2006 edition]

Board and Care Occupancies See *Residential Board and Care Occupancy.*

Bonding The permanent joining of metallic parts to form an electrically conductive path that will ensure electrical continuity and the capacity to conduct safely any current likely to be imposed. [NFPA 70, 2005 edition]

Branch Circuit The circuit conductors between the final overcurrent device protecting the circuit and the outlet(s). [NFPA 70, 2005 edition]

Building Any structure used or intended for supporting or sheltering any use or occupancy. [NFPA *101,* 2006 edition]

Building, Existing A building erected or officially authorized prior to the effective date of the adoption of this edition of the given code by the agency or jurisdiction. [NFPA *101,* 2006 edition]

Bulk Nitrous Oxide System An assembly of equipment as described in the definition of bulk oxygen systems that has a storage capacity of more than 3200 lb (1452 kg) [approximately 28,000 ft^3 (793 m^3) (at normal temperature and pressure)] of nitrous oxide. [NFPA 99, 2005 edition]

Bulk Oxygen System An assembly of equipment such as oxygen storage containers, pressure regulators, pressure relief devices, vaporizers, manifolds, and interconnecting piping that has a storage capacity of more than 566 m^3 (20,000 ft^3) of oxygen (at normal temperature and pressure) including unconnected reserves on hand at the site. [NFPA 99, 2005 edition]

Bulk System An assembly of equipment, such as storage containers, pressure regulators, pressure relief devices, vaporizers, manifolds, and interconnecting piping, that terminates at the source valve of oxygen or 1452 kg (3200 lb) of nitrous oxide including unconnected reserves on the site. [NFPA 99, 2005 edition]

Business Occupancy An occupancy used for the transaction of business other than mercantile. [NFPA *101,* 2006 edition]

Bypass-Isolation Switch A manually operated device used in conjunction with an automatic transfer switch to provide a means of directly connecting load conductors to a power source and disconnecting the automatic transfer switch. [NFPA 110, 2005 edition]

Common Path of Travel The portion of exit access that must be traversed before two separate and distinct paths of travel to two exits are available. [Note: Paths that merge are common paths of travel.] [NFPA *101*, 2006 edition]

Container Any vessel of 450 L (119 gal) or less capacity used for transporting or storing liquids. [NFPA 30, 2003 edition]

Critical Branch A subsystem of the emergency system consisting of feeders and branch circuits supplying energy to task illumination, special power circuits, and selected receptacles serving areas and functions related to patient care that are connected to alternate power sources by one or more transfer switches during interruption of the normal power source. [NFPA 99, 2005 edition]

Critical Radiant Flux The level of incident radiant heat energy on a floor-covering system at the most distant flameout point. [NFPA *101*, 2006 edition]

Critical System A system of feeders and branch circuits in nursing homes and custodial care facilities arranged for connection to the alternate power source to restore service to certain critical receptacles, task illumination, and equipment. [NFPA 99, 2005 edition]

Cylinder A supply tank containing high-pressure gases or gas mixtures at pressures that can be in excess of 2000 psi gauge (13.8 kPa gauge). [NFPA 99, 2005 edition]

Day Care Occupancies An occupancy in which four or more clients receive care, maintenance, and supervision, by other than their relatives or legal guardians, for fewer than 24 hours per day. [NFPA *101*, 2006 edition]

Detention and Correctional Occupancies An occupancy used to house one or more persons under varied degrees of restraint or security where such occupants are mostly incapable of self-preservation because of security measures not under the occupants' control. [NFPA *101*, 2006 edition]

Disaster Within the context of NFPA 99, *Standard for Health Care Facilities,* a disaster is defined as any unusual occurrence or unforeseen situation that seriously overtaxes or threatens to seriously overtax the routine capabilities of a health care facility. [NFPA 99, 2005 edition]

Disconnecting Means A device or group of devices or other means by which the conductors of a circuit can be disconnected from their source of supply. [NFPA 70, 2005 edition]

Dormitory A building or a space in a building in which group sleeping accommodations are provided for more than 16 persons, who are not members of the same family, in one room or a series of closely associated rooms under joint occupancy and single management, with or without meals but without individual cooking facilities. [NFPA *101*, 2006 edition]

Double-Insulated Appliances Appliances where the primary means of protection against electrical shock is not grounding. The primary means is by the use of combinations of insulation and separation spacings in accordance with an approved standard. [NFPA 99, 2005 edition]

Educational Occupancy An occupancy used for educational purposes through the twelfth grade by six or more persons for 4 or more hours per day or more than 12 hours per week. [NFPA *101*, 2006 edition]

Electrical Life Support Equipment Electrically powered equipment whose continuous operation is necessary to maintain a patient's life. [NFPA 99, 2005 edition]

Emergency Power Supply (EPS) The source of electric power of the required capacity and quality for an emergency power supply system (EPSS). [NFPA 110, 2005 edition]

Emergency Power Supply System (EPSS) A complete, functioning EPS system coupled to a system of conductors, disconnecting means and overcurrent protective devices, transfer switches, and all control, supervisory, and support devices up to and including the load terminals of the transfer equipment needed for the system to operate as a safe and reliable source of electric power. [NFPA 110, 2005 edition]

Emergency System A system of circuits and equipment intended to supply alternate power to a limited number of prescribed functions vital to the protection of life and safety. [NFPA 99, 2005 edition]

Equipment Grounding Bus A grounding terminal bus in the feeder circuit of the branch circuit distribution panel that serves a particular area. [NFPA 99, 2005 edition]

Equipment System A system of feeders and branch circuits arranged for delayed, automatic, or manual connection to the alternate power source and that serves primarily three-phase power equipment. [NFPA 99, 2005 edition]

Essential Electrical System A system composed of alternate sources of power and all connected distribution systems and ancillary equipment, designed to ensure continuity of electrical power to designated areas and functions of a health care facility during disruption of normal power sources and also to minimize disruption within internal wiring systems. [NFPA 99, 2005 edition]

Existing That which is already in existence on the date a given code or standard goes into effect. [NFPA *101*, 2006 edition]

Exit That portion of a means of egress that is separated from all other spaces of a building or structure by construction or equipment as required to provide a protected way of travel to the exit discharge. [Note: Exits include exterior exit doors, exit passageways, horizontal exits, exit stairs, and exit ramps.] [NFPA *101*, 2006 edition]

Exit Access That portion of a means of egress that leads to an exit. [NFPA *101*, 2006 edition]

Exit Discharge That portion of a means of egress between the termination of an exit and a public way. [NFPA *101*, 2006 edition]

Exposed Conductive Surfaces Those surfaces that are capable of carrying electric current and that are unprotected, uninsulated, unenclosed, or unguarded, permitting personal contact. [NFPA 99, 2005 edition]

Fault Current A current in an accidental connection between an energized and a grounded or other conductive element resulting from a failure of insulation, spacing, or containment of conductors. [NFPA 99, 2005 edition]

Selected NFPA Definitions

Feeder All circuit conductors between the service equipment, the source of a separately derived system, or other power supply source and the final branch-circuit overcurrent device. [NFPA 70, 2005 edition]

Fire Alarm Control Unit (Panel) A system component that receives inputs from automatic and manual fire alarm devices and might supply power to detection devices and to a transponder(s) or off-premises transmitter(s). The control unit might also provide transfer of power to the notification appliances and transfer of condition to relays or devices connected to the control unit. The fire alarm control unit can be a local fire alarm control unit or a master control unit. [NFPA 72, 2002 edition]

Fire Alarm System A system or portion of a combination system that consists of components and circuits arranged to monitor and annunciate the status of fire alarm or supervisory signal-initiating devices and to initiate the appropriate response to those signals. [*NFPA 72, 2002 edition*]

Fire Barrier A continuous membrane or a membrane with discontinuities created by protected openings with a specified fire protection rating, where such membrane is designed and constructed with a specified fire resistance rating to limit the spread of fire and that also restricts the movement of smoke. [NFPA *101,* 2006 edition]

Fire Barrier Wall A wall, other than a fire wall, that has a fire resistance rating. [NFPA *101,* 2006 edition]

Fire Compartment A space within a building that is enclosed by fire barriers on all sides, including the top and bottom. [NFPA *101,* 2006 edition]

Fire Protection Rating The time, in minutes or hours, that materials and assemblies used as opening protection have withstood a fire exposure as established in accordance with test procedures of NFPA 252, *Standard Methods of Fire Tests of Door Assemblies,* and NFPA 257, *Standard on Fire Test for Window and Glass Block Assemblies,* as applicable. [NFPA 850, 2005 edition]

Fire Resistance Rating The time, in minutes or hours, that materials or assemblies have withstood a fire exposure as established in accordance with the test procedures of NFPA 251, *Standard Methods of Tests of Fire Resistance of Building Construction and Materials.* [NFPA 850, 2005 edition]

Flame Spread The propagation of flame over a surface. [NFPA *101,* 2006 edition]

Governing Body The person or persons who have the overall legal responsibility for the operation of a health care facility. [NFPA 99, 2005 edition]

Ground A conducting connection, whether intentional or accidental, between an electrical circuit or equipment and the earth or to some conducting body that serves in place of the earth. [NFPA 70, 2005 edition]

Ground Fault Circuit Interrupter (GFCI) A device intended for the protection of personnel that functions to de-energize a circuit or portion thereof within an established period of time when a current to ground exceeds the value established for a Class A device. [NFPA 70, 2005 edition]

Ground Fault Protection of Equipment A system intended to provide protection of equipment from damaging line-to-ground fault currents by operating to cause a disconnecting means to open all ungrounded conductors of the faulted circuit. This protection is provided at current levels lower than those required to protect conductors from damage through the operation of a supply circuit overcurrent device. [NFPA 70, 2005 edition]

Grounding System A system of conductors that provides a low-impedance return path for leakage and fault currents. [NFPA 99, 2005 edition]

Hazard Current For a given set of connections in an isolated power system, the total current that would flow through a low impedance if it were connected between either isolated conductor and ground. The various hazard currents are as follows:
  (a) *Fault Hazard Current.* The hazard current of a given isolated power system with all devices connected except the line isolation monitor.
  (b) *Monitor Hazard Current.* The hazard current of the line isolation monitor alone.
  (c) *Total Hazard Current.* The hazard current of a given isolated system with all devices, including the line isolation monitor, connected. [NFPA 99, 2005 edition]

Hazardous Area An area of a structure or building that poses a degree of hazard greater than that normal to the general occupancy of the building or structure. [Note: Hazardous areas include areas for the storage or use of combustibles or flammables; toxic, noxious, or corrosive materials; or heat-producing appliances.] [NFPA 101, 2006 edition]

Health Care Facilities Buildings or portions of buildings in which medical, dental, psychiatric, nursing, obstetrical, and/or surgical care are provided. [NFPA 99, 2005 edition]

Health Care Occupancy An occupancy used for purposes of medical or other treatment or care of four or more persons where such occupants are mostly incapable of self-preservation because of age, physical or mental disability, or security measures not under the occupants' control. [NFPA 101, 2006 edition]

High-Rise Building A building where the floor of an occupiable story is greater than 75 ft (23 m) above the lowest level of fire department vehicle access. [NFPA 101, 2006 edition]

Horizontal Exit A way of passage from one building to an area of refuge in another building on approximately the same level, or a way of passage through or around a fire barrier to an area of refuge on approximately the same level in the same building, that affords safety from fire and smoke originating from the area of incidence and areas communicating therewith. [NFPA 101, 2006 edition]

Hospital A building or part thereof used on a 24-hour basis for the medical, psychiatric, obstetrical, or surgical care of four or more inpatients. [NFPA 101, 2006 edition]

Hotel A building or groups of buildings under the same management in which there are sleeping accommodations for more than 16 persons and that are primarily used by transients for lodging with or without meals. [NFPA 101, 2006 edition]

Impedance Impedance is the ratio of the voltage drop across a circuit element to the current flowing through the same circuit element. The unit of impedance is the ohm. [NFPA 99, 2005 edition]

Industrial Occupancy An occupancy in which products are manufactured or in which processing, assembling, mixing, packaging, finishing, decorating, or repair operations are conducted. [NFPA *101*, 2006 edition]

Invasive Procedure Any procedure that penetrates the protective surfaces of a patient's body (i.e., skin, mucous membrane, cornea) and that is performed with an aseptic field (procedural site). [Not included in this category are placement of peripheral intravenous needles or catheters used to administer fluids and/or medications, gastrointestinal endoscopies (i.e., sigmoidoscopies), insertion of urethral catheters, and other similar procedures.] [NFPA 99, 2005 edition]

Isolated Patient Lead A patient lead whose impedance to ground or to a power line is sufficiently high that connecting the lead to ground, or to either conductor of the power line, results in current flow below a hazardous limit in the lead. [NFPA 99, 2005 edition]

Isolated Power System A system comprising an isolating transformer or its equivalent, a line isolation monitor, and its ungrounded circuit conductors. [NFPA 99, 2005 edition]

Isolation Transformer A transformer of the multiple-winding type, with the primary and secondary windings physically separated, that inductively couples its secondary winding to the grounded feeder system that energizes its primary winding. [NFPA 99, 2005 edition]

Labeled Equipment or materials to which has been attached a label, symbol, or other identifying mark of an organization that is acceptable to the authority having jurisdiction and concerned with product evaluation, that maintains periodic inspection of production of labeled equipment or materials, and by whose labeling the manufacturer indicates compliance with appropriate standards or performance in a specified manner. [NFPA Official Definition]

Life Safety Branch A subsystem of the emergency system consisting of feeders and branch circuits, meeting the requirements of Article 700 of NFPA 70, *National Electrical Code*, and intended to provide adequate power needs to ensure safety to patients and personnel and that is automatically connected to alternate power sources during interruption of the normal power source. [NFPA 99, 2005 edition]

Limited Care Facility A building or portion of a building used on a 24-hour basis for the housing of four or more persons who are incapable of self-preservation because of age, physical limitations due to accident or illness, or limitations such as mental retardation/developmental disability, mental illness, or chemical dependency. [NFPA *101*, 2006 edition]

Limited-Combustible Refers to a building construction material not complying with the definition of noncombustible (see Section 3.3.150.3 of NFPA *101*, 2006 edition) that, in the form in which it is used, has a potential heat value not exceeding 3500 Btu/lb (8141 kJ/kg), where tested in accordance with NFPA 259, *Standard Test Method for Potential Heat of Building Materials,* and includes either of the following: (1) materials having a structural base of noncombustible material, with a surfacing not exceeding a thickness of 1/8 in. (3.2 mm) that has a flame spread index not greater than 50; (2) materials, in the form and thickness used, having neither a flame spread index greater than 25 nor evidence of continued progressive combustion, and of such composition that surfaces that would be exposed by cutting through the material on any plane would have neither a flame spread index greater than 25 nor evidence of continued progressive combustion. [NFPA *101,* 2006 edition]

Line Isolation Monitor A test instrument designed to continually check the balanced and unbalanced impedance from each line of an isolated circuit to ground and equipped with a built-in test circuit to exercise the alarm without adding to the leakage current hazard. [NFPA 99, 2005 edition]

Listed Equipment, materials, or services included in a list published by an organization that is acceptable to the authority having jurisdiction and concerned with evaluation of products or services, that maintains periodic inspection of production of listed equipment or materials or periodic evaluation of services, and whose listing states that either the equipment, material, or service meets appropriate designated standards or has been tested and found suitable for a specified purpose [Note: The means for identifying listed equipment may vary for each organization concerned with product evaluation; some organizations do not recognize equipment as listed unless it is also labeled. The authority having jurisdiction should utilize the system employed by the listing organization to identify a listed product.] [NFPA Official Definition]

Manifold A device for connecting the outlets of one or more gas cylinders to the central piping system for that specific gas. [NFPA 99, 2005 edition]

Manufactured Assembly A factory-assembled product designed for aesthetics or convenience that contains medical gas or vacuum outlets, piping, or other devices related to medical gases. [NFPA 99, 2005 edition]

Medical Air For purposes of NFPA 99, medical air is air that is supplied from cylinders, bulk containers, or medical air compressors or has been reconstituted from oxygen USP and oil-free, dry nitrogen NF. [Note: Air supplied from on-site compressor and associated air treatment systems (as opposed to medical air USP supplied in cylinders) that complies with the specified limits is considered medical air. Hydrocarbon carryover from the compressor into the pipeline distribution system could be detrimental to the safety of the end user and to the integrity of the piping system. Mixing of air and oxygen is a common clinical practice, and the hazards of fire are increased if the air is thus contaminated. Compliance with these limits is thus considered important to fire and patient safety. The quality of local ambient air should be determined prior to its selection for compressors and air treatment equipment.] [NFPA 99, 2005 edition]

Medical Air Compressor A compressor that is designed to exclude oil from the air stream and compression chamber and that does not under normal operating conditions or any single fault add any toxic or flammable contaminants to the compressed air. [NFPA 99, 2005 edition]

Medical/Dental Office A building or part thereof in which the following occur: (1) Examinations and minor treatment/procedures are performed under continuous supervision of a medical/dental professional; (2) only sedation or local anesthesia is involved and treatment or procedures do not render the patient incapable of self-preservation under emergency conditions; and (3) overnight stays for patients or 24-hour operation are not provided. [NFPA 99, 2005 edition]

Medical Gas A patient medical gas or medical support gas. (See also definitions for *Patient Medical Gas* and *Medical Support Gas.*) [NFPA 99, 2005 edition]

Medical Gas System An assembly of equipment and piping for the distribution of nonflammable medical gases such as oxygen, nitrous oxide, compressed air, carbon dioxide, and helium. [NFPA 99, 2005 edition]

Medical Support Gas Piped gases such as nitrogen and instrument air that are used to support medical procedures by operating medical and surgical tools, equipment booms, pendants, and similar medical support applications. [NFPA 99, 2005 edition]

Mercantile Occupancy An occupancy used for the display and sale of merchandise. [NFPA *101*, 2006 edition]

Mixed Occupancy A multiple occupancy where the occupancies are intermingled. [Note: See NFPA *101* for details on incidental occupancies.] [NFPA *101*, 2006 edition]

Motor Control Center An assembly of one or more enclosed sections having a common power bus and principally containing motor control units. [NFPA 70, 2005 edition]

Multiple Occupancy A building or structure in which two or more classes of occupancy exist. [NFPA *101*, 2006 edition]

Noncombustible (Material) A material that, in the form in which it is used and under the conditions anticipated, does not ignite, burn, support combustion, or release flammable vapors when subjected to fire or heat. Materials that are reported as passing ASTM E 136, *Standard Test Method for Behavior of Materials in a Vertical Tube Furnace at 750°C,* shall be considered noncombustible materials. [NFPA *101*, 2006 edition]

Nonflammable Medical Gas System See *Medical Gas System.*

Nursing Home A building or portion of a building used on a 24-hour basis for the housing and nursing care of four or more persons who, because of mental or physical incapacity, might be unable to provide for their own needs and safety without the assistance of another person. [NFPA *101*, 2006 edition]

Occupancy The purpose for which a building or other structure, or part thereof, is used or intended to be used. [NFPA *101*, 2006 edition]

Operating Supply See *Piped Distribution System.*

Overcurrent Any current in excess of the rated current of equipment or the ampacity of a conductor. It may result from overload, short circuit, or ground fault. [Note: A current in excess of rating may be accommodated by certain equipment and conductors for a given set of conditions. Therefore, the rules for overcurrent protection are specific for particular situations.] [NFPA 70, 2005 edition]

Panelboard A single panel or group of panel units designed for assembly in the form of a single panel, including buses and automatic overcurrent devices, and equipped with or without switches for the control of light, heat, or power circuits, designed to be placed in a cabinet or cutout box placed in or against a wall, partition, or other support and accessible only from the front. [NFPA 70, 2005 edition]

Patient Bed Location The location of a patient sleeping bed, or the bed or procedure table of a critical care area. [NFPA 99, 2005 edition]

Patient Care Area Any portion of a health care facility wherein patients are intended to be examined or treated.
  (a) *General Care Areas.* Patient bedrooms, examining rooms, treatment rooms, clinics, and similar areas in which it is intended that the patient will come in contact with ordinary appliances such as a nurse-call system, electric beds, examining lamps, telephones, and entertainment devices.
  (b) *Critical Care Areas.* Those special care units, intensive care units, coronary care units, angiography laboratories, cardiac catheterization laboratories, delivery rooms, operating rooms, postanesthesia recovery rooms, emergency departments, and similar areas in which patients are intended to be subjected to invasive procedures and connected to line-operated, patient care–related electrical appliances. [Note: For the purpose of NFPA 99, the use of intravenous needles or catheters used to administer fluids and/or medications, endoscopes, colonoscopes, sigmoidoscopes, and urinary catheters are not considered invasive.] (See also definition of *Invasive Procedure*.) [NFPA 99, 2005 edition]

Patient Care Vicinity A space, within a location intended for the examination and treatment of patients, extending 6 ft (1.8 m) beyond the normal location of the bed, chair, table, treadmill, or other device that supports the patient during examination and treatment and extending vertically to 7 ft 6 in. (2.3 m) above the floor. [NFPA 99, 2005 edition]

Patient Equipment Grounding Point A jack or terminal that serves as the collection point for redundant grounding of electric appliances serving a patient care vicinity or for grounding other items in order to eliminate electromagnetic interference problems. [NFPA 99, 2005 edition]

Patient Lead Any deliberate electrical connection that can carry current between an appliance and a patient. [NFPA 99, 2005 edition]

Patient Medical Gas Piped gases such as oxygen, nitrous oxide, helium, carbon dioxide, and medical air that are used in the application of human respiration and the calibration of medical devices used for human respiration. [NFPA 99, 2005 edition]

Piped Distribution System A pipeline network assembly of equipment that starts at and includes the source valve, warning systems (master, area, and local alarms), bulk gas system signal actuating switch wiring, interconnecting piping, and all other components up to and including the station outlets/inlets. [NFPA 99, 2005 edition]

Piping The tubing or conduit of the system. The three general classes of piping are main lines, risers, and branch (lateral) lines.
  (a) *Main Lines.* The piping that connects the source (pumps, receivers, etc.) to the risers or branches or both.
  (b) *Risers.* The vertical pipes connecting the system main line(s) with the branch lines on the various levels of the facility.
  (c) *Branch (Lateral) Lines.* Those sections or portions of the piping system that serve a room or group of rooms on the same story of the facility. [NFPA 99, 2005 edition]

Presignal Feature (Fire Alarm) A feature that allows initial fire alarm signals to sound only in department offices, control rooms, fire brigade stations, or other constantly attended central locations and for which human action is subsequently required to activate a general alarm, or a feature that allows the control equipment to delay the general alarm by more than 1 minute after the start of the alarm processing. If there is a connection to a remote location, the transmission of the alarm signal to the supervising station shall activate upon the initial alarm signal. [*NFPA 72,* 2002 edition]

Primary Supply See *Piped Distribution System.*

Public Way A street, alley, or other similar parcel of land essentially open to the outside air and deeded, dedicated, or otherwise permanently appropriated to the public for public use and having a clear width and height of not less than 10 ft (3050 mm). [NFPA *101,* 2006 edition]

Quiet Ground A system of grounding conductors, insulated from portions of the conventional grounding of the power system, that interconnects the grounds of electric appliances for the purpose of improving immunity to electromagnetic noise. [NFPA 99, 2005 edition]

Raceway An enclosed channel of metal or nonmetallic materials designed expressly for holding wires, cables, or busbars, with additional functions as permitted in NFPA 70, *National Electrical Code®.* Raceways include, but are not limited to, rigid metal conduit, rigid nonmetallic conduit, intermediate metal conduit, liquid-tight flexible conduit, flexible metallic tubing, flexible metal conduit, electrical nonmetallic tubing, electrical metallic tubing, underfloor raceways, cellular concrete floor raceways, cellular metal floor raceways, surface raceways, wireways, and busways. [NFPA 70, 2005 edition]

Reactance The component of impedance contributed by inductance or capacitance. The unit of reactance is the ohm. [NFPA 99, 2005 edition]

Reference Grounding Point The ground bus of the panelboard or isolated power system panel supplying the patient care area. [NFPA 99, 2005 edition]

Reserve Supply See *Piped Distribution System.*

Residential Board and Care Occupancy A building or portion thereof that is used for lodging and boarding of four or more residents, not related by blood or marriage to the owners or operators, for the purpose of providing personal care services. [NFPA 101, 2006 edition]

Residential Occupancy An occupancy that provides sleeping accommodations for purposes other than health care or detention and correction. [NFPA 101, 2006 edition]

Secondary Supply See *Piped Distribution System*.

Selected Receptacles A minimal number of receptacles selected by the governing body of a facility as necessary to provide essential patient care and facility services during loss of normal power. [NFPA 99, 2005 edition]

Self-Closing Equipped with an approved device that ensures closing after opening. [NFPA 101, 2006 edition]

Service The conductors and equipment for delivering electric energy from the serving utility to the wiring system of the premises served. [NFPA 70, 2005 edition]

Service Equipment The necessary equipment, usually consisting of a circuit breaker(s) or switch(es) and fuse(s) and their accessories, connected to the load end of service conductors to a building or other structure or an otherwise designated area and intended to constitute the main control and cutoff of the supply. [NFPA 70, 2005 edition]

Shall Indicates a mandatory requirement. [NFPA Official Definition]

Should Indicates a recommendation or that which is advised but not required. [NFPA Official Definition]

Single Treatment Facility A diagnostic or treatment complex under a single management comprising a number of use points but confined to a single contiguous grouping of use points (i.e., does not involve widely separated locations or separate distinct practices). [NFPA 99, 2005 edition]

Site of Intentional Expulsion All points within 0.3 m (1 ft) of a point at which an oxygen-enriched atmosphere is intentionally vented to the atmosphere. [Note: For example, for a patient receiving oxygen via a nasal cannula or face mask, the site of expulsion normally surrounds the mask or cannula; for a patient receiving oxygen while enclosed in a canopy or incubator, the site of intentional expulsion normally surrounds the openings to the canopy or incubator; for a patient receiving oxygen while on a ventilator, the site of intentional expulsion normally surrounds the venting port on the ventilator.] [NFPA 99, 2005 edition]

Smoke Barrier A continuous membrane, or a membrane with discontinuities created by protected openings, where such membrane is designed and constructed to restrict the movement of smoke. [NFPA 101, 2006 edition]

Smoke Compartment A space within a building enclosed by smoke barriers on all sides, including the top and bottom. [NFPA 101, 2006 edition]

Sprinkler System For fire protection purposes, an integrated system of underground and overhead piping designed in accordance with fire protection engineering standards. The installation includes one or more automatic water supplies. The portion of the sprinkler system aboveground is a network of specially sized or hydraulically designed piping installed in a building, structure, or area, generally overhead, and to which sprinklers are attached in a systematic pattern. The valve controlling each system riser is located in the system riser or its supply piping. Each sprinkler system riser includes a device for actuating an alarm when the system is in operation. The system is usually activated by heat from a fire and discharges water over the fire area. [NFPA 13, 2002 edition]

Station Inlet An inlet point in a medical/surgical piped vacuum distribution system at which the user makes connections and disconnections. [NFPA 99, 2005 edition]

Station Outlet An outlet point in a piped medical gas distribution system at which the user makes connections and disconnections. [NFPA 99, 2005 edition]

Storage Occupancy An occupancy used primarily for the storage or sheltering of goods, merchandise, products, vehicles, or animals. [NFPA *101*, 2006 edition]

Story The portion of a building located between the upper surface of a floor and the upper surface of the floor or roof next above it. [NFPA *101*, 2006 edition]

Structure That which is built or constructed. [NFPA *101*, 2006 edition]

Supply Source
 (a) *Operating Supply.* The portion of the supply system that normally supplies the piping systems. The operating supply consists of a primary supply or a primary and secondary supply.
 (b) *Primary Supply.* That portion of the source equipment that actually supplies the system.
 (c) *Reserve Supply.* Where provided, that portion of the source equipment that automatically supplies the system in the event of failure of the primary and secondary operating supplies.
 (d) *Secondary Supply.* Where provided, that portion of the source equipment that automatically supplies the system when the primary supply becomes exhausted. [NFPA 99, 2005 edition]

Surface-Mounted Medical Gas Rail Systems A surface-mounted gas delivery system intended to provide ready access for two or more gases through a common delivery system to provide multiple gas station outlet locations within a single patient room or critical care area. [NFPA 99, 2005 edition]

Switchboard A large single panel, frame, or assembly of panels on which are mounted, on the face or back or both, switches, overcurrent and other protective devices, buses, and usually instruments. Switchboards are generally accessible from the rear as well as from the front and are not intended to be installed in cabinets. [NFPA 70, 2005 edition]

Task Illumination Provisions for the minimum lighting required to carry out the necessary tasks in the areas described in Chapter 4 of NFPA 99, including safe access to supplies and equipment and access to exits. [NFPA 99, 2005 edition]

Waste Anesthetic Gas Disposal (WAGD) The process of capturing and carrying away gases vented from the patient breathing circuit during the normal operation of gas anesthetic or analgesia equipment. [NFPA 99, 2005 edition]

Wet Location The area in a patient care area where a procedure is performed that is normally subject to wet conditions while patients are present, including standing fluids on the floor or drenching of the work area, either of which condition is intimate to the patient or staff. [NFPA 99, 2005 edition]

Zone A defined area within the protected premises. A zone can define an area from which a signal can be received, an area to which a signal can be sent, or an area in which a form of control can be executed. [*NFPA 72, 2002 edition*]

References

NFPA Publications

National Fire Protection Association, 1 Batterymarch Park, Quincy, MA 02169-7471.
NFPA 1, *Uniform Fire Code*™, 2006 edition.
NFPA 10, *Standard for Portable Fire Extinguishers*, 2002 edition.
NFPA 12, *Standard on Carbon Dioxide Extinguishing Systems*, 2005 edition.
NFPA 13, *Standard for the Installation of Sprinkler Systems*, 2002 edition.
NFPA 14, *Standard for the Installation of Standpipe and Hose Systems*, 2003 edition.
NFPA 17, *Standard for Dry Chemical Extinguishing Systems*, 2002 edition.
NFPA 17A, *Standard for Wet Chemical Extinguishing Systems*, 2002 edition.
NFPA 25, *Standard for the Inspection, Testing, and Maintenance of Water-Based Fire Protection Systems*, 2002 edition.
NFPA 30, *Flammable and Combustible Liquids Code*, 2003 edition.
NFPA 31, *Standard for the Installation of Oil-Burning Equipment*, 2001 edition.
NFPA 33, *Standard for Spray Application Using Flammable or Combustible Materials*, 2003 edition.
NFPA 37, *Standard for Installation and Use of Stationary Combustion Engines and Gas Turbines*, 2002 edition.
NFPA 45, *Standard on Fire Protection for Laboratories Using Chemicals*, 2004 edition.
NFPA 51, *Standard for the Design and Installation of Oxygen–Fuel Gas Systems for Welding, Cutting, and Allied Processes*, 2002 edition.
NFPA 54, *National Fuel Gas Code*, 2006 edition.
NFPA 55, *Standard for the Storage, Use, and Handling of Compressed Gases and Cryogenic Fluids in Portable and Stationary Containers, Cylinders, and Tanks*, 2005 edition
NFPA 58, *Liquefied Petroleum Gas Code*, 2004 edition.
NFPA 70, *National Electrical Code*®, 2005 edition.
NFPA 72®, *National Fire Alarm Code*®, 2002 edition.
NFPA 80, *Standard for Fire Doors and Fire Windows*, 1999 edition.
NFPA 82, *Standard on Incinerators and Waste and Linen Handling Systems and Equipment*, 2004 edition.
NFPA 90A, *Standard for the Installation of Air-Conditioning and Ventilating Systems*, 2002 edition.
NFPA 91, *Standard for Exhaust Systems for Air Conveying of Vapors, Gases, Mists, and Noncombustible Particulate Solids*, 2004 edition.

References

NFPA 92A, *Standard for Smoke-Control Systems Utilizing Barriers and Pressure Differences,* 2006 edition.
NFPA 92B, *Standard for Smoke Management Systems in Malls, Atria, and Large Spaces,* 2005 edition.
NFPA 96, *Standard for Ventilation Control and Fire Protection of Commercial Cooking Operations,* 2004 edition.
NFPA 97, *Standard Glossary of Terms Relating to Chimneys, Vents, and Heat-Producing Appliances,* 2003 edition.
NFPA 99, *Standard for Health Care Facilities,* 2005 edition.
NFPA 99B, *Standard for Hypobaric Facilities,* 2005 edition.
NFPA 101®, *Life Safety Code®*, 2006 edition.
NFPA 101A, *Guide on Alternative Approaches to Life Safety,* 2004 edition.
NFPA 105, *Standard for the Installation of Smoke Door Assemblies,* 2003 edition.
NFPA 110, *Standard for Emergency and Standby Power Systems,* 2005 edition.
NFPA 111, *Standard on Stored Electrical Energy Emergency and Standby Power Systems,* 2005 edition.
NFPA 204, *Standard for Smoke and Heat Venting,* 2002 edition.
NFPA 211, *Standard for Chimneys, Fireplaces, Vents, and Solid Fuel–Burning Appliances,* 2003 edition.
NFPA 220, *Standard on Types of Building Construction,* 2006 edition.
NFPA 221, *Standard for High Challenge Fire Walls, Fire Walls, and Fire Barrier Walls,* 2006 edition.
NFPA 251, *Standard Methods of Tests of Fire Resistance of Building Construction and Materials,* 2006 edition.
NFPA 252, *Standard Methods of Fire Tests of Door Assemblies,* 2003 edition.
NFPA 253, *Standard Method of Test for Critical Radiant Flux of Floor Covering Systems Using a Radiant Heat Energy Source,* 2006 edition.
NFPA 255, *Standard Method of Test of Surface Burning Characteristics of Building Materials,* 2006 edition.
NFPA 257, *Standard on Fire Test for Window and Glass Block Assemblies,* 2000 edition.
NFPA 259, *Standard Test Method for Potential Heat of Building Materials,* 2003 edition.
NFPA 265, *Standard Methods of Fire Tests for Evaluating Room Fire Growth Contribution of Textile Wall Coverings on Full Height Panels and Walls,* 2002 edition.
NFPA 286, *Standard Methods of Fire Tests for Evaluating Contribution of Wall and Ceiling Interior Finish to Room Fire Growth,* 2006 edition.
NFPA 418, *Standard for Heliports,* 2001 edition.
NFPA 701, *Standard Methods of Fire Tests for Flame Propagation of Textiles and Films,* 2004 edition.
NFPA 750, *Standard on Water Mist Fire Protection Systems,* 2003 edition.
NFPA 2001, *Standard on Clean Agent Fire Extinguishing Systems,* 2004 edition.
Brannigan, Vincent, J. D., "Record of Appellate Courts on Retrospective Firesafety Codes," *NFPA Fire Journal* (November 1981).

Fire Alarm Signaling Systems, 2003 (3rd edition), Richard W. Bukowski and Wayne D. Moore, editors, NFPA and Society of Fire Protection Engineers.
Fire Protection Handbook, 2003 (19th edition), Arthur E. Cote, editor.
Fire Protection Systems: Inspection, Test & Maintenance Manual, 2000 (3rd edition), Wayne G. Carson and Richard L. Klinker.
"Halon and Beyond: Developing New Alternatives," *NFPA Journal* (November/December 1994), p. 41.
Health Care Facilities Handbook, 2005 (8th edition), Richard P. Bielen, editor.
Life Safety Code® Handbook, 2003 (9th edition), Ron Cote and Gregory Harrington, editors.
National Electrical Code Handbook, 2005 (10th edition), Mark W. Earley, Jeffrey S. Sargent, Joseph V. Sheehan, and John M. Caloggero, editors.
Rohr, K. D., "U.S. Experience with Sprinklers: Who Has Them? How Well Do They Work?" Research Report (1999).
The SFPE Handbook of Fire Protection Engineering, 2002 (3rd edition), Philip J. DiNenno, editor, NFPA and Society of Fire Protection Engineers.

AISI Publications

American Iron and Steel Institute, 1140 Connecticut Avenue, Suite 705, Washington, DC 20036.
Designing Fire Protection for Steel Beams (1984).
Designing Fire Protection for Steel Columns (1980).
Designing Fire Protection for Steel Trusses (1991).

ANSI Publications

American National Standards Institute, Inc., 25 West 43rd Street, 4th Floor, New York, NY 10036.
ANSI C84.1, *Electric Power Systems and Equipment—Voltage Ratings* (1995).
ANSI website, various pages, © 2000, www.ansi.org (retrieved August 2000).
ANSI Z49.1, *Safety in Welding and Cutting* (1999).
ANSI Z79.11, *Standard for Anesthetic Equipment—Scavenging Systems for Excess Anesthetic Gases* (1982).

ASA Publications

American Society of Anesthesiologists, 520 N. Northwest Highway, Park Ridge, IL 60068.
Waste Anesthetic Gases: Information for Management in Anesthetizing Areas and the Postanesthesia Care Unit (PACU) (1999).

ASME Publications

American Society of Mechanical Engineers, Three Park Avenue, New York, NY 10016-5990.
ASME/ANSI A17.1, *Safety Code for Elevators and Escalators* (2004).
ASME/ANSI A17.3, *Safety Code for Existing Elevators and Escalators* (2002).

ASME B31.1, *Power Piping* (2004).
ASME B31.3, *Process Piping* (2002).

ASSE Publications

American Society of Sanitary Engineering, 901 Canterbury Road (Suite A), Westlake, OH 44145.

ASSE 6010, *Professional Qualifications Standard for Medical Gas Systems Installers* (2001).

ASSE 6030, *Professional Qualifications Standard for Medical Gas Systems Verifiers* (2001).

ASTM Publications

American Society for Testing and Materials, 100 Barr Harbor Drive, West Conshohocken, PA 19428-2959.

ASTM B 819, *Standard Specification for Seamless Copper Tube for Medical Gas Systems* (2000).

ASTM E 119a, *Standard Test Methods for Fire Tests of Building Construction and Materials* (2000).

ASTM E 814, *Standard Test Methods for Fire Tests of Through-Penetration Fire Stops* (2002).

ASTM F 1343, *Standard Specification for Anesthetic Gas Scavenging Systems—Transfer and Receiving Systems* (2002).

ASTM F 1850, *Standard Specification for Particular Requirements for Anesthesia Workstations and Their Components* (2000, revised 2005).

ASTM website, various pages, © 2000, www.astm.org (retrieved August 2000).

CGA Publications

Compressed Gas Association, 4221 Walney Road, 5th Floor, Chantilly, VA 20151-2923.

CGA C-7, *Guide to the Preparation of Precautionary Labeling and Marking of Compressed Gas Containers* (2004).

CGA G-1, *Acetylene* (2003).

CGA G-4, *Oxygen* (1996, reaffirmed 2002).

CGA G-4.1, *Cleaning Equipment for Oxygen Service* (2004).

CGA G-4.3, *Commodity Specification for Oxygen* (2000).

CGA G-5.3, *Commodity Specification for Hydrogen* (2004).

CGA G-6, *Carbon Dioxide* (2003).

CGA G-6.5, *Standard for Small, Stationary Insulated Carbon Dioxide Supply Systems* (2001).

CGA G-6.2, *Commodity Specification for Carbon Dioxide* (2004).

CGA G-7.1, *Commodity Specification for Air* (2004).

CGA G-8.1, *Standard for the Installation of Nitrous Oxide Systems at Consumer Sites* (1990, reaffirmed 2002).

CGA G-8.2, *Commodity Specification for Nitrous Oxide* (2000).

CGA G-9.1, *Commodity Specification for Helium* (2004).

CGA G-10.1, *Commodity Specification for Nitrogen* (2004).
CGA G-11.1, *Commodity Specification for Argon* (2004).
CGA M-1, *Guide for Medical Gas Installations at Consumer Sites* (2003).
CGA O2-DIR, *Directory of Cleaning Agents for Oxygen Service* (2000)
CGA P-1, *Safe Handling of Compressed Gases in Containers* (2000).
CGA P-2.5, *Transfilling of High Pressure Gaseous Oxygen to Be Used for Respiration* (2000).
CGA P-2.6, *Transfilling of Liquid Oxygen Used for Respiration* (2005).
CGA P-2.7, *Guide for the Safe Storage, Handling, and Use of Portable Liquid Oxygen Systems in Healthcare Facilities* (2000).
CGA P-6, *Standard Density Data, Atmospheric Gases and Hydrogen* (2005).
CGA P-9, *The Inert Gases: Argon, Nitrogen and Helium* (2001).
CGA P-23, *Standard for Categorizing Gas Mixtures Containing Flammable and Nonflammable Components* (2003).
CGA P-24, *Guide to the Preparation of Material Safety Data Sheets, MSDS* (2000, reaffirmed 2002).
CGA P-24D, *Material Safety Data Sheets on Disk* (2000).
CGA V-1, *Compressed Gas Cylinder Valve Outlet and Inlet Connections* (ANSI B57.1) (2003).
CGA V-5, *Diameter Index Safety System (Non-Interchangeable Low-Pressure Connections for Medical Gas Applications)* (2005).
CGA V-6, *Standard Cryogenic Liquid Transfer Connection* (2000).
CGA website, various pages (n.d.), www.cganet.com (retrieved August 2000).

ICC Publications

International Code Council, 5203 Leesburg Pike, Suite 600, Falls Church, VA 22041.
International Building Code (2006).
International Fire Code (2006).
International Fuel Gas Code (2006).
International Plumbing Code (2006).
International Mechanical Code (2006).

IEEE Publication

Institute of Electrical and Electronics Engineers, Three Park Avenue, 17th Floor, New York, NY 10016-5997.
IEEE Std 602, *Recommended Practice for Electric Systems in Health Care Facilities—White Book* (1996).

JCAHO Publications

Joint Commission on Accreditation of Healthcare Organizations, One Renaissance Boulevard, Oakbrook Terrace, IL 60181.
Comprehensive Accreditation Manual for Hospitals: The Official Handbook (2006).
Environment of Care Handbook (2004, 2nd edition).

JCAHO website, various pages, © 2004, 2005, www.jcaho.org (retrieved December 2004; January 2005).

MSS Publications

Manufacturers Standardization Society of the Valve and Fittings Industry Inc., 127 Park Street NE, Vienna, VA 22180.
SP-58, *Pipe Hangers and Supports—Materials, Design, and Manufacture* (2002).
SP-69, *Pipe Hangers and Supports—Selection and Application* (2003).

Underwriters Laboratories Publications

Underwriters Laboratories Inc., 333 Pfingsten Road, Northbrook, IL 60062-2096.
UL 9, *Standard for Fire Tests of Window Assemblies* (2000).
UL 10B, *Standard for Fire Tests of Door Assemblies* (1997).
UL 197, *Standard for Commercial Electric Cooking Appliances* (2003).
UL 263, *Standard for Fire Tests of Building Construction and Materials* (2003).
UL 300, *Standard for Fire Testing of Fire Extinguishing Systems for Protection of Commercial Cooking Equipment* (1996).
UL 498, *Standard for Attachment Plugs and Receptacles* (2001).
UL Fire Resistance Directory (2005).
UL website, various pages, © 2000, www.ul.com (retrieved August 2000).

Other Publications

ACI 216R, *Guide for Determining the Fire Endurance of Concrete Elements,* Farmington Hills, MI: American Concrete Institute, 1989, reapproved 2001.
ADA, Americans with Disabilities Act, *Accessibility Guidelines for Buildings and Facilities* (ADAAG), Washington, DC: U.S. Department of Justice, as published in the *Federal Register,* July 26, 1991.
American Electricians' Handbook, 9th edition, Terrell Croft, Clifford C. Carr, John H. Watt, editors, New York: McGraw-Hill, 1969.
American Electricians' Handbook, 14th edition, Wilford I. Summers and Terrell Croft, editors, New York: McGraw-Hill/TAB Electronics, 2002.
American Institute of Architects (AIA), *Guidelines for Design and Construction of Hospital and Health Care Facilities*, Washington, DC: The American Institute of Architects Academy of Architecture with assistance from the U.S. Department of Health and Human Services, 2001.
American Institute of Architects website, various pages, © 1999, www. aiaonline.com (retrieved August 2000).
Analytical Methods of Determining Fire Endurance of Concrete and Masonry Members—Model Code-Approved Procedures, Skokie, IL: Concrete and Masonry Industry Firesafety Committee, 1985.
Design for Fire Resistance of Precast Prestressed Concrete, A. H. Gustaferro and L. D. Martin, Chicago: Precast/Prestressed Concrete Institute (PCI), 1977.
Design of Fire-Resistive Exposed Wood Members, Washington, DC: American Forest and Paper Association, 2001.

Engineering Manual for Wet-Pipe Kitchen Hood Protection, Form No. QU5-85, Wilsonville, OR: Gaylord Industries, Inc., 1985.

Fire, Electrical & Life Safety Compendium, Chicago, IL: American Society for Healthcare Engineering, 2002.

Fire Resistance Design Manual, Washington, DC: Gypsum Association, 2000.

Gypsum Association website, various pages, www.gypsum.org (retrieved August 2000).

Hughes Associates, Inc., Published report Baltimore, MD: January, 1995.

IESNA Lighting Handbook: Reference and Application, Mark S. Rea, editor-in-chief, New York: Illuminating Engineering Society of North America, 1999/2000.

2000 Life Safety Code Workbook and Study Guide for Health Care Facilities, James K. Lathrop and Jennifer Holloman, Marblehead, MA: Opus Communications, 2003.

OSHA Fact Sheet 91-38, *Waste Anesthetic Gas,* Washington, DC: Occupational Safety and Health Administration, 1991.

Reinforced Concrete Fire Resistance, Schaumberg, IL: Concrete Reinforcing Steel Institute (CRSI), 1980.

Standard Handbook for Electrical Engineers, 10th edition, Donald G. Fink, John M. Carroll, editors, New York: McGraw-Hill, 1969.

Standard Handbook for Electrical Engineers, 14th edition, Donald G. Fink, editor, New York: McGraw-Hill, 2001.

Index

AAMI. *See* Association for the Advancement of Medical Instrumentation
ABHR. *See* Alcohol-based hand rub cleaners
Above ceiling permits, 222
Acceptance testing
   AHJ providing, 446, 612, 613
   chassis leakage current with, 152
   sprinkler systems with, 403
   standpipe systems having, 422–424
Access control systems, 263
ADA. *See* Americans with Disabilities Act
Addressable fire alarm systems, 493–494
Administration departments, 27
Administrators, communicating with, 598–600
Agencies, outside, 604
Agent distribution, 461
AHJ. *See* Authorities having jurisdiction
AIA. *See* American Institute of Architects
Air compressor, 533–534
Air connectors, 343
Air duct
   coverings for, 343–344
   detector placement in, 322f
   inlet tube orientation in, 323f
   pendant-mounted, 322f
Air filters, 536
Air movement, 353
Air sampling system, 496
Air-atomizing nozzle, 472
Air-conditioning and ventilating systems, 9, 203
   fire safety from, 343
   maintenance provisions by, 345–346
Airports/community emergency planning, 10
Alarms
   check valves for, 382, 383f
   device testing for, 429
   facility engineer responsible for, 121–123
   for narcotics, 122
   indicators for, 123

inspecting, 401
kitchen fire-extinguishing systems with, 457–458
master, 553–554
requirements of, 109–110
security, 122–123
testing of, 408–409
Alcohol-based hand rub cleaners (ABHR), 285
Alerting system, 121
Alternating-tread device, 276f
Aluminum, 61
Ambient-compensated circuit breaker, 74
Ambulatory health care
   requirements for, 216
   travel distance in, 290
Ambulatory health care occupancies
   defined, 189–190
   outpatient treatment provided by, 20
Ambulatory patients, 205
American Institute of Architects (AIA)
   guidelines from, 11–12
   guidelines prepared by, 4, 187
   receptacles required by, 129
American National Standards Institute (ANSI), 5
American Society for Healthcare Engineering (ASHE), 31
American Society for Testing and Materials (ASTM), 5
Americans with Disabilities Act (ADA), 494
Ammeter, 57f
Ampacity, 61
Anesthetizing locations
   defined, 24
   electrical requirements for, 131
ANSI. *See* American National Standards Institute
Antifreeze loop, 386
Antifreeze sprinkler systems
   proper solution for, 409
   wet-pipe system similar to, 385–386

641

Antivibration mounting, 535
Apartment buildings, 192
Area alarm panel, 554f
Area separation wall, 224
Areas of refuge, 277
As-built drawings, 169
ASHE. *See* American Society for Healthcare Engineering
ASME/ANSI A17-1, *Safety Code for Elevators and Escalators*, 348–349
ASME/ANSI A17-3, *Safety Code for Existing Elevators and Escalators*, 348–349
Assembly occupancies
  cafeteria as, 21
  defined, 188
  egress doors in, 196
Association for the Advancement of Medical Instrumentation (AAMI), 142
ASTM. *See* American Society for Testing and Materials
ASTM E 84, *Standard Test Method for Surface Burning Characteristics of Building Materials*, 204
ASTM E 119, *Standard Methods for Fire Tests of Building Construction Materials*, 204
ASTM E 814, *Standard Test Method for Fire Tests of Through-Penetration Fire Stops*, 204
ASTM F 1343, *Standard Specification for Anesthetic Equipment-Scavenging Systems for Anesthetic Gases*, 576
ASTM F 1850, *Standard Specification for Particular Requirements for Anesthesia Workstations and Their Components*, 576
Astragal, 260
Atrium
  features of, 312t
  mini, 311, 312t
  protection requirements for, 312f
Authorities having jurisdiction (AHJ), 3, 12–13, 339
  acceptance testing by, 446, 612, 613
  facilities determining, 15
  standpipe systems installation witnessed by, 423–424
Automatic door closers, 264
Automatic sprinklers
  as heat activated, 375
  buildings protected by, 316
  gravity chutes protected by, 350f
  history of, 377–378
  inspecting, 511
  installation standards for, 454t
  misconceptions about, 376–377
  new buildings requiring, 375
  retrofitting cost for, 510
  trash/laundry chutes protected by, 350
  water flow from, 378
Automatic starting operation, 103
Automatic transfer switch, 104–105, 448

Bacteria filters, 521
Balanced doors, 266, 267f
Barriers. *See also* Fire barrier walls; Smoke barriers
  continuity of, 218–219
  fire resistance of, 223
  fire resistance rating for, 220–221
  openings/penetrations in, 229–238, 230t
  requirements for, 218, 231
  sealing penetrations in, 238–241
Bathrooms, receptacles in, 85
Battery units, 112
BCMC. *See* Board for the Coordination of Model Codes
Birthing centers
  birthing rooms in, 25
  electrical system in, 128
  variations of, 18–19
Blocking, 316
Blood bank alarms, 122
Board and care facilities, 192–193
Board for the Coordination of Model Codes (BCMC), 183
Boiler rooms, 300–302
Branch circuit conductors, 133
Branch circuits, 81
Building codes, 181–184
  building separation walls in, 224
  compartmentation types in, 218
  facility engineers determining, 184
  family of, 183–184, 184t
  fire-resistive structures and, 208–209
  history of, 181–183
  mixed occupancies with, 198
  new v. existing structures for, 199–202
  occupancy classification important to, 187–188
  other handbooks with, 203–204
  penetration protection addressed by, 240
  regionalized, 182–183
  table of heights and areas for, 206
  uniformity in, 183

Index **643**

Building construction (and safety code), 9, 10, 184, 188, 198, 202, 204, 210, 213, 214, 416
Building materials, 10, 204
Buildings, 388
  automatic sprinklers protecting, 316
  automatic sprinklers required for, 375
  compartmentation in, 218
  construction checklist for, 361–368
  conversion problems for, 359
  corridors in, 359–360
  exterior fire exposure protection for, 9
  fire evacuating, 509
  occupant load determined for, 279–285
  service equipment categories, 339
  smoke compartment in, 328
Business occupancies, 21, 193
  fire alarm required in, 338f
  hazardous area in, 329, 330f
Bypass isolation switch, 173

Cable, 59–63
  information systems requiring, 140–141
  type AC, 63f
  type MC, 64f
  types of, 59–60
Capacitive coupling, 116–117
Capacitors
  as surge protector, 55–56
  power factor correction, 56f
Capacity factors, 282t
Carbon dioxide
  advantage of, 456
  Inergen using, 466–467
Carbon dioxide extinguishing systems, 8, 372, 484
Carbon dioxide kitchen hood systems, 455
CCU. *See* Coronary care unit
Ceiling finish, 333–335
Ceiling-cavity plenums, 344
Centers for Medicare and Medicaid Services (CMS), 396
Central dictation systems, 124–125
Central vacuum cleaning systems, 582
CGA. *See* Compressed Gas Association
Chassis leakage current
  acceptance tests for, 152
  test circuit for, 152f
Check valves
  alarms with, 382
  patient gas equipment with, 583

Circuit breakers
  as switches, 88
  defined, 74
  GFCIs installed on, 90
Circuiting, 79–82
Class A hazards, 482t
Class B hazards, 483t
Class I systems, 418
Class II systems, 419
Class K fires, 458
Clean agent fire-extinguishing systems, 10, 373, 463
Clean agents, 468t
Clean fire suppression agents, 463–464
Clear width, 281–282
Clinical departments
  communications with, 603
  role of, 27
Clinical engineer
  medical equipment tested by, 143, 152
  role of, 28
Close-coupled single stage in-line pump, 433f
CMS. *See* Centers for Medicare and Medicaid Services
Code provisions, 304t
Code-making organizations
  conflicting requirements for, 111
  fire resistance determined by, 203
Codes
  combustible materials and, 214
  facilities meeting, 605
  fire barrier wall complying with, 225
  health care facilities with, 3, 4
  laboratory determining, 25–26
  NEC conflicting with, 14
  NFPA writing, 6
  organizations volunteering, 4
  resolving conflicts with, 14–15
  state v. local, 14
  wood frame construction and, 212
  X-ray equipment and, 175
Codes and standards
  conflicts/duplication among, 13–14
  creating/revising, 10–11
  cross-referencing, 11–12
  essential electrical systems with, 97–98
  fire protection systems with, 372–374
  flammable piped gas system with, 579
  health care facilities with, 7–10
  IPS with, 118
  lighting with, 92

Codes and standards (*continued*)
NFPA writing, 6
pressurized gas with, 533
training with, 615
Cogeneration plants, 48
Combined dry-pipe preaction sprinkler system, 387
Combustible cooking media, 458
Combustible materials, 214
Commercial cooking equipment, 451
Commercial cooking operations
fire protection for, 9, 331, 353–354, 373
fire safety requirements for, 351
Committees, 11
Common path of travel, 293–302, 293f, 295f
limits of, 294t
requirements for, 301, 302f
Communication systems, 123–126. *See also* Information systems
backup of, 126
installation of, 125
paging, 124
types of, 123
Communications, 595
administration and, 598–600
clinical departments and, 603
methods of, 600–603
outside agencies and, 604
safety committee and, 605
safety improved with, 135
support services and, 603–604
utility company and, 50
Compartmentation
building safety determined by, 218
fire limited by, 219f
Compressed air system
air monitored from, 537
dewpoint of, 558f
nonflammable gas systems as, 580
oxygen concentration in, 559f
Compressed Gas Association (CGA)
documents published by, 520
reference documents from, 579
standards developed by, 5
Compressed gases/cryogenic fluids, 9, 519, 524, 539, 579
Concealed spaces
combustible, 316
protecting, 315–316
Conductive flooring
flammable anesthetic agents and, 132
operating suites with, 173

Conductors
branch circuit, 133
marking, 62
operating temperatures for, 61
raceways with, 68
size of, 60
ungrounded, 75
Conduit, 68–69
Construction
communications assisting, 614
design considerations for, 387–396
sprinkler protection for, 509
Construction projects, 610
Construction types, 208–212
building height determining, 206–208
cross-reference of, 213t
determining, 214–217
fire resistance ratings and, 209t
limitations, 207t
multiple, 216–217
occupancies classifications with, 215–216
Containers, 530
Contamination, 570–571
patient harm from, 524
piping network with, 544
sources of, 536
Continuing education, 617, 618
Contractor's Material and Test Certificate, 423f–427f
Control valves, 394–395
fire department accessing, 394–395
fire-rated enclosure location for, 395f
gas flow controlled by, 544–545
sprinkler systems with, 379–380
supervisory switches attached to, 403
water damage and, 380
Convenience openings
defined, 313
two-story, 313f
Cooking areas, 451
Cooking equipment
exhaust system required by, 352
standards for, 351–354
Cooking operations, 351–352
Cooling systems, 108–109
Coronary care unit (CCU), 78, 297
Correctional/detention occupancies, 20, 190
Corridor
dead-end, 294, 295f, 296f
existing buildings with, 359–360
exit remoteness from, 293f

Index **645**

exits access from, 291
fire barriers separating, 228
fire resistance rating and, 249–250
habitable room accessing, 297, 298f
health care occupancy and, 284–285
lighting in, 93
minimum widths for, 284t
protection of, 247, 253–254, 255t
spaces open to, 249f
waiting spaces open to, 250f
Corridor doors
as self-closing/self-latching, 328
latch for, 250
requirements for, 251f, 253f
Corridor walls
drop tile ceilings and, 250
fire resistance rating for, 229
inspecting, 367
new facilities with, 251f
requirement exceptions for, 248t
unsprinklered compartment permitting, 252f
Corrosion, 399
Coupling mechanism, 541–542
Critical care areas
defined, 23
grounding systems in, 77
receptacles in, 129–130
Cross main, 411f
Cross-corridor doors, 264
Cryogenics, 530
Cultural resources, 204
Current
conditioners for, 55
LIMs monitoring, 119
Current leakage tests
medical equipment requiring, 145
NFPA 99 covering, 153
patient leads with, 146
test circuit for, 145f
test circuit (isolated) for, 147f, 153f
test circuit (isolated/nonisolated) for, 148f, 154f
test circuit (nonisolated) for, 146f, 153f
Cylinders, 518
as high pressure tanks, 528
management of, 533
NFPA 99 requirement for, 590
storage for, 586

Dampers
installation of, 456
NFPA 96 requirements for, 353–354

Day care occupancies, 21, 189
Dead-end corridors
common examples of, 295f
limitations of, 296f
rules for, 294
Deaths, 309
Deep-fat fryers, 456
Defend-in-place
evacuation v., 204–205
valve placement for, 546
Defend-in-place occupancies, 203
Delayed egress lock
egress arrangement using, 262f
panic bar used for, 261
Delivery rooms
physiological monitoring equipment in, 132–133
receptacles in, 132
Deluge sprinkler systems, 387
Design considerations
new construction with, 387–396
standpipe systems with, 417–419
Design engineers
building use information for, 388
fire system integration by, 394
Design stage, 592
Diagonal rule, 292
Diameter Index Safety System (DISS), 548
Dielectric fittings, 541
Diesel drive controller, 439f
Diesel engine drive, 437f
Diesel generator, 172
Diesel-driven pumps, 437–438
Differential clapper valve, 383
Disaster plan
health care facilities with, 43
JCAHO requiring, 616
Disconnecting devices, 72–73, 162
Disconnects, 51–52
DISS. See Diameter Index Safety System
Distribution systems
cable/wires in, 59–63
circuiting in, 79–82
one-switch feeding for, 82f
piped gas system with, 539–540
Documentation
as operation/testing log, 113f
EPSS with, 112–113
facility engineer with, 14
medical equipment repair manuals as, 146–147
Door assemblies, 10, 204, 205, 232

Door width
   minimum clearance for, 257f, 258f
   minimum requirements by occupancy for, 258t
Doors
   as common egress element, 256
   assembly occupancy egress with, 196
   balanced, 267
   clearance requirements for, 237
   compliance problems with, 244–245
   exceptions to, 259t
   fire protection rating for, 328
   horizontal sliding, 260f, 266, 267f
   inspecting, 362
   means of egress as, 261
   minimum widths for, 285, 285t
   revolving, 266
   self-closing required for, 264
   smoke barriers with, 318–319
   smoke passage resisted by, 327
   swing direction for, 259–260
   types of, 258–259
   width measured for, 256–258
Dormitories, 191–192
Double-cylinder locks, 261
Double-interlock preaction sprinkler system, 386–387
Draft stopping, 316
Drift compensation, 494
Drop tile ceilings, 250
Dry chemical extinguishing systems, 8, 373
Dry chemical kitchen hood system, 454
Dry-pipe sprinkler systems
   freezing areas using, 383–385
   hydraulic calculations for, 385
Dry-pipe valve, 385f
Ducts, 69
Dutch doors, 253, 254f

Educational occupancies, 22, 188
EES. *See* Essential electrical systems
Egress. *See also* Delayed egress lock; Means of egress
   assembly occupancy doors for, 196
   delayed egress lock for, 262f
   door width for, 257f
   doors for, 256
   exits and, 291f
   illumination of, 303–305
   impediments to, 294–296
   minimum widths for, 283t
   multistory building requiring, 277
   occupant load and, 281
   stairs for, 266–267
Egress capacity
   evaluating, 279–285
   sufficiency determined for, 283
Egress components
   capacity determined for, 282
   clear width determined for, 281–282
   minimum dimensions for, 283
Elderly care areas, 93
Electric closet, 54f
Electric motors, 434
Electrical design, 41
Electrical hazards
   avoiding, 39–43
   grounding reducing, 76
   health care facilities with, 37–39
   potential for, 38–39
   safeguarding from, 41–42
Electrical isolation
   manufacturer's test circuit measuring, 147f, 154f
   patient leads testing, 154
Electrical metallic tubing (EMT), 65
Electrical safety requirements, 9
Electrical service
   continuity lacking for, 38
   disruptions in, 49
   jurisdictions and, 51
   types of, 45–48
   workspace requirements for, 52
Electrical switches, 460
Electrical systems. *See also* Electrical hazards; Essential electrical systems; Raceways; Receptacles
   as-built drawings for, 169
   birthing centers with, 128
   capacitive coupling in, 116–117
   disconnecting devices in, 162
   disconnects in, 51–52
   faults isolated in, 41
   grounded, 116f
   licensed electricians working on, 164
   main service switch testing for, 161
   maintaining, 163–168
   power amounts delivered by, 40
   protection of, 70–76
   riser diagram for, 169
   service switch for, 82f
   system design of, 39
   testing, 161–163
   ungrounded, 117f

Index

Electrical systems and equipment, 35–178
Electrical Testing Laboratories (ETL), 7
Electricity, 524
Electronic computer/data processing equipment, 9
Electronic supervisory switches, 380
Elevators, 348–349
   communication system on, 349
   fire alarm systems in, 358
   inspecting, 363
   maintenance contractors for, 348
Emergencies
   hazardous materials and, 10
   piped gas system with, 559–560
   valve closing in, 545
Emergency departments
   categories for, 26
   electrical requirements of, 138
Emergency generators, 101–103, 448
   equipment used for, 107f
   outdoor engine for, 109f
   overloading of, 102
   records for, 175–176
   voltage requirements of, 106
Emergency lighting, 95, 303–305
   code provisions for, 304t
   inspecting, 365
   problems with, 305
   thresholds for, 304f
Emergency personnel
   fire alarm control panels alerting, 497–498
   zones and, 491–492
Emergency power supply system (EPSS)
   documentation for, 112–113
   maintenance log for, 114
Emergency response notification, 498
Emergency room, 589
Emergency/standby power systems, 9, 97–99, 106, 110–111, 112–113, 339, 343
   arrangement made for, 48
   branch circuits in, 81
   psychiatric areas using, 137
   transfer switches connecting, 103
EMT. *See* Electrical metallic tubing
Enclosed exits stairs, 344
Enclosure-compensated thermal-magnetic circuit breakers, 74
End shutoff valve, 552
End-cap, 552
End-stage filter, 537
Energy conservation, 95
Energy converters, 106–109

Engineer. *See* Design engineers; Facility engineer
Engineering department. *See* Facility engineer
Engineering staff, 617
Environmental Protection Agency (EPA), 538
EPA. *See* Environmental Protection Agency
EPSS. *See* Emergency power supply system
Equipment. *See also* Cooking equipment; Fire-extinguishing equipment; Medical equipment
   as emergency generators, 107f
   commercial cooking, 451
   cooking, 351–354, 352
   electrical supply location of, 70
   facility engineer requesting, 599–600
   fire-extinguishing, 338, 451, 474–475
   ground-fault protection for, 53
   grounding, 76
   health care facilities with, 1, 50–51, 607
   history records for, 608
   information systems, 141
   inspecting, 177–178
   lighting influencing, 134
   maintenance of, 617
   manuals for, 607
   manufacturers information for, 133
   medical, 142–159, 175
   NEC requirements for, 50–51
   new v. repair of, 599–600
   operating suites with, 132–133
   operators for, 617
   patient gas, 583
   patient piped vacuum systems with, 562–563
   physiological monitoring, 132–133
   preventive maintenance records, 608–609
   radiology having, 134
   records for, 607–608
   room for, 53f
   service, 339
   suppression systems and, 463
   testing records for, 609
   X-ray, 175
Escalators. *See* Elevators
Essential electrical systems (EESs)
   battery units powering, 112
   bypass isolation switch for, 173
   codes and standards for, 97–98
   power restored by, 555–556
   testing/maintenance for, 110–114
   types of, 98–101
   wiring of, 101

648 Index

ETL. *See* Electrical Testing Laboratories
Evacuation, 204–205
Examination rooms, 94
Exhaust systems, 9
 cleaning, 460–461
 cooking equipment requiring, 352
 inspection schedule for, 352
 requirements for, 353
Exit access corridors
 protection of, 229
 protective features of, 247f
Exit discharge, 364–365
Exit enclosures, 226f
 fire barrier created by, 226
 fire resistance rating required for, 227, 245t
 HVAC penetration prohibited for, 344
 means of egress and, 361–362
 penetrations allowed in, 246
 protecting, 244–254
 requirements for, 310
 use of, 246–247
Exit passageway, 344
 as horizontal means of egress, 273
 outside connected by, 274f
 travel distance reduced by, 274f
Exit signs, 306
 inspecting, 365
 orientation of, 307f
 specific color not required for, 308
Exit stairway, 278f
Exits, 245f
 as enclosed exits stairs, 344
 arrangement of, 286f
 corridor accessing, 291
 corridor remote from, 293f
 defined, 244
 diagonal rule for, 292
 means of egress as, 291f
 mezzanine with, 315
 number of, 285–287, 301
 protecting, 244–254
 room with single, 288f
 smoke barriers and, 287f
 smoke compartment access to, 287
 travel distance to, 288–289, 289f
Exterior exit access, 296, 297f
Exterior walls, 210–211
External disaster, 616

Facilities
 codes met by, 605
 gas outlet protection in, 549
 owners of, 459–460
 survey preparation by, 31–32
 upgrades for, 598–599
Facility engineer
 building codes determined by, 184
 clinical departments and, 603
 communications by, 597, 604
 criticism received by, 614
 documentation for, 14
 electrical loads anticipated by, 48
 facility needs communicated by, 598–599
 fire department interfacing with, 372
 fire protection systems selected by, 369
 GFCI requirement known by, 91
 intercompany relationships with, 600–601
 maintenance scheduled by, 137
 new equipment requests from, 599–600
 polarity checks by, 86
 progress report by, 611–612
 rail system installation by, 550
 readings taken by, 112
 receptacles needed by, 130
 records maintained by, 110, 119–120
 role of, 27–28
 spare parts supplied by, 165–167
 system design assisted by, 490
Factory Mutual (FM), 7
Family dwelling, 191
Faulty hazard current, 118
FE-13, 464–465
FE-227
 as liquefied compressed gas, 465–466
 safety considerations for, 466
Federal government, 11–12
Field inspection, 215
Filters, 353
Finishes, 335
Fire
 building evacuation for, 509
 fire-extinguishing equipment for, 474–475
 flame spread classifications for, 333t
 interior finish indices for, 333–334
 interior finish spreading, 332–333
 personnel response to, 507
 portable fire extinguishers suppressing, 474
 procedures for, 501–508
 sprinkler systems controlling, 379t
 systems protecting against, 369–370
Fire alarm control panels, 492f
 emergency signaling components in, 497–498
 fire detection and alarm systems, 491–494

Index   649

Fire alarm systems, 492f. *See also* Fire detection and alarm systems
  business occupancies requiring, 338f
  elevators with, 358
  installation requirements for, 121
  integrity monitored for, 491
  most occupancies requiring, 337
  requirements for, 337t
  sprinkler system interfacing with, 393–394
  visual alert systems for, 498
  water motor gong used by, 384f
Fire alarm-indicating devices, 402
Fire barrier walls, 9, 204, 216, 220f
  building separation walls as, 224
  codes complied with by, 225
  corridor separation with, 228
  exit enclosures creating, 226
  horizontal continuity and, 219
  sleeve penetrating, 240
  smoke barriers v., 318
  types of, 223–229
  vertical continuity for, 221f
Fire codes, 186–187, 372
Fire compartment, 271–272
Fire dampers, 345
Fire department
  control valves accessed by, 394–395
  health care facilities notifying, 498–499
Fire detection and alarm systems, 488
  addressable fire alarm systems as, 493–494
  alarm devices, 494–498
  alarm systems, 488
  automatic systems, 490
  fire alarm control panels, 491–494
  inspecting, 501, 513–514
  manual systems, 489, 494–495
  testing for, 499–500
Fire doors, 9, 203, 237, 318
  as horizontal sliding door, 233f
  bottom clearances for, 237t
  clearances, 236–237
  double egress, 232f
  fire protection ratings for, 231–232
  label painted over for, 235–236
  labeling, 234–236
  letter designation for, 236
  problems with, 237
  self-closing self-latching, 233–234
  single swinging, 231f
  single swinging hardware on, 232f
  surface-mounted rolling steel, 235f
  vertically sliding, 234f
  wired glass glazing limitations on, 236t
Fire escape
  outside stairs v., 275f
  restricted use for, 275–276
  stairs/ladder inspection for, 363
Fire-extinguishing equipment, 8, 338, 373, 457. *See also* Portable fire extinguishers
  fire classes for, 474–475
  fire suppressed by, 451
Fire fighters
  piped medical gas system and, 589
  standpipe assisting, 413
  two phases for, 349
Fire hazard, 332
Fire hoses, 10, 415f
Fire prevention codes, 186–187
Fire protection
  checklist for, 511–514
  vertical openings and, 310t
  vertical openings sealed for, 309–315
Fire Protection Handbook (NFPA), 242
Fire protection ratings
  doors with, 328
  fire doors with, 231–232
  labels providing, 234–235
Fire protection systems
  codes and standards for, 372–374
  components for, 371
  facility engineer selecting, 369
Fire pump controller, 435f, 436f
Fire pumps
  design considerations for, 438–440
  electrical wiring for, 436
  inspecting, 445–446, 512
  inspection/maintenance summary for, 441t–444t
  inspection/testing/maintenance for, 441–449
  location requirements for, 431–432
  maintaining, 449
  pump drivers for, 433–438
  room ventilation for, 438
  testing, 446–448
  types of, 431–433
Fire resistance
  barrier for, 224
  code-making organizations determining, 203
  field inspections determining, 215
  fire exposure withstood for, 205
  heavy timber construction and, 211

650 Index

Fire resistance (*continued*)
   test determining, 223
   time-temperature curve for, 206f
Fire resistance rating
   corridor walls with, 229
   corridors and, 249–250
   exit enclosures requiring, 245t
   self-closing door required for, 264
   smoke barriers with, 318
Fire safety plan, 507–508
Fire safety symbols, 9
Fire windows, 9, 203, 237, 238, 318
Fire resistance ratings, 210
   barrier requirements for, 221
   construction types and, 209t
   conversions for, 213
Fire-resistive structures, 208–209
Fittings, 398–400, 411f, 541–542, 565
Fixed-baffle hood assemblies, 457
Fixed-temperature devices, 495
Fixed-temperature line-type detector, 495
Flame spread classifications, 333, 334
Flammable anesthetizing agents, 131–132
Flammable gas systems
   codes and standards for, 579
   laboratories with, 580
Flammable/combustible liquids, 8
Flexible metallic conduit (FMC), 65–66
Flexible nonmetallic tubing, 66
Floors
   covering systems for, 10, 237, 336
   finish limitations for, 336t
   vertical openings in, 310f
Flow requirements, 587
Flow testing device, 428f
Fluorescent lamp interference, 94
FM. *See* Factory Mutual
FMC. *See* Flexible metallic conduit
Foam blanket, 451
Food preparation facilities, 352
Furnishings
   curtains/drapes regulated, 336–337
   nonsprinklered facilities and, 337
Fuse-puller-type switch, 72
Fuses, 71–73
Fusible link, 233–234, 454, 461

GA. *See* Gypsum Association
Gas, pressurized
   codes and standards for, 533
   different pressures for, 543
   hazards from, 515
   patient bedrooms with, 549
   performance criteria for, 532–533
   services from, 515
   supplying, 528
Gas source room, 537–538
Gaseous fire suppression agents, 464
Gaseous systems, 468–469
Gauges, 400, 408
General care areas
   defined, 23
   receptacles in, 128–129
Generator room, 102
Generator sets, 111–112
Generators. *See* Emergency generators
GFCI. *See* Ground-fault circuit interrupters
Gift shops, 331–332
Gooseneck piping, 391
Gravity chutes, 350f
Grease buildup, 460t
Grease fires
   king hood extinguishing systems suppressing, 450
   massive cooling for, 455
Gross area, 281
Gross/net area, 281f
Ground-fault circuit interrupters (GFCIs), 71, 75
   duplex receptacle with, 90f
   personnel protected by, 89–91
   testing of, 91
Ground-fault current, 131
Ground-fault protection systems
   equipment protected by, 53
   main service switch and, 174
   NEC articles for, 50–51
   NFPA 70 requiring, 162
   ungrounded conductors opened with, 75
Grounding, 76–79
   information systems equipment require, 141
   low-impedance return for, 41–42
   patient care areas testing, 162–163
   patient care requiring, 77
   quiet, 78–79
   reliability, 566
   requirements for, 76–77
Guards, 268, 269
Guidelines, 4
Guidelines for the Design and Construction of Hospital and Health Care Facilities (AIA), 17, 84
Gypsum Association (GA), 5, 221
Gypsum board, 214

Index **651**

Habitable room, 297, 298f
Halocarbon clean agent, 468–469
Halogenated agent extinguishers, 485
Halon 1301, 8, 464
Handrails
   ramps requiring, 272–273
   requirements for, 268
Harmonic waveforms, 57
Hazard currents, 120–121
Hazardous area, 331t
   as moving targets, 332
   business occupancies with, 330f
   defined, 327
   gift shops as, 331–332
   locker rooms not considered as, 358–359
   new v. existing facility, 331
   protecting, 327–332, 328–329, 330–332
   storage areas as, 284
   trash/laundry rooms considered, 331–332, 350
Hazardous locations, 26
   defined, 227–228
   electrical requirements in, 135
   exceptions for, 228
   flammable anesthetizing agents in, 131–132
   occupancy influencing, 194
   protection of, 227f
   requirements for, 200–201
Hazardous materials, 10, 195
Hazardous occupancies, 194–195
Hazards
   clean fire suppression agents for, 463–464
   design engineer considering, 391–392
   hospitals with, 519
   inspecting, 366
   minimizing, 540
Headwall, 78f
Health care facilities, 112, 118, 127, 131, 138, 144, 331, 524, 537, 544, 618. *See also* Limited care facilities; Nursing homes
   AHJ determined by, 15
   AIA *Guidelines* for, 187
   alerting systems specified in, 121
   areas categorized in, 22–27
   building occupancies in, 20–22
   categories of, 16–20
   classification by, 99
   codes specific to, 3, 4
   current leakage tests from, 153
   cylinder requirements from, 590

   departments in, 27–28
   disaster plan for, 43
   education not typical in, 188
   electrical hazards in, 37–39
   emergency generator requirements by, 102
   equipment layout/service for, 50–51
   equipment manuals kept by, 607
   essential electrical system requirements from, 98
   fire department notification for, 498–499
   horizontal exit in, 272f, 273f
   hospital defined in, 17
   hyperbaric defined by, 19
   independent, power requirements, 101
   inspecting, 29
   JCAHO standards for, 6
   Life Safety Code® compliance for, 185
   multiple occupancies for, 196–197
   NFPA codes and standards for, 7–10
   occupant load factors for, 280t
   personnel training from, 615
   project planning for, 610–611
   smoke compartment in, 326f
   system/equipment requirements from, 8
   systems/equipment in, 1
   travel distance in, 290t
Health care occupancies
   defined, 189
   mini-atrium not permitted in, 311
   self-preservation incapability for, 20
   shafts required by, 310
   suites in, 297–300
Heat exchange-cooled diesel-driven pump, 438f
Heating, ventilation, and air-conditioning equipment (HVAC), 343–347
   penetrations for, 320–321
   shafts for, 345
   systems, 239f
Heavy timber structures
   components of, 211f
   exterior walls in, 210–211
   fire resistance and, 211
Heliports, 10, 331
Hemodialysis unit, 357
High-reflectance reflectors, 95
High-rise buildings, 338
Home care, 19–20
Hood lights, 354
Horizontal barrier, 216
Horizontal continuity, 219

652 Index

Horizontal exits, 225f, 226f
   areas required for, 325t
   benefits of, 271
   capacity of, 325
   fire compartment with, 271–272
   health care facilities with, 272f, 273f
   inspecting, 362–363
   requirements for, 271f
   story subdivided by, 278f
   walls, 225–226
Horizontal means of egress, 273
Horizontal sliding door, 266, 267f
Horizontal-shaft, single-stage centrifugal pump, 432f
Hose cabinets, 510
Hose connections
   inspecting, 402
   routine testing of, 409
Hose valves
   locations of, 418–419
   stairwell housing, 418f
Hospital. *See also* Health care facilities
   defined, 17
   electrical safety in, 127
   type 1 EES required for, 100
Hospital grade receptacles, 129
Hot spots, 161
Hotels, 191–192
HVAC. *See* Heating, ventilation, and air-conditioning equipment
Hydrants, 10
Hydraulic nameplate, 401, 401f
Hydrogen gas, 136
Hydrostatic testing, 486t
   piping network integrity from, 402
   pressurized water used for, 428
   suppression-agent storage cylinders requiring, 462
   training for, 462, 485–486
Hyperbaric areas
   defined, 19
   electrical requirements of, 139
Hypobaric chambers
   defined, 19
   no longer used, 140
Hypobaric facilities, 9, 140

ICBO. *See* International Conference of Building Officials
ICC. *See* International Code Council
ICU. *See* Intensive care unit

IEEE. *See* Institute of Electrical and Electronic Engineers
IESNA. *See* Illuminating Engineering Society of North America
Illuminating Engineering Society of North America (IESNA), 92
Illumination, 365
IMC. *See* Intermediate metal conduit
IMO. *See* International Marine Organization
Impingement nozzles, 471–472
Incipient stage, 497
Indicating valves
   sprinkler systems with, 381f
   water flow with, 379–380
Indicators, 552
Inductive power, 55
Industrial fire brigades, 10
Industrial occupancies, 21, 193–194
Inergen
   as gaseous fire-extinguishing agent, 466
   oxygen content lowered by, 467
Information
   engineering department disseminating, 601
   equipment history as, 608
   major repair, 599
   record keeping providing, 605
   safety communicated as, 598
Information systems, 140–141
   data processing in, 26, 38
   UPS required for, 141
Inhalation anesthetizing locations, 131–132
Initiating device circuit technology, 493
In-line shutoff valve, 552
Inspection(s), 171
   alarms requiring, 401
   automatic sprinklers requiring, 511
   buildings requiring, 367–368
   construction stage requiring, 593
   corridor walls requiring, 367
   design stage requiring, 592
   doors requiring, 362
   electrical systems and equipment with, 176–177
   emergency lighting requiring, 365
   equipment requiring, 176–177
   escalators requiring, 363
   exhaust system scheduled for, 352
   exit discharge requiring, 364–365
   exit signage requiring, 365
   facility owners performing, 459–460

fire detection and alarm systems requiring, 513–514
fire escape ladders requiring, 363
fire escape stairs requiring, 363
fire pumps requiring, 512
fire resistance requiring, 215
frequency of, 472t–473t
grease buildup requiring, 460t
health care facilities with, 29
horizontal exit requiring, 362–363
hose connections requiring, 401
inspectors on-site for, 32–34
interior finishes requiring, 367
kitchen hood extinguishing systems requiring, 512–513
laundry chutes requiring, 366
operating features requiring, 368
OSHA performing, 29
piping network requiring, 398–400
portable fire extinguishers requiring, 483–484, 513
postconstruction stage requiring, 593
preparing for, 30–32, 605
ramps requiring, 363
sprinkler system requiring, 397–398
stairs requiring, 362
suppression systems requiring, 513
trash chutes requiring, 366
travel distance requiring, 364
vertical opening requiring, 366
Institute of Electrical and Electronic Engineers (IEEE), 4, 45
industry practices recommended by, 6
medical equipment recommendations from, 142
Instrument air piped gas systems, 576
Insulation, 60
Intensive care unit (ICU), 78, 78f, 297
Interior finishes, 10, 332–336
flame spread classification for, 334t
flame spread indices for, 333–334
inspecting, 367
materials for, 334–335
Interior floor finish, 10
classes of, 336t
research of, 335–336
Interior walls, 333–335
Intermediate metal conduit (IMC), 65
Internal disaster, 616
International Building Code, 224
International Code Council (ICC), 183

International Conference of Building Officials (ICBO), 182
International Marine Organization (IMO), 471
Invasive procedure, 23
IPS. *See* Isolated power systems
Isolated input, 146
Isolated power systems (IPS), 54, 94, 115–121
codes and standards for, 118
ground-fault current limited by, 131
installation/maintenance of, 119–120
LIMs in, 117
location requirements for, 173
operating suites with, 132
transformer isolation for, 118–119

JCAHO. *See* Joint Commission on Accreditation of Healthcare Organizations
Joint Commission on Accreditation of Healthcare Organizations (JCAHO), 29, 69, 85
disaster plan required by, 616
health care facility standards from, 6
inspector/surveyor on-site from, 32–34
medical equipment safety promoted by, 155
meeting standards of, 605
standards/accreditation by, 30, 31
surveyors authority important for, 13

Kitchen hood extinguishing systems
grease fires suppressed by, 450
inspecting, 512–513
inspection/maintenance of, 459–462
maintenance of, 461–462
schematic of, 453f
system design for, 456–459
types of, 453–456
Knife-blade switch, 72

Labeling, 585f
fire door, 234–236
fire extinguishers with, 486
system maintenance with, 461
Laboratories, 578
codes difficult for, 25–26
fire protection for, 9, 134
non-patient piped gas systems in, 576–577
receptacles in, 134–135
survey preparation by, 32

Index

Laundry chutes
  as hazardous area, 331–332, 352
  as vertical openings, 349–351
  inspecting, 366
Level 1 gas systems, 525–526, 534
  criteria for, 536
  installation requirements for, 557
  medical compressed-air source for, 535f
  oxygen container for, 532f
  walls required in, 589
Level 1 vacuum systems, 561
  pumps required for, 562–563
  receivers used in, 564
  sensors used in, 568–569
Level 2 gas systems, 525–526, 534
  criteria for, 536
  exceptions for, 554
  installation requirements for, 557
  walls required in, 589
Level 2 vacuum systems, 561
  receivers used in, 564
  sensors used in, 568–569
Level 3 gas systems, 525–526, 534, 555, 556
  criteria for, 530
  moisture indicator for, 536–537
Level 3 vacuum systems, 561, 569
Life Safety Code®, 8, 69, 189, 194, 203, 206, 242, 373, 474, 488–489
  building construction requirements from, 185
  compliance reasons for, 355
Lighting, 91–96. *See also* Emergency lighting
  artificial, 91
  codes and standards for, 92
  controls, 88
  corridors requiring, 93
  emergency, 95, 303–305
  examination rooms with, 94
  maintenance of, 96
  psychiatric areas with, 137
  radiology equipment influenced by, 134
  safeguards for, 42
Lightning protection systems, 10
Limited care facilities
  defined, 17–18
  electrical requirements for, 128
Limited combustible material, 214
LIMs. *See* Line isolation monitors
Line isolation monitors (LIMs), 117
  current monitored by, 119
  operating suites with, 118
Line-of-sight devices, 497

Line-voltage variation, 144
Liquefied compressed gas, 465–466
Liquefied Petroleum Gas Code, 9, 339, 341–342, 579
Load-shedding circuits, 103–106
  emergency power source from, 105–106
  restrictions on, 102
Locker rooms, 358–359
Locking arrangements, 261
Locks, 263
Locks and latches, 260–261
Low-Expansion Foam, 8

Magnetic Resonance Imaging (MRI), 38
Main drain test, 427
Main line valve, 545
Main service switch
  electrical systems with, 161
  ground-fault protection and, 174
Maintenance. *See also* Preventive maintenance
  communication systems needing, 126
  contractors for, 348
  EPSS log for, 114
  frequency of, 473
  lighting requiring, 96
  NFPA 25 requiring, 409–410
  personnel, 120
  qualified contractor for, 398
  sprinkler systems requiring, 409–412, 411–412
  types of, 410
Major rehabilitation, 357
Major repairs, 599
Manifold, 528–529
  gas supplied through, 529f
  nitrogen connected to, 530f
  nitrous oxide connected to, 531
Manual break-glass station, 489f
Manual release stations, 458
Manufactured assemblies
  external devices on, 551f
  installation easier from, 551
  vacuum station inlets in, 568
Manufacturer's test circuit
  chassis leakage current measured by, 152f
  current leakage tests with, 145f
  current leakage tests (isolated) with, 147f, 153f
  current leakage tests (isolated/nonisolated) with, 148f, 154f

Index **655**

current leakage tests (nonisolated) with, 146f, 153f
electrical isolation measured by, 147f, 154f
Marking system, 476f
Materials, 340–341
Means of egress, 199
   accessible, 277–278
   arrangement of, 289–302
   components of, 254–278
   defined, 243–244
   doors without keys for, 261
   exit access arrangements for, 291f
   exit enclosures and, 361–362
   horizontal exits and, 271
   lighting required for, 94
   marking, 305–308
   minimum number of, 286
   mobility impaired access to, 302–303
   number of, 286t
   provisions for, 256
   ramps as, 272–273
   reference for, 242
Measurements, 256
Mechanical defects
   avoiding, 42–43
   defined, 37–38
Mechanical rooms
   common path of travel and, 300–302
   single exit requirements from, 302f
Mechanical safety measures, 586
Medical air, 550
Medical appliances
   certification for, 149
   cord-connected test for, 152
   line-voltage variation requirements for, 144
Medical equipment, 142–159
   clinical engineer testing, 143, 152
   code requirements for, 149
   electrical power requirements for, 150
   facility testing requirements for, 151–156
   guideline requirements for, 142–143
   installation/grounding for, 156–157
   JCAHO safety promoted for, 155
   manufacturer's documentation for, repair manuals, 146–147
   manufacturer's requirements for, 143–148
     current leakage tests, 145
     power cords, 143
     wiring/switches, 144
   purchase order specifications for, 148–151
   receptacles compatible with, 150
   specifications for, 149
   testing intervals for, 175
   training for, 157–159
Medical gas, 121–122
Medical gas concentration test, 557
Medical office buildings (MOB), 193
Medical-surgical vacuum systems, 521, 580–581
   connection to, 575
   vacuum level measured for, 591
Meetings, planned
   agenda of, 601
   frequency/location of, 602–603
   project progress reports in, 600–601
Megger test, 162
Memory-metal fittings, 541
Mercantile occupancies, 22, 193
Metallic gas tube fittings, 541
Mezzanines
   as intermediate level, 313
   limitations to, 314
   opened/closed, 315f
MIC. *See* Microbiologically influenced corrosion
Microbiologically influenced corrosion (MIC), 399–400
Mini-atrium
   defined, 311
   features of, 312t
   health care occupancies not permitting, 311
Mixed occupancies
   building codes for, 198
   stringent provisions for, 195–196
Mixed-load service, 45
MOB. *See* Medical office buildings
Mobility impaired, 302–303
Model building code
   ceiling termination for, 222
   educational occupancies defined by, 188
   formulating, 183
   National Board of Fire Underwriters developing, 182
   occupancy and, 198
   occupancy separations from, 228
Model fire codes, 186–187
Moisture, 535
Moisture indicator, 536–537
Molded-case thermal magnetic circuit breaker, 74
Monitoring systems, 555–556

Index

Motor control centers
  accessible location for, 70
  information for, 170
  power bus common for, 69
MRI. *See* Magnetic Resonance Imaging
Multigenerator installation, 108
Multioutlet assemblies, 67
Multiple occupancy building, 195, 196–197, 196f
Multistory building
  egress requirements for, 277
  vertical opening fire problems for, 348

National Board of Fire Underwriters, 182
National Electrical Code® *(NEC*®*)*
  as most widely adopted code, 342
  codes conflict and, 14
  electrical hazard safeguards from, 7
  equipment requirements from, 50–51
  ground-fault protection from, 53
  ground-fault protection systems required by, 162
  receptacles addressed by, 128–129
  receptacles defined by, 83
  utilities standards from, 339
National Fire Alarm Code®, 8, 121, 186–187, 320, 373, 489
National Fire Protection Association (NFPA)
  code defined by, 3
  codes and standards written by, 6
National Fuel Gas Code, 9, 339, 340, 579
National Institute of Occupational Safety and Health (NIOSH), 522
NEC®. *See National Electrical Code*®
Net area, 281
Net/gross area, 281f
Network supplies, 46–48
NFPA. *See* National Fire Protection Association
NFPA 1, *Uniform Fire Code*™, 7, 186–187, 372
NFPA 10, *Standard for Portable Fire Extinguishers*, 8, 372
  fire classification by, 474–475
  model fire codes referencing, 186–187
NFPA 11, *Standard for Low-Expansion Foam*, 8
NFPA 12, *Standard on Carbon Dioxide Extinguishing Systems*, 8, 372
NFPA 12A, *Standard on Halon 1301 Fire Extinguishing Systems*, 8

NFPA 13, *Standard for the Installation of Sprinkler Systems*, 372, 404–407
  design/installation of, 7
  model fire codes referencing, 186–187
  privacy curtains tested for, 395
  sprinkler activation calculated by, 379
  sprinkler spacing requirements from, 389
NFPA 14, *Standard for the Installation of Standpipe and Hose Systems*, 8, 373
NFPA 17, *Standard for Dry Chemical Extinguishing Systems*, 8, 373
NFPA 17A, *Standard for Wet Chemical Extinguishing Systems*, 8, 373, 457
NFPA 20, *Standard for the Installation of Stationary Pumps for Fire Protection*, 8, 373
NFPA 22, *Standard for Water Tanks for Private Fire Protection*, 8
NFPA 24, *Standard for the Installation of Private Fire Service Mains and Their Appurtenances*, 8
NFPA 25, *Standard for the Inspection, Testing, and Maintenance of Water-Based Fire Protection Systems*, 373
  maintenance types from, 409–410
  routine testing from, 403
  system impairment addressed by, 7
NFPA 30, *Flammable and Combustible Liquids Code*, 8
NFPA 31, *Standard for the Installation of Oil-Burning Equipment*, 8, 347
NFPA 33, *Standard for Spray Application Using Flammable or Combustible Materials*, 8
NFPA 37, *Standard for the Installation and Use of Stationary Combustion Engines and Gas Turbines*, 8
NFPA 45, *Standard on Fire Protection for Laboratories Using Chemicals*, 8, 134
NFPA 50, 538–539
NFPA 51, *Standard for the Design and Installation of Oxygen–Fuel Gas Systems for Welding, Cutting, and Allied Processes*, 579
NFPA 51B, *Standard for Fire Prevention during Welding, Cutting and Other Hot Work*, 9
NFPA 53, *Recommended Practice on Materials, Equipment, and Systems Used in Oxygen-Enriched Atmospheres*, 9
NFPA 54, *National Fuel Gas Code*, 9, 339, 340, 579

Index

NFPA 55, *Standard for the Storage, Use, and Handling of Compressed Gases and Cryogenic Fluids in Portable and Stationary Containers, Cylinders, and Tanks*, 9, 519, 524, 539, 579
NFPA 56K, *Recommended Practice on Medical Surgical Vacuum Systems*, 521
NFPA 58, *Liquefied Petroleum Gas Code*, 9, 339, 341–342, 579
NFPA 70, *National Electrical Code® (NEC®)*
  as most widely adopted code, 342
  codes conflict and, 14
  electrical hazard safeguards from, 7
  equipment requirements from, 50–51
  ground-fault protection from, 53
  ground-fault protection systems required by, 162
  receptacles addressed by, 128–129
  receptacles defined by, 83
  utilities standards from, 339
NFPA 70E, *Standard for Electrical Safety Requirements for Employee Workplaces*, 9
NFPA 72, *National Fire Alarm Code®*, 8, 121, 186–187, 320, 373, 489
NFPA 75, *Standard for the Protection of Electronic Computer/Data Processing Equipment*, 9, 140
NFPA 76B, *Standard on the Safe Use of Electricity in Patient Areas of Health Care Facilities*, 1980 edition, 142
NFPA 80, *Standard for Fire Doors and Fire Windows*, 9, 203, 237, 318
NFPA 80A, *Recommended Practice for Protection of Buildings from Exterior Fire Exposures*, 9
NFPA 82, *Standard on Incinerators and Waste and Linen Handling Systems and Equipment*, 9, 332
NFPA 85, *Boiler and Combustion Systems Hazards Code*, 9
NFPA 88A, *Standard for Parking Structures*, 9
NFPA 90A, *Standard for the Installation of Air-Conditioning and Ventilating Systems*, 9, 203
  fire safety from, 343
  maintenance provisions by, 345–346
NFPA 90B, *Standard for the Installation of Warm Air Heating and Air-Conditioning Systems*, 9
NFPA 91, *Standard for Exhaust Systems for Air Conveying of Vapors, Gases, Mists, and Noncombustible Particulate Solids*, 9, 346

NFPA 92A, *Standard for Smoke-Control Systems Utilizing Barriers and Pressure Differences*, 9
NFPA 92B, *Standard for Smoke Management Systems in Malls, Atria, and Large Spaces*, 9
NFPA 96, *Standard for Ventilation Control and Fire Protection of Commercial Cooking Operations*, 9, 331, 351, 353–354, 373
NFPA 97, *Standard Glossary of Terms Relating to Chimneys, Vents, and Heat-Producing Appliances*, 9
NFPA 99, *Standard for Health Care Facilities*, 112, 118, 127, 131, 138, 144, 331, 524, 537, 544, 618
  alerting systems specified in, 121
  classification by, 99
  current leakage tests from, 153
  cylinder requirements from, 590
  emergency generator requirements by, 102
  essential electrical system requirements from, 98
  hospital defined in, 17
  hyperbaric defined by, 19
  system/equipment requirements from, 8
NFPA 99B, *Standard for Hypobaric Facilities*, 9, 140
NFPA 101®, *Life Safety Code®*, 8, 69, 189, 194, 203, 206, 242, 373, 474, 488–489
  building construction requirements from, 185
  reasons for compliance, 355
NFPA 101A, *Guide on Alternative Approaches to Life Safety*, 9
NFPA 105, *Standard for the Installation of Smoke Door Assemblies*, 9, 318
NFPA 110, *Standard for Emergency and Standby Power Systems*, 9, 97, 106, 110–111, 112–113, 343
NFPA 111, *Standard on Stored Electrical Energy Emergency and Standby Power Systems*, 9, 97, 339, 343
NFPA 115, *Recommended Practice on Laser Fire Protection*, 9
NFPA 170, *Standard for Fire Safety Symbols*, 9
NFPA 204, *Guide for Smoke and Heat Venting*, 9
NFPA 211, *Standard for Chimneys, Fireplaces, Vents, and Solid Fuel-Burning Appliances*, 9, 347

Index

NFPA 214, *Standard on Water-Cooling Towers,* 9
NFPA 220, *Standard on Types of Building Construction,* 9, 204, 210, 213, 214
NFPA 221, *Standard for High Challenge Fire Walls, Fire Walls, and Fire Barrier Walls,* 9, 204, 216, 223–224
NFPA 232, *Standard for the Protection of Records,* 10
NFPA 241, *Standard for Safeguarding Construction, Alteration, and Demolition Operations,* 10
NFPA 251, *Standard Methods of Tests of Fire Resistance of Building Construction and Materials,* 10, 204, 205
NFPA 252, *Standard Methods of Fire Tests of Door Assemblies,* 204, 232
NFPA 253, *Standard Method of Test for Critical Radiant Flux of Floor Covering Systems Using a Radiant Heat Energy Source,* 10, 237, 336
NFPA 255, *Standard Method of Test of Surface Burning Characteristics of Building Materials,* 10
NFPA 257, *Standard on Fire Test for Window and Glass Block Assemblies,* 10, 204, 232
NFPA 259, *Standard Test Method for Potential Heat of Building Materials,* 10, 204
NFPA 265, *Standard Methods of Fire Tests for Evaluating Room Fire Growth Contribution of Textile Wall Coverings on Full Height Panels and Walls,* 10, 335
NFPA 286, *Standard Methods of Fire Tests for Evaluating Contributions of Wall and Ceiling Interior Finish to Room Fire Growth,* 10, 335
NFPA 291, *Recommended Practice for Fire Flow Testing and Marking of Hydrants,* 10
NFPA 418, *Standard for Heliports,* 10, 331
NFPA 424, *Guide for Airports/Community Emergency Planning,* 10
NFPA 600, *Standard on Industrial Fire Brigades,* 10
NFPA 701, *Standard Methods of Fire Tests for Flame Propagation of Textiles and Films,* 10
NFPA 704, *Standard System for the Identification of the Hazards of Materials for Emergency Response,* 10
NFPA 750, *Standard on Water Mist Fire Protection Systems,* 470–471
NFPA 780, *Standard for the Installation of Lightning Protection Systems,* 10
NFPA 801, *Standard for Fire Protection for Facilities Handling Radioactive Materials,* 10
NFPA 909, *Code for the Protection of Cultural Resources,* 204
NFPA 1600, *Standard on Disaster/Emergency Management and Business Continuity Programs,* 10
NFPA 1961, *Standard on Fire Hoses,* 10
NFPA 1962, *Standard for the Care, Use and Service Testing of Fire Hose Including Couplings and Nozzles,* 10
NFPA 2001, *Standard on Clean Agent Fire Extinguishing Systems,* 10, 373, 463
NFPA 5000®, *Building Construction and Safety Code®,* 10, 184, 188, 198, 202, 214, 416
NFPA regulations, 11
NIOSH. *See* National Institute of Occupational Safety and Health
Nitrogen, 530f
Nitrous oxide, 531
Noncombustible materials
  defined, 214
  ordinary structures with, 210
Noncombustible structures, 209–210
Nonflammable gas systems
  compressed air most common for, 580
  laboratories with, 578
Nonisolated input, 146
Non-patient piped gas systems, 517–518
  laboratories with, 576–577
  limited use for, 522
  requirements for, 520
Non-patient piped vacuum systems, 580–582
Nonrechargeable extinguishers, 484
Non-sleeping suites
  as unsupervised area, 298, 299
  size limits and, 360
Nonsprinklered construction
  as facilities, 337
  corridor protection in, 250–251
  door requirement in, 252
  smoke barriers in, 323
Novec 1230, 467–468
Nurse call systems, 124
Nursery, 345
Nursing homes
  defined, 17
  electrical requirements for, 128

power requirements for, 100
supervised automatic sprinkler system
 required by, 338
Nursing suite, 299f

Obstructions
 investigating, 411
 sprinkler systems having, 410–411
Occupancies, 20–22, 187–195, 201. *See also*
 Business occupancies; Mixed
 occupancies
 construction type for, 215–216
 converting, 202
 educational, 22, 188
 fire barriers separating, 228
 industrial, 21, 193–194
 mixed, 195–196, 198
 residential, 21, 190–191, 197
 separated, 195, 198t
 separation wall influencing, 219
 travel distance limits in, 289t
Occupant load
 buildings with, 279–285
 egress sufficient for, 281
 health care facilities and, 280t
Occupant-use hose, 414–415, 415
Occupational Safety and Health
 Administration (OSHA)
 inspections by, 29
 "Waste Anesthetic Gas" issued by, 573–574
Oil-burning equipment, 8, 347
Oil-free compressor, 534
One-half diagonal rule, 292
One-third rule, 314f
Openings
 barriers with, 229–238
 fire doors, 231
 fire protection ratings for, 233t
 HVAC ducts as, 238
 penetrations differing from, 230
Operating features, 368
Operating room
 area alarm panel for, 554f
 conductive flooring in, 173
 flammable vapors in, 590
 hazard currents in, 120–121
 isolated power supply in, 132
 labor and delivery, 25
 line isolation monitor in, 118
 physiological monitoring equipment in,
 132–133
 utility wall in, 548f

Operating system, 494
Ordinary structures, 210
Organizations, 4
OSHA. *See* Occupational Safety and Health
 Administration
OS&Y. *See* Outside stem and yoke
Outdoor areas, 94
Outside stem and yoke (OS&Y), 380
Overcurrent protection, 52, 75
Overload, 74
Owner's information certificate, 388f
Oxygen, 455
 compressed air system with, 559f
 container for, 585f
 container of, 532f
 cylinder for, 584f, 585f
 for medical purposes, 528
 Inergen lowering, 467
 storage of, 584–585
 uses of, 584
Oxygen–fuel gas systems, 579

Paging systems, 124
Panelboard, 77
 circuit breaker in, 90
 defined, 69
 devices for, 69–70
 information for, 170
Panic bar, 261
Parking structures, 9, 324
Path of least deviation, 547
Patient(s)
 ambulatory, 205
 contaminants harming, 524
 gas ingested by, 532–533
 home care requirements and, 19–20
 piped air system not for, 577f
 proper treatment of, 597
 separation from, 577–578
Patient bedrooms
 gas outlet for, 549
 intervening room to, 298f
 lighting of, 92
 over-the-bed lamps in, 93f
 quick response sprinklers for, 375
Patient care areas
 circuiting requirements for, 80–81
 defined, 22–23
 grounding systems tested in, 162–163
 lighting required for, 91
 NFPA grounding standards for, 77
 receptacle grounding in, 174

Patient care areas (*continued*)
  receptacle testing in, 175
  receptacle wiring in, 130f
  shutoff valves accessible in, 545
  wet locations in, 24
Patient gas equipment, 583
Patient leads, 154
Patient physiological monitoring systems, 124
Patient piped gas systems
  background of, 517–518
  components of, 524, 525f
  incident containment for, 519
  levels of, 525–526
  requirements for, 518–520
  restrictions to, 527
Patient piped vacuum systems
  components of, 562f
  distribution system for, 565–567
  equipment creating, 562–563
  precursors to, 520–521
  pumps for, 564
  requirements for, 521–522
  vacuum sources for, 563f
Patient vacuum equipment, 588
Pediatric areas, 521
Penetration(s)
  areas evaluated for, 240–241
  barriers with, 229–238
  exit enclosures allowing, 246
  HVAC ducts as, 238
  HVAC system with, 239f
  openings differing from, 230
  partitions sealed from, 68–69
  problems with, 241
  smoke barriers and, 321–322, 510
Penetration protection
  barrier sealing and, 238–241
  building codes addressing, 240
Performance testing, 612
Personnel. *See* Emergency personnel
PET. *See* Positron Emission Tomography
Phase converter, 55
Physical properties, 470
Physiological monitoring equipment, 132–133
Piped air system, 577f
Piped gas system. *See also* Flammable gas systems; Level 1 gas systems; Non-patient piped gas systems; Patient piped gas systems
  checklist for, 592–593
  design considerations for, 526–527
  distribution system for, 539–540
  gas mix-ups for, 548
  gas source room for, 537–538
  installation monitoring of, 556
  laboratories using, 578
  master alarm panel for, 553f
  master alarms in, 553–554
  monitoring of, 552–553
  outdoor source supply for, 538–539
  "patient use" separate in, 537
  piped vacuum system differing from, 561
  primary supply for, 531
  quality factors to, 528
  sensors in, 551–552
  source shutoff valve for, 545
  staff training for, 558–560
  verification of, 557
  walls required in, 589
  water (preventing) in, 589
Piped medical gas system, 589
Piped vacuum system. *See also* Level 1 vacuum systems; Medical-surgical vacuum systems; Patient piped vacuum systems
  contamination of, 570–571
  design criteria for, 570
  devices connected to, 588
  grounding reliability in, 566
  monitoring, 569
  performance criteria for, 569–570
  piped gas system differing from, 561
  renovating, 571
  sensors in, 568–569
  shutoff valves for, 567
  system verifiers for, 572
  termination points in, 568
  WAGD systems differing from, 574
Piping materials
  gas system, 540–541
  vacuum system, 565–566
Piping network
  cleanliness of, 558f
  concealing, 391–392
  considerations for, 540
  contamination of, 544
  coupling mechanism for, 541–542
  gooseneck for, 391
  inspecting, 398–400
  marking methods for, 542–543
  methods for, 566
  path of least deviation for, 547
  pipe identification for, 542–543
  terminating methods for, 547–552
Plans, 215

Index **661**

Plastics, foamed, 335
Plenums, 69
Pneumatic testing, 402, 429
Pneumatic tube system, 581, 582f
Polarity checks, 86
Polytetrafluoro-ethylene (PTFE), 536
Pop-off device, 575
Pop-off valve, 522
Portable fire extinguishers, 8, 372. *See also* Fire extinguishing equipment; Standpipe systems
   characteristics of, 478–481
   class A hazards using, 482t
   class B hazards using, 483t
   code requirements for, 477
   compatibility with, 458–459
   hazard placement for, 482–483
   hydrostatic testing for, 486t
   inspecting, 483–484, 513
   lettershaped symbol marking for, 477f
   location of, 475–483
   maintenance of, 484
   marking system for, 476f
   NFPA 10 requirements for, 474–475
   rating, 475
   recharging, 485
Positron Emission Tomography (PET), 38
Postconstruction stage, 593
Power. *See also* Emergency power supply system (EPSS); Isolated power systems; Uninterrupted power supply
   electrical systems delivering, 40
   factor, 55, 56f
   portable tool analyzing, 58f
   rewiring for additional, 68
   utility company supplying, 44
Power conditioners, 57
Power transfer switch, 436f
Power-operated doors, 266
Preaction sprinkler systems, 386, 387
Pressure, 528
Pressure jet nozzle, 472
Pressure reducing valves, 427
Pressure switches, 385
Preventive maintenance
   accountability for, 167
   logs kept for, 171
   procedures for, 163–167
   record keeping for, 168–171, 608–609
   work order for, 156f–157f, 158f–159f, 165f–167f
Primary voltage, 48–49

Privacy curtains, 396
Private fire service mains, 8
Product safety, 6–7
Projected beam unit, 496
Psychiatric areas, 25, 136–138
   electrical requirements for, 136
   hospital locations for, 18
   lighting requirements for, 137
   security issues in, 138
PTFE. *See* Polytetrafluoro-ethylene
Public way
   area of refuge access to, 277
   defined, 243
Pump drivers, 433–438
Pump room, 564–565
Pumps, 562–563
Punch list, 613
Purchase order, 148–151

Quick response sprinklers, 319
   life safety increased by, 378
   patient sleeping rooms requiring, 375
Quiet grounding, 78
Quiet-type switches, 88

Raceways, 64–67
   as cable/wire enclosures, 64
   conductors installed in, 68
   flexible metallic, 65–66
   nonmetallic, 66
   rigid metallic, 65
   underfloor, 67
Radial system
   double-ended in, 47f
   single radial distribution in, 46f
   two feeder lines in, 46f
Radioactive materials, 10
Radio-frequency interference filters (RFI), 94
Radiology areas, 133–134
   as imaging centers, 25
   electrical requirements for, 133
Radiology equipment, 134
Rail system, 550
Ramps
   as means of egress, 272–273
   inspecting, 363
Rate-of-rise detector, 495
Reactive power, 55
Receivers, 564
Receptacles, 83–87
   additional, 85
   critical care areas with, 129–130

Receptacles (*continued*)
  delivery rooms with, 132
  duplex, 86–87
  general care areas with, 128–129
  GFCIs in, 89–90, 90f
  laboratories with, 134–135
  locations, 84–85
  medical equipment compatible with, 150
  NFPA 70 addressing, 128–129
  patient care areas grounding, 174
  patient care areas testing, 162–163, 175
  patient care areas with, 130f
  precautions for, 85–86
  psychiatric areas with, 136–137
  surface mounted, 86f
  types, 83–84
Reconstruction, 202
Record keeping, 10
  emergency generators with, 175
  facility requirements for, 34
  information exchanged by, 605
  preventive maintenance with, 168–171
  records stored for, 609–610
  requirements for, 606–607
  system loads and, 111–112
Refrigeration alarms, 122
Relationships, 597
Relays, 88–89
Renovations
  existing services relocated for, 611
  special provisions for, 201–202
Residential occupancies, 21, 190–191, 197
Residential range hoods, 352
Resistors, discharge, 56f
Respirators
  cross-connection prevention for, 588
  power to, 587–588
Restaurant system, 454
Revolving doors, 266
RFI. *See* Radio-frequency interference filters
Rigid metal conduit, 65
Rigid nonmetallic raceways, 66
Riser check valve, 384f, 545
Riser diagram, 169, 170
Riser-type control valves, 382
Rooming house, 191
Rotating electric generators, 108

Safety
  committee important for, 605
  devices for, 109–110
  information adding to, 598

Saponification, 451
SBCC. *See* Southern Building Code Congress
Scavenging, 573
Secondary network supply, 46–47, 47f
Secondary voltage, 48–49
Security alarms, 122–123
Security issues, 138
Selective coordination, 75
Self-preservation, 203
Separated occupancies
  fire resistance requirements not specified for, 195
  smoke detectors and, 198f
Separation walls, 219, 224
SEPSS. *See* Stored emergency power supply system
Service valve, 545
Shafts, 310
Shakedown period, 613–614
Shutoff valves, 559
  piped vacuum system with, 567
  strategic location for, 527
  zone, 546f
Single-phase power, 55
Sleeping suites, 298
Sleeping-type occupancy, 288
Smoke barriers, 316–327
  areas required for, 325t
  door problems and, 319
  doors for, 318–319
  existing facilities with, 325–327
  exit arrangements and, 287f
  fire barriers v., 318
  fire resistance rating for, 318, 324f
  improper, 220
  new/existing construction, 323, 327t
  nonsprinklered construction with, 323
  opening protection, 326f, 329f
  penetrations sealed in, 321–322, 510
  removing, 356
  requirement exceptions for, 324
  requirements for, 200, 271f, 323–324
  smoke partitions v., 317–318, 317t
Smoke compartment
  detector location in, 321f
  existing building with, 328
  exit access for, 287
  new health care facility with, 326f
  size limitations for, 325
  sprinklers in, 319
Smoke control, 347

Index **663**

Smoke dampers, 317
   air openings required in, 319
   issues regarding, 356
   large capacity systems with, 344
   smoke detection required for, 320
Smoke detection, 320
   incipient stage detection by, 497
   installation of, 496f
   location of, 320f, 321f
   location requirements of, 265f
   locations for, 345
   sensitivity of, 500–501
   smoke dampers requiring, 320
Smoke door assemblies, 9, 318
Smoke management systems, 9
Smoke partitions, 317–318, 317t
Smoke-control systems, 9
Smokeproof enclosures, 269f
Snap switches, 88
SOC. *See* Statement of Conditions
Soffits, 392
Source shutoff valve, 545
Southern Building Code Congress (SBCC), 182
Spark-ignited generator, 172
Special projects, 601–602
Special-hazard systems, 454
Specification grade receptacles, 129
Spot networks, 47
Spot-type detectors, 495, 495f
Sprinkler protection
   atria requiring, 312
   mini-atria and, 311
   new construction requiring, 207–208, 509
Sprinkler systems, 372. *See also* Antifreeze sprinkler systems; Automatic sprinklers; Dry-pipe sprinkler systems; Suppression systems; Water mist extinguishing systems; Wet-pipe sprinkler systems
   acceptance testing of, 403
   aesthetics/location of, 389–391
   as link-and-lever system, 377–378
   compartment requirements for, 249–250
   concealed, 392f
   control panel for, 458
   control valves in, 379–380
   design/installation of, 7
   facts about, 376
   fire alarm systems interfacing with, 393–394
   fire controlled by, 379t
   furnishings not interfering with, 336
   high-rise buildings requiring, 338
   indicating valves for, 381f
   inspecting, 396–401, 397t
   inspection/testing/maintenance of, 397t
   maintenance of, 409–412
   model fire codes referencing, 186–187
   models of, 390f
   modifying, 388–389
   NFPA 13 calculating for, 379
   NFPA 13 spacing requirements for, 389
   obstructions in, 410–411
   piping concealed for, 391–392
   privacy curtains interfering with, 395–396
   privacy curtains tested for, 395
   quick response, 319
   return bend arrangement for, 391f
   sprinkler activation calculated by, 379
   sprinkler spacing requirements from, 389
   testing, 402–409
   types, 380–387
   water supply and, 392–393
   wet-pipe, 381, 382, 382f, 383f, 400
   zones for, 378–379
Staff members, 120
Stairs
   as egress element, 266–267
   identification for, 270
   inspecting, 362
   outside v. fire escape, 275f
   sign placement for, 270f
   signs for, 270f
   storage prohibited under, 247
   vertical opening for, 268
Standards, 3
Standpipe systems, 8, 373
   acceptance testing for, 422–424
   classes of, 417
   design considerations for, 417–419
   fire hose assembly for, 415f
   history of, 413–416
   inspecting, 511
   inspection/maintenance summary for, 420t–422t
   inspection/testing/maintenance of, 419–430
   maintenance contracted for, 429
   record keeping for, 429–430
   requirements for, 416
   types of, 416–417
   valving for, 414f
Statement of Conditions (SOC), 323

Station inlets, 567
  manufactured assemblies with, 568
  surface-mounted rail systems with, 567
Station outlets
  gas flowing out of, 547
  malfunctions to, 549
Stationary combustion engines/gas turbines, 8
Stationary pumps, 8, 373
Steam, 470
Storage areas
  as hazardous area, 284
  cylinders in, 586
  under stairs prohibited as, 247
Storage occupancies, 21, 194
Stored emergency power supply system (SEPSS), 97
Stored energy systems, 107
Stored-pressure extinguishers, 484
Stored-pressure types, 484
Structures, 199–202
Supervised automatic sprinkler system, 338
Supervisory switches, 403
Support services, 603–604
Suppression systems
  discharging, 500
  equipment damage avoided with, 463
  inspecting, 513
Suppression-agent storage cylinders, 469–473
  hydrostatic pressure testing for, 462
  location of, 458
Surface metal raceways, 66
Surface nonmetallic raceway, 67f
Surface-mounted rail systems
  equipment connection easier with, 550
  vacuum station inlets installed in, 567
Surge protector, 55–56
Surgical suites, 266
Surveys, 30–32
Switchboards, 69, 70
  information for, 170
  main switch in, 72f
Switches
  automatic transfer, 104–105
  electrical, 460
  load disconnect by, 71–72
  main service, 161
  supervisory, 403
  three-way, 88
  transfer, 104f
  types of, 87–89
  vane-type flow, 382
  water flow alarm, 490f

Switching systems, 88–89
System design
  facility engineer assisting, 490
  kitchen hood extinguishing systems with, 456–459
System design characteristics, 471–472
Systems, 1

Table of heights and areas, 206
Temperature-controlled switches, 88
Terminating devices, 60
Test circuit
  chassis leakage current from, 152f
  current leakage tests on, 145f, 146f, 147f, 148f, 153f, 154f
Test standards, 5
Testing, 503–507
Thermal testing, 61
Three-story communicating space, 311f
Three-way switches, 88
Time-delay, 73, 103
Time-temperature curve, 206f
Tools, 58f
Torque wrenches, 60
Total hazard current, 118
Training
  codes and standards for, 615
  cost cutting influencing, 618
  engineering staff receiving, 617
  fire extinguishers requiring, 486–487
  hydrostatic testing with, 462, 485–486
  medical equipment with, 157–159
  piped gas system requiring, 558–560
Transfer switches
  emergency power source connected by, 103, 105
  uncovered, 104f
Transformers, 53–55
Transient voltage, 38
  grounding system for, 78
  preventing, 80
Trapping systems, 520
Trash chutes, 331–332, 349–351
  inspecting, 366
Travel distance
  ambulatory health care with, 290t
  exit reached as, 288–289
  exits measuring, 289f
  inspecting, 364
  limitations, 290f
  occupancies having, 289t
  smoke barrier door and, 325

Index **665**

Treatment areas, 38
Treatment suite
  intervening room to, 300f
  two intervening rooms to, 301f
True root-mean-square (rms) meter, 58
Tubing, 542f
Type 1 essential electrical system, 98–99
Type 2 essential electrical system, 99–100
Type 3 essential electrical system, 100
Type AC cable, 62, 63f
Type FCC cable, 62
Type M1 cable, 62
Type MC cable, 63, 64f
Type NM, 63
Type NMC cable, 63
Type NMS cable, 63
Type SE cable, 63
Type UF cable, 63
Type USE cable, 63

UL. *See* Underwriters Laboratories
UL 9, *Fire Tests of Window Assemblies,* 204
UL 10B, *Fire Tests of Door Assemblies,* 204
UL 263, *Fire Tests of Building Construction and Materials,* 204
UL 300-compliant suppression systems, 452–453
Uncompensated circuit breakers, 74
Underwriters Laboratories (UL)
  product safety from, 6–7
  testing procedures changed by, 87
Uniform Fire Code™. *See* NFPA 1, *Uniform Fire Code*™
Uninterrupted power supply (UPS), 135
  information systems requiring, 141
  respirators/ventilators with, 587
United States (U.S.), 181
UPS. *See* Uninterrupted power supply
U.S. *See* United States
Utility company
  arrangements with, 45–50
  communications with, 50
  NFPA 70 standards for, 339
  power supply from, 44
  service record of, 49
Utility wall, 548f

Vacuum alarms, 121–122
Vacuum sources
  patient piped vacuum systems with, 563f
  pumps for, 565f
Vacuum station inlets, 567

Vacuum system, 565–566. *See also* Piped vacuum system
Valves. *See* Check valves; Control valves; Hose valves; Indicating valves; Shutoff valves
Vane-type flow switches, 382
Vent-cleaning service, 460
Ventilation, 135
Ventilators, 587–588
Vertical continuity
  barriers with, 220–221
  fire barrier with, 221f
  problem with, 222
Vertical openings, 309–315
  fire protection requirements for, 310t
  floors with, 310f
  inspecting, 366
  laundry chutes as, 349–351
  multistory buildings with, 348
  sealed/enclosed for, 309
  trash/laundry chutes protected as, 349–351
  unprotected, 313
Vertical shaft turbine-type pump, 434f
Vision panels, 238
Visual inspections, 167–168, 502–503
Voltage
  communication system's, 125
  conditioners, 55
  emergency generators and, 106
  laboratory requiring, 135
  primary supplies of, 48–49
  regulators, 57
  secondary supplies of, 48–49
  variations in, 40
Voluntary codes, 4
Voluntary standardization system, 5

WAGD. *See* Waste anesthetic gas disposal
Walk-throughs, 613
Warm air heating/air-conditioning systems, 9
"Waste Anesthetic Gas" (OSHA), 573–574
Waste anesthetic gas disposal (WAGD), 522–523, 573–576
  methods of, 574–575
  performance criteria for, 575–576
  testing, 576
Water, 470
  damage from, 380
  flow of, 378
  mist systems using, 472–473
  motor gong, 382, 384f
  supply, 392–393

Water flow alarm switches, 403, 490f
Water mist extinguishing systems
  applications for, 470–471
  as suppression agent, 469–473
Water tanks, 8
Water-assisted wet chemical kitchen hood systems, 455
Water-based fire protection systems, 373
  maintenance types from, 409–410
  routine testing from, 403
  system impairment addressed by, 7
Water-cooling towers, 9
Watts, 55
Wet chemical extinguishing systems, 8, 373, 457
Wet chemical kitchen hood systems, 455
Wet locations
  defined, 23–24
  electrical hazards in, 39, 42
  electrical installations in, 173
  electrical requirements for, 130–131
  GFCIs in, 89
  patient care areas as, 24
  receptacles in, 85
  receptacles tested in, 163
Wet-pipe sprinkler kitchen hood systems, 455
Wet-pipe sprinkler systems
  alarm check valve for, 382, 383f
  antifreeze sprinkler systems similar to, 385–386
  as reliable inexpensive systems, 381, 382f
  gauge inspection for, 400
Wiring methods, 68–69
Wood frame construction
  codes referring to, 212
  components of, 212f
Work orders, 156f–157f, 158f–159f, 165f–167f, 609

X-ray equipment, 175

Zone shutoff valves, 546f
Zones, 491–492
Zone-type arrangements, 491